D1387899

Applied Bayesian Hierarchical Methods

Applied Bayesian Hierarchical Methods

Peter D. Congdon

University of London
England, UK

CRC Press is an imprint of the
Taylor & Francis Group an **informa** business
A CHAPMAN & HALL BOOK

Chapman & Hall/CRC
Taylor & Francis Group
6000 Broken Sound Parkway NW, Suite 300
Boca Raton, FL 33487-2742

Printed in the United States of America on acid-free paper
10 9 8 7 6 5 4 3 2 1

International Standard Book Number: 978-1-58488-720-1 (Hardback)

Library of Congress Cataloging-in-Publication Data

Congdon, P.
Applied Bayesian hierarchical methods / Peter D. Congdon.
p. cm.
Includes bibliographical references and index.
ISBN 978-1-58488-720-1 (hardcover : alk. paper)
1. Multilevel models (Statistics) 2. Bayesian statistical decision theory. I. Title.

QA279.5.C66 2010
519.5'42--dc22 2010008252

Visit the Taylor & Francis Web site at
http://www.taylorandfrancis.com

and the CRC Press Web site at
http://www.crcpress.com

Contents

Preface

The use of Markov Chain Monte Carlo (MCMC) methods for estimating hierarchical models, often involving complex data structures, is sometimes described as a revolutionary development, and has arguably facilitated the fitting of such models. This book is intended to provide an intermediate level treatment of Bayesian hierarchical models and their applications. In a sense, all Bayesian models are hierarchical but the present volume seeks to demonstrate the advantages of a Bayesian approach to datasets involving inferences for collections of related units or variables, or in methods (e.g., nonlinear regression) for which parameters can be treated as random collections. It is in such applications that the Bayesian approach based on MCMC techniques has provided particular benefits.

Examples of the application settings that occur are provided by meta-analysis, data structured in space or time, multilevel and longitudinal data, multivariate data, and survival time data. Such settings form the subject matter of different chapters in the book and an applied focus, with attention to computational issues, is uppermost. The main package used for worked examples is WinBUGS, which allows the analyst to explore alternative likelihood assumptions, regression structures or assumptions on prior densities, while parameter sampling mechanisms rely on the package's inbuilt updating schemes. WinBUGS can be obtained from http://www.mrc-bsu.cam.ac.uk/bugs and its successor, OpenBUGS, is available at http://mathstat.helsinki.fi/openbugs/. The same flexibility, though in a narrower range of modeling settings, applies to the BayesX package. BayesX can be obtained at www.stat.uni-muenchen.de/~bayesx, and is particularly useful in nonlinear regression, as shown in Chapter 10. To demonstrate MCMC sampling from first principles, Chapter 1 also includes worked examples using the R package.

My acknowledgments are first to Chapman & Hall/CRC for giving me the opportunity to develop the book and for their encouragement in progressing it to finality. Particular thanks are due to Sarah Morris at C&H/CRC. For comments on particular chapters or advice on methods used in particular sections of the book, my thanks are due to Sid Chib, Valen Johnson, Kalyan Das, and Peter Hooper. A set of programs linked to the worked examples in the book is available from "Downloads and Updates" at the CRC Web site for this book, namely, http://www.crcpress.com/product/isbn/9781584887201. Comments on the book's content, or on these programs, would be appreciated and can be sent to me at p.congdon@qmul.ac.uk.

Peter Congdon
Geography & Centre for Statistics, QMUL

Author

Peter Congdon is a health statistician, originally qualified at the London School of Economics, with particular interest in spatial variations in health, spatial statistical techniques and Bayesian methods in general. He is affiliated with the QMUL Centre for Statistics and the QMUL Department of Geography, and since 2001 has been a research professor at QMUL. He is the author of a wide range of articles and books, in both statistics and applications areas, and is an elected member of the International Statistical Institute and a Chartered Statistician.

1

Bayesian Methods for Complex Data: Estimation and Inference

1.1 Introduction

The Bayesian approach to inference focuses on updating knowledge about the unknowns, θ, in a statistical model on the basis of observations y, with revised knowledge expressed in the posterior density, $p(\theta|y)$. The sample of observations y being analyzed provides new information about the unknowns, while the prior density $p(\theta)$ of the unknowns represents accumulated knowledge about them before observing or analyzing the data y. Hypotheses on parameters are similarly based on posterior probabilities conditional on the observed data.

Compared to frequentist approaches, there is considerable flexibility with which prior evidence about parameters can be incorporated in an analysis, and use of informative priors (to express accumulated knowledge) can reduce the possibility of confounding and provides a natural basis for evidence synthesis (Dunson, 2001; Shoemaker et al., 1999). The Bayes approach provides uncertainty intervals on parameters that are consonant with everyday interpretations (Willink and Lira, 2005), and has no problems in comparing the fit of non-nested models, such as a nonlinear model and its linearized version.

Furthermore, Bayesian estimation and inference has a number of advantages in terms of its relevance to the types of data and problems tackled by modern scientific research. These are a primary focus later in the book. For example, much of the data in social and health research has a complex structure, involving hierarchical nesting of subjects (e.g., pupils within schools), crossed classifications (e.g., pupils classified by school and by homeplace), spatially configured data, or repeated measures on subjects (MacNab et al., 2004). The Bayesian approach naturally adapts to hierarchically or spatio-temporally correlated effects via conditionally specified hierarchical priors under a three-stage prior (Gustafson et al., 2006; Lindley and Smith, 1972), with the first stage specifying the likelihood of the data given unknown random individual or cluster effects, the second stage specifying the density of the population of random effects, and the third stage providing priors (or hyperpriors) on the parameters of the population density. Bayesian estimation via repeated sampling from posterior densities facilitates modeling of complex data with random effects treated as unknowns, and not integrated out as often in frequentist approaches (Davidian and Giltinan, 2003). The integrated likelihood approach may become infeasible or unreliable in complex varying coefficient models (Tutz and Kauermann, 2003), and different parameter estimates may be obtained according to the maximization methods used (Molenberghs and Verbeke, 2004).

While Bayesian inference may have benefits, until relatively recently its practical implementation was impeded by computational restrictions. The increased application of Bayesian methods has owed much to the development of Markov Chain Monte Carlo (MCMC) algorithms for estimation (Gelfand and Smith, 1990; Gilks et al., 1996), which draw repeated parameter samples from the posterior distributions of statistical models, including complex models (e.g., models with multiple or nested random effects). Sampling-based parameter estimation via MCMC provides a full posterior density of a parameter so that any clear non-normality is apparent (Dellaportas and Smith, 1993), and hypotheses about parameters or interval estimates can be assessed from the MCMC samples without assumptions of asymptotic normality that typically underlie frequentist estimation.

As mentioned in the preface, a substantial emphasis in the book is on practical implementation for tutorial purposes, via illustrative data analysis and attention to statistical computing. Accordingly, the worked examples in the rest of the chapter illustrate MCMC sampling and Bayesian posterior inference from first principles, mostly using R code. In subsequent chapters, WinBUGS (which incorporates inbuilt MCMC algorithms) and to a lesser extent BayesX are used for computation. To retain some of the advantages of working in R (e.g., graphical presentations) and to access already developed inference tools, one may use[1] the R2WinBUGS facility (Sturtz et al., 2005), which links R to WinBUGS1.4.

As just mentioned, Bayesian modeling of hierarchical and random effect models via MCMC techniques has extended the scope for modern data analysis. Despite this, application of Bayesian techniques in such models also raises particular issues, discussed below or in later chapters. These include:

1. propriety and identifiability issues when diffuse priors are applied to variance or dispersion parameters for collections of random effects (Hadjicostas and Berry, 1999; Hobert and Casella, 1996; Palmer and Pettit, 1996);

2. selecting the most suitable form of prior for variance or dispersion parameters, such as inverse gamma, lognormal, or uniform priors on variances (Gelman, 2006a);

3. appropriate priors for models with several sources of random variation, e.g., separate conjugate priors, as against shrinkage priors, which ensure propriety and express each variance as a component of total random variation (Daniels, 1999; Natarajan and Kass, 2000; Wakefield, 2007);

4. potential bias in shrinkage estimation, such as oversmoothing in the presence of genuine outliers in spatial applications (Conlon and Louis, 1999);
5. the scope for specification bias in hierarchical models for complex data structures where a range of plausible model structures are possible (Chiang et al., 1999).

1.2 Posterior Inference from Bayes Formula

Statistical analysis uses probability models to summarize a set of observations, $y = (y_1, \ldots, y_n)$, by a collection of unknown parameters of a particular dimension (say d), $\theta = (\theta_1, \ldots, \theta_d)$. Consider the joint density $p(y, \theta) = p(y|\theta)p(\theta)$, where $p(y|\theta)$ is the sampling model or likelihood, and $p(\theta)$ defines existing knowledge, or expresses assumptions regarding the unknowns that can be justified by the nature of the application (e.g., that random effects are spatially distributed in an area application). The analysis seeks to update knowledge about the unknowns, θ, using the data, y, and so interest focuses on the posterior density $p(\theta|y)$ of the unknowns. Since $p(y, \theta)$ also equals $p(y)p(\theta|y)$, where $p(y)$ is the unconditional density of the data (also known as the marginal likelihood), one may obtain,

$$p(y, \theta) = p(y|\theta)p(\theta) = p(y)p(\theta|y). \tag{1.1}$$

This can be rearranged to provide the required posterior density as,

$$p(\theta|y) = \frac{p(y|\theta)p(\theta)}{p(y)}. \tag{1.2}$$

The marginal likelihood $p(y)$ may be obtained by integrating the numerator on the right side of Equation 1.2 over the support for θ, namely,

$$p(y) = \int p(y|\theta)p(\theta)d\theta.$$

From Equation 1.2, the term $p(y)$ acts as a normalizing constant necessary to ensure $p(\theta|y)$ integrates to 1, and so one may write,

$$p(\theta|y) \propto p(y|\theta)p(\theta), \tag{1.3}$$

namely, that the posterior density (updated evidence) is proportional to the likelihood (data evidence) times the prior (historic evidence or elicited model assumptions).

In some cases when the prior on θ is conjugate with the posterior on θ (i.e., has the same density form), the posterior density and marginal likelihood can be obtained analytically. When θ is low-dimensional, numerical integration is an alternative, and approximations to the required integrals can be used, such as the Laplace approximation (Raftery, 1996). In more complex applications, such approximations are not feasible and the integration to obtain $p(y)$ is intractable, so that direct sampling from $p(\theta|y)$ is not feasible. In such situations, MCMC methods provide a way to sample from $p(\theta|y)$ without necessarily knowing its analytic form. They create a Markov chain of sampled values, $\theta^{(1)}, \ldots, \theta^{(T)}$, with transition kernel, $K(\theta_{\text{cand}}|\theta_{\text{curr}})$, which have $p(\theta|y)$ as their limiting distribution. Using the large samples from the posterior distribution obtained by MCMC, one can estimate posterior quantities of interest, such as posterior means, medians, and highest density regions (Chen and Shao, 1999; Hyndman, 1996).

1.3 Markov Chain Monte Carlo Sampling in Relation to Monte Carlo Methods: Obtaining Posterior Inferences

MCMC methods are iterative sampling methods that can be encompassed within the broad class of Monte Carlo methods. However, MCMC methods must be distinguished from conventional Monte Carlo methods that generate independent simulations, $\{u^{(1)}, u^{(2)}, \ldots, u^{(T)}\}$, from a target density, $\pi(u)$. From such simulations the expectation of a function $g(u)$ under $\pi(u)$, namely,

$$E_\pi[g(u)] = \int g(u)\pi(u)\,du,$$

is estimated as,

$$\bar{g} = \sum_{t=1}^{T} g(u^{(t)}),$$

and, under independent sampling from $\pi(u)$, $\bar{g}$ tends to $E_\pi[g(u)]$ as $T \to \infty$. However, independent sampling from the posterior density, $p(\theta|y)$, is not usually feasible.

When suitably implemented, MCMC methods offer an effective alternative way to generate samples from the joint posterior distribution, $p(\theta|y)$, but differ from conventional Monte Carlo methods in that successive sampled parameters are dependent or autocorrelated. The target density for MCMC samples is therefore the posterior density, $\pi(\theta) = p(\theta|y)$, and MCMC sampling is especially relevant when the posterior cannot be stated exactly in analytic form, e.g., when the prior density assumed for θ is not conjugate with the likelihood $p(y|\theta)$. The fact that successive sampled values are dependent means that larger samples are needed for equivalent precision, and the effective number of samples is less than the nominal number. MCMC techniques can also be used to generate random samples from particular statistical distributions, so offering an alternative to techniques such as rejection sampling and inverse transform sampling.

For the parameter sampling case, assume a preset initial parameter value, $\theta^{(0)}$. Then MCMC methods involve generating a correlated sequence of sampled values, $\theta^{(t)}$ $(t = 1, 2, 3, \ldots)$, where updated values, $\theta^{(t)}$, are drawn from a transition distribution,

$$K\left(\theta^{(t)}|\theta^{(0)}, \ldots, \theta^{(t-1)}\right) = K\left(\theta^{(t)}|\theta^{(t-1)}\right),$$

that is Markovian in the sense of depending only on $\theta^{(t-1)}$. The transition distribution, $K(\theta^{(t)}|\theta^{(t-1)})$, is chosen to satisfy additional conditions, ensuring that the sequence has the joint posterior density, $p(\theta|y)$, as its stationary distribution. These conditions typically reduce to requirements on the proposal and acceptance procedure used to generate new parameter samples. The proposal density and acceptance rule must be specified in a way that guarantees irreducibility and positive recurrence; see, e.g., Andrieu and Moulines (2006). Under such conditions, the sampled parameters, $\theta^{(t)}$ $\{t = B, B + 1, \ldots, T\}$, beyond a certain burn-in phase in the sampling (of length B) can be viewed as a random sample from $p(\theta|y)$ (Roberts and Rosenthal, 2004).

In practice, MCMC methods are applied separately to individual parameters or groups ("blocks") of more than one parameter (Roberts and Sahu, 1997). In fact, different MCMC methods may be applied to different parameters. So, assuming θ contains more than one parameter, and consists of C components or "blocks," $\{\theta_1, \ldots, \theta_C\}$, different updating methods may be used for each component. If there are d parameters, the number of blocks may well be less than d, since several parameters in a single block may be updated jointly (a so-called "block update"). Note that for simplicity, we will often represent the MCMC algorithms as involving a generic parameter θ.

There is no limit to the number of samples T of θ that may be taken from the posterior density $p(\theta|y)$. Estimates of the marginal posterior densities for each parameter can be made from the MCMC samples, including estimates of location parameters (e.g., posterior means, modes or medians), together with the estimated certainty or precision of these parameters in terms of posterior standard deviations, credible intervals, or highest posterior density intervals. Such samples can also be used to provide probabilities on hypotheses relating to the parameters (Smith and Gelfand, 1992). For example, one form of 95% credible interval for θ_h may be estimated using the 0.025 and 0.975 quantiles of the sampled output $\{\theta_h^{(t)}, t = B + 1, \ldots, T\}$. To reduce irregularities in the histogram of sampled values for a particular parameter, a smooth form of the posterior density can be approximated by applying kernel density methods to the sampled values.

Monte Carlo posterior summaries typically include estimated posterior means and variances of the parameters, obtainable as moment estimates from the MCMC output, namely,

$$\hat{E}(\theta_h) = \bar{\theta}_h = \sum_{t=B+1}^{T} \theta_h^{(t)}/(T-B),$$

$$\hat{V}(\theta_h) = \sum_{t=B+1}^{T} \left(\theta_h^{(t)} - \bar{\theta}_h\right)^2/(T-B).$$

This is equivalent to estimating the integrals,

$$E(\theta_h|y) = \int \theta_h p(\theta_h|y)d\theta_h,$$

$$V(\theta_h|y) = \int \theta_h^2 p(\theta_h|y)d\theta_h - [E(\theta_h|y)]^2 = E(\theta_h^2|y) - [E(\theta_h|y)]^2.$$

One may also use the MCMC output to obtain posterior means, variances, and credible intervals for functions, $\Delta = \Delta(\theta)$, of the parameters (van Dyk, 2003, 150). These are estimates of the integrals,

$$E[\Delta(\theta)|y] = \int \Delta(\theta)p(\theta|y)d\theta,$$

$$V[\Delta(\theta)|y] = \int \Delta^2 p(\theta|y)d\theta - [E(\Delta|y)]^2 = E(\Delta^2|y) - [E(\Delta|y)]^2.$$

For $\Delta(\theta)$, its posterior mean is obtained by calculating $\Delta^{(t)}$ at every MCMC iteration from the sampled values $\theta^{(t)}$. The theoretical justification for such estimates is provided by the MCMC version of the law of large numbers (Tierney, 1994), namely, that,

$$\sum_{t=B+1}^{T} \frac{\Delta[\theta^{(t)}]}{T-B} \rightarrow E_\pi[\Delta(\theta)],$$

as $T \rightarrow \infty$, provided that the expectation of $\Delta(\theta)$ under $\pi(\theta) = p(\theta|y)$, denoted $E_\pi[\Delta(\theta)]$, exists. MCMC methods also allow inferences on parameter comparisons (e.g., ranks of parameters or contrasts between them) (Marshall and Spiegelhalter, 1998).

1.4 Hierarchical Bayes Applications

The paradigm in Section 1.2 is appropriate to many problems, where uncertainty is limited to a few fundamental parameters, the number of which is independent of the sample size n—this is the case, for example, in a normal linear regression when the independent variables are known without error and the units are not hierarchically structured. However, in more complex data sets or with more complex forms of model or response, a more general perspective than that implied by Equations 1.1 through 1.3 is available and also implementable using MCMC methods.

Thus, a class of hierarchical Bayesian models are defined by latent data (Paap, 2002) intermediate between the observed data and the underlying parameters driving the process. A terminology useful for relating hierarchical models to substantive issues is proposed by Wikle (2003) in which y defines the data stage, latent effects b defines the process stage, and θ defines the parameter stage. For example, the observations, $i = 1, \ldots, n$, may be arranged in clusters, $j = 1, \ldots, J$, so that the observations can no longer be regarded as independent. Rather, subjects from the same cluster will tend to be more alike than individuals from different clusters, reflecting latent variables that induce dependence within clusters.

Modeling such dependencies involves a three stage hierarchical Bayes (HB) prior set up for the joint density,

$$p(y, b, \theta) = p(y|b, \theta)p(b|\theta)p(\theta), \tag{1.4}$$

with a first stage likelihood, $p(y|b, \theta)$, and second stage density, $p(b|\theta)$, for the latent data conditional on the higher stage parameters θ. The first stage density, $p(y|b, \theta)$, in Equation 1.4 is a conditional likelihood, conditioning on b as well as θ, sometimes also called the complete data or augmented data likelihood. The application of Bayes theorem now specifies,

$$p(\theta, b|y) = \frac{p(y|b, \theta)p(b|\theta)p(\theta)}{p(y)},$$

and the marginal posterior for θ may now be represented as,

$$p(\theta|y) = \frac{p(\theta)p(y|\theta)}{p(y)} = \frac{p(\theta) \int p(y|b, \theta)p(b|\theta)\, db}{p(y)}.$$

where

$$p(y|\theta) = \int p(y, b|\theta)\, db = \int p(y|b, \theta)p(b|\theta)\, db,$$

is the observed data likelihood, namely, the complete data likelihood with b integrated out; $p(y|\theta)$ is sometimes also known as the integrated likelihood.

Often the latent data exist for every observation, or they may exist for each cluster in which the observations are structured (e.g., a school-specific effect, b_j, for multilevel data, y_{ij}, on pupils i nested in schools j). Let $\theta = [\theta_L, \theta_b]$ consist of parameter subsets relevant to the likelihood and to the latent data density, respectively. Then the data are generally taken as independent of θ_b given b, so that Equation 1.4 becomes

$$p(y, b, \theta) = p(y|b, \theta_L)p(b|\theta_b)p(\theta_b, \theta_L).$$

The latent variables b can be seen as a population of values from an underlying density (e.g., varying log odds of disease) and the θ_b are then population hyperparameters (e.g., mean and variance of the log odds) (Dunson, 2001). As examples, Paap (2002) mentions unobserved states describing the business cycle, Johannes and Polson (2006) mention unobserved volatilities in stochastic volatility models, while Albert and Chib (1993) consider the missing or latent continuous data $\{b_1, \ldots, b_n\}$, which underlie binary observations $\{y_1, \ldots, y_n\}$. The subject-specific latent traits in psychometric or educational item analysis can also be considered this way (Johnson and Albert, 1999), as can the variance scaling factors in the robust Student t errors version of linear regression (Geweke, 1993), or the subject-specific slopes in a growth curve analysis of panel data on a collection of subjects (Lindstrom and Bates, 1990).

Typically, the integrated likelihood, $p(y|\theta)$, cannot be stated in closed form and classical likelihood estimation relies on numerical integration or simulation (Paap, 2002, 15). By contrast, MCMC methods can be used generate random samples indirectly from the posterior distribution $p(\theta, b|y)$ of parameters and latent data given the observations. This requires only that the augmented data likelihood be known in closed form, without needing to obtain the integrated likelihood $p(y|\theta)$. To see why, note that the marginal posterior of the parameter set, θ, may alternatively be derived as,

$$p(\theta|y) = \int p(\theta, b|y)\, db = \int p(\theta|y, b) p(b|y)\, db,$$

with marginal densities for component parameters θ_h of the form (Paap, 2002, 5),

$$\begin{aligned} p(\theta_h|y) &= \int_{\theta_{[h]}} \int_b p(\theta, b|y)\, db d\theta_{[h]}, \\ &\propto \int_{\theta_{[h]}} p(\theta|y) p(\theta) d\theta_{[h]} = \int_{\theta_{[h]}} \int_b p(\theta) p(y|b, \theta) p(b|\theta)\, db d\theta_{[h]}, \end{aligned}$$

where $\theta_{[h]}$ consists of all parameters in θ with the exception of θ_h. The derivation of suitable MCMC algorithms to sample from $p(\theta, b|y)$ is based on the Clifford–Hammersley theorem, namely, that any joint distribution can be fully characterized by its complete conditional distributions. In the HB context, this implies that the conditionals $p(b|\theta, y)$ and $p(\theta|b, y)$ characterize the joint distribution $p(\theta, b|y)$ from which samples are sought, and so MCMC sampling can alternate between updates $p(b^{(t)}|\theta^{(t-1)}, y)$ and $p(\theta^{(t)}|b^{(t)}, y)$ on conditional densities, which are usually of a simpler form than $p(\theta, b|y)$. The imputation of latent data in this way is sometimes known as data augmentation (van Dyk, 2003).

To illustrate application of MCMC methods to parameter comparisons and hypothesis tests in a HB setting, Shen and Louis (1998) consider hierarchical models with unit or cluster specific parameters, b_j, and show that if such parameters are the focus of interest, their posterior means are the optimal estimates. Suppose instead that the ranks of the unit or cluster parameters, namely,

$$R_j = \text{rank}(b_j) = \sum_k 1(b_j \geq b_k),$$

(where $1(A)$ is an indicator function that equals 1 when A is true, 0 otherwise) are required for deriving "league tables." Then the conditional expected ranks are optimal, and obtained by ranking b_j at each MCMC iteration and taking their means over all samples. By contrast, ranking posterior means of b_j themselves can perform poorly (Goldstein and Spiegelhalter, 1996; Laird and Louis, 1989). Similarly, when the empirical distribution function (EDF) of the unit parameters (e.g., to be used to obtain the fraction of parameters above a threshold) is required, the conditional expected EDF is optimal.

A posterior probability estimate that a particular b_j exceeds a threshold τ, namely, of the integral $Pr(b_j > \tau|y) = \int_\tau^\infty p(b_j|y)db_j$, is provided by the proportion of iterations where $b_j^{(t)}$ exceeds τ, namely,

$$Pr(b_j > \tau|y) = \sum_{t=B+1}^{T} 1(b_j^{(t)} > \tau)/(T-B).$$

Thus, one might in an epidemiological application wish to obtain the posterior probability that an area's smoothed relative mortality risk, b_j, exceeds unity, and so count iterations where this condition holds. If this probability exceeds .95 then a significant excess risk is indicated, whereas a low probability (the sampled relative risk rarely exceeds 1) would indicate a significantly low mortality level in the area.

In fact the significance of individual random effects is one aspect in assessing the gain of a random effects model over a model involving only fixed effects, or of assessing whether a more complex random effects model offers a benefit over a simpler one (Knorr-Held and Rainer, 2001, 116). Since the variance can be defined in terms of differences between elements of the vector $(b_1, \ldots, b_J)$ as opposed to deviations from a central value (Kendall, 1943, 47), one may also consider which contrasts between pairs of b values are significant. Thus, Deely and Smith (1998) suggest evaluating probabilities, $Pr(b_j \leq \tau b_k|k \neq j, y)$, where $0 < \tau \leq 1$, namely, the posterior probability that any one hierarchical effect is smaller by a factor τ than all the others.

1.5 Metropolis Sampling

The earliest scheme for posterior density estimation using a variant of MCMC was developed by Metropolis et al. (1953). The Metropolis sampling algorithm is a special case of Metropolis–Hastings considered in Section 1.8. Reverting to the canonical framework in Section 1.2, let $p(y|\theta)$ denote the likelihood, and $p(\theta)$ denote the prior density for θ, or more specifically the prior densities $p(\theta_1), \ldots, p(\theta_C)$ on the components of θ. Then the Metropolis algorithm involves a symmetric proposal density (e.g., a normal, Student t, or uniform density) $q(\theta_{\text{cand}}|\theta^{(t)})$ for generating candidate parameter values, θ_{cand}.

Under the Metropolis algorithm, the acceptance probability for potential candidate values may be obtained as,

$$\alpha = \min\left(1, \frac{\pi(\theta_{\text{cand}})}{\pi(\theta^{(t)})}\right) = \min\left(1, \frac{p(\theta_{\text{cand}}|y)}{p(\theta^{(t)}|y)}\right) = \min\left(1, \frac{p(y|\theta_{\text{cand}})p(\theta_{\text{cand}})}{p(y|\theta^{(t)})p(\theta^{(t)})}\right). \tag{1.5}$$

So one compares the (likelihood * prior), namely, $p(y|\theta)p(\theta)$, for the candidate and existing parameter values. If the (likelihood * prior) is higher for the candidate value, it is automatically accepted and $\theta^{(t+1)} = \theta_{\text{cand}}$. However, even if the (likelihood * prior) is lower for the candidate value, such that α is less than 1, the candidate value may still be accepted. This is decided by random sampling from a uniform density, $U^{(t)}$, and the candidate value is accepted if $\alpha \geq U^{(t)}$.

The third equality in Equation 1.5 follows because the marginal likelihood, $1/M = p(y)$, in the Bayesian formula,

$$p(\theta|y) = p(y|\theta)p(\theta)/p(y) = Mp(y|\theta)p(\theta),$$

cancels out, as it is a constant. Stated more completely, to sample parameters under the Metropolis algorithm, it is not necessary to know the normalized target distribution, namely, the posterior density, $\pi(\theta|y)$; it is enough to know it up to a constant factor.

So the Metropolis algorithm *can* be implemented by using the full posterior distribution,

$$\pi(\theta) = p(\theta|y) = Mp(y|\theta)p(\theta),$$

as the target distribution—which in practice involves comparisons of the un-normalized posterior, $p(y|\theta)p(\theta)$. However, for updating values on a particular parameter, θ_h, it is not just M that cancels out in the ratio,

$$\pi(\theta_{\text{cand}})/\pi(\theta^{(t)}) = \frac{p(y|\theta_{\text{cand}})p(\theta_{\text{cand}})}{p(y|\theta^{(t)})p(\theta^{(t)})},$$

but any parts of the likelihood or prior not involving θ_h (these parts can be viewed as constant when θ_h is being updated).

When those parts of the likelihood or prior not relevant to θ_h are abstracted out, the remaining part of $p(\theta|y) = Mp(y|\theta)p(\theta)$, the part relevant to updating θ_h, is known as the full conditional density for θ_h (Gilks, 1996). One may denote the full conditional density for θ_h as,

$$\pi_h(\theta_h|\theta_{[h]}) \propto p(y|\theta_h)p(\theta_h),$$

where $\theta_{[h]}$ denotes the parameter set excluding θ_h. So the probability for updating θ_h can be obtained *either* by comparing the full posterior (known up to a constant M), namely,

$$\alpha = \min\left(1, \frac{\pi\left(\theta_{h,\text{cand}}\right)}{\pi\left(\theta_h^{(t)}\right)}\right) = \min\left(1, \frac{p(y|\theta_{h,\text{cand}})p(\theta_{h,\text{cand}})}{p\left(y|\theta_h^{(t)}\right)p\left(\theta_h^{(t)}\right)}\right),$$

or by using the full conditional for the hth parameter, namely,

$$\alpha = \min\left(1, \frac{\pi_h\left(\theta_{h,\text{cand}}|\theta_{[h]}^{(t)}\right)}{\pi_h\left(\theta_h^{(t)}|\theta_{[h]}^{(t)}\right)}\right).$$

Then, one sets $\theta_h^{(t+1)} = \theta_{h,\text{cand}}$ with probability α, and $\theta_h^{(t+1)} = \theta_h^{(t)}$ otherwise.

1.6 Choice of Proposal Density

There is some flexibility in the choice of proposal density q for generating candidate values in the Metropolis and other MCMC algorithms, but the chosen density and the parameters incorporated in it are relevant to successful MCMC updating and convergence (Altaleb and Chauveau, 2002). A standard recommendation is that the proposal density for a particular parameter, θ_h, should approximate the posterior density $p(\theta_h|y)$ of that parameter. In some cases, one may have an idea (e.g., from a classical analysis) of what the posterior density is, or what its main defining parameters are. A normal proposal is often justified as many posterior densities do approximate normality. For example, Albert (2007) applies a Laplace approximation technique to estimate the posterior mode, and uses the mean and variance parameters to define the proposal densities used in a subsequent stage of Metropolis–Hastings sampling.

The rate at which a proposal generated by q is accepted (the acceptance rate) depends on how close θ_{cand} is to $\theta^{(t)}$, and this in turn depends on the variance, σ_q^2, of the proposal density. A higher acceptance rate would typically follow from reducing σ_q^2, but with the risk that the posterior density will take longer to explore. If the acceptance rate is too high, then autocorrelation in sampled values will be excessive (since the chain tends to move in a restricted space), while a too low acceptance rate leads to the same problem, since the chain then gets locked at particular values. One possibility is to use a variance or dispersion estimate, σ_m^2 or Σ_m, from a maximum likelihood or other mode finding analysis (which approximates the posterior variance) and then scale this by a constant, $c > 1$, so that the proposal density variance is $\sigma_q^2 = c\sigma_m^2$. Values of c in the range 2–10 are typical. For θ_h of dimension d_h with covariance Σ_m, a proposal density dispersion $2.38^2\ \Sigma_m/d_h$ is shown as optimal in random walk schemes (Roberts et al., 1997). Working rules are for an acceptance rate of 0.4 when a parameter is updated singly (e.g., by separate univariate normal proposals), and 0.2 when a group of parameters are updated simultaneously as a block (e.g., by a multivariate normal proposal). Geyer and Thompson (1995) suggest acceptance rates should be between 0.2 and 0.4.

Typical Metropolis updating schemes use uniform, standard normal, or standard Student t variables W_t. A normal proposal density $q(\theta_{\text{cand}}|\theta^{(t)})$ involves samples $W_t \sim N(0,1)$, with candidate values,

$$\theta_{\text{cand}} = \theta^{(t)} + \sigma_q W_t,$$

where σ_q determines the size of the jump (and the acceptance rate). A uniform random walk samples $W_t \sim \text{Unif}(-1,1)$ and scales this to form a proposal, $\theta_{\text{cand}} = \theta^{(t)} + \kappa W_t$, with the value of κ determining the acceptance rate. As noted above, it is desirable that the proposal density approximately matches the shape of the target density, $p(\theta|y)$. The Langevin random walk scheme is an example of a scheme including information about the shape of $p(\theta|y)$ in the proposal, namely, $\theta_{\text{cand}} = \theta^{(t)} + \sigma_q[W_t + 0.5\nabla \log(p(\theta^{(t)}|y)]$, where ∇ denotes the gradient function (Roberts and Tweedie, 1996).

Sometimes candidate parameter values are sampled using a transformed version of a parameter, for example, normal sampling of a log variance rather than sampling of a variance (which has to be restricted to positive values). In this case an appropriate Jacobean adjustment must be included in the likelihood. Example 1.2 illustrates this.

1.7 Obtaining Full Conditional Densities

As noted above, Metropolis sampling may be based on the full conditional density when a particular parameter, θ_h, is being updated. These full conditionals are particularly central in Gibbs sampling (see below). The full conditional densities may be obtained from the joint density, $p(\theta, y) = p(y|\theta)p(\theta)$, and in many cases reduce to standard densities (normal, exponential, gamma, etc.) from which direct sampling is straightforward. Full conditional densities are derived by abstracting out from the joint model density, $p(y|\theta)p(\theta)$ (likelihood times prior), only those elements including θ_h and treating other components as constants (George et al., 1993; Gilks, 1996).

Consider a conjugate model for Poisson count data, y_i, with means, μ_i, that are themselves gamma distributed; this is a model appropriate for overdispersed count data with actual variability, $\text{var}(y)$, exceeding that under the Poisson model (Molenberghs et al., 2007). Suppose the second stage prior is $\mu_i \sim Ga(\alpha, \beta)$, namely,

$$p(\mu_i|\alpha,\beta) = \mu_i^{\alpha-1} e^{-\beta\mu_i}\beta^{\alpha}/\Gamma(\alpha),$$

and further that $\alpha \sim E(A)$ (namely α is exponential with parameter A), and $\beta \sim Ga(B, C)$, where A, B, and C are preset constants. So the posterior density, $p(\theta|y)$ of $\theta = (\mu_1, \ldots, \mu_n, \alpha, \beta)$, given y is proportional to,

$$e^{-A\alpha}\beta^{B-1}e^{-C\beta}\left\{\prod_i e^{-\mu_i}\mu_i^{y_i}\right\}[\beta^\alpha/\Gamma(\alpha)]^n\left\{\prod_i \mu_i^{\alpha-1}e^{-\beta\mu_i}\right\}, \tag{1.6}$$

where all constants (such as the denominator $y_i!$ in the Poisson likelihood, as well as the inverse marginal likelihood M) are combined in the proportionality constant.

It is apparent from inspecting Equation 1.6 that the full conditional densities of μ_i and β are also gamma, namely,

$$\mu_i \sim Ga(y_i + \alpha, \beta + 1),$$

and

$$\beta \sim Ga(B + n\alpha, C + \sum_i \mu_i),$$

respectively. The full conditional density of α, also obtained from inspecting Equation 1.6, is

$$p(\alpha|y, \beta, \mu) \propto e^{-A\alpha}[\beta^\alpha/\Gamma(\alpha)]^n\left\{\prod_i \mu_i^{\alpha-1}\right\}.$$

This density is nonstandard and cannot be sampled directly (as can the gamma densities for μ_i and β). Hence, a Metropolis or Metropolis–Hastings step can be used for updating it.

Example 1.1. Estimating Normal Parameters via Metropolis To illustrate Metropolis sampling in practice using symmetric proposal densities, consider $n = 1000$ values y_i generated randomly from a $N(3, 25)$ distribution, namely, a normal with mean $\mu = 3$ and variance $\sigma^2 = 25$. Note that the average sampled y_i is 3.13. Using the generated y, we seek to estimate the mean and variance, now treating them as unknowns. Setting $\theta = (\mu, \sigma^2)$, the likelihood is

$$p(y|\theta) = \prod_{i=1}^{n} \frac{1}{\sigma\sqrt{2\pi}} \exp\left(-\frac{(y_i - \mu)^2}{2\sigma^2}\right).$$

Assume a flat prior for μ, and a prior, $p(\sigma) \propto 1/\sigma$, on σ; this is a form of noninformative prior (see Albert, 2007, 109). Then the posterior density, $p(\theta|y) = Mp(y|\theta)p(\theta)$, is proportional to,

$$\frac{1}{\sigma^{n+1}}\prod_{i=1}^{n} \exp\left(-\frac{(y_i - \mu)^2}{2\sigma^2}\right),$$

with the marginal likelihood and other constants incorporated in the proportionality constant.

Parameter sampling via Metropolis involves σ rather than σ^2, and uniform proposals. Thus, assume uniform $U(-\kappa,\kappa)$ proposal densities around the current parameter values $\mu^{(t)}$ and $\sigma^{(t)}$, with $\kappa = 0.5$ for both parameters. The absolute value of $\sigma^{(t)} + U(-\kappa,\kappa)$ is used to generate σ_{cand}. Note that varying the lower and upper limit of the uniform sampling (e.g., taking $\kappa = 1$ or $\kappa = 0.25$) may considerably affect the acceptance rates.

An R code for $\kappa = 0.5$ is presented in the Appendix to this chapter[2], and uses the full posterior density (rather than the full conditional for each parameter) as the target density for assessing candidate values. In the acceptance step, the log of the ratio $p(y|\theta_{\text{cand}})p(\theta_{\text{cand}})/p(y|\theta^{(t)})p(\theta^{(t)})$ is compared to the log of a random uniform value to avoid computer over/underflow. With $T = 10{,}000$, acceptance rates for the proposals of μ and σ are 48 and 34%, respectively, with posterior means 3.13 and 4.99. Other posterior summary tools (e.g., kernel density plots, effective sample sizes) are included in the R code. Also included is a posterior probability calculation to assess $Pr(\mu < 3|y)$, with a result of 0.18, and a command for a plot of the changing posterior expectation for μ over the iterations. The code uses the full normal likelihood, via the dnorm function in R; a suggested exercise is to recode in terms of the log of un-normalized posterior density $(1/\sigma^{n+1})\prod_{i=1}^{n}\exp(-(y_i-\mu)^2/2\sigma^2)$.

Example 1.2. Extended Logistic with Metropolis Sampling Following Morgan (1988) and Carlin & Louis (2000), consider an extended logistic model for the Bliss mortality data, involving death rates, p_i, at dose, w_i. Thus, for deaths, y_i, at six dose points, one has

$$\begin{aligned} y_i &\sim Bin(n_i, p_i), \\ p_i &= h(w_i) = [\exp(z_i)/(1+\exp(z_i))]^m, \\ z_i &= (w_i - \mu)/\sigma, \end{aligned}$$

where m and σ are both positive. To simplify notation, write $V = \sigma^2$.

Consider Metropolis sampling involving log transforms of m and V, and separate univariate normal proposals in a Metropolis scheme. Jacobian adjustments are needed in the posterior density to account for the two transformed parameters. The full posterior, $p(\mu, m, V|y)$, is proportional to,

$$p(m)p(\mu)p(V)\prod_i [h(w_i)]^{y_i}[1-h(w_i)]^{n_i-y_i},$$

where $p(\mu)$, $p(m)$, and $p(V)$ are priors for μ, m, and V. Suppose the priors $p(m)$ and $p(\mu)$ are as follows:

$$\begin{aligned} m &\sim Ga(a_0, b_0), \\ \mu &\sim N(c_0, d_0^2), \end{aligned}$$

where the gamma has the form, $Ga(x|\alpha, \beta) = (\beta^{\alpha}/\Gamma(\alpha))x^{\alpha-1}e^{-\beta x}$. Also for $p(V)$ assume,

$$V \sim IG(e_0, f_0),$$

where the inverse gamma has the form, $IG(x|\alpha, \beta) = (\beta^{\alpha}/\Gamma(\alpha))x^{-(\alpha+1)}e^{-\beta/x}$. The parameters $(a_0, b_0, c_0, d_0, e_0, f_0)$ are preset. The posterior is then proportional to

$$\left(m^{a_0-1}e^{-b_0 m}\right) \exp\left(-0.5\left[\frac{\mu - c_0}{d_0}\right]^2\right) V^{-(e_0+1)}e^{-f_0/V} \times \prod_i [h(w_i)]^{y_i}[1 - h(w_i)]^{n_i - y_i}.$$

Suppose the likelihood is respecified in terms of parameters, $\theta_1 = \mu, \theta_2 = \log(m)$, and $\theta_3 = \log(V)$. Then the full posterior in terms of the transformed parameters is proportional to,

$$\left(\frac{\partial m}{\partial \theta_2}\right)\left(\frac{\partial V}{\partial \theta_3}\right) p(\mu)p(m)p(V) \prod_i [h(w_i)]^{y_i}[1 - h(w_i)]^{n_i - y_i}.$$

One has $(\partial m/\partial\theta_2) = e^{\theta_2} = m$ and $(\partial V/\partial\theta_3) = e^{\theta_3} = V$. So taking account of the parameterization $(\theta_1, \theta_2, \theta_3)$, the posterior density is proportional to,

$$\left(m^{a_0}e^{-b_0 m}\right) \exp\left(-0.5\left[\frac{\mu - c_0}{d_0}\right]^2\right) V^{-e_0}e^{-f_0/V} \prod_i [h(w_i)]^{y_i}[1 - h(w_i)]^{n_i - y_i}.$$

The R code[3] assumes initial values for $\mu = \theta_1$ of 1.8, for $\theta_2 = \log(m)$ of 0, and for $\theta_3 = \log(V)$ of 1. Assume preset parameters ($a_0 = 0.25, b_0 = 0.25, c_0 = 2, d_0 = 10, e_0 = 2.000004, f_0 = 0.001$), number of iterations, $T = 50{,}000$, and standard deviations in the respective normal proposal densities of 0.01, 0.1, and 0.1. Metropolis updates involve comparisons of the log (un-normalized) posterior and logs of uniform random variables $\{U_h^{(t)}, h = 1, \ldots, 3\}$. Posterior medians (and 95% intervals) for $\{\mu, m, V\}$ are obtained as 1.81 (1.79, 1.83), 0.36 (0.20, 0.73), 0.00033 (0.00017, 0.00072) with acceptance rates of 0.41, 0.65, and 0.81. A suggested exercise is to calibrate the standard deviations in the proposal densities so that the acceptance rates for $\log(m)$ and $\log(V)$ are reduced to around 0.4.

1.8 Metropolis–Hastings Sampling

The Metropolis–Hastings (M–H) algorithm is the overarching algorithm for MCMC schemes that simulate a Markov chain, $\theta^{(t)}$, with $p(\theta|y)$ as its stationary distribution. Following Hastings (1970), the chain is updated from $\theta^{(t)}$ to θ_{cand} with probability,

$$\alpha(\theta_{\text{cand}}|\theta^{(t)}) = \min\left(1, \frac{p(\theta_{\text{cand}}|y)q(\theta^{(t)}|\theta_{\text{cand}})}{p(\theta^{(t)}|y)q(\theta_{\text{cand}}|\theta^{(t)})}\right),$$

where the proposal density q (Chib and Greenberg, 1995) may be nonsymmetric, so that $q(\theta_{\text{cand}}|\theta^{(t)})$ does not equal $q(\theta^{(t)}|\theta_{\text{cand}})$. $q(\theta_{\text{cand}}|\theta^{(t)})$ is the probability (or density ordinate) of θ_{cand} for a density centered at $\theta^{(t)}$, while $q(\theta^{(t)}|\theta_{\text{cand}})$ is the probability of moving back from θ_{cand} to the current value. If the proposal density is symmetric, with $q(\theta_{\text{cand}}|\theta^{(t)}) = q(\theta^{(t)}|\theta_{\text{cand}})$, then the M–H algorithm reduces to the Metropolis algorithm discussed above. The M–H transition kernel is

$$K(\theta_{\text{cand}}|\theta^{(t)}) = \alpha(\theta_{\text{cand}}|\theta^{(t)})q(\theta_{\text{cand}}|\theta^{(t)}),$$

for $\theta_{\text{cand}} \neq \theta^{(t)}$, with a nonzero probability of staying in the current state, namely,

$$K(\theta^{(t)}|\theta^{(t)}) = 1 - \int \alpha(\theta_{\text{cand}}|\theta^{(t)})q(\theta_{\text{cand}}|\theta^{(t)})d\theta_{\text{cand}}.$$

Conformity of M–H sampling to the requirement that the Markov chain eventually samples from $\pi(\theta)$ is considered by Mengersen and Tweedie (1996) and Roberts and Rosenthal (2004).

If the proposed new value, θ_{cand}, is accepted, then $\theta^{(t+1)} = \theta_{\text{cand}}$, while if it is rejected the next state is the same as the current state, i.e., $\theta^{(t+1)} = \theta^{(t)}$. As mentioned above, since the target density $p(\theta|y)$ appears in ratio form, it is not necessary to know the normalizing constant, $M = 1/p(y)$.

If the proposal density has the form,

$$q(\theta_{\text{cand}}|\theta^{(t)}) = q(\theta^{(t)} - \theta_{\text{cand}}),$$

then a random walk Metropolis scheme is obtained (Albert, 2007, 105; Gelman et al., 2004). Another option is independence sampling, when the density $q(\theta_{\text{cand}})$ for sampling candidate values is independent of the current value $\theta^{(t)}$.

While it is possible for the target density to relate to the entire parameter set, it is typically computationally simpler in multiparameter problems to divide θ into C blocks or components, and use the full conditional densities in componentwise updating. Consider the update for the hth parameter or parameter block. At step h of iteration $t+1$, the preceding $h-1$ parameter blocks are already updated via the M–H algorithm, while $\theta_{h+1}, \ldots, \theta_C$ are still at their iteration t values (Chib and Greenberg, 1995). Let the vector of partially updated parameters apart from θ_h be denoted,

$$\theta_{[h]}^{(t)} = \left(\theta_1^{(t+1)}, \theta_2^{(t+1)}, \ldots, \theta_{h-1}^{(t+1)}, \theta_{h+1}^{(t)}, \ldots, \theta_C^{(t)}\right).$$

The candidate value for θ_h is generated from the hth proposal density, denoted $q_h(\theta_{h,\text{cand}}|\theta_h^{(t)})$. Also governing the acceptance of a proposal are full conditional densities, $\pi_h(\theta_h^{(t)}|\theta_{[h]}^{(t)}) \propto p(y|\theta_h^{(t)})p(\theta_h^{(t)})$, specifying the density of θ_h conditional on known values of other parameters, $\theta_{[h]}$. The candidate value $\theta_{h,\text{cand}}$ is then accepted with probability,

$$\alpha = \min\left(1, \frac{p(y|\theta_{h,\text{cand}})p(\theta_{\text{cand}})q(\theta_h^{(t)}|\theta_{\text{cand}})}{p(y|\theta_h^{(t)})p(\theta_h^{(t)})q(\theta_{\text{cand}}|\theta_h^{(t)})}\right). \tag{1.7}$$

Example 1.3. Extended Logistic with Metropolis–Hastings Sampling (continued) We again consider an extended logistic model for the Bliss mortality data. However, the analysis now involves a gamma M–H proposal scheme for the positive parameters m and σ^2, so that there is no need for a Jacobian adjustments in the posterior density. Specifically, the proposal,

$$q(y|x) = Ga(\kappa, \kappa/x) = \frac{\kappa^\kappa}{x^\kappa\Gamma(\kappa)} y^{\kappa-1}e^{-\kappa y/x},$$

is used. Note that κ has a role as a precision parameter that can be tuned to provide improved acceptance rates: larger values of κ mean that the sampled candidate values,

$$x_{\text{cand}} \sim Ga(\kappa, \kappa/x_{\text{curr}}),$$

will be more closely clustered about the current parameter value, x_{curr}. Here κ is set to 10 for sampling candidate values of m and σ^2. The last 49,000 iterations of an MCMC chain[4] of $T = 50{,}000$ iterations give acceptance rates for μ, m, and σ^2 of 0.41, 0.30, and 0.50. The posterior medians (and 95% credible intervals) for μ, m, and σ are, respectively, 1.81 (1.79, 1.83), 0.36 (0.20, 0.70), and 0.0186 (0.013, 0.027). A suggested exercise is to adopt different κ values for m and σ^2 to achieve acceptance rates closer to 0.4.

Example 1.4. Normal Random Effects in a Hierarchical Binary Regression To exemplify a HB model involving a three stage prior, consider binary data, $y_i \sim Bern(p_i)$, from Sinharay and Stern (2005) on the survival or otherwise of $n = 244$ newborn turtles arranged in $J = 31$ clutches, numbered in increasing order of the average birthweight of the turtles. A known predictor is turtle birthweight, x_i. Let C_i denote the clutch that turtle i belongs to. Then to allow for varying clutch effects, one may specify, for cluster $j = C_i$, a probit regression with

$$p_i|b_j = \Phi(\beta_1 + \beta_2 x_i + b_j),$$

where $\{b_j \backsim N(0, 1/\tau_b), j = 1, \ldots, J\}$. It is assumed that $\beta_k \sim N(0, 10)$ and $\tau_b \sim Ga(1, 0.001)$.

An M–H step involving a gamma proposal is used for τ_b, and Metropolis updates for other parameters.[5] Trial runs suggest τ_b is approximately between

5 and 10, and a gamma proposal, $Gam(\kappa, \kappa/\tau_{b,\mathrm{curr}})$, with $\kappa = 100$ is adopted. A run of $T = 5000$, iterations with $B = 500$ provides posterior medians (95% intervals) for $\{\beta_1, \beta_2, \sigma_b = 1/\sqrt{\tau_b}\}$ of −2.66 (−3.80, −1.27), 0.36 (0.15, 0.54), and 0.24 (0.15, 0.46), and acceptance rates for $\{\beta_1, \beta_2, \tau_b\}$ of 0.31, 0.44, and 0.25. Acceptance rates for the clutch random effects (using normal proposals with standard deviation 0.5) are between 0.42 and 0.51. However, none of the clutch effects appear to be significant, in the sense of entirely positive or negative 95% credible intervals (see Table 1.1). The first effect b_1 (for the clutch with lowest average birth-weight) has median and 95% interval, 0.35 (−0.07, 0.97), and is the closest to being significant, while for b_J the median is −0.28 (−0.83, 0.14).

TABLE 1.1
Random effects posterior summary.

Clutch	2.5%	Median	97.5%
1	−0.07	0.35	0.97
2	−0.69	−0.12	0.43
3	−0.52	−0.01	0.54
4	−0.68	−0.15	0.32
5	−0.55	−0.06	0.43
6	−0.43	0.07	0.59
7	−0.60	−0.10	0.40
8	−0.58	−0.05	0.38
9	−0.21	0.25	0.82
10	−0.50	−0.02	0.42
11	−0.21	0.22	0.78
12	−0.36	0.14	0.80
13	−0.56	−0.01	0.55
14	−0.66	−0.06	0.46
15	−0.33	0.16	0.83
16	−0.68	−0.05	0.50
17	−0.47	−0.04	0.36
18	−0.60	−0.07	0.39
19	−0.57	−0.10	0.32
20	−0.39	0.09	0.74
21	−0.39	0.08	0.59
22	−0.45	0.04	0.54
23	−0.58	−0.12	0.35
24	−0.69	−0.07	0.47
25	−0.71	−0.15	0.28
26	−0.52	−0.04	0.50
27	−0.47	0.05	0.57
28	−0.58	−0.07	0.42
29	−0.30	0.16	0.72
30	−0.58	−0.09	0.32
31	−0.83	−0.28	0.14

1.9 Gibbs Sampling

The Gibbs sampler (Casella and George, 1992; Gelfand and Smith, 1990; Gilks et al., 1993) is a special componentwise M–H algorithm, whereby the proposal density, q, for updating θ_h equals the full conditional $\pi_h(\theta_h|\theta_{[h]}) \propto p(y|\theta_h)p(\theta_h)$. It follows from Equation 1.7 that proposals are accepted with probability 1. If it is possible to update all blocks this way, then the Gibbs sampler involves parameter block by parameter block updating, which when completed forms the transition from $\theta^{(t)} = (\theta_1^{(t)}, \ldots, \theta_C^{(t)})$ to $\theta^{(t+1)} = (\theta_1^{(t+1)}, \ldots, \theta_C^{(t+1)})$. The most common sequence used is

1. $\theta_1^{(t+1)} \sim f_1\left(\theta_1|\theta_2^{(t)}, \theta_3^{(t)}, \ldots, \theta_C^{(t)}\right)$;
2. $\theta_2^{(t+1)} \sim f_2\left(\theta_2|\theta_1^{(t+1)}, \theta_3^{(t)}, \ldots, \theta_C^{(t)}\right)$;
3. $\theta_C^{(t+1)} \sim f_C\left(\theta_C|\theta_1^{(t+1)}, \theta_2^{(t+1)}, \ldots, \theta_{C-1}^{(t+1)}\right)$.

While this scanning scheme is the usual one for Gibbs sampling, there are other options, such as the random permutation scan (Roberts and Sahu, 1997) and the reversible Gibbs sampler, which updates blocks 1 to C and then updates in reverse order.

Example 1.5. Gibbs Sampling Example: Schools Data Meta Analysis Consider the schools data from Gelman et al. (2004, 138), consisting of point estimates y_j ($j = 1, \ldots, J$) of unknown effects θ_j, where each y_j has a known design variance σ_j^2 (though the listed data provides σ_j not σ_j^2). The first stage of a hierarchical normal model assumes,

$$y_j \sim N(\theta_j, \sigma_j^2),$$

and the second stage specifies a normal model for the latent θ_j,

$$\theta_j \sim N(\mu, \tau^2).$$

The full conditionals for the latent effects, θ_j, namely, $p(\theta_j|y, \mu, \tau^2)$ are as specified by Gelman et al. (2004, 135). Assuming a flat prior on μ, and that the precision $1/\tau^2$ has a $Ga(a, b)$ gamma prior, the full conditional for μ is $N(\bar{\theta}, \tau^2/J)$, and that for $1/\tau^2$ is gamma with parameters $(J/2 + a, 0.5\sum_j(\theta_j - \mu)^2 + b)$.

For the R application, the setting $a = b = 0.1$ is used in the prior for $1/\tau^2$. Starting values for μ and τ^2 in the MCMC analysis are provided by the mean of the y_j and the median of the σ_j^2. A single run in R of T = 20,000 samples[6] provides the posterior means and standard deviations shown in Table 1.2.

TABLE 1.2
Schools normal meta-analysis posterior summary.

	μ	τ	θ_1	θ_2	θ_3	θ_4	θ_5	θ_6	θ_7	θ_8
Mean	8.0	2.5	9.0	8.0	7.6	8.0	7.1	7.5	8.8	8.1
Standard deviation	4.4	2.8	5.6	4.9	5.4	5.1	5.0	5.2	5.2	5.4

1.10 Assessing Efficiency and Convergence: Ways of Improving Convergence

It is necessary to decide how many iterations to use to accurately represent the posterior density and to ensure that the sampling process has converged. Nonvanishing autocorrelations at high lags mean that less information about the posterior distribution is provided by each iterate and a higher sample size is necessary to cover the parameter space. Autocorrelation will be reduced by "thinning," namely, retaining only samples that are $S > 1$ steps apart $\{\theta_h^{(t)}, \theta_h^{(t+S)}, \theta_h^{(t+2S)}, \ldots, \}$ that more closely approximate independent samples; however, this results in a loss of precision. The autocorrelation present in MCMC samples may depend on the form of parameterization, the complexity of the model, and the form of sampling (e.g., block or univariate sampling for collections of random effects). Autocorrelation will reduce the effective sample size, $T_{\text{eff},h}$, for parameter samples $\{\theta_h^{(t)}, t = B+1, \ldots, B+T\}$ below T. The effective number of samples (Kass et al., 1998) may be estimated[7] as

$$T_{\text{eff},h} = T \bigg/ \left[1 + 2\sum_{k=0}^{\infty} \rho_{hk}\right],$$

where,

$$\rho_{hk} = \gamma_{hk}/\gamma_{h0},$$

is the kth lag autocorrelation, γ_{h0} is the posterior variance $V(\theta_h|y)$, and γ_{hk} is the kth lag autocovariance, $\text{cov}[\theta_h^{(t)}, \theta_h^{(t+k)}|y]$. In practice, one may estimate $T_{\text{eff},h}$ by dividing T by $1+2\sum_{k=0}^{K^*} \rho_{hk}$, where K^* is the first lag value for which $\rho_{hk} < 0.1$ or $\rho_{hk} < 0.05$ (Browne et al., 2009).

Also useful for assessing efficiency is the Monte Carlo standard error, which is an estimate of the standard deviation of the difference between the true posterior mean, $E(\theta_h|y) = \int \theta_h p(\theta_h|y)d\theta_h$, and the simulation-based estimate,

$$\bar{\theta}_h = \frac{1}{T}\sum_{t=B+1}^{T+B} \theta_h^{(t)}.$$

A simple estimator of the Monte Carlo variance is

$$\frac{1}{T}\left[\frac{1}{T-1}\sum_{t=1}^{T}(\theta_h^{(t)}-\bar{\theta}_h)^2\right],$$

though this may be distorted by extreme sampled values; an alternative batch means method is described by Roberts (1996). The ratio of the posterior variance in a parameter to its Monte Carlo variance is a measure of the efficiency of the Markov chain sampling (Roberts, 1996), and it is sometimes suggested that the Monte Carlo standard error[8] should be less than 5% of the posterior standard deviation of a parameter (Toft et al., 2007).

The effective sample size is mentioned above, while Raftery and Lewis (1992, 1996) estimate the iterations required to estimate posterior summary statistics to a given accuracy. Suppose the following posterior probability,

$$Pr[\Delta(\theta|y) < b] = p,$$

is required. Raftery and Lewis seek estimates of the burn-in iterations, B, to be discarded, and the required further iterations, T_{req}, in order to estimate p to within r with probability s; typical quantities might be $p = 0.025$, $r = 0.005$, and $s = 0.95$. The selected values of $\{p, r, s\}$ can also be used to derive an estimate of the required minimum iterations, T_{min}, if autocorrelation were absent, with the ratio,

$$I = T_{\text{req}}/T_{\text{min}},$$

providing a measure of additional sampling required due to autocorrelation. Use of the R gibbsit function (available from STATLIB) is illustrated for obtaining these quantities in the Appendix to this chapter.

As to the second issue mentioned above, there is no guarantee that sampling from an MCMC algorithm will converge to the posterior distribution, despite obtaining a high number of iterations. Convergence can be informally assessed by examining the time series or trace plots of parameters. Ideally, the MCMC sampling is exploring the posterior distribution quickly enough to produce good estimates (this property is often called "good mixing"). Some techniques for assessing convergence (as against estimates of required sample sizes) consider samples $\theta^{(t)}$ from only a single long chain, possibly after excluding initial $t = 1, \ldots, B$ burn-in iterations. These include the spectral density diagnostic of Geweke (1992), the CUSUM method of Yu and Mykland (1998), and a quantitative measure of the "hairiness" of the CUSUM plot (Brooks and Roberts, 1998).

Slow convergence (usually combined with poor mixing and high autocorrelation in sampled values) will show in trace plots that wander, and that exhibit short-term trends, rather than fluctuating rapidly around a stable mean. Failure to converge is typically a feature of only some model parameters; for example, fixed regression effects in a general linear mixed model may show convergence, but not the parameters relating to the random components. Often measures of overall fit (e.g., model deviance) converge, while component parameters do not.

Problems of convergence in MCMC sampling may reflect problems in model identifiability, either formal nonidentification as in multiple random effects models, or poor empirical identifiability when an overly complex model is applied to a small sample ("overfitting"). Choice of diffuse priors tends to increase the chance that models are poorly identified, especially in complex hierarchical models for small data samples (Gelfand and Sahu, 1999). Elicitation of more informative priors and/or application of parameter constraints may assist identification and convergence.

Alternatively, a parameter expansion strategy may also improve MCMC performance (Browne et al., 2009; Gelman et al., 2008; Ghosh, 2008). For example, in a normal meta-analysis model (Chapter 3) with

$$y_j \sim N(\mu + \theta_j, \sigma_y^2); \quad \theta_j \sim N(0, \sigma_\theta^2), \qquad j = 1, \ldots, J,$$

conventional sampling approaches may become trapped near $\sigma_\theta = 0$, whereas improved convergence and effective sample sizes may be achieved by introducing a redundant scale parameter, $\lambda \sim N(0, V_\lambda)$,

$$y_j \sim N(\mu + \lambda\xi_j, \sigma_y^2); \quad \xi_j \sim N(0, \sigma_\xi^2).$$

The expanded model priors induce priors on the original model parameters, namely,

$$\theta_j = \lambda\xi_j; \quad \sigma_\theta = |\lambda|\sigma_\xi.$$

The setting for V_λ is important: too much diffuseness may lead to effective impropriety.

Another source of poor convergence is suboptimal parameterization or data form; for example, convergence is improved by centering independent variables in regression applications (Roberts and Sahu, 2001; Zuur et al., 2002). Similarly, delayed convergence in random effects models may be lessened by sum to zero or corner constraints (Clayton, 1996; Vines et al., 1996), or by a centered hierarchical prior (Gelfand et al., 1995, 1996), in which the prior on each stochastic variable is centered at a higher level stochastic mean—see the next section. However, the most effective parameterization may also depend on the balance in the data between different sources of variation. In fact, noncentered parameterizations, with latent data independent from hyperparameters, may be preferable in terms of MCMC convergence in some settings (Papaspiliopoulos et al., 2003).

1.10.1 Hierarchical model parameterization to improve convergence

While priors for unstructured random effects may include a nominal mean of zero, in practice a posterior mean of zero for such a set of effects may not be achieved during MCMC sampling. For example, the mean of the random effects can be confounded with the intercept. One may apply a corner constraint by setting a particular random effect (say the first) to a known value, usually zero (Scollnik, 2002). An empirical sum to zero constraint may be achieved by centering the sampled random effects, say,

$$u_i \sim N(0, \sigma_u^2), \quad i = 1, \ldots, n,$$

at each iteration (sometimes known as "centering on the fly"), so that,

$$u_i^* = u_i - \bar{u},$$

and inserting u_i^* rather than u_i in the model defining the likelihood. Another option (Scollink, 2002; Vines et al., 1996) is to define an auxiliary effect, $u_i^a \sim N(0, \sigma_u^2)$, and obtain u_i, following the same prior $N(0, \sigma_u^2)$, but now with a guaranteed mean of zero, by the transformation,[9]

$$u_i = \sqrt{\frac{n}{n-1}}(u_i^a - \bar{u}^a).$$

To illustrate a centered hierarchical prior (Gelfand et al., 1995; Browne et al., 2009), consider two-way nested data, with $j = 1, \ldots, J$ repetitions over subjects $i = 1, \ldots, n$,

$$y_{ij} = \mu + \alpha_i + u_{ij},$$

with $\alpha_i \backsim N(0, \sigma_\alpha^2)$ and $u_{ij} \backsim N(0, \sigma_u^2)$. The centered version defines $\kappa_i = \mu + \alpha_i$ with $y_{ij} = \kappa_i + u_{ij}$, so that,

$$y_{ij} \backsim N(\kappa_i, \sigma_u^2),$$
$$\kappa_i \backsim N(\mu, \sigma_\alpha^2).$$

For three-way nested data, the standard model form is

$$y_{ijk} = \mu + \alpha_i + \beta_{ij} + u_{ijk},$$

with $\alpha_i \backsim N(0, \sigma_\alpha^2)$ and $\beta_{ij} \backsim N(0, \sigma_\beta^2)$. The hierarchically centered version defines $\zeta_{ij} = \mu + \alpha_i + \beta_{ij}$ and $\kappa_i = \mu + \alpha_i$, so that,

$$y_{ijk} \backsim N(\zeta_{ij}, \sigma_u^2),$$
$$\zeta_{ij} \backsim N(\kappa_i, \sigma_\beta^2),$$

and

$$\kappa_i \backsim N(\mu, \sigma_\alpha^2).$$

Roberts and Sahu (1997, 309) set out the contrasting sets of full conditional densities under the standard and centered representations, and compare Gibbs sampling scanning schemes.

Papaspiliopoulos et al. (2003) compare MCMC convergence for centered, noncentered, and partially noncentered hierarchical model parameterizations according to the amount of information the data contain about the latent effects, $\kappa_i = \mu + \alpha_i$. Thus for two-way nested data, the (fully) noncentered parameterization (NCP), involves new random effects, $\widetilde{\kappa}_i$, with,

$$
\begin{aligned}
y_{ij} &= \widetilde{\kappa}_i + \mu + \sigma_u e_{ij}, \\
\widetilde{\kappa}_i &= \sigma_\alpha z_i,
\end{aligned}
$$

where e_{ij} and z_i are standard normal variables. In this form, the latent data, $\widetilde{\kappa}_i$, and hyperparameter, μ, are independent *a priori*, and so the NCP may give better convergence when the latent effects, κ_i, are not well identified by the observed data, y. A partially noncentered form is obtained using a number, $w \in [0, 1]$, and

$$
\begin{aligned}
y_{ij} &= \widetilde{\kappa}_i^w + w\mu + u_{ij}, \\
\widetilde{\kappa}_i^w &= (1 - w)\mu + \sigma_\alpha z_i,
\end{aligned}
$$

or equivalently,

$$
\widetilde{\kappa}_i^w = (1 - w)\kappa_i + w\widetilde{\kappa}_i.
$$

Thus, $w = 0$ gives the centered representation and $w = 1$ gives the noncentered parameterization. The optimal w for convergence depends on the ratio σ_u/σ_α. The centered representation performs best when σ_u/σ_α tends to zero, while the noncentered representation is optimal when σ_u/σ_α is large.

1.10.2 Multiple chain methods

Many practitioners prefer to use two or more parallel chains with diverse starting values to ensure full coverage of the sample space of the parameters (Gelman and Rubin, 1996; Toft et al., 2007). Diverse starting values may be based on default values for parameters (e.g., precisions set at different default values such as 1, 5, 10 and regression coefficients set at zero), or on the extreme quantiles of posterior densities from exploratory model runs. Online monitoring of sampled parameter values, $\{\theta_k^{(t)}, t = 1, \ldots, T\}$, from multiple chains, $k = 1, \ldots, K$, assists in diagnosing lack of model identifiability. Examples might be models with multiple random effects, or when the mean of the random effects is not specified within the prior, as under difference priors over time or space that are considered in Chapter 4 (Besag et al., 1995). Another example is factor and structural equation models where the loadings are not specified so as to anchor the factor scores in a consistent direction, so that the "name" of the common factor may switch during MCMC updating

(Congdon, 2003). Single runs may still be adequate for straightforward problems, and single chain convergence diagnostics (Geweke, 1992) may be applied in this case. Single runs are often useful for exploring the posterior density and as a preliminary to obtain inputs to multiple chains.

Convergence for multiple chains may be assessed using Gelman–Rubin scale reduction factors that measure the convergence of the between-chain variance in $\theta_k^{(t)} = \left(\theta_{1k}^{(t)}, \ldots, \theta_{dk}^{(t)}\right)$ to the variance over all chains. These factors converge to 1 if all chains are sampling identical distributions, whereas for poorly identified models, variability of sampled parameter values between chains will considerably exceed the variability within any one chain. To apply these criteria, one typically allows a burn in of B samples while the sampling moves away from the initial values to the region of the posterior. For iterations, $t = B+1, \ldots, T+B$, a pooled estimate of the posterior variance, $\sigma^2_{\theta_h|y}$, of θ_h is

$$\sigma_{\theta_h|y} = V_h/T + TW_h/(T-1),$$

where variability within chains W_h is defined as,

$$W_h = \frac{1}{(T-1)K} \sum_{k=1}^{K} \sum_{t=B+1}^{B+T} \left(\theta_{hk}^{(t)} - \bar{\theta}_{hk}\right)^2,$$

with $\bar{\theta}_{hk}$ being the posterior mean of θ_h in samples from the kth chain, and where,

$$V_h = \frac{T}{K-1} \sum_{k=1}^{K} (\bar{\theta}_{hk} - \bar{\theta}_{h\cdot})^2,$$

denotes between-chain variability in θ_h, with $\bar{\theta}_{h\cdot}$ denoting the pooled average of the $\bar{\theta}_{hk}$. The potential scale reduction factor (PSRF) compares $\sigma^2_{\theta_h|y}$ with the within-sample estimate, W_h. Specifically, the scale factor is $R_h = (\sigma^2_{\theta_h|y}/W_h)^{0.5}$ with values under 1.2 indicating convergence. A multivariate version of the PSRF for vector θ mentioned by Brooks and Gelman (1998) and Brooks and Roberts (1998) involves between- and within-chain covariances V_θ and W_θ, and pooled posterior covariance $\Sigma_{\theta|y}$. The scale factor is defined by,

$$R_\theta = \max_b \frac{b'\Sigma_{\theta|y} b}{b' W_\theta b} = \frac{T-1}{T} + \left(1 + \frac{1}{K}\right) \lambda_1,$$

where λ_1 is the maximum eigenvalue of $W_\theta^{-1} V_\theta / T$.

An alternative multiple chain convergence criterion, also proposed by Brooks and Gelman (1998), avoids reliance on the implicit normality assumptions in the Gelman–Rubin scale reduction factors based on analysis of variance over chains. Normality approximation may be improved by parameter transformation (e.g., log or logit), but problems may still be encountered when posterior densities are skewed or possibly multimodal (Toft et al., 2007).

The alternative criterion uses a ratio of parameter interval lengths: for each chain the length of the $100(1-\alpha)\%$ interval for a parameter is obtained, namely, the gap between 0.5α and $(1-0.5\alpha)$ points from T simulated values. This provides K within-chain interval lengths, with mean L_U. From the pooled output of TK samples, an analogous interval, L_P, is also obtained. The ratio, L_P/L_U, should converge to 1 if there is convergent mixing over the K chains. The corresponding $\hat{R}$ statistics are available both in WinBUGS and as part of the R2WinBUGS output (see note 1 in the Appendix).

1.11 Choice of Prior Density

Choice of an appropriate prior density, and preferably a sensitivity analysis over alternative priors, is fundamental in the Bayesian approach; for example, see Daniels (1999), Gelman (2006), and Gustafson et al. (2006) on priors for random effect variances. Before the advent of MCMC methods, conjugate priors were often used in order to reduce the burden of numeric integration. Now nonconjugate priors (e.g., finite range uniform priors on standard deviation parameters) are widely used. There may be questions of sensitivity of posterior inference to the choice of prior, especially for smaller datasets, or for certain forms of models; examples are the priors used for variance components in random effects models, the priors used for collections of correlated effects, for example, in hierarchical spatial models (Bernardinelli et al., 1995), priors in nonlinear models (Millar, 2004), and priors in discrete mixture models (Green and Richardson, 1997).

In many situations, existing knowledge may be difficult to summarize or elicit in the form of an "informative prior" and to express prior ignorance one may use "default" or "noninformative" priors. This is typically less problematic—in terms of posterior sensitivity—for fixed effects, such as regression coefficients (when taken to be homogenous over cases) than for variance parameters. Since the classical maximum likelihood estimate is obtained without considering priors on the parameters, a possible heuristic is that a noninformative prior leads to a Bayesian posterior estimate close to the maximum likelihood estimate. It might appear that a maximum likelihood analysis would therefore necessarily be approximated by flat or improper priors, but such priors may actually be unexpectedly informative about different parameter values (Zhu and Lu, 2004).

A flat or uniform prior distribution on θ, expressible as $p(\theta) = 1$, is often adopted on fixed regression effects, but is not invariant under reparameterization. For example, it is not true for $\phi = 1/\theta$ that $p(\phi) = 1$, as the prior for a function $\phi = g(\theta)$, namely,

$$p(\phi) = \left| \frac{d}{d\phi} g^{-1}(\phi) \right|,$$

demonstrates. By contrast, on invariance grounds, Jeffreys (1961) recommended the prior, $p(\sigma) = (1/\sigma)$, for a standard deviation, as for $\phi = g(\sigma) = \sigma^2$, one obtains $p(\phi) = (1/\phi)$. More general analytic rules for deriving noninformative priors include reference prior schemes (Berger and Bernardo, 1992), and Jeffreys prior,

$$p(\theta) \propto |I(\theta)|^{0.5},$$

where $I(\theta)$ is the information matrix, namely, $I(\theta) = -E\left(\partial^2 l(\theta)/\partial l(\theta_g)\partial l(\theta_h)\right)$, and $l(\theta) = \log(L(\theta|y))$ is the log-likelihood. Unlike uniform priors, a Jeffreys prior is invariant under transformation of scale since $I(\theta) = I(g(\theta))(g'(\theta))^2$ and $p(\theta) \propto I(g(\theta))^{0.5}g'(\theta) = p(g(\theta))g'(\theta)$ (Kass and Wasserman, 1996, 1345).

1.11.1 Including evidence

Especially for establishing the intercept (e.g., the average level of a disease), or regression effects (e.g., the impact of risk factors on disease), or variability in such impacts, it may be possible to base the prior density on cumulative evidence via meta-analysis of existing studies, or via elicitation techniques aimed at developing informative priors. This is well established in engineering risk and reliability assessment, where systematic elicitation approaches, such as maximum-entropy priors, are used (Hodge et al., 2001; Siu and Kelly, 1998). Thus, known constraints for a variable identify a class of possible distributions, and the distribution with the greatest Shannon–Weaver entropy is selected as the prior. Examples are $\theta \backsim N(m, V)$ if estimates m and V of the mean and variance are available, or an exponential with parameter $-q/\log(1-p)$ if a positive variable has an estimated pth quantile of q.

Simple approximate elicitation methods include the histogram technique, which divides the domain of an unknown θ into a set of bins and elicits prior probabilities that θ is located in each bin; then $p(\theta)$ may be represented as a discrete prior or converted to a smooth density. Prior elicitation may be aided if a prior is reparameterized in the form of a mean and prior sample size; for example, beta priors, $Be(a, b)$, for probabilities can be expressed as $Be(m\tau, (1-m)\tau)$, where $m = a/(a+b)$ and $\tau = a+b$ are elicited estimates of the mean probability and prior sample size. This principle is extended in data augmentation priors (Greenland and Christensen, 2001), while Greenland (2007) uses the device of a prior data stratum (equivalent to data augmentation) to represent the effect of binary risk factors in logistic regressions in epidemiology.

If a set of existing studies is available, providing evidence on the likely density of a parameter, these may be used in a form of preliminary meta-analysis to set up an informative prior for the current study. However, there may be limits to the applicability of existing studies to the current data, and so pooled information from previous studies may be down-weighted. For example, the precision of the pooled estimate from previous studies may be scaled downwards, with the scaling factor possibly an extra unknown. When a maximum likelihood (ML) analysis is simple to apply, one option is to adopt the ML mean as a prior mean, but with the ML precision matrix downweighted (Birkes and Dodge, 1993).

More comprehensive ways of downweighting historical/prior evidence have been proposed, such as power prior models (Chen et al., 2000; Ibrahim and Chen, 2000). Let $0 \leq \delta \leq 1$ be a scale parameter with beta prior that weights the likelihood of historical data, y_h, relative to the likelihood of the current study data, y. Following Chen et al. (2000, 124), a power prior has the form,

$$p(\theta, \delta|y_h) \propto [p(y_h|\theta)]^{\delta}[\delta^{a_\delta - 1}(1-\delta)^{b_\delta - 1}]p(\theta),$$

where $p(y_h|\theta)$ is the likelihood for the historical data, and (a_δ, b_δ) are prespecified beta density hyperparameters. The joint posterior density for (θ, δ) is then,

$$p(\theta, \delta|y, y_h) \propto p(y|\theta)[p(y_h|\theta)]^{\delta}[\delta^{a_\delta - 1}(1-\delta)^{b_\delta - 1}]p(\theta).$$

Chen and Ibrahim (2006) demonstrate connections between the power prior and conventional priors for hierarchical models.

1.11.2 Assessing posterior sensitivity: Robust priors

To assess sensitivity to prior assumptions, the analysis may be repeated over a limited range of alternative priors. Thus, Sargent (1997) and Fahrmeir and Knorr-Held (1997) suggest a gamma prior on inverse precisions, $1/\tau^2$, governing random walk effects (e.g., baseline hazard rates in survival analysis), namely, $1/\tau^2 \sim Ga(a, b)$, where a is set at 1, but b is varied over choices such as 0.05 or 0.0005. One possible strategy involves a consideration of both optimistic and conservative priors, with regard, say, to a treatment effect or the presence of significant random effect variation (Gustafson et al., 2006; Spiegelhalter, 2004).

Another relevant principle in multiple effect models is that of uniform shrinkage governing the proportion of total random variation to be assigned to each source of variation (Daniels, 1999; Natarajan and Kass, 2000). So for a two level normal linear model with,

$$y_{ij} = x_{ij}\beta + \eta_j + e_{ij},$$

with $e_{ij} \sim N(0, \sigma^2)$ and $\eta_j \sim N(0, \tau^2)$, one prior (e.g., inverse gamma) might relate to the residual variance, σ^2, and a second conditional $U(0, 1)$ prior relates to the ratio $(\tau^2/\tau^2 + \sigma^2)$ of cluster to total variance. A similar effect is achieved in structural time series models (Harvey, 1989) by considering different forms of signal to noise ratios in state–space models, including several forms of random effect (e.g., changing levels and slopes, as well as season effects). Gustafson et al. (2006) propose a conservative prior for the one level linear mixed model,

$$y_i \sim N(\eta_i, \sigma^2),$$
$$\eta_i \sim N(\mu, \tau^2),$$

namely, a conditional prior, $p(\tau^2|\sigma^2)$, aiming to prevent overestimation of τ^2. Thus in full,

$$p(\sigma^2, \tau^2) = p(\sigma^2)p(\tau^2|\sigma^2),$$

where $\sigma^2 \sim IG(e, e)$ for some small $e > 0$, and

$$p(\tau^2|\sigma^2) = \frac{a}{\sigma^2}[1 + \tau^2/\sigma^2]^{-(a+1)}.$$

The case, $a = 1$, corresponds to the uniform shrinkage prior of Daniels (1999), where,

$$p(\tau^2|\sigma^2) = \frac{\sigma^2}{[\sigma^2 + \tau^2]^2},$$

while larger values of a (e.g., $a = 5$) are found to be relatively conservative.

For covariance matrices, Σ, between random effects of dimension k, the emphasis in recent research has been on more flexible priors than afforded by the inverse Wishart (or Wishart priors for precision matrices). Barnard et al. (2000) and Liechty et al. (2004) consider a separation strategy whereby,

$$\Sigma = \text{diag}(S).R.\text{diag}(S),$$

where S is a $k \times 1$ vector of standard deviations, and R is a $k \times k$ correlation matrix. With the prior sequence, $p(R, S) = p(R|S)p(S)$, Barnard et al. suggest $\log(S) \sim N_k(\xi, \Lambda)$ where Λ is usually diagonal. For the elements r_{ij} of R, constrained beta sampling on $[-1, 1]$ can be used subject to positive definitiveness constraints on Σ. Daniels and Kass (1999) consider the transformation, $\eta_{ij} = 0.5 \log(1 - r_{ij}/1 + r_{ij})$, and suggest an exchangeable hierarchical shrinkage prior, $\eta_{ij} \sim N(0, \tau^2)$, where,

$$p(\tau^2) \propto (c + \tau^2)^{-2};$$
$$c = 1/(k - 3).$$

This is an example of an exchangeable hierarchical prior considered more fully in Chapter 3. Daniels and Kass (2001) consider other options for hierarchical modeling of the correlations in Σ. While a full covariance prior (e.g., assuming random slopes on all k predictors in a multilevel model) can be applied from the outset, MacNab et al. (2004) propose an incremental model strategy, starting with random intercepts and slopes but without covariation between them, in order to assess for which predictors there is significant slope variation. The next step applies a full covariance model only for the predictors showing significant slope variation.

Formal approaches to prior robustness may be based on "contamination" priors. For instance, one might assume a two group mixture with larger probability $1 - r$ on the "main" prior, $p_1(\theta)$, and a smaller probability, such as $r = 0.1$, on a contaminating density, $p_2(\theta)$, which may be any density (Gustafson, 1996). More generally, a sensitivity analysis may involve some form of mixture of priors, for example a discrete mixture over a few alternatives, a fully nonparametric approach (see Chapter 3), or a Dirichlet weight mixture over a small range of alternatives (e.g., Jullion and Lambert, 2007). For an example of such an analysis for a random effect variance, see Example 1.11.6.

A mixture prior can include the option that the parameter is not present (e.g., that a variance or regression effect is zero). A mixture prior methodology of this kind for regression effects is presented by George and McCulloch (1993). Increasingly also random effects models are selective, including a default allowing for random effects to be unnecessary (Albert and Chib, 1997; Cai and Dunson, 2006; Fruhwirth-Schnatter and Tuchler, 2008).

In hierarchical models the prior specifies both the form of the random effects (fully exchangeable over units or spatially/temporally structured), the density of the random effects (normal, mixture of normals, etc.), and the third stage hyperparameters. The form of the second stage prior $p(b|\theta_b)$ amounts to a hypothesis about the nature and form of the random effects. Thus, a hierarchical model for small area mortality may include spatially structured random effects, exchangeable random effects with no spatial pattern, or both, as under the convolution prior of Besag et al. (1991); it also may assume normality in the different random effects, as against heavier-tailed alternatives. A prior specifying the errors as spatially correlated and normal is likely to be a working model assumption, rather than a true cumulation of knowledge, and one may have several models for $p(b|\theta_b)$ being compared (Disease Mapping Collaborative Group, 2000), with sensitivity not just being assessed on the hyperparameters.

Random effect models often start with a normal hyperdensity, and so posterior inferences may be sensitive to outliers or multiple modes, as well as to the prior used on the hyperparameters. Indications of lack of fit (e.g., low conditional predictive ordinates for particular cases) may suggest robustification of the random effects prior. Robust hierarchical models are adapted to pooling inferences and/or smoothing in data subject to outliers or other irregularities; for example, Jonsen et al. (2006) consider robust space-time state–space models with Student t rather than normal errors in an analysis of travel rates of migrating leatherback turtles. Other forms of robust analysis involve discrete mixtures of random effects (e.g., Lenk and Desarbo, 2000), possibly under Dirichlet or Polya process models (e.g., Kleinman and Ibrahim, 1998). Robustification of hierarchical models reduces the chance of incorrect inferences on individual effects, important when random effects approaches are used to identify excess risk or poor outcomes (Conlon and Louis, 1999; Marshall et al., 2004).

1.11.3 Problems in prior selection in hierarchical Bayes models

For the third stage parameters (the hyperparameters) in hierarchical models, choice of a diffuse noninformative prior may be problematic as improper priors may induce improper posteriors that prevent MCMC convergence, since conditions necessary for convergence (e.g., positive recurrence) may be violated (Berger et al., 2005). This may apply even if conditional densities are proper, and Gibbs or other MCMC sampling proceeds apparently straightforwardly. A simple example is provided by the normal two level model with subjects, $i = 1, \ldots, n$, nested in clusters, $j = 1, \ldots, J$,

$$y_{ij} = \mu + \theta_j + u_{ij},$$

where $\theta_j \backsim N(0, \tau^2)$ and $u_{ij} \backsim N(0, \sigma^2)$. Hobert and Casella (1996) show that the posterior distribution is improper under the prior $p(\mu, \tau, \sigma) = (1/\sigma^2\tau^2)$, even though the full conditionals have standard forms, namely,

$$p\left(\theta_j|y, \mu, \sigma^2, \tau^2\right) = N\left(\frac{n(\bar{y}_j - \mu)}{n + \frac{\sigma^2}{\tau^2}}, \frac{1}{\frac{n}{\sigma^2} + \frac{1}{\tau^2}}\right),$$

$$p\left(\mu|y, \sigma^2, \tau^2, \theta\right) = N\left(\bar{y} - \bar{\theta}, \frac{\sigma^2}{nJ}\right),$$

$$p\left(1/\tau^2|y, \mu, \sigma^2, \theta\right) = Ga\left(\frac{J}{2}, 0.5\sum_j \theta_j^2\right),$$

$$p\left(1/\sigma^2|y, \mu, \tau^2, \theta\right) = Ga\left(\frac{nJ}{2}, 0.5\sum_{ij}(y_{ij} - \mu - \theta_j)^2\right),$$

so that Gibbs sampling could in principle proceed.

Whether posterior propriety holds may also depend on the level of information in the data, whether additional constraints are applied to parameters in MCMC updating, and the nature of the improper prior used. For example, Rodrigues and Assuncao (2008) demonstrate propriety in the posterior of spatially varying regression parameter models under a class of improper priors. More generally, Markov random field (MRF) priors, such as random walks in time or spatial conditional autoregressive priors (Chapter 4), may have joint forms that are improper, with a singular covariance matrix—see for example, the discussion by Sun et al. (2000, 28–30). The joint prior only identifies differences between pairs of effects and unless additional constraints are applied to the random effects, this may cause issues with posterior propriety.

It is possible to define proper priors in these cases by introducing autoregression parameters (Sun et al., 1999), but Besag et al. (1995, 11) mention that "the sole impropriety in such [MRF] priors is that of an arbitrary level and is removed from the corresponding posterior distribution by the presence of any informative data." The indeterminacy in the level is usually resolved by applying "centering on the fly" (at each MCMC iteration) within each set of random effects and under such a linear constraint, MRF priors become proper (Rodrigues and Assunção, 2008, 2409). Alternatively, "corner" constraints on particular effects, namely, setting them to fixed values (usually zero), may be applied (Clayton, 1996; Koop, 2003, 248), while Chib and Jeliazkov (2006) suggest an approach to obtaining propriety in random walk priors.

Datta and Smith (2003) consider propriety in multilevel normal linear models,

$$y_{ij} = x_{ij}\beta + z_{ij}b_j + u_{ij},$$

with repetitions $i = 1, \ldots, n_j$ in clusters $j = 1, \ldots, J$ and demonstrate propriety in terms of the ranks of the predictor matrices Z and X as compared to J and $n = \sum n_j$. For example, consider the Fay–Herriott small income model with,

$$y_i = x_i\beta + b_i + u_i,$$

where $b_i \backsim N(0, \psi_b)$, but variances, $\text{var}(u_i) = D_i$, are known (as in meta-regressions discussed in Chapter 3). Then, in the case $D_i = D$ and under the Jeffreys prior,

$$p(\beta, \psi_b) \propto \frac{1}{D + \psi_b},$$

the posterior is proper if $J \geq \text{rank}(X) + 3$. Similarly, Hadjicostas and Berry (1999) consider a conjugate hierarchical model with $y_j \backsim Po(o_j\lambda_j)$ and $\lambda_j \backsim Ga(\alpha, \beta)$ along with the prior,

$$p(\alpha, \beta) \propto \alpha^{k_1}(\alpha + s_1)^{k_2}\beta^{k_3}(\beta + s_2)^{k_4},$$

where $\{s_1, s_2\}$ are positive leading to a joint posterior,

$$p(\lambda_1, \ldots, \lambda_J, \alpha, \beta|y)$$
$$\propto \frac{1}{[\Gamma(\alpha)\beta^{\alpha'}]^J} \prod_{j=1}^{J} \left[\lambda_j^{\alpha+y_j-1} e^{-\lambda_j(o_j+1/\beta)}\right] \alpha^{k_1}(\alpha + s_1)^{k_2}\beta^{k_3}(\beta + s_2)^{k_4}.$$

The prior is improper for certain k values, but proper posteriors can still be obtained provided certain conditions are satisfied by the data.

Priors that are just proper mathematically (e.g., gamma priors on $1/\tau^2$ with small scale and shape parameters) are often used on the grounds of expediency, and justified as letting the data speak for themselves. However, such priors may cause identifiability problems as the posteriors are close to being empirically improper. This impedes MCMC convergence (Gelfand and Sahu, 1999; Kass and Wasserman, 1996, 1361). Furthermore, just proper priors on variance parameters in fact favour particular values, despite being supposedly only weakly informative. Gelman (2006a) suggests possible (less problematic) options including a finite range uniform prior on the standard deviation (rather than variance), and a positive truncated t density.

Example 1.6. Seeds Data: Mixture Prior for Random Effects Variance To illustrate a simple discrete mixture to the prior for a random effects variance, consider the seeds data from Crowder (1978), with model,

$$\begin{aligned} y_i &\sim \text{Binomial}(S_i, p_i), \\ \text{logit}(p_i) &= \beta_1 + \beta_2 x_{1i} + \beta_3 x_{2i} + \beta_4 x_{1i} x_{2i} + u_i, \\ u_i &\sim N(0, \sigma_u^2). \end{aligned}$$

The program in the WinBUGS14 examples (Volume I) considers two mutually exclusive alternatives for the random effects variance, σ_u^2, a uniform $U(0,\ 1000)$ prior on σ_u, and a gamma $Ga(0.001, 0.001)$ prior on the precision $\tau = 1/\sigma_u^2$.

However, instead of a single form of prior for σ_u^2, one may instead average over two or more plausible alternatives, either plausible mathematically, or in terms of the level of evidence they include (their informativeness using, say, previous studies). For example, one may use a discrete mixture[10] with equal prior probabilities for,

$$\text{M1: } \sigma_{1u} \backsim U(0, 100), \eta_1 = 1/\sigma_{1u}^2,$$

and

$$\text{M2: } \eta_2 = 1/\sigma_{2u}^2 \backsim Ga(0.1, 0.1).$$

The latter option is slightly less diffuse than the $1/\sigma_u^2 \backsim Ga(0.001, 0.001)$ option used in the WinBUGS14 examples. Thus, with $\pi = (\pi_1, \pi_2) = (0.5, 0.5)$, the realized precision is

$$\begin{aligned} \tau_u &= \eta_\delta, \\ \delta &\sim \text{Mult}(1, \pi), \\ \sigma_{1u} &\sim U(0, 100); \quad \eta_1 = 1/\sigma_{1u}^2, \\ \eta_2 &\sim Ga(0.1, 0.1). \end{aligned}$$

A two chain run of this model ($B = 1000$, $T = 10{,}000$) provides a posterior median (with 95% interval) for σ_υ of 0.37 (0.21, 0.67), higher than the median of 0.27 obtained under $1/\sigma_u^2 \backsim Ga(0.001, 0.001)$. The uniform option is chosen in under 2% of iterations.

The Dirichlet weight approach of Jullion and Lambert (2007) is here applied to average the precision over two priors. Again, the first is $\sigma_{1u} \backsim U(0, 100)$, with precision $\eta_1 = 1/\sigma_{1u}^2$, and the second is $\eta_2 \backsim Ga(0.1, 0.1)$. Thus, with τ_u denoting the overall weighted precision obtained by averaging over the two options, one has,

$$\begin{aligned} \tau_u | w &= w_1 \eta_1 + w_2 \eta_2, \\ w &\sim \text{Dirichlet}(\phi), \end{aligned}$$

where (ϕ_1, ϕ_2) are prior weights, each set to 0.5. This approach is less likely than the one just considered to be heavily weighted to one or other option.

This approach produces a posterior median for σ_u of 0.39 (0.20, 0.71) with posterior weights (w_1, w_2) of 0.27 and 0.73 on the two alternative priors.

It is also possible to include an option $\sigma_u^2 = 0$ in the mixture prior, via the model,

$$
\begin{aligned}
y_i &\sim \text{Binomial}(S_i, p_i), \\
\text{logit}(p_i) &= \beta_1 + \beta_2 x_{1i} + \beta_3 x_{2i} + \beta_4 x_{1i} x_{2i} + \kappa \sigma_m u_i^*, \\
u_i^* &\sim N(0, 1),
\end{aligned}
$$

where $\kappa \sim Bern(\pi_\kappa)$ is a binary inclusion indicator, and π_k can be taken as known or assigned a beta prior. Here we take $\pi_\kappa \sim Be(1,1)$. Additionally, when $\kappa = 1$, a discrete mixture prior, as in the first model, is adopted with equal prior probabilities for the two nonzero options, $\sigma_{1m} \backsim U(0, 100)$, $\eta_1 = 1/\sigma_{1m}^2$ and $\eta_2 = 1/\sigma_{2m}^2 \backsim Ga(0.1, 0.1)$. In effect, there is averaging over three models, with the realized standard deviation, $\sigma_u = \kappa\sigma_m$, averaging over iterations when $\kappa = 0$ as well as when $\kappa = 1$. A two chain run of 100,000 iterations (with 10,000 burn in) shows that, in fact, the posterior probability that $\kappa = 1$ is below 0.5, namely, 0.32. The posterior mean for the realized random effect standard deviation, namely, $\kappa\sigma_m$, is 0.13, with the density now showing a spike at zero.

Appendix: Computational Notes

1. The interface to WinBUGS from R involves specifying data, inits, parameters, and model, as well as number of chains, total iterations, and burn-in. One option is to save the WinBUGS code and data (with extensions .bug and .dat, respectively) to the R working directory. Another option is to specify the WinBUGS code in R using the cat command (see below for an example). For guidance, one may refer to http://www.stat.columbia.edu/~gelman/bugsR/runningbugs.html.

To illustrate, consider data relating male life expectancy y_i (in 2003–2005) to a health and disability score, x_i, for 353 English local authorities. The expectancy data are from http://www.nchod.nhs.uk/ and the health scores are from http://www.communities.gov.uk/communities/neighbourhoodrenewal/deprivation/deprivation07. Consider the quadratic model, $y_i \sim N(\mu_i, 1/\tau)$, $\mu_i = b_1 + b_2 x_i + b_3 x_i^2$, with $N(0,\ 1000)$ priors for b_j and $\tau \sim Ga(1, 0.01)$. One may specify initial values by generating them randomly or by presetting them.

Using the cat function, and with three chains, a possible command sequence is

```
library(R2WinBUGS)
expecs <- read.table ("c:\\Rdata\\Ch1_expecs.txt", header=TRUE)
y <- expecs$y; x <- expecs$x
data <- list("x","y")
```

```
cat("model { for (i in 1:353) {y[i] ~dnorm(mu[i],tau)
mu[i] <- b[1]+b[2]*x[i]+b[3]*x[i]*x[i]}
for (j in 1:3) {b[j] ~dnorm(0,0.001)}
tau ~dgamma(1,0.01)}", file="expecs.bug")
# initialise unknowns randomly
# inits <- function(){list(b = rnorm(3, 0, 100), tau=rgamma(1,1)) }
# initialise unknowns via preset values
inits1 <- list(b=c(70,0,0),tau=1); inits2 <- list(b=c(80,0,0),tau=2)
inits3 <- list(b=c(90,0,0),tau=3); inits <- list(inits1, inits2,inits3)
# name parameters
parameters=c("b","tau"); T <- 10000; B <- 1000
# interface to bugs
expecs.sim <- bugs(data,inits, parameters,
model="expecs.bug",n.chains=3,n.iter=T,n.burnin=B,n.thin=1)
# posterior summary
expecs.sim
# plots, etc
attach.all(expecs.sim$sims.list)
summary(b[,1]); summary(b[,2]); summary(b[,3])
# graphical summary of convergence and posterior density
plot(expecs.sim)
```

2. In Example 1.1, the data are generated ($n = 1000$ values) and then the underlying parameters are estimated as follows:

```
# generate data
y = rnorm(1000,3,5);
# initial vector setting and parameter values
mu <- sig <- numeric(T); mu[1] <- 3; sig[1] <- 5
T <- 10000; B <- T/10; u.mu <- runif(T); u.sig <- runif(T)
REJmu <- 0; REJsig <- 0
# log posterior density (up to a constant)
logpost = function(mu,sig){
loglike = sum(dnorm(y,mu,sig,log=TRUE))
return(loglike - log(sig))}
# MCMC sampling loop
for (t in 2:T) { print(t)
    mut <- mu[t-1]; sigt <- sig[t-1]
# uniform proposals with kappa=0.5
     mucand <- mut + runif(1,-0.5,0.5)
     sigcand <- abs(sigt + runif(1,-0.5,0.5))
alph.mu = logpost(mucand,sigt)-logpost(mut,sigt)
if (log(u.mu[t]) <= alph.mu) mu[t] <- mucand
else { mu[t] <- mut; REJmu <- REJmu+1 }
alph.sig = logpost(mu[t],sigcand)-logpost(mu[t],sigt)
if (log(u.sig[t]) <= alph.sig) sig[t] <- sigcand
```

```
else { sig[t] <- sigt; REJsig <- REJsig+1 }}
# MCMC Rejection rates
REJratemu <- REJmu/T; REJratesig <- REJsig/T;
# posterior plots and summaries
hist(mu,50) # mu posterior marginal
plot(mu) # mu sequence of sampled values
plot(mu,sig) # covariance plot of sampled values
cat("mubar = ",mean(mu),"\n")
cat("sdbar = ",mean(sig),"\n")
cat("Rejection Rate mu = ",REJratemu,"\n")
cat("Rejection Rate sigma = ",REJratesig,"\n")
# Kernel density plots
plot(density(mu),main="Density plot for mu posterior")
plot(density(sig),main="Density plot for sigma posterior")
# Monte Carlo Standard Errors
source("http://www.stat.psu.edu/~mharan/batchmeans.R")
bm1 <- bm(mu); bm2 <- bm(sig)
# Effective sample sizes and ACF plots
library(R2WinBUGS); effectiveSize(mu); effectiveSize(sig)
acf(mu,main="acf plot, mu"); acf(sig,main="acf plot, sig")
# posterior probability on hypothesis μ < 3
sum(mu < 3)/T
# evolution plot for posterior mean estimate over iterations
estvssamp(mu)
# Raftery-Lewis required iterations
source("C:\\Documents and Settings\\peter congdon\\gibbsit.R")
gibbsit(cbind(mu,sig),q=0.025,r=0.005,s=0.95)
```

Taking $T = 10,000$ leads to acceptance rates of 48% and 34% on μ and σ, respectively.

3. An R code for Metropolis sampling of the extended logistic model is

```
f = function(mu,th2,th3) {V <- exp(th3); m1 <- exp(th2); sig <- sqrt(V)
x <- (w-mu)/sig; xt <- exp(x)/(1+exp(x)); h <- xt^m1;
loglike <- y*log(h)+(n-y)*log(1-h)
logpriorm1 <- a0*th2-m1*b0
logpriorV <- -e0*th3-f0/V
logpriormu <- -0.5*((mu-c0)/d0)^2-log(d0)
logprior <- logpriormu+logpriorV+logpriorm1
f <- sum(loglike)+logprior}
# Read in data and set initial values and vectors
w = c(1.6907, 1.7242, 1.7552, 1.7842, 1.8113, 1.8369, 1.8610, 1.8839)
n = c(59, 60, 62, 56, 63, 59, 62, 60); y = c(6, 13, 18, 28, 52, 53, 61, 60)
T = 50000; mu <- numeric(T); th3 <- numeric(T); th2 <- numeric(T);
V<- numeric(T); m1<- numeric(T); samp<- matrix(0, nrow=T, ncol=3);
pm <- numeric(3)
```

```
MCMCsamp <- matrix(0, nrow=T, ncol=3)
# initial parameter values
mu[1] <- 1.8; th2[1] <- 0; th3[1] <- 1;
k1 = 0; k2 = 0; k3 = 0; u1 <- runif(T); u2 <- runif(T); u3 <- runif(T)
a0=0.25; b0=0.25; c0=2; d0=10; e0=2.004; f0=0.001
# metropolis proposal standard devn's
sd1 <- 0.01; sd2 <- 0.1; sd3 <- 0.1
# main MCMC loop
for (i in 2:T) {mucand <- mu[i-1]+sd1*rnorm(1,0,1)
tcand <- f(mucand,th2[i-1],th3[i-1])
tcurr <- f(mu[i-1],th2[i-1],th3[i-1])
if (log(u1[i]) <= tcand-tcurr) mu[i] <- mucand else
{mu[i] <- mu[i-1]; k1 <- k1+1 }
th2cand <- th2[i-1]+sd2*rnorm(1,0,1)
tcand <- f(mu[i],th2cand,th3[i-1])
tcurr <- f(mu[i],th2[i-1],th3[i-1])
if (log(u2[i]) <= tcand-tcurr) th2[i] <- th2cand else
{th2[i] <- th2[i-1]; k2 <- k2+1 }
m1[i] <- exp(th2[i])
th3cand <- th3[i-1]+sd3*rnorm(1,0,1)
tcand <- f(mu[i],th2[i],th3cand)
tcurr <- f(mu[i],th2[i],th3[i-1])
if (log(u3[i]) <= tcand-tcurr) th3[i] <- th3cand else
{th3[i] <- th3[i-1]; k3 <- k3+1}
V[i] <- exp(th3[i])
samp[i-1,1] <- mu[i]; samp[i-1,2] <- m1[i]; samp[i-1,3] <- V[i]}
# output for posterior data analysis
write.table(samp, file="MCMCsamp.txt")
# posterior summary (iterations 1000 to T=50,000)
quantile(mu[1000:T], probs=c(.025,0.5,0.975))
quantile(m1[1000:T], probs=c(.025,0.5,0.975))
quantile(V[1000:T], probs=c(.025,0.5,0.975))
# acceptance rates
1-k1/T; 1-k2/T; 1-k3/T
```

4. An R code for M–H estimation of the extended logistic for the Bliss mortality analysis is

```
f = function(mu,m1,sig2) { x <- (w-mu)/sqrt(sig2); xt <- exp(x)/(1+exp(x))
h <- xt^m1; loglike <- y*log(h)+(n-y)*log(1-h)
logpriorm1 <- (a0-1)*log(m1)-m1/b0
logpriorsig2 <- -(e0+1)*log(sig2)-1/(f0*sig2)
logpriormu <- -0.5*((mu-c0)/d0)^2-log(d0)
logprior <- logpriormu+logpriorsig2+logpriorm1
f <- sum(loglike)+logprior}
w = c(1.6907, 1.7242, 1.7552, 1.7842, 1.8113, 1.8369, 1.8610, 1.8839)
```

```
n = c(59, 60, 62, 56, 63, 59, 62, 60); y = c(6, 13, 18, 28, 52, 53, 61, 60)
k = 8; T = 50000
mu <- numeric(T); sig2 <- numeric(T); m1 <- numeric(T);
sig <- numeric(T); pm <- numeric(3)
mu[1] <- 1.8; m1[1] <- 0.5; sig2[1] <- 0.001
k1 = 0; k2 = 0; k3 = 0; u1 <- runif(T); u2 <- runif(T); u3 <- runif(T)
a0=0.25; b0=4; c0=2; d0=10; e0=2.004; f0=1000
kapm1=10; kapsig2=10
# Main MCMC loop
for (t in 2:T) { mu.n <- mu[t-1]+0.01*rnorm(1,0,1)
tn <- f(mu.n,m1[t-1],sig2[t-1])
tc <- f(mu[t-1],m1[t-1],sig2[t-1])
if (log(u1[t]) <= tn-tc) mu[t] <- mu.n else
{mu[t] <- mu[t-1]; k1 <- k1+1 }
m1.n <- rgamma(1,kapm1,kapm1/m1[t-1])
fn <- f(mu[t],m1.n,sig2[t-1])
fc <- f(mu[t],m1[t-1],sig2[t-1])
tn <- fn+log(dgamma(m1[t-1],kapm1,kapm1/m1.n))
tc <- fc+log(dgamma(m1.n,kapm1,kapm1/m1[t-1]))
if (log(u2[t]) <= tn-tc) m1[t] <- m1.n else
{m1[t] <- m1[t-1]; k2 <- k2+1 }
sig2.n <- rgamma(1,kapsig2, kapsig2/sig2[t-1]);
fn <- f(mu[t],m1[t],sig2.n)
fc <- f(mu[t],m1[t],sig2[t-1])
tn <- fn+log(dgamma(sig2[t-1], kapsig2,kapsig2/sig2.n))
tc <- fc+log(dgamma(sig2.n, kapsig2,kapsig2/sig2[t-1]))
if (log(u3[t]) <= tn-tc) sig2[t] <- sig2.n else
{sig2[t] <- sig2[t-1]; k3 <- k3+1 }
sig[t] <- sqrt(sig2[t]) }
# posterior quantiles
quantile(mu[1000:T], probs=c(.025,0.5,0.975))
quantile(m1[1000:T], probs=c(.025,0.5,0.975))
quantile(sig[1000:T], probs=c(.025,0.5,0.975))
# acceptance rates
1-k1/T; 1-k2/T; 1-k3/T
```

5. An R code for the turtle survival data is

```
# un-normalized posterior density
f = function(b,a,tau,e) { sig <- 1/sqrt(tau)
for (i in 1:N){ p[i] <- pnorm(a+b*x[i]+e[C[i]])
LL[i] <- y[i]*log(p[i])+(1-y[i])*log(1-p[i])}
logpr[1] <- -0.5*a^2/10
logpr[2] <- -0.5*b^2/10
logpr[3] <- -0.001*tau
for (j in 1:J){ LLr[j] <- -0.5*e[j]^2/sig^2-log(sig)}
```

```
f <- sum(LL[1:N])+sum(LLr[1:J])+sum(logpr[1:3])}
# settings
T = 5000; B =T/10; N <- 244; J <- 31; k1 = 0; k2 = 0; k3 =0; kap=100
U1 <- log(runif(T)); U2 <- log(runif(T)); U3 <- log(runif(T))
a <- numeric(T); b <- numeric(T); tau <- numeric(T); logpr <- numeric(3)
s <- numeric(T); p <- numeric(N); e <- numeric(J); LL <- numeric(N);
LLr <- numeric(J); ec <- matrix(0,T,J); en <- matrix(0,T,J); kran <-
numeric(J)
# initial values
b[1] <- 0.35; a[1] <- -2.6; tau[1] <- 5; for (j in 1:J) {ec[1,j] <- 0; kran[j] <- 0}
# Main loop
for (t in 2:T) { print (t)
bn <- b[t-1]+0.05*rnorm(1,0,1)
tn <- f(bn,a[t-1],tau[t-1],ec[t-1,]); tf <- f(b[t-1],a[t-1],tau[t-1],ec[t-1,])
if (U1[t] <= tn-tf) b[t] <- bn
else {b[t] <- b[t-1]; k1 <- k1+1 }
# update intercept
an <- a[t-1]+0.2*rnorm(1,0,1)
tn <- f(b[t],an,tau[t-1],ec[t-1,]); tf <- f(b[t],a[t-1],tau[t-1] ,ec[t-1,])
if (U2[t] <= tn-tf) a[t] <- an
else {a[t] <- a[t-1]; k2 <- k2+1}
# update precision
taun <- rgamma(1,kap,kap/tau[t-1])
s[t-1] <- 1/sqrt(tau[t-1])
tn <- f(b[t],a[t],taun,ec[t-1,])+log(dgamma(tau[t-1],kap,kap/taun))
tc <- f(b[t],a[t],tau[t-1],ec[t-1,])+log(dgamma(taun,kap,kap/tau[t-1]))
if (U3[t] <= tn-tf) tau[t] <- taun
else {tau[t] <- tau[t-1]; k3 <- k3+1}
# update cluster effects
for (j in 1:J) { en[j] <- ec[t-1,j]; ec[t,j] <- ec[t-1,j]}
for (j in 1:J) { en[j] <- ec[t-1,j]+0.5*rnorm(1,0,1)
tn <- f(b[t],a[t],tau[t],en[]); tf <- f(b[t],a[t],tau[t],ec[t,])
if (log(runif(1)) <= tn-tf) ec[t,j] <- en[j]
else { en[j] <- ec[t-1,j]
kran[j] <- kran[j]+1}}}
# summaries
quantile(a[B:T], probs=c(.025,0.5,0.975))
quantile(b[B:T], probs=c(.025,0.5,0.975))
quantile(tau[B:T], probs=c(.025,0.5,0.975))
quantile(s[B:T], probs=c(.025,0.5,0.975))
for (j in 1:J) {print(quantile(ec[B:T,j], probs=c(.025,0.5,0.975)))}
# acceptance rates
1-k1/T; 1-k2/T; 1-k3/T
for (j in 1:J) {print(1-kran[j]/T)}
```

6. There are $J + 2$ unknowns in the R code (NB the σ_j^2 are not unknowns) for implementing these Gibbs updates. There are $T = 20{,}000$ MCMC samples to be accumulated in the array MCMCsamp. With $a = b = 0.1$ in the prior for $1/\tau^2$, and remembering that the normal density in R uses the standard deviation, one then has

```
y=c(28,8,-3,7,-1,1,18,12); sigma=c(15,10,16,11,9,11,10,18)
J <- 8; T <- 20000; sigma2 <- sigma^2
MCMCsamp <- matrix(0, nrow=T, ncol=J+2)
# starting values
mu=mean(y); tau2=median(sigma2)
# main sampling loop
for (t in 1:T) {th.mean=(y/sigma2+mu/tau2)/(1/sigma2+1/tau2)
th.sd=sqrt(1/(1/sigma2+1/tau2))
theta=rnorm(J,th.mean,th.sd)
mu=rnorm(1,mean(theta),sqrt(tau2/J))
invtau2=rgamma(1,J/2+0.1,sum((theta-mu)^2)/2+0.1)
tau2 <- 1/invtau2; tau <- sqrt(tau2)
MCMCsamp[t,3:(2+J)] = theta;
MCMCsamp[t,1] =mu; MCMCsamp[t,2] =tau}
write.table(MCMCsamp, file="MCMCsamp.txt")
```

7. The effective sample size can be obtained in R by loading the R2WinBUGS library (Sturtz et al., 2005), and then using the effectiveSize(theta.h) command, where theta.h is the name for the sampled values of a particular parameter.

8. In the R program, one may use a program developed by Jones et al. (2006) for estimating the MC standard error, which can be downloaded via the command

```
source("http://www.stat.psu.edu/~mharan/batchmeans.R")
```

and then simply specifying

```
bmse <- bm(theta.h).
```

9. For example, consider a random effects logistic model (one allowing for overdispersion) for seeds, y_i, germinating among a total, S_i, on plate, i, as in Crowder (1978). Denote the probability of germination by p_i, $y_i \sim$ Binomial (S_i, p_i), with,

$$\text{logit}(p_i) = \beta_1 + \beta_2 x_{1i} + \beta_3 x_{2i} + \beta_4 x_{1i} x_{2i} + u_i,$$
$$u_i \sim N(0, \sigma_u^2),$$

where x_{1i} and x_{2i} are, respectively, the seed type and root extract of plate i. Adopting priors $\sigma_u \sim U(0, 10)$, and $\{\beta_h \sim N(0, 1000), h = 1, 4\}$, the auxiliary effect method may be coded in WinBUGS as

```
model { for( i in 1 : n ) {y[i] ~dbin(p[i],S[i])
        u.a[i] ~dnorm(0,inv.sig2.u); u[i] <- sqrt(n/(n-1))*(u.a[i]-mean(u.a[]))
        logit(p[i]) <- beta[1]+beta[2]*x1[i]+beta[3]*x2[i]+beta[4]*x1[i]*x2[i]+u[i]}
```

```
for (h in 1:4) {beta[h]~dnorm(0,1.0E-3)}
sig.u~dunif(0,10);inv.sig2.u <-1/(sig.u*sig.u)}
```

10. For the discrete mixture over two options for the variance, the relevant WinBUGS code is

```
model {for( i in 1 : n ) {r[i] ~dbin(p[i],S[i]); u[i] ~dnorm(0,tau)
logit(p[i]) <- beta[1]+beta[2]*x1[i]+beta[3]*x2[i]+beta[4]*x1[i]*x2[i]+u[i]}
for (j in 1:4) {beta[j]~dnorm(0,1.0E-3)}
# mixture over priors and selected precision and s.d. of random effects
tau <- taumx[del]; sig <- 1/sqrt(tau); del ~dcat(pi[1:2])
# Prior 1: uniform on SD
    sig1 ~dunif(0,100); taumx[1] <- 1/(sig1*sig1)
#Prior 2: gamma on precision
    taumx[2] ~dgamma(0.1, 0.1)}
```

For the Dirchlet weight mixture, a sequence of gamma priors is used to reproduce the Dirichlet. Thus

```
model { for( i in 1 : n) {r[i] ~dbin(p[i],S[i]); u[i] ~dnorm(0,tau)
logit(p[i]) <- beta[1]+beta[2]*x1[i]+beta[3]*x2[i]+beta[4]*x1[i]*x2[i]+u[i]}
for (j in 1:4) {beta[j]~dnorm(0,1.0E-3)}
# precision and s.d. of random effects
    tau <- w[1]*taumx[1]+w[2]*taumx[2];      sig <- 1/sqrt(tau)
    for (k in 1:2) {w[k] <- W[k]/sum(W[]); W[k] ~dgamma(phi[k],1)}
# Prior 1: uniform on SD
    sig1 ~dunif(0,100); taumx[1] <- 1/(sig1*sig1)
#Prior 2: gamma on precision
    taumx[2] ~dgamma(0.1, 0.1)}
```

When the option of a zero variance is added, the code is

```
model {for( i in 1 : n) {r[i] ~dbin(p[i],S[i])
ustar[i] ~dnorm(0,1)
logit(p[i]) <- beta[1]+beta[2]*x1[i]+beta[3]*x2[i]+beta[4]*x1[i]*x2[i]
+kap*sigm*ustar[i]}
for (j in 1:4) {beta[j]~dnorm(0,1.0E-3)}
kap ~dbern(pi.kap); pi.kap ~dbeta(1,1)
sig <- kap*sigm
del ~dcat(pi.m[1:2]); taum <- taumx[del]; sigm <- 1/sqrt(taum)
# Prior 1: uniform on SD
    sig1 ~dunif(0,100); taumx[1] <- 1/(sig1*sig1)
#Prior 2: gamma on precision
    taumx[2] ~dgamma(0.1, 0.1)}
```

2

Model Fit, Comparison, and Checking

2.1 Introduction

Model assessment involves both choice between competing models in terms of best fit, and checks to ensure model adequacy. For example, even if one model has superior fit, it still needs to be established whether predictions from the model check with, namely, reproduce satisfactorily, the observed data. Checking may also seek to establish whether model assumptions (e.g., normality of random effects) are justified, and whether particular observations are poorly fit (Berkhof et al., 2000; Sinharay and Stern, 2003).

Once adequacy is established for a set of candidate models, one may seek to choose a particular best fitting model to base inferences on, or average over two or more adequate models with closely competing fit. This chapter focuses on three main strategies to assessing model fit and carrying out model checks, the formal approach, approaches based on posterior analysis of the deviance, and predictive methods based on samples of replicate data. Particular emphasis is placed on their application in hierarchical models.

What is termed the formal approach to Bayes model selection is based on integration over the model parameter space to estimate marginal likelihoods and posterior model probabilities, leading to possible model averaging. The canonical situation is provided by a "model closed" or M-closed scenario (Bernardo and Smith, 1994; Key et al., 1999) where the set of models under consideration are judged to include the correct model. Then formal model choice strategies are directed toward finding which model is most likely given the data.

Let prior model probabilities be denoted $p(m = k)$, where $m \in (1, \ldots, K)$ is a model indicator. Then posterior model probabilities are obtained as,

$$p(m = k|y) = \frac{p(y|m = k)p(k)}{p(y)},$$

where,

$$p(y|m = k) = \int p(y|\theta_k)p(\theta_k)d\theta_k,$$

is the marginal likelihood for model k, with parameter θ_k of dimension d_k. Section 2.2 considers approximations to the marginal likelihood and to Bayes factors that compare such likelihoods. In simple models, such as normal linear regressions with regression coefficients and the residual variance the only unknowns, the formal approach is relatively simple to implement, and marginal likelihoods are available analytically under certain priors (Bos, 2002).

Approximate methods (Tierney and Kadane, 1986) for obtaining summary fit measures (e.g., marginal likelihoods) or posterior densities of parameters are also reliable in simple models. A large sample approximation for the log marginal likelihood is provided by the Bayesian Information Criterion (Schwarz, 1978) defined as,

$$\text{BIC} = \log[p(y|\hat{\theta}_k)] - 0.5d_k \log(n),$$

where $\hat{\theta}_k$ is the maximum likelihood estimator.

Posterior model probabilities on nested models are also obtainable by adding model selection indicators, as illustrated by Bayesian variable selection algorithms (Fernandez et al., 2001; Mitchell and Beauchamp, 1988) for choosing predictors in regression. Such selection has recently been extended to variance hyperparameters in hierarchical models (e.g., Cai and Dunson, 2006; Chen and Dunson, 2003; Fruhwirth-Schnatter and Tuchler, 2008; Kinney and Dunson, 2008) and avoids the sometimes complex issues involved in estimating marginal likelihoods of different models. Section 2.4 considers variance selection in hierarchical models.

However, in more complex random effect applications with discrete responses or hierarchically structured data, there remain issues that impede straightforward application of the formal approach (Han and Carlin, 2001). For example, in approximating marginal likelihoods, there is a choice whether or not to integrate over random effects (Sinharay and Stern, 2005). The more commonly advocated approach of integrating out random effects becomes impractical when there are multiple possibly correlated random effects. The formal approach is also sensitive to priors adopted on parameters, which in the case of random effect models include the form of prior on variance components (e.g., inverse gamma or uniform) as well as the degree of prior informativeness. As priors become more diffuse, the formal approach tends to select the simplest least parameterized models, in line with the so-called Lindley or Bartlett paradox (Bartlett, 1957). Finally, the formal approach to model averaging requires both posterior densities, $p(\theta_k|y, m = k)$, and posterior model probabilities, $p(m = k|y)$. Estimates of posterior densities, $p(\theta_k|y, m = k)$, may be difficult to obtain in complex random effects models with large numbers of parameters.

However, straightforward and pragmatic approaches to model comparison, applicable to complex hierarchical models, are available as alternatives to formal methods. The two main approaches are based on posterior densities of fit measures (e.g., log-likelihood, deviance) and on predictive asssessment using samples of replicate data. Section 2.4 considers the posterior deviance as a fit measure, and the related measure of model complexity (effective dimension) that is of considerable utility in comparing hierarchical models. Bayesian fit measures such as the deviance information criterion (DIC) are analogous to information theoretic approaches in frequentist statistics (Burnham and Anderson, 2002), but more widely applicable (e.g., to non-nested models). The components of the overall fit deriving from each observation (e.g., the deviance contributions from particular observations) may be used in model checking (Plummer, 2008).

The predictive approach to model choice and diagnosis (Section 2.5) has also been simplified by Markov Chain Monte Carlo (MCMC) (Gelfand, 1996). Predictive methods shift the focus onto observables away from parameters (Geisser and Eddy, 1979) and seek to alleviate the impact on model comparison of factors such as specification of priors. The predictive approach is particularly advantageous in model checking, namely, ensuring that a model actually reproduces the data satisfactorily (e.g., Kacker et al., 2008), but is also applied to model choice, for example under posterior predictive loss criteria (Gelfand and Ghosh, 1998).

Predictive model checking typically involves repeated sampling of replicate data, y_{new}, from a model's parameters at each MCMC iteration (Gelfand et al., 1992). For a satisfactory model, this process generates data like the observed data such that (y, y_{new}) are exchangeable draws from the joint density (Stern and Sinharay, 2005, 176–77),

$$p(y_{\text{new}}, y, \theta) = p(y_{\text{new}}|\theta, y)p(y|\theta)p(\theta) = p(y_{\text{new}}|\theta)p(y|\theta)p(\theta).$$

When all the data are used in model estimation, such sampling provides estimates of the posterior predictive density of model k, $p(y_{\text{new}}|y, m = k)$. However, predictive comparisons based on models using all the data in estimation may be overly favorable to the model being fitted (i.e., be conservative in terms of detecting model discrepancies) (Bayarri and Berger, 1999). An alternative involves cross-validation (Alqalaff and Gutafson, 2001), where the model predicts values for certain observations (the test sample) on the basis of a model estimated using the remaining observations (the learning sample). Key et al. (1999) argue that cross-validation is approximately optimal in an M-open scenario, where none of the models being considered is believed to be the true model.

2.2 Formal Methods: Approximating Marginal Likelihoods

As mentioned above, the global fit of a model with parameter vector θ under the formal Bayes paradigm is provided by the marginal likelihood, $p(y|m = k)$, obtained by integrating the likelihood,

$$p(y) = \int p(y|\theta)p(\theta)d\theta.$$

The marginal likelihood is also a component in Bayes formula, such that at any parameter value θ,

$$p(\theta|y) = \frac{p(y|\theta)p(\theta)}{p(y)}.$$

Consider models 1 and 2 with equal prior model probabilities, $p(m = 1) = p(m = 2) = 0.5$. Then the ratio of posterior model probabilities is obtained as,

$$\frac{p(m = 2|y)}{p(m = 1|y)} = \frac{p(y|m = 2)}{p(y|m = 1)} = B_{21},$$

where B_{21} is the Bayes factor. Kass and Raftery (1995) provide guidelines for interpreting B_{21}. If $2\log_e B_{21}$ is larger than 10 the evidence for model 2 is very strong, while values of $2\log_e B_{21} < 2$ are inconclusive as evidence in favor of one or other model. Note that such criteria are influenced by the prior adopted. In general, diffuse priors (whether on fixed effect parameters or variances) are to be avoided as they tend to favor the selection of the simpler model.

Estimating the marginal likelihood by direct integration is generally infeasible in multiparameter applications. Hence, a range of approximations have been proposed for estimating marginal likelihoods or associated model choice criteria, such as the Bayes factor. For example, on suitable rearrangement, the Bayes formula implies that the marginal likelihood may be approximated by estimating the posterior ordinate, $p(\theta|y)$, in the relation,

$$\log[p(y)] = \log[p(y|\theta_h)] + \log[p(\theta_h)] - \log[p(\theta_h|y)],$$

where θ_h is a point with high posterior density (e.g., posterior mean or median). One may estimate $p(\theta|y)$ by kernel density methods or by moment approximations based on MCMC output—see Lenk and DeSarbo (2000) for a discussion of such estimates. Let $g(\theta)$ denote an estimated density that approximates $p(\theta|y)$. One may then evaluate $g(\theta)$ at θ_h (Bos, 2002; Sinharay and Stern, 2005), so providing an estimate of the log marginal likelihood as,

$$\log[p(y|\theta_h)] + \log[p(\theta_h)] - \log[g(\theta_h)].$$

The relation, $\log[p(y)] = \log[p(y|\theta_h)] + \log[p(\theta_h)] - \log[p(\theta_h|y)]$, also implies a sampling-based estimator of the log marginal likelihood. Since this relation applies for all samples, $\theta^{(r)}$, one may average over values,

$$H^{(r)} = \log\left[p(y|\theta^{(r)})\right] + \log\left[p(\theta^{(r)})\right] - \log\left[g(\theta^{(r)})\right],$$

to estimate the log of the marginal likelihood, $\log[p(y)]$. This is likely to be the most suitable approach for larger samples, to avoid numeric overflow. For small samples, one may set $L^{(r)} = p(y|\theta^{(r)})$, $\pi^{(r)} = p(\theta^{(r)})$, and $g^{(r)} = g(\theta^{(r)})$. Then an estimator of the marginal likelihood is provided by the simple average of the ratios $(L^{(r)}\pi^{(r)}/g^{(r)})$.

Alternatively, suppose θ contains B parameter sub-blocks. When the full conditionals of each sub-block are available in closed form, Chib (1995) considers a marginal/conditional decomposition of $p(\theta_h|y)$ as follows,

$$p(\theta_h|y) = p(\theta_{1h}|y)p(\theta_{2h}|\theta_{1h}, y)p(\theta_{3h}|\theta_{1h}, \theta_{2h}, y), \ldots, p(\theta_{Bh}|\theta_{1h}, \ldots, \theta_{B-1,h}, y),$$

with $p(\theta_h|y)$, and thus $p(y)$, estimated by using $B-1$ sampling sequences subsidiary to the main scheme. If $B = 2$, namely, $\theta_h = (\theta_{1h}, \theta_{2h})$, the posterior ordinate $p(\theta_h|y)$ is then $p(\theta_{1h}|y)p(\theta_{2h}|y, \theta_{1h})$, where $p(\theta_{1h}|y)$ is estimated from the output of the main sample, e.g., as,

$$p(\theta_{1h}|y) = \sum_{r=1}^{R} p(\theta_{1h}|y, \theta_2^{(r)}),$$

or by an approximation technique (e.g., assuming univariate/multivariate posterior normality of θ_1 or a kernel method). The second ordinate is available by inserting θ_{1h} and θ_{2h} in the relevant full conditional density. Chib and Jeliazkov (2001) extend this method to cases where full conditionals do not have a known normalizing constant and have to be updated by Metropolis–Hastings steps.

2.2.1 Importance and bridge sampling estimates

Let θ_k be the parameter vector for model k and the marginal likelihoods, $p(y|m = k)$, be denoted M_k, and let,

$$p^*(\theta_k|y, m = k) = p(y|\theta_k, m = k)p(\theta_k|m = k),$$

denote the un-normalized posterior density of θ_k with,

$$p^*(\theta_k|y, m = k)/c_k = p(\theta_k|y).$$

Then by definition,

$$p(y|m = k) = \int p^*(\theta_k|y, m = k)d\theta.$$

Consider a function $g(\theta)$ with known normalizing constants, often termed an importance function, and one that should ideally approximate the posterior $p(\theta|y)$. Then one has,

$$p(y|m = k) = \int p^*(\theta_k|y, m = k)d\theta = \int \frac{p^*(\theta_k|y, m = k)}{g(\theta_k)} g(\theta_k)d\theta_k.$$

This suggests that an estimator for the marginal likelihood may be obtained using samples $\tilde{\theta}_k^{(r)} (r = 1, \ldots, R)$ from $g(\theta_k)$, namely,

$$M_k = \sum_r \frac{p^*(\tilde{\theta}_k^{(r)}|y, m = k)}{g(\tilde{\theta}_k^{(r)})}.$$

Let $\tilde{L}_k^{(r)} = p(y|\tilde{\theta}_k^{(r)})$, $\tilde{\pi}_k^{(r)} = p(\tilde{\theta}_k^{(r)})$, and $\tilde{g}_k^{(r)} = g(\tilde{\theta}_k^{(r)})$. Then the importance sample estimator may be written in terms of weights, $w_k^{(r)} = \tilde{\pi}_k^{(r)}/\tilde{g}_k^{(r)}$, comparing the prior and importance function, namely,

$$M_k = \sum_r \tilde{L}_k^{(r)} w_k^{(r)}.$$

Bridge sampling estimators of marginal likelihoods use the fact that the marginal likelihood of model k is the normalizing constant, $c_k = p(y|m = k)$, in the relation,

$$p(\theta_k|y, m = k) = \frac{p(y|\theta_k, m = k)p(\theta_k|m = k)}{p(y|m = k)} = \frac{p^*(\theta_k|y, m = k)}{c_k}.$$

The Bayes factor, $B_{jk} = (p(y|m = j)/p(y|m = k))$, is then a ratio, c_j/c_k, of normalizing constants. Let $g(\theta)$ be an approximation to $p(\theta|y)$ with known normalizing constant (e.g., suppose g consists of a multivariate normal density and a gamma density). Then one has

$$1 = \frac{\int \alpha(\theta_k)p(\theta_k|y)g(\theta_k)d\theta_k}{\int \alpha(\theta_k)g(\theta_k)p(\theta_k|y)d\theta_k} = \frac{E_g[\alpha(\theta_k)p(\theta_k|y)]}{E_p[\alpha(\theta_k)g(\theta_k)]},$$

where $\alpha(\theta)$ is a bridge function linking the densities $g(\theta)$ and $p(\theta|y)$ (see Meng and Wong, 1996), $E_g[]$ denotes expectation with regard to the density $g(\theta)$, and $E_p[]$ denotes expectation with regard to the density $p(\theta|y)$. Substituting $p^*(\theta_k|y, m = k)/c_k$ for $p(\theta|y)$ in $1 = E_g[\alpha(\theta_k)p(\theta_k|y)]/E_p[\alpha(\theta_k)g(\theta_k)]$ gives the result,

$$c_k = \frac{E_g[\alpha(\theta_k)p^*(\theta_k|y, m = k)]}{E_p[\alpha(\theta_k)g(\theta_k)]}.$$

For simplicitly omit conditioning on model k. Then with samples $\theta^{(r)}(r = 1, \ldots, S)$ and $\tilde{\theta}^{(r)}(r = 1, \ldots, R)$ from $p(\theta|y)$ and $g(\theta)$, respectively, one may estimate the marginal likelihood $p(y)$ of a particular model as,

$$\left\{\frac{1}{R}\sum_{r=1}^{R}\left[\alpha(\tilde{\theta}^{(r)})p^*(\tilde{\theta}^{(r)}|y)\right]\right\} \bigg/ \left\{\frac{1}{S}\sum_{r=1}^{S}\left[\alpha(\theta^{(r)})g(\theta^{(r)})\right]\right\}.$$

Setting $\alpha(\theta) = 1/g(\theta)$ then gives a marginal likelihood estimator,

$$M = \frac{1}{R}\sum_{r=1}^{R}\frac{p^*(\tilde{\theta}^{(r)}|y)}{g(\tilde{\theta}^{(r)})},$$

that uses only samples from the approximate posterior (or importance) density $g(\theta)$.

Setting $\alpha(\theta) = (1/p^*(\theta|y))$ gives an estimator based on the harmonic mean of the ratios $(p^*(\theta^{(r)}|y)/g(\theta^{(r)}))$, and using parameters sampled from $p(\theta|y)$ rather than $g(\theta)$ (Gelfand and Dey, 1994). So

$$\frac{1}{M} = \frac{1}{S}\sum_{r=1}^{S} \frac{g(\theta^{(r)})}{p^*(\theta^{(r)}|y)}.$$

The choice, $\alpha(\theta) = (1/g(\theta)p^*(\theta|y))$, leads to the geometric estimator of Lopes and West (2004), namely,

$$M = \frac{\frac{1}{R}\sum_{r=1}^{R}\left[p^*(\tilde{\theta}^{(r)}|y)/g(\tilde{\theta}^{(r)}|y)\right]^{0.5}}{\frac{1}{S}\sum_{r=1}^{S}\left[g(\theta^{(r)}|y)/p^*(\theta^{(r)}|y)\right]^{0.5}}.$$

A recursive scheme for obtaining an optimal estimate of $\alpha(\theta)$ is also available, and mentioned by Lopes and West (2004, 54) and Frühwirth-Schnatter (2004, equation 8). This simplifies if $R = S$, as in the first illustrative worked application below. With $R \neq S$, $s_1 = S/(S+R)$, and $s_2 = 1 - s_1$, one has an updated estimate for M at recursion j,

$$M_j = A(M_{j-1})/B(M_{j-1}),$$

where,

$$A(u) = \sum_r W_{2r}/(s_1 W_{2r} + s_2 u), \quad B(u) = \sum_s 1/(s_1 W_{1s} + s_2 u),$$
$$W_{2r} = p(y|\tilde{\theta}^{(r)})p(\tilde{\theta}^{(r)})/g(\tilde{\theta}^{(r)}),$$

and

$$W_{1s} = p(y|\theta^{(s)})p(\theta^{(s)})/g(\theta^{(s)}).$$

2.2.2 Path sampling

Another approximation may be obtained by a technique known as path sampling (Gelman and Meng, 1998). Consider a path variable t ranging from 0 to 1, and define the power posterior based on various levels of weighted likelihood, namely,

$$p_t(\theta|y) \propto [p(y|\theta)]^t p(\theta).$$

Define the posterior expectation,

$$z(y|t) = \int [p(y|\theta)]^t p(\theta)d\theta,$$

so that $z(y|t=0)$ is the integral of the prior, namely, 1 for proper priors, while $z(y|t=1)$ is the marginal likelihood, $p(y) = \int p(y|\theta)p(\theta)d\theta$.

To derive an estimate of $z(y|t=1)$, one may use the identity,

$$\log(p(y)) = \log\left(\frac{z(y|t=1)}{z(y|t=0)}\right) = \int_0^1 E_{\theta|y,t}\log[p(y|\theta)]\,dt,$$

which states that the log marginal likelihood is the expected log likelihood with respect to the power posterior at temperature t, with t ranging from 0 to 1. This follows (Friel and Pettitt, 2008) because,

$$\begin{aligned}\frac{d}{dt}\log[z(y|t)] = \frac{1}{z(y|t)}\frac{d}{dt}z(y|t) &= \frac{1}{z(y|t)}\frac{d}{dt}\left[\int\{p(y|\theta)\}^t p(\theta)d\theta\right]\\ &= \frac{1}{z(y|t)}\int\{p(y|\theta)\}^t\log[p(y|\theta)]p(\theta)d\theta\\ &= \int\frac{\{p(y|\theta)\}^t p(\theta)}{z(y|t)}\log[p(y|\theta)]d\theta\\ &= E_{\theta|y,t}\log[p(y|\theta)].\end{aligned}$$

One may numerically evaluate the integral over t using the trapezoid rule over T intervals defined using $T+1$ temperature functions, $q_s = a_s^c$, defined at cutpoints $\{a_0,\ldots,a_L\}$ in [0,1], where c is a specified positive power. So the estimate $\log(M_c)$ of the log marginal likelihood at that power is obtained by summing over T grid points that combine information from successive expected log likelihoods,

$$\log(M_c) = \sum_{s=0}^{T-1}(q_{s+1}-q_s)\frac{1}{2}\left[E_{\theta|y,q_{s+1}}\log[p(y|\theta)] + E_{\theta|y,q_s}\log[p(y|\theta)]\right].$$

Friel and Pettitt (2008) take $c=4$, while Song and Lee (2004) take $c=1$. So with $T=40$ intervals, equally spaced cutpoints $\{a_0=0, a_1=0.025, a_2=0.05,\ldots,a_{40}=1\}$, and setting $c=4$, one has $q_0=0$, $q_1=(0.025)^4,\ldots,q_{39}=0.975^4$, $q_{40}=1$. The Monte Carlo standard error of $\log(M_c)$ is obtained as the square root of the summed variances of the contributions to $\log(M_c)$ at each of T grid points. Thus, let $\delta_s = 1/2(q_{s+1}-q_s)$ and let ν_s be the Monte Carlo variance of $E_{\theta|y,q_{s+1}}\log[p(y|\theta)]$. Then the variance at each grid point is $\delta_s^2\nu_s$ and the Monte Carlo variance of $\log(M_c)$ is $\sum_{s=0}^{T-1}\delta_s^2\nu_s$.

2.2.3 Marginal likelihood for hierarchical models

For conjugate hierarchical models (e.g., Poisson-gamma mixtures), the marginal likelihood can be obtained analytically (Albert, 1999). However, general linear mixed models (Clayton, 1996) are widely used for handling multiple random effects, with regression terms,

$$\eta_i = X_i\beta + W_i b_i,$$

where X_i and W_i are predictors, and b_i are latent data. For such nonconjugate schemes, the marginal likelihood is not obtainable analytically, and one possible approach to evaluating marginal likelihoods is to work with the integrated likelihood,

$$p(y|\theta) = \int p(y, b|\theta)\, db = \int p(y|b, \theta)p(b|\theta)\, db,$$

where the random effects or latent data have been integrated out, and where θ includes hyperparameters ψ (e.g., covariances) governing b, as well as parameters φ (e.g., fixed regression effects) not relevant to the random effect hyperdensity (Fruhwirth-Schnatter, 1999; Sinharay and Stern, 2005). This can be done in practice in MCMC sampling by applying importance sampling, the Laplace approximation, or numeric integration methods to the complete data likelihood, $p(y, b|\theta)$.

However, it may be argued that under a Bayesian approach the distinction between fixed and random regression coefficients is less relevant, and so use of the integrated likelihood and implied numerical complexity may be avoided. For example, one may (e.g., Clayton, 1996) adopt a unified perspective on the parameters in the joint precision matrix for the fixed effects (and other parameters not in the hyperdensity) φ, and the random effects hyperparameters, ψ. Sinharay and Stern (2005) also mention obtaining the marginal likelihood by considering the expanded parameter set, $\omega = (b, \varphi, \psi)$, so that,

$$\begin{aligned} p(y) &= \iint p(y|\varphi, \psi)p(\varphi, \psi)d\varphi d\psi, \\ &= \iiint p(y|\varphi, b)p(b|\psi)p(\psi, \varphi)\, dbd\psi d\varphi. \end{aligned}$$

The advantage of working with the expanded likelihood, $p(y|b, \varphi)$, is the avoidance of repeated integration, but this comes at the expense of an often considerably increased dimension of the parameter space (namely, by the number of components in b). Marginal likelihood approximation retaining the expanded likelihood is considered in real examples by Nandram and Kim (2002) and Gelfand and Vlachos (2003).

Let $g(\theta|b)$ be a density subject to $\int g(\theta|b)d\theta = 1$, where $\theta = (\psi, \varphi)$, and let θ^* be an appropriate fixed point (e.g., a posterior mean). Chen (2005) mentions an estimator for the log marginal likelihood, $M = p(y)$, in a hierarchical modeling situation based on the identity,

$$\begin{aligned} p(y|\theta^*) &= \int p(y|\theta^*, b)p(b|\theta^*)g(\theta|b)d\theta db, \\ &= \int \frac{g(\theta|b)}{p(\theta)} \frac{p(y|\theta^*, b)p(b|\theta^*)}{p(y|\theta, b)p(b|\theta)} p(y|\theta, b)p(b|\theta)p(\theta)d\theta db, \\ &= p(y)E\left[\frac{g(\theta|b)}{p(\theta)} \frac{p(y|\theta^*, b)p(b|\theta^*)}{p(y|\theta, b)p(b|\theta)} \middle| y\right], \end{aligned}$$

where the expectation is over samples from $p(\theta, b|y)$. Taking logarithms provides,

$$\log[M] = \log[p(y|\theta^*)] - \log\left[E\left[\frac{g(\theta|b)}{p(\theta)}\frac{p(y|\theta^*, b)p(b|\theta^*)}{p(y|\theta, b)p(b|\theta)}\bigg| y\right]\right].$$

So with samples $\{\theta^{(r)}, b^{(r)}\}$ from $p(\theta, b|y)$, an estimator for $\log[M]$ is

$$\log[M] = \log[p(y|\theta^*)] - \log\left[\frac{1}{R}\sum_{r=1}^{R}\frac{g(\theta^{(r)}|b^{(r)})}{p(\theta^{(r)})}\frac{p(y|\theta^*, b^{(r)})p(b^{(r)}|\theta^*)}{p(y|\theta^{(r)}, b^{(r)})p(b^{(r)}|\theta^{(r)})}\right].$$

One option is then to set $g(\theta|b) = p(\theta)$, leading to,

$$\log[M] = \log[p(y|\theta^*)] - \log\left[\frac{1}{R}\sum_{r=1}^{R}\frac{p(y|\theta^*, b^{(r)})p(b^{(r)}|\theta^*)}{p(y|\theta^{(r)}, b^{(r)})p(b^{(r)}|\theta^{(r)})}\right],$$

The component, $\log[p(y|\theta^*)] = \log(L^*)$, may be estimated from the Monte Carlo average,

$$L^* = \frac{1}{R}\sum_{r=1}^{R} p\left(y|\theta^*, b^{(r)}\right).$$

Chen (2005) shows that a variance minimizing estimator is, however, obtained by setting $g(\theta|b) = p(\theta|b, y)$, namely, the conditional posterior density of θ given b.

Example 2.1. Marginal Likelihood Estimates for Turtle Mortality Data This example uses approximations to the marginal likelihood in data from Sinharay and Stern (2005) under the complete likelihood perspective just mentioned. It compares a simple fixed effect model against a random effects alternative for nested binary data, y_{ij}, on $n = 244$ newborn turtles, $i = 1, \ldots, m_j$, clustered into clutches, $j = 1, \ldots, J$, with responses $y_{ij} = 1$ or 0 according to survival or death. The known predictor is turtle birthweight, x_{ij}, so there are $p = 2$ regression parameters including an intercept. The $J = 31$ clutches are numbered in increasing order of the average birthweight of the turtles in the clutch. Graphical analysis suggests that heavier turtles have better survival chances, but also suggests extraneous variability in survival rates across clutches.

Two alternative models are to be evaluated for the probability, $\pi_{ij} = Pr(y_{ij} = 1)$. One involves a fixed effects only regression on birthweight with a probit link. The other assumes additional random effects based on clutch membership. So, model 1 specifies $\pi_{ij1} = \Phi(\beta_1 + \beta_2 x_{ij})$, while model 2 has

$$\pi_{ij2} = \Phi(\beta_1 + \beta_2 x_{ij} + b_j),$$

where $b_j \backsim N(0, \sigma_b^2)$. Sinharay and Stern (2005) compare several methods of deriving formal model fit measures, namely, marginal likelihoods or Bayes factors. Here the marginal likelihood of the alternative models is initially estimated using the temperature path approach of Friel and Pettitt (2008), and with $N(0, 10)$ priors on fixed effects $\{\beta_1, \beta_2\}$ as in Sinharay and Stern (2005).

There are several possible sources of sensitivity. The formal model measures will depend not only on the degree of informativeness in the priors (e.g., the prior variance on the fixed effects), but may depend on the form of prior, for example, the prior density adopted on the random effects variance, σ_b^2, or precision, $\tau_b = 1/\sigma_b^2$. Possibilities include gamma, lognormal, or uniform priors. Sinharay and Stern (2005) consider a form of shrinkage prior,

$$p\left(\sigma_b^2\right) \propto \frac{1}{1+\sigma_b^2},$$

that tends to downweight larger values of σ_b^2. For example, the value $\sigma_b^2 = 1$ has half the prior weight of $\sigma_b^2 = 0$. With this prior, Sinharay and Stern obtain an inconclusive Bayes factor of 1.31 in favor of the simpler fixed effects only model.

To illustrate sensitivity to the prior on the variance, a gamma prior on τ_b is initially assumed, namely, $\tau_b \sim Ga(1, 1)$. The path sampling approach[1] of Friel and Pettitt (2008) and Song and Lee (2004) is applied, with $q_s = a_s^4$ and $T = 20$, and with $\{a_0, \ldots, a_{T+1}\}$ then having elements $\{0.00001, 0.05, 0.10, \ldots, 0.95, 1\}$. For numeric stability, the first element in a_s is taken as 0.00001 rather than 0, so that $q_0 = 1E-20$. This device could possibly be avoided with more informative priors, e.g., a $N(0, 1)$ prior on β_2. Although formally the estimate of $\log[p(y)]$ is obtained by piecing together the separate posterior estimates, $E_{\theta_k|y,t_s}p(y|\theta_k)$, an essentially identical estimate is obtained by applying the trapezoid rule at each iteration and monitoring the composite log marginal likelihood node.

The marginal likelihood estimate for model 1 is obtained as -155.29 compared to -155.53 for the random effects extension giving $B_{12} = 1.27$. These estimates are based on the last 90,000 iterations of single chain runs of 100,000 iterations[2], and have respective Monte Carlo standard errors of 0.030 and 0.117. Relatively large clutch effects (mean with posterior sd) are obtained under model 2 for clutches 1, 9, and 11, namely, 0.82 (0.34), 0.65 (0.42), and 0.60 (0.39). Posterior plots of the effects indicate normality, so the first effect might be judged significant or necessary. To assess significance, an alternative device, namely, monitoring the probability that b_j is positive, may be applied—see Example 2.2, also using the turtles mortality data. With a slightly more diffuse $Ga(0.1, 0.1)$ prior on τ_b, the marginal likelihood for the random effects model is raised to -155.15 and B_{12} reduced to 0.8.

The bridge sampling method is also applied under the prior $\tau_b \sim Ga(1, 1)$, and with importance sample estimates of $\{\beta_1, \beta_2, \tau_b, b_1, \ldots, b_J\}$ based on a preliminary run. A gamma approximation is used for $p(\tau_b|y)$ and normal approximations for the remaining parameters. Thus the preliminary run provided a posterior mean and standard deviation for τ_b of 3.74 and 1.45 and this translates to a gamma $Ga(6.65,\ 1.78)$ approximation for $p(\tau_b|y)$.

A WinBUGS code[3] to provide the quantities required for the iterative scheme (in MATLAB®) to obtain the optimal α function involves sampling $\{\beta_1, \beta_2, \tau_b, b_1, \ldots, b_J\}$ both from the posterior and the importance functions; this is followed (for both sets of parameter samples) by log likelihood calculations, and by normal and gamma function evaluations over both log prior and log importance densities.

$R = 5000$ samples of the four required statistics under the prior, $\tau_b \sim Ga(1, 1)$, are obtained and the recursions run 500 times. The bridge samping method gives an estimate for $\log(M)$ of -156.5 for the fixed effects model, and -160.2 for the random effects model; so $B_{12} = 46.5$. Although both path and bridge sampling reach the same conclusion regarding the better model, the variation in marginal likelihood estimates between them reflects features of the approximations used, such as the arbitrariness of the power in the temperature function (path method), and imperfections in the importance sample densities as approximations to the posterior density (bridge method).

2.3 Effective Model Dimension and Deviance Information Criterion

Classical approaches to model choice methods are frequently based on penalized likelihood criteria, such as the Akaike information criterion (AIC) (Akaike, 1973), and the Bayesian information criterion (BIC) (Schwarz, 1978). Such criteria are applicable in comparing fixed effects models with known dimension d, and with models assumed nested within one another. With L denoting a log likelihood, and $D = -2L$ denoting the deviance, log likelihood ratio tests comparing maximized log likelihoods of models 1 and 2 are obtained with,

$$C = -2(\log L_1 - \log L_2) = D_1 - D_2,$$

where C is approximately chi-square, with degrees of freedom $d_2 - d_1$ equal to the number of additional parameters in the more complex model 2. The AIC is defined as $2d - 2L = D + 2d$ and the difference in AICs between models 1 and 2 is $\Delta AIC = C + 2(d_1 - d_2)$. However, classical likelihood ratio testing is not possible in random effects models or models with parameter constraints (e.g., order of size constraints) that make the effective number of estimated parameters itself a random variable so that the asymptotic distribution of the log likelihood ratio is unknown.

Spiegelhalter et al. (2002) provide a penalized fit criterion analogous to the AIC and BIC, called the deviance information criterion (DIC). This is applicable to comparing non-nested models and also to models including random effects where the true model dimension is another unknown. The DIC is based on the posterior distribution of the deviance statistic,

$$D(\theta|y) = -2\log[p(y|\theta)] + 2\log[h(y)],$$

where $p(y|\theta)$ is the likelihood of data y given parameters θ, and $h(y)$ is a standardizing function of the data only (and so does not affect model choice).

Suppose the deviance is monitored as an extra node in an MCMC run, providing samples $\{D^{(1)}, \ldots, D^{(R)}\}$. The overall fit of a model is measured by the posterior expected deviance obtained by averaging over the posterior density of the parameters,

$$\bar{D} = E_{\theta|y}[D],$$

while the effective model dimension, d_e, is estimated as,

$$d_e = E_{\theta|y}[D] - D(E_{\theta|y}[\theta]) = \bar{D} - D(\bar{\theta}),$$

namely, the expected deviance minus the deviance at the posterior means of the parameters; the latter is also known as the plug-in deviance (Plummer, 2008). In hierarchical random effects models, the effective number of parameters total is typically lower than the nominal number of parameters, due to borrowing of strength under the hyperdensity (e.g., Buenconsejo et al., 2008; Zhu et al., 2006).

The DIC is then obtainable as the expected deviance plus the effective model dimension,

$$\text{DIC} = \bar{D} + d_e = D(\bar{\theta}) + 2d_e.$$

So the DIC will prefer models with lower values of $\bar{D}$, combined with small values of d_e (which indicate a relatively parsimonious model). A possible disadvantage with the DIC is that it can be affected by reparameterization of θ, or by the form of link in general linear models, with this applying in particular to the "plug-in" deviance, $D(\bar{\theta})$; hence the value of d_e may be sensitive to parameterization.

The DIC and d_e can be disaggregated to individual observations, and provide a measure of local complexity, namely, of observations that are more problematic under the model relative to others. Spiegelhalter et al. (2002, 602) mention that the local complexity measures,

$$d_{ei} = \bar{D}_i - D_i(\bar{\theta}),$$

measure the leverage of observation i, defined as the relative influence that each observation has on its own fitted value. Unusually large observation-specific DIC measures, namely,

$$\text{DIC}_i = \bar{D}_i + d_{ei},$$

are used by Spiegelhalter et al. (2002) as indicators of outlier status—observations inconsistent with the model.

The DIC can be seen as a Bayesian version of AIC and may underpenalize model complexity, as pointed out by discussants to Spiegelhalter et al. (2002). Plummer (2008) confirms that the DIC underpenalizes complex models, particularly when the ratio of the sample size to the effective number of parameters is relatively low. By contrast, it is well established (Burnham and Anderson, 2002) that the BIC tends to select overly parsimonious models. A fit criterion analogous to the BIC may be defined as,

$$\mathrm{DIC}^* = D(\bar{\theta}) + d_e \log(n),$$

and was used by Pourahmadi and Daniels (2002, 228) for panel data with repeated observations over n subjects. Note that the model with the lowest DIC or DIC* will not necessarily be a suitable model if it does not reproduce the data adequately. Hence, model checks are required to assess consistency of predictions from the model with the actual observations.

Just as there are alternative approaches to marginal likelihood derivation in hierarchical models, Spiegelhalter et al. (2002) point out that for such models, one cannot uniquely define the likelihood or model complexity without specifying the level of the hierarchy that is the model focus. Thus, one might analyze count data using a complete data likelihood (with unknown latent data b as well as hyperparameters θ) using a Poisson-gamma or Poisson-lognormal model, or alternatively apply a negative binomial likelihood with the random effects integrated out (Fahrmeir and Osuna, 2003), and the complexity measures will obviously differ. Model choice may be affected by the focus, as shown by Plummer (2008, 530) in an analysis of a discrete mixture model, with one approach considering a complete data likelihood, $p_C(y|b, \theta)$ (with the parameters including missing component indicators), and the other considering the integrated likelihood, $p_I(y|\theta)$. Ando (2007) considered DICs based on both conditional and integrated likelihoods, namely, DIC^C and DIC^I, and showed that both tend to select overfitted (i.e., nonparsimonious) models.

Issues of focus as well as the derivation of the complexity measure, d_e, are considered by Celeux et al. (2006). In general terms, a complexity measure or effective parameter count is obtained by comparing the mean deviance with the deviance at the pseudo-true parameter values θ^t (Spiegelhalter et al., 2002, Section 2.2). There are various possible estimators, $\tilde{\theta}$, of the pseudo-true parameter values, θ^t, apart from the elementwise posterior means. Another possibility is to consider the posterior mode value, $\hat{\theta}$, which generates the maximum posterior density, $p(\theta|y) \propto p(y|\theta)p(\theta)$ (Celeux et al., 2006, 654), namely,

$$\hat{\theta} = \arg\max_{\theta} p(\theta|y).$$

In missing data applications (e.g., discrete mixture models and random effect models) with missing data b, this extends to considering the pair $(\hat{\theta}, \hat{b})$ that generates the maximum posterior density (Celeux et al., 2006, 656). Celeux et al. (2006) mention other possibilities for $\tilde{\theta}$, such as the EM (expectation-maximization) maximum likelihood estimate.

Celeux et al. state different DIC definitions under three alternative foci (observed data likelihood, complete data likelihood, and conditional likelihood) and under different options for $\tilde{\theta}$. For the observed data focus with likelihood $p(y|\theta)$, obtained possibly after integrating out random effects, one has,

$$\text{DIC} = \bar{D} + d_e = D(\tilde{\theta}) + 2d_e = 2\bar{D} - D(\tilde{\theta}) = -4E_\theta[\log\{p(y|\theta)\}|y] + 2\log[p(y|\tilde{\theta})].$$

It can be seen that taking $\widetilde{\theta}$ as the posterior mean amounts to assuming,

$$\text{DIC} = -4E_\theta[\log\{p(y|\theta)\}|y] + 2\log[p\{y|E_\theta(\theta|y)\}],$$

whereas taking $\tilde{\theta}$ as the posterior mode $\hat{\theta}$ amounts to an alternative DIC definition, denoted DIC_2 by Celeux et al., namely,

$$\text{DIC} = -4E_\theta[\log\{p(y|\theta)\}|y] + 2\log[p(y|\hat{\theta})].$$

For a complete data focus, with likelihood $p(y,b|\theta) = p(y|b,\theta)p(b|\theta)$ including the second stage likelihood model, $p(b|\theta)$, for the missing data (e.g., Kuhn and Lavielle, 2005), one obtains,

$$\bar{D} = -2E_{\theta,b}[\log\{p(y,b|\theta)\}|y].$$

Taking b as additional parameters, one may define $\tilde{\theta}$ on the basis of joint modal or maximum a posteriori parameters, $(\hat{\theta}, \hat{b})$, so that d_e is obtained by comparing the average deviance $\bar{D}$ with,

$$D(\tilde{\theta}) = -2\log[p(y, \hat{b}|\hat{\theta})].$$

The joint mode $(\hat{\theta}, \hat{b})$ may be estimated by monitoring the posterior density over an MCMC sequence, and finding that set of values $\{\theta^{(r)}, b^{(r)}\}$ associated with the maximum value, $p_{\max}(\theta|y)$, of the posterior density. The DIC may then be defined as,

$$\text{DIC} = -4E_{\theta,b}[\log\{p(y,b|\theta)\}|y] + 2\log[p(y,\hat{b}|\hat{\theta})],$$

with complexity estimated as,

$$d_e = -2E_{\theta,b}[\log\{p(y,b|\theta)\}|y] + 2\log[p(y,\hat{b}|\hat{\theta})].$$

This procedure has the disadvantage that ordinates of prior densities (including densities for random effects) have to be evaluated at each iteration. To ensure satisfactory estimation of $\log[p(y,\hat{b}|\hat{\theta})]$, one may consider values obtained from different batches of MCMC sequences and/or different chains, and ensure that they show consistency in the estimated $\log[p(y,\hat{b}|\hat{\theta})]$. Similarly, under a conditional likelihood approach, the DIC may be defined as,

$$\text{DIC} = -4E_{\theta,b}[\log\{p(y|\theta,b)\}|y] + 2\log[p(y|\hat{b},\hat{\theta})],$$

with complexity estimated as,

$$d_e = -2E_{\theta,b}[\log\{p(y|b,\theta)\}|y] + 2\log[p(y|\hat{b},\hat{\theta})].$$

The deviance $D(\bar{\theta}|y)$ at the posterior mean $\bar{\theta}$ of the parameters may also be estimated using posterior means of quantities involved in defining the deviance, such as case means (Poisson likelihood), means and overdispersion parameter (negative binomial likelihood), means and variance (normal likelihood), and so on (Spiegelhalter et al., 2002, 596). Thus, let μ_i denote case-specific means and ξ denote any another parameters needed to derive the deviance. Then the estimate, $D(\bar{\mu}, \bar{\xi}|y)$, may be more easily obtainable than $D(\bar{\theta}|y)$ in complex (e.g., discrete mixture) models, or in models with many random effects, where the number of nominal parameters may considerably exceed the number of cases. For an illustration, see an application of the Potts spatial model in Chapter 4 (Example 4.9). This type of procedure is also mentioned by Spiegelhalter (2006) in terms of monitoring the "direct parameters" that appear in the distributional syntax and plugging these into the deviance; it was adopted in the paper by Ohlssen et al. (2006, Section 2).

Example 2.2. Turtle Survival This example seeks to demonstrate possible issues in establishing whether one model improves over another under the posterior deviance approach, and how inferences may be affected by alternative priors on hyperparameters. Simulations are also used to show how the effective parameter count reflects the amount of random heterogeneity in the data.

As in Example 2.1, two models are considered for the observed survival data. Normal $N(0, 10)$ priors are adopted on the fixed effects and as the first of three options, a $Ga(0.1, 0.1)$ prior is initially assumed on the random effects precision, $\tau_b = 1/\sigma_b^2$, in model 2 that includes clutch effects. The fixed effects priors follow Sinharay and Stern (2005), whereas the prior on the precision follows the commonly adopted diffuse setting, $Ga(\varepsilon, \varepsilon)$, with ε small (e.g., Green and Richardson, 2000; Lambert et al., 2005). A possible alternative is $Ga(1, \varepsilon)$ as in Besag et al. (1995) and Fahrmeir and Lang (1997). These options correspond in the limit as $\varepsilon \to 0$ to $p(\tau_b) \propto 1/\tau_b$ and to a flat prior on τ_b.

With two chain runs[4] involving R = 10,000 iterations (B = 1000 iterations for burn in), the DICs of models 1 and 2 are obtained as 301.6 and 298.1, with $d_{e1} = 1.98$, $d_{e2} = 12.8$, $D(\bar{\theta}_1) = 299.7$, and $D(\bar{\theta}_2) = 272.5$. The effective parameter total, d_{e2}, for model 2 is about 11 higher than model 1, whereas there are 32 extra nominal parameters (the random effects variance and each of the 31 cluster effects), illustrating the impact of borrowing strength on the parameter total.

Comparing the DICs suggests a small advantage for the random effects model, though the small DIC difference might not be judged significant according to the rule of thumb in Spiegelhalter et al. (2002, Section 9.2.4). A more specific focus on the parameter estimates from model 2 shows the density of σ_b to be relatively symmetric and to have its mass away from zero

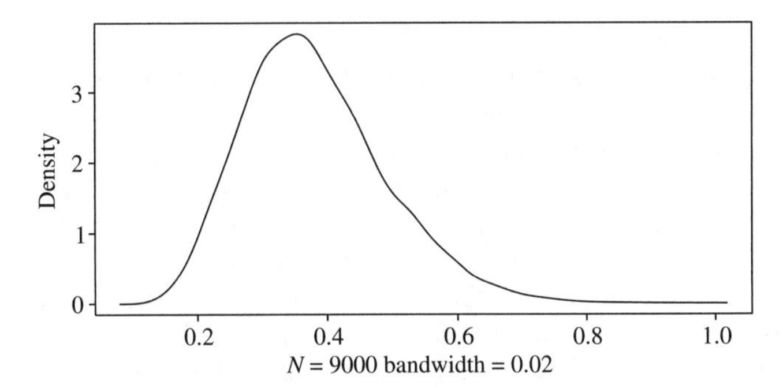

FIGURE 2.1
Kernel density clutch effects standard deviation.

(Figure 2.1), supporting the preference by the DIC for this model. In this connection, McNab et al. (2004) illustrate how—when a form of random variation is not supported by the data—the density of the random effects standard deviation can be heavily skewed to the left or "spiked," with the posterior mass piled up against zero.

It is also relevant to consider the significance of individual random effects. In fact, only the first cluster effect seems to be clearly significant (with a .97 chance of exceeding zero), but the last has a posterior probability of .06 of exceeding 0, so it is close to significance. It is also relevant (in assessing the gain from adding random effects) to consider effect contrasts $\{b_i - b_j\}$ and whether these are significantly different from zero.

Note that BIC analogues DIC* for models 1 and 2 are obtained as 310.7 and 316.5, with implicit marginal likelihood approximations of -155.3 and -158.2. This is calculated—in line with Pourahmadi and Daniels (2002)—by taking the number of observations as the cluster total of $J = 31$, so that $\text{DIC}^* = D(\bar{\theta}) + d_e \log(J)$. So the BIC analog, like the formal criteria considered in Example 2.1, tend to find evidence against the more heavily parameterized model 2, in contrast to the DIC.

The alternative $Ga(1, 1)$ prior for $\tau_b = 1/\sigma_b^2$ used in Example 2.1 is also applied. This prior is less likely to generate very small near zero precisions (i.e., less likely to generate large random effect variances). By contrast, small ε values in the $Ga(\varepsilon, \varepsilon)$ prior may lead to a spike at (virtually) zero values (Lambert, 2006). The DIC for model 2 is then 299.7, with the effective parameter count about 4 higher, namely, $d_{e2} = 17.1$, and $D(\bar{\theta}_2) = 282.6$. It is apparent that the effective parameter count, d_e, may be sensitive to the prior adopted for the inverse variance. Unlike the formal fit criteria in Example 2.1, the DIC still prefers the random effect model.

A further option is a bounded uniform prior on the standard deviation of the clutch effects, for example, $\sigma_b \sim U(0, 10)$. This provides $\text{DIC}_2 = 299$, with d_{e2} lower at 11.8, but the deviance at the parameter means higher at

$D(\bar{\theta}_2) = 287.3$. The mean for σ_b is 0.33, with a posterior density bounded away from zero. This option again confirms that the effective parameter count, d_e, may be sensitive to the prior adopted for the variance or precision, and the number of individually significant random effects may also be sensitive to variance priors. Under this bounded uniform prior on σ_b, no clutch effects are significant in the sense of $Pr(b_j > 0|y)$ being over 0.95 or under 0.05, whereas with $\tau_b \sim Ga(0.1, 0.1)$ there were significant effects.

To illustrate the impact of the level of random variation in the data on the effective parameter estimate, the birthweight and clutch membership data are retained, but survival indicators, y_{ij}, are simulated[5] under various preset levels of σ_b^2. Three different preset values σ_b^2, namely, $\sigma_b^2 = 1$, $\sigma_b^2 = 0.01$, and $\sigma_b^2 = 0.001$ are considered.

One may then estimate d_e for the values of y sampled under the three options on σ_b^2. The prior, $\sigma_b \sim U(0, 10)$, is adopted. With a high level of random variation ($\sigma_b^2 = 1$), we obtain $d_e = 24.7$, while low random variation ($\sigma_b^2 = 0.01$) gives $d_e = 6.9$, and a situation with limited random variation ($\sigma_b^2 = 0.001$) leads to $d_e = 6.3$. Therefore, the effective parameters associated with a random effects density increase with the level of heterogeneity present. The analysis is complicated by overestimation of σ_b^2 against its true value (i.e., the one on which the simulated y are based), with posterior means of 1.55, 0.06, and 0.04 under the three scenarios.

2.4 Variance Component Choice and Model Averaging

A considerable amount of research has been devoted to MCMC selection of significant fixed effects, namely, significant predictors in regression—sometimes called Bayesian "variable selection" or predictor selection. Such techniques are consistent with a formal Bayes approach, but less constrained by the complex integration issues that may be involved in obtaining marginal likelihoods. Different possible approaches to predictor selection are considered by Fernandez et al. (2001), George and McCulloch (1997), and Sala-i-Martin et al. (2004). Let J_j be a binary indicator for retaining or excluding the jth regression coefficient β_j, George and McCullough (1993, 1997) develop stochastic search variable selection (SSVS) using a mixture prior,

$$p(\beta_j|J_j) = 1(J_j = 1)p(\beta_j|J_j = 1) + 1(J_j = 0)p(\beta_j|J_j = 0),$$

in which the "inclusion prior," $p(\beta_j|J_j = 1)$, is a diffuse or possibly informative prior, but one that allows realistic search for the parameter value. By contrast, the "exclusion prior," $p(\beta_j|J_j = 0)$, is centered at zero with high precision. For example, one might have,

$$p(\beta_j|J_j = 1) \sim N(0, V_j),$$

with V_j large, but,

$$p(\beta_j|J_j = 0) \sim N(0, V_j/K_j) \qquad K_j \gg 1,$$

with K_j chosen so that the sampling from the prior is constrained to values around zero, that is, to substantively insignificant values. If all p predictors apart from the intercept are open to inclusion or exclusion, then MCMC sampling over parameters β_j and indicators J_j is averaging over 2^p possible models (Fernandez et al., 2001).

By contrast, Kuo and Mallick (1998) and Smith and Kohn (1996) take the selection indicators, J_j, and coefficients, β_j, to be independent rather than being governed by mixture priors. Assuming normal priors, one has $\beta_j = 0$ if $J_j = 0$, but $p(\beta_j) \sim N(0, V_j)$ if $J_j = 1$. Following Zellner (1986), the prior on $(\beta_0, \beta_1, \ldots, \beta_p)$ when $J_j = 1$ may be specified as a g-prior, namely,

$$(\beta_0, \beta_1, \ldots, \beta_p|\sigma^2) \sim N_{p+1}(B, g\sigma^2(XX)^{-1}),$$

where g is a known constant and B is typically a vector of zeroes (Vannucci, 2000).

Model indicator selection ideas has also been applied to the parameters governing random effects so that only genuine sources of heterogeneity are retained. Such covariance selection helps ensure sparse structure in the covariance matrix of the selected (retained) random effects (Frühwirth-Schnatter and Tüchler, 2008). Selection may relate to the retention or otherwise of univariate random effects—for example, a multilevel model with a random intercept (as in the turtle survival analysis) or the convolution model of Besag et al. (1991) for area count data. For multivariate random effects, such as random cluster intercepts, b_{0j}, and slopes $\{b_{1j}, \ldots, b_{pj}\}$ in a multilevel analysis,

$$y_{ij} = b_{0j} + b_{1j}x_{1ij} + \cdots + b_{pj}x_{pij} + u_{ij},$$

one can consider retaining covariances, Σ_{bgh}, subject to variances in both effects, b_{gj} and b_{hj}, being retained. Thus, Smith and Kohn (2002) identify zero off-diagonal elements in the inverse, $\Pi_b = \Sigma_b^{-1}$, of the variance–covariance matrix. Alternatively, one may allow exclusion also of variance components (diagonal terms in Σ_b), which necessarily leads to exclusion of associated covariances.

Shively et al. (1999) suggest variance component selection based on informative priors in the sense of using an exploratory run of the data to provide a sensible prior on the variance component. In general, noninformative priors are to be avoided in model selection applications as they bias selection toward null models (Ghosh and Dunson, 2008). Shively et al. (1999) suggest initial runs of random effects models based on the usual choice for variances, such as a diffuse inverse gamma or uniform over a wide interval (Yau and Kohn, 2003, 193), or an inverse Wishart prior for covariance matrices. Suppose σ_b^2 is a single variance component (e.g., an intercept variance) as in the turtle data example above. Long run of samples of $h^{(r)} = \log(\sigma_b^{2(r)})$ would then be obtained in a first stage analysis using standard diffuse priors, providing the basis for lognormal priors on σ_b^2,

$$\sigma_b^2 \sim LN(M_\sigma, V_\sigma),$$

at the second stage. Specifically, as in Yau et al. (2003, 34), the median of $h^{(r)}$ provides the mean, M_σ, for a second stage lognormal prior, while the variance, V_σ, is provided by n times the variance, $V_h = \text{var}(h^{(r)})$, where n is the total sample size, or the number of units involved in defining the particular random effect. Shively et al. (1999, 779–80) argue that scaling the variance in this way leads to model selection that approximately replicates selection via the BIC. Variance inflation by a factor of n is a particular setting in a g-prior adjustment as discussed by Liang et al. (2008, 413), and one might instead choose a data-based approach with $V_\sigma = gV_h$ involving a uniform prior on the shrinkage factor $g/(1+g)$.

Consider a model with K variance components. For example, the convolution model of Besag et al. (1991) widely used in disease mapping has $K = 2$ components. Then the variance component selection at the second stage involves binary indicators, $J_k \sim Bern(\pi_k)$, where (at any particular MCMC iteration) a sampled value, $J_k = 1$, corresponds to retaining the random effects b_{kj} (for areas j) with a nonzero variance, while $J_k = 0$ corresponds to $\sigma_{bk}^2 = 0$, and amounts to setting all $b_{kj} = 0$. The π_k could be preset or taken as extra unknowns. If π_k are preset (e.g., $\pi_k = 0.5$), then the posterior probability, $Pr(J_k = 1|y)$, would be obtained from the MCMC run, and the odds ratio $Pr(J_k = 1|y)/[1 - Pr(J_k = 1|y)]$, compared to its prior value (which is 1 when $\pi_k = 0.5$).

One may also follow an analog to the George and McCulloch (1997) approach in predictor selection, whereby $J_k = 1$ corresponds to a realistic variance prior, while $J_k = 0$ corresponds to choosing an informative prior giving a very low variance, which is substantively negligible. The setting for the latter would typically be context specific and depend on the scale of the predictors (including intercepts) to which random effects are applied. For example, if b_j were random cluster effects (variable intercepts) in a two level normal model for responses y_{ij},

$$y_{ij} = b_j + X_i\beta + e_{ij},$$

with intercept variance anticipated to be around 5, then the "no effective variance" option could be centered at a variance of 0.05, for instance via an informative $Ga(20, 1)$ prior on the precision.

Selection schemes applicable to both diagonal and off-diagonal elements in covariance matrices for random effects have been developed by Cai and Dunson (2006), Chen and Dunson (2003), and Fruhwirth-Schnatter and Tuchler (2008). Thus consider a general linear mixed model for nested responses y_{ij} (as in longitudinal data with repetitions i over subjects j) with means μ_{ij}.

These means are linked to a $P \times 1$ vector of regressors, X_{ij}, and $Q \times 1$ vector of regressors, Z_{ij}, via the model,

$$g(\mu_{ij}) = X'_{ij}\beta + Z'_{ij}b_j,$$

where g is an appropriate link, $\beta = (\beta_1, \ldots, \beta_Q)'$ denotes the central fixed effects, and $b_j = (b_{j1}, \ldots, b_{jQ})'$ are zero mean random effects with covariance $\sum_b = \{\sigma_{bkl}\}$. For continuous data—and discrete outcomes subject to overdispersion—an observation level residual is also present, so that,

$$g(\mu_{ij}) = X'_{ij}\beta + Z'_{ij}b_j + u_{ij},$$

with u_{ij} usually taken as i.i.d. errors.

Following Cai and Dunson (2006), one possible Cholesky decomposition of the covariance matrix for $b_j = (b_{j1}, \ldots, b_{jQ})'$ has the form,

$$\Sigma_b = \Lambda\Gamma\Gamma'\Lambda,$$

where $\Lambda = \mathrm{diag}(\lambda_1, \ldots, \lambda_Q)$ and Γ is a lower triangular matrix,

$$\Gamma = \begin{pmatrix} 1 & 0 & \cdots & 0 \\ \gamma_{21} & 1 & \cdots & 0 \\ \cdots & \cdots & \ddots & 0 \\ \gamma_{Q1} & \gamma_{Q2} & \cdots & 1 \end{pmatrix},$$

implying,

$$\sigma_{bkl} = \lambda_k\lambda_l\left(\gamma_{r_2 r_1} + \sum_{s=1}^{r_1-1}\gamma_{ks}\gamma_{ls}\right),$$

where $r_2 = \max(k, l)$, $r_1 = \min(k, l)$. Then one has,

$$g(\mu_{ij}) = X'_{ij}\beta + Z'_{ij}\Lambda\Gamma c_j + u_{ij},$$

where $\{c_{jq} \sim N(0, 1),\ q = 1, \ldots, Q\}$ are uncorrelated standard normal variables.

The selection indicators for retaining variances and covariances are $J_q \sim Bern(\pi_\Lambda)$, governing the diagonal terms in Λ, and $H_{kl} \sim Bern(\pi_\Gamma)$ governing the terms in Γ. Note that retaining γ_{kl} requires not only $H_{kl} = 1$, but $J_k = J_l = 1$. If either J_k or J_l is zero then γ_{kl} is necessarily excluded. Cai and Dunson (2006) suggest positive truncated normal priors with variance 10 for the diagonal terms λ_q, namely,

$$\begin{aligned} \lambda_q &\sim N(0, 10)I(0,) \qquad && \text{if} \quad J_q = 1, \\ \lambda_q &= 0 && \text{if} \quad J_q = 0, \end{aligned}$$

but one may also use the data-based approach of Shively et al. (1999). Diffuse priors are not recommended (Cai and Dunson, 2008, 72) as they may favor the null model. There may also be a case for interlinked priors for λ_q and the variances of the u_{ij} effects (if present).

Smith and Kohn (2002) parameterize the covariance matrix through a Cholesky decomposition of its inverse. Thus,

$$\Sigma_b^{-1} = HDH',$$

where H is lower triangular with ones on the diagonal, while D is diagonal, $D = \text{diag}(d_1, \ldots, d_Q)$. So for $Q = 2$, one obtains,

$$\Sigma_b^{-1} = \begin{bmatrix} h_1 & d_1 h_{21} \\ d_1 h_{21} & h_{21}^2 d_1 + d_2 \end{bmatrix},$$

and for $Q = 3$,

$$\Sigma_b^{-1} = \begin{bmatrix} d_1 & d_1 h_{21} & d_1 h_{31} \\ d_1 h_{21} & h_{21}^2 d_1 + d_2 & h_{21} h_{31} d_1 + h_{32} d_2 \\ d_1 h_{31} & h_{21} h_{31} d_1 + h_{32} d_2 & h_{31}^2 d_1 + h_{32}^2 d_2 + d_3 \end{bmatrix}.$$

Kohn and Smith apply selection to the elements of H, so that, potentially, H could be reduced to an identity matrix. They take $J_{jk}^h \sim Bern(\pi_H)$, where π_H is an extra unknown and,

$$h_{jk} \neq 0 \text{ if } J_{jk}^h = 1, \; j \geq k,$$
$$h_{jk} = 0 \text{ if } J_{jk}^h = 0.$$

If all selection indicators are zero, Σ_b is shrunk toward the diagonal matrix D^{-1}.

Selection can be extended to the diagonal terms in D also, with the "variance absent" option corresponding to infinite precision, $d_q \to \infty$. The George–McCulloch strategy of effectively zero variance would correspond to a very high precision chosen for each variation source, q, according to context. Thus, $J_q^d = 1$ corresponds to selecting d_q from priors that allow realistic exploration, while $J_q^d = 0$ corresponds to setting d_q to large values, taken so large as to be effectively equivalent to random variation being negligible.

Fruhwirth-Schnatter and Tuchler (2008) consider the covariance matrix decomposition,

$$\Sigma_b = CC',$$

with C a lower triangular matrix of dimension Q including unknown diagonal terms C_{qq}. To illustrate the resulting covariance selection procedure, a hierarchical linear normal model with varying cluster regression effects, $\beta_j = \beta + b_j$, of dimension Q, would be reframed as,

$$y_{ij} = X_{ij}(\beta + b_j) + u_{ij} = X_{ij}\beta + X_{ij} C z_j + u_{ij},$$

where $u_{ij} \sim N(0, 1/\tau_u)$ and $z_j = (z_{j1}, z_{j2}, \ldots, z_{jQ})'$ is a $Q \times 1$ vector distributed as $N_Q(0, I)$. Consider binary indicators, J_{kl}, for retention or otherwise of each of the $Q(Q+1)/2$ elements of C. Then,

$$C_{kl} \neq 0 \text{ if } J_{kl} = 1 \qquad (\text{for } k \geq l),$$
$$C_{kl} = 0 \text{ if } J_{kl} = 0,$$

and b_{jk} is 0 at a particular iteration if all C_{kl} in the kth row of C are zero. A possible prior for the J_{kl} indicators is Bernoulli with probability π_J, where π_J follows a beta density,

$$\pi_J \sim Be(T_J + 1, Q(Q+1)/2 - T_J + 1),$$

based on the total free covariance parameters and the number T_J of J_{kl} taking the value 1. For $Q = 1$ in a model where a cluster level random intercept is to be tested for inclusion, one would have,

$$\mu_{ij} = \beta_0 + X_{ij}\beta + b_j + u_{ij} = \beta_0 + X_{ij}\beta + cz_j + u_{ij},$$

where $z_j \backsim N(0, 1)$, and $c \neq 0$ if $J = 1$ and $c = 0$ if $J = 0$. The (model averaged) estimate of the covariance matrix, Σ_b, of b_j over $r = 1, \ldots, R$ iterations of a chain is obtained as,

$$\Sigma_b = \frac{1}{R}\sum_{r=1}^{R} C^{(r)}(C')^{(r)}.$$

Example 2.3. Hypertension Trial To illustrate covariance selection for potentially correlated multiple random effects, this example considers clinicial trial data from Brown and Prescott (1999). In the trial concerned, 288 patients are randomly assigned to one of three drug treatments for hypertension, 1 = Carvedilol, 2 = Nifedipine, and 3 = Atenolol. The data consist of a baseline reading, BS_i, of diastolic blood pressure, and four post-treatment blood pressure readings, y_{it}, at two weekly intervals (weeks 3, 5, 7, and 9 after treatment). Some patients are lost to follow up (there are 1092 observations rather than $4 \times 288 = 1152$), but for simplicity their means are modeled for all $T = 4$ periods.

The baseline analysis (model 1) includes random patient intercepts and random slopes on the baseline reading. Additionally, the new treatment, Carvelidol, is referenced in the fixed effects comparison vector, $\eta = (\eta_1, \eta_2, \eta_3)$, leading to the corner constraint, $\eta_1 = 0$. Then for patients, $i = 1, \ldots, 288$, with treatments, Tr_i, and waves $t = 1, \ldots, 4$,

$$y_{it} = \beta_1 + b_{1i} + (\beta_2 + b_{2i})BS_i + \eta_{Tr_i} + u_{it},$$

with $u_{it} \sim N(0, 1/\tau_u)$ taken to be uncorrelated through time. In line with a commonly adopted methodology, the b_{qi} are taken to be bivariate normal with mean zero and covariance Σ_b. The precision matrix, Σ_b^{-1}, is assumed to be Wishart with 2 degrees of freedom and identity scale matrix, S. The observation level precision is taken to have a gamma prior, $\tau_u \backsim Ga(1, 0.001)$.

The last 4500 iterations from a two chain run of 5000 iterations give posterior means (and 95% intervals) for $\beta = (\beta_1, \beta_2)$ of 52.8 (43.6, 75.2) and 0.39 (0.09, 0.48). Posterior means (and 95% intervals) for the random effect standard deviations, $\sigma_{bj} = \sqrt{\Sigma_{bjj}}$, of $\{b_{1i}, b_{2i}\}$ are 0.98 (0.37, 2.7) and 0.094 (0.084, 0.103). Although the posterior density of σ_{b1} is bounded away from zero, none of the ratios $|b_{1i}/\text{sd}(b_{1i})|$ of posterior means to standard deviations of the varying intercepts exceed 0.04. By contrast, 74 of 288 ratios $|b_{2i}/\text{sd}(b_{2i})|$ exceed 2. Correlation between the effects does not seem to be apparent, with σ_{b12} having a 95% interval straddling zero.

So, absence of significant intercept effects at the individual patient level contrasts with widely significant slope variations. This suggests possible redundancy in the full bivariate random effects approach. To implement the Shively et al. (1999) method, the logs of the variances Σ_{bjj} are monitored and have posterior medians (−0.35, −4.8) and variances (0.94, 0.0084). Following Yau et al. (2003, 567), these variances are multiplied by the number of patients (288). This procedure provides parameters for lognormal priors on σ^2_{bj} which are those used for covariance selection. Independent rather than correlated effects $\{b_{1i}, b_{2i}\}$ are now assumed.

Following the principle suggested by George and McCulloch (1993), an alternative to the "variance present" option (where the variances are those of the random intercepts and slopes) is not to set the variance to zero, but to a variance that is inconsequential in the application. These are set here[6] at $\sigma^2_{b1} = 0.001$ and $\sigma^2_{b2} = 0.0001$, so that the precisions τ_{b1} and τ_{b2} are 1000 and 10,000; this specification forms model 2.

After 10,000 iterations in a two chain run of 50,000, both chains overwhelmingly choose to exclude the first variance component, namely, the one associated with variations b_{1i} around the intercept β_1. Prior probabilities for retaining each component are set at 0.5, but the posterior probability for retaining σ^2_{b1} is under 0.01, while the second component (for varying slopes on baseline blood pressure) has a posterior retention probability of 1.

Finally in model 3, full covariance selection is considered via the approach of Fruhwirth-Schnatter and Tuchler (2008). Context-based informative priors for the diagonal elements of C are assumed, namely, $C_{11} \sim E(1)$ and $C_{22} \sim E(10)$, based on the posterior means 0.98 and 0.09 for the random effects standard deviations from the baseline analysis. For the lower diagonal term a normal prior, $C_{21} \sim N(0, 1)$, is assumed. These options are preferred to, say, adopting diffuse priors on the C_{jk} terms, in order to stabilize the covariance selection analysis. Such a strategy may be compared to the Shively et al. (1999) use of lognormal priors for variances based on results of earlier analysis. Note that the covariance term, Σ_{21}, is nonzero only when both C_{11} and C_{21} are retained[7].

The last 9000 iterations of a two chain run of 10,000 iterations give a posterior probability of 1 for retaining slope variation, while the posterior probability for intercept variation is 0.6. Posterior means (and 95% intervals) for the random effect standard deviations $\sigma_{bj} = \sqrt{\Sigma_{bjj}}$ of $\{b_{1i}, b_{2i}\}$ are 0.68 (0, 3.8), and 0.065 (0.050, 0.097), with a clear spike at zero in the density for σ_{b1}. Analogous posterior results for β_q $(q = 1, 2)$ are 49.2 (43.2, 55.1) and 0.43 (0.37, 0.49), so that the average baseline effect, β_2, is both higher and more precisely identified than in the conventional bivariate random effects model (model 1).

Finally, it may be noted that using diffuse $Ga(1, 0.001)$ priors on C_{11} and C_{22}, and a $N(0, 1000)$ prior on C_{12}, did not produce a posterior analysis resembling the one using more informative priors. A 20,000 iteration two chain run in fact shows a zero posterior probability for retaining varying slopes. This is in line with a general principle that model selection tends to choose the null model if diffuse priors are taken on the parameter(s) subject to inclusion or rejection (Cai and Dunson, 2008).

2.5 Predictive Methods for Model Choice and Checking

A number of studies have pointed to drawbacks in focusing solely on the marginal likelihood or Bayes factor as a single global assessment measure of the performance of complex models, and also mentioned computational and inferential difficulties with the Bayes factor when priors are diffuse, as well as the need to examine fit for individual observations to make sense of global criteria (e.g., Gelfand, 1996; Johnson, 2004). While formal Bayes methods can be extended to assessing the fit of single observations (Pettit and Young, 1990), it may be argued that predictive likelihood methods offer a more flexible approach to assessing the role of individual observations. In fact, predictive methods have a role both in model choice and model checking.

2.5.1 Predictive model checking and choice

The formal predictive likelihood approach assumes only part of the observations are used in estimating a model. On this basis, one may obtain cross-validation predictive densities (Vehtari and Lampinen, 2002), $p(y_s|y_{[s]})$, where y_s denotes a subset of y (the "validation data"), and $y_{[s]}$ is the complementary "test data" formed by excluding y_s from y. If $[i]$ is defined to contain all the data $\{y_1, \ldots, y_{i-1}, y_{i+1}, \ldots, y_n\}$ except for a single observation i, then the densities,

$$p(y_i|y_{[i]}) = \int p(y_i|\theta, y_{[i]})p(\theta|y_{[i]})d\theta,$$

are called conditional predictive ordinates (CPOs) (e.g., Chaloner and Brant, 1988; Geisser and Eddy, 1979), and sampling from them shows what values of y_i are likely when a model is applied to all the data points except the ith, namely, to the data $y_{[i]}$. The predictive distribution, $p(y_i|y_{[i]})$, can be compared to the actual observation in various ways (Gelfand et al., 1992).

For example, to assess whether the observation is extreme (not well fitted) in terms of the model being applied, replicate data, $y_{i,\text{rep}}$, may be sampled from $p(y_i|y_{[i]})$ and their concordance with the data may be represented by probabilities (Marshall and Spiegelhalter, 2003),

$$Pr(y_{i,\text{rep}} \leq y_i|y_{[i]}).$$

These are estimated in practice by counting iterations r where the constraint $y_{i,\text{rep}}^{(r)} \leq y_i$ holds. For discrete data, this assessment is based on the probability,

$$Pr(y_{i,\text{rep}} < y_i|y_{[i]}) + 0.5Pr(y_{i,\text{rep}} = y_i|y_{[i]}).$$

Gelfand (1996) recommends assessing concordance between predictions and actual data by a tally of how many actual observations, y_i, are located within the 95% interval of the corresponding model prediction, $y_{i,\text{rep}}$. For example, if 95% or more of all the observations are within 95% posterior intervals of the predictions $y_{i,\text{rep}}$, then the model is judged to be reproducing the observations satisfactorily.

The collection of predictive ordinates $\{p(y_i|y_{[i]}), i = 1, n\}$ is equivalent to the marginal likelihood $p(y)$ when $p(y)$ is proper, in that each uniquely determines the other. A pseudo Bayes factor is obtained as a ratio of products of leave one out cross-validation predictive densities (Vehtari and Lampinen, 2002) under models M_1 and M_2, namely,

$$\text{PsBF}(M_1, M_2) = \prod_{i=1}^{n} \{p(y_i|y_{[i]}, M_1)/p(y_i|y_{[i]}, M_2)\}.$$

In practical data analysis, one typically uses logs of CPO estimates and totals the log(CPO) to derive log pseudo marginal likelihoods and log pseudo Bayes factors (Sinha et al., 1999, 588).

Monte Carlo estimates of conditional predictive ordinates, $p(y_i|y_{[i]})$, may be obtained without actually omitting cases, so formal cross validation based on n separate estimations (the first omitting case 1, the second omitting case 2, etc.) may be approximated by using a single estimation run. For parameter samples $\{\theta^{(1)}, \ldots, \theta^{(R)}\}$ from an MCMC chain, an estimator for the CPO, $p(y_i|y_{[i]})$, is

$$\frac{1}{p(y_i|y_{[i]})} = \frac{1}{R}\sum_{r=1}^{R} \frac{1}{p(y_i|\theta^{(r)})},$$

namely, the harmonic mean of the likelihoods for each observation (Aslanidou et al., 1998; Silva et al., 2006). In computing terms, an inverse likelihood needs to be calculated for each case at each iteration, the posterior means of these inverse likelihoods obtained, and the CPOs are the inverse of those posterior mean inverse likelihoods. Denoting the inverse likelihoods as $H_i^{(r)} = (1/p(y_i|\theta^{(r)}))$, one would in practice take minus the logarithms of the posterior means of H_i, for example in a spreadsheet environment since small numbers may be involved. The sum over all cases of these estimates of $\log(\text{CPO}_i)$ provides a simple estimate of the log pseudo marginal likelihood. In the turtle data example, the fixed effects only model 1 has a PsBF of -151.8, while the random effects model 2 has a Pseudo Bayes Factor (PsBF) of -149.6 under a $Ga(0.1, 0.1)$ prior for τ_b, and under a $Ga(1, 1)$ prior has a PsBF $= -150.8$. So the pseudo Bayes factors tends to support the random effects option.

Model fit (and hence choice) may also be assessed by comparing samples y_{rep} from the posterior predictive density based on all observations, though such procedures may be conservative since the presence of y_i influences the sampled $y_{i,\text{rep}}$ (Marshall and Spiegelhalter, 2003). Laud and Ibrahim (1995) and Meyer and Laud (2002) propose model choice based on minimization of the criterion,

$$C = E[c(y_{\text{rep}}, y)|y] = \sum_{i=1}^{n} \{\text{var}(y_{i,\text{rep}}) + [y_i - E(y_{i,\text{rep}})]^2\},$$

where for y continuous, $c(y_{\text{rep}}, y)$ is the predictive error sum of squares,

$$c(y_{\text{rep}}, y) = (y_{\text{rep}} - y)'(y_{\text{rep}} - y).$$

The C measure can be obtained from the posterior means and variances of sampled $y_{i,\text{rep}}^{(r)}$ or from the posterior average of $\sum_{i=1}^{n} \left(y_{i,\text{rep}}^{(r)} - y_i\right)^2$. Carlin and Louis (2000) and Buck and Sahu (2000) propose related model fit criteria appropriate to both metric and discrete outcomes.

Posterior predictive loss (PPL) model choice criteria allow varying trade-offs in the balance between bias in predictions and their precision (Gelfand and Ghosh, 1998; Ibrahim et al., 2001). Thus, for k positive and y continuous, one possible criterion has the form,

$$\text{PPL}(k) = \sum_{i=1}^{n} \left\{\text{var}(y_{i,\text{rep}}) + \left(\frac{k}{k+1}\right) [y_i - E(y_{i,\text{rep}})]^2 \right\}.$$

This criterion would be compared between models at selected values of k, typical values being $k = 0$, $k = 1$, and $k = 10{,}000$, where higher k values put greater stress on accuracy in predictions and less on precision. One may consider calibration of such measures, namely, expressing the uncertainty of C or PPL in a variance measure (Ibrahim et al., 2001; Laud and Ibrahim, 1995). De la Horra and Rodrıguez-Bernal (2005) suggest predictive model choice based on measures of distance between the two densities that can potentially be used for predicting future observations, namely, sampling densities and posterior predictive densities.

To assess poorly fitted cases, the CPO values may be scaled (dividing by their maximum) and low values for particular observations (e.g., under 0.001) will then show observations that the model does not reproduce effectively (Weiss, 1994). If there are no very small scaled CPOs then a relatively good fit of the model to all data points is suggested and is likely to be confirmed by other forms of predictive check. The ratio of extreme percentiles of the CPOs is useful as an indicator of a good fitting model, e.g., the ratio of the 99th to the 1st percentile.

An improved estimate of the CPO may be obtained by weighted resampling from $p(\theta|y)$ (Marshall and Spiegelhalter, 2003; Smith and Gelfand, 1992). Samples $\theta^{(r)}$ from $p(\theta|y)$ can be converted (approximately) to samples from $p(\theta|y_{[i]})$ by resampling the $\theta^{(r)}$ with weights,

$$w_i^{(r)} = G\big(y_i|\theta^{(r)}\big) \Big/ \sum_{r=1}^{R} G\big(y_i|\theta^{(r)}\big),$$

where,

$$G\big(y_i|\theta^{(r)}\big) = 1/p\big(y_i|\theta^{(r)}\big),$$

is the inverse likelihood of case i at iteration r. Using the resulting resampled values, $\tilde{\theta}^{(r)}$, corresponding predictions, $\tilde{y}_{\text{rep}}$, can be obtained, which are a sample from $p(y_i|y_{[i]})$.

2.5.2 Posterior predictive model checks

A range of model checks can also be applied using samples from the posterior predictive density without actual case omission. To assess predictive performance, samples of replicate data y_{rep} from,

$$p(y_{\text{rep}}|y) = \int p(y_{\text{rep}}|\theta)p(\theta|y)d\theta,$$

may be taken, and checks made against the data, for example, whether the actual observations y are within 95% credible intervals of y_{rep}. Formally, such samples are obtained by the method of composition (Chib, 2008), whereby if $\theta^{(r)}$ is a draw from $p(\theta|y)$, then $y_{\text{rep}}^{(r)}$ drawn from $p(y_{\text{rep}}|\theta^{(r)})$ is a draw from $p(y_{\text{rep}}|y)$. In a satisfactory model, namely, one that adequately reproduces the data being modeled, predictive concordance (accurate reproduction of the actual data by replicate data) is at least 95% (Gelfand, 1996, 158).

Other comparisons of actual and predicted data can be made, for example by a chi-square comparison (Gosonuiou et al., 2006). Johnson (2004) proposes a Bayesian chi-square approach based on partitioning the cumulative distribution into K bins, usually of equal probability. Thus, one chooses quantiles,

$$0 \equiv a_0 < a_1 < \cdots < a_{K-1} < a_K \equiv 1,$$

with corresponding bin probabilities

$$p_k = a_k - a_{k-1}, \quad k = 1, \ldots, K.$$

Then using model means μ_i for subject $i \in (1, \ldots, n)$, one obtains the implied cumulative density, q_i, say $a_{k^*-1} < q_i < a_{k^*}$, and allocates the fitted point to a bin randomly chosen from bins $1, \ldots, k^*$. For example, under a Poisson likelihood, suppose there are $K = 5$ equally probable intervals, with $p_k = 0.2$. If $\mu_i = 1.4$, the probability assigned to an observation $y_i = 1$ by the cumulative density function falls in the interval (0.247, 0.592), which straddles bins 2 and 3. To allocate a bin, a $U(0.247, 0.592)$ variable is sampled, and the predicted bin is 2 or 3 according to whether the sampled uniform variable falls within (0.247, 0.4) or (0.4, 0.592). The totals so obtained accumulating over all subjects define predicted counts, $m_k(\tilde{\theta})$, which are compared (at each MCMC iteration) to actual counts, np_k, as in formula (3) in Johnson (2004). This provides the Bayesian chi-square criterion,

$$R^B(\tilde{\theta}) = \sum_{k=1}^{K} \frac{[m_k(\tilde{\theta}) - np_k]^2}{np_k},$$

where $R^B(\tilde{\theta})$ is asymptotically χ^2_{K-1}, regardless of the parameter dimension of the model being fitted[8]. One can assess the posterior probability that $R^B(\tilde{\theta})$ exceeds the 95th percentile of the χ^2_{K-1} density. Poor fit will show in probabilities considerably exceeding 0.05.

Analogs of classical significance tests are obtained using the posterior predictive p value (Kato and Hoijtink, 2004). This was originally defined (Meng, 1994) as the probability that a test statistic, $T(y_{\text{rep}})$, of future observations; y_{rep}, is larger than or equal to the observed value of $T(y)$, given the adopted model M, the response data y and any ancilliary data x,

$$p_{\text{post}} = Pr[T(y_{\text{rep}}, x) \geq T(y, x)|y, x, M],$$

where x would typically be predictors measured without error. The probability is calculated over the posterior predictive distribution of y_{rep} conditional on M and x. By contrast, the classical p-test integrates over y, as in,

$$p_c = Pr[T(y_{\text{rep}}, x) \geq T(y, x)|x, M].$$

The formulation of Meng (1994) is extended by Gelman et al. (1996) to apply to discrepancy criteria, $D(y, \theta)$, based on data and parameters, as well as to observation-based functions $T(y)$. So,

$$p_{\text{post}} = Pr[D(y_{\text{rep}}, x, \theta) \geq D(y, x, \theta)|y, x, M],$$

where the probability is taken over the joint posterior distribution of y_{rep} and θ given M and x. In estimating the corresponding p_{post}, the discrepancy is calculated at each MCMC iteration. This is done both for the observations, giving a value $D(y, x, \theta^{(r)})$, and for the replicate data, $y^{(r)}_{\text{rep}}$, sampled from $p(y^{(r)}_{\text{rep}}|\theta^{(r)}, x)$, resulting in a value $D(y^{(r)}_{\text{rep}}, x, \theta^{(r)})$ for each sampled parameter $\theta^{(r)}$. The proportion of samples where $D(y^{(r)}_{\text{rep}}, x, \theta^{(r)})$ exceeds $D(y, x, \theta^{(r)})$

is then the Monte Carlo estimate of p_{post}. For example, Kato and Hoijtink (2004) show good performance of p_{post} using both statistics T and discrepancies D in a normal multilevel model context with subjects $i = 1, \ldots, m_j$ in clusters $j = 1, \ldots, J$,

$$y_{ij} = b_{1j} + b_{2j}x_{ij} + u_{ij},$$

where $u_{ij} \backsim N(0, \sigma_j^2)$. The hypotheses considered (i.e., in the form of reduced models) are $b_{kj} = \beta_k$ and $\sigma_j^2 = \sigma^2$.

Posterior predictive checks may be used to assess model assumptions. For instance, in multilevel and general linear mixed models, assumptions of normality regarding random effects are often made by default, and a posterior check against such assumptions is sensible. A number of classical tests have been proposed, such as the Shapiro–Wilk W statistic (Royston, 1993) and the Jarque–Bera test (Bera and Jarque, 1980). These statistics can be derived at each iteration for actual and replicate data, and the comparison, $D(y_{\text{rep}}, x, \theta) \geq D(y, x, \theta)$, applied over MCMC iterations to provide a posterior predictive p value.

2.5.3 Mixed predictive checks

The posterior predictive check makes double use of the data and so may be conservative as a test (Bayarri and Berger, 1999, 2000), since the observation y_i has a strong influence on the replicate $y_{i,\text{rep}}$. For example, Sinharay and Stern (2003) show that posterior predictive checks may fail to detect departures from normality in random effects models. However, also in the context of random effects and hierarchical models, Marshall and Spiegelhalter (2003, 2007) mention a mixed predictive scheme, which uses a predictive prior distribution, $p(y_{i,\text{rep}}|b_{i,\text{rep}}, \theta)$, for a new set of random effects. The associated model check is called a mixed predictive p test, whereas the (conservative) option of sampling from $p(y_{i,\text{rep}}|b_i, \theta)$ results in what Marshall and Spiegelhalter (2007, 424) term full-data posterior predictive p values. Mixed predictive replicates for each case seek to reduce dependence on the observation for that case, as the replicate data is sampled conditional only on global hyperparameters. Therefore, mixed predictive p values are expected to be less conservative than posterior predictive p values.

Let b denote random effects for cases $i = 1, \ldots, n$, or for clusters j in which individual cases are nested. To generate a replicate $y_{i,\text{rep}}$ for the ith case under the mixed scheme involves sampling (θ, b) from the usual posterior $p(\theta, b|y)$ conditional on all observations, but the sampled b are ignored and instead replicate b_{rep} values taken. A fully cross-validatory method would require that $b_{i,\text{rep}}$ be obtained by sampling from $p(b_{i,\text{rep}}|y_{[i]})$, or $b_{j,\text{rep}}$ sampled from $p(b_{j,\text{rep}}|y_{[j]})$; in fact, Green et al. (2009) compare mixed predictive assessment schemes with full cross-validation based on omitting single observations.

As full cross-validation is computationally demanding when using MCMC methods, approximate cross-validatory procedures are proposed by Marshall and Spiegelhalter (2007), in which the replicate random effect is sampled from $p(b_{i,\mathrm{rep}}|y)$, followed by a step sampling $y_{i,\mathrm{rep}}$ from $p(y_{i,\mathrm{rep}}|b_{i,\mathrm{rep}}, \theta)$. A discrepancy measure T^{obs} based on the observed data is then compared to its reference distribution,

$$p(T|y) = \int p(T|b)p^M(b|y)\,db,$$

where $p^M(b|y) = \int p(b|\theta)p(\theta|y)d\theta$ may be termed the "predictive prior" for b (Marshall and Spiegelhalter, 2007, 413). This contrasts with more conservative posterior predictive checks based on replicate sampling from $p(y_{i,\mathrm{rep}}|b_i, \theta)$, under which T^{obs} is compared to the reference distribution,

$$p(T|y) = \int p(T|b)p(b|y)d\theta.$$

Marshall and Spiegelhalter (2003) confirm that a mixed predictive procedure reduces the conservatism of posterior predictive checks in relatively simple random effects models, and is more effective in reproducing $p(y_i|y_{[i]})$ than weighted importance sampling. However, this procedure may be influenced by the informativeness of the priors on the hyperparameters θ, and also by the presence of multiple random effects.

Marshall and Spiegelhalter (2007) also consider full cross-validatory mixed predictive checks to assess conflict in evidence regarding random effects b between the likelihood and the second stage prior; see also Bayarri and Castellanos (2007). Consider nested data $\{y_{ij}, i = 1, \ldots, n_j;\ j = 1, \ldots, J\}$ with likelihood,

$$y_{ij} \sim N(b_j, \sigma^2),$$

and second stage prior on random cluster effects,

$$b_j \sim N(\mu, \tau^2).$$

Under a cross-validatory approach the discrepancy measure T_j^{obs} for cluster j would be based on the remaining data $y_{[j]}$ with cluster j excluded, and its reference distribution is then,

$$p(T_j^{\mathrm{rep}}|y_{[j]}) = \int p(T_j^{\mathrm{rep}}|b_j, \sigma^2)p(b_j|\mu, \tau^2)p(\sigma^2, \tau^2, \mu|y_{[j]})\,db_j d\sigma^2 d\tau^2 d\mu.$$

Marshall and Spiegelhalter (2007) also propose a conflict p test based on comparing a predictive prior replicate, $b_{j,\mathrm{rep}}|y_{[j]}$, with a fixed effect estimate or "likelihood replicate" $b_{j,\mathrm{fix}}$ for b_j based only on the data. The latter is obtained using a highly diffuse fixed effects prior on the b_j, rather than a borrowing strength hierarchical prior, for example $b_j \sim Be(1,1)$ or $b_j \sim Be(0.5, 0.5)$. Defining,

$$b_{j,\mathrm{diff}} = b_{j,\mathrm{rep}} - b_{j,\mathrm{fix}},$$

the conflict p value for cluster j is obtained as,

$$p_{j,\text{conf}} = Pr(b_{j,\text{diff}} \leq 0|y).$$

This can be compared to a mixed predictive p value, based on sampling $y_{j,\text{rep}}$ from a cross-validatory model using only the remaining cases $y_{[j]}$ to estimate parameters, and then comparing $y_{j,\text{rep}}$, or some function $T_j^{\text{rep}} = T(y_{j,\text{rep}})$, with $y_{j,\text{obs}}$ or with $T_j^{\text{obs}} = T(y_{j,\text{obs}})$. Thus, depending on the substantive application, one may define lower or upper tail mixed p values,

$$p_{j,\text{mix}} = Pr\left(T_j^{\text{rep}} \leq T_j^{\text{obs}}|y\right),$$

or

$$p_{j,\text{mix}} = Pr\left(T_j^{\text{rep}} \geq T_j^{\text{obs}}|y\right),$$

with the latter being relevant in (say) assessing outliers in hospital mortality comparisons. If $T(y) = y$ and y is a count, then a mid p value is relevant instead with the upper tail test being,

$$p_{j,\text{mix}} = Pr(y_{j,\text{rep}} > y_{j,\text{obs}}|y_{[j]}) + 0.5Pr(y_{j,\text{rep}} = y_{j,\text{obs}}|y_{[j]}).$$

Example 2.4. Regional Mortality in China This example considers predictive checks in a model for mortality contrasts between the 31 provinces of China. Regional life table analysis often proceeds by using a fixed effects model independently in each region, which takes no account of spatial structure in mortality risk, and also does not model correlation in mortality between adjacent age groups. Here a random effects model pools strength over ages and regions. The analysis relates to deaths in 2000 in $i = 1, \ldots, n$ Chinese provinces with $n = 31$, with deaths also differentiated by age x (for $X = 21$ groups from 0 to 4, 5 to 9 through to 100+), by gender s, and by an urban–rural subdivision r within each province ($r = 1$ for urban, $r = 2$ for rural). Populations are from the 2000 China Census. Whereas a fixed effects model involves $31 \times 21 \times 2 \times 2 = 2604$ parameters, the approach developed here is expected to be more parsimonious while also reproducing the data satisfactorily. A Bayesian approach enables stochastic variation in life table parameters to be fully assessed (e.g., via posterior densities for province life expectancies).

The deaths data, $y_{\text{ris}x}$, collected in conjunction with the 2000 China Census, are subject to undercounting. Bannister and Hill (2004) obtain correction factors using the general growth balance method, and here these factors are applied to correct for mortality under-recording. Specifically, populations at risk, $P_{\text{ris}x}$, are reduced to account for deflation in the death recording; the male mortality adjustment factor was 1.113, so recorded deaths are retained as the response variables, but analyzed in relation to populations scaled by 1/1.113=0.898; for females the adjustment factor is 1.181, so female populations are scaled by 0.847. Due to relatively large death counts, the $y_{\text{ris}x}$ are assumed to follow a negative binomial density, namely,

$$p(y_{\text{ris}x}|\xi_{\text{ris}x},\alpha) = \frac{\Gamma(\alpha + y_{\text{ris}x})}{\Gamma(\alpha)\Gamma(y_{\text{ris}x}+1)}\left(\frac{\alpha}{\alpha+\xi_{\text{ris}x}}\right)^{\alpha}\left(\frac{\xi_{\text{ris}x}}{\alpha+\xi_{\text{ris}x}}\right)^{y_{\text{ris}x}},$$

with means $\xi_{\text{ris}x}$. Equivalently, $y_{\text{ris}x} \backsim Po(\mu_{\text{ris}x})$, with gamma means,

$$\mu_{\text{ris}x} \backsim G(\alpha, \alpha/\xi_{\text{ris}x}),$$

whereby $E(\mu_{\text{ris}x}) = \xi_{\text{ris}x}$, $\text{Var}(\mu_{\text{ris}x}) = \xi^2_{\text{ris}x}/\alpha$, and,

$$\text{Var}(y_{\text{ris}x}) = E[\text{Var}(y_{\text{ris}x}|\mu_{\text{ris}x})] + \text{Var}[E(y_{\text{ris}x}|\mu_{\text{ris}x})] = \xi_{\text{ris}x} + \xi^2_{\text{ris}x}/\alpha.$$

A multiplicative model for mortality parameters, $\xi_{\text{ris}x}$, involves:

1. a population at risk ($P_{\text{ris}x}$) offset;
2. overall mortality level (fixed effect) parameters by gender s, κ_s;
3. parameters η_{xs} to represent age-sex mortality differentials and following a first order random walk prior reflecting correlation in rates between successive ages. These age parameters are taken to apply uniformly across all provinces in line with the multiplicative model, though in fact, regional variation in age effects may be present in the Chinese mortality data;
4. fixed effect parameters, γ_s, to represent a gender-specific rural mortality differential and present only in the model for ξ_{2isx};
5. division effects, λ_{is}, reflecting unmeasured risks likely to be spatially patterned (e.g., differences in environment, health care). They are assigned a multivariate conditional autoregressive prior (see Chapters 4 and 7) with Wishart distributed precision matrix.

Then the multiplicative model under (1)–(5) is

$$\log(\xi_{1isx}) = \log(P_{1isx}) + \kappa_s + \eta_{xs} + \lambda_{is},$$
$$\log(\xi_{2isx}) = \log(P_{2isx}) + \kappa_s + \gamma_s + \eta_{xs} + \lambda_{is},$$

with model checks based on full data posterior replicates and on mixed predictions using replicate samples from the random effects $\{\eta_{xs}, \lambda_{is}\}$. The analysis[9] also includes derivation of Monte Carlo estimates of $\log(\text{CPO}_{\text{ris}x})$ statistics.

The second half of a two chain run of 3000 iterations shows that just over 5% of the cases are not included in the 95% intervals of full data posterior predictive data, $y_{\text{rep,ris}x}$. These condition on random effects sampled from the posterior rather than on replicate random effects. Using this potentially conservative method suggests that in overall terms the model's predictions seem broadly concordant with the data. However, examination of

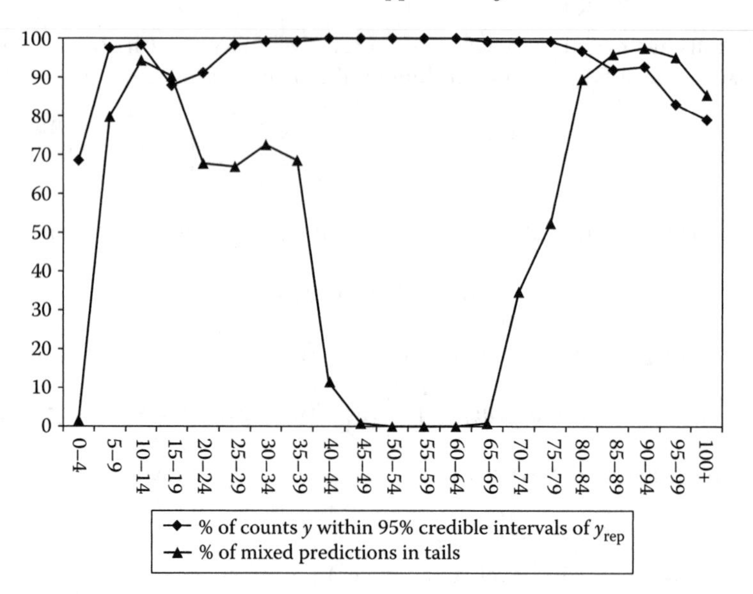

FIGURE 2.2
Predictive discrepancies by age band.

model discrepancies at the age group level shows a relatively high number of death counts for ages 0–4 and the oldest ages are not predicted satisfactorily. There are $2604/21 = 124$ observations in each age group, and 39 of the 124 death counts in the age group 0–4 are not satisfactorily predicted (see Figure 2.2). The log(CPO) statistics also show deficient fit for the youngest age group. The mixed predictions are assessed by sampling $\{\eta_{\text{rep}}, \lambda_{\text{rep}}\}$ and $\{y_{\text{rep}}|\eta_{\text{rep}}, \lambda_{\text{rep}}, \kappa, \gamma\}$ and deriving p values

$$Pr[y_{h,\text{rep}} < y_h|y) + 0.5Pr[y_{h,\text{rep}} = y_h|y),$$

where $h = (r, i, s, x)$. These mixed p tests show a broadly similar pattern (also in Figure 2.2), with high probabilities at the oldest ages, though they also highlight discrepancies in the model for young adult mortality. An alternative approach is therefore indicated, such as to allow spatially varying age effects, for at least some age groups.

2.6 Estimating Posterior Model Probabilities

A Monte Carlo method for estimating posterior model probabilities based on independent MCMC sampling of two or more different models is presented in Congdon (2007a), building on work by Carlin and Chib (1995), and Scott

(2002). This method allows iteration-specific model averaging as compared to approaches that undertake averaging over posterior parameter densities. Let $\theta = (\theta_1, \ldots, \theta_K)$ be parameters for models $(1, \ldots, K)$, with dimension $(d_1, \ldots, d_K)$, and define a model indicator $m \in (1, \ldots, K)$. Conditional on the model $m = j$, the parameter vector, θ_j, defines the likelihood for y, and y is independent of parameters in other models, $\theta_k, k \neq j$. As above, the marginal likelihood given $m = j$ is (Carlin and Chib, 1995),

$$\begin{aligned} p(y|m = j) &= \int p(y|\theta, m = j)p(\theta|m = j)d\theta \\ &= \int p(y|\theta_j, m = j)p(\theta|m = j)d\theta_j. \end{aligned}$$

The term $p(\theta|m = j)$ includes an "own model" prior $p(\theta_j|m = j)$, and $K-1$ "cross-model" priors $p(\theta_k|m = j, k \neq j)$; the latter are termed pseudo priors by Carlin and Chib (1995). Proper densities are required for the cross-model priors, $p(\theta_k|m = j, k \neq j)$, in order that $p(\theta|m = j)$ integrate to 1. With samples from the own model prior, cross-model priors, and posterior densities, $p(\theta_j|y, m = j)$, of all K models, one may estimate model weights at each iteration and so form model averaged parameters at each iteration. The posterior average of the weights may be used to derive posterior probabilities for each model.

From the posterior model probability formula,

$$p(m = j|y) = \int p(m = j, \theta|y)d\theta = \int p(m = j|y, \theta)p(\theta|y)d\theta,$$

it is apparent that a Monte Carlo estimate of $Pr(m = j|y)$ over T samples is provided by,

$$\bar{w}_j = \sum_{t=1}^{T} p\left(m = j|y, \theta^{(t)}\right)/T,$$

where $\{\theta^{(t)} = (\theta_1^{(t)}, \ldots, \theta_2^{(t)}, \ldots, \theta_K^{(t)}), t = 1, T\}$ are samples of parameters from the K posterior densities. The weights for model j at iteration t are,

$$\begin{aligned} w_j^{(t)} = p(m = j|y, \theta^{(t)}) &= \frac{p(m = j, y, \theta^{(t)})}{p(y, \theta^{(t)})} \\ &= \frac{p(y|m = j, \theta^{(t)})p(\theta^{(t)}|m = j)p(m = j)}{p(y, \theta^{(t)})}. \end{aligned}$$

The numerator in the preceding equation contains the term $p(\theta^{(t)}|m = j)$ involving a mix of own model and cross-model priors, namely,

$$p(\theta|m = j) = p(\theta_1|m = j)p(\theta_2|m = j) \cdots p(\theta_j|m = j) \cdots p(\theta_K|m = j).$$

The choice for the cross-model prior is arbitrary though a proper prior is preferred (Carlin and Chib, 1995). The simplification,

$$p(\theta_k|m = j) = g_k \qquad (\text{all } j \text{ when } k \neq j),$$

is therefore admissible, with g_k a proper density. So,

$$\begin{aligned} p(\theta_k|m = 1) &= \cdots = p(\theta_k|m = k-1) = p(\theta_k|m = k+1) \\ &= \cdots = p(\theta_k|m = K) = g_k, \end{aligned}$$

and there are K cross-model priors $\{g_1, \ldots, g_K\}$. For example, one may set g_k to be an estimate of $p(\theta_k|m = k, y)$, namely, the "own model" posterior density of θ_k given y and $m = k$.

One then has,

$$\begin{aligned} p(\theta|m = j) &= p(\theta_j|m = j) \prod_{k \neq j}^{K} p(\theta_k|m = j) \\ &= p(\theta_j|m = j) g_1 g_2 \cdots g_{j-1} g_{j+1} \cdots g_K. \end{aligned}$$

The model weights at each iteration become,

$$w_j^{(t)} = \frac{p\left(y|m = j, \theta_j^{(t)}\right) p\left(\theta_j^{(t)}|m = j\right) \left[\prod_{h \neq j} g_h^{(t)}\right] p(m = j)}{p(y, \theta^{(t)})},$$

where the denominator is,

$$\begin{aligned} p\left(y, \theta^{(t)}\right) &= \sum_{k=1}^{K} p\left(y, \theta^{(t)}, m = k\right) \\ &= \sum_{k=1}^{K} \left\{ p\left(y|\theta^{(t)}, m = k\right) p\left(\theta_k^{(t)}|m = k\right) \left[\prod_{h \neq k} g_h^{(t)}\right] p(m = k) \right\} \\ &= \sum_{k=1}^{K} p\left(y|\theta_k^{(t)}, m = k\right) p\left(\theta_k^{(t)}|m = k\right) \left[\prod_{h \neq k} g_h^{(t)}\right] p(m = k). \end{aligned}$$

Then

$$w_j^{(t)} = \frac{p(y|m = j, \theta_j^{(t)}) p(\theta_j^{(t)}|m = j) [\prod_{h \neq j} g_h^{(t)}] p(m = j)}{\sum_{k=1}^{K} p(y|\theta_k^{(t)}, m = k) p(\theta_k^{(t)}|m = k) [\prod_{h \neq k} g_h^{(t)}] p(m = k)},$$

and dividing through by the product of the K cross-model priors $(g_1 g_2 \cdots g_K)$ gives

$$w_j^{(t)} = \frac{\left\{ \frac{p(y|m=j, \theta_j^{(t)}) p(\theta_j^{(t)}|m=j) p(m=j)}{g_j^{(t)}} \right\}}{\sum_{k=1}^{K} \left\{ \frac{p(y|\theta_k^{(t)}, m=k) p(\theta_k^{(t)}|m=k) p(m=k)}{g_k^{(t)}} \right\}}.$$

For the common case when $K = 2$, one has,

$$w_1^{(t)} = \frac{\frac{p(y|\theta_k^{(t)},m=1)p(\theta_k^{(t)}|m=1)p(m=1)}{g_1^{(t)}}}{\frac{p(y|\theta_k^{(t)},m=1)p(\theta_k^{(t)}|m=1)p(m=1)}{g_1^{(t)}} + \frac{p(y|\theta_k^{(t)},m=2)p(\theta_k^{(t)}|m=2)p(m=2)}{g_2^{(t)}}}. \quad (2.1)$$

Letting $H_j^{(t)} = \left(P(y|m = j, \theta_j^{(t)})P(\theta_j^{(t)}|m = j)/g_j^{(t)}\right)$, one has,

$$w_j^{(t)} = \frac{H_j^{(t)} P(m = j)}{\sum_{k=1}^{K}[H_k^{(t)} P(m = k)]}.$$

In the $H_j^{(t)}$ term, g_j have a role parallel to that of an importance function (Geweke, 1989), and so densities, g_j, which are heavy tailed relative to the posterior may be preferred in order to minimize the variance of the components $H_j^{(t)}$ of $w_j^{(t)}$ (Yuan and Druzdzel, 2005). This might mean, for example, using a heavy tail modification Student $g_j = t(m_j, s_j^2, \nu)$ with low degrees of freedom ν, even if the own model posterior density for θ_j with mean and variance (m_j, s_j^2) is approximately normal (Congdon, 2007a).

This approach to posterior model probability estimation avoids problems involved in tuning or "jump" proposal densities to ensure that models are visited sufficiently often (Friel and Pettitt, 2008; Green and O'Hagan, 1998). Moreover, model averaging is on the basis of continuous quantities, namely, the $w_k^{(t)}$ obtained for all models and so is more efficient. It is convenient to work in the log scale and define the quantities,

$$q_k^{(t)} = \log\left\{p\left(y|\theta_k^{(t)}, m = k\right) p\left(\theta_k^{(t)}|m = k\right)\left[\prod_{a\neq k} g_a\right] p(m = k)\right\},$$

calculate deviations $\Delta q_k^{(t)} = q_k^{(t)} - \max_k(q_k^{(t)})$, with $\mathrm{w}_k^{(t)}$ then obtained by exponentiating,

$$w_k^{(t)} = \frac{\exp\left(\Delta q_k^{(t)}\right)}{\sum_k \exp\left(\Delta q_k^{(t)}\right)}.$$

As a toy example*, consider comparing two beta priors, $\pi \sim Be(\alpha, \beta)$, for a binomial probability, when the data are $x = 1$ sucesses in $n = 10$ trials. Two models are compared, differing in the priors adopted for the unknown probability, namely, a diffuse prior,

$$M_1 : \; x \sim Bin(n, \pi); \pi \sim Be(1, 1),$$

*This example is based on discussion with Samu Mäntyniemi [samu.mantyniemi@helsinki.fi].

and an informative prior,

$$M_2: \ x \sim Bin(n, \pi); \pi \sim Be(30, 30).$$

This choice can be solved analytically to give $P(M_1|x) = 0.8634$. In particular, integrating π out from the binomial-beta mixture model, $x \backsim Bin(n, \pi)$, $\pi \backsim Be(\alpha, \beta)$, gives a beta-binomial distribution, $x \sim Beta-Bin(n, \alpha, \beta)$, with density,

$$p(x|n, \alpha, \beta) = \frac{Be(x+\alpha, n-x+\beta)}{Be(\alpha, \beta)} \frac{n!}{(n-x!)x!},$$

which can be evaluated at $p(x|10, 1, 1)$ and $p(x|10, 30, 30)$.

To apply the method in Congdon (2007a), as in Equation 2.1 particularly, the posterior densities of π under the two models (which may be taken to provide cross-model priors g_1 and g_2) are obtainable analytically as $Be(2, 10)$ and $Be(31, 39)$, because $\pi|x \sim Be(\alpha+x; \beta+n-x)$. Alternatively, an initial run provides posterior means (sd) for π, denoted $m_j(s_j)$, under the two models of 0.1656 (0.1025), and 0.4427 (0.0586), respectively. One may then use the fact that beta density, $Be(\alpha, \beta)$, parameters α and β are obtainable as,

$$\alpha = -m(m^2 - m + s^2)/s^2, \ \beta = (m-1)(m^2 - m + s^2)/s^2.$$

Using the latter results provides beta cross-model priors with parameters (2, 10.1) and (31.3, 39.5). A subsequent MCMC run of 100,000 iterations assumes equal prior model probabilities $Pr(M_1) = Pr(M_2) = 0.5$. This run[10] obtains likelihoods, prior ordinates, and importance sample ordinates at each iteration, giving iteration-specific weights, and a posterior mean weight for model 1 equal to the analytic probability to four significant places, namely 0.8634, with a Monte Carlo standard error of 7.5E-6.

2.6.1 Random effects models

For highly parameterized random effect models, or models with multiple random effects, the same principle applies except that posterior densities are usually not available analytically. There is also an issue of "focus" to consider for such models (Spiegelhalter et al., 2002). Model selection for random effects models may be based on integrated likelihoods, with random effects integrated out analytically or numerically—see Congdon (2007a). However, an alternative focus is to retain random effects $b_j^{(t)}$ in model j in an expanded parameter collection (Spiegelhalter et al., 2002). Then the numerator in $w_j^{(t)}$ becomes,

$$p\left(y|m=j, b_j^{(t)}, \theta_j^{(t)}\right) p\left(b_j^{(t)}|\theta_j^{(t)}, m=j\right) p\left(\theta_j^{(t)}|m=j\right) \left[\prod_{h \neq j} g_h^{(t)}\right] p(m=j).$$

Additionally the cross-model prior g_h becomes $p(b_h|\theta_h, m = j)p(\theta_h|m = j)$. Choice of g_h is again arbitrary, but a reasonable default is an estimate of the own model posterior density for the parameters, namely,

$$p(b_h|\theta_h, m = h, y)p(\theta_h|m = h, y),$$

the density of $\{b_h, \theta_h\}$ given y and $m = h$.

Example 2.5. Turtle Mortality This example reworks the turtle mortality model comparison, with a lognormal option for the precisions τ_b in the random effects model 2, namely, $\tau_b \sim LN(0, 10)$, but with $N(0, 10)$ priors retained on the fixed effects $\{\beta_1, \beta_2\}$. According to the method of Friel and Pettitt (2008), this model has a log marginal likelihood of -155.03 compared to -155.29 for the fixed effects model 1 (from iterations 5,000 to 50,000 of single chain runs using OpenBUGS version 2.2). This leads to respective posterior probabilities of 0.443 (model 1) and 0.557 (model 2)[11].

To apply the method of Congdon (2007a), separate exploratory runs of models 1 and 2 are taken to provide estimates of the components of the cross-model priors g_1 and g_2, namely, 34 parameters in model 2, and two in model 1. For the random effects model 2, the cross-model prior component for τ_b, namely, $\log(\tau_b)$, is monitored in this exploratory run and $\log(\tau_b)$ is found to have a posterior mean and standard deviation of 2.8 and 1.27, respectively. This provides parameters for a lognormal density on τ_b in the cross-model prior g_2. Normal densities for the random effects b_j and regression parameters $\{\beta_1, \beta_2\}$ are assumed for the remaining components of g_2, and in g_1 also. From iterations 5000 to 50,000 of a single chain run, respective posterior probabilities[12] on the two models are obtained. Like those derived from the marginal likelihood estimates, these are inconclusive for model choice, but very similar numerically, namely, 0.57 for the hierarchical model 2 and 0.43 for the simpler model.

Appendix: Computational Notes

1. A WinBUGS code is based on a stacked rather than nested form of input data, so that $h = 1, \ldots, m_1$ for cases in the 1st cluster, $h = m_1 + 1, \ldots, m_2$ for cases in the 2nd cluster, and so on. A code with each temperature point, q_s, taken singly uses synthetic data $w_h = 1$ to define a nonstandard likelihood and adopts the device suggested by Barry (2006). Thus,

```
model {for (h in 1:n) {w[h] <- 1; L.tem[h] <- pow(L[h],q.s)
w[h]~dunif(a1[h],b1[h]); a1[h] <- -1/L.tem[h]; b1[h] <- 1/L.tem[h]
LL[h] <- log(L[h]); pi[h] <- phi(beta[1]+beta[2]*x[h]+b[clutch[h]])
log(L[h]) <- y[h]*log(pi[h])+(1-y[h])*log(1-pi[h])}
for (j in 1:p) { beta[j]~dnorm(0,0.1)}
for (j in 1:J) { b[j] ~dnorm(0,tau)}
tau ~dgamma(1,1); tLL <- sum(LL[1:n])}.
```

The marginal likelihood is obtained by piecing together the separate posterior likelihood estimates at each grid point (tLL in the above code, obtained with q.s=a_s^4) according to their individual Monte Carlo variances.

Alternatively an inclusive code spanning all $T+1$ cutpoints is as follows

```
model {for (h in 1:n) {for (s in 1:T+1) {w[h,s] <- 1; L.tem[h,s] <- pow
(L[h,s],q[s])
w[h,s] ~dunif(a1[h,s],b1[h,s])
a1[h,s] <- -1/L.tem[h,s]; b1[h,s] <- 1/L.tem[h,s]
LL[h,s] <- log(L[h,s])
pi[h,s] <- phi(beta[1,s]+beta[2,s]*x[h]+b[clutch[h],s])
log(L[h,s]) <- y[h]*log(pi[h,s])+(1-y[h])*log(1-pi[h,s])}}
for (s in 1:T+1) { for (j in 1:p) { beta[j,s] ~dnorm(0,0.1)}
for (j in 1:J) { b[j,s] ~dnorm(0,tau.b[s])}
tau.b[s] ~dgamma(1,1); q[s] <- pow(a[s],4)
tLL[s] <- sum(LL[1:n,s]) }
a[1] <- 0.00001; for (s in 1:T) {a[s+1] <- s/T}
for (s in 1:T) {mc[s] <- (q[s+1]-q[s])*(tLL[s+1]+tLL[s])*0.5 }
logML <- sum(mc[])}
```

where the $tLL[1:T]$ correspond to $E_{\theta_k|y,t_s}p(y|\theta_k)$.

2. Slightly different results for the posterior mean log marginal likelihood are obtained using WinBUGS14 and OpenBUGS. The mean of -155.53 for the random effects model is obtained using WinBUGS14.

3. The four relevant series are monitored in the following WinBUGS code and then input to a matlab program, which is also listed. Thus, the random effects model code in WinBUGS is

```
model {for (h in 1:n) { y[h] ~dbern(pi[h]); pi[h] <- phi(nu[h])
nu[h] <- beta[1] + beta[2]*x[h] + b[Clutch[h]]
nu.g[h] <- beta.g[1] + beta.g[2]*x[h] + b.g[Clutch[h]]
# log-likelihoods, posterior and importance sampled parameters
LL[h] <- y[h]*log(phi(nu[h])) + (1-y[h])*log(1-phi(nu[h]))
LL.g[h] <- y[h]*log(phi(nu.g[h])) + (1-y[h])*log(1-phi(nu.g[h]))}
# quantities for calculating optimal function
log.pstar <- sum(LL[])+sum(pr[]); log.imp <- sum(imp[])
log.pstar.g <- sum(LL.g[])+sum(pr.g[]); log.imp.g <- sum(imp.g[])
# matlab inputs
mon[1] <- log.pstar;      mon[2] <- log.imp;
mon[3] <- log.pstar.g;    mon[4] <- log.imp.g;
# sample beta[1:p] from posterior and imp densities, and obtain
# prior and importance ordinates for both samples
for (j in 1:p) {beta[j] ~dnorm(M[j],P[j]);
beta.g[j] ~dnorm(beta.m[j],beta.pr[j]); beta.pr[j] <- 1/pow(beta.se[j],2)
pr.g[j] <- 0.5*log(P[j]/6.28)-0.5*P[j]*pow(beta.g[j]-M[j],2)
imp.g[j] <- 0.5*log(beta.pr[j]/6.28)
```

```
-0.5*beta.pr[j]*pow(beta.g[j]-beta.m[j],2)
pr[j] <- 0.5*log(P[j]/6.28)-0.5*P[j]*pow(beta[j]-M[j],2)
imp[j] <- 0.5*log(beta.pr[j]/6.28)-0.5*beta.pr[j]*pow(beta[j]-beta.m[j],2)}
# sample b[1:J] from posterior and imp densities
for (j in 1:J) {b[j] ~dnorm(0,tau); b.g[j] ~dnorm(b.m[j],b.pr[j])
b.pr[j] <- 1/(b.sd[j]*b.sd[j])
pr.g[p+j] <- 0.5*log(tau/6.28)-0.5*tau*b.g[j]*b.g[j]
imp.g[p+j] <- 0.5*log(b.pr[j]/6.28)-0.5*b.pr[j]*pow(b.g[j]-b.m[j],2)
pr[p+j] <- 0.5*log(tau/6.28)-0.5*tau*b[j]*b[j]
imp[p+j] <- 0.5*log(b.pr[j]/6.28)-0.5*b.pr[j]*pow(b[j]-b.m[j],2)}
# importance and posterior samples of precision
tau.g ~dgamma(r.m,s.m); tau ~dgamma(r,s)
pr.g[p+J+1] <- (r-1)* log(tau.g)-s*tau.g +r*log(s.m)-loggam(r.m)
imp.g[p+J+1] <- (r.m-1)*log(tau.g)-s.m*tau.g+r.m*log(s.m)-loggam(r.m)
pr[p+J+1] <- (r-1)*log(tau) - s*tau + r*log(s) - loggam(r)
imp[p+J+1] <- (r.m-1)*log(tau) - s.m*tau+ r.m*log(s.m) - loggam(r.m)}
```

The subsequent matlab function code carries out Nrecurse recursions to obtain the Meng-Wong optimal estimator for R=S samples from posterior and importance densities. It also obtains the simple importance sample estimator.

```
function [logML]=MLrecurse(Nrecurse,R,log_pstar, log_imp, log_pstar_g,
log_imp_g)
% initial estimate of Marg LKD
r(1) = 1;
for t=1:R W1(t) = exp(log_pstar(t)-log_imp(t));
W2(t) = exp(log_pstar_g(t)-log_imp_g(t));
ISratio(t)=exp(log_pstar_g(t)-log_imp_g(t));
end
logML_IS=log(mean(ISratio(1:R)))
% revised estimates of optimal marg LKD estimator
for j=2:Nrecurse A(j)=0; B(j)=0;
for t=1:R A(j)=A(j)+W2(t)/(0.5*W2(t)+0.5*r(j-1));
B(j)=B(j)+1/(0.5*W1(t)+0.5*r(j-1));
end
% revised estimate of marg LKD and log ML at recursion j
r(j) = A(j)/B(j); logML= log(r(j));
end
```

4. The WinBUGS coding of the random effects option (model 2) is again based on a stacked rather than nested form of input data, so that $h = 1, \ldots, m_1$ for cases in the 1st cluster, $h = m_1 + 1, \ldots, m_2$ for cases in the 2nd cluster, and so on. Thus,

```
model {for (h in 1:n) {y[h]~dbern(pi[h]);
pi[i] <- phi(beta[1] + beta[2]*x[h] + b[Clutch[h]])
# log-likelihood
LL[h] <- y[i]*log(pi[h]) + (1-y[i])*log(1-pi[h])
```

```
# CPO estimate is inverse of posterior mean of g[h] (section 2.5)
g[h] <- 1/exp(LL[h])}
# deviance node
Dv <- -2*sum(LL[])
for (j in 1:p) {beta[j] ~dnorm(M[j],P[j])}
# alternative prior on random effects precision/variance
tau~dgamma(r,s); sig2 <- 1/tau
# sig2~dunif(r,s); tau <- 1/sig2
for (j in 1:J) { b[j] ~dnorm(0,tau.b); p.sig[j] <- step(b[j])}}
```

5. This is achieved with the WinBUGS code (for stacked data)

```
model {for (i in 1:n) { y[i] ~dbern(pi[i]);
pi[i] <- phi(-2.9 + 0.4*x[i] + b[Clutch[i]])}
tau.b <- 1/sig2b
for (j in 1:J) { b[j]~dnorm(0,tau.b)}}
```

with the input data containing the preset σ_b^2, together with clutch and predictor values, but not the actually observed y values. Instead, simulated y_{ij} are obtained by using the "gen inits" and "info/node" commands.

6. In Example 2.3, the WinBUGS code uses normal priors on $\log(\sigma_{bj}^2)$, equivalent to lognormal priors on σ_{bj}^2. Additionally, the code assumes y in stacked form indexed by patient and reading, so that

```
model {for (i in 1:1092) {y[i] ~dnorm(mu[pat[i],rdg[i]],tau.u) }
for (j in 1:N) {for (q in 1:q) {b[q,j] ~dnorm(0,tau.b[q])}
for (t in 1:T) {mu[j,t]<- beta[1]+b[1,j]+(beta[2]+b[2,j])*Bs[j] + eta[Tr[j]]}}
# variance component selection
for (q in 1:2) {J[q] ~dbern(0.5); sig2b[q] <- exp(logsig2b[q])
sigb[q] <- sqrt(1/tau.b[q])}
tau.b[1] <- 1000*(1-J[1])+J[1]/sig2b[1]
tau.b[2] <- 10000*(1-J[2])+J[2]/sig2b[2]
logsig2b[1]~dnorm(-0.35,0.0035); logsig2b[2]~dnorm(-4.8,0.4)
# other priors
for (q in 1:Q) {beta[q] ~dnorm(M.beta[q],T.beta[q])}
tau.u ~dgamma(1,0.001); eta[1] <- 0; for (k in 2:3) {eta[k] ~dnorm
(0,0.001)}}
```

7. The distinctive code elements governing the random effects, and their retention or otherwise under a selection mechanism, are

```
for (i in 1:N) {for (j in 1:Q){z[i,j] ~dnorm(0,1)}
b[1,i] <- J[1,1]*C[1,1]*z[i,1];
b[2,i] <- J[2,1]*C[2,1]*z[i,1]+J[2,2]*C[2,2]*z[i,2]}
Retain[1] <- J[1,1]; Retain[2] <- max(equals(J[2,1],1),equals(J[2,2],1))
C[1,1] ~dexp(1);C[1,2] <- 0; J[1,2] <- 0
C[2,1] ~dnorm(0,1); C[2,2] ~dexp(10)
```

```
for (j in 1:2) { sd.b[j] <- sqrt(Sig.b[j,j]); for (k in 1:2) {Csamp[j,k] <-
J[j,k]*C[j,k]
Ctran[j,k] <- Csamp[k,j]; Sig.b[j,k] <- inprod(Csamp[j,], Ctran[,k])}}
Qr <- Q*(Q+1)/2; aJ <- T.J+1; bJ <- Qr-T.J+1;
T.J <- sum(tJ[]); p.J~dbeta(aJ,bJ)
for (j in 1:Q) { tJ[j] <- sum(J[j,1:j]); for (k in 1:j) {J[j,k] ~dbern(p.J)}}
```

8. The Bayesian chi-square method is illustrated using model 5 for the Scottish lip cancer incidence, also considered in Johnson (2004, 2374–76). Thus, with E_i denoting expected incidence counts,

$$y_i \backsim Po(E_i \rho_i)$$

where ρ_i are modeled as diffuse fixed effects. The code is as follows:

```
model { for (i in 1:n) {y[i] ~dpois(mu[i]);
log(mu[i]) <- log(E[i]) + b[i]; b[i] ~dnorm(0,0.001)
# Poisson probs (up to maximum count 50), ym[i]=y[i]-1 unless y=0
for (j in 1:51) {cdf[i,j] <- exp(-mu[i])*pow(mu[i],j-1)/exp(logfact(j-1))
*step(y[i]-j+1)}
for (j in 1:51) {cdfm[i,j] <- exp(-mu[i])*pow(mu[i],j-1)/exp(logfact(j-1))
*step(ym[i]-j+1)}
# cdf probs for y[i] and (y[i]-1)
t[i] <- sum(cdf[i,1:51]); tm[i] <- sum(cdfm[i,1:51])
# lower limit of interval from which bin randomly chosen
s[i] <- (1-equals(y[i],0))*tm[i]
u[i] ~dunif(0,1); a[i] <- s[i]+u[i]*(t[i]-s[i])
ybin[i,1] <- step(0.2-a[i]); ybin[i,5] <- step(a[i]-0.8)
ybin[i,2] <- step(a[i]-0.2)*step(0.4-a[i])
ybin[i,3] <- step(a[i]-0.4)*step(0.6-a[i])
ybin[i,4] <- step(a[i]-0.6)*step(0.8-a[i])}
for (k in 1:K) {mhat[k] <- sum(ybin[,k]); m[k] <- n*p[k]
r.B[k] <- pow(mhat[k]-m[k],2)/m[k]}
# compare R.B with 95th quantile of the chi2 distribution for K-1 df
R.B <- sum(r.B[]); P <- step(R.B-9.49)}
```

From iterations 5–100 thousand of a single chain run the probability that R^B exceeds the 95% point of a χ^2_4 is 0.157 and the posterior means of the number (mhat[] in the code) of the $n = 56$ counts assigned to the five bins are (8.6, 9.9, 10.9, 12.1, 14.5)

9. The WinBUGS code for Example 2.4 is

```
model { for (i in 1:n) { for (s in 1:2) { for (x in 1:X) {
for (r in 1:2) {y[r,i,s,x] ~dnegbin(pi[r,i,s,x],alp)
yrep[r,i,s,x] ~dnegbin(pi[r,i,s,x],alp)
yrep.mx[r,i,s,x] ~dnegbin(pi.rep[r,i,s,x],alp)
pi[r,i,s,x] <- alp/(alp+xi[r,i,s,x])
pi.rep[r,i,s,x] <- alp/(alp+xi.rep[r,i,s,x])
```

```
d.extreme[r,i,s,x] <- step(y[r,i,s,x]-yrep.mx[r,i,s,x]-0.001)
+0.5*equals(y[r,i,s,x],yrep.mx[r,i,s,x])
log(xi[r,i,s,x]) <- log(P[r,i,s,x])+gm[r,s]+eta[s,x]+lam[s,i]
log(xi.rep[r,i,s,x]) <- log(P[r,i,s,x])+gm[r,s]+eta.rep[s,x]+lam.rep[s,i]
LL[r,i,s,x] <- loggam(a1[r,i,s,x])-logfact(y[r,i,s,x])-loggam(alp)
+alp*log(pi[r,i,s,x])+y[r,i,s,x]*log(pi.m[r,i,s,x])
# log(CPO) are minus log(posterior mean G)
G[r,i,s,x] <- 1/exp(LL[r,i,s,x])
a1[r,i,s,x] <- y[r,i,s,x]+alp; pi.m[r,i,s,x] <- 1-pi[r,i,s,x]}}}}
for (j in 1:2) { for (k in 1:2) {gm[j,k]~dnorm(-5,0.001)}}
alp~dunif(0,1000);
# age-sex RW component
w[1] <- 1;          adj[1] <- 2;          num[1] <- 1
w[(X-2)*2 + 2] <- 1;     adj[(X-2)*2 + 2] <- X-1;     num[X] <- 1
for (x in 2:X-1) {w[2+(x-2)*2] <- 1; adj[2+(x-2)*2] <- x-1
w[3+(x-2)*2] <- 1;adj[3+(x-2)*2] <- x+1;num[x] <- 2}
for (j in 1:2) {tau[j] ~dgamma(1,0.001)
eta[j,1:X] ~car.normal(adj[],w[],num[],tau[j])
eta.rep[j,1:X] ~car.normal(adj[],w[],num[],tau[j])}
# spatial component
lam[1:2,1:n] ~mv.car(mapnei[], wnei[], numnei[], omeg.lam[,])
lam.rep[1:2,1:n] ~mv.car(mapnei[], wnei[], numnei[], omeg.lam[,])
omeg.lam[1:2, 1:2] ~dwish(Q[ , ], 2)
Sig2.lam[1:2,1:2] <- inverse(omeg.lam[,])
corr.lam <- Sig2.lam[1,2]/sqrt(Sig2.lam[1,1]*Sig2.lam[2,2])
for (i in 1 : Nnei) { wnei[i] <- 1}}
```

10. The code for the beta-binomial toy example, with the mean of w[1:2] being the relevant output, and with the same data for both models, namely, list(x=c(1,1)), is

```
model {x[1]~dbin(p[1],10); p[1]~dbeta(1,1)
        x[2]~dbin(p[2],10); p[2]~dbeta(30,30)
# Likelihood terms for models
LL[1]<-logfact(10)-logfact(10-x[1])-logfact(x[1])+x[1]*log(p[1])+(10-x[1])
*log(1-p[1])
LL[2]<-logfact(10)-logfact(10-x[2])-logfact(x[2])+x[2]*log(p[2])+(10-x[2])
*log(1-p[2])
# prior ordinates for models
LP[1]<-loggam(2)-loggam(1)-loggam(1)+0*log(p[1])+0*log(1-p[1])
LP[2]<-loggam(60)-loggam(30)-loggam(30)+29*log(p[2])+29*log(1-p[2])
# importance sample ordinates for models
g[1]<-loggam(12.1)-loggam(2)-loggam(10.1)+1*log(p[1])+9.1*log(1-p[1])
g[2]<-loggam(70.8)-loggam(31.3)-loggam(39.5)+30.3*log(p[2])+38.5*log(1-
p[2])
# Combining likelihoods, priors and importance samples
```

```
L[1]<-LL[1]+LP[1]+log(0.5)-g[1]; L[2]<-LL[2]+LP[2]+log(0.5)-g[2]
# Scaling with the largest likelihood, and protecting for underflow
maxL<-ranked(L[],2); SL[1]<-max(L[1]-maxL,-700); SL[2]<-max(L[2]-maxL,-
700)
# exponentiating and normalizing
expSL[1]<- exp(SL[1]); expSL[2]<- exp(SL[2]);
# model weights
w[1]<-expSL[1]/sum(expSL[]); w[2]<-expSL[2]/sum(expSL[])}.
```

11. The coding of the Friel and Pettitt method for model 2 in Example 2.5 is (with T=20 grid points)

```
model {for (h in 1:n) {for (s in 1:T+1) { w[h,s] <- 1; L.tem[h,s] <-
pow(L[h,s],q[s])
w[h,s] ~dunif(a1[h,s],b1[h,s])
a1[h,s] <- -1/L.tem[h,s]; b1[h,s] <- 1/L.tem[h,s]
LL[h,s] <- log(L[h,s])
pi[h,s] <- phi(beta[1,s]+beta[2,s]*x[h]+b[clutch[h],s])
log(L[h,s]) <- y[h]*log(pi[h,s])+(1-y[h])*log(1-pi[h,s])}}
for (s in 1:T+1) { for (j in 1:p) { beta[j,s] ~dnorm(0,0.1)}
for (j in 1:J) { b[j,s] ~dnorm(0,tau.b[s])}
log.tau.b[s] ~dnorm(0,0.1);
tau.b[s] <- exp(log.tau.b[s])
q[s] <- pow(a[s],4); expLL[s] <- sum(LL[1:n,s]) }
a[1] <- 0.01; for (s in 1:T) {a[s+1] <- s/T}
for (s in 1:T) {mc[s] <- (q[s+1]-q[s])*(expLL[s+1]+expLL[s])*0.5 }
logML <- sum(mc[])}
```

12. The code for the method of Congdon (2007a) applied in Example 2.5 with a lognormal prior on τ_b is

```
model {for (i in 1:N){Y1[i] ~dbern(p1[i]); Y2[i] ~dbern(p2[i])
Y1[i] <- y[i]; Y2[i] <- y[i]
p1[i] <- phi(b1[1]+b1[2]*x[i])
p2[i] <- phi(b2[1]+b2[2]*x[i]+e[clutch[i]])
LL[i,2] <- y[i]*log(p2[i])+(1-y[i])*log(1-p2[i])
LL[i,1] <- y[i]*log(p1[i])+(1-y[i])*log(1-p1[i])}
# priors
for (j in 1:J) {e[j] ~dnorm(0,tau)}
tau ~dlnorm(0,pr.tau); log.tau <- log(tau)
for (j in 1:p) { b2[j] ~dnorm(M,T); b1[j] ~dnorm(M,T)}
# prior ordinates, reg coeffs
for (j in 1:p) {prior1[j] <- 0.5*log(T/6.28)-0.5*T*pow(b1[j]-M,2)
prior2[j] <- 0.5*log(T/6.28)-0.5*T*pow(b2[j]-M,2)}
# cross prior ordinates, reg coeffs
for (j in 1:p) { pr.b2[j] <- 1/(psd.b2[j]*psd.b2[j]);
pr.b1[j] <- 1/(psd.b1[j]*psd.b1[j])
```

```
g1[j] <- 0.5*log(pr.b2[j]/6.28)-0.5*pr.b2[j]*pow(b2[j]-pm.b2[j],2)
g2[j] <- 0.5*log(pr.b1[j]/6.28)-0.5*pr.b1[j]*pow(b1[j]-pm.b1[j],2)}
# prior ordinate precision (tau) model 2
pr.logtau <- 1/(psd.logtau*psd.logtau)
prior2[p+J+1] <- 0.5*log(pr.tau/6.28)-log.tau-0.5*log.tau*log.tau
# cross prior ordinate, precision
g2[p+J+1] <- 0.5*log(pr.logtau/6.28)-log.tau
-0.5*pr.logtau*pow(log.tau-pm.logtau,2)
# prior ordinates, random effects model 2
for (j in 1:J) { prior2[p+j] <- 0.5*log(tau/6.28)-0.5*tau*e[j]*e[j]}
for (j in 1:J) {pr.e[j] <- 1/(psd.e[j]*psd.e[j])
# cross prior ordinates, random effects model 2
g2[p+j] <- 0.5*log(pr.e[j]/6.28)-0.5*pr.e[j]*pow(e[j]-pm.e[j],2)}
# model comparison
for (j in 1:2) { TL[j] <- sum(LL[,j]); H[j] <- exp(q[j])
q[j] <- max(logh[j]-maxh,-500)
# model probability
w[j] <- H[j]/sum(H[1:2])}
logh[2] <- TL[2]+sum(prior2[1:34])-sum(g2[1:34])
logh[1] <- TL[1]+sum(prior1[1:2])-sum(g1[1:2])
maxh <- ranked(logh[1:2],2) }
```

3

Hierarchical Estimation for Exchangeable Units: Continuous and Discrete Mixture Approaches

3.1 Introduction

What is sometimes termed ensemble estimation or "borrowing strength from the ensemble" refers to inferences for collections of similar units, $i = 1, \ldots, n$ (schools, health agencies, etc.) (Burr and Doss, 2005; George et al., 1993; Morris, 1983; Rao, 1975). Among possible examples are collections of clinical trials, surgical death rates for hospitals, exam pass rates for schools, goal averages for basketball players (Hsiao, 1997), or teenage pregnancy rates in health areas (Deely and Smith, 1998). Fixed effects models for such collections are problematic (Marshall and Spiegelhalter, 1998), whereas hierarchical random effects approaches pool information across units to obtain more reliable estimates for each unit, and help identify units with unusually high or low values.

Bayesian hierarchical models for modeling collections of similar units typically adopt unit or cluster-specific models with conditional means, $\{y_i|b_i\}$, depending on unobserved effects, b_i, with density, $p(b_i|\psi)$, known variously as the mixing distribution, hyperdensity, or higher stage density with hyperparameter(s) ψ. To reflect the conditioning on this density, estimation for sets of similar units is known as hierarchical modeling (Kass and Steffey, 1989; Lee, 2004). For instance, in the first stage of the conjugate Poisson-gamma model considered below, the observed counts are conditionally independent given the unknown means that are taken to have generated them. At the second stage, these means are themselves determined by the gamma density parameters, while the density for the gamma parameters forms the third stage. The goals of such analysis may vary (Shen and Louis, 1998), but often the aim is to provide conditional estimates of outcome rates or effects in each unit given the parameters ψ of the common density. Alternatively, rankings of the units may be required or probabilities of significant difference between units or against a threshold (Deely and Smith, 1998).

The procedures considered in this chapter are typically based on an exchangeability principle—that units are similar enough to justify being modeled by a common density and that the units are not configured in ways (e.g., over time or space) that imply higher correlations between some units than others (Lindley and Smith, 1972, 4; Spiegelhalter et al., 2004). Structuring of units in space, time, or other forms of nonexchangeability does not preclude borrowing strength, but a prior relating to that structuring is required (see Chapter 4). Exchangeability means that there is no prior basis for supposing some units have higher true effects than others, or that certain subgroups of units are more similar between themselves than other subgroups (e.g., that mortality in hospitals i and j is more similar than between hospitals i and k). For units of the same type and observations generated under similar conditions, exchangeability means all possible permutations of the sequence of units have the same probability: random variables $\{y_1, \ldots, y_n\}$ are exchangeable if their joint distribution $P(y_1, \ldots, y_n)$ is invariant under permutation of its arguments, so that,

$$P(y_1^*, \ldots, y_n^*) = P(y_1, \ldots, y_n),$$

where $\{y_1^*, \ldots, y_n^*\}$ is any permutation of $\{y_1, \ldots, y_n\}$ (Greenland and Draper, 1998).

Sometimes units are better considered exchangeable within subgroups of the data. A UK example relates to mortality in cardiac surgery units, with exchangeability within "closed" procedures involving no use of heart bypass during anesthesia, and "open" procedures where the heart is stopped and a heart bypass is needed (Spiegelhalter, 1999). Sometimes exchangeability can only be supported for "residual effects," b_i, obtained after controlling for known differences in the denominator population (e.g., the patient casemix underlying different hospital death rates).

Hierarchical smoothing methods result in shrinkage of estimates for each unit toward the average outcome rate in the population within which exchangeability is assumed; shrinkage will be greater for units with observations based on small samples. When the single population hierarchical model is appropriate, pooling of strength results in more precise estimates, and may provide better out-of-sample predictive performance—see Deely and Smith (1998) for an application of such predictions to "performance indicators." The estimated locations of individual units are typically pulled toward the population average, so that pooling strength may increase the risk of bias, as compared to unadjusted fixed effect estimates. The increase in precision but possible bias inherent in hierarchical estimation provides a dilemma known as the bias-variance trade-off. In some applications, inferences are over more than one variable as well as over a collection of similar units (Everson and Morris, 2000; van Houwelingen et al., 2002). Inferences will typically be improved for related outcomes over similar units (e.g., exam success rates in different pupils, or surgical and nonsurgical mortality in different hospitals).

While smoothing is the leading motivation for hierarchical models, a related theme is to achieve smoothed estimates that allow appropriately for heterogeneity between sample units—that is, they do not "oversmooth" and show some robustness or flexibility to individual units, or to clusters of units, which are somewhat discrepant or outlying from the rest of the population. Such heterogeneity will often be associated with overdispersion in Poisson or binomial data, or with heavy-tailed data in the case of symmetric departures from normality in continuous variables. One way to modify the standard densities (e.g., binomial, Poisson, normal) to take account of heterogeneity greater than postulated under that density is to allow adaptive continuous mixing at unit level. Examples of such mixing are the scale mixture approach to the t-density discussed in Section 3.2, or in the Poisson-gamma mixture—see Hsiao (1997) for a binomial-beta example. Another option is discrete mixing (see Sections 3.6 et seq), in which a single population assumption is replaced by an assumption of two or more latent subpopulations. Shrinkage will then be toward the subgroup characteristics that each unit has the highest posterior probability of belonging to.

Undershrinkage (undersmoothing) also raises issues: this will lead to overestimation of random effect variability and is to be avoided when a type I error has worse consequences than a type II error (Gustafson et al., 2006). Similarly, Spiegelhalter (2005) points out that there is a danger in performance indicator analysis that the units (e.g., institutions) that one is trying to detect could be "accommodated" by a random effects approach, and it is therefore important that robust methods are used to estimate the standard deviation of the random effects distribution.

3.2 Hierarchical Priors for Ensemble Estimation using Continuous Mixtures

Observations for related units are often available in aggregate form, such as means, y_i, for a metric variable or numbers of successes for a binomial variable, though originally collected in disaggregated form for repetitions, j, within each unit of observation i. Consider the first stage sampling density, $p(y|b)$, for a set of n observations, $\{y_i, i = 1, \ldots, n\}$, conditional on the parameter vector, $b = \{b_1, \ldots, b_n\}$. For example, consider a Poisson model, $y_i \backsim Po(b_i)$, where y_i is the number of accidents at different road sites in a fixed period (e.g., a year), and b_i is a measure of accident proneness. Instead of assuming all units have the same proneness, it may be more realistic to allow b_i to vary over units according to a stage 2 density, $p(b_i|\psi)$, for instance a gamma density to allow for positive skew in proneness. Then at stage 3 the population parameters $\psi = (\psi_1, \ldots, \psi_Q)$ (hyperparameters) that generate b_i are specified.

Of central interest are the posterior densities, $p(b_i|y)$, and $p(\psi|y)$, and the probabilities under these densities that b_i and ψ are in specified intervals, such as the probability of an overall positive effect, $Pr(\psi > 0|y)$, when y_i are measures of (say) a clinical treatment benefit. One may also be interested in predictions for hypothetical future units (e.g., for a new clinical trial or for the next year in a performance ranking application), $p(y_{\text{new}}|y)$. If $p(\psi|y)$ can be obtained analytically, or samples $\psi^{(1)}, \psi^{(2)}, \ldots, \psi^{(t)}$ obtained directly, then the posteriors $p(b_i|y)$, $p(\psi|y)$, and $p(y_{\text{new}}|y)$ can be obtained by Monte Carlo simulation, as in,

$$p(b_i|y) = \int p(b_i|y, \psi)p(\psi|y)d\psi,$$

leading to the estimate $p(b_i|y) = \sum_{t=1}^{T} p(b_i|y, \psi^{(t)})$.

An alternative to direct simulation is to simulate the full posterior $p(\psi, b|y)$ using Markov Chain Monte Carlo (MCMC) methods (Marshall and Spiegelhalter, 1998), for example by obtaining samples $\{b^{(t)}, \psi^{(t)}\}$ from the full conditional posteriors $p(b_i|b_{[i]}, \psi, y)$ and $p(\psi_q|\psi_{[q]}, b, y)$. Often the first stage density, $p(y|b)$, is in the full exponential family, so that,

$$p(y_i|b_i) = \exp\left(\frac{y_i b_i - B(b_i)}{A(\phi_i)} + C(y_i, \phi_i)\right), \tag{3.1}$$

where ϕ_i is a scale parameter. Assuming a conjugate second stage prior, the conditional posterior of each b_i follows the same density. For example, assume (Das and Dey, 2006, 2007; Diaconis and Ylvisaker, 1979; Ferreira and Gamerman, 2000; Frees, 2004) that,

$$p(b_i|\psi) = k_1 \exp(b_i g_1(\psi) - B(b_i)g_2(\psi)), \tag{3.2}$$

where k_1 is a normalizing constant. Then the posterior density of b_i and ψ given y is of exponential form,

$$p(b_i, \psi|y) = k_2 \exp\left(\left[g_1(\psi) + \frac{y_i}{A(\phi_i)}\right] b_i - B(b_i)\left[g_2(\psi) + \frac{1}{A(\phi_i)}\right]\right). \tag{3.3}$$

With proper log-concave priors, $p(\psi)$, the full conditionals, $p(\psi_q|\psi_{[q]}, b, y)$, are log-concave and can be sampled using methods such as those of Gilks and Wild (1992) and a convergent MCMC sequence generally obtained.

By contrast, if improper priors are assumed on $\{\psi_1, \ldots, \psi_Q\}$, then the full posterior, $p(b, \psi|y)$, is not necessarily proper (Browne and Draper, 2006; George and Zhang, 2001; George et al., 1993), and empirical convergence of the MCMC sequence, $\{b^{(t)}, \psi^{(t)}\}$, may be problematic even if the posterior is proper analytically. George and Zhang (2001) consider posterior propriety results for the Poisson-gamma, the binomial-beta, and multinomial-Dirichlet models in terms of conditions on the hyperparameter prior tail behavior. For the latter two hierarchical model schemes, no improper prior can guarantee a proper posterior. Similar convergence and identification issues apply to the general linear mixed model formulation.

3.3 The Normal-Normal Hierarchical Model and Its Applications

A widely applied conjugate hierarchical scheme assumes normal sampling and normally distributed latent effects. A typical borrowing strength template is for continuous observations for unit level effects, y_i, and intra-unit variation, $a(\phi_i) = s_i^2$, even though the underlying data might have involved two-way nesting with $j = 1, \ldots, J_i$ replications for units $i = 1, \ldots, n$. This is often the case in clinical meta-analysis where patient-level results are summarized as treatment or risk factor effect measures (e.g., change in a clinical measure between treatment and control groups, or the slope of a dose-response curve) along with moment estimates of sampling variances. Assuming the observed summary measures are exchangeable in terms of being obtained from similar study designs and relating to similar types of unit (Spiegelhalter et al., 2004, 92), they may be regarded as draws from an underlying common density for the unknown true means, b_i.

Often the normal-normal model is applied to ostensibly binomial data using normal approximations for the effect measures (e.g., Albert, 1996a; Carlin, 1992). Suppose $r_{i\text{T}}$ of $N_{i\text{T}}$ treated subjects in trial i exhibit a particular response (e.g., disease or death) as compared to $r_{i\text{C}}$ of $N_{i\text{C}}$ control subjects. Define log odds,

$$\omega_{i\text{T}} = \log(r_{i\text{T}}/(N_{i\text{T}} - r_{i\text{T}}),$$

and

$$\omega_{i\text{C}} = \log(r_{i\text{C}}/(N_{i\text{C}} - r_{i\text{C}}),$$

in the treated and control arms in each trial. Then the log of the odds ratio forms the unit level response,

$$y_i = \omega_{i\text{T}} - \omega_{i\text{C}},$$

assumed approximately normal with variance,

$$s_i^2 = \frac{1}{r_{i\text{T}}} + \frac{1}{N_{i\text{T}} - r_{i\text{T}}} + \frac{1}{r_{i\text{C}}} + \frac{1}{N_{i\text{C}} - r_{i\text{C}}},$$

(see Example 3.1). It is also possible to take y_i as a log relative risk between treatment and control groups, namely,

$$y_i = \log\left(\frac{r_{i\text{T}}}{N_{i\text{T}}}\right) - \log\left(\frac{r_{i\text{C}}}{N_{i\text{C}}}\right),$$

with variance,

$$\frac{1}{r_{i\text{T}}} + \frac{1}{r_{i\text{T}}} - \frac{1}{N_{i\text{T}}} - \frac{1}{N_{i\text{C}}}.$$

Another option is to take the risk difference,

$$y_i = \frac{r_{i\mathrm{T}}}{N_{i\mathrm{T}}} - \frac{r_{i\mathrm{C}}}{N_{i\mathrm{C}}},$$

as approximately normal with variance,

$$\frac{r_{i\mathrm{T}}(N_{i\mathrm{T}} - r_{i\mathrm{T}})}{N_{i\mathrm{T}}^3} + \frac{r_{i\mathrm{C}}(N_{i\mathrm{C}} - r_{i\mathrm{C}})}{N_{i\mathrm{C}}^3}.$$

Unless heavy tails, skewness, or multiple modes are suspected, an appropriate hierarchical model then has a first stage specifying a normal density for the observations y_i (with variances allowed to differ by units) and a second stage normal density for the b_i with variance constant over groups. So for a univariate outcome,

$$y_i \sim N(b_i, s_i^2), \tag{3.4}$$

and

$$b_i \sim N(\mu, \tau^2). \tag{3.5}$$

Integrating out the b_i, the marginal likelihood for y_i is then,

$$y_i|\mu, \tau^2 \sim N\left(\mu, s_i^2 + \tau^2\right).$$

Often the summary measures are unit or trial means and different observational variances are associated with differing sample sizes, N_i, so that $s_i^2 = \sigma^2/N_i$, where σ^2 is an additional unknown. While clinical meta-analysis applications are common, a similar scenario occurs in small area estimation from multiple surveys where s_i^2 are sampling variances obtained according to the survey design.

More complex situations can be fitted into this framework. For example, Abrams et al. (2000) consider the effect of testing positive or negative in a screening test on subsequent levels of anxiety; see also Abrams et al. (2005). Let x_{ik} be baseline anxiety in study i, with $k = 1$ (tested positive) and $k = 2$ (tested negative), and with N_{i1} and N_{i2} subjects in different arms. Let z_{ik} be follow-up anxiety according to screening result, and let $d_{ik} = z_{ik} - x_{ik}$ denote change in anxiety. Then the measure of interest is the contrast between anxiety growth according to screening result, namely, $y_i = d_{i1} - d_{i2}$, with variance

$$s_i^2 = \frac{(N_{i1} - 1)V(d_{i1}) + (N_{i2} - 1)V(d_{i2})}{(N_{i1} + N_{i2} - 2)},$$

where

$$V(d_{ik}) = V(x_{ik}) + V(z_{ik}) - 2\rho\sqrt{V(x_{ik})V(z_{ik})},$$

and ρ is a within-subject correlation taken constant across studies and arms. Studies may not report all the relevant statistics: they may report the d_{ik} and their variances, or the separate baseline and follow-up measures in each arm, $\{x_{ik}, z_{ik}\}$, and their variances. In either case, meta-analysis requires a prior on ρ.

In Equation 3.4, assume independent priors on the hyperparameters,

$$p(\tau^2, \mu) = p(\tau^2)p(\mu),$$

with a commonly adopted option being,

$$\tau^2 \sim IG\left(\frac{\nu}{2}, \frac{\nu\lambda}{2}\right),$$
$$\mu \sim N\left(m_\mu, V_\mu\right),$$

where $\nu, \lambda, m_\mu, V_\mu$ are known. The full posterior conditional for b_i is then (Browne and Draper, 2006; George et al., 1993; Silliman, 1997, 927),

$$p(b_i|b_{[i]}, \mu, \tau, y) = p(b_i|\mu, \tau, y) = N([1 - w_i]y_i + w_i\mu, D_i),$$

where

$$D_i = \left(\frac{1}{S_i^2} + \frac{1}{\tau^2}\right)^{-1} = \frac{\tau^2 S_i^2}{\tau^2 + S_i^2},$$
$$w_i = \frac{s_i^2}{s_i^2 + \tau^2},$$

and the first equality is by virtue of conditional independence of b_i. The full conditional for τ^2 is

$$\tau^2 \backsim IG\left(0.5[n + \nu], 0.5\left[\nu\lambda + \sum_{i=1}^{n}(y_i - \mu)^2\right]\right),$$

while that for μ involves a precision weighted average of m_μ, and the average of b_i, namely,

$$\mu \backsim N\left(\bar{b}\left(\frac{nV_\mu}{nV_\mu + \tau^2}\right) + m_\mu\left(\frac{\tau^2}{nV_\mu + \tau^2}\right), \frac{\tau^2 V_\mu}{nV_\mu + \tau^2}\right).$$

Allowing interrelatedness between units leads to inferences about underlying unit means that are different from those obtained under alternative scenarios sometimes used, namely, (a) the "independent units" case, with b_i taken as unknown and mutually unrelated fixed effects, with $\tau^2 \to \infty$; and (b) the complete pooling model of classical meta-analysis where the studies are regarded as effectively interchangeable and $\tau^2 = 0$.

By contrast, the intermediate "exchangeable units" Bayes model leads to a posterior mean for b_i,

$$E[b_i|y] = w_i\mu + [1 - w_i]y_i,$$

that averages over the prior mean μ and the data mean y_i with weights $w_i = (s_i^2/s_i^2 + \tau^2)$ and $1 - w_i = (\tau^2/s_i^2 + \tau^2)$, respectively, as is apparent from the Gibbs sampling full conditionals. The b_i under an exchangeability scenario have narrower posterior intervals than under an independent units assumption, with precision related to the confidence about the prior mean and the prior assumed for τ^2 (see also Section 3.4). Assume the intra-study variances can be expressed as $s_i^2 = \sigma^2/N_i$ and then set $\tau^2 = \sigma^2/N_\mu$, where N_μ is the sample size assigned to the prior mean. Then the weights, w_i, become $N_\mu/(N_i + N_\mu)$, demonstrating that shrinkage to the prior mean increases as the confidence about the prior mean increases.

Sometimes it is necessary to control explicitly for trial design, study location, and other design features in order to justify an exchangeability assumption (Marshall and Spiegelhalter, 1998; Pauler and Wakefield, 2000; Prevost et al., 2000). Similarly, in survey-based small area estimation, the estimate of b_i may incorporate information from administrative area data X_i (Fay and Herriott, 1979; Rao, 2003). So with centered predictors X_i of dimension p (excluding a constant term) the above model becomes,

$$y_i \sim N(b_i, s_i^2),$$
$$b_i \sim N(\mu + X_i\beta, \tau^2),$$

with marginal likelihood then (DuMouchel, 1996),

$$y_i|\beta, \tau^2 \sim N(\mu + X_i\beta, s_i^2 + \tau^2).$$

This is sometimes known as meta-regression, with recent Bayesian applications including Levy et al. (2005) and Batterham (2005). Writing the model as

$$y_i = \mu + X_i\beta + \delta_i + \varepsilon_i,$$
$$\delta_i \sim N(0, \tau^2),$$
$$\varepsilon_i \sim N(0, s_i^2),$$

the true effect for unit i is then $\mu + X_i\beta + \delta_i$.

The normal-normal model may be robustified against skewness or heavy tails in either the sampling density or the latent effects density. If non-normality is suspected at the second stage, a heavy-tailed prior can be used to accommodate possibly outlying studies. This may be achieved by introducing study-specific scale parameters at the second stage (West, 1984), discounting the influence of atypical studies on posterior estimates of the overall effect μ, and avoiding overshrinkage of individual study effects b_i. These scale factors are positive and most commonly taken as gamma with scale and shape $\nu/2$, providing a scale mixture version of the Student t density. So,

$$b_i \backsim N(\mu, \tau^2/\lambda_i),$$
$$\lambda_i \backsim Ga\left(\frac{\nu}{2}, \frac{\nu}{2}\right).$$

Skewness in the observed data can often be reduced or eliminated by transformation. However, continuous data (e.g., cost data or data resulting from psychometric tests) will sometimes have more unusual departures from normality that render transformation inapplicable, such as clumping of zero values as well as positive skewness in positive responses (Delucchi and Bostrom, 2004). Skewness in the latent effects, b_i, may also be handled by more specific parametric adaptations (e.g., Fernandez and Steele, 1998; Sahu et al., 2003), multivariate versions of which are considered in Section 3.5. Another robust option adapted to skewness and/or multiple modes is a discrete mixture over two or more normal densities (Marshall and Spiegelhalter, 1998).

3.4 Prior for Second Stage Variance

The prior assumed for τ^2 plays an important role in governing the degree of shrinkage or pooling strength (Lambert et al., 2005), with diffuse priors leading to lesser shrinkage (Conlon et al., 2007). As discussed in Chapter 1, improper or highly diffuse priors may also lead to identification or propriety problems. For example, the prior

$$p(\tau^2) \propto \frac{1}{\tau^2},$$

equivalent to taking $\tau^2 \backsim IG(0,0)$ and to a flat prior on $\log(\tau)$ over $(0,\infty)$, can lead to improper posteriors in random effects models (DuMouchel and Waternaux, 1992). A just proper alternative, such as $\tau^2 \backsim IG(c,c)$ with c small is often used. However, this prior has a spike near zero (Browne and Draper, 2006), and different values of c can influence posterior influences despite the supposedly diffuse nature of the prior (Gelman, 2006a).

One might carry out a sensitivity analysis over a range of proper but diffuse $Ga(c,d)$ priors for $1/\tau^2$, such as $\{c = 0.1, d = 0.001\}$ or $c = d = 0.0001$ (Fahrmeir and Lang, 2001; van Dongen, 2006). An alternative scheme is to compare alternative values of c in $Ga(1,c)$ priors for $1/\tau^2$ (Besag et al., 1995), possibly using a mixture prior over M possible values for c_m in the prior $1/\tau^2 \sim Ga(1,c_m)$, such as $c_m = 1, 0.1, 0.01$, and 0.001 (Jullion and Lambert, 2007). Then for $c = (c_1,\ldots,c_m)$, and $p = (p_1,\ldots,p_m)$,

$$c|p \sim \sum_{m=1}^{M} p_m Ga(1,c_m),$$

$$p \sim \text{Dirichlet}(\omega),$$

where $(\omega_1,\ldots,\omega_M)$ are prior weights.

Introducing some degree of prior information may be relevant, and is natural under the inverse chi-squared density (sometimes called the scaled inverse chi-squared) with parameters $\{\nu,\lambda\}$. For τ^2 a variance, taking

$$\tau^2 \backsim \chi^{-2}(\nu, \lambda),$$

is equivalent to assuming $\tau^2 \sim IG(\nu/2, \nu\lambda/2)$, where λ is a prior guess at the mean variance and ν is a prior sample size (or level of confidence) parameter. Conlon et al. (2007) consider informative inverse gamma priors on τ^2 for interstudy variability in log-expression ratios in a microarray data application; for example, they use relatively large prior sample sizes ν.

Smith et al. (1995) discuss elicitation of informative inverse gamma priors for τ^2 based on anticipated variation in the underlying rates b_i, and the fact that assuming normality, 95% of b_i will lie between $\mu - 1.96\tau$ and $\mu + 1.96\tau$. Assume b_i are measured on a log scale (e.g., log relative risks or log odds) and suppose the expected ratio of the 97.5th and 2.5th percentiles of risks (or odds) between centers or studies is 5, then the gap between the 97.5th and 2.5th percentiles for b_i is $\log(5) = 1.61$. For normal b_i, the prior mean for τ^2 is then $(0.5 \times 1.61/1.96)^2 = 0.17$, and the prior mean for $1/\tau^2$ is 5.93. If the upper limit for the ratio of the 97.5th and 2.5th percentile of risks or odds is set at 10, this defines the 97.5th percentile of τ^2, namely, $(0.5 \times 2.3/1.96)^2 = 0.34$. The expectation and variability is then used to define an inverse gamma prior on τ^2 or a gamma prior on $1/\tau^2$. Another procedure based on expected contrasts in relative risk (RR) or relative odds (ROs) is mentioned by Marshall and Spiegelhalter (2007, 422): 95% of units will have RRs or ROs in the range $\exp(\pm 1.96\tau)$, and an expectation of reasonable homogeneity might correspond to values of τ less than $\tau_h = 0.2$. Setting $\psi = 0.5\tau_h = 0.1$, these expectations are expressed via a half normal prior on τ, with $\tau = |T|$ where,

$$T \sim N(0, \psi^2),$$

with prior 95% point at $1.96 \times \psi = 0.2$.

As another way to use prior evidence on variability, Marshall and Spiegelhalter (1998) mention a hyperprior for the scale parameter, ϕ, in a gamma prior for $1/\tau^2$, namely,

$$1/\tau^2 \sim Ga(\gamma, \phi),$$
$$\phi \sim Ga(c, d),$$

where d is a small multiple of $1/R^2$ and R is the range of the observed center effects, with γ and c constrained according to $\gamma > 1 > c$. When the first stage sampling density involves an unknown variance, Gustafson et al. (2006) suggest a conditional prior sequence adapted to avoiding undersmoothing, namely,

$$p\left(\sigma^2, \tau^2\right) = p\left(\sigma^2\right) p\left(\tau^2|\sigma^2\right),$$

where $\sigma^2 \sim IG(e, e)$ for some small $e > 0$, and

$$p\left(\tau^2|\sigma^2\right) \propto \left[\frac{1}{\tau^2 + \sigma^2}\right]^{a+1} \exp\left[-\frac{b}{\tau^2 + \sigma^2}\right].$$

This corresponds to a truncated inverse Gaussian prior on τ^2, with $Z \sim IG(a,b)$ or $1/Z \sim Ga(a,b)$, where $Z = \tau^2 + \sigma^2$. The case $\{a = 1, b = 0\}$ corresponds to the uniform shrinkage prior, while larger values of a (e.g., $a = 5$) are "conservative" in the sense of guarding against overestimation of τ^2.

3.4.1 Nonconjugate priors

Among nonconjugate strategies (for normal-normal meta-analysis) an effective choice in terms of being genuinely noninformative (Gelman, 2006a) is a bounded uniform prior on the random effects standard deviation, $\tau \sim U(0, H)$, with H large. However, this prior may be biased toward relatively large variances when the number of units (trials, studies, etc.) is small (van Dongen, 2006, 92).

Variations on the uniform shrinkage prior, suggested by Christiansen and Morris (1997) and Daniels (1999), may also be used. One is a uniform prior on the shrinkage weights, $w_i = (s_i^2/s_i^2 + \tau^2)$, or on the shrinkage weight, $w = (\sigma^2/\sigma^2 + \tau^2)$, when σ^2 is unknown. Alternatively, one might represent different shades of opinion (sceptical, neutral, enthusiastic with regard to meta-analytic shrinkage) via the shrinkage weight. One might set a prior probability of 1/3 on the value $w = 0.9$, or on values $w > 0.9$, corresponding to nearly complete shrinkage to μ as under classical meta-analysis. A prior probability of 1/3 would also be set on $w = 0.1$ or values $w < 0.1$, corresponding to a sceptical view on exchangeability. Finally, a prior probability of 1/3 could be set on neutral values, $w \sim U(0.1, 0.9)$.

Another possibility when the s_i^2 are provided as part of study summaries, is a uniform prior on the average shrinkage (Spiegelhalter et al., 2004, Chapter 5), namely,

$$w = \frac{s_0^2}{s_0^2 + \tau^2},$$

where

$$\frac{1}{s_0^2} = \frac{1}{n}\sum_{i=1}^{n}\frac{1}{s_i^2},$$

is the harmonic mean of the study sampling variances. DuMouchel (1996) proposes a uniform prior on $s_0/(s_0+\tau)$ that is equivalent to a Pareto prior, namely,

$$p(\tau) = \frac{s_0}{(s_0 + \tau)^2}.$$

This prior is proper but with $E(\tau) = \infty$, and with (0.01, 0.25, 0.5, 0.75, 0.99) percentile points at $(s_0/99, s_0/3, s_0, 3s_0, 99s_0)$. Note that the Pareto can also be parameterized as

$$p(u) = bs_0^b u^{-b-1},$$

with $\tau = u - s_0$ when $b = 1$.

An increasingly popular option is half-normal, half-Student t, or half-Cauchy priors on the second stage standard deviation τ. If $T \sim N(0, V)$ and $\tau = |T|$, then τ is half-normal with variance V (Spiegelhalter et al., 2004). One then has $E(\tau|V) = \sqrt{2V/\pi}$ and $\text{var}(\tau|V) = V(1-2/\pi)$. If τ_U represents a likely upper value for τ, then one may take $V = (\tau_U/1.96)^2$ as in Pauler and Wakefield (2000). Note that if $T \sim N(m, V)$ (i.e., the normal has an unknown mean), then $\tau = |T|$ is folded-normal with

$$E(\tau|V) = \sqrt{2V/\pi}\exp(-m^2/2V) - m[1 - 2\Phi(m/V^{0.5})],$$

and variance $m^2 + V$. Gelman (2006a) and Zhao et al. (2006) adopt folded noncentral t-densities for τ, obtained by dividing the absolute value of a normal variable by the square root of a gamma variable. If the normal variable has mean zero then the folded noncentral t becomes a half-t variable.

In particular, half-t and half-Cauchy priors for the second stage parameter τ may be achieved by a reparameterization of the second stage prior on the latent trial means that strictly involves parameter redundancy. Such overparameterization may improve MCMC convergence (Gelman, 2006a). With preset parameters ν and A (degrees of freedom and prior scale, respectively) one has,

$$\begin{aligned} b_i &= \mu + \xi\eta_i, \\ \xi &\sim N(0, A), \\ \eta_i &\sim N\left(0, \sigma_\eta^2\right), \\ 1/\sigma_\eta^2 &\sim \chi_\nu^2, \end{aligned}$$

with the standard deviation of b_i then obtained as $\tau = |\xi|\sigma_\eta$; see van Dongen (2006) for an application. Setting $\nu = 1$ leads to a half-Cauchy prior,

$$p(\sigma_b) \propto (\tau^2 + A)^{-1},$$

where Gelman (2006a, 524) uses a value $A = 25$ in a meta-analysis with small n, based on a prior belief that τ was well below 100.

3.5 Multivariate Meta-Analysis

A multivariate analysis for metric outcomes may arise in two main ways. The first occurs in clinical applications involving treatment and control arms. Often the event rate in the control arm is taken as indicating the baseline risk in the patient population being studied and there is interest in whether the treatment effect is related in any way to baseline risk (Arends, 2006). Suppose $r_{i\text{T}}$ of $N_{i\text{T}}$ treated subjects in trial i exhibit a particular response (e.g., disease or death) as compared to $r_{i\text{C}}$ of $N_{i\text{C}}$ control subjects; also define log odds $y_{i\text{T}} = \log(r_{i\text{T}}/(N_{i\text{T}} - r_{i\text{T}}))$ and $y_{i\text{C}} = \log(r_{i\text{C}}/(N_{i\text{C}} - r_{i\text{C}}))$. Often the analysis focuses on the log of the odds ratio obtained as $y_{iT} - y_{iC}$, and treats it as approximately normal (see Example 3.1).

However, to separate out baseline risk, one may model $\{y_{iT}, y_{iC}\}$ as (approximately) bivariate normal. If the trial is randomized it is legitimate to assume that $\{y_{iT}, y_{iC}\}$ are independent at the first stage (van Houwelingen et al., 2002). So,

$$\begin{pmatrix} y_{iT} \\ y_{iC} \end{pmatrix} \backsim N\left(\begin{pmatrix} b_{iT} \\ b_{iC} \end{pmatrix}, \begin{pmatrix} s_{iT}^2 & 0 \\ 0 & s_{iC}^2 \end{pmatrix}\right),$$
$$\begin{pmatrix} b_{iT} \\ b_{iC} \end{pmatrix} \backsim N\left(\begin{pmatrix} \mu_T \\ \mu_C \end{pmatrix}, \Upsilon\right),$$

where $\Upsilon = \begin{pmatrix} \tau_T^2 & \tau_{TC} \\ \tau_{TC} & \tau_C^2 \end{pmatrix}$, with diagonal terms τ_T^2 and τ_C^2 representing variability in the true treatment and control event rates, and with $\gamma = \mu_T - \mu_C$ defining the underlying treatment effect with variance $\tau_T^2 + \tau_C^2 - 2\tau_{TC}$. The conditional variance of the treatment effect given the true control group rate is $\tau_T^2 - (\tau_{TC}^2/\tau_C^2)$. So baseline risk explains a portion $\tau_C^2 - 2\tau_{TC} + (\tau_{TC}^2/\tau_C^2)/\tau_T^2 + \tau_C^2 - 2\tau_{TC}$ of the treatment effect variance.

It is also possible, of course, to have a multivariate analysis generated when more than one outcome is associated with a specific unit, such as surgical and nonsurgical mortality in each hospital in the case study of Everson and Morris (2000). In this case, suppose there are K outcomes, then,

$$\begin{pmatrix} y_{i1} \\ y_{i2} \\ \cdot \\ y_{iK} \end{pmatrix} \backsim N\left(\begin{pmatrix} b_{i1} \\ b_{i2} \\ \cdot \\ b_{iK} \end{pmatrix}, S_i\right),$$

where

$$S_i = \begin{pmatrix} s_{i1}^2 & s_{i12} & \cdot & s_{i1K} \\ s_{i21} & s_{i2}^2 & \cdot & s_{i2K} \\ \cdot & \cdot & \cdot & \cdot \\ s_{iK1} & s_{iK2} & \cdot & s_{iK}^2 \end{pmatrix},$$

is the known covariance matrix between outcomes for trial i. A multivariate normal second level prior for $(b_{i1}, \ldots, b_{iK})$ involves means $\{\mu_1, \ldots, \mu_K\}$ and $K \times K$ covariance matrix Υ.

Multivariate normality is often a simplification and one may wish to allow both for heavier tails, skewness, or multimodality; see Genton (2004) and Lee and Thompson (2008) for reviews of recent developments in skew-elliptical densities, which are one possible avenue to greater robustness in these directions. These models build on the principle suggested by Azzalini (1985) that if f and g are symmetric densities with parameters μ and σ, with G the cumulative density corresponding to g, then the new density defined by,

$$h(x|\mu,\sigma,\delta) = \frac{2}{\sigma} f\left(\frac{x-\mu}{\sigma}\right) G\left(\delta\frac{x-\mu}{\sigma}\right),$$

is skew for nonzero δ.

Following Sahu et al. (2003), a multivariate skew-normal model is a particular type of skew-elliptical model (of dimension K) obtained by considering errors $\varepsilon_{K\times 1} \backsim N_K(0,\Sigma)$, positive variables $Z_{K\times 1} \backsim N_K(0,I)$, and taking $y = DZ + \varepsilon$ where D is a diagonal matrix, $\text{diag}(\delta_1,\ldots,\delta_K)$. In a regression setting with a K dimensional mean μ, one has,

$$y|Z = z \backsim N_K(\mu + Dz, \Sigma).$$

Values $\delta_k > 0$ correspond to positive skew in the kth outcome, while a negative δ_k arises from negative skew. A multivariate skew-t model (allowing for both heavier tails than the normal, and also for skewness) is obtained by sampling $Z_{K\times 1} \backsim t_{K,\nu}(0,I)$, where ν is a degrees of freedom parameter, then,

$$y|Z = z \backsim t_{K,\nu+K}\left(\mu + Dz, \frac{\nu + z^T z}{\nu + K}\Sigma\right).$$

Example 3.1. Nicotine Replacement Therapies To illustrate approximately normal responses based on discrete (binomial) data, this example considers $n = 90$ studies[1] of the benefits of nicotine replacement therapy (NRT) (Cepeda-Benito et al., 2004). The data are supplied as the numbers, $r_{i\mathrm{T}}$, quitting smoking among those under therapy, $N_{i\mathrm{T}}$, and numbers of quitters, $r_{i\mathrm{C}}$, in control or placebo groups of size $N_{i\mathrm{C}}$. Then the empirical log odds ratios measuring treatment effects, namely,

$$y_i = \log\left(\frac{r_{i\mathrm{T}}}{N_{i\mathrm{T}} - r_{i\mathrm{T}}}\right) - \log\left(\frac{r_{i\mathrm{C}}}{N_{i\mathrm{C}} - r_{i\mathrm{C}}}\right),$$

are approximately normal with variances,

$$s_i^2 = \frac{1}{r_{i\mathrm{T}}} + \frac{1}{N_{i\mathrm{T}} - r_{i\mathrm{T}}} + \frac{1}{y_{i\mathrm{C}}} + \frac{1}{N_{i\mathrm{C}} - r_{i\mathrm{C}}}.$$

A normal higher stage is assumed with $y_i \backsim N(b_i, s_i^2)$ and $b_i \backsim N(\mu, \tau^2)$. A uniform shrinkage prior on

$$w = \frac{s_0^2}{s_0^2 + \tau^2},$$

as considered above, is assumed for the second stage variance. Additionally a $N(0,\ 100)$ prior on μ is adopted. Various kinds of predictions may be considered. Here the predicted treatment effect in a new trial is sampled according to,

$$b_{\text{new}} \backsim N(\mu, \tau^2),$$
$$y_{\text{new}} \backsim N(b_{\text{new}}, s_0^2).$$

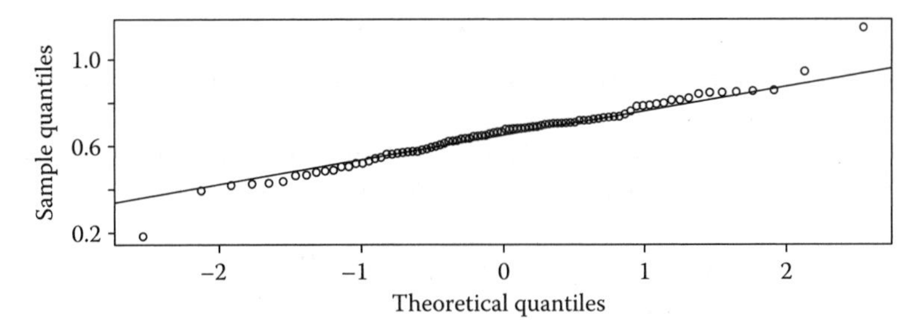

FIGURE 3.1
Normal Q-Q plot.

Early convergence in a two chain run of 5000 iterations is obtained. Using the last 4000 for inference, a clear benefit of NRT is indicated, with the odds ratio $\exp(\mu)$ having a posterior mean (and 95% credible interval) of 1.92 (1.73, 2.14). τ^2 is estimated as 0.082 (mean) and 0.076 (median). A Q-Q plot for the posterior mean, b_i, does not suggest marked departures from normality[2] except possibly in the more extreme b_i (Figure 3.1), while a Shapiro–Wilk normality test has a p-value just over 0.05. However, a Jarque–Bera test (Jarque and Bera, 1980) suggests a significant departure from normality, with p-value of 0.0018. Although the analysis would appear conclusive in terms of treatment benefit, the predicted odds ratio for a new trial includes null values for the benefit from NRT, having mean (95% CI) of 2.16 (0.7, 5.1).

To allow for possibly outlying trials and downweight their effect, an alternative analysis adopts a second stage Student density with,

$$b_i \backsim N(\mu, \tau^2/\lambda_i),$$
$$\lambda_i \backsim G\left(\frac{\nu}{2}, \frac{\nu}{2}\right).$$

Less typical trial results will have values of λ_i considerably under 1 and a test for the posterior probability that λ_i is less than 1 can be included. The prior on ν is specified in two steps as $\nu \backsim E(\kappa)$ and $\kappa \backsim U(0.01, 0.5)$. Evidence in support of a heavy-tailed second stage is equivocal. From the last 9000 of a two chain run of 10,000 iterations, ν is estimated at 9.5, suggesting departure from normality. The posterior mean and median for τ^2 are reduced to 0.051 and 0.046, respectively. On the other hand, only two trials (4 and 36) have posterior probability that $\lambda_i < 1$ in excess of 0.7, namely 0.87 for trial 4 and 0.83 for trial 36. Trial 4 has an exceptionally high empirical log odds ratio in support of NRT.

Example 3.2. Bacillus Calmette–Guérin Vaccine Trials: Bivariate Model Following van Houwelingen et al. (2002), an example of a bivariate meta-analysis involves data from 13 trials regarding the effectiveness of Bacillus Calmette–Guérin (BCG) vaccine against tuberculosis. Each trial compares vaccinated and nonvaccinated groups of size $\{N_{\mathrm{T}}, N_{\mathrm{C}}\}$ with the outcome being counts of tuberculosis $\{r_{\mathrm{T}}, r_{\mathrm{C}}\}$, and with the infection rate in the control arm taken as indicating the baseline risk. The response variables are the log odds in each trial arm, $y_{i\mathrm{T}} = \log(r_{i\mathrm{T}}/(N_{i\mathrm{T}} - r_{i\mathrm{T}}))$ and $y_{i\mathrm{C}} = \log(r_{i\mathrm{C}}/(N_{i\mathrm{C}} - r_{i\mathrm{C}}))$.

Here the analysis assumes normality at both levels[3], with

$$\begin{pmatrix} y_{i\mathrm{T}} \\ y_{i\mathrm{C}} \end{pmatrix} \backsim N\left(\begin{pmatrix} b_{i\mathrm{T}} \\ b_{i\mathrm{C}} \end{pmatrix}, \begin{pmatrix} s_{i\mathrm{T}}^2 & 0 \\ 0 & s_{i\mathrm{C}}^2 \end{pmatrix}\right),$$
$$\begin{pmatrix} b_{i\mathrm{T}} \\ b_{i\mathrm{C}} \end{pmatrix} \backsim N\left(\begin{pmatrix} \mu_T \\ \mu_{\mathrm{C}} \end{pmatrix}, \Sigma_b\right),$$

where $s_{i\mathrm{T}}^2 = 1/r_{i\mathrm{T}} + 1/(N_{i\mathrm{T}} - r_{i\mathrm{T}})$, $s_{i\mathrm{C}}^2 = 1/r_{i\mathrm{C}} + 1/(N_{i\mathrm{C}} - r_{i\mathrm{C}})$. It is assumed that the precision matrix Σ_b^{-1} of the latent effects is Wishart with identity scale matrix and 2 degrees of freedom, while the $\{\mu_{\mathrm{T}}, \mu_{\mathrm{C}}\}$ parameters have $N(0,\ 1000)$ priors.

Posterior means for $(\mu_{\mathrm{T}}, \mu_{\mathrm{C}})$ are estimated as $(-4.83, -4.02)$, with mean vaccination effect of -0.80 $(-1.3, -0.3)$, slightly more negative than the estimate of -0.74 found by van Houwelingen et al. (2002) using classical methods (in the SAS package). The posterior mean for Σ_b is $\left(\begin{smallmatrix} 1.86 & 2.24 \\ 2.24 & 3.32 \end{smallmatrix}\right)$, with correlation between treatment and control effects (where effects are log odds) obtained from monitoring the components of Σ_b as 0.90. Similarly, the slope of the regression to predict the vaccination group log odds from the control group log odds, obtained by averaging $\Sigma_{b12}^{(t)}/\Sigma_{b22}^{(t)}$ over iterations t, is 0.67. The variance of the true treatment effects, $b_{i\mathrm{T}} - b_{i\mathrm{C}}$, is obtained by monitoring $V_{\mathrm{t}} = \Sigma_{b,11} + \Sigma_{b,22} - 2\Sigma_{b,12}$, while the conditional variance of the vaccination log odds effects $b_{i\mathrm{T}}$ given $b_{i\mathrm{C}}$ (and hence the variance of $b_{i\mathrm{T}} - b_{i\mathrm{C}}$ given $b_{i\mathrm{C}}$) is obtained by monitoring $V_{\mathrm{c}} = \Sigma_{b,11} - \Sigma_{b,12}^2/\Sigma_{b,22}$. Finally, the proportion of treatment effect variation explained by baseline risk (i.e., the true log odds in the control group), obtained by monitoring $1 - V_{\mathrm{c}}/V_{\mathrm{t}}$, has a posterior mean of 0.51.

3.6 Heterogeneity in Count Data: Hierarchical Poisson Models

The adoption of higher stage densities for count data is often linked to apparent departures from the Poisson mean-variance assumption. The most common departure is that count data show more variability than expected under the Poisson, so that the coefficient of variation, $(V(y)/\bar{y})$, exceeds 1. Overdispersion may reflect unobserved subject frailties, multiple modes, nonrandom sampling (Efron, 1986), or widely different exposures, o_i (e.g., when a count outcome y_i is surgical deaths for hospitals, with means $\mu_i o_i$ where o_i are patient totals). The conjugate continuous mixture models in the presence of excess heterogeneity is the Poisson-gamma, though greater flexibility in more complex models (e.g., multilevel or multivariate) is generally obtained by mixing with nonconjugate links.

The Poisson-gamma model allows for unit mean rates, μ_i, to vary according to a gamma density, $\mu_i \backsim Ga(\alpha, \beta)$, which is unimodal but flexibly shaped. Thus for count data y_i assumed Poisson with means μ_i, set,

$$B(b_i) = e^{b_i},$$

in Equation 3.1, where $\mu_i = e^{b_i}, a(\phi_i) = 1$, and $c(y_i, \phi_i) = \log y_i!$ Then Equation 3.2 has the form,

$$\begin{aligned} p(b_i|\psi) &= k_1 \exp\left(b_i g_1(\psi) - e^{b_i} g_2(\psi)\right) \\ &= k_1 \left(e^{b_i}\right)^{g_1(\psi)} \exp\left(-e^{b_i} g_2(\psi)\right), \end{aligned}$$

namely, a gamma density for $\mu_i = e^{b_i}$ with parameters $\alpha = g_1(\psi) + 1$ and $\beta = g_2(\psi)$. The conditional posterior is

$$p(b_i|y, \alpha, \beta) = k_2 \left(e^{b_i}\right)^{\alpha + y_i} \exp\left(-e^{b_i}[\beta + 1]\right),$$

namely, a gamma for μ_i with parameters $\alpha + y_i$ and $\beta + 1$. Denoting the mean of the μ_i as $\xi = \alpha/\beta$, one obtains $V(\mu_i) = \alpha/\beta^2 = \xi^2/\alpha$. Then,

$$V(y_i) = E[V(y_i|\mu_i)] + V[E(y_i|\mu_i)] = \xi + \xi^2/\alpha,$$

so that overdispersion is present when $\phi > 0$, where $\phi = 1/\alpha$.

Different parameterizations of the Poisson-gamma mixture can be used. For example, one may set $\mu_i = \xi\omega_i$, with overall mean parameter ξ, and multiplicative random effects, ω_i, having mean 1 for identifiability, namely, $\omega_i \backsim G(\alpha, \alpha)$ with $V(\omega_i) = 1/\alpha$. Integrating the ω_i out, as in

$$p(y_i|\xi, \alpha) = \int p(y_i|\omega_i, \xi) p(\omega_i|\alpha) d\omega_i,$$

leads to a marginal negative binomial density for the y_i, namely,

$$p(y_i|\xi, \alpha) = \frac{\Gamma(\alpha + y_i)}{\Gamma(\alpha)\Gamma(y_i + 1)} \left(\frac{\alpha}{\alpha + \xi}\right)^{\alpha} \left(\frac{\xi}{\alpha + \xi}\right)^{y_i}.$$

If predictors X_i are present, negative binomial regression is obtained with $\xi_i = \exp(X_i\beta)$ (see Chapter 5). Note that in many applications there may be forms of truncation, as when zero counts do not enter the analysis (Eaton, 1974; Larson and Soule, 2006). So a zero truncated negative binomial has

$$p(y_i|\xi, \alpha, y_i > 0) = \frac{\frac{\Gamma(\alpha + y_i)}{\Gamma(\alpha)\Gamma(y_i + 1)} \left(\frac{\alpha}{\alpha + \xi}\right)^{\alpha} \left(\frac{\xi}{\alpha + \xi}\right)^{y_i}}{1 - \left(\frac{\alpha}{\alpha + \xi}\right)^{\alpha}}.$$

Alternatively, one may assume $y_i \backsim Po(\mu_i), \mu_i \backsim G(\alpha, \beta)$, with $E(\mu_i) = m = \alpha/\beta$ and $\text{var}(\mu_i) = V_\mu = \alpha/\beta^2$ (e.g., Clayton and Kaldor, 1987). When this parameterization includes offsets, o_i, the posterior $p(\mu_i, \alpha, \beta|y)$ is proportional to,

$$L(\alpha, \beta, \mu|y)p(\alpha, \beta) =$$

$$\left[\prod_{i=1}^{n} \frac{\exp(-\mu_i o_i)(\mu_i o_i)^{y_i}}{y_i!}\right]\left[\left(\frac{\beta^\alpha}{\Gamma(\alpha)}\right)^n \left(\prod_{i=1}^{n} \mu_i\right)^{\alpha-1} \exp\left(-\beta \sum_{i=1}^{n} \mu_i\right)\right] p(\alpha, \beta),$$

with conditional posterior for μ_i now $G(\alpha + y_i, \beta + o_i)$. Hence the posterior mean is

$$E(\mu_i|y_i, \alpha, \beta) = \frac{y_i + \alpha}{o_i + \beta}.$$

The conditional likelihoods (George et al., 1993, 191) for α and β under this structure are obtained from $L(\alpha, \beta, \mu|y)$, namely,

$$L(\alpha|\beta, \mu) = k_\alpha \left(\frac{\beta^\alpha}{\Gamma(\alpha)}\right)^n \left(\prod_{i=1}^{n} \mu_i\right)^{\alpha-1},$$

and

$$L(\beta|\alpha, \mu) = k_\beta \beta^{n\alpha} \exp\left(-\beta \sum_{i=1}^{n} \mu_i\right),$$

where k_α and k_β are normalizing constants. Hence $L(\beta|\alpha, \mu)$ is gamma with parameters $n\alpha + 1$ and $\sum_{i=1}^{n} \mu_i$. The conditional posteriors, $p(\alpha|\beta, \mu) = L(\alpha|\beta, \mu)p(\alpha)$ and $p(\beta|\alpha, \mu) = L(\beta|\alpha, \mu)p(\beta)$, are log-concave when the priors $p(\alpha)$ and $p(\beta)$ are log-concave. Assuming a gamma prior, $p(\beta) = G(c, d)$, the full conditional for β is $G(n\alpha + 1 + c, \sum_{i=1}^{n} \mu_i + d)$. However, the full conditional for α is nonstandard whatever form for $p(\alpha)$ is adopted.

Another Poisson-gamma mixture formulation (e.g., Albert, 1999; Christiansen and Morris, 1996) assumes,

$$y_i|\lambda_i \backsim Po(o_i \lambda_i),$$

$$\lambda_i \backsim Ga\left(\zeta, \frac{\zeta}{\mu_i}\right),$$

where $V(\lambda_i) = \mu_i^2/\zeta$ and the Poisson corresponds to $\zeta \to \infty$. If $\mu_i = \mu$ and a gamma prior is assumed for μ, then the posterior mean for λ_i conditional on μ and ζ is

$$E(\lambda_i|y, \zeta, \mu) = \frac{y_i + \zeta}{o_i + \zeta/\mu} = B_i \mu + (1 - B_i)\frac{y_i}{o_i},$$

where

$$B_i = \frac{\zeta}{\zeta + o_i \mu},$$

measures the level of shrinkage toward the overall mean μ. Thus, shrinkage will be greater when o_i (e.g., the population at risk in a mortality application) is small, or when ζ is large. As for the second stage variance in the normal-normal model, the prior on ζ influences the degree of shrinkage that is obtained. Let $r_i = y_i/o_i$. Then Christiansen and Morris (1996) suggest a uniform prior based on the average shrinkage factor,

$$B_0 = \frac{\zeta}{\zeta + \min(o_i)\bar{r}} \sim U(0,1),$$

with the prior value of ζ then obtained as $B_0 \min(o_i)\bar{r}/(1 - B_0)$.

Extended parameterizations of the negative binomial have been suggested (Liu and Dey, 2007). Winkelmann and Zimmermann (1991) suggest a variance function,

$$V(y_i) = E[V(y_i|\mu_i)] + V[E(y_i|\mu_i)] = \xi + \phi\xi^{k+1},$$

with $k \geq -1$, and obtained by taking $\mu_i \backsim G(\xi^{1-k}/\phi, \xi^{-k}/\phi)$. Setting $k = 0$ and $k = 1$ leads to what are called NB1 and NB2 forms of the negative binomial, under which the variances are linear and quadratic in ξ, namely, $V(y_i) = \xi + \phi\xi$ and $V(y_i) = \xi + \phi\xi^2$, respectively.

3.6.1 Nonconjugate Poisson mixing

Alternatives to the conjugate model are the Poisson lognormal model, and models such as the generalized Poisson density, zero inflated Poisson (ZIP), and hurdle model adapted to different types of departure from the typical Poisson frequency pattern. The Poisson lognormal model has been suggested as a more appropriate model, than the conjugate mixture in certain applications such as species abundance—see Bulmer (1974) and Diserud and Engen (2000). It is obtained for $y_i \sim Po(\mu_i)$ when μ_i are lognormally distributed, or equivalently when the logarithms $w_i = \log(\mu_i)$ of the Poisson means are assumed normal with mean M and variance V (Aitchison and Ho, 1989). The marginal density under lognormal mixing is obtained by integrating the sampling density over the domain of the log mean, namely,

$$p(y_i|M,V) = \frac{(2\pi V)^{-0.5}}{y_i!} \int_0^\infty \mu_i^{y_i - 1} e^{-\mu_i} \exp\left[\frac{-(\log \mu_i - M)^2}{2V}\right] d\mu_i,$$

with marginal mean and variance, respectively, $e^{M+V/2}$ and $e^{2M+V}[e^V - 1]$. As $V \to 0$, this reduces to a Poisson density. An alternative parameterization (Weems and Smith, 2004) has $y_i \sim Po(\mu_i U_i)$ with $\log(\mu_i) = \beta_0 + \beta_1 x_{1i} + \cdots + \beta_p x_{pi}$, and $\log(U_i) \sim N(1, V)$.

The Poisson-lognormal generalizes readily to multivariate count data (Chib and Winkelmann, 2001) or to mixing with heavier tails than available under the lognormal; for example, the log Student t with a low degrees of freedom parameter for a heavy-tailed albeit symmetric mixing density. Skew normal and skew Student t mixing can also be used since in some applications extremes of frailty tend to be above rather than below the center of density (Sahu et al., 2003).

The exchangeable Poisson lognormal model is quite widely applied to pooling inferences over sets of units (e.g., hospitals) when health event totals, y_i, such as surgical deaths are obtained and there are o_i expected events; the Poisson lognormal is also widely applied in modeling for spatially structured disease count data (Chapter 4). The o_i might be based on multiplying the patient total for hospital i by an average event rate and are usually assumed known (i.e., not to be subject to measurement error). If the average rate is based on the total set of n hospitals, then one has $\Sigma y_i = \Sigma o_i$, and with $\mu_i = o_i \rho_i$ one has,

$$y_i \backsim Po(o_i \rho_i),$$

with the ρ_i interpretable as relative risks averaging 1 over all units. However, this feature is not always present, and allowing for mean risk other than 1 (e.g., if a national surgical mortality rate is applied to a particular set of hospitals) the Poisson-lognormal then assumes,

$$\log(\rho_i) = \beta_0 + w_i,$$

where $w_i \backsim N(0, V_w)$ are exchangeable normal random effects, with relative risks, ρ_i, pooled toward a global average rate, $\exp(\beta_0)$, according to the size of V_w. Equivalently, $v_i = \exp(w_i)$ are lognormal with mean $\mu = \exp(0.5\, V_w)$ and variance $\mu^2(\exp^{V_w} - 1)$.

Generalized Poisson and Poisson process models are also often useful in particular settings, including underdispersion (Consul, 1989; Podlich et al., 2004; Scollnik, 1995). The generalized Poisson density (Consul, 1989) specifies[4]

$$p(y|\lambda, \rho) = \frac{\lambda(\lambda + y\rho)^{y-1}}{y!} e^{-\lambda - \rho y},$$

with mean $\lambda/(1 - \rho)$, variance $\lambda/(1 - \rho)^3$, and hence coefficient of variation $1/(1 - \rho)^2 \geq 1$. This reduces to a Poisson density as $\rho \to 0$.

Example 3.3. Hospital Mortality To exemplify the Poisson-gamma methodology, consider counts of patient deaths following heart transplant surgery in 131 hospitals in the United States between October 1987 and December 1989. These were analyzed by Christiansen and Morris (1996, 1997). Let o_i be expected deaths (calculated by a logit regression on patient characteristics). Then,

$$y_i|\lambda_i \backsim Po(o_i \lambda_i),$$
$$\lambda_i \backsim Ga\left(\zeta, \frac{\zeta}{\mu}\right),$$

with shrinkage factors,

$$B_i = \frac{\zeta}{\zeta + o_i \mu}.$$

The prior on ζ is indirect, via a uniform prior on $B_0 = \zeta/(\zeta + \min(o_i)\bar{r})$.

A two chain run of 10,000 iterations[5] provides a deviance information criterion (DIC) of 456 ($d_e = 32$) and high values for both B_0 and ζ, namely, 0.987 and 10.9. This might be taken as possibly indicating that a Poisson-gamma mixture is not needed (see Albert, 1999). A possible predictive check is that the 95% intervals of $y_{\text{rep},i}$ include at least 95% of the data points y_i (Gelfand, 1996). In fact, all the observations meet this criterion.

Christiansen and Morris (1996) argue that exchangeability between all 131 units might not be applicable since hospitals with larger patient totals have lower crude death rates. As one remedy for such a pattern, one might take $y_i \backsim Po(\nu_i)$, $\nu_i \backsim Ga(\zeta, \zeta/\mu_i)$ where,

$$\log(\mu_i) = \beta_1 + \beta_2 \log(o_i),$$

now includes a regression on $\log(o_i)$. So expected deaths is no longer an offset with implicit coefficient $\beta_2 = 1$. Here we instead split the hospitals into two groups with indicator G_i, one group (with $G_i = 1$) containing 37 hospitals with under 10 patients, the other (with $G_i = 2$) containing the remaining 94 hospitals. Different means and variance parameters are assumed in the two groups. So,

$$y_i|\lambda_i \backsim Po(o_i\lambda_i),$$
$$\lambda_i \backsim Ga\left(\zeta_{G_i}, \frac{\zeta_{G_i}}{\mu_{G_i}}\right),$$

with uniform priors on group-specific average shrinkage factors ($k = 1, 2$),

$$B_{0k} = \frac{\zeta_k}{\zeta_k + \min(o_i; G_i = k)\bar{r}_k}.$$

This extension to partial exchangeability produces a deviance reduction to 447 with the effective dimension still about 32. The mean mortality RR (with 95% interval) is found to be 2.1 (1.4, 2.9) in the low workload hospitals, but lower, namely, 0.95 (0.82, 1.1), in the higher workload hospitals. The variance factor, ζ_k, is higher in the low workload hospitals, but the average shrinkages, B_{0k}, are similar, at 0.95 and 0.93, respectively.

3.7 Binomial and Multinomial Heterogeneity

Heterogeneity in binary and categoric outcomes is commonly found in consumer and demographic data. Among possible approaches are the beta-binomial, the logistic-normal, and generalizations of the binomial (e.g., Alanko and Duffy, 1996); analogous methods apply for categoric data with the conjugate model being the multinomial-Dirichlet. Although the Poisson-gamma mixture is widely applied to health and disease events, the beta-binomial may also be used if populations are relatively small, and has different implications for shrinkage: shrinkage is greater under the Poisson-gamma (Howley and Gibberd, 2003). Binomial and multinomial mixture methods have recently become popular in the analysis of ecologic problems where marginals of a contingency table are available, often from different sources, such as census and voting data, but the internal cells are unobserved (King, 1997; King et al., 2004).

For binomial data, $y_i \backsim Bin(N_i, \pi_i), i = 1, \ldots, n$, the exponential family parameterization sets,

$$B(b_i) = N_i \ \log(1 + e^{b_i}),$$

in Equation 3.1, where $\pi_i = e^{b_i}/(1 + e^{b_i}), a(\phi_i) = 1$, and $c(y_i, \phi_i) = \log \binom{N_i}{y_i}$. Then Equation (3.2) has the form,

$$\begin{aligned} p(b_i|\psi) &= k_1 \exp(b_i g_1(\psi) - N_i \log(1 + e^{b_i}) g_2(\psi)) \\ &= k_1 \left(\frac{e^{b_i}}{1 + e^{b_i}}\right)^{g_1(\psi)} (1 + e^{b_i})^{-N_i g_2(\psi) + g_1(\psi)}, \end{aligned}$$

namely, a beta density for π_i with parameters $g_1(\psi)$ and $N_i g_2(\psi) - g_1(\psi)$. The conditional posterior of π_i is then also a beta density with parameters $g_1(\psi) + y_i$ and $N_i[g_2(\psi) + 1] - g_1(\psi) - y_i$. The marginal density is known as the beta-binomial with,

$$p(y_i|g_1, g_2) = \binom{N_i}{y_i} \frac{Be(g_1 + y_i, N_i(g_2 + 1) - (g_1 + y_i))}{Be(g_1, N_i g_2 - g_1)}.$$

Shrinkage effects are apparent under the beta mixing parameterization,

$$\pi_i \backsim Be(\gamma\rho, \gamma(1 - \rho)),$$

where $\rho \in (0, 1)$, and where $\gamma > 0$, termed the spread parameter by Howley and Gibberd (2003), is inversely related to the prior variance of the proportions $\rho(1 - \rho)/(1 + \gamma)$. The conditional posterior for π_i is $\pi_i \backsim Be(\gamma\rho + y_i, \gamma(1 - \rho) + N_i - y_i)$, and the posterior mean is

$$E(\pi_i|y, \gamma, \rho) = \frac{\gamma}{\gamma + N_i} \rho + \frac{N_i}{\gamma + N_i} \left(\frac{y_i}{N_i}\right),$$

namely, a weighted average of the observed rate and the prior mean rate. Shrinkage to the prior mean is greater when γ is large and for small populations, N_i. The marginal density is

$$\begin{aligned} p(y_i|\gamma, \rho) &= \binom{N_i}{y_i} \frac{Be(\gamma\rho + y_i, \gamma(1 - \rho) + N_i - y_i)}{Be(\gamma\rho, \gamma(1 - \rho))} \\ &= \binom{N_i}{y_i} \frac{\Gamma(\gamma\rho + y_i)\Gamma(\gamma(1 - \rho) + N_i - y_i)\Gamma(\gamma)}{\Gamma(\gamma\rho)\Gamma(\gamma(1 - \rho))\Gamma(\gamma + N_i)}, \end{aligned}$$

with expectation $E(y_i) = E[E(y_i|\pi_i)] = E(N_i\pi_i) = N_i\rho$, and variance,

$$V(y_i) = V[E(y_i|\pi_i)] + E[V(y_i|\pi_i)] = \rho(1-\rho)\left(\frac{\gamma+N_i}{\gamma+1}\right),$$

so that $\gamma \to \infty$ corresponds to the binomial density.

Quintana and Tam (1996) consider both marginal and conditional likelihood MCMC estimation approaches to the beta-binomial. With beta mixing according to $\pi_i \backsim Be(a,b)$, and prior $p(a,b)$, they apply Hastings sampling to the joint marginal likelihood (with π_i integrated out)

$$L(a,b,y) \propto \left[\frac{\Gamma(a+b)}{\Gamma(a)\Gamma(b)}\right]^n \left[\prod_i \frac{\Gamma(a+y_i)\Gamma(b+N_i-y_i)}{\Gamma(a+b+N_i)}\right] p(a,b),$$

and mixed Gibbs–Hastings sampling to the joint conditional likelihood,

$$L(a,b,\pi,y) \propto \left[\frac{\Gamma(a+b)}{\Gamma(a)\Gamma(b)}\right]^n \left[\prod_i \pi_i^{a+y_i-1}(1-\pi_i)^{b+N_i-y_i+1}\right] p(a,b).$$

They also consider implications for posterior parameter correlation of the reparameterization (Lee and Sabavala, 1987),

$$\pi_i \sim Be(\mu,\eta),$$
$$\mu = a/(a+b); \eta = 1/(1+a+b),$$

where η is a measure of heterogeneity.

3.7.1 Nonconjugate priors for binomial mixing

Alternatives to conjugate beta mixing are the binomial with normal errors in the link, generalized binomial models (Makuch et al., 1989), generalized beta-binomial models (Rodriguez-Avi et al., 2007), and models adapted to departures from the typical binomial frequency pattern, such as zero-inflated binomial models. The logistic-normal model with normal random effects in the logit link specifies,

$$y_i|\pi_i \backsim Bin(N_i,\pi_i),$$
$$\text{logit}(\pi_i) = b_i,$$
$$b_i|\mu,\tau \backsim N(\mu,\tau^2).$$

Here, π_i then follows a logistic-normal density,

$$p(\pi_i|\mu,\tau^2) = \frac{1}{\tau\sqrt{2\pi}} \exp\left(-\frac{1}{2\tau^2}\left[\log\frac{\pi_i}{1-\pi_i} - \mu\right]^2\right) \frac{1}{\pi_i(1-\pi_i)}.$$

The logistic-normal prior with $\tau = 2.67$ and $\mu = 0$ matches a Jeffreys prior on π_i in the first two moments, and setting $\tau = 1.69$ matches the uniform prior in the first two moments (Agresti and Hitchcock, 2005). As for the Poisson-lognormal, one may generalize to heavier tailed or skewed mixing densities. Teather (1984) proposes a family of symmetric prior densities for $\text{logit}(\pi_i)$ that includes the normal and double exponential as special cases. Alternative links (e.g., probit) or mixing over links are possible.

In many applications (e.g., studies with patients allocated to multiple treatment), the random effect variation represents differential frailty in the patient population of the study, so that for studies, $i = 1, \ldots, n$ with $k = 1, \ldots, K$ treatment categories,

$$\begin{aligned} y_{ik} &\backsim Bin(N_{ik}, \pi_{ik}), \\ \text{logit}(\pi_{ik}) &= b_i + \beta_k, \\ b_i &\backsim N(0, \tau^2), \end{aligned}$$

where β_k are fixed treatment effects, while b_i can be interpreted as between study variation in treatment effects. For example, Gao (2004) considers this structure for data on a meta-analysis of eight randomized clinical trials comparing healing rates in duodenal ulcer patients. For trials with treatment and control arms only, with patient totals $\{N_{i\text{T}}, N_{i\text{C}}\}$, the logistic-normal model is often applied in meta-analysis when trial totals are small, rather than adopting a normal approximation (Parmigiani, 2002; Warn et al., 2002). In fact, other links (combined with binomial sampling) may be more useful in clinical interpretability.

The prior structure often focuses on the control arm probabilities, $\pi_{i\text{C}}$, and on differences between trial and control group probabilities. Thus, assume,

$$\begin{aligned} y_{i\text{T}} &\backsim Bin(N_{i\text{T}}, \pi_{i\text{T}}), \\ y_{i\text{C}} &\backsim Bin(N_{i\text{C}}, \pi_{i\text{C}}). \end{aligned}$$

Then analysis of treatment–control differences, δ_i, on the log odds ratio scale would involve transforms $\omega_{i\text{T}} = \text{logit}(\pi_{i\text{T}})$, and $\omega_{i\text{C}} = \text{logit}(\pi_{i\text{C}})$, and taking,

$$\delta_i = \omega_{i\text{T}} - \omega_{i\text{C}},$$

one might assume,

$$\delta_i \sim N(\Delta, \sigma_\delta^2).$$

For the $\pi_{i\text{C}}$, random effect options might be to take $\omega_{i\text{C}} \backsim N(\mu_\text{C}, \tau_\text{C}^2)$, with $\{\mu_\text{C}, \tau_\text{C}^2\}$ as additional unknowns, or $\pi_{i\text{C}} \backsim Be(a_\text{C}, b_\text{C})$ with $\{a_\text{C}, b_\text{C}\}$ additional unknowns.

Consider instead a log link, so that $\omega_{i\text{T}} = \log(\pi_{i\text{T}})$, and $\omega_{i\text{C}} = \log(\pi_{i\text{C}})$, again with $\delta_i \backsim N(\Delta, \sigma_\delta^2)$. The δ_i now measure log relative risks, which are often more clinically useful than log odds ratios, and $\exp(\Delta)$ will measure the RR of (say) recurrence or mortality under the treatment. In practice, sampling has to be constrained to ensure δ_i is less than $-\log(\pi_{i\text{C}})$, so that,

$$\omega_{i\mathrm{T}} = \omega_{i\mathrm{C}} + \min(\delta_i, -\log(\pi_{i\mathrm{C}})),$$
$$\delta_i \backsim N(\Delta, \sigma_\delta^2).$$

Similarly, for a risk difference analysis, $\omega_{i\mathrm{C}} = \pi_{i\mathrm{C}}, \pi_{i\mathrm{T}} = \omega_{i\mathrm{T}} = \delta_i + \omega_{i\mathrm{C}}$, with $\delta_i \backsim N(\Delta, \sigma_\delta^2)$, but sampling has to be constrained to ensure that $\pi_{i\mathrm{T}} \in [0, 1]$. This involves confining δ_i to the interval $[-\pi_{i\mathrm{C}}, 1-\pi_{i\mathrm{C}}]$ with the actually sampled model specifying,

$$\omega_{i\mathrm{T}} = \omega_{i\mathrm{C}} + \min(\max(\delta_i, -\pi_{i\mathrm{C}}), 1 - \pi_{i\mathrm{C}}).$$

If the control group probabilities are regarded as proxies for the underlying risk of subjects in a study, then the model involves a regression on centered control group effects, namely,

$$\omega_{i\mathrm{T}} = \omega_{i\mathrm{C}} + \delta_i + \beta(\omega_{i\mathrm{C}} - \bar{\omega}_\mathrm{C}),$$
$$\delta_i \backsim N(\Delta, \sigma_\delta^2),$$

where $\bar{\omega}_\mathrm{C}$ is the average of the control arm effects (calculated at each iteration), and β is an extra unknown.

3.7.2 Multinomial mixtures

For representing overdispersion in multinomial data with M categories,

$$(y_{i1}, \ldots, y_{iM}) \backsim Mult(N_i, [\pi_{i1}, \ldots, \pi_{iM}]),$$
$$N_i = \sum_m y_{im},$$

the beta prior generalizes to a Dirichlet prior with parameters $(\alpha_{i1}, \ldots, \alpha_{iM})$. With $\pi_i = [\pi_{i1}, \ldots, \pi_{iM}]$, $\alpha_{im} = \alpha_m$, and $A = \sum_m \alpha_m$, one has,

$$p(\pi_i|\alpha) = \frac{\Gamma(A)}{\prod_{m=1}^M \Gamma(\alpha_m)} \prod_{m=1}^M \pi_{im}^{\alpha_m - 1},$$

so that prior means for π_{im} are α_m/A, with variances $\alpha_m(K - \alpha_m)/A^2(A+1)$. The posterior density for $[\pi_{i1}, \ldots, \pi_{iM}]$ is Dirichlet with parameters $(y_{i1} + \alpha_1, \ldots, y_{iM} + \alpha_M)$. Assuming equal prior mass is assigned to all categories, namely, $\alpha_1 = \alpha_2 = \cdots = \alpha_M$, there is greater shrinkage or flattening toward an equal prior cell probability across the M categories as A increases.

Greater flexibility may be provided by a multivariate generalization of the logistic-normal prior (Aitchison and Shen, 1980; Hoff, 2003). Thus, with $(y_{i1}, \ldots, y_{iM}) \backsim Mult(N_i, [\pi_{i1}, \ldots, \pi_{iM}])$,

$$\pi_{ij} = \frac{\exp(b_{ij})}{\sum_{m=1}^M \exp(b_{im})},$$

where the vector $(b_{i1}, \ldots, b_{i,M-1})$ of the first $M - 1$ effects is multivariate normal with mean $\mu_i = (\mu_{i1}, \ldots, \mu_{i,M-1})$ and covariance matrix Σ of dimension $M - 1$. For the reference category, one sets $b_{iM} = 0$. If the categories are ordered and similarity of probabilities in adjacent categories is expected on substantive grounds, the covariance matrix or its inverse may be stipulated in line with a low order autoregressive form; this is known as histogram smoothing (Leonard, 1973).

Another generalization is to add a higher stage prior on the Dirichlet parameters, for example on the total mass A. Thus, Albert and Gupta (1982) consider a two-stage prior in multinomial-Dirichlet analysis of contingency tables. With the reparameterization, $\alpha_i = A\rho_i$, where $\sum_m \rho_m = 1$, one possible hierarchical prior generalizes the binomial-beta with,

$$\begin{aligned}
\pi_i = [\pi_{i1}, \ldots, \pi_{iM}] &\backsim Dir(A\rho_1, \ldots, A\rho_M), \\
A &\backsim Ga(a_A, b_A), \\
(\rho_1, \ldots, \rho_M) &\backsim Dir(w_1, \ldots, w_M),
\end{aligned}$$

where w_m and $\{a_A, b_A\}$ are known.

3.7.3 Ecological inference using mixture models

Binomial-beta and multinomial-Dirichlet models (or nonconjugate alternatives) have recently found wide application in ecological inference. Much of the impetus for this research has come from political science, and may involve counts of a behavior or event for unit i (e.g., constituency) with M outcomes (e.g., party voting affiliation) by demographic attribute with c levels (e.g., social class, ethnic group). The underlying data are the totals N_{imc}. What is observed in practice are the marginals N_{im+} (e.g., constituency voting data by party voted for), and information from another source (e.g., from the census) on the relative distribution of the voting age population across levels of the demographic attribute. This is proxy information regarding the ratios $x_{ic} = N_{i+c}/N_{i++}$ (which might be census-based percentages of the voting population in different ethnic groups).

Consider the simplest case, ecological inference in 2×2 tables. Suppose the observations are the total electorate, N_i, the number who turn out, V_i, and (from census data) the proportion, x_i, of the voting age population who are black. Given this information, the goal of ecological inference is to estimate parameters governing the internal table cells, namely, the proportions, r_{i1} and r_{i2}, of black and white voters who turned out. Since $M = 2$, the data are binomial, and the overall turnout rate in area i is modeled as $V_i \sim Bin(N_i, p_i)$. Modeling of the turnout rates in terms of ethnic-specific voting rates proceeds using the probabilistic statement,

$$\begin{aligned}
Pr(\text{Turnout}) = {} & Pr(\text{Turnout}|\text{Black})Pr(\text{Black}) \\
& + Pr(\text{Turnout}|\text{White})Pr(\text{White}),
\end{aligned}$$

with the corresponding relation in area i being,

$$p_i = r_{i1}x_i + r_{i2}(1 - x_i).$$

Among possible priors for the unknown r_{i1} and r_{i2} in a 2×2 ecological problem are:

1. independent beta densities $r_{i1} \sim Be(a_1, b_1), r_{i2} \sim Be(a_2, b_2)$;
2. a bivariate normal for $w_{ij} = \text{logit}(r_{ij})$, with mean $\mu = (\mu_1, \mu_2)$ and covariance Σ, allowing $\{r_{i1}, r_{i2}\}$ to be correlated;
3. a trivariate normal for $w_{i1} = \text{logit}(r_{i1})$, $w_{i2} = \text{logit}(r_{i2})$, and $w_{i3} = \text{logit}(x_i)$.

Imai et al. (2010) typify ecological missing data as data "coarsening" and the first two priors above are consistent with coarsening at random. By contrast, the final option amounts to modeling the joint density $p(x, r)$ of racial composition x and turnout behavior $r = (r_1, r_2)$ via the sequence $p(x|r)p(r)$. This is similar to joint modeling of missingness and observed data in non-random models for missing data (Pastor, 2003) and hence may be termed coarsening not at random. If predictors of turnout rates are available, then the means μ_{i1} and μ_{i2} include regression terms.

Example 3.4. Breast Cancer Recurrence: Binomial Meta-Analysis under Different Treatment Scales Parmigiani (2002, 127) considers 14 trials concerning the impact of tamoxifen on breast cancer recurrence rates. The trials are mostly large and a normal approximation might well be applied, though one trial involved only 20 patients. A binomial analysis is adopted with $y_{i\text{T}} \backsim Bin(N_{i\text{T}}, \pi_{i\text{T}})$, and $y_{i\text{C}} \backsim Bin(N_{i\text{C}}, \pi_{i\text{C}})$. Defining $\omega_{i\text{T}} = \text{logit}(\pi_{i\text{T}})$, $\omega_{i\text{C}} = \text{logit}(\pi_{i\text{C}})$, and $\delta_i = \omega_{i\text{T}} - \omega_{i\text{C}}$, the logit scale treatment effects are assumed normal,

$$\delta_i \backsim N(\Delta, \sigma_\delta^2),$$

with diffuse normal and inverse gamma priors on Δ and σ_δ^2, respectively. A random effects beta density is assumed for the control group rates, namely, $\pi_{i\text{C}} \backsim Be(a_\text{C}, b_\text{C})$ with uniform priors on the unknowns, $a_\text{C} \backsim U(1, 100)$ and $b_\text{C} \backsim U(1, 100)$. To provide a summary index of treatment benefit, the treatment gain, δ_{new}, for a hypothetical new trial is sampled and added to a predicted baseline recurrence rate, $\pi_{\text{new,C}}$ (transformed on the appropriate scale) to give a predicted new trial treatment rate, $\pi_{\text{new,T}}$. Then the probability that the predictive relative risk, $\text{RR}_{\text{new}} = (\pi_{\text{new,T}}/\pi_{\text{new,C}})$, exceeds 1 is obtained.

Treatment and placebo groups are compared on three different effect scales, namely, the log-odds ratio (LOR), the log-relative risk (LRR), and the absolute risk difference (ARD). On the LRR scale, the predictive density[6] for RR_{new} has 95% interval (0.78, 1.01) with a 3.5% chance that RR_{new} exceeds 1. The ARD scale admits a larger element of doubt, with a 95% interval

(0.64, 1.05) and $Pr(\text{RR}_{\text{new}} > 1|y) = 0.075$. By contrast, under an LOR scale, RR_{new} has 95% interval (0.78, 0.98) with only a 1% chance of exceeding 1.

Example 3.5. Voter Registration This example considers the race and literacy data from King (1997) for 1040 US counties. For county i, x_i is the county proportion of blacks, and p_i is the overall literacy rate in a binomial model, $L_i \sim Bin(N_i, p_i)$, with literate and population totals, L_i and N_i. Then,

$$p_i = r_{i1}x_i + r_{i2}(1 - x_i),$$

where r_{i1} and r_{i2} are the literacy rates of blacks and whites, respectively, taken to be unknown for modeling purposes. In fact, the true values of r_{i1} and r_{i2} are known (denoted r^*_{i1} and r^*_{i2}), so that the performance of alternative models can be assessed. The first model applied is the "coarsening at random" model of Imai et al. (2010) with a bivariate normal on the logits of r_{i1} and r_{i2}, with $N(0, 1)$ priors on μ_i, and a Wishart prior on Σ with identity scale matrix. An absolute error performance measure, $E_r = \Sigma_i \Sigma_j |r_{ij} - r^*_{ij}|$, is applied. In a two chain run[7] in OpenBUGS, using initial values based on exploratory analysis, convergence is obtained by iteration 10,000 in a 20,000 iteration two-chain run. With inferences based on the remaining iterations, the correlation ρ between logits of black and white literacy rates has a mean (95% interval) of 0.66 (0.53, 0.78), a value higher than the 0.27 in a frequentist analysis reported by Imai et al. (2010) in the manual for the R package eco (for ecological inference in 2×2 tables). Other parameter means are $\mu = (0.49, 3.0)$, $\sigma_{11} = 0.17$, and $\sigma_{22} = 0.95$, with mean $E_r = 113.8$.

Taking a trivariate normal for $w_{i1} = \text{logit}(r_{i1})$, $w_{i2} = \text{logit}(r_{i2})$, and $w_{i3} = \text{logit}(x_i)$ involves a different data input scheme, with $\text{logit}(x_i)$ as the third column in a 1040×3 matrix, with the first two columns containing missing data. The Wishart scale matrix has degrees of freedom, $\nu = 3$, and scale matrix with diagonal elements equal to νV_{ii}, where V_{ii} $(i = 1, 2)$ are based on the preceding model and V_{33} is based on the empirical variance of w_{i3}. From the second half of a two-chain run of 50,000 iterations, the absolute error criterion now averages 104, thereby assuming nonrandom data coarsening improves fit. The correlation between logit literacy rates is lower than under the bivariate model (and more precisely estimated) at 0.17 (0.13, 0.21). The element σ_{31} of the covariance matrix is significantly negative, so that the registration rate for blacks is inversely related to the percent of voters who are black.

3.8 Discrete Mixtures and Nonparametric Smoothing Methods

Hierarchical models for pooled inferences or density estimation based on a single underlying population with a specific parametric form are often a simplification. Pooling strength applications such as meta-analysis and density estimation are often seeking to identify the main features of the data, or to predict further observations y_{new} via the predictive distribution $p(y_{\text{new}}|y)$, and a single population model may not be appropriate for data exhibiting asymmetry, multiple modes, isolated outliers, or outlier clusters (Mohr, 2006). While the standard densities can be extended (e.g., to reflect asymmetry), mixtures of standard densities (normal and t densities) can be used to represent a wide variety of density shapes (Everitt and Hand, 1981). Use of a single population density model in such circumstances will provide improper pooling and poor predictions for a new unit (Hoff, 2003). For example, a normal random-effects analysis of hospital mortality rates may shrink extreme rates considerably, and this might mask potentially unusual results for units with smaller totals of patients at risk (Ohlssen et al., 2007).

Among the principles that govern robust smoothing and regression methods for nonstandard densities are discrete mixing of densities over $K > 1$ subpopulations (Bohning, 1999) and various types of local regression based on kernel or smoothness priors (Muller et al., 1996). In this chapter, the focus is on discrete mixture modeling, where the Bayesian approach has been coupled with many recent advances. These include the Bayesian analogue to nonparametric maximum likelihood estimation, with MCMC implementation as set out by Diebolt and Robert (1994), Richardson and Green (1997), and Robert (1996) and numerous developments of the Dirichlet process (DP) methodology as reviewed by Hanson et al. (2005). The Bayesian approach is flexible in terms of prior structures that can be imposed in estimation, either grounded in substantive theory, or to improve definition of the subgroups (e.g., Robert and Mengersen, 1999). On the other hand, repeated sampling without appropriate parameter constraints is subject to "label switching," since labeling of the subgroups is arbitrary (Chung et al., 2004; Fruhwirth-Schattner, 2001).

3.8.1 Finite mixtures of parametric densities

In a discrete parametric mixture model, a single parametric density is typically assumed in each subpopulation, $k \in (1, \ldots, K)$, but a different hyperparameter, ψ_k, so that within this subpopulation, $y \backsim p(y|\psi_k)$. Unobserved subgroup or allocation indicators $S_i \in (1, \ldots, K)$ describe how the units are distributed over subpopulations. These are also known as configuration indicators (Gopalan and Berry, 1998). The joint or complete data density $p(y, S)$ can be written,

$$p(y_i, S_i) = p(y_i|S_i)p(S_i) = p(y_i|\psi_{S_i})\pi_{S_i},$$

where $p(y|S) = p(y|\psi_S)$ is the density for y_i conditional on S_i, and $\{\pi_1, \ldots, \pi_K\}$ are the prior subgroup probabilities, with $\sum_{k=1}^{K} \pi_k = 1$. The unconditional or marginal density for a single y_i is

$$p(y_i|\pi_1,\ldots,\pi_K,\psi_1,\ldots,\psi_K)=\sum_{k=1}^{K}\pi_k p(y_i|\psi_k),$$

with the total likelihood (Diebolt and Robert, 1994) being

$$p(y|\pi,\psi)=\prod_{i=1}^{n}\sum_{k=1}^{K}\pi_k p(y_i|\psi_k).$$

Classical analysis via nonparametric maximum likelihood estimation involves maximization of the log of this marginal density—for example, see Rattanasiri et al. (2004) for a disease mapping application, where y_i are malaria counts and $p(y|\psi)$ is a Poisson density.

In MCMC applications, discrete mixture models can be represented hierarchically using the latent subpopulation indicators (Marin et al., 2005, 462). Thus at the highest stage or level are the parameters $\varphi=(\pi_1,\ldots,\pi_K,$ $\psi_1,\ldots,\psi_K)$, then the missing configuration data, the distribution of which depends on φ,

$$S_i\sim P(S_i|\varphi),$$

and at the lowest (first) stage the distribution of the observations $p(y|\varphi,S)$ depends on both φ and $S=(S_1,\ldots,S_n)$. The joint distribution is therefore,

$$p(y,S,\varphi)=p(y|S,\varphi)p(S|\varphi)p(\varphi).$$

3.8.2 Finite mixtures of standard densities

There is a considerable literature on univariate and multivariate normal mixtures for continuous data, and on Poisson and binomial mixtures for discrete data, with Bayesian references including Hurn et al. (2003), Militino et al. (2001), Richardson and Green (1997), and Roberts et al. (1998). Overdispersed or skew alternatives to the major densities can be used in discrete mixtures instead: for continuous data, the Student t distribution involves an additional tuning parameter useful for outlier accommodation, and greater robustness to such points may be obtained by discrete mixtures over univariate and multivariate Student t densities with varying degrees of freedom (Lin et al., 2004). Discrete mixtures of skew normal and skew Student t densities are considered by Lin et al. (2007a, 2007b). Lin et al. (2007a) argue that a simple normal discrete mixture model tends to overfit when additional components are added to capture skewness in continuous data.

Parameter sampling via MCMC is facilitated by conjugate prior choices for the mixing density. For example, consider a univariate normal mixture with $\psi_k=(\mu_k,\sigma_k^2)$, and

$$p(y|\pi,\mu,\sigma)=\sum_{k=1}^{K}\pi_k\phi(y|\mu_k,\sigma_k),$$

where $\phi(y|\mu,\sigma)$ is the normal density, $N(\mu,\sigma^2)$. The conjugate prior for $\psi_k = (\mu_k,\sigma_k)$ takes $\sigma_k^2 \backsim IG(\nu_k/2, V_k/2)$, namely,

$$p(\sigma_k^2) \propto \sigma_k^{-\nu_k-1} \exp\left(-V_k/2\sigma_k^2\right),$$

and

$$p\left(\mu_k|\sigma_k^2\right) = N\left(\xi_k, \frac{\sigma_k^2}{\kappa_k}\right).$$

Also assume a Dirichlet prior for the unknown mixture probabilities $(\pi_1,\ldots,\pi_K) \backsim Dir(\alpha,\ldots,\alpha)$, with α preset or possibly an extra unknown. Gibbs sampling then samples the missing data (the allocation indicators) according to a multinomial density with probabilities at iteration t,

$$p(S_i^{(t)} = k|\pi^{(t)},\mu^{(t)},\sigma^{(t)}) = \rho_{ik}^{(t)} = \frac{\pi_k^{(t)}\phi(y_i|\mu_k^{(t)},\sigma_k^{(t)})}{\sum_{k=1}^K \pi_k^{(t)}\phi(y_i|\mu_k^{(t)},\sigma_k^{(t)})}.$$

Let $d_{ik}^{(t)} = 1$ if $S_i^{(t)} = k$ and $d_{ik}^{(t)} = 0$ otherwise. Suppose $N_k^{(t)} = \#\{S_i^{(t)} = k\}$ is the total number of cases with $S_i^{(t)} = k$, that $m_k^{(t)} = \sum d_{ik}^{(t)} y_i/N_k^{(t)}$ is the average response for these cases, and that $E_k^{(t)} = \sum d_{ik}^{(t)}(y_i - m_k^{(t)})^2$ is the sum of squared errors for this subgroup. Then, with conditioning on remaining parameters understood, the π_k are updated according to a Dirichlet with

$$\left(\pi_1^{(t)},\pi_2^{(t)},\ldots,\pi_K^{(t)}\right) \backsim D(\alpha + N_k^{(t)}, \alpha + N_2^{(t)},\ldots,\alpha + N_K^{(t)}),$$

the subgroup variances are sampled from an updated inverse gamma,

$$\sigma_k^{2(t)} \backsim IG\left(0.5\left[\nu_k + N_k^{(t)}\right], 0.5\left[V_k + E_k^{(t)} + \frac{N_k^{(t)}\kappa_k}{\kappa_k + N_k^{(t)}}\left(\xi_k - m_k^{(t)}\right)\right]\right),$$

and the subgroup means are updated according to

$$\mu_k^{(t)} \backsim N\left(\frac{\kappa_k\xi_k + N_k^{(t)} m_k^{(t)}}{\kappa_k + N_k^{(t)}}, \frac{\sigma_k^{2(t)}}{\kappa_k + N_k^{(t)}}\right).$$

Diebolt and Robert (1994) suggest stabilizing adjustments to these updates to improve convergence. A refinement is to take the mixture proportions as subject-specific as in $(\pi_{i1},\ldots,\pi_{iK}) \backsim Dir(\alpha,\ldots,\alpha)$, and in the updates for $\pi_{ik}^{(t)}$, the $N_k^{(t)}$ are replaced by binary indicators according to which class subject i is allocated to at a particular iteration.

3.8.3 Inference in mixture models

Parametric mixture models such as the univariate normal just considered are subject to identification issues due to the arbitrariness of the subpopulation labels. Another form of identifiability relates to potential overfitting (e.g., K taken too large) (Frühwirth-Schnatter, 2006, 107). To illustrate label identifiability, in the absence of parameter constraints or other prior information to distinguish the components, the likelihood is invariant under permutation of the components and there are $K!$ possible labeling schemes. It is essential to produce MCMC draws with a unique labeling if interest lies in the estimation of group-specific parameters or classification probabilities, π_k (Frühwirth-Schnatter et al., 2004). Note though that inferences on some aspects of the model are unaffected by group labeling—for example, the unit means that pool over the population-wide category means. Cluster labeling issues are also not generally considered in the Dirichlet process approach (Section 3.9), where the emphasis is on the smoothed unit means.

Identifying (usually ordering) constraints may be imposed on parameters to avoid label switching (Richardson and Green, 1997; Roeder and Wasserman, 1997). Label switching refers to permuting the mixture component subscripts without altering the likelihood (Redner and Walker, 1984). However, Celeux et al. (2000), Geweke (2007), and Marin et al. (2005) consider drawbacks to such identifiability constraints (e.g., distortions of the posterior distribution of the parameters). For example, in a normal mixture, constraints may be imposed on prior masses, π_k (e.g., $\pi_1 > \pi_2 > \cdots > \pi_K$), or on the subpopulation parameters, μ_k, or on the scale parameters, σ_k. A preliminary MCMC sampling analysis without parameter constraints may be used to assess the most suitable form of constraint (Fruhwirth-Schattner, 2001). Another possibility is to use maximum likelihood solutions (e.g., using the R package flexmix) to set constraints and/or relatively informative priors that are sensible for the dataset. Reanalysis of the posterior output to impose a consistent labeling is another possibility (Frühwirth-Schnatter, 2001), as are data-based priors, albeit not fully Bayesian (Wasserman, 2000). For example, in a two group model without regression on predictors, the unit with the maximum y value could be prelabeled as belonging to one or other subpopulation. Diebolt and Robert (1994) suggest excluding MCMC samples where the configuration indicators all fall into one group.

Particular types of parameterization may be used to improve identification. Robert (1996) and Robert and Mengersen (1999) suggest introducing dependence between the parameters, ψ_k, in different components such that they are perturbations of one another. For example, a normal mixture model with $\psi_k = (\mu_k, \sigma_k^2)$ would be based on taking $\{\theta_1, \sigma_1^2\}$ as reference parameters and adopting the parameterization,

$$\begin{aligned}
\sigma_2 &= \sigma_1\omega_1,\\
\sigma_3 &= \sigma_2\omega_2,\\
\sigma_4 &= \sigma_3\omega_3,\\
&\ldots\\
\sigma_K &= \sigma_{K-1}\omega_{K-1} = \sigma_1\omega_1\omega_2\cdots\omega_{K-1},
\end{aligned}$$

where $\omega_k \backsim U(0,1)$. With $\theta_1 = \mu_1$, the prior on the series of normal means takes a perturbation form,

$$\mu_2 = \theta_1 + \sigma_1\theta_2,$$
$$\mu_3 = \theta_1 + \sigma_1\theta_2 + \sigma_1\sigma_2\theta_3,$$
$$\dots$$
$$\mu_K = \theta_1 + \sigma_1\theta_2 + \sigma_1\sigma_2\theta_3 + \cdots + (\sigma_1\sigma_2\cdots\sigma_{K-1})\theta_K.$$

The mixture weights have the form,

$$\pi_1 = p_1,$$
$$\pi_2 = (1-p_1)p_2,$$
$$\pi_3 = (1-p_1)(1-p_2)p_3,$$
$$\dots$$
$$\pi_{K-1} = (1-p_1)(1-p_2)\dots(1-p_{K-2})p_{K-1},$$
$$\pi_K = (1-p_1)(1-p_2)\dots(1-p_{K-1}),$$

with $p_k \backsim U(0,1)$. This prior is still invariant under permutation of the cluster indices and an indentifying constraint is placed on the variances by taking $1 \geq \omega_1 \geq \cdots \geq \omega_{K-1}$. An advantage of this representation is that an improper prior on $\{\mu_1, \sigma_1^2\}$ can be used (Robert and Titterington, 1998). For the two group case, Basu (1996) presents the parameterization $\nu = \sigma_1^2/\sigma_2^2$ and $\Delta = (\mu_2 - \mu_1/\sigma_1)$ to test for normal or Student t unimodality as against bimodality; posterior probabilities of unimodality are obtained using the results of Robertson and Fryer (1968).

Celeux et al. (2000) and others apply postprocessing to the MCMC output resulting from a discrete mixture analysis without parameter constraints; the goal is to reconfigure the output with a consistent labeling. Suppose there are p parameters in any subpopulation. If MCMC convergence is assumed, one may select a short run of iterations (say $S = 100$ iterations) where there is no label switching to provide a reference labeling. The initial run of parameter samples provides a base reference label sequence, $1, 2, \dots, K$ (one among the $K!$ possible), and K means of dimension, p, $\bar{\theta}_k = \{\bar{\theta}_{1k}, \bar{\theta}_{2k}, \dots, \bar{\theta}_{pK}\}$, that can be permuted to include all other remaining $K!-1$ possible labeling schemes. In a subsequent run of R iterations where label switching might occur, iteration r is assigned to that scheme (among the $K!$) closest to it in distance terms and a relabeling applied if there has been a switch away from the base reference label. Additionally, the means under the schemes are recalculated at each iteration, $S + r$ (Celeux et al., 2000, 965).

3.8.4 Particular types of discrete mixture model

Heterogeneity within classes can be accommodated using discrete mixtures for unit level conjugate or nonconjugate random effects (Fruhwirth-Schnatter et al., 2004; Lenk and DeSarbo, 2000). For example, the standard discrete mixture to account for heterogeneity in count data involves $K < n$ homogenous subpopulations with means $\mu_1, \ldots, \mu_K$,

$$y_i \sim \sum_{k=1}^{K} \pi_k Po(\mu_k),$$

where π_k is the prior probability that a unit belongs to subpopulation k, with $\sum_{k=1}^{K} \pi_k = 1$. Alternatively, accounting for heterogeneity within subpopulations would involve K Poisson-gamma subgroups,

$$y_i \sim Po(\mu_i),$$
$$\mu_i \sim \sum_{k=1}^{K} \pi_k Ga(a_k, b_k),$$

or K Poisson-lognormal subgroups,

$$y_i \sim Po(\mu_i),$$
$$\mu_i \sim \sum_{k=1}^{K} \pi_k LN(\mu_k, \sigma_k^2),$$

where $LN(m, V)$ denotes a lognormal density with mean m and variance V.

Discrete mixtures can also be used to modify the shape of standard densities such as the Poisson or binomial. For example, a manufacturing process may move between different regimes, one where faults are essentially unknown and another where they occur according to a Poisson process. This will generate excess zeroes as compared to the standard Poisson, leading to a zero inflated Poisson (ZIP). One may introduce a binary regime indicator, S_i, with marginal probability $\pi = Pr(S_i = 1)$ that the fault-free regime applies, and $(1 - \pi)$ that sampling is from a Poisson density with mean μ. In more generality, with $p(y|\psi)$ as a density for count data (e.g., Poisson, negative binomial, binomial), the corresponding zero-inflated density is

$$p(y = 0|\pi, \psi) = \pi + (1 - \pi)p(y = 0|\psi) \quad y = 0,$$
$$p(y|\pi, \psi) = (1 - \pi)p(y|\psi) \quad y > 0.$$

Conditionally, $Pr(S_i = 1|y > 0) = 0$, while

$$Pr(S_i = 1|y = 0) = \frac{\pi}{\pi + (1 - \pi)p(y = 0|\psi)}.$$

The process generating the S_i needs only to be considered for zero observations, $y_i = 0$, and the complete data likelihood (assuming S_i to be given) is

$$L(\pi, \psi|y, S) = \prod_{y_i > 0} (1 - \pi_i)p(y_i|\psi) \prod_{y_i = 0} \pi_i^{S_i} [(1 - \pi_i)p(0|\psi_i)]^{1 - S_i}.$$

For example, if $p(y|\psi)$ is taken to be Poisson with mean $\psi = \mu$, then $E(y|\pi, \mu) = (1-\pi)\mu$ and

$$V(y|\pi, \mu) = (1-\pi)\mu(1+\pi\mu) > E(y|\pi, \mu),$$

so that the ZIP model is necessarily overdispersed[8].

3.8.5 The logistic-normal alternative to the Dirichlet prior

A generalization of the logistic-normal to multivariate contexts has been applied to nonparametric analysis and by authors such as Aitchison and Shen (1980), Hoff (2003), and Lenk (1988). The goal is to replace the restrictive Dirichlet prior for the unknown mixture probabilities, π_k, with a multinomial logistic framework. Consider the case where units are exchangeable, and there are no covariates relevant to allocation between subpopulations. Then for subjects or units $i = 1, \ldots, n$, and assuming,

$$y_i \backsim \sum_{k=1}^{K} \pi_{ik} p_k(y_i|\psi_k),$$

the mixing probabilities are obtained as,

$$\pi_{ik} = \frac{e^{z_{ik}}}{1 + \sum_{k=1}^{K} e^{z_{ik}}}, \qquad k = 1, \ldots, K-1,$$

$$\pi_{iK} = \frac{1}{1 + \sum_{k=1}^{K} e^{z_{ik}}},$$

where the $\{z_{ik}, k = 1, \ldots, K-1\}$ are multivariate normal with mean ν and variance Σ_z. For example, Hoff (2003) argues for the use of normal mixtures in density smoothing and in this case the $p_k(y|\psi_k)$ would be univariate or multivariate normal themselves. This approach generalizes to multivariate skew normal or multivariate Student t densities, and can be adapted to allow nonexchangeable mixture priors, as in histogram smoothing (Leonard, 1973).

Instead of subject-specific z_{ik}, one may also assume a single vector $\{z_1, \ldots, z_{K-1}\}$ to be multivariate normal. For unique identification of the subgroups, one may impose order constraints on the parameters in ψ_k or on those underlying $\{z_1, \ldots, z_{K-1}\}$. In the univariate normal case with $\psi_k = \{\mu_k, \sigma_k^2\}$, one might assume an ordering either on the means μ_k, or on the means ν_k of the z_{ik}.

Example 3.6. Galaxy Data The number of clusters detected in the much analysed galaxy data has varied over different studies, under the model,

$$y_i \backsim \sum_{k=1}^{K} \pi_k N\left(\mu_k, \sigma_k^2\right),$$

and y being measured in thousands of kilometers per second. A classical analysis using the flexmix package (Leisch, 2004) in R shows no gain in moving beyond four clusters, whereas Ishwaran and James (2002) find at least five to six clusters with a Dirichlet process approach and under an inverse gamma prior for the σ_k^2. They do, however, find only four clusters when a uniform prior $U(0,\ 20.83)$ is used for σ_k^2 with 20.83 being the observed variance, $V(y)$. Ando (2007) reports six clusters (assuming a monotonic constraint on the μ_k) via several model fit criteria.

Here, K is taken to be 6 with an identifiability constraint on the normal means applied, together with additional data-based features to ensure sensible inferences. Thus, μ_1 is taken to be normal with a minimum of 9.176, namely, to be at least as large as the minimum y value. Then $\mu_k = \mu_{k-1}+\delta_k$, where the increments are distributed $Ga(1,\ 0.001)$ with a maximum of 20. Without the latter constraint, implausibly large μ_6 values were sometimes sampled. Diffuse $Ga(1,\ 0.001)$ priors are assumed on $1/\sigma_k^2$. Initial values for a two chain run were based on an initial single chain run with trial values for parameters.

Using these initial values, convergence[9] is attained by iteration 2,000 in a run of 10,000 iterations and the clusters means (with their posterior standard deviations) are obtained as {9.7 (0.16), 16.1 (0.05), 19.8 (0.17), 22.7 (0.43), 28.2 (2.5), 33.9 (3.0)}, while the mean cluster probabilities are {0.092, 0.034, 0.34, 0.46, 0.034, 0.04}. The mean BIC is 461, obtained using the posterior mean of minus twice the likelihood of the marginal density,

$$p(y|\pi,\mu,\sigma) = \sum_{k=1}^{K} \pi_k \phi(y|\mu_k,\sigma_k),$$

and adding a penalty based on the known parameter total. While an apparently sensible model, replicate samples y_{new} for some data points are biased away from the observations; for example, both $y_6 = 10.23$ and $y_7 = 10.41$ have posterior predictive means and 95% intervals of 9.7 $\{8.7, 10.6\}$, while observations y_{48} to y_{77} vary from 21.8 to 25.6, but have posterior predicted means between 22.5 and 22.8.

To illustrate the logistic-normal approach, the same number of subgroups is assumed, with the same priors on the component group normal parameters $\{\mu_k, \sigma_k^2\}$. $N(0,\ 1000)$ priors are adopted on the ν_k parameters and a Wishart with 5 degrees of freedom and identity scale matrix assumed for Σ_z^{-1}. The fit obtained is very similar to that under a Dirichlet prior. Replicate data, y_{new}, sampled from the model still do not reproduce observations y_6 and y_7 very closely, and show a flat profile for observations 48 to 77, with y_{new} around 22.6.

3.9 Nonparametric Mixing via Dirichlet Process and Polya Tree Priors

In applications of hierarchical models, inferences may depend on the assumed forms (e.g., normal, gamma) for higher stage priors, and will be distorted if there are unrecognized features such as multiple modes in the underlying second stage effects. Instead of assuming a known prior distribution, G, for second stage latent effects, such as b_i in the normal-normal model of Section 3.3, the Dirichlet process (DP) prior involves a distribution on G itself, so acknowledging uncertainty about its form (Gill and Casella, 2009). The DP prior involves a baseline or base prior G_0, the expectation of G, and a precision or mass parameter, α, governing the concentration of the prior for G about its mean G_0. For any partition $A_1, \ldots, A_M$ on the support of G_0, the vector $\{G(A_1), \ldots, G(A_M)\}$ of probabilities $G(A_m)$ contained in the set $\{A_m, m = 1, \ldots, M\}$ follows a Dirichlet distribution $D(\alpha G_0(A_1), \ldots, \alpha G_M(A_M))$.

Original forms of the DP prior assumed G_0 to be known (fixed). One problem with a DP when G_0 is known is that it assigns a probability of 1 to the space of discrete probability measures (Hanson et al., 2005, 249). An alternative is to take the parameters in G_0 to be unknown, and to follow a set of parametric distributions, with possibly unknown hyperparameters, resulting in a mixture of Dirichlet process or MDP model (Walker et al., 1999, 489). General computational procedures for such models are discussed by Jara (2007) and Ohlssen et al. (2007).

Following West et al. (1994), assume conventional first stage sampling densities $y_i \backsim p(y_i|b_i, \psi)$, with distributions $P(y_i|b_i, \psi)$. The uncertainty about the appropriate form of prior arises about the distribution G for the latent effects b_i. Under a DP prior, any set of unit-specific parameters $\{b_1, \ldots, b_n\}$ generated from G lies in a set of $K \leq n$ distinct values $\{\zeta_1, \ldots, \zeta_K\}$, which are sampled from G_0. The concentration parameter α governing the closeness of G to G_0 can be taken as an unknown, or assigned a preset value (e.g., $\alpha = 1$). The number of distinct values or clusters K is stochastic, with an implicit prior determined by α, with limiting mean $\alpha \log(1 + n/\alpha)$. Note that the posterior mean of K is not necessarily a reliable guide to the number of components in the data or effects (e.g., components with substantive meaning), though it can be interpreted as an upper bound on the number of components (Ishwaran and Zarepour, 2000, 381–82). Related algorithms include the Chinese Restaurant process (Ishwaran and James, 2003), a method for randomly assigning objects to groups that results in samples having the equivalent sampling distribution as that obtained by the Dirichlet Process.

Given the realised number of clusters K (at any particular MCMC iteration), b_i are sampled from the set $\{\zeta_1, \ldots, \zeta_K\}$ according to a multinomial distribution. Define cluster indicators $S = \{S_1, \ldots, S_n\}$, where $S_i = k$ if $b_i = \zeta_k$ and denote $N_k = \#\{S_i = k\}$ as the total number of units with $S_i = k$ (i.e., units in the same cluster with a common value ζ_k for the second stage latent effect). If α is taken as unknown, its prior is important in determining the number of clusters. Taking $\alpha \backsim Ga(\eta_1, \eta_2)$ where η_1 and η_2 are relatively large will tend to discourage unduly small or large values for α. Typical values are $\eta_1 = \eta_2 = 1$ or $\eta_1 = \eta_2 = 2$, though taking $\eta_2 > \eta_1$ as in $\{\eta_1 = 2, \eta_2 = 4\}$ tends to encourage repetitions in ζ_k, and can be used to assess the number of components present in the data (Ishwaran and Zarepour, 2000, 377). It is clear that the parameters used in the prior for α may affect the number of components, but typically there is less concern with this aspect in nonparametric mixture modeling (Leslie et al., 2007).

Consider the assignment of a latent effect b_i to a particular unit, given that the remaining $n-1$ latent effects $b_{[i]} = \{b_1, \ldots, b_{i-1}, b_{i+1}, \ldots, b_n\}$ are already assigned. Also let $S_{[i]}$ be a particular configuration of the remaining $n-1$ effects in $b_{[i]}$ into $K_{[i]}$ distinct values, with $N_{[i]k} = \#\{S_i = k,\ k \neq i\}$ denoting the total of those $n-1$ units having a common value $\zeta_{[i]k}$. Then the conditional prior for b_i follows a Polya urn scheme (Dunson et al., 2007, 165; Hanson et al., 2005, 252; West et al., 1994),

$$\begin{aligned}(b_i|b_{[i]}, S_{[i]}, K_{[i]}, \alpha) &\backsim \frac{\alpha}{\alpha+n-1}G_0 + \frac{1}{\alpha+n-1}\sum_{k\neq i}\delta(b_k)\\ &\backsim \frac{\alpha}{\alpha+n-1}G_0 + \frac{1}{\alpha+n-1}\sum_{k=1}^{K_{[i]}} N_{[i]k}\delta(\zeta_{[i]k}),\end{aligned} \tag{3.5}$$

where $\delta(u)$ denotes a degenerate distribution having a single value at u. So b_i is distinct from the remaining latent values with probability $(\alpha/\alpha+n-1)$, in which case it is drawn from the base prior G_0. Alternatively it is selected from the existing distinct effects, $\zeta_{[i]k}$, according to a multinomial with probabilities proportional to $(N_{[i]k}/\alpha+n-1)$. This selection scheme extends to the predictive scenario, i.e., to the latent effect for a hypothetical new unit $n+1$, with

$$(b_{n+1}|b, S, K, \alpha) \backsim \frac{\alpha}{\alpha+n}G_0 + \frac{1}{\alpha+n}\sum_{k=1}^{K} N_k\delta(\zeta_k).$$

Predictions of the first stage response for unit $n+1$ are obtained as,

$$(y_{n+1}|b, S, K, \alpha) \backsim \frac{\alpha}{\alpha+n}P_{n+1}(|\zeta_{n+1}) + \frac{1}{\alpha+n}\sum_{k=1}^{K} N_k P_{n+1}(|\zeta_k),$$

where ζ_{n+1} is an extra draw from G_0. Predictions beyond $n+1$ may be relevant in panel or time series applications (Hirano, 1998).

In terms of Gibbs sampling, Equation 3.5 implies conditional posteriors (Ishwaran and James, 2001, 166; Neal, 2000; West et al., 1994, 367),

$$(b_i|y, b_{[i]}, S_{[i]}, K_{[i]}, \alpha) \backsim \alpha q_{i0} g_0(b_i|y)p(y_i|b_i) + \sum_{k=1}^{K_{[i]}} q_{ik}\delta(\zeta_{[i]k}),$$

where $g_0(b_i|y)$ is the density corresponding to G_0 evaluated at b_i, and where

$$\begin{aligned}q_{i0} &= \int p(y_i|b_i)g_0(b_i)\,db_i\\ q_{ik} &= N_{[i]k}p(y_i|\zeta_{[i]k}) \qquad k > 0.\end{aligned} \tag{3.6}$$

Normalizing the values αq_{i0} and q_{ik} to probabilities $\{r_{i0}, r_{i1}, \ldots, r_{iK_{[i]}}\}$ summing to 1, the conditional posteriors for the subgroup indicators are then,

$$Pr\left(S_i = k | y, b_{[i]}, S_{[i]}, K_{[i]}\right) = r_{ik},$$

where $S_i = 0$ corresponds to drawing a new sample from G_0 under the Polya urn scheme.

3.9.1 Specifying the baseline density

An important aspect of the MDP framework is the specification of G_0. Assume there are p parameters, $(\psi_1, \ldots, \psi_p)$, in G_0, then one has

$$\begin{aligned} y_i | b_i &\backsim p(y_i | b_i), \\ b_1, \ldots, b_n &| G, \\ G | \alpha, G_0 &\backsim DP(\alpha G_0), \\ G_0 &= \{p_{01}(\psi_1 | \xi_1), \ldots, p_{0p}(\psi_p | \xi_p)\}, \end{aligned}$$

where $\psi_1, \ldots, \psi_p$ are unknown, and also possibly some of the defining ξ parameters. Consider a normal mixture with both means and variances possibly differing for each unit (Cao and West, 1996; Hirano, 2002), namely,

$$y_i \backsim N(\mu_i, \sigma_i^2).$$

The appropriate prior G for $b_i = (\mu_i, \sigma_i^2)$ is not certain, therefore,

$$\begin{aligned} (\mu_i, \sigma_i^2) &\backsim G, \\ G &\backsim DP(\alpha G_0), \end{aligned}$$

where G_0 involves the priors $\mu_i \backsim p_{01}(\mu_i | \xi_1), \sigma_i^2 \backsim p_{02}(\sigma_i^2 | \xi_2)$, with ξ_1 and ξ_2 possibly including further unknowns. For example, Hirano (2002) takes,

$$1/\sigma_i^2 \sim \chi^2(s)/(sQ),$$

and

$$\mu_i \sim N\left(m, c\sigma_i^2\right),$$

where $s, Q, m,$ and c are specified but may be varied in a sensitivity analysis.

The marginal distribution of y_i (averaged over all possible G) in this case is a mixture of normal distributions, with the number of subgroups, K, randomly varying between 1 and n. The n unit-specific parameter pairs, $b_i = (\mu_i, \sigma_i^2)$, are selected under G from the set of $K_{[i]}$ possible values, $\zeta_k = (\mu_k, \sigma_k^2)$, already drawn from G_0, or by fresh sampling from G_0. The q_{ih} in Equation 3.6 are then obtained as

$$\begin{aligned} q_{i0} &= \int \frac{1}{\sigma_i \sqrt{2\pi}} e^{-(y_i - \mu_i)^2 / 2\sigma_i^2} g_0(\mu_i, \sigma_i^2) d\mu_i d\sigma_i^2, \\ q_{ik} &= N_{[i]k} \frac{1}{\sigma_k \sqrt{2\pi}} e^{-(y_i - \mu_k)^2 / 2\sigma_k^2} \qquad k > 0. \end{aligned}$$

As another example, Kleinman and Ibrahim (1998) consider Gibbs updates in an MDP framework for parameters in general linear mixed models for nested data. For example, let X_i and Z_i be predictors of dimension q and r (possibly overlapping) and consider repeated data, y_{it}, over subjects i, with observation vectors $y_i = (y_{i1}, \ldots, y_{iT})$ and first stage model,

$$y_i \sim N\left(X_i\beta + Z_i b_i, \sigma^2\right),$$

where one may assume conventional normal and inverse gamma priors for β and σ^2. However, for $b_i = (b_{i1}, \ldots, b_{ir})$, greater flexibility is obtained by taking,

$$b_i \backsim G,$$
$$G \backsim DP(\alpha, G_0),$$

where G_0 is the multivariate normal of dimension r, with mean 0 but unknown covariance D. The Wishart distribution in the Gibbs update for D^{-1} is modified for clustering of values among the sampled b_i (Kleinman and Ibrahim, 1998, 94).

3.9.2 Truncated Dirichlet processes and stick-breaking priors

Implementation may be simplified if an alternative way to generate the DP prior is adopted. The basis of this alternative scheme is to regard the density of the unit level effects, b_i, as an infinite mixture of point masses or continuous densities (Hirano, 1998; Ohlssen et al., 2007), with

$$b_i \sim \sum_{k=1}^{\infty} \pi_k h(b_i|\psi_k).$$

This approach is called a Dirichlet process mixture by Hanson et al. (2005, 250) and a dependent Dirichlet process by Dunson et al. (2007, 164). For practical application, Ishwaran and James (2002) and Ishwaran and Zarepour (2000) suggest the infinite representation be approximated by one truncated at $M \leq n$ components with

$$g(b) = \sum_{m=1}^{M} \pi_m h(b|\psi_m),$$

where π_m are sampled by introducing $M-1$ beta distributed random variables,

$$V_m \backsim Be(c_m, d_m),$$

with $V_M = 1$ to ensure the random weights, π_m, sum to 1 (Ishwaran and James, 2001; Sethuraman, 1994). Then $\pi_1 = V_1$ and

$$\pi_m = (1 - V_1)(1 - V_2)\cdots(1 - V_{m-1})V_m \qquad m > 1.$$

This method of generation is known as stick breaking, since at each stage, the procedure randomly breaks what is left of a stick of unit length and assigns the length of the break to the current π_m.

Following Pitman and Yor (1997), the beta parameters $\{c_m, d_m\}$ in the prior for V_m can be written as $c_m = 1 - c, d_m = d + mc$, where $c \in [0, 1)$ and $d > -c$. For an infinite dimensional mixture, the Dirichlet process is obtained by taking $c = 0$ and $d = \alpha$, so that $V_m \backsim Be(1, \alpha)$. When a finite (truncated) mixture is used, setting $c_m = 1 + c/M$ and $d_m = \alpha - m\alpha/M = \alpha(1 - m/M)$ is asymptotically equivalent to the DP process (Ishwaran and James, 2001; Ishwaran and Zarepour, 2002).

However, using an approximate DP scheme with $V_m \backsim Be(1, \alpha)$ and M large is equivalent to the infinite DP process for practical purposes (Ishwaran and James, 2002; Ishwaran and Zarepour, 2000, 383). If a $Ga(\eta_1, \eta_2)$ prior is used for α, its full conditional is $\alpha \backsim Ga(M + \eta_1 - 1, \eta_2 - \log(\pi_M))$ (Ishwaran and Zarepour, 2000, 387). The realized number of clusters is $K \leq M$ as above, and Ishwaran and James (2002) suggest AIC and BIC penalties based on K that can be used for model selection.

Taking $V_m \backsim Be(\alpha, 1)$ rather than $V_m \backsim Be(1, \alpha)$ in the truncated stick-breaking scheme means that larger values of α now imply greater clustering into a few subpopulations. This is an example of the beta process priors considered by Ishwaran and Zarepour (2000). Other truncated mixture sampling schemes that start with a prior on α to give an implicit prior on a stochastic K are available. For example, Ishwaran and Zarepour (2000, 376) consider taking α as an unknown in,

$$(\pi_1, \ldots, \pi_M) \backsim D\left(\frac{\alpha}{M}, \frac{\alpha}{M}, \ldots, \frac{\alpha}{M}\right).$$

Alternatively, Green and Richardson (2001, 357) start off with a prior on K and then select the cluster indicators from a multinomial vector with probabilities $p(S_i = k) = \pi_i$, where $(\pi_1, \ldots, \pi_K)$ follow a Dirichlet density $D(\delta, \ldots, \delta)$. They refer to this as an explicit allocation prior and show how the DP prior is obtained as $K \to \infty$ and $\delta \to 0$ in such a way that $K\delta \to \alpha > 0$.

3.9.3 Polya Tree priors

The Polya Tree (PT) is a more general class than the Dirichlet process and has the benefit that it can place probability 1 on the space of continuous densities (Hanson et al., 2005; Walker et al., 1999). In essence, if the support of a parameter ω is denoted Γ then the PT prior chooses the most appropriate value for ω by successive binary partitioning of Γ. The first partition splits Γ into two disjoint sets $\{B_0, B_1\}$; the probabilities of moving into B_0 and B_1 are C_{00} and $C_{01} = 1 - C_{00}$, with C_{00} set to 0.5. At the second partition, B_0 is split into $\{B_{00}, B_{01}\}$ and B_1 is split into $\{B_{10}, B_{11}\}$, so there are 2^2 sets. At the third partition, B_{00} is split into $\{B_{000}, B_{001}\}$, B_{01} into $\{B_{010}, B_{011}\}$, B_{10} into $\{B_{100}, B_{101}\}$, and B_{11} into $\{B_{110}, B_{111}\}$, so there are 2^3 sets. Generally, the number of sets at the mth partition is 2^m.

The partition probabilities at second and subsequent stages are unknown. Let ε denote a sequence of 0s and 1s. For example, suppose B_1 is selected at step 1 and B_{11} is selected at step 2, then $\varepsilon = [1, 1]$. The choice at the next stage between sets $B_{\varepsilon 0}$ and $B_{\varepsilon 1}$ (i.e., between B_{110} and B_{111}) is governed by probabilities $(C_{\varepsilon 0}, C_{\varepsilon 1})$, with a beta prior for $C_{\varepsilon 0}$, and $C_{\varepsilon 1} = 1 - C_{\varepsilon 0}$. The canonical form for the prior on the partition probabilities at partition m is

$$C_{\varepsilon 0} \backsim Be(c_m, c_m),$$
$$c_m = dm^2,$$

where d may be taken as an extra unknown. The Dirichlet process occurs when $c_m = d/2^m$, so that $c_m \to 0$ as $m \to \infty$, whereas $c_m \to \infty$ as $m \to \infty$ is appropriate if the underlying distribution G is expected to be continuous.

While, theoretically, the completely continuous case corresponds to $m \to \infty$, in practice the partitioning is truncated at a finite value M. Hanson and Johnson recommend $M = \log_2(n)$, where n is the sample size. The partitions can be taken to coincide with percentiles of G_0, so for example,

$$B_0 = (-\infty, G_0^{-1}(0.5)], \qquad B_1 = [G_0^{-1}(0.5), \infty);$$
$$B_{00} = (-\infty, G_0^{-1}(0.25)], \qquad B_{01} = [G_0^{-1}(0.25), G_0^{-1}(0.5)],$$
$$B_{10} = [G_0^{-1}(0.5), G_0^{-1}(0.75)], \qquad B_{11} = [G_0^{-1}(0.75), \infty);$$

and so on. Let d_{ki} at partition k and option i be a re-expression of the B_ε (e.g., for $k = 3$, $d_{31} = B_{000}$, $d_{32} = B_{001}$, $d_{33} = B_{010}$, $d_{34} = B_{011}$, $d_{35} = B_{100}$, $d_{36} = B_{101}$, $d_{37} = B_{110}$, $d_{38} = B_{111}$). Then at partition k, for $i = 1, \ldots, 2^k$, the interval boundaries are

$$d_{ki} = \left[G_0^{-1}\left(\frac{i-1}{2^k}\right), G_0^{-1}\left(\frac{i}{2^k}\right)\right],$$

with appropriate modifications for the extreme tails.

For example, consider a PT prior on unstructured errors in a Poisson lognormal mixture, with

$$y_i \backsim Po(\mu_i),$$
$$\log(\mu_i) = \beta + \sigma b_i.$$

Then G_0 for $v_i = \sigma b_i$ is a $N(0, \sigma^2)$ density, with G_0 for b_i being a $N(0, 1)$ density. So with $M = 3$ levels, the relevant ordinates from G_0 for defining the eight intervals are (−1.15, −0.67, −0.32, 0, 0.32, 0.67, 1.15).

Example 3.7. Nicotine Replacement Therapy Xia et al. (2005) analyzed the NRT trials data of Example 3.1 and detected $K = 2$ subpopulations using a discrete mixture of normals model with $y_i \sim N(b_i, s_i^2)$, with s_i^2 known, and

$$b_i \sim \sum_{k=1}^{K} \pi_k N\left(\mu_k, \tau_k^2\right).$$

They found one subgroup to have a nonsignificant treatment effect, with its μ parameter straddling zero. Here we consider both a conventional discrete normal mixture with $K = 2$, and a DP-based mixture. In the former, a monotonicity constraint is applied to the μ_k. The mean likelihood of this model was improved by taking relatively informative $N(0, 1)$ priors on μ_k and $Ga(1, 1)$ priors on $1/\tau_k^2$; these may be judged reasonable in terms of the likely range of treatment-control log odds ratios typically encountered in trials.

With inferences based on the second half of a two chain run of 10,000 iterations, the average likelihood stands at -56 compared to -62 for a standard normal-normal model with $K = 1$. Posterior means for μ_k (and 95% credible intervals) are found to be 0.38 (-0.83, 0.76) and 0.93 (0.55, 2.19), with respective component probabilities 0.48 and 0.52. Sampling replicate data, $y_{\text{rep},i}$, shows the observations to be reproduced effectively with no posterior probabilities, $Pr(y_{\text{rep},i} > y_i|y)$, exceeding 0.05 or 0.95, and in fact varying between 0.11 and 0.88.

In the DP model, conventional priors for $\{b_i, \mu_k, \tau_k^2\}$ are replaced by a DPM structure (Section 3.9) with $b_i = \zeta_k$ when $S_i = k$ and where the realised number of clusters is $K \leq M$, where a maximum of $M = 20$ possible clusters is assumed. The M potential values $\{\mu_m, \tau_m^2\}$ are sampled from normal densities with means $\mu_m \backsim N(m_\mu, 1)$, where m_μ is itself unknown, and with $1/\tau_m^2 \backsim G(1, 1)$. A $Ga(3, 3)$ prior is assumed on the Dirichlet concentration parameter α.

A two chain run[10] of 5000 iterations shows convergence in α, K, and the latent effects b after around 1000 iterations. The posterior mean and median of K are, respectively, 3.5 and 3, supporting a relatively small number of components in the second stage prior of NRT effects; α has a posterior mean of 0.78. A plot of the posterior means of the b_i does not show sharply distinct subgroups (Figure 3.2), though outliers can be seen, such as trial 36, a trial with a relatively large number of subjects in the "women, long term follow-up category" in Cepeda-Benito et al. (2004). However, the effects show more peakedness than under a normal density (superimposed plot).

Sampling replicate data, $y_{\text{rep},i}$, shows the observations to be reproduced effectively with no posterior probabilities, $Pr(y_{\text{rep},i} > y_i|y)$, exceeding 0.05 or 0.95, and in fact varying between 0.09 and 0.93. The latter value for $Pr(y_{\text{rep},i} > y_i|y)$ is for trial 66 where r_{T} is only 2 (from $N_{\text{T}} = 86$ in the treatment group).

Example 3.8. Eye-Tracking Data Escobar and West (1998) present count data on eye-tracking anomalies in schizophrenic patients ($n = 101$). The data are overdispersed and the first analysis assumes a DP Poisson-gamma mixture with G_0 being a gamma density with unknown shape and scale parameters. So,

$$\begin{aligned} y_i &\backsim Po(b_i), \\ b_i &\backsim G, \\ G &\backsim DP(\alpha G_0), \\ G_0 &= Ga(c_g, d_g). \end{aligned}$$

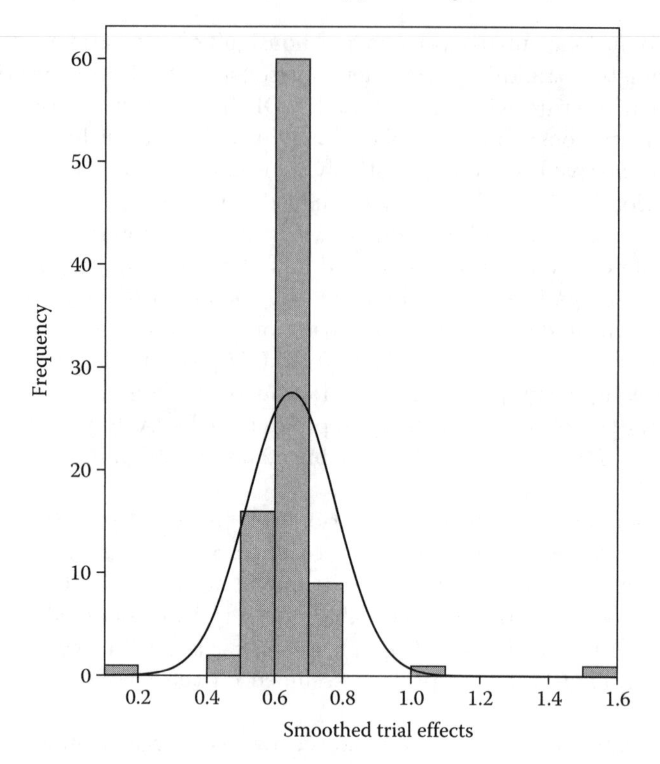

FIGURE 3.2
Histogram of smoothed trial effects.

Taking c_g and d_g to be unknowns results in an MDP prior, which is implemented using the Polya urn prior (Equation 3.5). A $Ga(1, 1)$ prior is assumed on α, and exponential $E(1)$ priors on the parameters (c_g, d_g) with a minimum of 0.5 on c_g for numerical stability.

In line with Marshall and Spiegelhalter (2003), the observed y_i are compared with replicates sampled from the predictive distribution $p(y_{\text{rep}}|y)$ to see if y_i are at odds with the model. Discrepancies could be due to genuine outlier status, or to model failures. For discrete data, the relevant p-value is

$$Pr(y_{\text{rep},i} < y_i) + 0.5Pr(y_{\text{rep},i} = y_i).$$

A related check is whether the 95% intervals for $y_{\text{rep},i}$ include y_i (Gelfand, 1996).

From the last 4000 iterations of a two chain run[11] of 5000 iterations in OpenBUGS, an average of $K = 12$ distinct clusters is obtained, with posterior mean (sd) for α, c_g, and d_g of 2.86 (1.5), 0.77 (0.27), and 0.13 (0.06). Figure 3.3 shows the prediction y_{new} for a new case, and demonstrates that the main source of overdispersion is skewness in the latent frailties, b_i, rather than multiple modes. The predictive checks based on replicate samples are

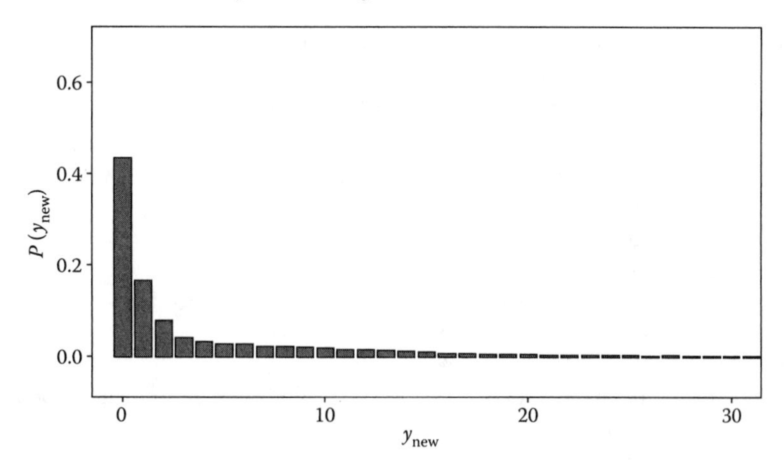

FIGURE 3.3
Prediction for new case.

satisfactory. Note that the same does not apply if the gamma mixing density parameters are set, e.g., $c_g = d_g = 1$. In this case, bimodal posteriors are obtained on some b_i (e.g., b_{92}) and predictive checks for $y_{101} = 34$ suggest it to be an extreme observation.

A second analysis involves a Polya Tree prior in a Poisson-lognormal model, namely, $y_i \backsim Po(\mu_i)$ with

$$\log(\mu_i) = \beta + \sigma b_i,$$

where G_0 for $v_i = \sigma b_i$ is an $N(0, \sigma^2)$ density. The number of stages is set at $M = 4$ and a $Ga(1, 1)$ prior is assumed on $1/\sigma^2$. Once an interval, $B_{\varepsilon m}$, is selected, uniform sampling to generate b_i takes place within the interval defined by G_0, except in the tails where the sampling is from a $N(0, 1)$.

As for the Polya urn model (with the same MCMC details), both types of predictive check indicate no major discrepancies. σ has posterior mean (and 95% interval) 2.0 (1.6, 2.5). If σ is taken to equal 1 so that G_0 is assumed known, then predictive discrepancies do occur. Taking $\sigma = 1$ also leads to bimodal posteriors for individual b_i indicating a clash between prior and data, such that the prior cannot accommodate certain values.

Appendix: Computational Notes

1. The code for the nicotine replacement example is

```
model {# predictions
ynew ~dnorm(b.new,inv.s20); b.new ~dnorm(mu,inv.tau2); OR.new <-
exp(ynew)
```

```
for (i in 1:n) {# empirical logits and sampling variances
y[i] <- log(rT[i]/(nT[i]-rT[i]))-log(rC[i]/(nC[i]-rC[i]))
s2[i] <- 1/rT[i] + 1/(nT[i]-rT[i]) + 1/rC[i] + 1/(nC[i]-rC[i]); inv.s2[i] <-
1/s2[i]
# observation model
y[i] ~dnorm(b[i],inv.s2[i])
# second stage
b[i] ~dnorm(mu,inv.tau2)}
inv.tau2 <- 1/tau2; s2.0 <- n/sum(inv.s2[]); inv.s20 <- 1/s2.0;
# uniform for average shrinkage
w ~dunif(0,1); tau2 <- (s2.0-s2.0*w)/w
# posterior mean of NRTpos estimates probability of treatment benefit
NRTpos <- step(mu);mu ~dnorm(0,0.01);
# overall odds-ratio
OR.NRT <- exp(mu)}
```

Under the second stage t-prior the code for the latent effects becomes

```
b[i] ~dnorm(mu,prec[i]);
prec[i] <- inv.tau2*lam[i]
lam[i] ~dgamma(nu.2,nu.2); step.lam[i] <- step(1-lam[i])}
```

2. A Q-Q plot may be obtained in R by pasting in the b[1:90] posterior means in Example 3.1, naming the vector b, and then using the commands

```
> qqnorm(b)
> qqline(b)
```

The Shapiro-Wilk test may be obtained by loading the stats package, and then using the command shapiro.test(b). The Jarque-Bera test may be obtained by loading the tseries package, and then using the command jarque.bera.test(b).

3. The WinBUGS code for this model is

```
model {for (i in 1:n) {# define response
y[i,1] <- log(rT[i]/(nT[i]-rT[i])); y[i,2] <- log(rC[i]/(nC[i]-rC[i]))
S2[i,1,1] <- 1/rT[i] + 1/(nT[i]-rT[i]); S2[i,1,2] <- 0;
S2[i,2,1] <- 0; S2[i,2,2] <- 1/rC[i] + 1/(nC[i]-rC[i]);
# define hierarchical model
y[i,1:2] ~dmnorm(b[i,1:2],Precy[i,1:2,1:2])
Precy[i,1:2,1:2] <- inverse(S2[i,1:2,1:2])
ynew[i,1:2] ~dmnorm(b[i,1:2],Precy[i,1:2,1:2])
b[i,1:2] ~dmnorm(mu[1:2],Precb[1:2,1:2])}
Sigma.b[1:2,1:2] <- inverse(Precb[,]);Precb[1:2,1:2] ~dwish(Q[,],2)
V.t <- Sigma.b[1,1]+Sigma.b[2,2]-2*Sigma.b[1,2]
V.c <- Sigma.b[1,1]-pow(Sigma.b[1,2],2)/Sigma.b[2,2]
# variation in treatment effects due to baseline risk
r2.base <- 1-V.c/V.t
corr.TC <- Sigma.b[1,2]/sqrt(Sigma.b[1,1]*Sigma.b[2,2]);
b.TC <- Sigma.b[1,2]/Sigma.b[2,2]
```

```
# gam is population wide vaccine effect
gam <- mu[1]-mu[2]; for (i in 1:2) {mu[i] ~dnorm(0,0.001)}}
```

4. As an illustration of applying these extended models, and to provide a template for assessing overdispersion in other count datasets, consider data on the distribution of 402 sow bugs beneath 122 boards from Scollnik (1995). The observed frequencies are for $0.1, \ldots, 17$ bugs found. The vector of predicted probabilities under the generalized Poisson density is contained in prob[1:n] in the following code, where prob[1] is the probability of a board having zero bugs, etc. So with n=122, the code is

```
model {for (i in 1:n) {log(p[i]) <-
log(lam)+(y[i]-1)*log(lam+y[i]*rho)-(lam+y[i]*rho)-loggam(y[i]+1)
z[i] <- 1; z[i] ~dunif(a[i],b[i]); a[i] <- -1/p[i]; b[i] <- 1/p[i]}
for (i in 1:n) {log(prob[i]) <- log(lam)+(i-2)*log(lam+(i-1)*rho)
-(lam+(i-1)*rho)-loggam(i); yhat[i] <- n*prob[i]}
lam ~dgamma(2,0.25); rho <- rho.val[k.rho]; k.rho ~dcat(p.rho[1:101])
for (i in 1:101){p.rho[i] <- 1/101; rho.val[i] <- (i-1)*0.01}}
```

The sample variance for these data is considerably in excess of the sample mean and counts of one are deflated by social behavior among the bugs. Because of the discrete prior on ρ, a Bayes factor on the generalized Poisson as against the usual Poisson model can in principle be obtained. In fact, the posterior mean (sd) of ρ is 0.56 (0.05), and values of ρ under 0.4 have zero posterior probability (in a two chain run of 10,000 iterations with 1,000 burn in).

5. The WinBUGS code for the hospital mortality example (exchangeable model) is

```
model { for (i in 1:131) {# checks using replicate samples
yrep[i] ~dpois(nu[i]); ch[i] <- step(y[i]-yrep[i]-0.001)+0.5*equals(yrep[i],y[i])
y[i] ~dpois(nu[i]); nu[i] <- o[i]*lam[i]; r[i] <- y[i]/o[i]
lam[i] ~dgamma(zeta,b)}
B0 ~dunif(0,1); zeta <- B0*ranked(o[],1)*mean(r[])/(1-B0)
b <- zeta/mu; mu ~dexp(1)}
```

while the two group model has code

```
model { for (i in 1:131) {# checks using replicate samples
yrep[i] ~dpois(m[i]); ch[i] <- step(y[i]-yrep[i]-0.001)+0.5*equals(yrep[i],y[i])
y[i] ~dpois(nu[i]); nu[i] <- o[i]*lam[i]; r[i] <- y[i]/o[i]
lam[i] ~dgamma(zeta[G[i]],b[G[i]])}
z0[1] <- ranked(o[1:37],1)*mean(r[1:37])
z0[2] <- ranked(o[38:131],1)*mean(r[38:131])
for (k in 1:2) {B0[k] ~dunif(0,1); zeta[k] <- B0[k]*z0[k]/(1-B0[k])
        b[k] <- zeta[k]/mu[k]; mu[k] ~dexp(1)}}
```

6. The code used in Example 3.4 (cancer recurrence) is as follows:

```
model { for (i in 1:n) { rT[i]~dbin(pT[i],nT[i]); rC[i] ~dbin(pC[i],nC[i]);
pC[i] ~dbeta(a.C,b.C); del[i] ~dnorm(mu.del,tau.del)}
```

```
# alternative scales
# log OR scale
# for (i in 1:n) {logit(pT[i]) <- logit(pC[i])+del[i]}
# logit(pT.new) <- logit(pC.new)+del.new
# log RR scale
# for (i in 1:n) {log(pT[i]) <- log(pC[i])+min(del[i],-log(pC[i]))}
# log(pT.new) <- log(pC.new)+del.new
# absolute risk difference scale
for (i in 1:n) {pT[i] <- pC[i]+min(max(del[i],-pC[i]),(1-pC[i]))}
pT.new <- pC.new+del.new
# predictive relative risk (all models)
del.new ~dnorm(mu.del,tau.del); pC.new ~dbeta(a.C,b.C);
RRnew <- pT.new/pC.new; RRnew.above.1 <- step(RRnew-1)
# hyperpriors
a.C ~dunif(1,100); b.C ~dunif(1,100)
mu.del ~dnorm(0,0.001); tau.del ~dgamma(1,0.001)}
```

7. The code for the bivariate normal model is

```
model {for (i in 1:1040){ L[i] ~dbin(p[i], N[i])
     p[i] <- x[i]*r[1,i]+(1-x[i])*r[2,i]
     for (j in 1:2) {r[j,i] <- 1/(1+exp(-w[i,j])); e[i,j] <- abs(r[j,i]-rstar[i,j])}
w[i,1:2] ~dmnorm(mu[1:2],T.w[,])}
T.w[1:2,1:2] ~dwish(Q[,],2); Sig[1:2,1:2] <- inverse(T.w[,])
rho <- Sig[1,2]/sqrt(Sig[1,1]*Sig[2,2])
for (j in 1:2) { mu[j] ~dnorm(0,1)
for (k in 1:2) {Q[j,k] <- equals(j,k)}}
E.r <- sum(e[,])}
```

with initial value files list(T.w = structure(.Data = c(5,-0.8,-0.8,1),.Dim = c(2,2)),mu = c(0.5,3)) and list(T.w = structure(.Data = c(4,-1,-1,0.9),.Dim = c(2,2)),mu = c(0.55,3.1)).

For the trivariate model the code is

```
model {for (i in 1:1040){ L[i] ~dbin(p[i], N[i])
# literacy rate
            p[i] <- x[i]*r[1,i]+(1-x[i])*r[2,i]
for (j in 1:2) {r[j,i] <- 1/(1+exp(-w[i,j])); e[i,j] <- abs(r[j,i]-rstar[i,j])}
            x[i] <- 1/(1+exp(-w[i,3]))
w[i,1:3] ~dmnorm(mu[1:3],T.w[,])}
T.w[1:3,1:3] ~dwish(Q[,],3); E.r <- sum(e[,])
Sig[1:3,1:3] <- inverse(T.w[,]); rho <- Sig[1,2]/sqrt(Sig[1,1]*Sig[2,2])
for (j in 1:3) { mu[j] ~dnorm(0,1)
for (k in 1:3) {Q[j,k] <- 3*V[j]*equals(j,k)}}}
```

with initial value files list(T.w = structure(.Data = c(6.9,-1.7,8.1,-1.7,1.3,-2.5,8.1,-2.5,13.7),.Dim = c(3,3)),mu = c(0.8,3.5,-0.4)) and list(T.w = structure(.Data = c(5.1,-2.7,5.7,-2.7,0.8,-3.9,5.7,-3.9,10.3),.Dim = c(3,3)),mu = c(0.7,3.8,-0.4)).

8. A BUGS implementation for a ZIP model may use the associated full conditionals for π and μ. With data arranged so that the n_0 subjects with $y_i = 0$ are placed first, and assuming priors $\pi \backsim Be(1,1)$, $\mu \backsim Ga(1, 0.001)$ the code is

```
model {for (i in 1:n0) {S0[i]~dbern(p.S)}
p.S <- pi/(pi + (1 - pi) * exp(-mu))
for (i in n0+1:n) { y[i]~dpois(mu)}
s0 <- sum(S0[]); pi~dbeta(aw,bw);
aw <- s0 + 1; bw <- n - s0 + 1
mu ~dgamma(a.mu,b.mu);
a.mu <- sum(y[]) + 1; b.mu <- n - s0 + 0.001}
```

For example, applying this code to the hard disk read-write error data from Xie et al. (2001) gives π = 0.862 and μ = 8.66, close to the maximum likelihood estimates, namely, π = 0.865 and μ = 8.64.

9. A code including both the conjugate and logistic-normal priors (with KM=K-1 and n=82) follows. One or other prior will need to be commented out in practice.

```
model { for (i in 1:n) {y[i] <- Y[i]/1000; y[i] ~dnorm(mu[G[i]],tau[G[i]])
ynew[i] ~dnorm(mu[G[i]],tau[G[i]]); ML[i] <- log(sum(c[i,]))}
for (k in 1:K) {tau[k] ~dgamma(1,0.001)}
# constrained prior on latent means
mu[1] ~dnorm(10,0.001) I(9.176,)
for (k in 1:KM) {del[k] ~dgamma(1,0.001) I(,20); mu[k+1] <- mu[k]
+del[k]}
# Fit
TLL <- sum(ML[]); pars <- (K-1)+2*K; BIC <- -2*TLL+pars*log(n)
#
# Mixing Prior 1
#
for (i in 1:n) {G[i] ~dcat(pi[1:K])
for (k in 1:K) {c[i,k] <- pi[k]*exp(0.5*log(tau[k]/6.28)
                -0.5*tau[k]*pow(y[i]-mu[k],2))}}
pi[1:K] ~ddirch(alph[1:K]); for (k in 1:K) {alph[k] <- 1}
#
# Mixing Prior 2
#
for (i in 1:n) {z[i,1:KM] ~dmnorm(nu[1:KM],T.z[1:KM,1:KM])
G[i]~dcat(p.z[i,1:K]); p.z[i,K] <- 1/(1+sum(exz[i,1:KM]))
for (k in 1:K) {c[i,k] <- p.z[i,k]*exp(0.5*log(tau[k]/6.28)
            -0.5*tau[k]*pow(y[i]-mu[k],2))}
for (k in 1:KM) {exz[i,k] <- exp(z[i,k])
p.z[i,k] <- exz[i,k]/(1+sum(exz[i,1:KM]))}}
for (k in 1:KM) {nu[k] ~dnorm(0,0.001);
for (m in 1:KM) {Q[k,m] <- equals(k,m)}}
T.z[1:KM,1:KM] ~dwish(Q[,],KM)}
```

10. The code for the NRT analysis is

```
model {for (i in 1:n) { y[i] <- log(rT[i]/(nT[i]-rT[i]))-log(rC[i]/(nC[i]-rC[i]))
s2[i] <- 1/rT[i] + 1/(nT[i]-rT[i]) + 1/rC[i] + 1/(nC[i]-rC[i]);
y[i] ~dnorm(b[i],inv.s2[i]); inv.s2[i] <- 1/s2[i]
yrep[i] ~dnorm(b[i],inv.s2[i]); testrep[i] <- step(yrep[i]-y[i])
# subgroup indicator
S[i] ~dcat(p[1:M]); b[i] <- phi[S[i]];
# log-likelihood
LL[i] <- 0.5*log(inv.s2[i]/6.28)-0.5*inv.s2[i]*pow(y[i]-b[i],2)
for (m in 1:M) {memb[i,m] <- equals(S[i],m)}}
TL <- sum(LL[])
# base prior
for (m in 1:M) { phi[m] ~dnorm(mu[m],inv.tau2[m])
realclus[m] <- step(sum(memb[,m])-1)
inv.tau2[m] ~dgamma(1,1); mu[m] ~dnorm(m.mu,1)}
m.mu ~dnorm(0,1)
# treatment benefit
p.ben <- step(mean(b[]))
# truncated Dirichlet process
alpha ~dgamma(a0,b0) I(0.1,); V[M] <- 1; p[1] <- V[1]
for (m in 1:M-1){
# c[m] <- 1+alpha/M; d[m] <- alpha*(1-m/M)
c[m] <- 1; d[m] <- alpha
V[m] ~dbeta(c[m],d[m]); p[m+1] <- V[m+1]*(1-V[m])*p[m]/V[m]}
# total clusters
    K <- sum(realclus[])}
```

11. The code for the Polya urn scheme for the DP Poisson-gamma model (with nP=n+1) is

```
model { # prediction
pnew[1] <- alph/(alph+n); bnew[1] ~dgamma(c.g,d.g)
for (k in 2:nP) {pnew[k] <- 1/(alph+n); bnew[k] <- b[k-1]}
Snew ~dcat(pnew[1:nP]); b.new <- bnew[Snew]
ynew ~dpois(b.new)
for (i in 1:n) {y[i] ~dpois(b[i]); yrep[i] ~dpois(b[i]);
# checks using replicate samples
ch1[i] <- step(y[i]-yrep[i]-0.001); ch2[i] <- equals(yrep[i],y[i])
ch[i] <- ch1[i]+0.5*ch2[i]}
# urn prior
b[1] ~dgamma(c.g,d.g); newclus[1] <- 1
for (i in 2:n) {bstar[i,1] ~dgamma(c.g,d.g)
p[i,1] <- alph/(alph+i-1); S[i] ~dcat(p[i,1:n])
newclus[i] <- equals(S[i],1); b[i] <- bstar[i,S[i]]
for (k in 1:i-1) {bstar[i,k+1] <- b[k]; p[i,k+1] <- 1/(alph+i-1)}
for (k in i:n-1) {bstar[i,k+1] <- 0; p[i,k+1] <- 0}}
```

```
K <- sum(newclus[]); alph ~dgamma(1,1); c.g ~dexp(1) I(0.5,); d.g
~dexp(1)}
```

The code for the Polya Tree prior in this data analysis is

```
model {for (i in 1:n) {y[i] ~dpois(mu[i]); log(mu[i]) <- beta+sig*b[i];
b[i] <- bstar[B[M,i]]
yrep[i] ~dpois(mu[i]); ch1[i] <- step(y[i]-yrep[i]-0.001);
ch2[i] <- equals(yrep[i],y[i]); ch[i] <- ch1[i]+0.5*ch2[i]}
beta ~dnorm(0,0.01); tau ~dgamma(1,1); sig <- 1/sqrt(tau)
# Polya Tree Process for b[]
for (m in 2:M) { c[m] <- 0.1*pow(m,2)}
for (i in 1:n) { V[1,i] ~dbern(0.5)
for (m in 2:M) { p[m,i] ~dbeta(c[m],c[m]); V[m,i] ~dbern(p[m,i])}
# level 1 choice (convert V=0,1 to B=1,2)
B[1,i] <- V[1,i]+1
# choices at level 2 and above
for (m in 2:M) { B[m,i] <- sum(BC[m,i,1:m-1])+V[m,i]+1
for (k in 1:m-1) {BC[m,i,k] <- V[m-k,i]*pow(2,k)}}}
# Sample within ordinates of base distribution
bstar[1] ~dnorm(0,1) I(,U[1]); bstar[M2] ~dnorm(0,1) I(L[M2],)
for (i in 2:M2-1) {w[i] ~dbeta(1,1); bstar[i] <- (1-w[i])*L[i]+w[i]*U[i]}}
```

4

Structured Priors Recognizing Similarity over Time and Space

4.1 Introduction

In the analysis of data over time or space, one often expects positive covariation between units that are close to each other in those domains, so that exchangeable priors are not necessarily appropriate. Consider health event counts for small geographical areas, or relatively rare diseases, when small event totals or small populations at risk lead to unstable estimates of rates or relative risks. One is then led to hierarchical methods for pooling strength over sets of areas to achieve more stable estimates (Riggan et al., 1991; Waller, 2002). An assumption of exchangeable random effects then implies global smoothing, with area rates or risks smoothed toward the overall mean. If there is spatial covariation (when contiguous areas have similar disease levels), a more appropriate smoothing mechanism would incorporate local smoothing toward the mean of adjacent areas (Clayton and Kaldor, 1987).

Related ideas in time series are the extraction of relatively smooth trends or regular seasonal effects from time series subject to random variation. For example, modern state–space models recognize the presence of multiple underlying components in time series (West and Harrison, 1997), with the priors governing the evolution of the components emphasizing an expectation of smoothness. An example of such smoothness priors involves kth order differences in successive latent parameters, θ_t, as in $\Delta^k\theta_t \sim N(0, W)$ for times $t = k+1, \ldots, T$ (Kashiwagi and Yanagimoto, 1992; Kitagawa and Gersch, 1996). While time series are sometimes analyzed exchangeably, at least within subgroups of the data, as in change point models (Mira and Petrone, 1996), in most applications there is a gain from modeling temporal covariation.

Priors for time or space covariance modeling are structured in the sense of explicitly recognizing adjacency or proximity, and use this structure as the basis for smoothing or prediction. Often, smoothing of a time series or of rates in small areas is an end in itself; for example, spatial smoothing of health data for administrative areas to reflect similarity of disease risks in nearby areas is a more reliable guide for health interventions (e.g., Zhu et al., 2006, 3). Similarly, Berzuini and Clayton (1994, 828) mention use of state–space priors for pooling strength in problems with several time scales (e.g., age, time, cohort), whereby random time parameters depend on each other in a way that reflects their proximity on each scale.

However, structured priors may also be more suitable when the goals of analysis include out-of-sample prediction in addition to description. For example, time series forecasts place higher weight on recent observations, as in first order autoregressions or Markov random walks, but the parameters for the second stage model use all the data. In spatial applications, such as geostatistics, a frequent goal is interpolation of a modeled surface to unsampled locations based on proximity to observed locations (Gotway and Wolfinger, 2003; Jiruse et al., 2004; Webster et al., 1994).

While there may be benefits from borrowing strength methods that take account of correlations between units, the use of multiple random effects to represent unobserved components in time or spatial series raises potential identification issues. For example, priors for correlated unit effects in time or space may consider differences between adjacent units without specifying the mean level of the effects. Markov Chain Monte Carlo (MCMC) methods may then require centering of the effects during sampling to ensure identification of other parameters. Methods for smoothing or interpolation in space or time may also need to retain robustness to take account of regime shifts, or to accommodate temporal or spatial outliers. Structured priors assume relatively smooth variation over adjacent units, and their parameters may be distorted if mechanisms are not incorporated for accommodating extreme points (see Sections 4.6 and 4.10 relating to time and spatial series, respectively).

To illustrate commonality in structured priors for temporal and spatial series, one may cite simultaneous and conditional autoregressive priors defined over observations, latent effects, or both. In particular, the pairwise difference or Markov random field (MRF) prior may be specified via conditional densities, which are naturally suited for Gibbs sampling (Finley et al., 2006). For univariate effects, $\theta = (\theta_1, \ldots, \theta_n)$, the conditional MRF prior takes the form (Besag et al., 1995, 11; Rue and Tjelmeland, 2002),

$$p(\theta_i|\theta_{[i]}) \propto \tau \exp\left[-\sum_{j\neq i} w_{ij}\Phi(\tau[\theta_i - \theta_j])\right],$$

where $\theta_{[i]}$ denotes values for cases other than i, w_{ij} are weights specifying dependence between units i and j, $\Phi(u)$ is an increasing function in u, subject to $\Phi(u) = \Phi(-u)$, and τ is a precision parameter. Under a neighborhood prior, where $w_{ij} = 1$ when units i and j are neighbors and $w_{ij} = 0$ otherwise, an equivalent representation is

$$p(\theta_i|\theta_{[i]}) \propto \tau \exp\left[-\sum_{j\in\partial_i} \Phi(\tau[\theta_i - \theta_j])\right],$$

where ∂_i is the set of areas or times adjacent to area or time i. The case $w_{ij} = 1$ if $|i - j| = 1$ and $w_{ij} = 0$ otherwise, leads to first order random walk priors relevant to modeling time ordered data. The MRF prior generalizes to variables θ_{ij} in two-dimensional lattices (e.g., areas i and times j), and a neighborhood might then be defined as $\partial_{ij} = [(i+1,j),(i-1,j),(i,j+1),(i,j-1)]$ (Lavine, 1999). Taking $\Phi(u) = u^2/2$ leads to a Gaussian or L2 norm conditional prior for θ_i (Waller, 2002),

$$\theta_i|\theta_{[i]} \backsim N\left(\sum_{j\neq i}\frac{w_{ij}\theta_j}{w_{i+}}, \frac{1}{\tau w_{i+}}\right),$$

whereas if $\phi(u) = |u|$, then,

$$p(\theta_i|\theta_{[i]}) \propto \tau \exp\left(-\tau\sum_{j\# i} w_{ij}|\theta_i - \theta_j|\right),$$

known as the L1 norm prior (Richardson et al., 2004). To achieve robust smoothing, the latter form may be better suited to spatial or temporal discontinuities since its mode is at the median rather than the mean. Other methods that reflect spatial or temporal outliers include convolution smoothing (Besag et al., 1991; Knorr-Held, 2000), based on an equal mixture of global and local smoothing, or discrete mixtures of spatially or temporally structured priors (Sections 4.8 and 4.10).

While the MRF prior can be used for simple time series model, a more general scheme for specifying priors for modeling time series data is provided by the state–space approach, considered in Sections 4.3 and 4.4 (Harvey et al., 2006). A linear state–space (or dynamic linear model) specification for the changing level of a metric response, y_t, has the form,

$$y_t = \beta_t X_t + u_t,$$
$$\beta_t = \beta_{t-1} G_t + w_t,$$

where $u_t \sim N(0, V_t)$ and $w_t \sim N(0, W_t)$ are white noise (unstructured) random variation, X_t is a predictor or design matrix, and G_t is a known matrix governing the evolution of the state vector β_t (Durbin, 2000; West and Harrison, 1997). The time-structured latent effects, β_t, may include levels, trends, seasonal, or cyclical effects. Taking u_t and w_t to be normal leads to the normal dynamic linear model (West, 1998), with extension to general linear model forms for discrete data leading to dynamic general linear models. State–space principles can also be applied to model stochastic variances, for example in stochastic volatility models, as opposed to changing levels (Section 4.5).

In terms of Bayesian computing options for temporal and spatial series, one may mention the development of conditional autoregression and spatial functions in WinBUGS, such as the car.normal, mv.car, and spatial.exp functions. As well as WinBUGS, a number of temporal and spatial analysis tools and models are available in R. Spatial modules, some with Bayesian options, include spdep (Bivand and Gebhardt, 2000), geoR and geoRglm (Ribeiro and

Diggle, 2001), and spBayes. Time series analysis functions in R are discussed by Shumway and Stoffer (2006), while the MSBVAR package specifically addresses Bayesian vector autoregressive models, with applications considered by Brandt and Freeman (2006) and Brandt and Williams (2007).

This chapter first considers schemes for modeling correlated observations and latent effects in time series. Sections 4.2 and 4.3 consider autoregressive and state–space priors for time series analysis, with Section 4.4 considering state–space methods for discrete time series, Section 4.5 considering Bayesian approaches to stochastic volatility, and Section 4.6 considering models adaptive to temporal discontinuities. The last four sections consider error structures for area and point data.

4.2 Modeling Temporal Structure: Autoregressive Models

A time series is a sequence of stochastic observations that are ordered in time, most often at equally spaced discrete times, $t = 1, \ldots, T$, though extensions to unequally spaced intervals are relatively straightforward (Lee and Nelder, 2001). Major goals of time series analysis include modeling the interrelationship of variables evolving jointly through time, as in econometric growth models (Paap and van Dijk, 2003), forecasting future values of time series variables (Beck, 2004), and identifying the structural components of the sequence of observations (Huerta and West, 1999). Areas where Bayesian perspectives have greatly influenced recent developments include dynamic general linear models (Carter and Kohn, 1994), time series model selection (Troughton and Godsill, 1998), and models with common latent variables such as factor stochastic volatility models. Below, several leading approaches to time series modeling are reviewed, beginning with established methods (e.g., Box–Jenkins), which help set a context for more recent techniques (e.g., for modeling stochastic volatility).

Many time series show evidence of serial dependence in the observations or error terms, leading to what are sometimes denoted as observation-and parameter-driven models, respectively (Oh and Lim, 2001); an analogous distinction is often made in analyzing spatial series (Larch and Walde, 2008). A widely used model for expressing such serial dependence is the lag p autoregressive or $AR(p)$ model. An $AR(p)$ scheme for dependent outcomes, y_t, in a normal linear framework is represented by

$$y_t = \phi_0 + \phi_1 y_{t-1} + \phi_2 y_{t-2} + \cdots + \phi_p y_{t-p} + u_t,$$

where the innovation errors, $u_t \sim N(0, \sigma^2)$, are homoscedastic white noise, independent of each other and lagged y values $\{y_{t-1}, \cdots, y_{t-p}\}$. So $E(u_t u_{t-s}) = E(u_{t-j} u_{t-j-s}) = 0$ for all s and j. Note that a full likelihood analysis will refer to p latent preseries values (Marriott et al., 1996), with Marriott et al. (2003) suggesting preseries values follow a heavy-tailed version of the density assumed for the observed series, for instance $(y_0, y_{-1}, \ldots, y_{1-p})$ as Student t with variance σ^2 and low degrees of freedom ν (e.g., $\nu = 2$). Autoregressive dependence may also be present in error terms, such that,

$$y_t = \phi_0 + \phi_1 y_{t-1} + \phi_2 y_{t-2} + \cdots + \phi_{p_1} y_{t-p_1} + \varepsilon_t,$$
$$\varepsilon_t = \rho_1 \varepsilon_{t-1} + \rho_2 \varepsilon_{t-2} + \cdots + \rho_{p_2} \varepsilon_{t-p_2} + u_t.$$

Furthermore, moving average effects may occur in the white noise errors, u_t, with an impact on y_t of lagged disturbances u_t. A lag q moving average effect, combined with a lag p effect in the y_t series, provides the $ARMA(p, q)$ model,

$$y_t = \phi_0 + \phi_1 y_{t-1} + \cdots + \phi_p y_{t-p} + u_t + \gamma_1 u_{t-1} + \gamma_2 u_{t-2} + \cdots + \gamma_q u_{t-q}.$$

Assuming the y-series is centered around its mean, and defining $By_t = y_t - y_{t-1}$, one has $y_t - \phi_1 y_{t-1} - \cdots \phi_p y_{t-p} = y_t(1 - \phi_1 B - \cdots \phi_p B^p) = \Phi(B) y_t$, and the $ARMA(p, q)$ model can be written,

$$\Phi(B) y_t = \Gamma(B) u_t.$$

Classical estimation methods typically require stationarity and constant variances in estimating such models. Stationarity is equivalent to the roots of $\Phi(B) = 1 - B - B^2 \, .. - B^p$ being outside the unit circle, and invertibility refers to the same condition on the roots of $\Gamma(B)$. This typically involves preliminary data differencing or transformation to gain stationarity, or regression to remove trend (e.g., Abraham and Ledolter, 1983, 225), with the actual model then applied to differenced data or to regression residuals. To assess whether stationarity has been achieved, one can consider the autocorrelation sequence of model residuals: a stationary process should show a sequence fading to zero at high lags, whereas significant values at high lags indicate nonstationarity. In Bayesian analyzes, it is common to estimate parameters without presuming stationarity (or invertibility), but obtain the posterior probabilities of stationarity via monitoring the sampled parameters (Marriott et al., 1996; McCulloch and Tsay, 1994).

4.2.1 Random coefficient autoregressive models

A hierarchical generalization of the $AR(p)$ prior allows the lag coefficient to vary over time, as in random coefficient AR or RCAR models—see, e.g., Berkes et al. (2009), Lee (1998), and Nicholls and Quinn (1980). These are also called time-varying autoregressive or TVAR models. Thus, for a centered and univariate y, an $RCAR(p)$ model in the observations specifies,

$$y_t = \sum \phi_{tj} y_{t-j} + u_t,$$
$$\phi_t = \mu_\phi + \Sigma_\phi^{0.5} e_t,$$

where $u_t \sim N(0, \sigma^2)$, $e_t \backsim N_p(0, I)$, $\phi_t = (\phi_{t1}, \ldots, \phi_{tp})$, and $\mu_\phi = (\phi_1, \ldots, \phi_p)$. Instead of a multivariate normal prior for the ϕ_t, sequential updating of the ϕ_t may be applied, for example via a multivariate random walk (Section 4.3), with,

$$\phi_t = \phi_{t-1} + w_t \qquad w_t \sim N_p(0, W_t).$$

Another possibility (Godsill et al., 2004) is to take both the AR coefficient vector and the innovation variance σ^2 to be time varying, for example by setting a random walk prior on $h_t = \log(\sigma_t)$, or by a second stage autoregression, such as,

$$h_t \sim N(\rho_h h_{t-1}, \sigma_h^2).$$

As in many Bayesian applications, typically, stationarity constraints are not necessarily placed on the ϕ_{tj} at each t (Prado et al., 2000). However, if the AR parameters lie in the stationary region then the series can be considered locally stationary. For example, for an $RCAR(1)$ model including a latent preseries value, y_0, a hierarchical scheme such as,

$$\begin{aligned} y_t &\backsim N(\phi_{t1} y_{t-1}, \sigma^2) \qquad t > 1, \\ y_0 &\backsim t_2(m_0, \sigma^2), \\ \phi_{t1} &\backsim N(\phi_1, \sigma_\phi^2) \qquad t > 1, \end{aligned}$$

may be applied. For this model, stationarity holds if $\phi_1^2 + \sigma_\phi^2 < 1$.

4.2.2 Low order autoregressive models

Simple dependence models for observations or latent effects are obtained via first or second order autoregression. In the $AR(1)$ observation model, one has,

$$y_t - \mu = \phi(y_{t-1} - \mu) + u_t,$$

or

$$y_t = \phi y_{t-1} + u_t,$$

for centered data, where under stationarity, $-1 < \phi < 1$, and y_r and y_s for $1 \leq r \leq s \leq T$ are conditionally independent, given $\{y_{r+1}, \ldots, y_{s-1}\}$ if $r - s > 1$ (Rue and Held, 2005). The $AR(2)$ model has,

$$y_t = \phi_1 y_{t-1} + \phi_2 y_{t-2} + u_t,$$

where stationarity requires $\phi_1 + \phi_2 < 1, \phi_2 - \phi_1 < 1$, and $|\phi_2| < 1$. An $AR(1)$ error sequence, $\varepsilon_t = \rho\varepsilon_{t-1} + u_t$ with $u_t \sim N(0, \sigma^2)$, similarly requires $-1 < \rho < 1$ for stationarity. The covariance for such a sequence has the form, $\text{Cov}(\varepsilon) = \sigma^2 C$, with (s,t)th element in the correlation matrix, $\text{corr}(\varepsilon_s, \varepsilon_t) = \rho^{|s-t|}/(1-\rho^2)$, so correlations decline as the gap between observations increases.

For the stationary $AR(1)$ observation model, $y_t = \phi y_{t-1} + u_t$, the marginal density of the first observation is $y_1 \backsim N(0, \sigma^2/(1-\phi^2))$, and the joint density can also be obtained by density decomposition as,

$$\begin{aligned} p(y_1, \dots, y_T) &= p(y_1)p(y_2|y_1)p(y_3|y_2)\dots p(y_T|y_{T-1}) \\ &\propto (1-\phi^2)^{0.5}\sigma^{-n}\exp[-0.5H/\sigma^2], \end{aligned}$$

where $H = (1-\phi^2)y_1^2 + \sum_{t=2}^{T}(y_t - \phi y_{t-1})^2$. The same sequence of marginal and conditional densities applies for $AR(1)$ autoregressive errors.

The precision (inverse covariance) matrix of autoregressive models has interesting theoretical properties demonstrating how conditional independence structures determine the precision matrix and vice versa (Rue and Held, 2005; Speed and Kiiveri, 1986). Specifically, zeros in the precision matrix define, and are defined by, conditional independencies in the joint density. Thus, for an $AR(1)$ prior on effects ε with lag coefficient ρ, the precision matrix, Π, is tridiagonal with (r, s)th cell equalling zero if, and only if, the complete conditional distribution of ε_r does not depend on ε_s, namely,

$$\Pi = \sigma^{-2}C^{-1} = \sigma^{-2}\begin{bmatrix} 1 & -\rho & 0 & & & & \\ -\rho & 1+\rho^2 & -\rho & & & & \\ 0 & -\rho & 1+\rho^2 & \cdots & & & \\ & & \cdots & \cdots & & & \\ & & & & -\rho & 1+\rho^2 & -\rho \\ & & & & 0 & -\rho & 1 \end{bmatrix}.$$

For an $AR(2)$ error sequence with lag parameters $\{\rho_1, \rho_2\}$, the precision matrix is

$$\Pi = \sigma^{-2}\begin{bmatrix} 1 & -\rho_1 & -\rho_2 & 0 & & & \\ -\rho_1 & 1+\rho_1^2 & -\rho_1(1-\rho_2) & -\rho_2 & & & \\ -\rho_2 & -\rho_1(1-\rho_2) & 1+\rho_1^2+\rho_2^2 & -\rho_1(1-\rho_2) & & & \\ 0 & -\rho_2 & -\rho_1(1-\rho_2) & 1+\rho_1^2+\rho_2^2 & \cdots & & \\ & & \cdots & \cdots & & & \\ & & & & -\rho_1(1-\rho_2) & 1+\rho_1^2 & -\rho_1 \\ & & & & -\rho_2 & -\rho_1 & 1 \end{bmatrix}.$$

Such simplifications in structure are useful in multidimensional applications involving spatio-temporal or multiple time scale errors. For example, if the covariance matrix of a spatio-temporal error, ε_{st}, is represented as a Kronecker product, $\Sigma_t \otimes \Sigma_s$, of a temporal covariance Σ_t and spatial covariance Σ_s, then the corresponding precision matrix is $\Pi_t \otimes \Pi_s$ (Bijma et al., 2005).

There is a considerable literature around the unit root and explosive root solutions of the $AR(1)$ model, $y_t = \phi y_{t-1} + u_t$. One may apply an autoregressive prior not constrained to stationarity, and a substantial posterior probability of nonstationarity would support using random walk priors (Section 4.3), as a parsimonious autoregressive prior that allows for potential nonstationarity. For example, Lubrano (1995) considers the alternative composite hypotheses,

$H_0 : \phi < 1$ and $H_1 : \phi \geq 1$. Schotman and van Dijk (1991) consider the autoregression plus trend observation model, $y_t = \phi_0 + \phi_1 y_{t-1} + \delta t + u_t$, and reframe it in equivalent $AR(1)$ error form as,

$$y_t = \delta_0 + \delta_1 t + \varepsilon_t,$$
$$\varepsilon_t = \phi\varepsilon_{t-1} + u_t,$$

while Chatuverdi and Kumar (2005) consider the unit root hypothesis under a more general polynomial trend, $y_t = \delta_0 + \Sigma_j \delta_j t^j + \varepsilon_t$.

4.2.3 Antedependence models

Structured antedependence models may offer flexibility in time series specification; they resemble autoregressions in entailing a regression over preceding observations or latent effects, but are specified in a way that avoids stationarity constraints (Nunez-Anton and Zimmerman, 2000; Pourahmadi, 2002). Observations $\{y_1, \ldots, y_T\}$ are antedependent of order s if y_t depends only on $\{y_{t-1}, \ldots, y_{t-s}\}$ for all $t \geq s$ (Gabriel, 1962). For example, Jaffrezic et al. (2003) consider a second order antedependence model for normal panel data of the form $y_{it} = \mu_{it} + g_{it} + u_{it}$, where μ_{it} models fixed effects, e.g., $\mu_{it} = x_{it}\beta$, u_{it} are unstructured white noise errors with fixed variance, and the genetic component, g_{it}, follows a second-order structured antedependence or $AD(2)$ scheme. The essence of this scheme for a pure time series is

$$g_1 = \eta_1,$$
$$g_2 = \phi_{12} g_1 + \eta_2,$$
$$g_t = \phi_{1t} g_{t-1} + \phi_{2t} g_{t-2} + \eta_t \qquad t > 2,$$

with $\eta_t \sim N(0, \omega_t)$. Due to the initial condition, $g_1 = \eta_1$, the antedependence parameters, such as $\{\phi_{1t}, \phi_{2t}\}$ in an $AD(2)$ model, are unconstrained, in contrast to the stationarity constraints needed for autoregressive models.

To reduce the number of parameters being estimated, changing variances, ω_t, may be modeled via a parametric function of time, for example,

$$\log(\omega_t) = \alpha_1 + \alpha_2 t + \alpha_3 t^2,$$

while the antedependence parameters can also be modeled using time functions. For example, a Box–Cox power law can be used to parameterize time-varying AD coefficients, ϕ_{kt}, namely,

$$\phi_{kt} = \phi_k^{r_t - r_{t-k}},$$

where $\{r_t = t^{\lambda_k - 1}/\lambda_k, r_{t-k} = (t-k)^{\lambda_k - 1}/\lambda_k\}$ if $\lambda_k \neq 0$, and $\{r_t = \log(t), r_{t-k} = \log(t-k)\}$ if $\lambda_k = 0$ (Nunez-Anton and Zimmerman, 2000). The ϕ and ω parameters may be adjusted to account for unevenly spaced times located at points $\{a_1, \ldots, a_T\}$. For example, an exponential parameterization is possible, such that an $AD(2)$ model would specify,

$$\phi_1(a_t, a_{t-1}) = \exp(-\lambda_1(a_t - a_{t-1})),$$

and

$$\phi_2(a_t, a_{t-2}) = \exp(-\lambda_2(a_t - a_{t-2})),$$

with $\lambda_j > 0$ corresponding to positive temporal dependence (Jaffrezic et al., 2004; Pourahmadi, 2002).

4.3 State–Space Priors for Metric Data

Nonstationary models based on state–space priors are widely used in applications where time series parameters are evolving through time, especially in analyzing multiple unobserved components due to trend, cyclical, or seasonal effects. The idea that a time series may best be viewed as being composed of several unobserved components contrasts with Box–Jenkins or ARMA methods that require differencing to eliminate trend or periodic effects and achieve stationary means and variances (Durbin, 2000, 2). ARMA models are selected using autocorrelation and partial and autocorrelation functions that are subject to sampling variability, and quite different models can provide similar fits for the same series. In fact, though ARMA sequences can be represented as particular instances of state–space models with implicit components. Among informative discussions on state–space vs. Box–Jenkins methods, see Durbin and Koopman (2001, 51) and Harvey and Todd (1983).

The normal linear state–space specification, or dynamic linear model, has the form,

$$y_t = \beta_t X_t + u_t,$$

where evolution of the p dimensional signal, β_t, is defined by a state equation,

$$\beta_t = \beta_{t-1} G_t + w_t,$$

with X_t being a $p \times 1$ design matrix (typically including an intercept), and G_t defining a $p \times p$ state evolution matrix. The normal errors u_t and w_t are independent of each other, with mean zero and variances V_t and W_t (or covariances for multivariate y). The initial state vector or initial condition has a separate (e.g., normal) prior such as $\beta_1 \sim N(m_1, W_1)$ (Strickland et al., 2008). Often, G_t has a simple form such as an identity matrix. For the case $G_t = G$, Gamerman (1998) mentions an inverse parameterization consequent on taking,

$$\delta_1 = \beta_1, \delta_t = \beta_t - G\beta_{t-1},$$

so that

$$\beta_t = \sum_{j=1}^{t} G^{t-j}\delta_j.$$

Algorithms using normal distribution properties can be applied to sequential updating (filtering), forward prediction, and retrospective smoothing of the state vector in the normal dynamic linear model. Letting $D_t = (y_t, y_{t-1}, \ldots, y_1)$, the prior, predictive, and posterior distributions of β_t are (Reis et al., 2006),

$$\begin{aligned} p(\beta_t|D_{t-1}) &= \int p(\beta_t|\beta_{t-1})p(\beta_{t-1}|D_{t-1})d\beta_{t-1}, \\ p(y_t|D_{t-1}) &= \int p(y_t|\beta_t)p(\beta_t|D_{t-1})d\beta_t, \\ p(\beta_t|D_t) &\propto p(\beta_t|D_{t-1})p(y_t|D_{t-1}). \end{aligned}$$

For the linear normal model with $V_t = V$, $W_t = W$, sequential updating provides posteriors,

$$\beta_t|D_t \backsim N(m_t, C_t),$$

where

$$\begin{aligned} a_t &= G_t m_{t-1}, \\ m_t &= a_t + A_t e_t, \\ C_t &= R_t - A_t A_t' q_t, \\ R_t &= G_t C_{t-1} G_t' + W, \\ q_t &= X_t' R_t X_t + V, \end{aligned}$$

with forecast errors,

$$e_t = y_t - X_t' a_t.$$

The one step ahead state and observation predictive densities are normal densities, namely,

$$\begin{aligned} (\beta_t|D_{t-1}) &\backsim N(a_t, R_t), \\ (y_t|D_{t-1}) &\backsim N(X_t' a_t, q_t). \end{aligned}$$

4.3.1 Simple signal models

As an illustration of a normal state–space or dynamic linear model, assume that observations are obtained with measurement error and in fact generated by a relatively smooth underlying signal, β_t. This is a hierarchical model—parallel to the normal-normal model of Chapter 3—with the first level being the observation equation, the second level being the state equation, and the priors on the variances and initial conditions defining hyperparameters at the third stage (Berliner, 1996). Assuming unstructured measurement errors, u_t, one has an observation or measurement equation,

$$y_t = \beta_t + u_t, \tag{4.1a}$$

for $t = 1, \ldots, T$ and a state equation defining the evolution of the signal,

$$\beta_t = \beta_{t-1} + w_t, \tag{4.1b}$$

for $t = 2, \ldots, T$. This is also known as a local level model (Durbin and Koopman, 2001), or random walk plus noise model (Durbin, 2000), and the second stage is a nonstationary first order random walk or $RW(1)$ prior, corresponding to the unit root case of an $AR(1)$ prior.

As for the $AR(1)$ prior, future values of the signal depend on $(\beta_t, \beta_{t-1}, \ldots, \beta_1)$ only through the current value β_t. The conditional form of the $RW(1)$ prior is

$$p(\beta_t|\beta_{[t]}, y) \propto \left\{ \begin{array}{ll} p(\beta_2|\beta_1)p(\beta_1)p(y_1|\beta_1) & t = 1 \\ p(\beta_{t+1}|\beta_t)p(\beta_t|\beta_{t-1})p(y_t|\beta_t) & t = 2, \ldots, T-1 \\ p(\beta_T|\beta_{T-1})p(y_T|\beta_T) & t = T \end{array} \right\},$$

so that for times $t = 2, \ldots, T-1$, there is averaging over preceding and following states. The first period signal (or "initial condition") β_1 is typically taken as an unknown fixed effect with large variance, while the observation error, u_t, and state error, w_t, are taken as, respectively, $N(0, V)$ and $N(0, W)$, and assumed uncorrelated in time, independent of one other, and also independent of the signal β_t. Assume $\beta_1 \sim N(b_1, S_1)$, $1/V \sim Ga(a_u, b_u)$, $1/W \sim Ga(a_w, b_w)$, then the full conditionals are,

$$\beta_1 \sim N\left(\left[\frac{\beta_2}{W} + \frac{b_1}{S_1} + \frac{y_1}{V}\right]\left[\frac{1}{W} + \frac{1}{S_1} + \frac{1}{V}\right]^{-1}, \left[\frac{1}{W} + \frac{1}{S_1} + \frac{1}{V}\right]^{-1}\right),$$

$$\beta_t \sim N\left(\left[\frac{(\beta_{t+1} + \beta_{t-1})}{W} + \frac{y_t}{V}\right]\left[\frac{2}{W} + \frac{1}{V}\right]^{-1}, \left[\frac{2}{W} + \frac{1}{V}\right]^{-1}\right) \qquad t = 2, \ldots, T-1,$$

$$\beta_T \sim N\left(\left[\frac{\beta_{T-1}}{W} + \frac{y_T}{V}\right]\left[\frac{1}{W} + \frac{1}{V}\right]^{-1}, \left[\frac{1}{W} + \frac{1}{V}\right]^{-1}\right),$$

$$1/V \sim Ga\left(a_u + \frac{T}{2}, b_u + 0.5\sum_{t=1}^{T}(y_t - \beta_t)^2\right),$$

$$1/W \sim Ga\left(a_w + \frac{(T-1)}{2}, b_w + 0.5\sum_{t=2}^{T}(\beta_t - \beta_{t-1})^2\right).$$

Higher order random walks in the signal are another possibility, with a kth order random walk having prior,

$$\Delta^k\beta_t \backsim N(0, W),$$

(Berliner, 1996; Fahrmeir and Lang, 2001; Kitagawa and Gersch, 1996). For example, a second difference random walk or $RW(2)$ prior specifies $y_t = \beta_t + u_t$ and state equation $\Delta^2\beta_t = w_t$. Hence,

$$\Delta(\Delta\beta_t) = \Delta(\beta_t - \beta_{t-1}) = \Delta\beta_t - \Delta\beta_{t-1} = (\beta_t - \beta_{t-1}) - (\beta_{t-1} - \beta_{t-2}) = w_t,$$

and the $RW(2)$ prior can be stated as,

$$\beta_t \sim N(2\beta_{t-1} - \beta_{t-2}, W).$$

Whereas first order random walks penalize abrupt jumps between successive values, the $RW(2)$ prior penalizes deviations from a linear trend. The $RW(2)$ and higher order RW priors therefore lead to a smoother evolution of β_t through time. This is relevant not just to time series, but to processes operating on other time scales (e.g., age, cohort), for example in survival analysis or in graduating (smoothing) demographic schedules (Carlin and Klugman, 1993).

4.3.2 Sampling schemes

Different MCMC sampling schemes have been proposed for state–space models according to the form of outcome (e.g., metric or discrete) and the form of observation-state equations (e.g., linear or nonlinear). Multistate or joint sampling of the state vectors, β_t, is generally more efficient than single-state sampling that updates one state parameter vector at a time (Knorr-Held, 1999). Carlin et al. (1992) in the context of non-normal/nonlinear dynamic models suggest single-state samples from $(\beta_1, \ldots, \beta_T)$. Joint sampling for β when y is metric is discussed by Carter and Kohn (1994) and Fruhwirth-Schnatter (1994), while de Jong and Shephard (1995) focus on sampling the u_t and w_t error series as opposed to the state effects β_t; a recent overview is provided by Reis et al. (2006). Gamerman (1998) proposes updating via the δ_t rather than the usually highly correlated β_t using the reparameterization mentioned above.

Knorr-Held (1999), Rue (2001), and Rue and Held (2005) use properties of the penalty (inverse covariance) matrix of the joint density for the state vectors as a basis for sampling sub-blocks of the elements $(\beta_1, \ldots, \beta_T)$. Thus, Gaussian state–space priors can be written in joint form as,

$$p(\beta_1, \ldots, \beta_T|W) \propto \exp\left(-\frac{\beta' K \beta}{2W}\right),$$

where the form of the penalty matrix K is determined by the form of autoregressive prior. For a first order random walk with $\beta_t \backsim N(\beta_{t-1}, W)$, the penalty matrix is

$$K = \begin{pmatrix} 1 & -1 & & & & & \\ -1 & 2 & -1 & & & & \\ & -1 & 2 & -1 & & & \\ & & .. & .. & .. & & \\ & & & -1 & 2 & -1 & \\ & & & & -1 & 2 & -1 \\ & & & & & -1 & 1 \end{pmatrix},$$

while for a second order random walk with $\beta_t \backsim N(2\beta_{t-1} - \beta_{t-2}, W)$, one has,

$$K = \begin{pmatrix} 1 & -2 & 1 & & & & & & & \\ -2 & 5 & -4 & 1 & & & & & & \\ 1 & -4 & 6 & -4 & 1 & & & & & \\ & 1 & -4 & 6 & -4 & 1 & & & & \\ & & .. & .. & .. & .. & & & & \\ & & & 1 & -4 & 6 & -4 & 1 & & \\ & & & & 1 & -4 & 6 & -4 & 1 \\ & & & & & 1 & -4 & 5 & -2 \\ & & & & & & 1 & -2 & 1 \end{pmatrix}.$$

For an $RW(p)$ prior at equally spaced time points, the elements of the matrix K (apart from edge effects) are expressible as,

$$k_{ij} = (-1)^{|i-j|} \binom{2p}{p - |i-j|} \text{ if } |i-j| \leq p,$$

and $k_{ij} = 0$ otherwise (Sun et al., 2006, Equation 10).

Let β_{ab} denote the subvector $(\beta_a, \beta_{a+1}, \ldots, \beta_b)$ of state effects, and K_{ab} denote the corresponding submatrix of K. Let $K_{1,a-1}$ and $K_{b+1,T}$ denote the submatrices to the left and right of K_{ab}, namely,

$$K = \begin{pmatrix} & K'_{1,a-1} & \\ K_{1,a-1} & K_{ab} & K_{b+1,T} \\ & K'_{b+1,T} & \end{pmatrix}.$$

Then the conditional density for β_{ab} given $\beta_{1,a-1}$, $\beta_{b+1,T}$ and W is normal, $\beta_{ab} \backsim N(\nu_{ab}, WK_{ab}^{-1})$, where

$$\nu_{ab} = \begin{array}{lr} -K_{ab}^{-1}K_{b+1,T}\beta_{b+1,T} & a=1 \\ -K_{ab}^{-1}[K_{1,a-1}\beta_{1,a-1} + K_{b+1,T}\beta_{b+1,T}] & a>1, b<T \\ -K_{ab}^{-1}K_{1,a-1}\beta_{1,a-1} & b=T. \end{array}$$

Using this density, a Metropolis–Hastings block sample may be used to update the full conditional,

$$p(\beta_{ab}|) \propto \prod_{t=a}^{b} p(y_t|\beta_t) p(\beta_{ab}|\beta_{b+1,T}, \beta_{1,a-1}, W).$$

This involves drawing a proposal, β_{ab}, from $N(\nu_{ab},\ WK_{ab}^{-1})$ with $\{\nu_{ab}, K_{ab}\}$ evaluated at the current sampled values β and W in a chain, with the proposal accepted or rejected according to a probability,

$$\min\left(1, \prod_{t=a}^{b} p(y_t|\beta_t^*) \Big/ \prod_{t=a}^{b} p(y_t|\beta_t)\right),$$

which may be calculated by comparing likelihoods only (Knorr-Held, 1999, 134).

4.3.3 Basic structural model

To allow for a trend in the mean level or signal, one may extend the state equation in Equation 4.1 to include a stochastic increment, so that,

$$y_t = \beta_t + u_t,$$
$$\beta_t = \beta_{t-1} + \Delta_t + w_{1t},$$
$$\Delta_t = \Delta_{t-1} + w_{2t},$$

where Δ_t represent the changing slope of the trend. This provides the local linear trend model or dynamic trend model (Fruhwirth-Schnatter, 1994). A constant parameter, Δ, models a linear trend, as in the Carter–Lee mortality forecasting model considered by Pedroza (2006); this is sometimes known as a random walk with drift. Other variations on the local linear model in Equation 4.1 include autoregressive rather than random walk state equations, such as,

$$\beta_t = \phi\beta_{t-1} + w_t,$$

as in Carlin et al. (1992, 496). An autoregression or random walk in y itself might be added, as in Ghosh and Tiwari (2007), who assume a local linear model for common cancer deaths of the form $y_{t+1} \backsim N(y_t + \beta_t, V)$.

The basic structural model or unobserved components model (Koopman, 1993; Koopman et al., 1999) adds seasonal effects, s_t, to the above local linear trend model, so that with μ_t representing the level of the series, one has,

$$y_t = \mu_t + s_t + u_t,$$
$$\mu_t = \mu_{t-1} + \Delta_t + w_{1t},$$
$$\Delta_t = \Delta_{t-1} + w_{2t},$$
$$s_t + s_{t-1} + \cdots + s_{t-S+1} = w_{3t},$$

where S is the number of seasons, and $w_{jt} \sim N(0, W_j)$. Fruhwirth-Schnatter (1994) sets out the full conditionals for this model under gamma priors for the precisions $1/W_j$. The last equation provides the time domain prior for seasonal effects, whereas a frequency domain prior specifies,

$$s_t = \sum_{j=1}^{[S/2]} s_{jt},$$
$$s_{jt} = s_{j,t-1}\cos(\lambda_j) + v_{j,t-1}\sin(\lambda_j) + w_{3t},$$
$$v_{jt} = -s_{j,t-1}\sin(\lambda_j) + v_{j,t-1}\cos(\lambda_j) + w_{4t},$$

where $\lambda_j = 2\pi j/S$ and $[S/2]$ denotes the integer part of $S/2$.

Certain series (e.g., natural phenomena) may show unknown periodicities, so cyclical components are added as well as, or instead of, seasonal components. For example, Piegorsch and Bailer (2005, 229) consider unknown frequencies in carbon dioxide concentrations from Mauna Loa volcano in

Hawaii. So for a local linear trend model with a single unknown cycle,

$$\begin{aligned}
y_t &= \mu_t + c_t + u_t,\\
\mu_t &= \mu_{t-1} + \Delta_t + w_{1t},\\
\Delta_t &= \Delta_{t-1} + w_{2t},\\
c_{t+1} &= c_t \cos(\lambda) + d_t \sin(\lambda) + w_{3t},\\
d_{t+1} &= -c_t \sin(\lambda) + d_t \cos(\lambda) + w_{4t},
\end{aligned}$$

where λ is an unknown frequency.

4.3.4 Identification questions

Identification issues in state–space random effect models occur for two main reasons. One is that the mean or level of the state effects is not specified (rather the mean of pairwise or higher differences is specified). The other is the presence of multiple confounded sources of random variation, as in the basic structural model with level and seasonal effects, whereas the data can only identify the sum of the random effects, $\mu_t + s_t$. These questions raise issues in MCMC sampling because an intercept (if included) will be confounded with the mean of at least one set of random effects.

To exemplify issues occurring due to the mean of the latent series, consider the measurement error with *RW*(1) signal model in Section 4.3.1. The state equation can be stated as,

$$\Delta\beta_t = \beta_t - \beta_{t-1} \backsim N(0, W),$$

so the prior only defines a level for differences in β_t, but the level of (undifferenced) β_t is not defined by the prior. If the model for y_t does not have a separate intercept parameter, the level of β_t will be identified by the level of y_t. Suppose though that the observation equation includes a separate constant with,

$$y_t = \mu + \beta_t + u_t.$$

Then μ and the mean of β_t are confounded and for identification, one may apply a centering or corner constraint to β_t. An identifying corner constraint involves setting a single β_t to a known value; taking the initial condition β_1 to have a known value, e.g., $\beta_1 = 0$, is one option (Clayton, 1996). By contrast, if the initial conditions (β_1 in an *RW*(1) prior, β_1 and β_2 in an *RW*(2) prior, etc.) are taken as unknowns, then a centering constraint may be applied at each MCMC iteration, so that the centered β_t satisfy $\sum_{t=1}^{T} \beta_t = 0$.

As in other models with multiple sources of random variation, priors on the variance components in state–space models may affect inferences. This is not simply a matter of scale and shape options for inverse gamma priors, but of how priors on variances (or precisions) influence the partitioning of total random variation between different sources (such as measurement error and state error variation in the simple signal model). One may recognize the interdependence between variance components using devices such as uniform priors on shrinkage ratios, $B = V/(V+W)$, combined with a prior on V or $V+W$ (Daniels, 1999). Alternatively, (V, W) may be reparameterized as (V, qV) where q is a signal to noise ratio. So with an inverse gamma prior on V, the prior on q might be centered on 1, in line with a prior belief that signal and observation variances are equal.

These approaches extend to models with competing sources of variation in the state equation. Consider the three errors, w_{jt} (for levels, slopes, and seasonals), in the basic structural model. Denoting $W_j = \text{Var}(w_{jt})$ and $V = \text{Var}(u_t)$, one may set $W_j = q_j V$ where q_j are signal to noise ratios (Harvey, 1989, 33; Koopman, 1993). One may then set priors on the q_j separately (e.g., separate gammas), or jointly; for example, via a multivariate normal on $\log(q_j)$. Another option is a prior on V and uniform priors on the ratios $V/(V+W_j)$. Such devices amount to assuming prior correlation between the respective variances.

An alternative approach to ensure stable identification is to set informative priors on the variance of each random walk, possibly based on expected stochastic variation around a deterministic trend. For example, following Berzuini and Clayton (1994), for $y_t \sim Po(\lambda_t)$, consider a second order random walk for $\beta_t = \log(\lambda_t)$,

$$\beta_t = 2\beta_{t-1} - \beta_{t-2} + w_t,$$

then the value, $W = 0$, for $\text{Var}(w_t)$ corresponds to a log-linear deterministic relationship between λ_t and time. To allow for stochastic variation, one may assume,

$$\frac{\nu W^*}{W} \backsim \chi^2_\nu,$$

or equivalently,

$$1/W \sim Ga\left(\frac{\nu}{2}, \frac{W^*\nu}{2}\right),$$

where W^* is a prior guess at W, and higher values of ν represent stronger degrees of belief in that guess. For example, taking $W^* = 0.01$ corresponds to assuming a 95% probability that λ_t will be within -18 and $+22\%$ of a log-linear extrapolation from β_{t-1}.

The single source of error approach (Ord et al., 2005) may also assist in achieving parsimony, and in resolving the partitioning of variance between multiple sources of variation in unobserved component models. Thus, the local linear trend model in multiple source of error (MSOE) form is

$$y_t = \mu_t + u_t,$$
$$\mu_t = \mu_{t-1} + \Delta_t + w_{1t},$$
$$\Delta_t = \Delta_{t-1} + w_{2t},$$

but in single source of error form is

$$y_t = \mu_t + u_t,$$
$$\mu_t = \mu_{t-1} + \Delta_t + \lambda_1 u_t,$$
$$\Delta_t = \Delta_{t-1} + \lambda_2 u_t,$$

where λ_1 and λ_2 are loadings. In contrast to the MSOE scheme, the state and observation errors are now correlated.

Example 4.1. Air Passenger Data As an example of the kind of issues that might occur in partitioning the sources of variation in a basic structural model, consider air passenger data analyzed in West and Harrison (1997) and elsewhere, namely, monthly totals of international airline passengers (for 1949–1961, so $T = 144$). A monthly seasonal effect is assumed, so there are $S - 1 = 11$ initial conditions for the s_t sequence. Then,

$$y_t = \mu_t + s_t + u_t,$$
$$\mu_t = \mu_{t-1} + \Delta_t + w_{1t},$$
$$\Delta_t = \Delta_{t-1} + w_{2t},$$
$$s_t + s_{t-1} + \cdots + s_{t-S+1} = w_{3t},$$

with $w_{jt} \sim N(0, W_j), u_t \sim N(0, V)$. A stable solution may depend on introducing some context-specific information and also inspecting the data to see what priors on hyperparameters are sensible, especially the variances $\{V, W_1, W_2, W_3\}$, without actually using data-based priors. Adopting vague priors on variance parameters may mean the trend and level components are not well identified empirically.

In this spirit, a diffuse $Ga(1, 0.001)$ prior is assumed on $1/V$, but priors on $\{q_j, j = 1, 3\}$ where $W_j = q_j V$, are set in line with contextual information and a general belief that "level and slope both change slowly over time" (Harvey and Todd, 1983, 300), such that the μ_t and Δ_t sequences should not show erratic fluctuations. On the other hand, seasonal fluctuations may be pronounced, as a plot of the data suggests. So $q_3 \sim Ga(1, 0.001)$, while q_1 and q_2 are assigned $E(10)$ priors, implying that W_1 and W_2 are expected to be small compared to V. $N(0, 1000)$ initial condition priors are assumed for the seasonal and growth effects.

The second half of a two chain run of 20,000 iterations is used for inferences[1]. Figure 4.1 shows a smoothly evolving mean and clear seasonal variations. Underlying growth (shown by the posterior means of the Δ_t parameters in the trend figure) is generally positive, though stalls between months 50–60 and 100–120. The posterior means (medians) of the W_j are 0.76 (0.74), 0.75 (0.70), and 172 (170), while those for q_j are 0.15 (0.11), 0.23 (0.20), and 58 (46).

An alternative prior sets $q_j \sim N(0, 1)\ I(0,)$, and also adopts less diffuse $N(0, 10)$ initial condition priors for the seasonal and growth effects. This

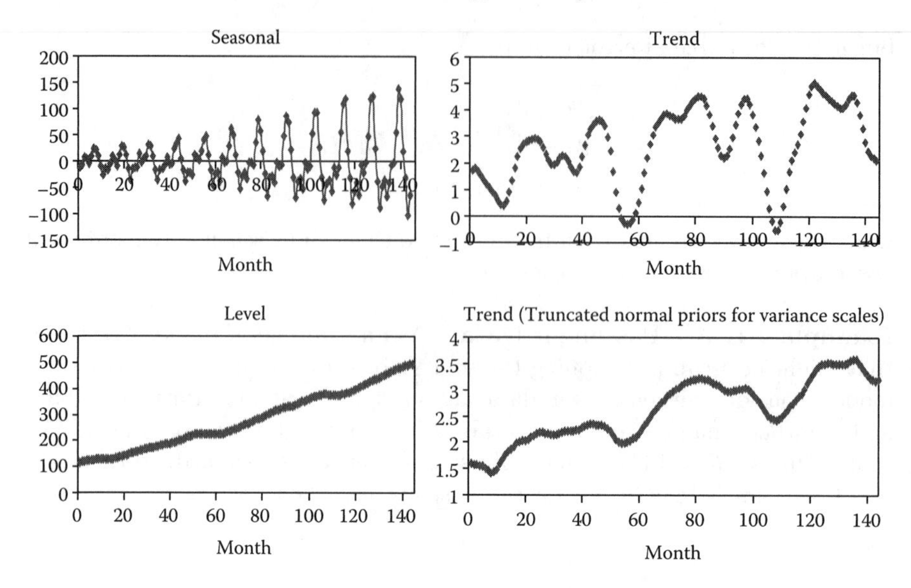

FIGURE 4.1
Air passenger BSM.

achieves early convergence in a two chain run of 20,000 iterations (5000 burn in). The posterior means on (q_1, q_2, q_3) are 0.46, 0.0035, and 2.87, respectively. There is a flatter trend effect (see Figure 4.1), though the level and seasonal plots are very similar to those of the first model.

Example 4.2. Global Sea Level Change This example compares in-sample predictions and out-of-sample forecasts from a local linear model with linear trend, and a simple hierarchical model involving unit level linear trends. A number of studies have analyzed local relative sea level records from tide gauge observations, and considered broader inferences regarding global mean sea level change. Local confounding factors may hinder quantifying a "global" signal from such data. However, Patwardhad and Small (1992) consider a set of stations from around the world with relatively long continuous records that were representative of other stations in the same region, and seemed relatively free of local confounding factors.

The analysis here follows them in using records for 1900–1980 from five stations, namely, San Francisco, Tonoura (Japan), Sydney, Bombay, and Cascais (Portugal), with out-of-sample forecasts to 2000. The data are in millimeters from the Permanent Service for Mean Sea Level website (http://www.pol.ac.uk/psmsl). For station j at time t, the model used by Patwardhad and Small (1992) involves a first order random walk in the mean global sea level, M_t, plus a homogenous linear trend (common coefficient b across sites j). So model 1 has,

$$y_{jt} = M_t + bt + u_{jt},$$

with $u_{jt} \backsim N(0, \sigma^2)$, and $M_t \backsim N(M_{t-1}, \sigma_M^2)$ for $t > 1$, and with the initial condition M_1 assigned a diffuse $N(6{,}900, 10{,}000)$ prior. A gamma prior is assumed for $\varsigma = (1/\sigma^2) + (1/\sigma_M^2)$ and with $u \backsim U(0,1)$, one obtains $(1/\sigma^2) = u\varsigma$. Patwardhad and Small mention that compilations of trends in relative sea level data suggest an upward trend of 0.5–3.0 mm/year, so a $N(0,1)$ prior on b seems reasonable. An alternative model (model 2) allowing site-specific linear trends is considered, namely,

$$y_{jt} = M_t + b_j t + u_{jt},$$

with $u_{jt} \backsim N(0, \sigma^2), b_j \backsim N(\mu_b, \sigma_b^2), M_t \backsim N(M_{t-1}, \sigma_M^2)$, and $\mu_b \backsim N(0, 1)$.

For model 1, a two chain run of 25,000 iterations converges early, and the last $R = 20{,}000$ iterations give a mean linear growth rate of 1.03 with 95% interval 0.45 to 1.49. Variation around the evolving sea level, M_t, is comparatively small, with the posterior median of σ_M^2 standing at 4, compared to a median 2010 for σ^2. Figure 4.2 plots the smoothed series, M_t, through to 1980, with forecasts thereafter. Let $y_{\text{rep},jt}$ be replicate data from the model. Then a posterior predictive loss (PPL) criterion is calculated (within the observed data period to 1980) as,

$$\text{PPL} = \sum_{t=1}^{80}\sum_{j=1}^{5} V(y_{\text{rep},jt}) + \frac{1}{R}\frac{k}{k+1}\sum_{t=1}^{80}\sum_{j=1}^{5}\sum_{r=1}^{R}(y_{\text{rep},jt}^{(r)} - y_{jt})^2,$$

and obtained (with $k = 1$) as 1830 in units of 1000. For model 2, a two chain analysis with the same details gives a mean (95% interval) for μ_b of 0.82 (0.13, 1.47). Variation around the evolving sea level, M_t, is raised, with the posterior median of σ_M^2 now 7.2, and the median σ^2 reduced to 1378. The PPL is much reduced, namely, 1273.

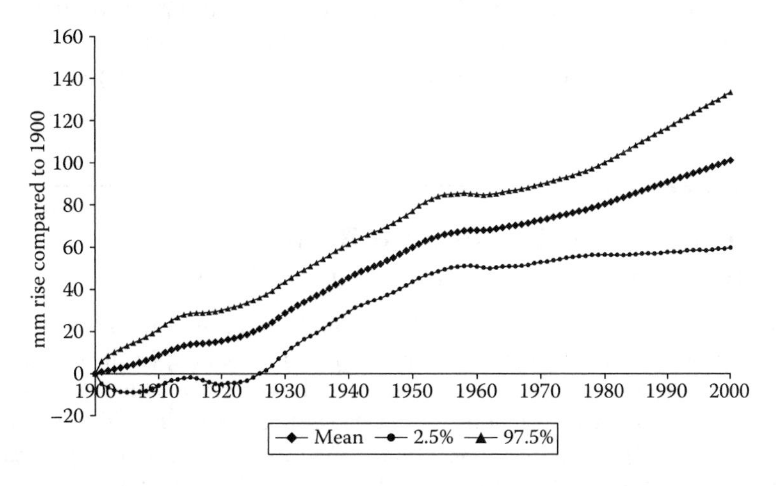

FIGURE 4.2
Change in global sea level (in mm relative to 1900).

Example 4.3. Luteinizing Hormone: State–Space Version of an ARMA Model Koopman et al. (1999) consider state–space versions of the $ARMA(p, q)$ model involving latent data, z_t, whereby for (for $p > q$),

$$y_t = z_t + u_t,$$
$$z_t = \phi_0 + \phi_1 z_{t-1} + \cdots + \phi_p z_{t-p} + (\gamma_1 + \phi_1)u_{t-1} + (\gamma_2 + \phi_2)u_{t-2} + \cdots + (\gamma_q + \phi_q)u_{t-q} + \phi_{q+1}u_{t-q-1} + \cdots + \phi_p u_{t-p},$$

with an analogous expression for $p \leq q$. For the $ARMA(1, 1)$ model, one obtains,

$$y_t = z_t + u_t,$$
$$z_{t+1} = \phi_0 + \phi_1 z_t + (\gamma_1 + \phi_1)u_t,$$

that is $z_{t+1} \backsim N(\phi_0+\phi_1 z_t, (\gamma_1+\phi_1)^2\sigma^2)$. Different forms for stating the state–space equivalent forms of an ARMA model for y_t are considered by authors such as De Jong and Penzer (2004) and Pearlman (1980).

Venables and Ripley (1994) analyze a series of 48 centered readings, y_t, of luteinizing hormone readings in a human female. They especially consider $AR(1)$, $ARMA(1, 1)$, and $AR(3)$ models. Here, a stationary $ARMA$(1,1) model for centered data, namely,

$$y_t = \phi y_{t-1} + u_t + \gamma u_{t-1},$$

is estimated using the latent data approach, with,

$$y_t \backsim N(z_t, \sigma^2),$$
$$z_t \backsim N(\phi z_{t-1}, (\gamma + \phi)^2\sigma^2).$$

A two chain run[2] of 10,000 iterations (1000 burn in) gives posterior means (and 95% intervals) for ϕ and γ of 0.68 (0.44, 0.91) and 0.72 (0.16, 0.99). The residual variance σ^2 is estimated at 0.07, lower than reported by Venables and Ripley. One step ahead forecasts are demonstrated in Figure 4.3.

4.4 Time Series for Discrete Responses: State–Space Priors and Alternatives

Dynamic generalized linear models (DGLMs) extend the Gaussian state–space representation to outcomes with density $p(y_t|\zeta_t)$ belonging to the exponential family of distributions, where ζ_t is the natural parameter. One may also condition on the history of previous observations plus previous and current predictors, $D_{t-1} = (y_{t-1}, y_{t-2}, \ldots, y_1, X_t, \ldots, X_1)$ to allow for observation-driven components in the model (Fahrmeier and Tutz, 2001, 242). So,

$$p(y_t|\zeta_t, D_{t-1}) = \exp[\phi_t\{y_t\zeta_t - b(\zeta_t)\}]c(y_t, \phi_t),$$

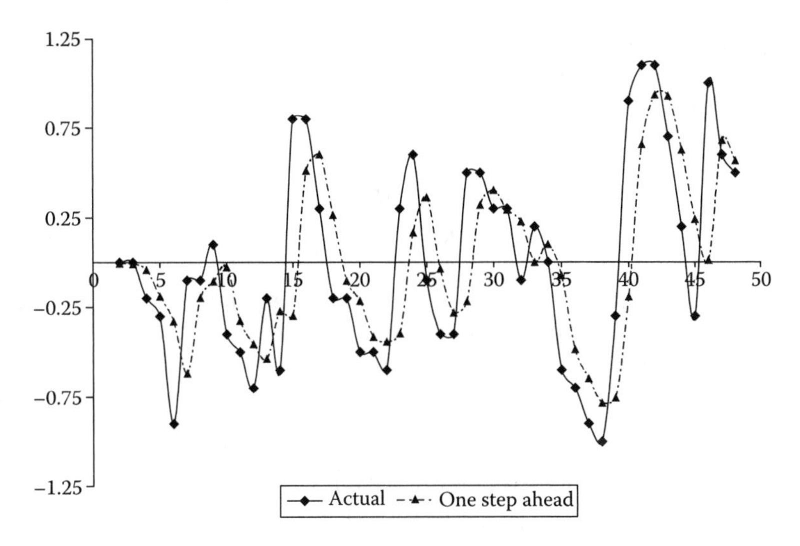

FIGURE 4.3
Luteinizing hormone.

with $\mu_t = E(y_t|\zeta_t) = b'(\zeta_t)$, and μ_t linked to a linear predictor, η_t, via a link function, g, namely, $g(\mu_t) = \eta_t$. Also, a known scale parameter, ϕ_t, defines the conditional variance, $\text{Var}(y_t|\zeta_t) = b''(\zeta_t)/\phi_t$. Then an observation equation for design matrix, X_t, of dimension p would typically be of the form,

$$g(\mu_t) = \eta_t = \beta_t X_t + u_t,$$

with state or system equation,

$$\beta_t = \beta_{t-1} G_t + w_t,$$

where $w_t \sim N_p(0, W)$, and $u_t \sim N(0, V)$ is not necessarily included for discrete responses, but may be necessary to represent unstructured (or structured) extra-variation.

An alternative state–space approach, sometimes termed a linear Bayes approach, involves conjugate priors for the natural parameters and a guide relationship,

$$h(\zeta_t) = \beta_t X_t,$$

linking the natural parameters to the state vector (West et al., 1985, 74; Ferreira and Gamerman, 2000, 60). So with time-specific parameters (g_t, h_t), the prior for the natural parameter at time t is

$$p(\zeta_t|D_{t-1}, g_t, h_t) = k(g_t, h_t) \exp[g_t\zeta_t - h_t b(\zeta_t)],$$

while the updated natural parameters have density,

$$p(\zeta_t|D_t, g_t, h_t) = k(g_t, h_t) \exp[(g_t + \phi_t y_t)\zeta_t - (h_t + \phi_t) b(\zeta_t)].$$

As for normal linear state–space models, the state vector may include level, trend, and seasonal effects. For an underlying signal model (X_t containing only an intercept), the observation and state equations become (Kitagawa and Gersch, 1996, Chapter 13),

$$g(\mu_t) = \beta_t + u_t,$$
$$\Delta^k \beta_t = w_t,$$

with $u_t \sim N(0,V), w_t \sim N(0,W)$. Thus, Kashiwagi and Yanagimoto (1992) consider Poisson data on disease counts, $y_t \sim Po(\mu_t)$, and take $k = 1$ in the signal equation. For binary data with $\pi_t = Pr(y_t = 1)$, a signal may be combined with Markov dependence on lagged responses (Cox, 1970). For example, a time-varying level and lag 1 effect could specify,

$$g(\pi_t) = \eta_t = \beta_{1t} + \beta_{2t} y_{t-1},$$
$$(\beta_{1t}, \beta_{2t}) \sim N_2([\beta_{1,t-1}, \beta_{2,t-1}], W).$$

Time series of categorical data vectors, namely, $y_t = (y_{t1}, y_{t2}, \ldots, y_{tJ})$ with only a single $y_{tj} = 1$ if (say) diagnosis j applies, or mutually exclusive choices j made at time t, are multinomial according to,

$$y_t = (y_{t1}, y_{t2}, \ldots, y_{tJ}) \backsim \text{Mult}(1, [p_{t1}, p_{t2}, \ldots, p_{tJ}]).$$

Typically, a multiple logit link is assumed for the unknown probabilities p_{tj} (Cargnoni et al., 1997; Fahrmeier and Tutz, 2001). A signal model would then involve a $(J-1)$ dimensional state vector, though by analogy to binary Markov dependence, the regression term, η_{tj}, for the jth choice may also involve lags on both the same response, $y_{t-k,j}$, and lagged cross-responses, $y_{t-k,m}$ $(m \neq j)$. For a general predictor, possibly varying by category, X_{tj}, one has,

$$p_{tj} = \exp(\beta_{tj} X_{tj}) \Bigg/ \sum_{j=1}^{J} \exp(\beta_{tj} X_{tj}),$$

where $\beta_{tJ} = 0$ for identifiability. Cross-series borrowing of strength via random walk priors may be applied for the $J-1$ category-specific state vectors, β_{tj}. Thus for the coefficient on predictor k, X_{tjk}, one might have,

$$\beta_{tk} \backsim N_{J-1}(\beta_{t-1,k}, \Sigma_k),$$

where $\beta_{tk} = (\beta_{t1k}, \ldots, \beta_{t,J-1,k})$, and Σ_k is of dimension $J-1$.

An alternative with binary and multinomial responses is to introduce the augmented metric data, y_t^*, which underlie the observed discrete responses. Thus, for binary data, consider the scheme,

$$y_t^* = \beta_t X_t + u_t,$$

where y_t^* is positive or negative according as $y_t = 1$ or $y_t = 0$, and the variance of u_t is assumed known for identifiability, usually with $\text{Var}(u_t) = 1$. A simple signal model with $X_t = 1$ may then be expressed (Carlin and Polson, 1992, 583) as,

$$\begin{aligned}
&y_t^* | W, y_t, \beta_t \propto N(\beta_t, 1)\ I(0, \infty) && \text{if} \quad y_t = 1, \\
&y_t^* | W, y_t, \beta_t \propto N(\beta_t, 1)\ I(-\infty, 0) && \text{if} \quad y_t = 0, \\
&\beta_t \sim N(\beta_{t-1}, W).
\end{aligned}$$

4.4.1 Other approaches

Another general scheme for modeling time series of exponential family data is the generalized autoregressive moving average (GARMA) representation (Benjamin et al., 2000; Li, 1994). This implies models with conditional means, μ_t, link function, $g(\mu_t) = \eta_t$, and regression term in the form,

$$\eta_t = \gamma_t X_t + \sum_{j=1}^{p} \phi_j [g(y_{t-j}) - \gamma_{t-j} X_{t-j}] + \sum_{k=1}^{q} \gamma_k [g(y_{t-k}) - \eta_{t-k}].$$

For example, for Poisson data, $y_t \sim Po(\mu_t)$, and $y_t^* = \max(y_t, m)$ for a small positive constant m, one has,

$$\log(\mu_t) = \gamma_t X_t + \sum_{j=1}^{p} \phi_j [\log(y_{t-j}^*) - \gamma_{t-j} X_{t-j}] + \sum_{k=1}^{q} \gamma_k [\log(y_{t-k}^* / \mu_{t-k})].$$

More general autoregression in the state vector (not limited to random walks) may be adopted. Thus, Chan and Ledolter (1995) and Oh and Lim (2001) adopt an autocorrelated error, θ_t, for count data with $y_t \sim Po(e^{\eta_t})$, such that,

$$\begin{aligned}
\eta_t &= \theta_t + X_t \gamma, \\
\theta_t &= \rho \theta_{t-1} + w_t,
\end{aligned}$$

with ρ constrained to stationarity. Dependence on lagged counts can also be achieved by binomial thinning, whereby,

$$\eta_t = \gamma_t X_t + \rho \circ y_{t-1},$$

is equivalent to $\eta_t = \gamma_t X_t + h_t$, $h_t \sim Bin(y_{t-1}, \rho)$.

Conjugate mixture schemes for time series counts are exemplified by Bockenholt (1999), and Jowaheer and Sutradhar (2002), with for instance,

$$\begin{aligned}
y_t &\sim Po(e^{\eta_t} \gamma_t), \\
\gamma_t &\sim Ga\left(\frac{1}{c}, \frac{1}{c}\right),
\end{aligned}$$

where η_t may be specified in state–space form, $\eta_t = \beta_t X_t, \beta_t = G\beta_{t-1} + \omega_t$. Marginally, $\text{Var}(y_t) = \exp(\eta_t) + c\exp(2\eta_t)$.

TABLE 4.1
Treated schizophrenia by age and sex.

	Cases		Population	
Age	**M**	**F**	**M**	**F**
0–15	0	0	123,512	117,775
16–24	75	25	62,808	59,439
25–34	235	93	90,722	89,149
35–44	299	186	90,625	87,755
45–54	298	238	84,956	82,847
55–64	171	217	63,229	62,147
65–74	129	207	47,180	53,337
75–84	58	119	25,762	39,653
85+	6	40	6,037	15,886
All	1271	1125	59,4831	607,988

Example 4.4. Treated Schizophrenia To illustrate dynamic general linear model (DGLM) approaches, consider data on treated schizophrenia prevalence totals, y_{ag}, by gender g and age a ($a = 1, \ldots, A$ for $A = 9$ age groups, 0–15, 16–24, 25–34, ..., 65–74, 85+) from the General Practice Research Database for England and Wales. This database covers around 1 in 40 of the population of England and Wales and includes totals of patients being treated for major chronic disease. The counts are zero for under 16s since schizophrenia is very rare at these ages (Table 4.1).

Under a state–space approach, first order random walks are assumed to pool strength in prevalence estimates over adjacent ages, a. So with a Poisson observation equation, one has,

$$y_{ag} \backsim Po(r_{ag}P_{ag}/1000),$$

where P_{ag} denote populations at risk by age and sex, and division by 1000 means that the r_{ag} will be rates per 1000 population. Then,

$$\log(r_{ag}) = \delta_{ag},$$

with gender-specific $RW(1)$ priors for δ_{ag},

$$\delta_{ag} \sim N(\delta_{a-1,g}, 1/\tau_g),$$

to model anticipated correlation in prevalence for neighboring age bands, and uniform priors on the standard deviations $1/\tau_g^{0.5}$. A first order antedependence (AD1) model (model 2) is also applied with,

$$\log(r_{ag}) = \lambda_{ag},$$

where,

$$\begin{aligned} \lambda_{1g} &= \omega_{1g}, \\ \lambda_{ag} &= \phi_g \lambda_{a-1,g} + \omega_{ag} \qquad\qquad a > 1, \end{aligned}$$

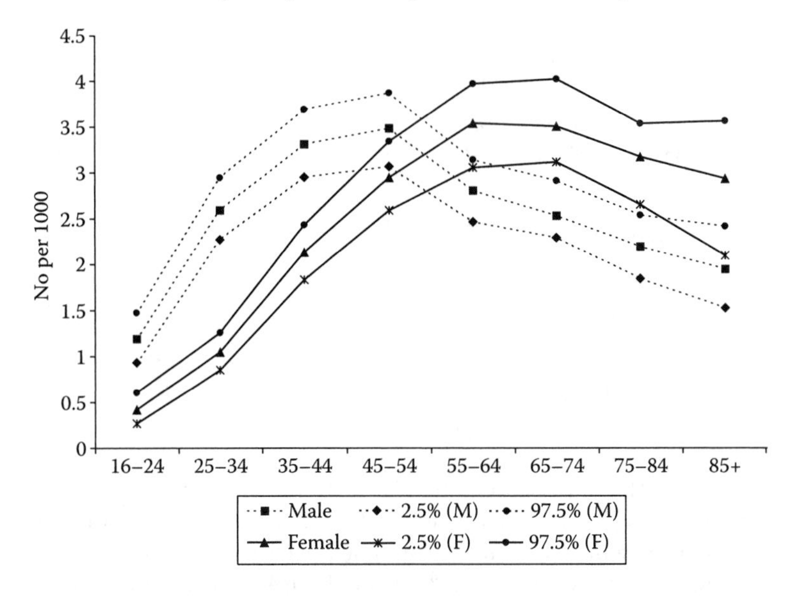

FIGURE 4.4
Fitted prevalence curves by age, antedependence model.

where $\omega_{ag} \backsim N(0, 1/\chi_a)$ and $\phi_g \backsim N(0, 1)$. A quadratic function in age is assumed for the precisions, χ_a.

Two chain runs of 100,000 iterations are applied[3] in OpenBUGS with inferences based on the second halves of the sampling. Both models provide a satisfactory predictive check (95% or over of observed data within 95% credible intervals of replicate data) and both models have a DIC around 144.6, though the AD1 model has fewer effective parameters (14.7 vs. 18.4). The ϕ coefficients in the latter have means (and 95% interval) of 0.82 (0.46, 0.93) and 0.94 (0.77, 1.02). Figure 4.4 plots the model 2 estimates of the male and female series, r_{ag}.

Example 4.5. Warwick Telephone Calls This example applies a dynamic GLM to half-hourly counts, y_t, of incoming telephone calls at the University of Warwick over a week (so $T = 7 \times 48 = 336$) (West et al., 1985). Assuming Poisson distributed counts, $y_t \sim Po(\mu_t)$, a log link model assumes evolving level and cyclical terms, the latter modeled using $K = 5$ harmonics of a Fourier representation with period $S = 48$. Then with,

$$\log(\mu_t) = \beta_t X_t,$$
$$\beta_t = \beta_{t-1} G_t + w_t,$$

one possible parameterization takes β_t of length 11, $X_t = (1, 1, 0, 1, 0, 1, 0, 1, 0, 1, 0)'$, $G_t = (H_0, H_1, \ldots, H_5)'$, and $H_0 = 1$, with,

$$H_k = \begin{bmatrix} \cos(\eta_k) & \sin(\eta_k) \\ -\sin(\eta_k) & \cos(\eta_k) \end{bmatrix},$$

for $k = 1, \ldots, K$, and $\eta_k = (2\pi k/S)$.

Here the parameterization adopted is

$$\log(\mu_t) = \lambda_t + \sum_{k=1}^{K} c_{kt},$$

$$\lambda_t = \lambda_{t-1} + w_{1t},$$

$$c_{kt} = c_{k,t-1}\cos(\eta_k) + d_{k,t-1}\sin(\eta_k) + w_{2t},$$

$$d_{kt} = -c_{k,t-1}\sin(\eta_k) + d_{k,t-1}\cos(\eta_k) + w_{3t},$$

$$w_{jt} \sim N(0, W_j).$$

To partition the variance among the components the square root of the total variance, $Z = \sum_j W_j$, is assigned a $U(0, 10)$ prior and the variances W_j obtained as $W_j = Zq_j$, where q_j are assigned a Dirichlet prior.

Convergence in a two chain run[4] with default starting values ($Z^{0.5} = 1$ in one chain, $Z^{0.5} = 0.25$ in the other, with the c, d, and λ parameters all set to 0) is obtained after 35,000 iterations, and a further 15,000 provide posterior medians for $\{W_1, W_2, W_3\}$ of 0.057, 6.1E-4, and 1.6E-4. The plot of one step ahead forecasts (Figure 4.5) matches the observations reasonably well: 80% of the observations (269/335) are included in the 95% credible intervals of the one step ahead predictions.

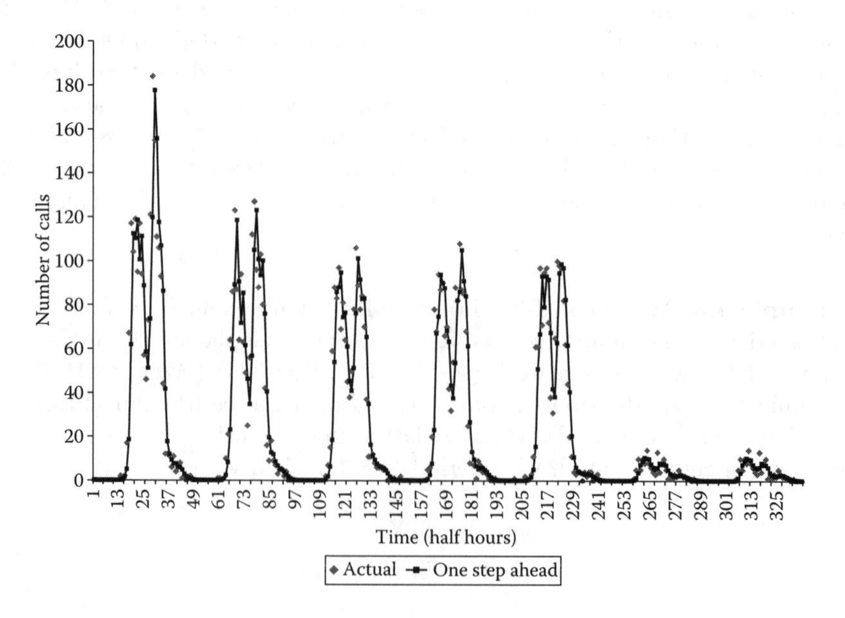

FIGURE 4.5
One step ahead predictions, telephone call data.

4.5 Stochastic Variances

Many state–space applications assume constant variances in the observation and state equations, but there is often nonstationarity in such variances (Broto and Ruiz, 2004). Certain types of data such as exchange rate and share price series, r_t, are particularly likely to demonstrate volatility clustering (Granger and Machina, 2006), with fluctuating variances, $\text{Var}(r_t)$. There are periods when volatility is relatively high and periods when volatility is relatively low, often with relatively smooth transition between high and low volatility regimes. In many applications the series is transformed to have an effectively zero mean (Meyer and Yu, 2000, 200); for example, the ratio of successive exchange rates, r_t/r_{t-1}, has approximate average 1, so that a response obtained as $y_t = \log(r_t/r_{t-1})$ can be taken to average zero. Hence, one may write a model without intercept (or predictor effects) as,

$$y_t = V_t^{0.5} u_t,$$

where $u_t \sim N(0,1)$, but the variances V_t are unknowns.

Stochastic volatility models may apply state–space techniques to model changing variances. A widely used template model in this class involves a stationary state equation (Harvey et al., 1994; Jacquier et al., 2004; Kim et al., 1998; Meyer and Yu, 2000). With $\theta_t = \log(V_t)$, one has,

$$\begin{aligned} y_t &= \sqrt{V_t} u_t = \exp\left(\frac{\theta_t}{2}\right) u_t, \qquad (4.2)\\ \theta_t &= \mu + \phi(\theta_{t-1} - \mu) + W^{0.5} w_t \qquad t > 1,\\ \theta_1 &\sim N\left(\mu, \frac{W}{1-\phi^2}\right),\\ \begin{pmatrix} u_t \\ w_t \end{pmatrix} &\sim N\left(\begin{pmatrix} 0 \\ 0 \end{pmatrix}, \begin{pmatrix} 1 & 0 \\ 0 & 1 \end{pmatrix}\right), \end{aligned}$$

where $|\phi| < 1$ measures persistence in the volatility, but the u_t and w_t series are uncorrelated. This scheme can be generalized to multivariate responses subject to volatility, such as a set of exchange rates—see Chapter 7, and Yu and Meyer (2006).

As a heavy-tailed alternative, one may consider a Student t for the y series as a scale mixture of normals (Jacquier et al., 2004). With ν degrees of freedom, one has,

$$\begin{aligned} y_t &= \sqrt{\lambda_t}\sqrt{V_t} u_t = \sqrt{\lambda_t} \exp\left(\frac{\theta_t}{2}\right) u_t,\\ 1/\lambda_t &\sim Ga\left(\frac{\nu}{2}, \frac{\nu}{2}\right), \end{aligned}$$

and other aspects as above. Although a uniform prior on ν is possible, for a recent alternative prior (applicable to other types of Student t regression) see

Fonseca et al. (2008). This model deals with isolated y outliers by introducing a large λ_t, and it requires a sequence of large $|y_t|$ before V_t is increased (Jacquier et al., 2004, 190).

By contrast, generalized autoregressive conditional heteroscedastic (GARCH) models involve autoregression in y_t^2 and/or V_t. A GARCH(p, q) model specifies,

$$V_t = \gamma + \sum_{j=1}^{p} \alpha_j y_{t-j}^2 + \sum_{j=1}^{q} \beta_j V_{t-j},$$

where coefficients $\{\gamma, \alpha_j, \beta_j\}$ are constrained to be positive, and setting $q = 0$ leads to the ARCH(p) model (Engle, 1982). Stationarity requires,

$$\sum_{j=1}^{p} \alpha_j + \sum_{j=1}^{q} \beta_j < 1,$$

though is not necessarily imposed a priori. Whichever approach is used, departures from normality are frequently relevant, such that $y_t/\sqrt{V_t}$ is non-Gaussian. Among heavy-tailed alternatives, one may consider a Student t, either $u_t \sim t(0, 1, \nu)$, or as a scale mixture of normals (Bauwens and Lubrano, 1998; Chib et al., 2002).

In case y has a nonzero mean or there are predictors, one may widen the model for y. For example, a model with a zero mean y and lag 1 effect in y would be

$$y_t = \rho y_{t-1} + \sqrt{V_t} u_t.$$

One variant is the doubly autoregressive model (Ling, 2004),

$$y_t = \rho y_{t-1} + u_t \sqrt{\gamma + \alpha y_{t-1}^2},$$

which can be shown equivalent to the random coefficient AR model,

$$y_t = (\rho + a_t) y_{t-1} + c_t,$$

where (a_t, c_t) are bivariate normal with mean 0 and covariance matrix, Diag(α, γ).

A generalization of the state–space approach is to introduce correlation between the u_t and w_t terms, and so reflect leverage effects. Positive and negative shocks then have different impacts on future volatility (Asai et al., 2006; Jacquier et al., 2004; Meyer and Yu, 2000; Chen and So, 2006). So one possible scheme has

$$y_t = \sqrt{V_t} u_t = \exp\left(\frac{\theta_t}{2}\right) u_t,$$

$$\theta_t = \mu + \phi(\theta_{t-1} - \mu) + \sqrt{W} w_t,$$

$$\begin{pmatrix} u_t \\ w_t \end{pmatrix} \sim N\left(\begin{pmatrix} 0 \\ 0 \end{pmatrix}, \begin{pmatrix} 1 & \varphi \\ \varphi & 1 \end{pmatrix}\right),$$

where φ is a correlation. A heavy-tailed version of the leverage model (Jacquier et al., 2004) may be obtained with,

$$y_t = \sqrt{\lambda_t}\exp\left(\frac{\theta_t}{2}\right)u_t,$$
$$1/\lambda_t \sim Ga\left(\frac{\nu}{2},\frac{\nu}{2}\right).$$

A GARCH model including leverage is obtained by setting $z_t = \sqrt{V_t}u_t$ in $y_t = \mu_y + \rho(y_{t-1} - \mu_y) + \sqrt{V_t}u_t$. Leverage is then obtained under the following asymmetric model (Glosten et al., 1993),

$$V_t = \gamma + \alpha_1 z_{t-1}^2 + \alpha_2 z_{t-1}^2 I(z_{t-1} > 0) + \beta V_{t-1}.$$

Under the model (Equation 4.2), assume priors $\mu \sim N(0, \sigma_\mu^2)$, $(\phi + 1)/2 \sim Be(r_\phi, s_\phi)$, and $W \sim IG(\kappa_W, \lambda_W)$, where $\{\sigma_\mu^2, r_\phi, s_\phi, \kappa_W, \lambda_W\}$ are known. Then, with $\psi = (\mu, \phi, W)$, the posterior is

$$p(\psi|y) \propto \left[\prod_{t=1}^{T}\exp\{-\theta_t/2\}\exp\left\{\frac{-y_t^2}{2e^{\theta_t}}\right\}\right]$$
$$\times\left[\left(\frac{1-\phi^2}{W}\right)^{0.5}\exp\left\{\frac{1-\phi^2}{2W}(\theta_1-\mu)^2\right\}\right]$$
$$\times\left[\prod_{t=2}^{T}\left(\frac{1}{W}\right)^{0.5}\exp\left\{-\frac{1}{2W}(\theta_t-\mu-\phi(\theta_{t-1}-\mu))^2\right\}\right]p(\mu)p(\phi)p(W),$$

and Gibbs sampling from full conditionals is obtained (Kim et al., 1998). The griddy-Gibbs technique may also enable Gibbs sampling of all parameters in a $GARCH(1,1)$ model, with normal or Student distributed u_t (Bauwens and Lubrano, 1998). Chib et al. (2002) consider more general Metropolis–Hastings techniques, including particle filtering, to sample from models with discontinuities in the observations.

Example 4.6. US 90-Day Treasury Bill Rate Consider 350 observations, r_t, of the US 90-day bill rate from July 1972 to August 2001. The data are obtained as $y_t = \log(r_t/r_{t-1})$—see Figure 4.6 for a plot of $\log(r_t)$, which shows several spells of high volatility. As one of several ways to represent the data, the double autoregressive model (Ling, 2004), namely,

$$y_t = \rho y_{t-1} + u_t\sqrt{\gamma + \alpha y_{t-1}^2},$$

is applied[5], with the constraint $\rho^2 + \alpha < 1$ sufficient to ensure $E(y_t^2 < \infty)$. The analysis conditions on the first observation. A two chain run of 5000 iterations with 500 burn-in gives posterior means (sd) for ρ and α of 0.43

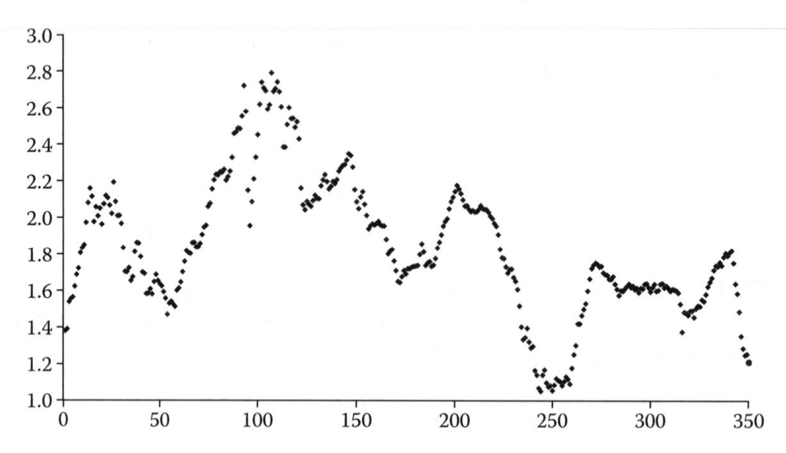

FIGURE 4.6
Treasury bill rate.

(0.07) and 0.50 (0.10), with γ estimated as 0.0018. Despite its low parameter count, sampling replicate data from this model provides 93.4% concordance with the actual observations; i.e., 93.4% of the data points are contained with 95% credible intervals of replicate data sampled from the model. A PPL criterion defined for iterations $r = 1, \ldots, R$ after the burn-in is obtained as

$$\mathrm{PPL} = \sum_{t=2}^{350} \mathrm{Var}(y_{\mathrm{rep},t}) + \frac{1}{4500} \frac{k}{k+1} \sum_{t=2}^{349} \sum_{r=1}^{R} (y_{\mathrm{rep},t}^{(r)} - y_t)^2,$$

and for $k = 1$ and $k = 100$ has values 2.59 and 3.83.

A second approach is based on a stationary state–space stochastic volatility model with,

$$y_t = \sqrt{V_t} u_t = \exp\left(\frac{\theta_t}{2}\right) u_t,$$
$$\theta_t = \mu + \phi(\theta_{t-1} - \mu) + W^{0.5} w_t \qquad t > 1,$$

where a Beta prior is adopted on ϕ^*, with ϕ obtained as $2\phi^* - 1$ (Meyer and Yu, 2000), and $1/W \sim Ga(2.5, 0.025)$. Starting values are important for early convergence of this model and are based on exploratory analysis showing $\mu \simeq -6$. With inferences from the second half of a two chain run of 100,000 iterations, this model produces PPL values of 2.71 and 4.04 at $k = 1$ and $k = 100$. Figure 4.7 plots the evolving variance, $V_t = \exp(\theta_t)$, under this model.

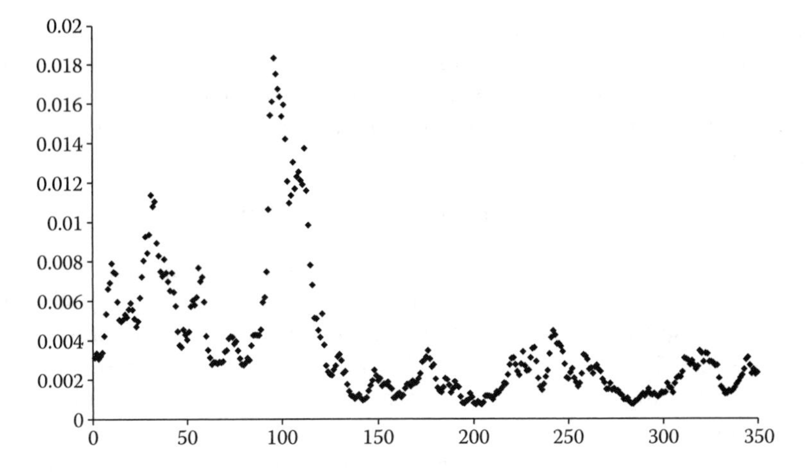

FIGURE 4.7
Plot of variances under state–space volatility.

4.6 Modeling Discontinuities in Time

Aberrant observations or shifts in a series can bias parameter estimates and other inferences in time series models (Chen and Liu, 1993; Hamilton, 2009; Tsay, 1986), and a variety of methods exist for modeling shifts or outliers in the observation, state, or error series. These extend to shifts in variance parameters also.

Robust versions of the priors for the component errors u_t and/or w_{jt} in dynamic models may be applied to allow flexibility in response to disparate observations. For example, a heavy-tailed alternative to Gaussian errors (Martin and Raftery, 1987) may be invoked by scale mixing at both levels in the local level model,

$$y_t = \beta_t + u_t, \beta_t = \beta_{t-1} + w_t,$$

with,

$$u_t \backsim N(0, V/\lambda_{1t}), w_t \backsim N(0, W/\lambda_{2t}),$$
$$\lambda_{1t} \backsim Ga\left(\frac{\nu_u}{2}, \frac{\nu_u}{2}\right), \lambda_{2t} \backsim Ga\left(\frac{\nu_w}{2}, \frac{\nu_w}{2}\right).$$

This generalization is adapted to detecting or accommodating additive outliers (outliers in the observation equation errors) and innovation outliers in the state equation errors. Geweke (1993) points out problems with adopting diffuse priors for ν, and possibilities include an exponential density such as $\nu \sim E(0.1)$ (Fernandez and Steel, 1998).

Many outlier mechanisms involve discrete mixing around default normal error assumptions, as in a contaminated normal density (Verdinelli and Wasserman, 1991). Thus, let π be a given prior probability of an outlier (e.g., $\pi = 0.05$). Then the observation error in a state–space model can be modified to allow innovation outliers,

$$u_t \backsim (1-\pi)N(0, W_1) + \pi N(0, W_2),$$

where $W_2 = KW_1$ with K large. A comprehensive generalization of the normal errors dynamic linear model is provided by taking y_t and β_t to follow the univariate or multivariate exponential power distribution (Gomez et al., 2002).

More specialized binary switching in observation error or state error processes may be applied (Diggle and Zeger, 1989), for example, adapted to positive pulses (e.g., periods with abnormally heavy rainfall). To illustrate switching in observation errors to accommodate positive pulses, consider the $AR(1)$ observation model,

$$y_t = \phi y_{t-1} + u_t,$$

such that usually $u_t = u_{1t}$, but exceptionally $u_t = u_{2t}$, where the latter error is necessarily positive, namely,

$$u_{1t} \sim N(0, \sigma^2),$$
$$u_{2t} \sim Ga(g_1, g_2),$$

where $\{g_1, g_2\}$ are preset. Define latent allocation indicators, $S_t \in (1, 2)$, as in Chapter 3. Then $u_t = u_{2t}$ with probabilities $\pi_t = Pr(S_t = 2)$, which might be defined by a separate model such as,

$$\text{logit}(\pi_t) = \eta_0 + \eta_1 y_{t-1}.$$

One may also distinguish innovation outliers from additive outliers corresponding to isolated shifts or "gross errors" in the observation series (Fox, 1972; Tsay, 1986). This involves separate binary indicators $\{S_{At}, S_{It}\}$, or a single multinomial indicator, S_t. For example, let π_A and π_I be prior probabilities of additive and innovative outliers, and consider an $AR(1)$ observation model with $AR(1)$ errors,

$$y_t = \phi_0 + \phi_1 y_{t-1} + a_t S_{At} + \varepsilon_t,$$
$$\varepsilon_t = \rho \varepsilon_{t-1} + u_t,$$

where $S_{At} \sim Bern(\pi_A)$, and $a_t \sim N(0, \sigma_a^2)$ represents the sizes of the additive outliers (McCulloch and Tsay, 1994). Innovation outliers are encompassed by a variance inflation mechanism with,

$$u_t \sim (1-\pi_I)N(0, V) + \pi_I N(0, KV),$$

with $K > 1$, as determined by latent indicators, $S_{It} \sim Bern(\pi_I)$.

The possibility of additive and innovative outliers coinciding at a single point may be discounted (Barnett et al., 1996; Gerlach et al., 1999). So with both additive and innovation outliers generated by variance inflation factors (respectively, K_{A} and K_{I}), one may have a single trinomial indicator, S_t, governing outlier occurrence, with $S_t = 1$ if neither type of outlier is present ($K_{\mathrm{A}} = 0, K_{\mathrm{I}} = 1$), $S_t = 2$ if an additive outlier is present ($K_{\mathrm{A}} = 10$, $K_{\mathrm{I}} = 0$), and $S_t = 3$ if an innovation outlier is present ($K_{\mathrm{A}} = 0, K_{\mathrm{I}} = 10$). Then,

$$S_t \sim \text{Mult}(1, [\pi_1, \pi_2, \pi_3]),$$

where π_2, and π_3 may be assigned preset values (e.g., $\pi_2 = \pi_3 = 0.025$), and,

$$\begin{aligned} y_t &= \phi_0 + \phi_1 y_{t-1} + a_{tS_t} + \varepsilon_t, \\ \varepsilon_t &= \rho\varepsilon_{t-1} + u_{tS_t}, \end{aligned}$$

where $a_{tS_t} \sim N(0, K_{\mathrm{A}}\sigma^2), u_{tS_t} \sim N(0, K_{\mathrm{I}}\sigma^2)$.

Enduring rather than temporary shifts in the mean or variance of a series require another approach. Models with a single or small number of enduring changes in the level of the series may be handled by extending conventional discrete mixture methods (e.g., Leonte et al., 2003; Mira and Petrone, 1996; Perreault et al., 2000). To illustrate binary switching in both levels and variances in an autoregressive error model (McCulloch and Tsay, 1993), determined by binary pairs (S_{1t}, S_{2t}), consider,

$$y_t = \mu_t + \varepsilon_t,$$

where a change in level is accomplished by letting,

$$\mu_t = \mu_{t-1} + S_{1t}\Delta_t,$$

when $S_{1t} = 1$, with $Pr(S_{1t} = 1) = \pi_1$, and Δ_t are random effects representing the shifts. The errors are $AR(p)$,

$$\varepsilon_t = \rho_1\varepsilon_{t-1} + \rho_2\varepsilon_{t-2} + \cdots + \rho_{t-p}\varepsilon_{t-p} + u_t,$$

where shifts in the variance of $u_t \sim N(0, V_t)$ occur when $S_{2t} = 1$ with $Pr(S_{2t} = 1) = \pi_2$. If there is conditioning on $(y_1, \ldots, y_p)$, then the variance sequence commences with $V_{p+1} = \sigma^2$, and subsequently,

$$\begin{aligned} V_t &= V_{t-1} &&\text{when} \quad S_{2t} = 0, \\ V_t &= \kappa_t V_{t-1} &&\text{when} \quad S_{2t} = 1, \end{aligned}$$

where κ_t are positive variables (e.g., gamma distributed) that model proportional shifts in the error variance.

Shocks in different components of the basic structural model can also be considered (De Jong and Penzer, 1998; Penzer, 2006). For example, in a three component local linear trend model, binary shock indicators (S_{1t}, S_{2t}, S_{3t}) are invoked such that,

$$\begin{aligned}
y_t &= \mu_t + S_{1t}\Delta_{1t} + u_t,\\
\mu_t &= \mu_{t-1} + S_{2t}\Delta_{2t} + \Delta_t + w_{1t},\\
\Delta_t &= \Delta_{t-1} + S_{3t}\Delta_{3t} + w_{2t},
\end{aligned}$$

where Δ_{1t} represent temporary additive shocks that occur when $S_{1t} = 1$, Δ_{2t} represent shifts in mean, and Δ_{3t} represent shifts in the slope.

Regime switching models (Geweke and Terui, 1993; Lubrano, 1995) typically involve discrete switching between two or more levels, regression regimes, or variances, though smooth transition mechanisms can also be used. The choice between regimes is governed by a binary switching function, S_t, or a continuous transition function, ϕ_t, with values between 0 and 1, such as the logit (Bauwens et al., 2000). A binary function, S_t, might be defined as 1 if time t exceeds a threshold κ and zero otherwise, as in change point models for the mean level of a series. In self-exciting threshold autoregressive (SETAR) models, the mechanism involves a lag in y; for example, $S_t = 1$ if $y_{t-1} > \kappa$. The continuous version in these two cases would be

$$\begin{aligned}
\phi_t &= \frac{\exp(\omega[t-\kappa])}{1+\exp(\omega[t-\kappa])},\\
\phi_t &= \frac{\exp(\omega[y_{t-1}-\kappa])}{1+\exp(\omega[y_{t-1}-\kappa])},
\end{aligned}$$

where ω is an extra unknown. Additionally, the lag r in the comparison $y_{t-r} > \kappa$ may be unknown (Geweke and Terui, 1993).

Example 4.7. Nile Discharges Data on Nile discharges for 1871–1970 ($T = 100$) have been analyzed by a variety of ARMA and other methods and illustrate possible identification issues associated with outlier and shift points. Balke (1993) compares an $AR(2)$ model for these data to one allowing for an intercept shift. To facilitate prior specification for latent preseries values, y_0 and y_{-1}, we center the original data, Y_t, by subtracting Y_1 from all points. So $y_t = Y_t - Y_1$.

Firstly, an $AR(2)$ model with no shift mechanism (and a heavy-tailed prior for the preseries points) is applied, namely,

$$\begin{aligned}
y_t &= \phi_0 + \phi_1 y_{t-1} + \phi_2 y_{t-2} + u_t \qquad t = 1, \ldots, T,\\
u_t &\sim N(0, \sigma^2),\\
y_t &\backsim t_2(0, \sigma^2) \qquad t = -1, 0,
\end{aligned}$$

with $N(0, 1)$ priors on $\{\phi_1, \phi_2\}$ so that nonstationarity is allowed. A two chain run[6] of 10,000 iterations (with 1000 burn-in) provides a DIC (and effective parameters d_e) of 1282.5 (4.3). The posterior means (and 95% intervals) on the AR parameters $\{\phi_1, \phi_2\}$ are obtained as 0.40 (0.21, 0.59), and 0.21 (0.02, 0.40).

Suppose, however, a shift in the series level is allowed: a series plot suggests such a shift around 1895. One may also allow for coefficient selection via binary variables, $d_j = 1$, if ϕ_j is to be retained, with prior probabilities $Pr(d_j = 1) = 0.5$. So,

$$y_t = \phi_{01} + \phi_{02}I(t > \kappa) + d_1\phi_1 y_{t-1} + d_2\phi_2 y_{t-2} + u_t,$$

where κ is taken to be uniform between 3 and T-3. Fitting this model indicates that the lag in y_{t-2} is now in doubt, with $Pr(d_2 = 1|y) = 0.09$, whereas $Pr(d_1 = 1|y) = 0.25$.

So an $AR(1)$ model with shift mechanism is applied, namely,

$$y_t = \phi_{01} + \phi_{02}I(t > \kappa) + \phi_1 y_{t-1} + u_t \qquad t = 1, \ldots, T.$$

With a two chain run of 10,000 iterations (and 1000 burn-in), the DIC is reduced to 1260.5 with $d_e = 5.5$, and κ is precisely identified with mean 28.2 (i.e., the year 1998) and 95% interval (26.1, 30.3). The lag 1 coefficient estimate is now in fact not strictly significant in the sense that the 95% interval (−0.04, 0.35) straddles zero. A suggested exercise is to fit a SETAR model,

$$y_t = \phi_{01} + \phi_{02}I(y_{t-1} > \kappa) + \phi_1 y_{t-1} + u_t \qquad t = 2, \ldots, T,$$

and assess the impact on ϕ_1.

Example 4.8. Box–Jenkins Series A This example involves the Box–Jenkins series A, and entails outlier modeling via variance inflation in the observation component of an autoregressive state–space model (cf. Gerlach et al., 1999). The observation model is

$$y_t \backsim N(\beta_0 + \theta_t, V_{S_t}),$$

where S_t is a trinomial indicator modeling the measurement error outlier mechanism. The state equation is

$$\theta_t = \phi\theta_{t-1} + w_t,$$

where $w_t \backsim N(0, W)$, and ϕ is constrained to stationarity.

As discussed above, outlier probabilities are often preset. However, if variance inflation factors are preset instead, then it is possible to take the outlier probabilities as unknowns. Thus, assume $\pi_1 = Pr(S_t = 1)$ is the unknown probability of a normal measurement error with variance V_1, while $\pi_2 = \pi_3$ are unknown probabilities of moderate and extreme outliers with variances $10V_1$ and $32V_1$, respectively. It is assumed that $V_1 \backsim Ga(1, 0.001)$, together with the parameterization,

$$\pi_1 = 1/(1 + r),$$
$$\pi_2 = \pi_3 = 0.5r/(1 + r),$$

where $r \backsim E(9)$. Additionally, the variances of the observation and state equations are linked by taking $W = qV_1$ with an $E(1)$ prior on q.

A two chain run[7] of 20,000 iterations (and 2000 burn-in) in OpenBUGS shows early convergence with estimated probability $\pi_1 = 0.93$ (and 95% interval from 0.77 to 0.99). The observation error variance, V_1, has a posterior mean of 0.031, while the state variance, W, has mean 0.036.

4.7 Spatial Smoothing and Prediction for Area Data

Whereas exchangeable hierarchical analysis is appropriate for independently generated area or point data, such data often cannot be regarded as independent because of the presence of similarities between neighboring areas or points (Anselin and Bera, 1998). Modeling area differences or point patterns with spatially structured effects reflects the empirical regularity that neighboring areas or points tend to be similar, and that similarity typically diminishes as distance increases. Even if some known predictors are available, it is likely that other relevant influences on the underlying process cannot be identified or measured, and this residual heterogeneity is likely (at least in part) to be spatially structured (Lawson, 2008, 94); for example, Gelfand et al. (2005a) consider spatial modeling of residuals in the analysis of species distributions, both for areas and points as the units, where unobserved influences might include habitat and interspecies competition. Bayesian techniques have played a central role in recent developments for analyzing spatial data, whether space is viewed from a discrete or continuous perspective, e.g., Banerjee et al. (2004) and Waller (2002).

In studies with a discrete framework, the data are typically aggregated, with observations consisting of counts (e.g., of diseased subjects in spatial epidemiology) or of regional indicators (e.g., average income per head or house prices in spatial econometrics). By contrast, in geostatistical models for geochemical readings, species distribution, or disease events in relation to a pollution source, a continuous spatial framework is more relevant (Section 4.11), allowing interpolation between observed point readings.

Consider metric responses y_i for areas i, or at sites specified by grid references $g_i = (g_{1i}, g_{2i})$. To allow greater flexibility, one may assume a "convolution" prior that compromises between structured and unstructured variation: so the model includes both a spatially structured random effect, s_i, and a fully exchangeable effect, u_i, with,

$$y_i = \alpha + u_i + s_i,$$

where $u_i \backsim N(0, \sigma_u^2)$, but s_i is spatially correlated. Alternatively, suppose y_i are counts, and that P_i are populations at risk with $y_i \backsim Bin(P_i, \pi_i)$. Then, one may specify,

$$\text{logit}(\pi_i) = \alpha + u_i + s_i,$$

where π_i are latent probabilities of the event. Alternatively, for rare events in relation to the risk population, a Poisson assumption is relevant with $y_i \backsim Po(P_i\lambda_i)$, and,

$$\log(\lambda_i) = \alpha + u_i + s_i,$$

where λ_i are latent event rates per unit of P_i. If the offsets to the Poisson mean are expected health events, E_i, such that $\sum_i y_i = \sum_i E_i$ with $y_i \backsim Po(E_i\lambda_i)$, then λ_i are interpretable as latent relative risks (Wakefield, 2007, 160).

One way to model the correlation in the elements of the vector $s = (s_1, \ldots, s_n)$ is to directly specify a joint multivariate prior with a covariance matrix that expresses spatial correlation between areas i and j or sites g_i and g_j (Richardson et al., 1992, 541; Wakefield, 2007). Typical assumptions in such models (also considered in Section 4.11) are of stationarity and isotropy, with the latter meaning the correlation is the same in all directions. For example, a multivariate normal prior would take,

$$(s_1, \ldots, s_n) \backsim N_n(0, \Sigma_s),$$

where,

$$\Sigma_s = \sigma_s^2 W = \sigma_s^2 \begin{bmatrix} 1 & w_{12} & \cdot & w_{1n} \\ w_{21} & 1 & \cdot & w_{2n} \\ \cdot & \cdot & \cdot & \cdot \\ w_{n1} & w_{n2} & \cdot & 1 \end{bmatrix},$$

and $w_{ij} = f(d_{ij})$ are correlation functions that decline as the spatial separation, d_{ij}, between areas i and j (or sites g_i and g_j) increases. For this model, one may decompose the total residual variation (metric response) or the total residual relative risk (count models with offset E_i) into spatial and nonspatial components, σ_s^2 and σ_u^2.

The correlation functions are defined to ensure that W is always non-negative definite (Mardia and Watkins, 1989). Cook and Pocock (1983) suggest,

$$w_{ij} = \exp(-\delta d_{ij}),$$

where $\delta > 0$, while Cliff and Ord (1973) propose a function combining inter-area distance, d_{ij}, and length b_{ij} of the common border between area i and j, namely,

$$w_{ij} = [d_{ij}]^{\beta_1}[b_{ij} + c]^{\beta_2},$$

where β_1 is negative and β_2 is positive. Another choice is the disc model with,

$$w_{ij} = \frac{2}{\pi}\left[\cos^{-1}\left(\frac{d_{ij}}{\kappa}\right) - \left\{\frac{d_{ij}}{\kappa}\left(1 - \frac{d_{ij}^2}{\kappa^2}\right)\right\}^{0.5}\right] \qquad d_{ij} \leq \kappa,$$

with $w_{ij} = 0$ for $d_{ij} > \kappa$, so that κ controls the decline in correlation with distance. Such choices are to some degree arbitrary, and inferences may be sensitive to the choice of spatial weights (e.g., Bhattacharjee and Jensen-Butler, 2006).

Another widely used scheme specifies the joint density via simultaneous autoregressive or SAR effects (Richardson et al., 1992). By analogy with ARMA time series models, the autoregression may operate both for (metric) responses, $y = (y_1, \ldots, y_n)'$, and for the error vector, $s = (s_1, \ldots, s_n)'$. Let $W = [w_{ij}]$ be a spatial dependence matrix as above, but with $w_{ii} = 0$ rather than $w_{ii} = 1$. One possible SAR scheme has the form,

$$y_i = \alpha + \rho_1 \sum_{h \neq i} w_{ih} y_h + s_i,$$

$$s_i = \rho_2 \sum_{h \neq i} w_{ih} s_h + u_i,$$

where ρ_1 and ρ_2 are coefficients of spatial autocorrelation, and $u = (u_1, \ldots, u_n)'$ are independently distributed, with diagonal covariance matrix, Σ_u. The covariance matrix, for $s = (s_1, \ldots, s_n)'$ is $(I - \rho_2 W)^{-1} \Sigma_u (I - \rho_2 W')^{-1}$. In matrix form,

$$y = 1\alpha + \rho_1 Wy + s,$$
$$s = \rho_2 Ws + u,$$

where 1 is a $n \times 1$ vector of ones. The simultaneous autoregressive prior parallels autoregressive models in time, such as one with $AR(1)$ dependence in both y and the errors,

$$y_t = \alpha + \rho_1 y_{t-1} + \varepsilon_t,$$
$$\varepsilon_i = \rho_2 \varepsilon_{t-1} + u_t.$$

Wall (2004) points out that SAR priors (and also CAR priors, as considered below) may generate implausible covariance patterns when considered in terms of the joint priors.

The ρ coefficients are constrained to lie between $(1/\eta_{\min})$ and $(1/\eta_{\max})$, where $\{\eta_1, \ldots, \eta_n\}$ are the eigenvalues of W, in order to ensure that $(I - \rho W)$ is invertible. This restriction can be implemented by choosing unconstrained priors for ρ_1 and ρ_2, but retaining only sampled values satisfying the criterion. If the spatial lag matrix in a SAR model is standardized to have row sums of unity, so that $w^*_{ij} = w_{ij}/\sum_h w_{ih}$, then the maximum eigenvalue of W^* is 1 and since negative spatial correlation is unlikely, one may specify uniform or beta priors on ρ coefficients in the interval $[0, 1]$.

Metropolis–Hastings updates are generally needed for the correlation parameters. Consider the spatial autoregressive model with $\rho_2 = 0$, namely,

$$y = 1\alpha + \rho Wy + u.$$

Assuming $\Sigma_u = \sigma^2 I$, the likelihood is (Lesage, 1997),

$$L\left(\alpha, \rho, \sigma^2 | y\right) \propto \frac{1}{\sigma^n} |I - \rho W| \exp\left[-\frac{1}{2\sigma^2}\sum_i u_i^2\right],$$

where $u_i = y_i - \alpha - \rho \sum_{h \neq i} w_{ih} y_h$. With normal and inverse gamma priors on α and σ^2, Gibbs updates may be used for these parameters, but Metropolis–Hastings sampling must be used for ρ, which has conditional posterior,

$$p\left(\rho | \alpha, \sigma^2, y\right) \propto |I - \rho W| \exp\left[-\frac{1}{2\sigma^2}\sum_i u_i^2\right].$$

4.8 Conditional Autoregressive Priors

In contrast to simultaneous autoregressive spatial priors, conditional autoregressive priors for $(s_1, \ldots, s_n)$ have the advantage of facilitating random effects analysis under a MCMC sampling approach. Such priors are often applied to discrete outcomes, such as health event counts, y_i, taken to be Poisson or binomial in relation to populations, P_i, or expected events, E_i (e.g., Besag et al., 1991; Norton and Niu, 2009). If there are no measured risk factors and the event is relatively infrequent (or populations at risk are small), one often seeks to estimate an underlying smooth pattern of relative risk by "borrowing strength" over areas taking account of spatial dependence (MacNab et al., 2006, 3967). Simple maximum likelihood estimates of event rates, y_i/P_i, or relative risks (e.g., by ratios $r_i = y_i/E_i$ of observed to expected events) assume a binomial or Poisson density with disease risk constant over areas and individuals within areas. In practice, individual risks vary within areas, and risks vary between areas, so that area counts are more variable than the standard density stipulates.

The extra-variation can be modeled by including random effects in a model for disease rates or risks, whether by conjugate (e.g., Poisson-gamma) or non-conjugate methods. Spatially correlated effects that vary smoothly over space may account for much extra-variation, and proxy unobserved risk factors that are also spatially correlated (Richardson and Monfort, 2000). These might be shared environmental or social capital factors in neighboring areas, or shared factors operating at a more aggregated regional scale, such as regional variation in cardiovascular mortality associated with differences in water hardness.

While spatially correlated random effects alone may be postulated, this is an informative prior assumption. For health data, there may be outlier areas (e.g., areas of deprived social renting surrounded by affluent areas), or localized policy interventions, or service variations, which affect the health event but cannot be regarded as spatially continuous. A more general and less informative approach allows adaptive downweighting of a spatial prior to allow the data to support an exchangeable unstructured prior for some areas.

4.8.1 Linking conditional and joint specifications

As discussed by Besag and Kooperberg (1995), one may use properties of the multivariate normal to obtain the conditional autoregressive prior from a joint spatial prior and vice versa. Thus consider the joint multivariate normal density for the effects, $s = (s_1, \dots, s_n)$, with mean zero and covariance Σ_s,

$$p(s) = \frac{1}{(2\pi)^{n/2}} |\Sigma_s|^{-0.5} \exp\left(-0.5 s' \Sigma_s^{-1} s\right).$$

Setting $Q = [q_{ij}] = \Sigma_s^{-1}$ to be the precision matrix and $s_{[i]} = (s_1, \dots, s_{i-1}, s_{i+1}, \dots, s_n)$, the conditional distributions for each s_i take a univariate normal form, corresponding to the pairwise interaction function, $\Phi(u) = u^2/2$ (Rue and Held, 2005, 22), namely,

$$s_i | s_{[i]} \backsim N\left(\sum_{j \neq i} \left[-\frac{q_{ij}}{q_{ii}}\right] s_j, \frac{1}{q_{ii}}\right),$$

with $\text{corr}(s_i, s_j | s_{[i,j]}) = (-q_{ij}/\sqrt{q_{ii} q_{jj}})$. Following Besag and Kooperberg (1995, 734), define $h_{ii} = 0$, and set,

$$h_{ij} = -q_{ij}/q_{ii} \qquad (i \neq j).$$

Also set $q_{ii} = a_i/\delta$ with variance parameter δ, so that,

$$h_{ij} = -q_{ij}\delta/a_i. \tag{4.3}$$

The above conditional density is then in the conditional autoregressive form specified by Besag (1974),

$$s_i | s_{[i]} \backsim N\left(\sum_{j \neq i} h_{ij} s_j, \delta/a_i\right). \tag{4.4}$$

To obtain the joint density from the conditional one, symmetry of Q means $-Q_{ij} = -Q_{ji}$, so that from Equation 4.3 the constraint,

$$h_{ij} a_i = h_{ji} a_j,$$

applies. Note that expressing $\delta/a_i = \tau_i^2$ or $a_i = \delta/\tau_i^2$, this constraint can also be stated (Cressie and Kapat, 2008) as,

$$h_{ij} \tau_j^2 = h_{ji} \tau_i^2.$$

Letting $R = A(I - H)$, where $A = \text{diag}(a_1, \dots, a_n)$, one has that R is symmetric with diagonal elements a_i and off-diagonal elements $-a_i h_{ij}$. So the joint density (Banerjee et al., 2004, 8; Besag and Green, 1993) implied by the conditional priors is

$$(s_1, \dots, s_n) \backsim N_n(0, \delta R^{-1}),$$

where $Q = \delta^{-1}R$. If R is positive definite as well as symmetric, the joint density of the spatial effects is proper. Positive definiteness of R holds under diagonal dominance (Besag and Kooperberg, 1995, 734; Rue and Held, 2005, 20), namely, that in at least one row (or column) of R, the diagonal element, r_{ii}, exceeds the absolute sum of the off-diagonal elements, $\left|\sum_{j \neq i} r_{ij}\right|$.

4.8.2 Alternative conditional priors

Various schemes for defining the h_{ij} and a_i in Equation 4.4 are possible, including options where R is not positive definite. Setting,

$$h_{ij} = \rho \frac{w_{ij}}{\sum_{k \neq i} w_{ik}}, \; a_i = \sum_{k \neq i} w_{ik},$$

where $0 \leq \rho \leq 1$, and taking $w_{ij} = w_{ji}$, with $w_{ii} = 0$, ensures the symmetry constraint is met, with $h_{ij}a_i = \rho w_{ij} = h_{ji}a_j$. The most commonly applied approach is to set $w_{ij} = 1$ for adjacent areas and $w_{ij} = 0$ otherwise, and let $a_i = d_i = \sum_{k \neq i} w_{ik}$, where d_i is then the number of areas adjacent to area i. For example, when a region is partitioned into grid cells, then each grid cell has eight (first order) neighbors (Gelfand et al., 2005a). However, distance or common boundary length-based forms for w_{ij} can be used.

In this case, $R = A(I - H)$ has diagonal elements a_i and off-diagonal elements $-\rho w_{ij}$. This provides the intrinsic conditional autoregression or $ICAR(\rho)$ prior, with,

$$s_i | s_{[i]} \backsim N\left(\rho \bar{A}_i, \frac{\delta}{d_i}\right),$$

where $\bar{A}_i$ is the average of the s_j in locality L_i of area i, i.e.,

$$\bar{A}_i = \frac{\sum_{j \in L_i} s_j}{d_i}.$$

Note that $R = A(I - H) = D - \rho W$ is positive definite, and the joint prior on $(s_1, \ldots, s_n)$ is proper, only when $|\rho| < 1$. Lower values of ρ imply lesser degrees of spatial dependence between s_i, though the limiting case when $\rho = 0$ has the disadvantage that the variance is not constant but depends on the number of neighbors, d_i.

Alternatively, in a $CAR(\rho)$ spatial prior, as distinct from the $ICAR(\rho)$ prior, one may set,

$$h_{ij} = \rho w_{ij}, \; a_i = 1,$$

so that

$$s_i | s_{[i]} \backsim N\left(\rho \sum_{j \neq i} w_{ij} s_j, \delta\right),$$

with a homogenous conditional variance (Cressie and Kapat, 2008, 729). In this case, $R = I - \rho W$ is positive definite, and so invertible (and the joint density is proper), when the correlation parameter is between $1/\eta_{\min}$ and $1/\eta_{\max}$, where $\eta_1, \ldots, \eta_n$ are the eigenvalues of W (Bell and Broemeling, 2000).

A compromise scheme for the variance deflators, a_i—see Leroux et al. (1999) and MacNab et al. (2006)—sets,

$$a_i = (1 - \lambda) + \lambda \sum_{j \neq i} w_{ij},$$

with $0 \leq \lambda \leq 1$ subject to a prior such as $\lambda \sim U(0, 1)$. The symmetry condition, $h_{ij} a_i = h_{ji} a_j$, is maintained by setting,

$$h_{ij} = \frac{\lambda w_{ij}}{1 - \lambda + \lambda \sum_{j \neq i} w_{ij}},$$

since $h_{ij} a_i = \lambda w_{ij} = \lambda w_{ji} = h_{ji} a_j$. So the joint density for $(s_1, \ldots, s_n)$ has covariance δR^{-1} where,

$$\begin{aligned} R &= \lambda F + (1 - \lambda) I, \\ f_{ii} &= \sum_{j \neq i} w_{ij}, \\ f_{ij} &= -w_{ij} \qquad i \neq j. \end{aligned}$$

The case $\lambda = 0$ corresponds to a lack of spatial interdependence, with R then reducing to an identity matrix, and borrowing strength confined to "global smoothing." By contrast, $\lambda = 1$ leads to the $ICAR(1)$ model (see Section 4.8.3). So,

$$s_i | s_{[i]} \backsim N\left(\frac{\lambda}{1 - \lambda + \lambda \sum_{j \neq i} w_{ij}} \sum_{j \neq i} w_{ij} s_j, \frac{\delta}{1 - \lambda + \lambda \sum_{j \neq i} w_{ij}} \right),$$

and when w_{ij} are defined by contiguity, one obtains,

$$s_i | s_{[i]} \backsim N\left(\frac{\lambda}{1 - \lambda + \lambda d_i} \sum_{j \in L_i} s_j, \frac{\delta}{1 - \lambda + \lambda d_i} \right).$$

The scheme of Leroux et al. (1999) can be generalized to allow greater spatial adaptivity with varying λ (Congdon, 2008a). The symmetry condition, $h_{ij} a_i = h_{ji} a_j$, is maintained by setting, $a_i = (1 - \lambda_i) + \lambda_i \sum_{j \neq i} w_{ij}$, and taking,

$$h_{ij} = \frac{\lambda_i \lambda_j w_{ij}}{1 - \lambda_i + \lambda_i \sum_{j \neq i} w_{ij}},$$

since this ensures the constraint,

$$h_{ij} a_i = h_{ji} a_j = \lambda_i \lambda_j w_{ij}.$$

A possible borrowing strength prior for these parameters is

$$\text{logit}(\lambda_i) \sim N(\lambda_\mu, 1/\tau_\lambda),$$

where the average, λ_μ, and precision, τ_λ, are extra unknowns. Setting $\Lambda = \text{diag}(\lambda_1, \ldots, \lambda_n)$, the covariance in the joint prior is then,

$$\delta[\Lambda F^* + (I - \Lambda)]^{-1},$$

where,

$$f^*_{ii} = \sum_{j \neq i} w_{ij},$$

$$f^*_{ij} = -w_{ij}\lambda_j \qquad i \neq j.$$

Pettitt et al. (2002) propose a scheme with,

$$h_{ij} = \frac{\phi w_{ij}}{1 + |\phi| \sum_{j \neq i} w_{ij}},$$

and

$$a_i = 1 + |\phi| \sum_{j \neq i} w_{ij},$$

where ϕ measures the strength of spatial dependency, and the case $\phi = 0$ corresponds to an absence of spatial interdependence, such that $R = I$ (see also Gschlößl and Czado, 2008). Gibbs updating for ϕ can be applied. So,

$$s_i | s_{[i]} \backsim N\left(\frac{\phi}{1 + |\phi| \sum_{j \neq i} w_{ij}} \sum_{j \neq i} w_{ij} s_j, \frac{\delta}{1 + |\phi| \sum_{j \neq i} w_{ij}} \right).$$

Under both the McNab et al. (2005) and Pettitt et al. (2002) schemes, the joint distribution of s is proper, ensuring a proper posterior when either is taken as the prior distribution. Retaining $h_{ij} = \phi w_{ij}/(1 + |\phi| \sum_{j \neq i} w_{ij})$, but setting $a_i = (1 + |\phi| \sum_{j \neq i} w_{ij})/(1 + |\phi|)$, means that $\phi \to \infty$ corresponds to the $ICAR(1)$ prior, with the conditional variance $(1 + |\phi|\delta)/(1 + |\phi| \sum_{j \neq i} w_{ij})$ tending to $\frac{\delta}{\sum_{j \neq i} w_{ij}}$.

4.8.3 ICAR(1) and convolution priors

The $ICAR(\rho)$ prior, when $\rho = 1$, is sometimes known as the $ICAR(1)$ model, when one has,

$$h_{ij} = \frac{w_{ij}}{\sum_{j \neq i} w_{ij}}, \; a_i = \sum_{j \neq i} w_{ij},$$

and for counts $y_i \sim Po(\lambda_i P_i)$, if one assumes,

$$\log(\lambda_i) = \alpha + s_i,$$

then borrowing of strength is purely spatial, with,

$$s_i | s_{[i]} \backsim N\left(\bar{A}_i, \frac{\delta}{\sum_{j \neq i} w_{ij}}\right),$$

where $\bar{A}_i = (\sum_{j \neq i} w_{ij} s_j / \sum_{j \neq i} w_{ij})$. The precision matrix of the joint prior is $\delta^{-1} R$ where,

$$r_{ii} = \sum_{j \neq i} w_{ij},$$

$$r_{ij} = -w_{ij} \qquad i \neq j.$$

When w_{ij} are binary indicators of adjacency ($w_{ij} = 1$ for areas i and j contiguous, $w_{ij} = 0$ otherwise), then $r_{ii} = d_i$ and the off-diagonal elements, r_{ij}, are -1 if i and j are neighbors, but zero otherwise. This case demonstrates most directly that conditional independence properties relating to the different spatial effects are stipulated by the matrix R and vice versa (Rue and Held, 2005, 4). Despite the relative simplicity of this form and the wide use of the $ICAR(1)$ conditional prior, R is not invertible under this model, and the joint prior is improper (Haran et al., 2003).

To see this in another way, consider the case where w_{ij} are binary, such that the joint prior can be specified in terms of pairwise comparisons between s_i (Knorr-Held and Becker, 2000). Let $i \backsim j$ denote that areas i and j are neighbors, then for a normal $ICAR(1)$ model, the joint prior in terms of differences, $s_i - s_j$, is (Hodges et al., 2003),

$$p(s_1, \ldots, s_n) \propto \delta^{-0.5(n-1)} \exp\left(-\frac{1}{2\delta} \sum_{i \backsim j} (s_i - s_j)^2\right).$$

Thus, the prior only specifies differences between spatial effects and not their overall level. However, all linear contrasts $c's$ with $c'1 = 0$ have proper distributions (Besag and Kooperberg, 1995, 740).

To tie down the effects and remove their locational invariance, one method involves centering the sampled values at every iteration to have mean zero. This is one form of linear constraint, and so the joint distribution becomes integrable and propriety is obtained (Rodrigues and Assuncao, 2008). Another possibility is a corner constraint, i.e., setting a particular effect to a known value, such as $s_1 = 0$ (Besag et al., 1995). Finally, one may omit the intercept so that s_i model the level of the data (Reich and Hodges, 2008). In this case, $y_i \backsim Po(P_i \exp(s_i))$ with s_i not constrained, rather than $y_i \backsim Po(P_i \exp(\alpha + s_i))$.

As mentioned above, a spatial effects only assumption is relatively informative, and the $ICAR(1)$ spatial prior is often combined with an exchangeable prior to form a convolution prior (Richardson et al., 2004). It may be argued that an exchangeable i.i.d effect should only be introduced in combination with an $ICAR(1)$ spatial prior, since conditional priors including a correlation parameter, such as the $ICAR(\rho)$, can adjust to varying mixtures of spatial and unstructured variation by varying the ρ parameter (Wakefield, 2007). Thus for a Poisson response, $y_i \backsim Po(\lambda_i P_i)$, one obtains the convolution prior of Besag et al. (1991), namely,

$$\log(\lambda_i) = \alpha + s_i + u_i,$$

with $s_i|s_{[i]} \backsim N(\bar{A}_i, \delta_s/d_i)$, and $u_i \backsim N(0, \delta_u)$ usually homoscedastic. However, heteroscedasticity or heavier tails than under the normal might be represented by taking $u_i \backsim N(0, \psi_i)$ where,

$$\psi_i = \delta_u/\kappa_i,$$

where κ_i are positive variables with mean 1 (Lesage, 1999a). Reich and Hodges (2008) allow spatial adaptivity by making the variance δ_s specific to area-pairs. While only the sum, $z_i = s_i + u_i$, is identifiable in this model, Norton and Niu (2009) show that the precisions, δ_s and δ_u, are identifiable from the distribution of z_i.

4.9 Priors on Variances in Conditional Spatial Models

As in the exchangeable hierarchical models considered in Chapter 3, the prior on δ, and on the pair $\{\delta_s, \delta_u\}$ in a convolution model, is important in governing the degree of smoothing toward the neighborhood or global mean. Although some applications of conditional autoregressive priors use vague priors for δ such as $p(\delta) \propto 1/\delta$ or just proper priors, with $1/\delta \backsim Ga(\varepsilon, \varepsilon)$ with ε small, these may lead to effective impropriety in the posterior such that MCMC convergence is impeded (Besag and Kooperberg, 1995, 741). The prior $1/\delta \backsim Ga(\varepsilon, \varepsilon)$ with ε small may also put undue weight on low variances.

Suppose the prior relates to a variance for unstructured effects in a log-linear model for relative risks, with $y_i \backsim Po(E_i\lambda_i)$. Wakefield (2007) mentions that a $Ga(0.001, 0.001)$ prior on δ_u in the model, $\log(\lambda_i) = \alpha + u_i$, is equivalent to assuming that the relative risks, $e^{\alpha + u_i}$, follow a log-t distribution with 0.002 degrees of freedom. To avoid such diffuse options, one may use a weakly data-based prior (technically amounting to an Empirical Bayes method) for this model (Mollie, 1996, 372). Thus, let $H = 1/V(q_i)$, where $q_i = \log(y_i/E_i)$ or $q_i = \log((y_i + 0.5)/(E_i + 0.5))$. Then in a convolution prior, one would take $1/\delta_s \backsim Ga(0.5bH/\bar{d}, b)$, and $1/\delta_u \backsim Ga(0.5bH, b)$, where $\bar{d}$ is the average number of neighbors, and $b < 1$ downweights information from the data.

Another option in the convolution prior is to link the two variances or precisions. Adapting the uniform shrinkage prior of Natarajan and Kass (2000) is not straighforward, since δ_s is a conditional variance, so that δ_s and δ_u cannot be regarded as components of the total residual relative risk. Wakefield

(2007, 167) suggests an approximate strategy based on the marginal variance of $w_i = s_i - s_n$. One may, however, monitor the ratio of spatial to total residual variation by obtaining an empirical estimate, $V_s = \text{Var}(s)$, of the marginal spatial variance (e.g., Eksler, 2008). It is then possible to set up a uniform prior on ratio, $r_u = (\delta_u/V_s + \delta_u)$ (see Example 4.9), resulting in a form of joint prior for δ_s and δ_u. One might also use a scaling factor, q, such that $\delta_u = qV_s$ with q centered at 1.

Example 4.9. Blood Lead in Children, Virginia Counties The data here, considered by Schabenberger and Gotway (2004), relate to elevated blood level readings, y_i, among n_i children (under 72 months) tested in the 133 counties of Virginia (including Independent Cities) in 2000. Recent data are described at http://www.vahealth.org/leadsafe/dataclpp.htm.

Numbers sampled, n_i, vary considerably (from 1 to 3808). Spatial proximity is binary, with $w_{ij}= 1$ for intercounty distances under 50 km and $w_{ij}= 0$ otherwise. Assuming binomial sampling with $y_i \sim Bin(n_i, \pi_i)$, one option considered is the convolution model of Besag et al. (1991), namely,

$$\text{logit}(\pi_i) = \alpha + s_i + u_i,$$

with conditional variance δ_s for $ICAR(1)$ spatial effect s_i, and variance δ_u for the unstructured effects. This is compared to the Leroux et al. (1999) model with $\text{logit}(\pi_i) = \alpha + s_i$ and,

$$s_i|s_{[i]} \backsim N\left(\frac{\lambda}{1-\lambda+\lambda d_i}\sum_{j\backsim i} s_j, \frac{\delta}{1-\lambda+\lambda d_i}\right).$$

For the convolution model, the priors on the precisions are

$$\delta_s^{-1} \sim Ga(1, 0.001),$$
$$r_u = \delta_u/(\delta_u + V_s) \sim U(0,1),$$

where $V_s = \text{Var}(s_i)$ is the marginal variance of the spatial effects calculated at each iteration. The uniform prior on the share, r_u, of unstructured in total variation ties the unknown variances together and governs the extent of global as against local shrinkage (cf Cohen et al., 1998; Daniels, 1999).

The posterior mean[8] for r_u based on iterations 1000–50,000 of a two chain run is 0.42. Fifteen (from 133) of the s_i parameters are judged significant in terms of posterior probabilities over 0.95 or under 0.05 that $s_i> 0$. By contrast, 40 composite terms, u_i+s_i, are significant. Such a pattern suggests that there may be no great advantage (in terms of substantive interpretation not just statistical terms) in having two distinct sets of effects. The average deviance, $\bar{D}$, is 501.4, and d_e is 77.6, giving a DIC of 579. The log pseudo marginal likelihood is −315.4, with the lowest log(CPO) being for Winchester (county 3), namely, −6.5.

Since the second model is less parameterized with a single set of effects, one might expect the average deviance to deteriorate. In fact, $\bar{D}$ is unchanged at

501.4, though a lower d_e of 73.6 leads to a reduced DIC of 575. The log psML is also slightly improved, at −313, giving a pseudo Bayes factor of 10; λ is estimated at 0.63, albeit with a wide 95% interval from 0.25 to 0.97; 34 effects s_i are now significant. One feature of note is the only slight increase in the effective parameter count in the convolution model as compared to this one; this may in part reflect the lack of significance of the majority of individual spatial effects in that model.

4.10 Spatial Discontinuity and Robust Smoothing

Spatial pooling assuming a smoothly varying outcome over contiguous areas may not be appropriate when there are clear discontinuities in the spatial pattern of events. For instance, a low mortality area surrounded by high mortality areas will have a distorted smoothed rate when heterogeneity is assumed to be entirely spatially structured. More generally, one may seek robustness against mis-specification of the distribution of latent event rates or risks; for example, virtually all applications of spatial conditional autoregression models assume normality by default. Finally, one may seek some degree of spatial adaptiveness. For example, under conditional autoregressive models, the conditional variance, δ, is constant across the region, whereas one might expect spatial correlation to be stronger in some subregions. In the convolution model, the variances δ_s and δ_u are global parameters, so that the relative amount of spatially structured and unstructured heterogeneity is constant across the study region (Congdon, 2007b; Knorr-Held and Becker, 2000).

Robustness against spatial outliers or non-normality may be important when event totals are small, since then the prior structure of the latent risks has a greater effect; this is the case with the much analyzed Scottish lip cancer data, where certain areas have elevated standard mortality ratios (SMRs) but small counts, y, and expected cases, E. A high relative risk apparent from a crude or moment estimate not based on a large y or E may be shrunk considerably under a spatial random effects approach, particularly if surrounded by lower morbidity areas, so that important excess risks may not be flagged up (Conlon and Louis, 1999).

Where extreme crude rates are observed, then a robust model is suggested even though the crude rates are unreliable estimators. One might adopt heavier-tailed alternatives such as the double exponential (Laplace) or L1-norm version of the $ICAR(1)$ prior, which Besag (1989, 399) mentions as preferable when s_i have discontinuities. For a connected graph (i.e., with no isolated areas in the region) this prior is

$$p(s_1, \ldots, s_n) \propto \frac{1}{\delta^{n-1}} \exp\left(-0.5\frac{1}{\delta}\sum_{j \neq i}|s_i - s_j|\right),$$

and has its posterior mode at the median rather than mean of the neighboring s_j. One might also apply Student t versions of the $ICAR(\rho)$, which if applied using scale mixtures give a natural measure of outlier status. Thus,

$$s_i|s_{[i]} \backsim N\left(\rho\bar{A}_i, \frac{\delta}{\gamma_i d_i}\right),$$

where $\gamma_i \backsim Ga(\nu/2, \nu/2)$, and very low values of γ_i correspond to spatial outliers.

Forms of discrete mixture have been proposed as particularly appropriate to modeling discontinuities in disease risk. Green and Richardson (2000) distinguish between clustering models and allocation models, while Knorr-Held and Rasser (2000) propose a scheme whereby at each MCMC iteration, areas are allocated to clusters of mutually contiguous areas, with identical risks within each cluster. Lawson and Clark (2002) propose a mixture of the $ICAR(1)$ and Laplace priors for the case $y_i \sim Po(E_i\lambda_i)$, with continuous (beta) weights, r_i, rather than binary mixture weights, namely,

$$\log(\lambda_i) = \alpha + r_i s_{1i} + (1 - r_i)s_{2i},$$

where $r_i \backsim Be(c, c)$, with c known, s_{1i} is an ICAR error, but s_{2i} follows a spatial Laplace prior. Other densities might be used for s_{2i} (e.g., ones allowing skewness). Following Congdon (2007b), analogous mixture forms can be applied to the errors in the convolution model itself, giving more emphasis to the unstructured term, u_i, in outlier areas,

$$\log(\lambda_i) = \alpha + r_i s_i + (1 - r_i)u_i.$$

This type of representation may also be useful for modeling edge effects, with u effects taking a greater role on the peripheral areas where neighbors are fewer. Another possibility is a discrete mixture in a "spatial switching" model (Congdon, 2007b), allowing an unstructured term only for areas where the pure spatial effects model is inappropriate. Thus for a count response,

$$\begin{aligned}
y_i &\backsim Po(E_i\lambda_{S_i,i}), \\
S_i &\backsim \text{Categoric}(\pi_1, \pi_2), \\
(\pi_1, \pi_2) &\backsim \text{Dirichlet}(\xi_1, \xi_2), \\
\log(\lambda_{1i}) &= \alpha + s_i, \\
\log(\lambda_{2i}) &= \alpha + s_i + u_i,
\end{aligned}$$

where ξ_j are extra unknowns, and s_i follow an $ICAR(1)$ prior. The posterior estimates for ξ_j provide overall weights of evidence in favor of a pure spatial model as compared to a convolution model, while high posterior probabilities, $Pr(S_i = 2|y)$, for particular areas indicate that pure spatial smoothing is inappropriate for them.

Fernandez and Green (2002) use a discrete mixture model generated via mixing over several spatial priors. Thus, for count data, assume K possible components with area-specific probabilities, π_{ik}, on each component,

$$y_i \backsim \sum_{k=1}^{K} \pi_{ik} Po(E_i \lambda_{ik}),$$

where $\log(\lambda_{ik}) = \alpha_k$ for a model without predictors. Then K sets of underlying spatial effects $\{s_{ik}\}$ are generated from separate conditional spatial priors, and used to estimate area-specific mixture weights,

$$\pi_{ik} = \exp(\chi s_{ik}) \Big/ \sum_{k=1}^{K} \exp(\chi s_{ik}),$$

where $\chi > 0$. As χ tends to 0, the π_{ik} tend to $1/K$ without spatial patterning, whereas large χ reduce overshrinkage.

Another discrete mixture model for robust spatial dependence modeling uses the Potts prior (Green and Richardson, 2002). Thus, let $S_i \in 1, \ldots, K$ be unknown allocation indicators with $y_i \backsim Po(E_i \mu_{S_i})$ where $\{\mu_1, \ldots, \mu_K\}$ are distinct cluster means. Also let $d_{ik} = 1$ if $S_i = k$. Then the joint prior for the allocation indicators incorporates spatial dependence with,

$$Pr(S_i = k) = \exp\left[\omega \sum_{j \sim i} I(d_{ik} = d_{jk})\right] \Big/ \sum_{h=1}^{K} \exp\left[\omega \sum_{j \sim i} I(d_{ih} = d_{jh})\right],$$

where $\omega > 0$ multiplies the number of same label neighbor pairs, so that lower values of ω indicate lesser spatial dependence. So pooling toward the local neighborhood average will tend not to occur if an area's latent risk is discrepant with those of its neighbors. Richardson et al. (2004) compare this model with the convolution model under various simulated scenarios for differentiated spatial risks. Additional effects can be included by multiplying μ_{S_i}; for example, a spatially unstructured multiplicative effect could be modeled as $\nu_i \backsim Ga(b_\nu, b_\nu)$, or a log-normal prior assumed with $\nu_i = \exp(u_i)$, and $u_i \backsim N(0, \delta_u)$. Then $y_i \backsim Po(E_i \mu_{S_i} \nu_i)$.

Assumptions such as normality in the spatial effects can be avoided by adapting the Dirichlet process stick-breaking prior of Sethuraman (1994) to spatial settings. The stick-breaking prior specifies an unknown distribution G by a mixture,

$$G = \sum_{m=1}^{M} p_m \delta(\rho_m),$$

where M may in principle be infinite, but in practical computing is taken as finite, the mixing probabilities satisfy $\sum_{m=1}^{M} p_m = 1$, and $\delta(\rho_m)$ has a point mass at ρ_m that may be scalar or vector values for areas (e.g., relative risks) or at grid locations. For example, the ρ_m may be draws from a baseline borrowing-strength prior G_0 such as a stationary Gaussian process in the case of continuous point-referenced spatial data, $y(g_i)$, at sites g_i. One may incorporate spatial information into either ρ_m, as in Gelfand et al. (2005b), or into the mixture probabilities, p_m, as in Griffin and Steele (2006). Such formulations are typically for point-referenced data, and allow for nonstationarity and non-Gaussian features in the response when the stationary Gaussian process (see Section 4.11) is not appropriate (Duan et al., 2007).

Example 4.10. Robust Priors for Attempted Suicides in NE London
This analysis compares the Potts prior and the convolution prior for modeling the distribution of attempted suicides in 133 wards (small administrative areas) in NE London over the two financial years 2002–2003 and 2003–2004. Attempted suicide (also called parasuicide) is deliberate self-injury with a nonfatal outcome where there is evidence that the patient intended to commit suicide (external ICD10 codes X60-X84,Y10-Y34). Expected parasuicides, E_i, are based on London as a whole, with the crude hospitalization risk ratios, $q_i = y_i/E_i$, averaging 154 (i.e., parasuicide is higher in this part of London than elsewhere).

A plot of q_i (Figure 4.8) shows discontinuities in the spatial patterning that may be linked to socioeconomic disparities between neighboring areas, when relatively deprived areas (which tend to have higher parasuicide levels) are surrounded by affluent areas, and vice versa. Hence, adopting a pure

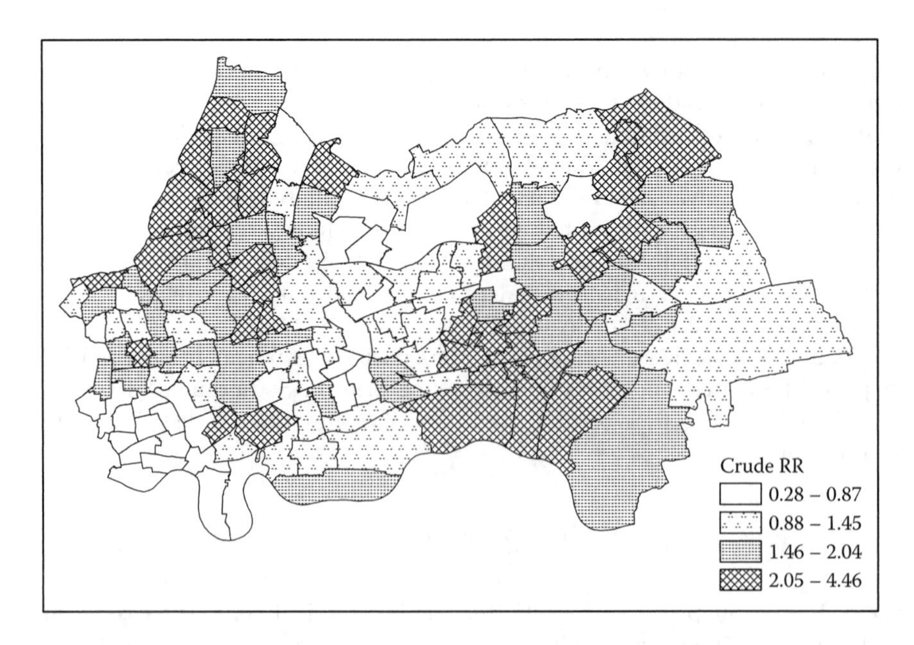

FIGURE 4.8
Unsmoothed curve relative risks.

spatial prior may be unduly informative and result in biased smoothing, as, for example, taking $y_i \backsim Po(E_i\lambda_i)$ with $\log(\lambda_i) = \alpha + s_i$, with s_i following an $ICAR(1)$ prior.

Instead, a Potts prior is applied with an exponential $E(1)$ prior on ω and with an ordering constraint on the latent cluster means, so $\mu_1 \leq \mu_2 \leq \cdots \leq \mu_K$, where K is set at 10. An order constraint on the cluster means is applied indirectly, so that for $k > 1$, $\mu_k = \mu_{k-1} + e^{\eta_k}$, where η_k are taken as random effects. The second half of a two chain run[9] of 10,000 iterations gives a mean scaled deviance,

$$2\sum_i \{y_i \log(y_i/(E_i\lambda_i)) - (y_i - E_i\lambda_i)\},$$

of 158, and DIC of 185 ($d_e = 26$), where the latter is obtained using posterior means of the Poisson parameters, $\nu_i = E_i\mu_{S_i}$. A predictive check using replicate data, $y_{\text{new},i}$, sampled from the model is satisfactory, with only one of the observations, y_i, not contained within 95% intervals of $y_{\text{new},i}$. The posterior mean (95% CI) of ω is 1.08 (0.85, 1.35) with $K = 10$ latent cluster means ranging from $\mu_1 = 0.52$ to $\mu_K = 4.02$, the latter being four times the expected parasuicide count based on London-wide parasuicide rates. The smoothed pattern (posterior means of $\lambda_i = \mu_{S_i}$ in Figure 4.9) shows fewer isolated high-risk areas than are apparent in the plot of unsmoothed risks.

The adaptive version of the convolution prior mentioned above, namely,

$$\log(\lambda_i) = r_i u_i + (1 - r_i)s_i,$$

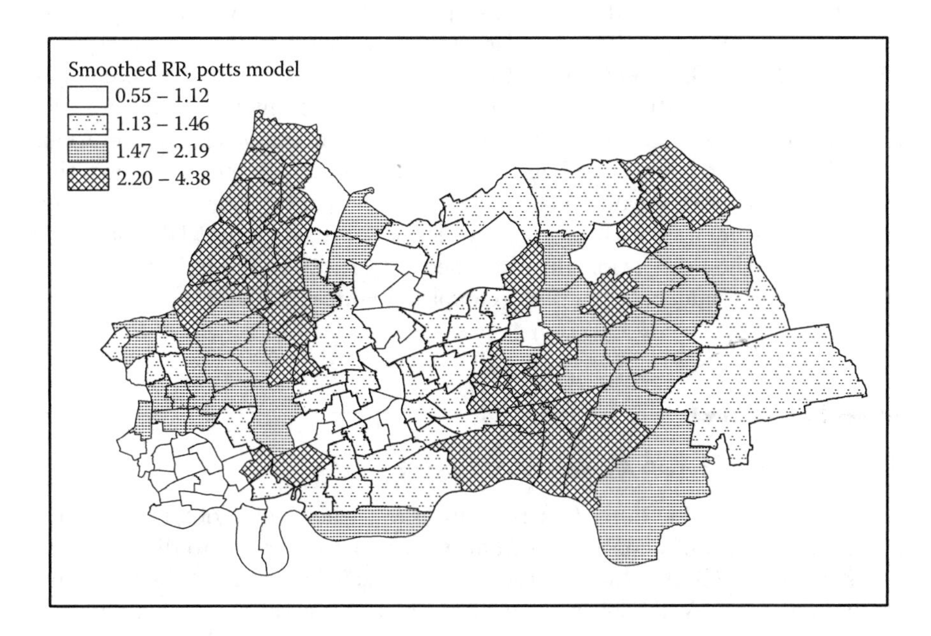

FIGURE 4.9
Smoothed relative risks, Potts Model.

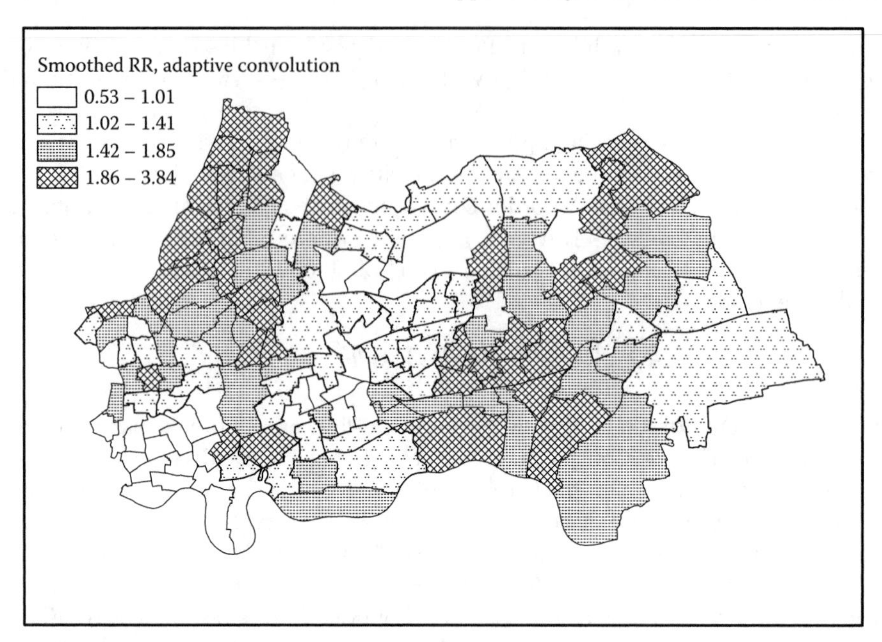

FIGURE 4.10
Smoothed relative risk, Adaptive Convolution Model.

is also run using the weakly data-based prior described above on the two precisions, and taking $r_i \sim Be(1,1)$. For these data, one has $H = \text{Var}(\log[q_i]) = 25.4$ and $\bar{d} = 5.3$. Taking $b = 0.01$, one has $1/\delta_s \backsim Ga(0.024, 0.01)$ and $1/\delta_u \backsim Ga(0.127, 0.01)$. The intercept is excluded from the model for the log relative risk, so s_i do not need to be centered at every MCMC iteration. Considering again the second half of a two chain run of 10,000 iterations, this option gives a lower mean deviance of 143.6 (comparing more closely to the $n = 133$ data points), with complexity estimated at 10.6, and DIC of 154.2. Compared to the Potts prior, the smoothed pattern is less dissimilar from that evident in the crude relative risks (compare Figures 4.10 and 4.8).

4.11 Models for Point Processes

A continuous spatial framework is appropriate when point observations are made. Nevertheless, a continuous framework is often applied to discrete area or lattice data (Berke, 2004; Kelsall and Wakefield, 2002; Webster et al., 1994; Yanli and Wall, 2004). Consider metric observations $(y_1, \ldots, y_n) = (y_1(g_1), \ldots, y_n(g_n))$ at points $\{g_1, g_2, \ldots, g_n\}$ in two-dimensional space G^2. To represent the spatially driven component in the variation of y, define a Gaussian spatial process $(s_1, \ldots, s_n) = (s(g_1), \ldots, s(g_n))$ with covariance matrix $\Sigma(d_{ij}) = \sigma_s^2 C(d_{ij})$, where the off-diagonal correlations depend on distances, $d_{ij} = \|g_i - g_j\|$, between points g_i and g_j, and $C(0) = 1$.

Such a process is ergodic if the off-diagonal elements in $\Sigma(d)$ tend to zero as $d \to \infty$ (so that covariance between values at two points vanishes for large enough distances), and isotropic if $\Sigma(d)$ depends only on the distance between g_i and g_j, and not on other features such as the direction from g_i to g_j or the coordinates of the g_i. The process is intrinsically stationary if $E[y(g+d) - y(g)] = 0$, namely, has a constant mean, and if the variance depends only on the lag, not on the point locations, namely,

$$E[y(g+d) - y(g)]^2 = V[y(g+d) - y(g)] = 2\gamma(d),$$

where $\gamma(d)$ is the semiovariogram (Waller and Gotway, 2004, 274). The covariance $\Sigma(d)$ and the semiovariogram are related via $\gamma(d) = \Sigma(0) - \Sigma(d)$ since,

$$\begin{aligned} 2\gamma(d) &= V[y(g+d) - y(g)] \\ &= V[y(g+d)] + V[y(g)] - 2\text{Cov}[y(g+d), y(g)] \\ &= \Sigma(0) + \Sigma(0) - 2\Sigma(d) = 2[\Sigma(0) - \Sigma(d)], \end{aligned}$$

so that $\gamma(0) = 0$.

A Gaussian spatial error, possibly together with an unstructured random effect, $u_i \backsim N(0, \sigma_u^2)$, and regressor effects may be used to define means, $\mu_i = E(y_i)$, in a normal linear model. However, this scheme generalizes to discrete responses using an appropriate link function (Diggle et al., 1998; Zhang, 2002). The variance parameter, σ_u^2, is sometimes called a nugget effect or nugget variance, defining measurement error or microscale spatial effects (spatial variation at lower scales than the smallest observed distance between sampled points). In Bayesian modeling, it is possible to take account of interplay between the nugget and the parameters of the spatial correlation function (Gramacy and Lee, 2008). Regressor effects might include a trend surface, $T(g_{1i}, g_{2i})$, defined by the coordinates of g_i (Diggle and Ribeiro, 2002, 133; O'Sullivan and Unwin, 2002), such as a quadratic polynomial with terms $(g_{1i}, g_{2i}, g_{1i}^2, g_{2i}^2, g_{1i}g_{2i})$. So for y continuous, one might have,

$$\begin{aligned} y_i &= \alpha + \beta.T(g_{1i}, g_{2i}) + s_i + u_i, \\ (s_1, \ldots, s_n) &\backsim N_n(0, \sigma_s^2 C), \\ u_i &\backsim N_n\left(0, \sigma_u^2\right). \end{aligned}$$

There are a number of isotropic schemes with $C(d_{ij})$, and hence $\gamma(d_{ij})$, parameterized to reflect anticipated distance decay in the correlation between points (e.g., Grunwald, 2005). For example, the exponential distance model has

$$C(d_{ij}) = \exp\left(-\frac{d_{ij}}{\phi}\right),$$

with range parameter $\phi > 0$, and smaller values of ϕ leading to more pronounced distance decay. The covariance function for $(e_1, \ldots, e_n)$, where $e_i = s_i + u_i$, is then,

$$\Sigma(d_{ij}) = \sigma_u^2 I(i = j) + \sigma_s^2 \exp\left(-\frac{d_{ij}}{\phi}\right),$$

while the semivariogram is

$$\gamma(d_{ij}) = \sigma_u^2 + \sigma_s^2 \left[1 - \exp\left(\frac{-d_{ij}}{\phi}\right)\right].$$

As d_{ij} tends to infinity, the semivariogram tends to an upper limit of $\sigma_u^2 + \sigma_s^2$, known as the sill. The powered exponential variant has

$$C(d_{ij}) = \exp\left[-\left(\frac{d_{ij}}{\phi}\right)^{\kappa}\right],$$

for $\phi > 0$ and $0 < \kappa \leq 2$.

The spherical model (Zhang, 2002) has nonzero covariance only within a certain range ϕ, namely,

$$C(d_{ij}) = 1 - \frac{3d_{ij}}{2\phi} + \frac{d_{ij}^3}{2\phi^3},$$

for $d < \phi$, whereas $C(d_{ij}) = 0$ for $d_{ij} \geq \phi$. Hence, the spherical function has covariance,

$$\Sigma(d_{ij}) = \sigma_u^2 I(i = j) + \sigma_s^2 \left[1 - \frac{3d_{ij}}{2\phi} + \frac{d_{ij}^3}{2\phi^3}\right] I(d_{ij} < \phi),$$

and semivariogram,

$$\gamma(d_{ij}) = \sigma_u^2 + \sigma^2 \left[\frac{3d_{ij}}{2\phi} - \frac{d_{ij}^3}{2\phi^3}\right] \text{ for } d_{ij} < \phi,$$

$$\gamma(d_{ij}) = \sigma_u^2 + \sigma_s^2 \text{ for } d \geq \phi.$$

Finally, Matern functions (Diggle et al., 2003) set,

$$C(d_{ij}) = \frac{2^{1-\kappa}}{\Gamma(\kappa)} \left(\frac{d_{ij}}{\phi}\right)^{\kappa} K_\kappa\left(\frac{d_{ij}}{\phi}\right),$$

where $K_\kappa(u) = \sum_{i=0}(-1)^i/i!\Gamma(\kappa + i + 1)(u/2)^{2i+\kappa}$ is a Bessel function of order κ.

Suppose continuous observations $y = (y_1, \ldots, y_n) = (y_1(g_1), \ldots, y_n(g_n))$ are made at locations $g = (g_1, \ldots, g_n)$, and that predictions $y_0 = (y_{01}, \ldots, y_{0k})$ are required at k new locations $g_0 = (g_{01}, \ldots, g_{0k})$. These are based on the posterior predictive density,

$$p(y_0|y) = \int p(y_0, \theta|y)d\theta = \int p(y_0|y, \theta)p(\theta|y)d\theta,$$

where θ is the vector of parameters involved in the model for y, namely, those defining its mean, and the covariance parameters for spatial and unstructured errors (Banerjee et al., 2004, 132). For example, Diggle et al. (2003, 51) consider a model, $y_i = \mu + s_i + u_i$, with $u_i \backsim N_n\left(0, \sigma_u^2\right)$, and spatial error process,

$$(s_1, \ldots, s_n) \backsim N_n(0, \sigma_s^2 C),$$

where prediction is required at a single new location, g_0. With d_0 denoting a $n \times 1$ vector of distances between g_0 and $g = (g_1, \ldots, g_n)$, and with $Q = \sigma_u^2 I + \sigma_s^2 C$, one has,

$$p(y_0|\theta, y) = N\left(\mu + \sigma_s^2 d_0' Q^{-1}(y - \mu 1_n), \sigma_s^2 - \sigma_s^2 d_0' Q^{-1} \sigma_s^2 d_0\right).$$

For $k > 1$, univariate predictions may be obtained at each new site, g_{01}, $g_{02}, \ldots, g_{0k}$, though multivariate predictions may be more precise. For non-normal z, one may apply the procedure of De Oliveira et al. (1997), where $y = G(z)$ is obtained by a normalizing transformation of z.

4.11.1 Discrete convolution models

Assuming a stationary Gaussian process described through its mean and covariance structure may result in slow estimation when there are a large number of points and is relatively inflexible when stationarity and isotropy assumptions are violated. An alternative representation, based on the Gaussian process but one that adapts to spatial nonstationarity and anisotropy is the process convolution approach (Higdon, 1998, 2007; Lee et al., 2005). This involves convolving a continuous white noise process, $w(g)$, with a symmetric smoothing kernel, $K(g)$, with the spatial effect obtained as,

$$s(g) = \int_G K(g - u) w(u)\, du,$$

where G is the region of interest. The spatial process might be combined with fixed effects defining level or regression impacts and with appropriate regression links for non-normal observations. For example, if $y(g)$ were binary, such as species presence or absence at site g (Gelfand et al., 2005a), then $y(g) \sim Bern(\pi(g))$ and,

$$\text{logit}[\pi(g)] = \beta_0 + s(g),$$

where β_0 defines the average intensity.

In practice, the continuous underlying process can be approximated by a discretized process (e.g., one defined on a regular lattice over G) provided the discretization is not too coarse relative to the smoothing kernel (Calder, 2003, 2007). So if there are $i = 1, \ldots, n$ observations at points $g_1, \ldots, g_n$ and grid locations $\{t_j, j = 1, \ldots, m\}$ with $t_j = (t_{1j}, t_{2j})$, over the region, one may define the discretized kernel smoother as,

$$s(g_i) = \sum_{j=1}^{m} K(g_i - t_j)w_j,$$

where for large m, the w_j can be taken as a collection of random effects (Higdon, 2007, 245). Lee et al. (2005) consider options for representing the kernel, possibly by a form with known variance (e.g., a standard normal), and consequent ways for modeling w_j. Note that if both the K function and w series have unknown variances, then there is nonidentifiability. Options for w_j include exchangeable effects or low order random walks, with unknown precision, τ_w. Assuming K is a normal kernel, one can, by varying τ_w, mimic the effect of the range parameter in a conventional Gaussian process model with a Gaussian variogram.

For example, Lee et al. (2005) consider $n = 12$ observations, y_i, in G^1 at equally spaced locations, g_i, between 0 and 10. These are generated according to a Gaussian process, $s(g)$, with mean 0 and covariance matrix,

$$C(d_{ij}) = \exp(-d_{ij}^2/25),$$

where d_{ij} relates to distances between points g_i and g_j on the line. A white noise error, u_i, with standard deviation 0.2 is also used to define y_i, so that $y_i = s(g_i) + u_i$. They then fit[10] a discrete convolution model to y_i so generated, using a grid with $m = 20$ points, t_j equally spaced between -2 and 12. They assume w_j follow a first order random walk, and assume the kernel is a normal density with standard deviation 0.6.

Best et al. (2000) consider a convolution model for health counts, $y_i \sim Po(P_i\lambda_i)$, observed for areas rather than points, where P_i are populations and λ_i are latent rates. In this case, a rectangular grid is defined over m points in the region, and an additive (rather than log link) regression is used for modeling the latent rates. So with a single predictor, x_i, taking positive values only, one has,

$$\lambda_i = \beta_0 + \beta_1 x_i + \beta_2 \sum_{j=1}^{m} K(g_i - t_j)w_j,$$

where w_j (and β parameters) are gamma distributed and the kernel function has a known variance. One can decompose the total risk parameter into three sources: one due to the background rate, β_0, one reflecting the known predictor, and one the latent spatially configured risk over the region.

Semiparametric approaches to spatial modeling based on the stick-breaking prior can also be related to this theme (Reich and Fuentes, 2007). Thus, there are kernel functions for each of m potential clusters, with the kernel centers, $t_j = (t_{1j}, t_{2j})$, being unknowns, and the cluster allocation probabilities for sites or areas i at location $g_i = (g_{1i}, g_{2i})$ incorporating spatial information. While the cluster effects, $w_j \sim N(0, 1/\tau_w)$, are unstructured, the cluster for area or point i is chosen using indicators,

$$S_i \sim \text{Categorical}(p_{i1}, \ldots, p_{im}),$$

with p_{ij} determined both via beta distributed $V_j \backsim Be(c,d)$, and by cluster-specific kernels, K_{ij}, constrained to lie in $[0,1]$. The realized spatial effect for area or point i is then w_{S_i}. Defining $R_{ij} = K_{ij}V_j$, one has,

$$\begin{aligned} p_{i1} &= R_{i1}, \\ p_{ij} &= R_{ij}(1-R_{i1})\cdots(1-R_{i,j-1}) \qquad j=2,\ldots,m-1, \\ p_{im} &= (1-R_{i1})\cdots(1-R_{i,m-1}), \end{aligned}$$

where (for example),

$$K_{ij} = \exp[-|g_i - t_j|/2\gamma_j],$$

defines a normal kernel with bandwidth γ_j. Bandwidths can be taken equal across kernel functions or vary across kernel functions according to a positive prior (e.g., inverse gamma).

Example 4.11. Trichloroethylene Groundwater Contamination This example considers spatial covariance modeling for continuous measures of groundwater contamination. The particular contamination involves concentrations (ppb) of trichloroethylene (TCE), a chemical used as a metal degreaser and in cleaning fluids that can enter ground and surface water from industrial discharges or from landfill disposal of industrial waste. Data from Kitanidis (1997, Table 2.2) relate to TCE readings in groundwater at 56 locations with predictions required for $k=3$ new points, namely, $g_{01} = (50,-50)$, $g_{02} = (100,-50)$, and $g_{03} = (170,-50)$.

Log transforms of the originally positive skewed readings are taken as the responses. The eastings and northings are scaled to give kilometre distances, so that the original four digit grid references are replaced by two digit grid references. A powered exponential distance model[11] is applied for spatial effects s_i, namely,

$$C_{ij}(d) = \exp\left[-\left(\frac{d}{\phi}\right)^{\kappa}\right],$$

with spatial error predictions $(s_{01},\ldots,s_{0k})$ for $k=3$ new points. A nugget term is included so that,

$$y_i = \alpha + s_i + u_i,$$

and the interplay between variances governed by a uniform prior on,

$$r_u = \frac{\sigma_s^2}{\sigma_s^2+\sigma_u^2},$$

with a $Ga(1, 0.001)$ prior adopted on $1/\sigma_s^2$.

A two chain run with of 10,000 iterations (2,500 burn-in) shows the $k=3$ predicted values close to the overall mean $\alpha = 4.3$, and the spatial errors (s_{01}, s_{02}, s_{03}) at the new locations all having 95% credible intervals straddling

TABLE 4.2
Response variable and grid reference data.

	$x1$	$x2$	y	E	Crude SMR
Barking and Dagenham	547.8	185.1	75	80.7	0.93
Barnet	524.3	191.7	145	169.8	0.85
Bexley	548.4	175.7	99	123.2	0.80
Brent	520.7	185.5	168	139.5	1.20
Bromley	541.8	167.6	152	169.1	0.90
Camden	527.9	184.3	173	107.2	1.61
Croydon	533.3	165.1	152	179.8	0.85
Ealing	515.9	181.4	169	160.4	1.05
Enfield	533.1	195.3	130	147.5	0.88
Greenwich	542.8	176.8	117	116.8	1.00
Hackney	534.2	185.5	124	102.8	1.21
Hammersmith and Fulham	523.8	178.5	119	91.8	1.30
Haringey	531.5	189.6	134	119.6	1.12
Harrow	515	189.5	90	114.8	0.78
Havering	553.1	188.2	98	131.1	0.75
Hillingdon	508.6	183.8	89	136.1	0.65
Hounslow	514	175.8	128	116.6	1.10
Islington	531.1	185.1	145	98.5	1.47
Kensington and Chelsea	525.6	179.5	130	88.8	1.46
Kingston upon Thames	519.4	167.5	69	79.8	0.86
Lambeth	530.8	174.6	246	144.9	1.70
Lewisham	537.5	174	166	134.7	1.23
Merton	525.8	169.3	95	98.9	0.96
Newham	541.2	183.6	135	118.6	1.14
Redbridge	543.8	188.9	98	130.6	0.75
Richmond upon Thames	517	173.4	97	96.1	1.01
Southwark	526.6	164.5	202	127.1	1.59
Sutton	533.6	177.1	75	97.7	0.77
Tower Hamlets	536.1	181.8	100	88.5	1.13
Waltham Forest	526.4	173.9	100	121.4	0.82
Wandsworth	527.2	181.1	153	156.8	0.98
Westminster, City of	537.9	189.6	194	114	1.70

zero. The spatial process has a low ϕ value, with posterior mean (95% interval) of 0.16 (0.1, 0.4), and so relatively strong distance decay, while κ has mean 1.3 (0.95, 1.6). Spatial variation outweighs nugget variation with respective posterior medians on σ_s^2 and σ_u^2 of 8.3 and 0.4, respectively. The DIC (and d_e) stands at 103 (22).

Example 4.12. Area Suicide Counts This example relates to a discrete area outcome, but subject to smoothing via geostatistical methods. Suicide mortality totals, y_i (male and female suicides combined over 1989–1993), in 32 London boroughs, expected deaths, E_i, and grid references (g_{i1}, g_{i2}) are shown in Table 4.2. The grid references are defined so that inter-area distances are in kilometres. To apply kernel smoothing, a two-way grid enclosing London is constructed; this has 13 equally spaced points (eastings from 500 to 560 at intervals of 5) on its east-west axis, and 11 north–south points (northings from 160 to 210). So there are $m = 143$ interior points, $t_j = (t_{j1}, t_{j2})$, defining the grid.

A discrete kernel approach is applied with a log link regression with $y_i \sim Po(\lambda_i)$ where,

$$\log(\lambda_i) = \log(E_i) + \beta_0 + \sum_j K_{ij} w_j,$$

and with a normal kernel,

$$K_{ij}(d_{ij}) = \frac{1}{\sigma_k \sqrt{2\pi}} \exp\left(-d_{ij}^2 / 2\sigma_k^2\right),$$

with distances $d_{ij} = \left[(g_{i1} - t_{j1})^2 + (g_{i2} - t_{j2})^2\right]^{0.5}$, and standard deviation σ_k preset at 8. This setting ensures a majority of K_{ij} are negligible (e.g., only 17/143 of K_{1j} are above 0.01, with 2 above 0.4). The w_j are assumed to be unstructured random effects with zero mean, and with standard deviation, σ_w, assigned a $U(0, 100)$ prior.

The second half of a two chain run[12] of 50,000 iterations gives a mean scaled deviance of 67, with σ_w having posterior mean 0.85. Predictive performance is satisfactory with observed counts for 31 of the 32 areas contained within 95% intervals for replicate data. Analogues to area-specific spatial effects, s_i, are obtained by centering the terms,

$$A_i = \sum_{j=1}^{m} K_{ij} w_j,$$

around their means. Then 24 of the 32 centered spatial effects, $s_i = A_i - \bar{A}$, are judged significant in that the posterior probability they exceed zero is above 0.95 or below 0.05.

To achieve a match between average deviance and the number of observations, one might simply add an unstructured error. An alternative that achieves the same goal of improved fit but preserves the majority of significant spatial effects is based on the switching spatial prior of Congdon (2007b). Thus,

$$\log(\mu_i) = \log(E_i) + \beta_0 + \sum_{j=1}^{m} K_{ij} w_j + \delta_i u_i,$$

where $\delta_i \sim Bern(0.05)$ and u_i is a zero mean unstructured normal effect. Three areas (21, 31, 32) have posterior probabilities, $Pr(\delta_i = 1|y)$, exceeding 0.9, but 16 of the s_i remain significant. The average deviance is reduced to 34.

Finally, a stick-breaking prior is applied with $m = 8$ potential clusters, and cluster centers (eastings and northings) taken to be uniform $t_{1j} \sim U(509, 533), t_{2j} \sim U(165, 195)$. The $\{w_j, j = 1, \dots, m\}$ are normal with precision $1/\sigma_w^2$ taken to be $Ga(1,\ 0.01)$, while the mixture parameters, V_j, are taken to be $Be(1, 2)$. The kernels are normal in form,

$$K_{ij} = \exp[-|g_i - t_j|/2\gamma_j],$$

with bandwidths γ_j taken to be $Ga(1.5, 40)$, and choice of scale parameter based on the maximum possible distance between areas, as in Reich and Fuentes (2007). Then,

$$\log(\mu_i) = \log(E_i) + \beta_0 + w_{S_i}.$$

A single chain run of 10,000 iterations produces an average deviance of 35.8, close to the number of areas, without a separate unstructured effect, u_i, being needed. Monitoring the realized spatial effects for borough i, namely w_{S_i}, 19 of the 32 effects are judged significant in that the posterior probability they exceed zero is above 0.95 or below 0.05.

Appendix: Computational Notes

1. In one chain the initial values for the μ_t are set equal to the y_t, and the first inits file is

```
list(invV=1,q=c(1,1,1),delta1=5,s1=c(0,0,0,0,0,0,0,0,0,0,0),
mu1=112,mu=c(NA,118,...,))
```

while the other inits file is based on an exploratory single chain run using these initial values. The code for the air passenger data analysis is then

```
model {for (t in 1:T) { y[t]~dnorm(m.y[t],invV); m.y[t] <- mu[t] +s[t]}
for (t in 2:T){ mu[t] ~dnorm(m.mu[t],invW[1])
m.mu[t] <- mu[t-1]+del[t-1]; del[t] ~dnorm(del[t-1],invW[2])}
for (t in 12:T) {s[t]~dnorm(m.s[t],invW[3]);
                  m.s[t] <- -sum(s[t-S+1:t-1])}
# initial conditions
for (j in 1:11) {s[j] <- s1[j]; s1[j]~dnorm(0,0.001)}
del[1] <- del1; mu[1] <- mu1
del1~dnorm(0,0.001); mu1~dnorm(0,0.00001)
# variances
invV~dgamma(1,0.001); q[3]~dgamma(1,0.001)
q[2]~dexp(10); q[1]~dexp(10)
for (j in 1:3) { invW[j] <- invV/q[j]; W[j] <- 1/invW[j]}}.
```

The code for the alternative prior is

```
model {for (t in 1:T) { y[t]~dnorm(m.y[t],tau[1]); m.y[t] <- mu[t] +s[t]}
```

```
for (j in 1:3) {q[j]~dnorm(0,1) I(0,); tau[j+1] <- tau[1]/q[j] }
for (t in 2:T){mu[t]~dnorm(m.mu[t],tau[2])
m.mu[t] <- mu[t-1]+del[t-1];del[t]~dnorm(del[t-1],tau[3])}
for (t in 12:T) {s[t]~dnorm(m.s[t],tau[4]);
m.s[t] <- -sum(s[t-S+1:t-1])}
# initial conditions
for (j in 1:11) {s[j] <- s1[j]; s1[j]~dnorm(0,0.1)}
del[1] <- del1; del1~dnorm(0,0.1)
mu[1] <- mu1; mu1 ~dnorm(0,0.00001); tau[1] ~dgamma(1,0.01)}
```

with initial values list(del1=0,mu1=0,s1=c(0,0,0,...,),tau=c(1,NA,NA,NA), q=c(1,1,1)) and list(del1=0,mu1=0,s1=c(0,0,0,...,),tau=c(10,NA,NA,NA), q=c(0.1,0.1,0.1)).

2. The code in Example 4.3 is

```
model { tau~dgamma(1,0.001); sig2 <- 1/tau
sig2z <- (gam+ph)*(gam+ph)*sig2; tau.z <- 1/sig2z
for (t in 1:T) { y[t] ~dnorm(z[t],tau)}
for (t in 2:T) { mu.z[t] <- ph*z[t-1]
# one step ahead predictions
yp.one[t]~dnorm(z[t-1],tau)
z[t] ~dnorm(mu.z[t],tau.z)}
z1 ~dnorm(0,0.001); z[1] <- z1
ph~dunif(-1,1); gam~dunif(-1,1)}
```

3. The codes for the two models in Example 4.4 are as follows

```
model { for (a in 1:9) {for (g in 1:2) {
y[a,g] ~dpois(m[a,g]); ynew[a,g] ~dpois(m[a,g])
m[a,g] <- r[a,g]*P[a,g]/1000; log(r[a,g]) <- del[g,a]}}
for(g in 1:2){del[g,1] ~dflat(); sig[g] ~dunif(0,100);
tau[g] <- pow(sig[g],-2)
for (a in 2:9){del[g,a] ~dnorm(del[g,a-1],tau[g])}}}
```

and

```
model { for (a in 1:9) {for (g in 1:2) {
y[a,g] ~dpois(m[a,g]); ynew[a,g] ~dpois(m[a,g])
m[a,g] <- r[a,g]*P[a,g]/1000;
log(r[a,g]) <- lam[a,g]}}
for (g in 1:2) {ph[g] ~dnorm(0,1); lam[1,g] <- omeg[1,g]
for (a in 1:9) {omeg[a,g] ~dnorm(0,chi[a])}
for (a in 2:9) {lam[a,g] <- ph[g]*lam[a-1,g]+omeg[a,g]}}
for (a in 1:9) {log(chi[a]) <- b[1]+b[2]*a+b[3]*a*a}
for (j in 1:3) {b[j] ~dnorm(0,1)}}.
```

4. The code for the telephone call data (Example 4.5) is

```
model { for (t in 1:T) {y[t] ~dpois(mu[t])
log(mu[t]) <- lam[t] + sum(c[1:K,t])}
```

```
for (k in 1:K) {c[k,1] ~dnorm(0,0.001); d[k,1] ~dnorm(0,0.001)}
Zr ~dunif(0,10); lam[1] ~dnorm(0,0.001)
for (j in 1:3) {v[j] ~dgamma(1,1); q[j] <- v[j]/sum(v[])
W[j] <- Zr*Zr*q[j]; tau[j] <- 1/W[j]}
for (t in 2:336) {yp[t] ~dpois(mu[t-1])
lam[t] ~dnorm(lam[t-1],tau[1])
for (k in 1:K) { c[k,t] ~dnorm(C[k,t],tau[2])
d[k,t] ~dnorm(D[k,t],tau[3])
C[k,t] <- c[k,t-1]*cos(2*k*pi/S)+d[k,t-1]*sin(2*k*pi/S)
D[k,t] <- -c[k,t-1]*sin(2*k*pi/S)+d[k,t-1]*cos(2*k*pi/S)}}}
```

5. The code for the double autoregression model of the treasury bill data, with $x_t = \log(r_t)$, is

```
model { for (t in 1:349) {y[t] <- x[t+1]-x[t]}
for (t in 2:349) {y[t]~dnorm(mu[t],tau[t]); mu[t] <- rho*y[t-1]
ynew[t] ~dnorm(mu[t],tau[t]); dnew[t] <- pow(y[t]-ynew[t],2)
var[t] <- gam + alph*y[t-1]*y[t-1]; tau[t] <- 1/var[t]}
rho <- sqrt(rho2); gam ~dgamma(1,1)
rho2 <- r[1]/sum(r[]); alph <- r[2]/sum(r[])
Dnew <- sum(dnew[2:349])
for (j in 1:3) {r[j]~dgamma(1,1)}}
```

The state–space model has code

```
model { for (t in 1:349) {y[t] <- x[t+1]-x[t]}
for (t in 2:349) {y[t] ~dnorm(m.y[t],tau[t]); ynew[t] ~dnorm(m.y[t],tau[t]);
tau[t] <- 1/exp(th[t]); var[t] <- 1/tau[t]; m.y[t] <- rho*y[t-1]
th[t] ~dnorm(m.th[t],invW); m.th[t] <- mu + ph*th[t-1]}
rho~dnorm(0,1); ph~dnorm(0,1); mu~dnorm(0,0.001)
invW~dgamma(1,0.001); th[1] <- 0}
```

6. The code for the first Nile discharge analysis is

```
model { for (t in 1:T) {y[t] <- Y[t]-Y[1];
                        y[t] ~dnorm(mu[t],tau)}
                        y0 ~dt(0,tau,2); ym1 ~dt(0,tau,2)
for (t in 3:T) {mu[t] <- ph0+ph[1]*y[t-1]+ph[2]*y[t-2]}
                mu[2] <- ph0+ph[1]*y[1]+ph[2]*y0;
                mu[1] <- ph0+ph[1]*y0+ph[2]*ym1
ph0 ~dflat(); for (j in 1:2) {ph[j]~dnorm(0,1)}
tau ~dgamma(1,0.001); sig <- sqrt(1/tau)}.
```

The code for the $AR(1)$ model with shift mechanism is

```
model { for (t in 1:T) {y[t] <- Y[t]-Y[1]; y[t]~dnorm(mu[t],tau)}
y0 ~dt(0,tau,2);
for (t in 2:T) {mu[t] <- ph0[2]*step(t-kap)+ph0[1]+ph*y[t-1]}
                mu[1] <- ph0[1]+ph*y0
ph ~dnorm(0,1); kap ~dunif(3,97)
for (j in 1:2) {ph0[j] ~dflat()}
tau ~dgamma(1,0.001); sig <- sqrt(1/tau)}
```

7. The code for the Box–Jenkins series A data is

```
model { for (t in 1:T) {y[t] ~dnorm(mu[t],tau[S[t]])
mu[t] <- beta0+th[t]; S[t] ~dcat(p[1:3])}
r ~dexp(9); p[1] <- 1/(1+r); p[2] <- 0.5*r/(1+r); p[3] <- 0.5*r/(1+r)
for (t in 2:T) {th[t] ~dnorm(m.th[t],tau.th); m.th[t] <- phi*th[t-1]}
th[1] <- 0; beta0 ~dnorm(0,0.00001)
tau.th <- tau[1]/q; tau[1] ~dgamma(1,0.001)
tau[2] <- tau[1]/10; tau[3] <- tau[1]/32
q ~dexp(1); phi ~dunif(-1,1)
sig2[1] <- 1/tau[1]; sig2[2] <- 1/tau.th}
```

8. The code excluding model fit elements for the blood lead example (Example 4.10) is

```
model {for (i in 1:133) {y[i] ~dbin(p[i],n[i])
u[i] ~dnorm(0,inv.delta.u); p.sig[i] <- step(s[i])
logit(p[i]) <- alph+s[i]+u[i]}
s[1:133] ~car.normal(map[],wei[],d[],inv.delta)
for (j in 1:1056) {wei[j] <- 1}
V.s <- pow(sd(s[]),2); r.u ~dunif(0,1);
delta.u <- r.u*V.s/(1-r.u); inv.delta.u <- 1/delta.u
inv.delta ~dgamma(1,0.001); alpha ~dflat()}
```

The Leroux et al. model has code

```
model {for (i in 1:N) {y[i] ~dbin(p[i],n[i])
p.sig[i] <- step(s[i]); logit(p[i]) <- alph+s[i]
s[i] ~dnorm(S[i],tau[i]); tau[i] <- inv.delta * (1-lam+lam*d[i])
S[i] <- (lam/(1-lam+lam*d[i]))*sum(Ws[C[i]+1:C[i+1] ])}
# sum weighted errors over neighbors
for (i in 1:NN) { Ws[i] <- s[map[i]] }
inv.delta ~dgamma(1,0.001); alph ~dflat(); lam ~dunif(0,1)}
```

9. The Potts prior model has code

```
model { for (i in 1:N) { y[i] ~dpois(nu[i]); ynew[i] ~dpois(nu[i])
nu[i] <- E[i]*lam[i]; lam[i] <- mu[S[i]]; S[i] ~dcat(p[i,1:K])
dv[i] <- y[i]*log(y[i]/nu[i])-(y[i]-nu[i])
for (k in 1:K) {J[i,k] <- equals(S[i],k);
# allocation probabilities
p[i,k] <- U[i,k] /sum(U[i,])
log(U[i,k]) <- omega*sum(wJ[cumnei[i] + 1 : cumnei[i + 1],k ])}}
for (i in 1 : NN ) {for (k in 1:K) { wJ[i,k] <- J[map[i],k] }}
for (k in 2:K) {eta[k-1] ~dnorm(0,tau.eta);
mu[k] <- mu[k-1]+exp(eta[k-1])}
D <- 2*sum(dv[]); mu1 ~dgamma(1,1); mu[1] <- mu1;
# priors
omega ~dexp(1); tau.eta ~dgamma(1,1)}
```

The code for the adaptive convolution prior is

```
model {for (i in 1:N) { y[i] ~dpois(nu[i]); ynew[i] ~dpois(nu[i])
```

```
nu[i] <- E[i]*lam[i]; log(lam[i]) <- r[i]*u[i]+(1-r[i])*s[i];
r[i] ~dbeta(1,1); u[i] ~dnorm(0,inv.delta.u);
s[i] ~dnorm(S[i],tau.s[i])
S[i] <- mean(Ws[ cumnei[i]+1 : cumnei[i+1] ])
tau.s[i] <- inv.delta.s*d[i]; d[i] <- cumnei[i+1]-cumnei[i]
dv[i] <- y[i]*log(y[i]/nu[i])-(y[i]-nu[i])}
for (i in 1 : NN ) {Ws[i] <- s[map[i]] }
D <- 2*sum(dv[]); h ~dexp(1)
# priors
inv.delta.u ~dgamma(0.127,0.01); inv.delta.s ~dgamma(0.024,0.01)}
```

10. The code for the Lee et al. (2005) line example, with precision τ_w for the w_j assigned a $Ga(1, 0.001)$ prior is

```
model { for (i in 1:n) {y[i] ~dnorm(mu[i],tau[1])
mu[i] <- beta+sum(Kw[i,])
for (j in 1:m) {K[i,j] <- phi((x[i]-t[j])/sd.kern)
Kw[i,j] <- K[i,j]*w[j]}}
w[1] ~dnorm(0,0.1); beta ~dflat()
for (j in 2:m) {w[j] ~dnorm(w[j-1],tau[2])}
for (j in 1:2) {tau[j] ~dgamma(1,0.001)}}
```

with data input

```
list(sd.kern=0.6,n=12,m=20,x=c(0,0.91,1.82,2.73,3.64,4.55,5.45,
6.36,7.27,8.18,9.09,10),y=c(1.19,0.80,1.07,0.45,0.22,0.11,
-0.06,0.13,-0.25,0.47,0.36,0.46),t=c(-2,-1.26,-0.53,0.21,0.94,1.68,2.42,3.15,
3.89,4.63,5.36,6.1,6.84,7.57,8.31,9.05,9.79,10.53,11.26,12))
```

11. The WinBUGS code for the TCE data is

```
model { for (i in 1 : 56) { y[i] <- log(Y[i]+1);  y[i] ~dnorm(mu[i],tau.u)
X1[i] <- x1[i]/100; X2[i] <- x2[i]/100; nought[i] <- 0; mu[i] <- alpha+s[i]}
# structured errors
        s[1:56] ~spatial.exp(nought[],X1[],X2[],tau.s,phi.inv,kappa)
for(i in 1:k) { X0east[i] <- x0east[i]/100; X0north[i] <- x0north[i]/100
# single site prediction
              s0[i] ~spatial.unipred(0,X0east[i], X0north[i],s[])
              y0[i] <- alpha+s0[i]}
# priors
alpha ~dnorm(0,0.001); phi.inv ~dunif(0,10); phi <- 1/phi.inv
kappa ~dunif(0,2); r.u ~dunif(0,1); sig2.s <- 1/tau.s; tau.s ~dgamma
(1,0.001);
sig2.u <- sig2.s*(1-r.u)/r.u; tau.u <- 1/sig2.u}
```

12. The code for Example 4.13 is

```
model { for (i in 1:n) { y[i] ~dpois(mu[i]); yrep[i] ~dpois(mu[i])
log(mu[i]) <- log(E[i])+beta0+sum(Kw[i,])
tkw[i] <- sum(Kw[i,]); s[i] <- tkw[i]-mean(tkw[])
```

```
for (j in 1:m) {d[i,j] <- sqrt(pow(x1[i]-t1[j],2)+pow(x2[i]-t2[j],2))
K[i,j] <- phi(-pow(d[i,j]/sig.k,2));
Kw[i,j] <- K[i,j]*w[j]}}
beta0 ~dflat()
for (j in 1:m) {w[j] ~dnorm(0,tau)}
sd.w ~dunif(0,100); tau <- pow(sd.w,-2)
for (i in 1 : n) { # probs of extreme values of s
p.sig[i] <- step(s[i])
# deviance
dev[i] <- y[i]*log(y[i]/mu[i])-(y[i]-mu[i])}
# Deviance
Dv <- 2*sum(dev[])}
```

The amended lines of code in the model extension are

```
log(mu[i]) <- log(E[i])+beta0+sum(Kw[i,])+del[i]*u[i]
u[i] ~dnorm(0,tau.u); del[i] ~dbern(0.05)
...
sd.u ~dunif(0,100); tau.u <- pow(sd.u,-2).
```

The code for the stick-breaking prior incorporating cluster specific kernels is

```
model {for(i in 1:n){y[i] ~dpois(mu[i])
log(mu[i]) <- log(E[i])+beta0+w[S[i]]
# realized spatial effect
s[i] <- w[S[i]]; pexc.s[i] <- step(s[i])
# deviance elements
dev[i] <- y[i]*log(y[i]/mu[i])-(y[i]-mu[i])
# cluster choice
S[i] ~dcat(p[i,1:m])
g1[i] <- eas[i]/10; g2[i] <- nor[i]/10}
beta0 ~dnorm(0,0.01); for (j in 1:m) {w[j] ~dnorm(0,tauw)}
tauw ~dgamma(1,0.01)
for (i in 1:n) {p[i,1] <- R[i,1]; R[i,m] <- 1
for (j in 2:m) {p[i,j] <- R[i,j]*prod(Rm[i,1:(j-1)])}}
for (j in 1:(m-1)) {v[j] ~dbeta(1,2)
# centres of kernel functions
t1[j] ~dunif(min1,max1); t2[j] ~dunif(min2,max2)
# bandwidth params
gam[j] ~dgamma(1.5,lambda)
for (i in 1:n) {d[i,j] <- pow(t1[j]-g1[i],2) + pow(t2[j]-g2[i],2)
R[i,j] <- exp(-0.5*d[i,j]/gam[j])*v[j]
Rm[i,j] <- 1-R[i,j]}}
# range parameter
lambda ~dunif(0,40)
# Deviance
Dv <- 2*sum(dev[])}
```

5

Regression Techniques Using Hierarchical Priors

5.1 Introduction

This chapter is concerned with the application of hierarchical priors to regression models for univariate metric and discrete responses, where the observation units are non-nested but may be spatially or temporally configured. Nested data applications are considered in Chapters 6 and 8. Particular applications involving latent responses or random effects are when such effects are used:

1. to improve model fit in general linear models in line with distributional assumptions;
2. to generate latent responses on a different scale to the observations;
3. to demonstrate heterogeneity in regression relationships or variance parameters over exchangeable sample units;
4. to represent random regression effects and correlated regression errors for responses structured in time or space.

Let $\{y_1, \ldots, y_n\}$ be observations from the exponential family density,

$$p(y_i|\theta_i, \phi) = \exp\left[\frac{y_i\theta_i - b(\theta_i)}{a_i(\phi)} + c(y_i, \phi)\right], \tag{5.1}$$

with canonical parameter, θ_i, dispersion function, $a_i(\phi) = \phi/w_i$, and w_i known. The mean of y is $\mu_i = E(y_i|\theta_i) = b'(\theta_i)$, linked to predictors X_i via a monotone link function, $g(\mu_i) = X_i\beta$, and the variance is

$$\text{Var}(y_i|\theta_i) = a_i(\phi)b''(\theta_i) = a_i(\phi)\text{Var}(\mu_i).$$

The exponential family includes as special cases the binomial, Poisson, exponential, gamma, and inverse Gaussian densities. The generalized linear model (Dey et al., 2000; McCullagh and Nelder, 1989) scheme extends normal linear regression concepts to such outcomes.

However, counts assumed to be Poisson or binomial often show a residual variance larger than expected under the exponential family models, due to unknown omitted covariates, clustering in the original units, or inter-subject variations in propensity (Albert and Pepple, 1989; Dey and Ravishanker, 2000; Gschlößl and Czado, 2006). Unless such excess dispersion is allowed for, standard errors are likely to be understated. The solution involves regression with conjugate or nonconjugate mixing for the residual variation (Section 5.2), and the focus in Markov Chain Monte Carlo (MCMC) is often on the complete data likelihood rather than the marginal model obtained by integrating over the random effects.

Binary and multinomial regression based on the generalized linear modeling principles is widely applied, with Bayesian strategies described in Dey et al. (2000). Sampling and inference in Bayesian general linear models is, however, complicated to the extent that conjugate priors are only available for normal regression (Holmes and Held, 2006). The auxiliary variable approach (Albert and Chib, 1993; van Dyk and Meng, 2001) circumvents this by introducing latent continuous responses underlying the binary or categorical observations, resulting in a specification (including priors) that effectively replicates normal regression (Section 5.3). This provides simplified MCMC sampling, improved residual tests, and facilitates multivariate analysis involving mixtures of continuous and discrete responses, as in the underlying variable approach in factor analysis (Bartholomew, 1987; Muthen, 1984).

For data assumed conditionally normal, one has $\theta_i = \mu_i = X_i\beta$ and $a_i(\phi) = \sigma^2$, and the canonical form of the normal linear regression model is

$$y_i = X_i\beta + \varepsilon_i, \tag{5.2}$$

where $\varepsilon_i \sim N(0, \sigma^2)$. However, the assumption of homoscedastic normal errors in Equation 5.2 may be restrictive in many modeling situations due to the relatively thin tails of the normal, particularly when unusual observations are present. Geweke (1993), Lange et al. (1989), and West (1984) consider regression based on a wider class of scale mixtures of normals, which leads to a varying scale parameter for each sample unit, and leads to heavier tails than the normal. Qin et al. (2000) propose an alternative scale mixture of uniforms method, which can lead to both heavier and lighter tails than the normal. Other approaches to heteroscedasticity are possible, including variance transformation and variance regression modeling (Cepeda and Gamerman, 2000); see Section 5.4. The other main limitation of Equation 5.2 is the assumption of identical regression effects for all cases. Alternatives are discrete mixture regressions, also known as regression regimes in time series applications (Hurn et al., 2003), and random coefficient models (Swamy and Mehta, 1975)—see Section 5.5.

For structured data observed over neighboring areas or periods, or points in time or space, a mis-specification of regression is likely to be apparent in correlated residuals. Either the error structure will need to accommodate such correlation or the design component of the regression model will need to be extended to reduce residual correlation. In time series regression, relatively simple autoregressive or moving average error structures are often sufficient to remove residual correlation, though time-varying regression effects are often relevant in particular applications—see Sections 5.6 and 5.7.

In spatial data modeling, correlated residuals may be attributable to omitted predictors, nonlinear effects, and spatial heterogeneity—see Section 5.8. In particular, varying regression effects over space may be indicated for some or all observed predictors and the prior governing the randomly varying coefficients will be spatially structured—see Section 5.9. While classical approaches center on geographically weighted regression, random spatially varying coefficient models based on Bayesian principles (Assunção, 2003; Gamerman et al., 2003) arguably provide greater inferential flexibility.

5.2 Regression for Overdispersed Discrete Data

For discrete data (Poisson, binomial, multinomial), overdispersion is typically due to unobserved variations between subjects (also called frailty effects) that are not represented by the observed covariates. For time series data, another possibility is contagion, violating the Poisson assumption that events occur randomly in time (Winkelmann and Zimmerman, 1995). Particular types of response pattern (e.g., an excess proportion of zero counts as compared to the expected Poisson frequency) may also cause overdispersion (Hall, 2000). Without correction for such extra-variability, regression parameter estimates may be biased, and their credible intervals will be too narrow, so that incorrect inferences about significance may be obtained (Cameron and Trivedi, 1998).

For example, the Poisson regression model for count data assumes that the mean and variance are equal, but overdispersion as compared to the Poisson assumption is routinely encountered. As discussed in Chapter 3, the conjugate mixture model for count data is the Poisson-gamma with,

$$y_i \backsim Po(\mu_i),$$
$$\mu_i \backsim Ga(\alpha_i, \eta_i).$$

Denoting the mean of μ_i as $\xi_i = \alpha_i/\eta_i$, one obtains $\text{Var}(\mu_i) = \alpha_i/\eta_i^2 = \xi_i^2/\alpha_i$ and

$$\text{Var}(y_i) = E[\text{Var}(y_i|\mu_i)] + \text{Var}[E(y_i|\mu_i)] = \xi_i + \xi_i^2/\alpha_i,$$

so providing overdispersion as α_i becomes smaller. The mean is modeled by regression, typically involving fixed effects only, with $\xi_i = \exp(\beta_0+\beta_1 x_{1i}+\cdots+\beta_p x_{pi})$. Identification requires constraints on the gamma mixture parameters, such as $\alpha_i = \alpha$ in the $\{\xi_i, \alpha_i\}$ parameterization, namely, $\mu_i \backsim Ga(\alpha, \alpha/\xi_i)$. Then with $\phi = 1/\alpha$, one has a quadratic variance function, denoted as NB2 (e.g., Cameron and Trivedi, 1998, 71) with,

$$\text{Var}(y_i) = E[\text{Var}(y_i|\mu_i)] + \text{Var}[E(y_i|\mu_i)] = \xi_i + \phi\xi_i^2. \tag{5.3}$$

Another possibility (Fahrmeir and Osuna, 2006; Greene, 2007) is to set $\mu_i = \xi_i\omega_i$, where $\omega_i \backsim Ga(\alpha, \alpha)$ so that the frailties average 1, with variance $\phi = 1/\alpha$. Integrating out ω_i leads to a marginal negative binomial density for y_i, namely,

$$p(y_i|\beta, \alpha) = \frac{\Gamma(\alpha + y_i)}{\Gamma(\alpha)\Gamma(y_i + 1)} \left(\frac{\alpha}{\alpha + \xi_i}\right)^{\alpha} \left(\frac{\xi}{\alpha + \xi_i}\right)^{y_i}.$$

Bayesian approaches to the negative binomial include Bradlow et al. (2002), Chun and Sumichrast (2007), and Dauxois et al. (2006). Dauxois et al. (2006) adopt prior distributions on the variance function coefficients to encompass Poisson, binomial, and negative binomial models simultaneously and decide which provides a better fit. Fahrmeir and Osuna (2006) adopt a $Ga(a, b)$ prior for the overdispersion parameter α, with $a = 1$, and with $b \sim Ga(1, 0.005)$ taken as an extra unknown. They show that the full conditional for this parameter then has no closed analytical form, so that a M-H algorithm is required to sample values with a random walk proposal.

A more general NBk form is described by Winkelmann and Zimmermann (1995), and involves a variance function,

$$\text{Var}(y_i) = E[\text{Var}(y_i|\mu_i)] + \text{Var}[E(y_i|\mu_i)] = \xi_i + \phi\xi_i^{k+1}, \tag{5.4}$$

with $k \geq -1$. This is obtained by an extra parameter in the mixing prior, namely,

$$\mu_i \backsim Ga\left(\frac{\xi_i^{1-k}}{\phi}, \frac{\xi_i^{-k}}{\phi}\right).$$

The values $k = 0$ and $k = 1$ lead to variance forms (NB1 and NB2) that are linear and quadratic in ξ_i, namely, $\text{Var}(y_i) = \xi_i + \phi\xi_i = (1 + \phi)\xi_i$ and $\text{Var}(y_i) = \xi_i + \phi\xi_i^2$, respectively.

Nonconjugate random mixture models are often adopted for count data, with normal or Student t errors in the log link (Kim et al., 2002). This is a more common approach when multiple or multilevel random effects are to be considered, with an example being the convolution prior (Besag et al., 1991) for area disease events, y_i, in population totals, P_i. Thus, for rare events, $y_i \backsim Po(P_i\mu_i)$, where,

$$\log(\mu_i) = X_i\beta + \varepsilon_i + s_i,$$

and both random effects $\{\varepsilon_i, s_i\}$ may account for overdispersion, but the ε_i are unstructured (exchangeable with regard to area identifiers) while the s_i are spatially structured. For count regressions only involving an unstructured error, one may specify,

$$\mu_i = E(y_i|X_i, \varepsilon_i) = \exp(X_i\beta + \sigma\varepsilon_i),$$

with $\varepsilon_i \sim N(0,1)$. Denoting $\nu_i = \exp(X_i\beta)$, the unconditional mean (Greene, 2007) is

$$E(y_i|X_i) = E_\varepsilon[E(y_i|X_i,\varepsilon_i)] = \nu_i \exp(\sigma^2/2),$$

and the unconditional variance is

$$\begin{aligned}\text{Var}(y_i|X_i) &= E_\varepsilon[\text{Var}(y_i|X_i,\varepsilon_i)] + \text{Var}_\varepsilon[E(y_i|X_i,\varepsilon_i)] \\ &= \nu_i \exp(\sigma^2/2)\{1 + \nu_i \exp(\sigma^2/2)[\exp(\sigma^2) - 1]\}.\end{aligned}$$

Taking $\phi = e^{\sigma^2} - 1$,

$$\text{Var}(y_i|X_i) = E(y_i|X_i,\varepsilon_i)[1 + \phi E(y_i|X_i,\varepsilon_i)],$$

showing that the variance has a quadratic form, as for the NB2 form of the negative binomial.

5.2.1 Overdispersed binomial and multinomial regression

Binomial regression with excess variation may occur when responses are arranged in clusters and responses from the same cluster are correlated: examples occur in teratological studies (e.g., when the observation unit is a litter of animals, and litters differ in terms of unknown genetic factors). Crowder (1978) assumed a conjugate approach with a beta distributed success probability, leading to a beta-binomial regression model—this form of overdispersion model is considered in Bayesian terms by Kahn and Raftery (1997). Thus, with $y_i \sim Bin(n_i, p_i)$, one assumes,

$$p_i \sim Beta(\gamma\pi_i, (1-\pi_i)\gamma),$$

with mean π_i and variance,

$$\pi_i(1-\pi_i)/(\gamma+1),$$

where $\gamma \geq 0$. Regression on known predictors involves a logit or other link,

$$g(\pi_i) = X_i\beta.$$

Setting $\varphi = (\gamma+1)^{-1}$, the unconditional variance of a beta-binomial response is of the form (Collett 2002, 201),

$$\text{Var}(y_i) = n_i p_i(1-p_i)[1 + (n_i - 1)\varphi].$$

Possible priors on the precision parameter γ include $P(\gamma) \propto 1/\gamma$, and (Albert, 1988),

$$P(\gamma) = 1/(1+\gamma)^2.$$

Nonconjugate random mixture models are often adopted for binomial data, with normal or Student t errors in the regression link (whether logit, probit, or complementary log-log). The presence of an error term permits regression variable selection using a g-prior approach (Zellner, 1983), which avoids specifying the prior covariances for the elements of β and instead assumes these covariances are a multiplier of those provided by the observations. Thus, Gerlach et al. (2002) and Kinney and Dunson (2007) propose variable and random effects selection in mixed logistic models, with g-priors on the fixed effects. A logit link example is

$$\begin{aligned} \text{logit}(\pi_i) &= X_i\beta + \varepsilon_i, \\ \varepsilon_i &\sim N(0, \sigma^2), \end{aligned}$$

with g-prior ($g > 1$, typically large),

$$\beta \sim N(B, \sigma^2 g(X'X)^{-1}).$$

For multinomial data (e.g., on voting patterns, y_{ij}, for parties, j, by constituency, i) overdispersion may occur when choice probabilities vary between N_i individuals in each observation unit, but clusters of individuals within each unit have similar probabilities. The individual level factors associated with such clustering are not observed, so a random effect will proxy such unobserved factors; for example, voters with different education levels may differ in their voting preferences, but only the average education in each constituency is observed. The raw percentages, y_{ij}/N_i, are also likely to show erratic features, whereas hierarchical models for pooling strength over units provide frequency smoothing and model interdependencies between categories.

This form of data may be modeled as a product multinomial likelihood conditioning on known $N_i = y_{i+}$. With probabilities, π_{ij}, of choices, $j = 1, \ldots, J$, the sampling model is

$$y_{ij} \sim M(N_i, [\pi_{ij}, \ldots, \pi_{iJ}]) \qquad i = 1, \ldots, n.$$

The conjugate approach for such heterogeneity is the multinomial-Dirichlet mixture, where the Dirichlet is the multivariate generalization of the beta density. However, the Dirichlet has a restricted covariance structure when there are dependencies between the response categories, j, within units, i. For example, for n constituencies and J political parties, one may expect both negative and positive correlations between π_{ij} for different parties. Greater flexibility is provided by modeling heterogeneity within the regression link, as in random effects multiple logit models (Hensher and Greene, 2003), or via multinomial probit models (Hausman and Wise, 1978).

Under the multiple logit form, define a $J - 1$ dimensional random effect, $\alpha_i = (\alpha_{i1}, \ldots, \alpha_{i,J-1})$, representing subject or unit level intercepts; these might be exchangeable or correlated (if, say, the units were areas and behaviors were spatially clustered). Then with X_i excluding an intercept,

$$\pi_{ij} = \exp(\alpha_{ij} + X_i\beta_j) \Big/ \sum_{k=1}^{J} \exp(\alpha_{ik} + X_i\beta_k),$$

with $\alpha_{iJ} = \beta_J = 0$ for identification. For example, one may assume,

$$(\alpha_{i1}, \ldots, \alpha_{i,J-1}) \sim N_{J-1}(A, D), \tag{5.5}$$

where D is an unknown covariance matrix, and A is the average intercept. This model may also be fitted by Poisson regression using the fact that the multinomial is equivalent to a Poisson distribution conditional on a fixed total; this involves defining n fixed effect predictors, a_i, to ensure the unit totals, N_i, are maintained. Thus, $y_{ij} \sim Po(\mu_{ij})$, with,

$$\log(\mu_{ij}) = a_i + \alpha_{ij} + X_i\beta_j,$$

for $j = 1, \ldots, J$, where a_i would typically be fixed effects assigned vague priors, e.g., $a_i \sim N(0, 1000)$.

Example 5.1. Crab Data This example considers Poisson overdispersion in terms of different negative binomial regression forms. The dataset is the crab data from Agresti (1996) with $n = 173$ observations relating the number of satellite males to the predictor x=carapace width. Overdispersion in the data is apparent with variance of 9.9 exceeding the mean of 2.9. A simple Poisson regression for these data assumes $\log(\mu_i) = \beta_1 + \beta_2 x_i$, and deviance (minus twice the posterior mean likelihood) is 925.2. By contrast, under a Poisson-gamma mixture with,

$$\begin{aligned} y_i &\backsim Po(\xi_i\omega_i), \\ \omega_i &\backsim Ga(\alpha, \alpha), \\ \log(\xi_i) &= \beta_0 + \beta_1 x_i, \end{aligned}$$

the Poisson deviance is 539 with posterior mean and 95% CI for α of 0.95 (0.65, 1.35). This can be estimated using the marginal negative binomial NB2 likelihood or the complete data Poisson-gamma likelihood, with the latter approach having the benefit of providing observation-specific frailties (Fahrmeir and Osuna, 2006). The coefficient β_2 under the NB2 model has mean 0.195 (0.10, 0.29), while ϕ is estimated as 1.1 (0.75, 1.55). This model may also be estimated in BayesX[1], under which the overdispersion parameter α has posterior mean 0.88, and mean deviance (minus twice the NB2 likelihood) of 754.4 and DIC of 757.4.

The suitability of the NB1 or NB2 form of negative binomial regression may be assessed using the NBk model, whereby for $y_i \sim Po(\mu_i)$ one has,

$$\mu_i \backsim Ga\left(\frac{\xi_i^{1-k}}{\phi}, \frac{\xi_i^{-k}}{\phi}\right),$$

and

$$\mathrm{Var}(y_i) = E[\mathrm{Var}(y_i|\mu_i)] + \mathrm{Var}[E(y_i|\mu_i)] = \xi_i + \phi\xi_i^{k+1}.$$

Gamma priors with scale and index parameters of 1 are adopted for ϕ and $k+1$. Since the predictor values are relatively large, a $N(0,1)$ prior is adopted on its regression coefficient to avoid numeric overflow. The second half of a two chain run of 10,000 iterations gives posterior means (95% interval) for k and ϕ of -0.39 (-0.93, 0.31) and 4.3 (2.0, 7.4). The posterior interval for k excludes 1, and so does not favor the NB2 parameterization. In fact, the overdispersion term in the variance function is close to involving a square root in the mean ξ_i. The mean deviance is reduced to 527 under the NBk model, and the coefficient on x is more precisely identified with mean and 95% CI of 0.19 (0.13, 0.25).

Example 5.2. Voting in Florida, 2000: Multinomial Overdispersion
This example considers normal random effects to model multinomial overdispersion via multiple logit links. The analysis relates to 2000 US presidential election voting data, y_{ij}, for $i = 1, \ldots, 67$ Florida polling districts and with N_i denoting total votes (Mebane and Sekhon, 2004). There are $J = 5$ choices (Buchanan, Nader, Gore, Bush, other) and three predictors:

1. x_1, the proportion of each county's votes for different presidential candidates in 1996;
2. x_2, changes between 1996 and 2000 in party registration;
3. x_3, percent of Census population Cuban in district i.

Specification of x_1 and x_2 (predictors specific for area and candidate) follows Mebane and Sekhon (2004), but x_3 differs from their variable. Mebane and Sekhon (2004) find substantial overdispersion in these data.

The sampling model is

$$y_{ij} \sim \mathrm{Mult}(N_i, [\pi_{i1}, \ldots, \pi_{i5}]) \qquad i = 1, \ldots, n,$$

$$\pi_{ij} = \phi_{ij} \Big/ \sum_{ij} \phi_{ij},$$

and to account for overdispersion, normal effects as in Equation 5.5 are included in multiple logit links. These are denoted α_{ij} and are of dimension $J-1$ with nonzero means A_j, namely, the intercepts for the first four choices. So,

$$\log(\phi_{ij}) = \alpha_{ij} + X_i\beta_j,$$
$$(\alpha_{i1}, \ldots, \alpha_{i,J-1}) \sim N_{J-1}(A, D),$$
$$\log(\phi_{iJ}) = 0,$$

with $\{\beta_{j,k}; j = 1,4, k = 1,3\}$ and A_j assigned flat priors, and the precision matrix, D^{-1}, assigned a Wishart prior with identity scale matrix and $J-1 = 4$ degrees of freedom[2].

Despite centering of predictors and taking mean random effects to be intercepts, convergence is delayed to over 50,000 iterations in certain β coefficients when widely separated initial values (for two chains) are chosen. The scale of the predictors is quite unusual in these data (evident in a very large coefficient for $\beta_{2,2}$), and possibly standardizing predictors would improve MCMC performance. Nevertheless, a posterior predictive check comparing chi-square values for replicate and observed data (Gelman et al., 1996) is satisfactory at around 0.30. This is not the case when a fixed effects only model is applied with,

$$\log(\phi_{ij}) = A_j + X_i\beta_j \qquad j = 1, \ldots, J-1.$$

There is then zero posterior probability that $\chi^2(y_{\text{rep}}, \theta) > \chi^2(y, \theta)$. Standard deviations of predictor effects are also considerably understated if allowance is not made for excess variation.

5.3 Latent Scales for Binary and Categorical Data

Prior specification and sampling in binary and categorical regression is often complicated by the nonconjugacy in the standard general linear model (GLM) approach (Holmes and Held, 2006). For example, sampling strategies for the logistic model include Metropolis–Hastings updates combined with an approximate posterior density for the regression coefficients obtained via iteratively reweighted least squares (Gamerman, 1997), and adaptive rejection sampling from the univariate conditional densities of the coefficients (Dellaportas and Smith, 1993). An alternative is to augment the observations with latent data on a metric scale.

Consider first binary responses. One may assume latent metric data, y^*, such that $y = 1$ when $y^* > 0$ and $y = 0$ when $y^* \leq 0$ (Albert and Chib, 1993). In economic choice applications (e.g., regarding economic participation or not), the latent scale y^* arises by comparing utilities U_{1i} and U_{0i} of options 1 and 0 with,

$$\begin{aligned} U_{ji} &= V_{ji} + \varepsilon_{ji} = X_i\beta_j^* + \varepsilon_{ji}, \\ y_i^* &= U_{1i} - U_{0i}. \end{aligned} \tag{5.6}$$

In other applications, y^* and U may be conceptualized differently; thus Heringstad et al. (2001) consider a threshold-liability model for the analysis of clinical mastitis as a binary response. Under the scheme (Equation 5.6) one has,

$$Pr(y_i = 1) = Pr(y_i^* > 0) = Pr(\varepsilon_{0i} - \varepsilon_{1i} < V_{1i} - V_{0i}) = Pr(\varepsilon_{0i} - \varepsilon_{1i} < X_i\beta),$$

where $\beta = \beta_1^* - \beta_0^*$. Alternative forms for ε lead to different links: taking ε_{ji} to be normal with mean zero and variance σ^2 leads to a probit link with

$Pr(y_i = 1) = \Phi(X_i\beta/\sigma)$. It is apparent that β and σ cannot be separately identified, and the commonest identifying device takes $\sigma^2 = 1$.

A probit regression with binary responses, y_i, may therefore be obtained by truncated normal sampling for y_i^* with the form of constraint determined by the observed y. Thus, if $y_i = 1$, y_i^* is constrained to be positive, and sampled from a normal with mean $X_i\beta$ (including an intercept in p-dimensional X_i) and variance 1. If $y_i = 0$, y_i^* is sampled from the same density but constrained to be negative. With a normal prior on the coefficients $\beta \sim N_p(B_0, V_0)$, the full conditional distribution of β is also normal, namely,

$$\beta|y^* \sim N(B, V),$$
$$B = V^{-1}(V_0^{-1}B_0 + X'y^*),$$
$$V = (V_0^{-1} + X'X)^{-1}.$$

Van Dyk and Meng (2001) use a "working parameter" sampling approach that retains σ^2 as an unknown, and amounts to the scheme,

$$y_i^* \sim N(X_i\beta\sigma, \sigma^2)I(0,) \qquad y_i = 1,$$
$$y_i^* \sim N(X_i\beta\sigma, \sigma^2)I(,0) \qquad y_i = 0.$$

Holmes and Held (2006) propose an alternative strategy to reduce autocorrelation and improve mixing in MCMC sampling by updating y^* and β jointly, and justified by the factorization,

$$p(\beta, y^*|y) = p(y^*|y)p(\beta|y^*),$$

where updating of β is as above, but y^* is updated from its marginal distribution integrated over β.

Heavier-tailed links are obtained by sampling y_i^* directly from a Student t with ν degrees of freedom, or by using the scale mixture version of the Student t density. This again involves constrained normal sampling but with gamma distributed subject-specific precisions $\lambda_i \sim Ga(\nu/2, \nu/2)$, so that,

$$y_i^* \sim N(X_i\beta, 1/\lambda_i)I(0, \infty) \text{ when } y_i = 1,$$
$$y_i^* \sim N(X_i\beta, 1/\lambda_i)I(-\infty, 0) \text{ when } y_i = 0.$$

For example, Chang et al. (2006) consider binary mastistis data in cattle, and argue that replacing a Gaussian model for the underlying liabilities with a heavy-tailed distribution for the underlying liabilities is expected to lead to more robust inferences. As well as a gamma density for λ_i, they consider a slash density, $p(\lambda_i|\nu) = \nu\lambda_i^{\nu-1}$. Skew densities for ε in Equation 5.6 have also been proposed. Thus, Bazan et al. (2006) mention a skew-probit link with augmentation scheme,

$$y_i^* = X_i\beta + \varepsilon_i,$$
$$\varepsilon_i = \sigma[-\delta V_i - (1 - \delta^2)W_i], \tag{5.7}$$

where V_i is half normal $V_i \sim HN(0,1)$, $W_i \sim N(0,1)$, $\delta \sim U(-1,1)$, and $\sigma = 1$ for identifiability. In hierarchical form, one has,

$$y_i^* \sim N(X_i\beta - \delta V_i, 1 - \delta^2).$$

Taking ε to be logistic, a logit regression is obtainable (e.g., Kinney and Dunson, 2007), by the augmentation scheme,

$$y_i^* \sim \text{logistic}(X_i\beta, 1)I(0,\infty) \text{ when } y_i = 1,$$
$$y_i^* \sim \text{logistic}(X_i\beta, 1)I(-\infty,0) \text{ when } y_i = 0,$$

where $y \sim \text{logistic}(\mu, \tau)$, when,

$$p(y|\tau,\mu) = \tau \exp(\tau[y-\mu])/\{1+\exp(\tau[y-\mu])\}^2,$$

with variance κ^2/τ^2, where $\kappa^2 = \pi^2/3$.

Several other approaches to generate the latent data underlying logit choice model have been suggested. Thus, the logit link can be obtained approximately by Student t sampling when $\nu = 8$, or equivalently by scale mixture normal sampling with $\lambda_i \sim Ga(\nu/2, \nu/2)$ (Albert and Chib, 1993), combined with constrained sampling according to the observed y values. Specifically, a t_8 variable is approximately 0.634 times a logistic variable, so that,

$$y_i^* \sim t_8\left(X_i\beta, \frac{1}{0.634^2}\right) I(0,\infty) \text{ when } y_i = 1,$$
$$y_i^* \sim t_8\left(X_i\beta, \frac{1}{0.634^2}\right) I(-\infty,0) \text{ when } y_i = 0.$$

Equivalently with $\lambda_i \sim Ga(4,4)$,

$$y_i^* \sim N\left(X_i\beta, \frac{1}{\lambda_i(0.634)^2}\right) I(0,\infty) \text{ when } y_i = 1,$$
$$y_i^* \sim N\left(X_i\beta, \frac{1}{\lambda_i(0.634)^2}\right) I(-\infty,0) \text{ when } y_i = 0.$$

Kinney and Dunson (2007) mention a slightly different approximation whereby a $t_{7.3}$ variable is approximately 0.647 times a logistic variable. So with $\lambda_i \sim Ga(\nu/2, \nu/2)$, where $\nu = 7.3$, and $\tilde{\sigma}^2 = \pi^2(\nu-2)/3\nu$,

$$y_i^* \sim N\left(X_i\beta, \tilde{\sigma}^2/\lambda_i\right) I(0,\infty) \text{ when } y_i = 1,$$
$$y_i^* \sim N\left(X_i\beta, \tilde{\sigma}^2/\lambda_i\right) I(-\infty,0) \text{ when } y_i = 0.$$

Logit models relate responses $y_i = 0$ or 1 to predictors X_i through proportional exponential functions of regressors,

$$Pr(y_i = k) \propto \exp\{\eta_k(X_i, \beta)\}$$

where η_k is a linear or nonlinear function. With $\eta_k(X_i\beta) = X_i\beta$, a latent exponential variable version (Scott, 2003) of the logit link involves sampling $\{z_{0i}, z_{1i}\}$ from exponential densities $E(\lambda)$, with parameters $\lambda_{0i} = 1$ and $\lambda_{1i} = \exp(X_i\beta)$. If $y_i = \text{argmin}(z_{0i}, z_{1i})$, then $Pr(y_i = k|X_i) \propto \lambda_{ki}$ as under a logit regression[3]. This principle extends to multiple logit regression with J categories, implemented by sampling $\{z_{0i}, z_{1i}, \ldots, z_{J-1,i}\}$.

In related work, Fruhwirth-Schnatter and Fruhwirth (2007) implement augmented data sampling for the logit model by using a discrete mixture approximation of the type 1 extreme value error in the McFadden (1974) formulation of the logit model. Thus with U_{0i} and U_{1i} as utilities of category 0 and 1, and

$$U_{1i} = X_i\beta + \varepsilon_i,$$

the binary logit is obtained when U_{0i} and ε_i follow type 1 extreme value distributions. Using the relation between the exponential and type 1 extreme distributions, and with $\nu_i = \exp(X_i\beta)$, one has,

$$\exp(-U_{0i}) \sim E(1), \exp(-U_{1i}) \sim E(\nu_i),$$

with the minimum of these variables also exponential,

$$\min[\exp(-U_{0i}), \exp(-U_{1i})] \sim E(1 + \nu_i). \qquad (5.8)$$

When $y_i = 1$, one has $U_{1i} > U_{0i}$, or equivalently $\exp(-U_{1i}) < -\exp(-U_{0i})$, so that from Equation 5.8,

$$\exp(-U_{1i}) \sim E(1 + \nu_i).$$

When $y_i = 0$, one has $U_{0i} > U_{1i}$, or equivalently $\exp(-U_{0i}) < -\exp(-U_{1i})$, so that,

$$\exp(-U_{0i}) \sim E(1 + \nu_i),$$
$$\exp(-U_{1i}) = \exp(-U_{0i}) + \delta_i,$$

where,

$$\delta_i \sim E(\nu_i).$$

A useful diagnostic feature resulting from the latent response approach is that the residuals, $y_i^* - X_i\beta$, are nominally a random sample from the assumed cumulative distribution for ε (Johnson and Albert, 1999). So for the latent data probit, the residual, $\varepsilon_i = y_i^* - X_i\beta$, is approximately $N(0, 1)$ if the model is appropriate for case i, whereas if the posterior distribution of ε_i is significantly different from $N(0, 1)$, then the model conflicts with the observed y. So one might obtain the probability $Pr(|\varepsilon_i| > 2|y)$ and compare it to its prior value, which is 0.045. For the latent data logit, one may obtain $Pr(|\varepsilon_i|/\kappa > 2|y)$, while for the logistic approximation, one monitors $Pr(|\varepsilon_i|\lambda_i > 2|y)$.

5.3.1 Augmentation for ordinal responses

Suppose that the categorical response, y_i, has J categories, with the observations measuring a latent response, y^*, according to the model,

$$y_i = j \text{ if } \alpha_{j-1} \leq y_i^* < \alpha_j.$$

The α_j are cut points dividing the values of y^* according to the observed y values. The model in the latent data (e.g., Spiess, 2006) is then,

$$y_i^* = X_i\beta_j + \varepsilon_{ji},$$

where ε_{ji} is usually either normally or logistically distributed. So $P(\varepsilon) = \Phi(\varepsilon)$, where Φ is the cumulative normal function, or $P(\varepsilon) = 1/(1+\exp(-\varepsilon))$.

The corresponding model for cumulative probabilities is

$$\begin{aligned} Pr(y_i^* \leq \alpha_j) &= Pr(X_i\beta_j + \varepsilon_{ji} \leq \alpha_j), \\ &= Pr(\varepsilon_{ji} \leq \alpha_j - X_i\beta_j). \end{aligned}$$

Thus,

$$Pr(y_i^* \leq \alpha_j) = \Phi(\alpha_j - X_i\beta_j),$$

or

$$Pr(y_i^* \leq \alpha_j) = 1/(1+\exp(-[\alpha_j - X_i\beta_j])),$$

according to the assumed form for ε_{ji}. Let $\gamma_{ji} = Pr(y_i^* \leq \alpha_j)$, then,

$$Pr(y_i = j) = Pr(\alpha_{j-1} \leq y_i^* < \alpha_j) = \gamma_{ji} - \gamma_{j-1,i}.$$

The probability that $y_i = 1$, namely,

$$Pr(y_i = 1) = Pr(\alpha_0 \leq y_i^* < \alpha_1) = \gamma_{1i},$$

is obtained by setting $\alpha_0 = -\infty$, while the probability that $y_i = J$, namely,

$$Pr(y_i = J) = Pr(\alpha_{j-1} \leq y_i^* < \alpha_J) = 1 - \gamma_{J-1,i},$$

is obtained by setting $\alpha_J = \infty$.

Assuming X_i excludes an intercept, the remaining $J-1$ cut points $\{\alpha_1, \alpha_2, \ldots, \alpha_{J-1}\}$ are unknowns subject to an order constrained prior $\alpha_1 \leq \alpha_2 \leq \cdots \leq \alpha_{J-1}$. By reparameterization,

$$\begin{aligned} \alpha_j &= \alpha_{j-1} + \exp(\Delta_j) \qquad (J > j > 1), \\ \alpha_1 &= \Delta_1, \end{aligned}$$

one may, however, specify unconstrained normal priors such as $\Delta_j \backsim N(0, V_\Delta)$, where V_Δ is preset or possibly itself unknown.

An equivalent specification of this model involves sets of $J - 1$ binary variables for each subject, namely, $z_{ji} = 1$ if $y_i \leq j$, and $z_{ji} = 0$ otherwise (e.g., Parsons et al., 2006, 512). So if $J = 3$, and if $y_i = 1$, then $z_{1i} = 1$, $z_{2i} = 1$; if $y = 2$, then $z_{1i} = 0$, $z_{2i} = 1$. So for ε normal,

$$Pr(y_i \leq j) = Pr(y_i^* \leq \alpha_j) = Pr(z_{ji} = 1) = \Phi(\alpha_j - X_i\beta_j).$$

Example 5.3. Irish Education Attainment This example involves a probit regression model for education data from Raftery and Hout (1993) relating to $n = 500$ Irish school pupils. A binary leaving certificate indicator is the response variable and explanatory variables are the Student's Verbal Reasoning Test Score, the Student's sex (1 = female, 0 = male), and father's occupational status. The third covariate has some missing values that are treated as missing at random.

To account for possible skewness in residuals the approach of Bazan et al. (2005) is applied, as in Equation 5.7, but modified to include a model selection indicator applied to the skewness parameter δ. Thus[4],

$$\begin{aligned} y_i^* &\sim N\left(\mu_i, \left[1 - k_\delta^2\delta^2\right]\right) I(A_i, B_i), \\ \mu_i &= X_i\beta - k_\delta\delta V_i, \\ V_i &\sim N(0,1), \end{aligned}$$

where the sampling intervals $\{A_i, B_i\}$ depend on the observed leaving certificate indicator, y. A uniform prior is adopted for the skew parameter δ, namely, $\delta \sim U(-1,1)$. The binary selection indicator has prior $k_\delta \sim Bern(\pi_\delta)$, where $\pi_\delta = 0.5$.

With inferences from iterations 5001 to 25,000 of a two chain run, all three predictors emerge as significant with means and 95% intervals of 0.034 (0.022, 0.044), 0.29 (0.05, 0.53), and 0.022 (0.013, 0.031). The posterior probability that $k_\delta = 1$ is 0.44, compared to a prior probability of 0.5, so no significant residual skew seems present.

Example 5.4. Delegation of Discretion in Trade Policy Epstein and O'Halloran (1996, 388) apply an ordered probit model to analyze changes in discretion in trade policy delegated to the US President by Congress between 1890 and 1990 (giving $T = 99$ observations). The response has $J = 3$ categories: 3 if the President's discretion is increased between successive years, 2 if it stays the same, and 1 if it is reduced. They relate changes in discretion to $p = 4$ predictors, namely, changes in log gross national product (x_1), changes in the log unemployment rate (x_2), changes in the log of the producer price index (x_3), and to a variable measuring *changes* in government disunity (x_4), where disunity in a particular year is measured by a trichotomy according to whether one or both chambers of Congress are in the same political party as the President. So x_4 can take values $\{-2, -1, 0, 1, 2\}$.

Following Epstein and O'Halloran, a proportional odds model[5] over responses j is assumed, namely, $\beta_{jk} = \beta_k$ $(k = 1, \ldots, p)$. Order constrained

$N(0,1)$ priors are assumed on the unknown cut points $\{\alpha_1, \alpha_2\}$ and a multivariate normal (MVN) prior on $\{\beta_1, \ldots, \beta_4\}$ with mean zero and diagonal precision matrix B_0, with prior variances of 1000.

A two chain run of 5000 iterations (with the last 4000 for inference) gives a nonsignificant posterior mean (95% CI) for the impact β_1 of x_1, namely, of 0.96 $(-5.0,\ 7.1)$. The 95% intervals for the impacts of x_2 and x_3 are inconclusive, though the posterior densities are biased to negative values. β_4 has mean -0.42 $(-0.82,\ -0.03)$, an estimate similar to the maximum likelihood estimate of -0.46 reported by Epstein and O'Halloran. The percentage of years accurately predicted by replicate responses is 62.6%. Similar coefficients are obtained from an augmented data approach[5] involving binary responses, $z_{tj} = 1$ if $y_t \leq j$, and $z_{tj} = 0$ otherwise.

5.4 Nonconstant Regression Relationships and Variance Heterogeneity

The canonical form of the linear normal regression model assumes constant regression relationships and a constant residual variance over all units. However, heteroskedasticity and varying regression effects over sample subsets are well-known problems in the regression literature. Apparent heteroscedasticity in residuals may in fact be due to varying effects of predictors: consider a linear model specified as $y_i = x_i\beta + w_i\gamma + v_i$, when in fact the true model is $y_i = x_i\beta + w_i(\gamma_\mu + c_i) + u_i$ with $\text{Var}(c_i) = \sigma_c^2$ and $\text{Var}(u_i) = \sigma_u^2$. The error v_i will then have nonconstant variance, $w_i^2\sigma_c^2 + \sigma_u^2$. In many applications, heteroscedasticity is apparent also in increasing dispersion of the residuals at higher fitted values. Nonconstant variance may be reduced or eliminated by making the variance a function of the predictors themselves, or by outcome and predictor term transform (Carroll and Ruppert, 1984). Both options may be combined with modeling of the variance.

Carroll and Ruppert (1982) consider the heteroscedastic linear model,

$$y_i = \eta_i + \sigma_i z_i,$$

with $\eta_i = X_i\beta$, z a zero mean symmetric density with known variance (e.g., normal with variance 1), and σ_i a possibly nonlinear function of η_i or X_i. Possibilities might be simply $\sigma_i = \exp(X_i\gamma)$, or $\sigma_i^2 = \exp(X_i\gamma)$ as in Cepeda and Gamerman (2000), where γ includes an intercept. This may be modified to $\sigma_i = \exp(W_i\gamma)$, where W_i may contain some or all of the variables in X_i. Alternatively, separate variance parameters may be specified, as in the forms,

$$\sigma_i = \sigma|\eta_i|^\lambda,$$
$$\sigma_i = \sigma\left(1 + \lambda\eta_i^2\right)^{0.5}.$$

The approach of Geweke (1993) involving a scale mixture distribution for regression errors can be used to deal with heteroscedasticity as well as with outliers; see also Fonseca et al. (2008). The class of scale mixtures of normals accommodates a wide variety of heavy-tailed distributions, as demonstrated by Fernandez and Steele (2000). In the linear model, $y_i = \eta_i + \sigma\varepsilon_i$, a scale mixture is generated by assuming the residuals are distributed as,

$$\varepsilon_i = z_i/\lambda_i^{0.5},$$

where z_i are $N(0,1)$, and λ_i are independent positive random variables. The t_ν distribution can be interpreted as a scale mixture of normal distributions by taking λ_i to be gamma with scale and shape $\nu/2$, with the Cauchy obtained by taking $\nu = 1$. Fonseca et al. (2008) consider alternative priors for ν, including the Jeffreys prior.

In a related approach, Qin et al. (2000) consider variance regression involving a scale mixture of uniforms. They show that $y_i \sim N(\mu_i, \sigma^2/\lambda)$ is equivalent to,

$$y_i|h_i \sim U\left(\mu_i - \sigma\sqrt{h_i}, \mu_i + \sigma\sqrt{h_i}\right),$$
$$h_i \sim Ga(1.5, \lambda/2),$$

with $h_i \sim Ga(1.5, 0.5)$ leading to the baseline normal $y_i \sim N(\mu_i, \sigma^2)$. Taking,

$$h_i \sim Ga(3\lambda/2, \lambda/2),$$

leads to alternative forms of kurtosis, namely, lighter tails than normal under $\lambda > 1$, and heavier tails than normal under $1 > \lambda > 0$. Variance regression may be achieved by the parameterization,

$$y_i|h_i \sim U\left(\mu_i - \frac{\sqrt{h_i}}{P_i}, \mu_i + \frac{\sqrt{h_i}}{P_i}\right),$$
$$P_i = \prod_{k=0}^{K} \gamma_k^{w_{ki}},$$

where $w_{0i} = 1$, $W_i = (1, w_{1i}, \ldots, w_{Ki})$ are positive covariates influencing the variance, and γ_k are positive coefficients. For example, one may adopt gamma priors, $\gamma_k \sim Ga(a_k, a_k)$, with a_k small.

5.5 Heterogenous Regression and Discrete Mixture Regressions

Heterogeneity in regression relationships is also a familiar question considered under Bayesian random effects perspectives (e.g., Smith, 1973a), following on earlier classical work (e.g., Hildreth and Houck, 1968). Thus, for a normal linear regression and R predictors, x_{ri}, apart from the intercept, a random regression effects model specifies,

$$y_i = \alpha + \sum_{r=1}^{R} x_{ri}(\beta_r + v_{ri}) + u_i,$$

where $\{v_{1i}, \ldots, v_{Ri}\}$ are independent or jointly dependent zero mean random effects. This approach is relevant to exchangeable, non-nested responses with metric predictors. Exchangeable random priors for effects of categorical predictors, x_{ri}, as opposed to metric ones, have been employed in log-linear model and analysis of variance applications. Smith (1973b) considers an analysis of variance application—see McCarthy (2007) for a recent discussion of random factors in analysis of variance—while Albert (1996b, 331) uses exchangeable random effect priors for interaction parameters in log-linear models.

A hierarchical regression approach especially relevant for metric predictors involves the Gaussian process, whereby for a metric response $y = (y_1, \ldots, y_n)'$, and predictor matrix $F = (1, X)$ of dimension $n \times (1 + R)$, one has (Gramacy, 2007; Gramacy and Lee, 2008),

$$\begin{aligned} y &\sim N_n(F\beta, \sigma^2 H), \\ \beta &\sim N_{R+1}(B, V), \\ H_{ij} &= K_{ij} + \gamma I(i = j), \end{aligned}$$

and the correlation matrix K_{ij} between observations i and j is defined in geostatistical terms (see Chapter 4) by distance functions $|x_i - x_j|$ involving the R predictors, excluding the intercept[6]. For example, a separable exponential power function specifies,

$$K_{ij} = \exp\left(-\sum_{r=1}^{R} (x_{ri} - x_{rj})^{p_r} \Big/ d_r \right),$$

where d_r is a range parameter, and $0 < p_r \leq 2$.

Variation in regression regimes between groups of observations is often approached using discrete regression mixtures (Section 5.5.1), while randomly varying regression effects applied to structured data (such as time series or spatially configured data) have received considerable attention also (see Sections 5.7 and 5.9).

5.5.1 Discrete mixture regressions

Discrete mixtures of regressions take the form,

$$p(y_i | x_{1i}, \ldots, x_{Ri}) = \sum_{k=1}^{K} \pi_k f_k(X_i, \beta_k, \phi_k),$$

where β_k denote component regression effects and ϕ_k any other parameters involved in defining densities f_k. They have found wide application, with Bayesian applications increasing with the advent of MCMC. An early example is the work on switching regression in time series by Goldfeld and Quandt (1975), with more recent applications including neural computing (Peng et al., 1996), marketing (Andrews and Currim, 2003), and health research (Yau et al., 2003). Examples of discrete mixture regressions include normal regression mixtures (Viele and Tong, 2002),

$$p(y_i|x_{1i}, \ldots, x_{Ri}) = \sum_{k=1}^{K} \pi_k N\left(\alpha_k + \sum_{r=1}^{R} x_{ri}\beta_{rk}, \sigma_k^2\right),$$

Poisson regression mixtures (Wedel et al., 1993),

$$p(y_i|x_{1i}, \ldots, x_{Ri}) = \sum_{k=1}^{K} \pi_k Po\left(\exp\left(\alpha_k + \sum_{r=1}^{R} x_{ri}\beta_{rk}\right)\right),$$

and logit regression mixtures (Hurn et al., 2003) for binary or binomial data,

$$p(y_i|x_{1i}, \ldots, x_{Ri}) = \sum_{k=1}^{K} \pi_k Bern\left(\frac{\exp\left(\alpha_k + \sum_{r=1}^{R} x_{ri}\beta_{rk}\right)}{1 + \exp\left(\alpha_k + \sum_{r=1}^{R} x_{ri}\beta_{rk}\right)}\right).$$

The probabilities, π_k, for the components may be predicted for each individual via regression (e.g., logit) also. In Bayesian applications, MCMC sampling is facilitated by the introduction of latent allocation indicators, $S_i \in (1, \ldots, K)$, for each case, which are sampled at each iteration from case-specific multinomial full conditionals,

$$\pi_k f_k(X_i, \beta_k, \phi_k) \Bigg/ \sum_{k=1}^{K} \pi_k f_k(X_i, \beta_k, \phi_k).$$

Discrete regression mixtures are useful for detecting subpopulations with different behaviors, while accounting for excess heterogeneity in part related to varying regression relationships (Wedel et al., 1993). They are also advocated as a natural solution to masked outliers, with one widely reproduced scenario (Rousseeuw, 1984) involving one subpopulation where a genuine regression exists, but another with (y_i, X_i) uncorrelated. A simple discrete mixture underlies the outlier accommodation method of Verdinelli and Wasserman (1991), which takes the form for normal data,

$$p(y_i|x_{1i}, \ldots, x_{Ri}) = \sum_{k=1}^{2} \pi_k N\left(\alpha + \sum_{r=1}^{R} x_{ri}\beta_r, \sigma_k^2\right),$$

where $\sigma_2^2 \gg \sigma_1^2$, and π_2 is taken small (e.g., $\pi_2 = 0.05$). This provides the scale contamination or variance-inflation mechanism for regression outliers, where outliers come from a normal distribution with the same mean as the remaining cases, but a higher variance. The Huber-M estimate when applied to normal regression is also expressible as a discrete mixture (Rice and Spiegelhalter, 2006). Viele and Tong (2002) argue that a full discrete mixture (including component varying regression effects) is more suitable to identify masked outliers, while Mohr (2007) regards the variance-inflation mechanism as better suited to modeling scattered outliers rather than outlier clusters.

Mohr (2007, 3958) advocates a two group model allowing for both clustered outliers (with clustering defined by similar values on predictors) and for scattered outliers, generated by a variance inflation mechanism. Thus, allocation indicators, S_{1i}, are used to identify cluster outliers (in group 1) with probability,

$$Pr(S_{1i} = 1) = \kappa_1 \exp[-\kappa_2(X_i^* - \psi)'(X_i^* - \psi)], \tag{5.9a}$$

where X_i^* is a subset of the full predictors (excluding the intercept), ψ is the group 1 mean on X^*, $0 \leq \kappa_1 \leq 1$, and $\kappa_2 > 0$. For a normal regression, cases in group 1 have density,

$$y_i|S_{1i} = 1 \sim N(X_i\beta_1, 1/\tau_1).$$

For cases not classified in the clustered outlier group, it is still possible to be a scattered outlier. So for the (majority of) cases with $S_{1i} = 0$, a second allocation indicator, $S_{2i} \sim Bern(\pi)$, has a value of 1 for scattered outliers, but is 0 for the main set of cases. For a normal regression, such cases have density,

$$y_i|S_{1i} = 0 \sim (1 - \pi)N(X_i\beta_2, 1/\tau_2) + \pi N(X_i\beta_2, D/\tau_2), \tag{5.9b}$$

where $D \gg 1$. The binary pair (S_{1i}, S_{2i}) can be replaced by a trinomial (see Example 5.6).

Certain identification and estimation issues apply to discrete regression mixtures, and a variety of sampling and post-processing methods, and priors to gain or improve identifiability, have been proposed. Firstly, improper priors are not suitable as they result in an improper posterior (Diebolt and Robert, 1994). Even with proper priors, a major issue revolves around component labeling, since different component labels cannot be distinguished during MCMC sampling unless some identifiability constraint is imposed a priori. Another issue involves small components (those with relatively low probabilities π_k), especially when combined with small samples (Wasserman, 2000), since at particular MCMC samples, no cases may be allocated to a particular group, k^*, so that the associated parameters are not updated. To deal with this problem, Viele and Tong (2002, 320) suggest Metropolis sampling of (β, σ^2, π) without introducing latent allocation indicators.

Sampling and estimation methods for discrete regression mixtures differ in whether they impose identifying constraints or allow switching between different numbers of components. Hurn et al. (2003) exemplify approaches that exclude identifying constraints on parameters, but rely on post-processing to identify clusters; they also employ a birth–death process to switch between different numbers of components K.

By contrast, Viele and Tong (2002, 317) advocate identifying restrictions in linear regression mixtures, for example on the variances of components, $\sigma_1^2 < \sigma_2^2 < \cdots < \sigma_K^2$. Ordering of variances may work better when the variances are well separated, whereas ordering of particular regression parameters works well when subpopulations are distinct in substantive terms. Similarly, Geweke and Keane (2000) consider a data augmented probit regression mixture and impose identifiability either by ordered regression intercepts, $\alpha_{k-1} < \alpha_k$, or by ordered precisions, $\tau_{k-1} < \tau_k$, with the latter option including the constraint that $\tau_{k^*} = 1$ for some $k^* \in (1, \ldots, K)$. A flexible error structure is then achieved with,

$$p(y_i^*|x_{1i}, \ldots, x_{Ri}) = \sum_{k=1}^{K} \pi_k N\left(\alpha_k + \sum_{r=1}^{R} x_{ri}\beta_r, 1 \Big/ \tau_k\right),$$

with straightforward extension to component-specific regression. Sampling of y_i^* is truncated to negative or positive values according to the observed binary response, y_i.

5.5.2 Zero-inflated mixture regression

Zero-inflated regression is a form of mixture regression applied to overdispersed or underdispersed data from Poisson, binomial, or negative binomial densities. For example, Poisson overdispersion may result from an excess number of zero counts, and this form of overdispersion occurs in applications in veterinary science (Rodrigues-Motta et al., 2007), sociology (MacAdam and Su, 2002), and quality control (Lambert, 1992). Under a zero inflated Poisson (ZIP) model, zero counts may result from either of two mechanisms: they may be true zeroes, as in quiescent periods for war protests (MacAdam and Su, 2002), or result from a stochastic mechanism, when the process is "active" but sometimes produces zero events. A distinction is therefore often made between structural and random zeroes (Martin et al., 2005).

The "active" stochastic mechanism, $f(y)$, may be described by any discrete density (Poisson, generalized Poisson, negative binomial, binomial, etc.). Let $d_i = 1$ for true zeroes as against stochastic zeroes when $d_i = 0$, with $Pr(d_i = 1) = \omega$. The inflation to the zero counts occurs under the degenerate option. Then,

$$P(y_i = 0) = Pr(d_i = 1) + f(y_i = 0|d_i = 0)Pr(d_i = 0)$$
$$P(y_i = j) = f(y_i = j|d_i = 0)Pr(d_i = 0) \qquad j = 1, 2, \ldots,$$

Regressors X_i may be relevant both to the binary inflation mechanism, and to the parameters defining the density $f(y_i = j)$ of the count data, such as a Poisson or negative binomial (Czado et al., 2007; Hall, 2000). So, under a zero-inflated Poisson regression, one might adopt a logit regression for the inflation process with,

$$\omega_i = Pr(d_i = 1|X_i) = \frac{\exp(X_i\gamma)}{1 + \exp(X_i\gamma)},$$

and a log link model, $\mu_i = \exp(X_i\beta)$, for the Poisson mean in $f(y_i = j|X_i, d_i = 0)$. The full ZIP model is

$$P(y_i = 0|X_i) = \omega_i + (1 - \omega_i)e^{-\mu_i},$$
$$P(y_i = j|X_i) = (1 - \omega_i)e^{-\mu_i}\mu_i^{y_i}/y_i! \qquad j = 1, 2, \ldots,$$

with variance then,

$$\text{Var}(y_i|\omega_i, \mu_i) = (1 - \omega_i)\left[\mu_i + \omega_i\mu_i^2\right] > \mu_i(1 - \omega_i) = E(y|\omega_i, \mu_i).$$

So modeling of excess zeros implies overdispersion. A useful representation for programming the zero-inflated Poisson involves the mixed scheme (Ghosh et al., 2006),

$$w_i \sim dbern(\omega),$$
$$y_i \sim Po(\mu_i(1 - w_i)).$$

The zero adjustment approach is also applicable to underdispersion (Ghosh and Kim, 2007). Czado et al. (2007) and Angers and Biswas (2003) consider Bayesian approaches to zero adjusted modeling using the generalized Poisson density (GPD), namely,

$$P(y_i|X_i) = \left(\frac{\mu_i}{1 + \alpha\mu_i}\right)^{y_i} \frac{(1 + \alpha y_i)^{y_i - 1}}{y_i!} \exp\left(\frac{-\mu_i(1 + \alpha y_i)}{1 + \alpha\mu_i}\right), \qquad (5.10)$$

which implies overdispersion for $\alpha > 0$, and underdispersion for $\alpha < 0$. A maximum likelihood GDP approach with adjustment for excess zeroes is considered by Bae et al. (2005), including a generalization to Poisson k-inflation (e.g., too many cases with $y_i = 1$ as compared to the Poisson if $k = 1$). For f a generalized Poisson (or any other relevant discrete density), the k-inflated model has the form,

$$P(y_i = k|X_i) = \omega_i + (1 - \omega_i)f(k; \mu_i, \alpha),$$
$$P(y_i = j|X_i) = (1 - \omega_i)f(j; \mu_i, \alpha) \qquad j \neq k.$$

Example 5.5. Radioimmunoassay and Esterase This example compares three heteroscedastic models for continuous radioimmunoassay observations (y) in relation to a single predictor, namely, esterase (x), as in Carroll and Ruppert (1982). A randomly varying coefficient model is also considered. Fit is based on sampling replicate data, with fit and penalty criteria derived as in Gelfand and Ghosh (1998).

So with,

$$y_i = \eta_i = \beta_0 + \beta_1 x_i + \sigma_i z_i,$$

the first model is a variance regression model with,

$$\sigma_i^2 = \exp(\gamma_0 + \gamma_1 x_i).$$

This yields a significant parameter, γ_1, with mean (95% interval) of 0.068 (0.044, 0.094), and such a significant relationship between the variance and predictor supports heteroscedasticity, whereas homoscedastic regression requires predictor effects on the variance to be nonsignificant. The penalty criterion, C_P (obtained by summing the posterior variances of replicates), is $1.42E+06$, while the predictive fit criterion, obtained as the posterior mean of $\sum(y_i - y_{\mathrm{rep},i})^2$, is $C_\mathrm{F} = 2.63E+06$.

The variance model used by Carroll and Ruppert (1982) is a power model in the absolute linear predictor, namely,

$$\sigma_i = \sigma(1 + |\eta_i|)^\lambda.$$

This is applied[7] with a $U(-2, 2)$ prior on λ, and with a $U(0,\ 250)$ prior on σ, which includes the observed standard deviation of 213. This gives an estimate (from OpenBUGS) for $\lambda = 0.63(0.41, 0.85)$, and provides improved fit criteria $(C_\mathrm{P}, C_\mathrm{F}) = (1.24E+06, 2.44E+06)$.

A Student t with ν degrees of freedom via normal scale mixing (centered on a single variance parameter σ^2) is then applied. A $U(0.01, 1)$ prior is applied on the inverse of the degrees of freedom $1/\nu$. This shows 17 data points with scale factors λ_i below 0.5, and ν estimated at 2.8. Although such estimates clearly show non-normality, the fit criteria deteriorate to $(C_\mathrm{P}, C_\mathrm{F}) = (1.72E+06, 2.94E+06)$.

Finally, a random regression coefficient model is applied, with a variance regression model additionally retained to see whether heteroscedasticity is reduced or eliminated. So,

$$\begin{aligned} y_i &= \alpha + \beta_i x_i + \sigma_i z_i, \\ \sigma_i^2 &= \exp(\gamma_0 + \gamma_1 x_i), \\ \beta_i &\sim N(\mu_\beta, \sigma_\beta^2), \end{aligned}$$

with priors,

$$\log(\sigma_b) \sim N(0, 1); \mu_\beta \sim N(0, 1000).$$

Two chains are run, with initial values for $\log(\sigma_b)$ of 1 and 2.

The second half of a run of 100,000 iterations shows a mean (95% interval) for $\log(\sigma_b)$ of 1.58 (0.13, 1.80), with γ_1 now having a 95% interval (−1.25, 0.55) straddling zero. The posterior means of the β_i range from 6.7 to 38, but with mean 17.1 close to that under other models. A histogram of the posterior mean β_i (in Figure 5.1, including a superimposed normal) shows more kurtosis than under a normal, with a few extreme highly positive coefficients. The fit criteria are markedly improved with $(C_\mathrm{P}, C_\mathrm{F}) = (150620, 154600)$.

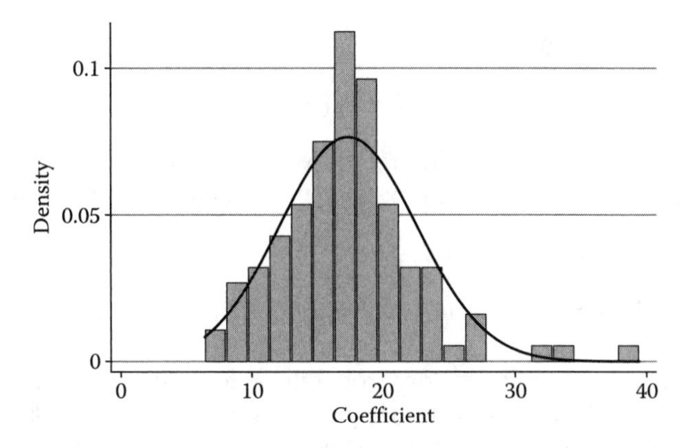

FIGURE 5.1
Varying esterase coefficients.

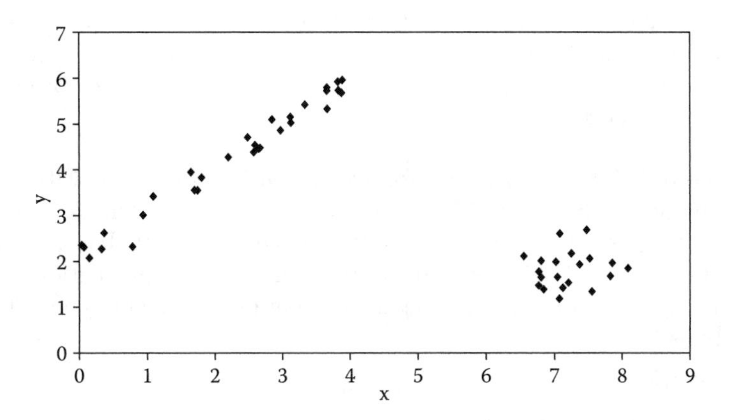

FIGURE 5.2
Clustered outliers.

Example 5.6. Clustered Outlier Data This example considers discrete mixture regression applied to simulated data that follow the scenario of Rousseeuw (1984). Thus, the first 30 points (the "main regression component") are generated as $y_i = 2 + x_i + u_i$, where u_i have standard deviation 0.2, while for cases 31–50, the bivariate pair (x_i, y_i), defining the "masked outlier component," are generated from a bivariate normal with mean (7,2) and diagonal covariance with diagonal terms 0.25—see Figure 5.2. To estimate the underlying mixture, the Mohr (2007) method is used, as in Equation 5.9, except that the cluster definition prior for the probability of belonging to group 1 (the masked outlier group) is simplified to,

$$Pr(S_{1i} = 1) = \exp[-\kappa(x_i - \psi)'(x_i - \psi)] = \exp(-h_i),$$

where the group 1 regression is

$$y_i|S_{1i} = 1 \sim N(X_i\beta_1, 1/\tau_1),$$

where $X_i = (1, x_i)$ includes an intercept. A value of 10 is taken for the variance inflation factor D in the mixture prior for the main observation group,

$$y_i \sim (1 - \pi)N(X_i\beta_2, 1/\tau_2) + \pi N(X_i\beta_2, D/\tau_2).$$

Then the three possible groups have probabilities $\{e^{-h_i}, (1-\pi)(1-e^{-h_i}), \pi(1-e^{-h_i})\}$.

A $Be(1, 9)$ prior on π favors low probabilities for scattered outliers, while an informative $U(0.04, 8.09)$ prior for the univariate cluster center, ψ, reflects the range of the observed (here simulated) covariate x_i. More diffuse priors on ψ were accompanied by failure to locate the regression parameters in the main regression component. A 10,000 iteration run[8] then shows early convergence with cases 1–30 having zero or effectively zero probabilities of belonging to group 1, defined by a posterior mean of 7.27 for ψ. The posterior mean for π is 0.067, with case 14 having a posterior probability of 0.47 of belonging to the scattered outlier group, but all other cases having posterior probabilities under 0.10 of being a scattered outlier.

Example 5.7. ZIP Regression for Decayed, Missing, and Filled Teeth
This example involves zero inflated regression analysis of counts of teeth decayed, missing, or filled (DMFT) from Bohning et al. (1999). Thus, 797 Brazilian children in six schools were subject to a dental health prevention trial; the school is in fact equivalent to a health prevention treatment type. The variables are as follows:

1. DMFT1—DMFT before intervention
2. DMFT2—DMFT at end of study
3. Gender (0—female ; 1—male)
4. Ethnicity: 1—dark ; 2—white ; 3—black)
5. School (kind of prevention):

   1—oral health education; 2—all four methods together;

   3—control school; 4—enriched diet with ricebran;

   5—mouthrinse with 0.2% NaF-solution; 6—oral hygiene.

The response y_i is DMFT2, with baseline DMFT scores included as a measure of severity of dental problems, and with school 3 as the reference prevention treatment.

As above, the model is

$$P(y_i = 0) = \omega + f(y_i = 0|d_i = 0)(1 - \omega),$$
$$P(y_i = j) = f(y_i = j|d_i = 0)(1 - \omega), \qquad j = 1, 2, \ldots$$

where $f(y_i = j|d_i = 0)$ is a Poisson likelihood with mean μ_i, and where $\log(\mu_i)$ is modeled as a function of gender, ethnicity, and school. Since the model is Poisson,

$$f(y_i = 0|d_i = 0) = \exp(-\mu_i).$$

One way of estimating the ZIP model involves a marginal rather than complete data likelihood, with WinBUGS coded for a nonstandard density[9]. A two chain run of 5000 iterations (convergent after 500) gives an estimate for ω of 0.047, and insignificant regression coefficients except for negative effects of schools (i.e., treatment types) 1, 2, and 5. The posterior means (sd) of the relevant coefficients are −0.24 (0.09), −0.33 (0.10), and −0.25 (0.08). A pseudo marginal likelihood estimate of −1243 is obtained from the posterior means of the inverse likelihoods. An alternative estimation method based on the mixed Poisson scheme of Ghosh et al. (2006) yields the same estimate of ω, and the same pattern of significant school effects.

For illustrative purposes, the zero inflated version of the generalized Poisson density,

$$P(y_i|X_i) = \left(\frac{\mu_i}{1+\alpha\mu_i}\right)^{y_i} \frac{(1+\alpha y_i)^{y_i-1}}{y_i!} \exp\left(\frac{-\mu_i(1+\alpha y_i)}{1+\alpha\mu_i}\right),$$

is also applied. A gamma prior is adopted on the overdispersion parameter, namely, $\alpha \sim Ga(1, 0.01)$. This model produces an improved pseudo marginal likelihood of −1229, with a lower posterior mean for ω of 0.017, but a posterior density for α, which is not clearly bounded away from zero—there is a spike at zero.

5.6 Time Series Regression: Correlated Errors and Time-Varying Regression Effects

The generalized linear model framework of Section 5.1 applies to time series regressions where responses may be binary, counts, or metric. Such time series regression may be characterized by serial correlation in regression residuals, overdispersion, or time-varying regression coefficients (Jung et al., 2006; Kedem and Fokianos, 2002). There may also be lagged dependence on observed or latent responses, or on predictors, with such dependence varying over time (Kitagawa and Gersch, 1985; Nicholls and Quinn, 1982). An autoregressive distributed lag or $ADL(p, q)$ regression includes lagged effects $\{y_{t-1}, \ldots, y_{t-p}; X_{t-1}, \ldots, X_{t-q}\}$ in both predictors and responses (Bauwens et al., 1999, 136).

If autocorrelation in the regression errors is suspected or postulated, as opposed to dependence on past responses or latent data, one option is parameter driven or latent process time series models involving autoregressive and moving average random effects (Chiogna and Gaetan, 2002; Cox, 1981). For y metric, with,

$$y_t = X_t\beta + \varepsilon_t, \qquad t = 1, \ldots, T,$$

where X_t is of dimension R, an $ARMA(p,q)$ error scheme specifies (Chib and Greenberg, 1994)

$$\varepsilon_t - \rho_1\varepsilon_{t-1} - \rho_2\varepsilon_{t-2}\cdots - \rho_p\varepsilon_{t-p} = u_t - \theta_1 u_{t-1} - \theta_2 u_{t-2}\cdots - \theta_q u_{t-q},$$

or $\rho_p(L)\varepsilon_t = \theta_q(L)u_t$, where $u_t \sim N(0,\sigma^2)$ are white noise errors.

For the normal linear regression case, $y_t = X_t\beta + \varepsilon_t$, where X_t does not include lagged y values, Chib (1993) sets out Gibbs sampling for the $AR(p)$ error model,

$$\rho_p(L)\varepsilon_t = u_t,$$

with $\rho(L) = 1 - \rho_1 L - \rho_2 L^2 \cdots - \rho_p L^p$. Assume conditioning on the initial p observations and priors,

$$\sigma^2 \sim IG\left(\frac{\nu_0}{2}, \frac{\delta_0}{2}\right), \rho \sim N_p\left(\rho_0, R_0^{-1}\right), \beta|\sigma^2 \sim N\left(\beta_0, \sigma^2 B_0^{-1}\right).$$

Also, restate the model as $\rho(L)y_t = \rho(L)X_t\beta + u_t$ or equivalently as $y_t^* = X_t^*\beta + u_t$, where $y_t^* = \rho(L)y_t$, and $X_t^* = \rho(L)X_t$. Finally, restate the error scheme, $\varepsilon_t = \rho_1\varepsilon_{t-1} + \rho_2\varepsilon_{t-2} + \cdots + \rho_p\varepsilon_{t-p} + u_t$, as $\varepsilon = E\rho + u$, where E is a $(T-p)\times p$ matrix with tth row $(\varepsilon_{t-1}, \ldots, \varepsilon_{t-p})$. Then with $\tau = \sigma^{-2}$, the full conditionals for the unknowns are

$$\begin{aligned}
&\beta|\sigma^2, \rho \sim N_k(\tilde{\beta}, \sigma^2\tilde{B}^{-1});\\
&\sigma^2|\beta, \rho \sim IG\left(\frac{n-p+\nu_0+k}{2}, \frac{\delta_0 + Q_\beta + d_\beta}{2}\right);\\
&\rho|\beta, \sigma^2 \sim N_p\left(\tilde{R}^{-1}[R_0\rho_0 + \tau E'\varepsilon]^{-1}, \tilde{R}^{-1}\right),
\end{aligned}$$

where $\tilde{\beta} = (B_0 + X^{*\prime}X^*)^{-1}(B_0\beta_0 + X^*y^*), \tilde{B} = (B_0 + X^{*\prime}X^*), Q_\beta = (\beta - \beta_0)'B_0(\beta - \beta_0), d_\beta = (y^* - X^*\beta)'(y^* - X^*\beta)$, and $\tilde{R} = (R_0 + \tau E'E)$. The prior and posterior for ρ can be constrained to a stationary region, though Chib (1993) suggests unconstrained sampling with draws in the nonstationary region rejected.

Widely applied options in practice for $ARMA$ error dependence in time series models for metric and discrete data are the simple $AR(1)$ and $MA(1)$ schemes. The $AR(1)$ model with,

$$\varepsilon_t = \rho\varepsilon_{t-1} + u_t, \tag{5.11a}$$

where $u_t \sim N(0,\sigma^2)$ are unstructured and independent of ε_t, is an effective scheme for controlling for temporal error dependence if (as often) most correlation from previous errors is transmitted through the impact of ε_{t-1}. This assumption is widely used in longitudinal models (e.g., Chi and Reinsel, 1989). As discussed in Chapter 4 whether the prior on ρ specifies stationarity is a central feature; for example, Palmer and Pettitt (1996) discuss the requirement for an informative prior on the regression intercept in an $AR(1)$ error model for a metric response, since in or near the unit root case ($\rho = 1$) the intercept is not defined. With $\sigma_\varepsilon^2 = \mathrm{Var}(\varepsilon_t)$, assuming stationarity with $|\rho| < 1$, $AR(1)$ error dependence means,

$$\mathrm{Var}(\varepsilon_t) = \rho^2 \mathrm{Var}(\varepsilon_{t-1}) + \sigma^2 + 2\rho cov(\varepsilon_{t-1}, u_t) = \rho^2\sigma_\varepsilon^2 + \sigma^2,$$

so that $\sigma_\varepsilon^2 = \sigma^2/(1-\rho^2)$, and the initial condition for the stationary case is

$$\varepsilon_1 \backsim N\left(0, \frac{\sigma^2}{1-\rho^2}\right). \tag{5.11b}$$

Also $cov(\varepsilon_t, \varepsilon_{t-1}) = \rho E(\varepsilon_{t-1}^2) + E(\varepsilon_{t-1}, u_t) = \rho\sigma_\varepsilon^2$ and $cov(\varepsilon_t, \varepsilon_{t-k}) = \rho^k \rho\sigma_\varepsilon^2 = \rho^k\sigma^2/(1-\rho^2)$.

$AR(1)$ error dependence for nonmetric responses is illustrated by the Poisson count outcomes case, $y_t \backsim Po(\mu_t)$, with (Chan and Ledolter, 1995; Nelson and Leroux, 2006),

$$\log(\mu_t) = X_t\beta + \varepsilon_t,$$
$$\varepsilon_t = \rho\varepsilon_{t-1} + u_t,$$

while the $MA(1)$ error model (Baltagi and Li, 1995) specifies $\varepsilon_t = u_t - \theta u_{t-1}$. Bayesian analysis of $AR(1)$ errors for count data is exemplified by Oh and Lim (2001) and Jung et al. (2006), who also consider augmented data sampling for count responses, while Chen and Ibrahim (2000) set out sampling algorithms under a power prior approach (that assumes historic data with the same form of design are available).

The Durbin–Watson statistic for $AR(1)$ error dependence, namely,

$$\mathrm{DW} = \frac{\sum(\varepsilon_t - \varepsilon_{t-1})^2}{\sum \varepsilon_t^2} = 2 - 2\frac{\sum(\varepsilon_t - \varepsilon_{t-1})}{\sum\varepsilon_t^2 \sum\varepsilon_{t-1}^2} = 2 - 2\rho,$$

is often used to test temporal autocorrelation (when predictors exclude lagged responses), and in a Bayesian context can be applied in a posterior predictive check. For example, Spiegelhalter (1998, 126) considers Poisson time series modeling of cancer cases, y_{ijt}, in age groups $i = 1, \ldots, I$, districts $j = 1, \ldots, J$, and years $t = 1, \ldots, T$, with μ_{ijt} being model means. At each iteration, deviance residuals, $d_{ijt} = -2\log\{p(y_{ijt}|\mu_{ijt})\}$, are obtained, and an average DW statistic derived for each age and district, namely,

$$\mathrm{DW}_{ij} = \frac{\sum_{t=2}^T (d_{ijt} - d_{ij,t-1})^2}{\sum_{t=1}^T (d_{ijt} - \overline{d_{ij.}})^2}.$$

A summary statistic for autocorrelation is then $\overline{\mathrm{DW}} = \sum\sum \mathrm{DW}_{ij}/IJ$, which can be obtained for both actual and replicate data and the procedure of Gelman et al. (1996) applied.

An alternative observation-driven approach to dependent errors is the generalized $ARMA(p,q)$ or $GARMA(p,q)$ scheme for discrete outcomes (Benjamin et al., 2003), based on re-expressing lagged residuals as differences between previous responses and regression terms. For count data with Poisson means, μ_t, an $AR(p)$ analog is

$$\log(\mu_t) = X_t\beta + \rho_1(\log y^*_{t-1} - X_{t-1}\beta) + \rho_2(\log y^*_{t-2} - X_{t-2}\beta) + \cdots + \rho_p(\log y^*_{t-p} - X_{t-p}\beta),$$

where lagged zero responses are handled by setting,

$$y^*_{t-k} = \max(c, y_{t-k}),$$

with $c \in (0,1]$. Moving average analogs in a $GARMA$ model compare $\log y^*_{t-j}$ with $\log \mu_{t-j}$, so a $GARMA(0,1)$ model would be

$$\log(\mu_t) = X_t\beta + \theta\left[\log\left(y^*_{t-1}/\mu_{t-1}\right)\right].$$

Davis et al. (2003) also compare previous responses with predictions via an observation-driven approach, namely,

$$\log(\mu_t) = X_t\beta + Z_t = X_t\beta + \sum_{k=1}^{\infty}\gamma_k e_{t-k} = X_t\beta + \sum_{k=1}^{\infty}\gamma_k\left(\frac{y_{t-k} - \mu_{t-k}}{\mu^{\lambda}_{t-k}}\right),$$

where $\lambda \in (0,1]$. The infinite moving average can be represented in the form,

$$Z_t = \phi_1(Z_{t-1} + e_{t-1}) + \cdots + \phi_p(Z_{t-p} + e_{t-p}) + \theta_1 e_{t-1} + \cdots + \theta_q e_{t-q},$$

with initial conditions $Z_t = e_t = 0$ for $t \leq 0$. So the choice $(p = 1, q = 0)$ gives $Z_t = \phi(Z_{t-1} + e_{t-1})$, while $(p = 0, q = 1)$ gives $Z_t = \theta e_{t-1}$.

The latent process driving autocorrelation may also be modeled using discrete mixture formulations. For example, Wang and Puterman (1999a) define Markov Poisson regression in which for each observed count, y_t, there corresponds an unobserved categorical variable, $S_t \in (1,\ldots,K)$, representing the state by which y_t is generated. The latent states are generated according to a stationary Markov chain with transition probabilities,

$$Pr(S_t = j|S_{t-1} = k) = \pi_{jk} \qquad \{j,k = 1,\ldots,K\}.$$

Conditional on $S_t = j$, the tth observation, y_t, is Poisson with mean $\mu_t = \exp(X_t\beta_j)$.

5.7 Time-Varying Regression Effects

Autocorrelated or heteroscedastic disturbances in time series regression may be caused by assuming the effects of regressors are constant across time, when in fact they are time varying. For example, Beck (1983) mentions how the ARCH model for volatility may be expressed as a varying regression impact. Thus, with $\varepsilon_t \sim N(0,1)$,

$$y_t = \alpha + \beta x_t + \delta_t h_t x_t + h_t \varepsilon_t = \alpha + \beta_t x_t + h_t \varepsilon_t,$$

where

$$\beta_t = \beta + \delta_t h_t,$$
$$h_t^2 = a_0 + \sum_{k=1}^{q} a_k \varepsilon_{t-k}^2.$$

Then, in periods when it is difficult to predict y_t (and so with larger h_t), the impact of x on y fluctuates according to the direction of δ_t and size of h_t.

Consider a dynamic linear model for metric responses with $\dim(X_t) = R$,

$$y_t = X_t \beta_t + \varepsilon_t.$$

A simple way to allow coefficient variation is to take,

$$\beta_t = \beta_\mu + u_t,$$

with u_t taken as exchangeable random effects, as in the random coefficient model of Swamy (1971). However, in time series contexts it is likely that deviations from the central coefficient effect, β_μ, will be correlated with nearby deviations in time. A flexible framework for time-varying parameter effects is provided by the linear Gaussian state–space model (West and Harrison, 1997, 73), involving first order random walks in scalar or vector coefficients, β_t,

$$\begin{aligned} y_t &= X_t \beta_t + \varepsilon_t & \varepsilon_t &\sim N_T(0, \Sigma_t) \\ \beta_t &= G_t \beta_{t-1} + \omega_t & \omega_t &\sim N_R(0, V_t). \end{aligned}$$

Often, $G_t = I$, $\Sigma_t = \Sigma$, and $V_t = V$, but if there is stochastic volatility, the variances (more particularly the log variances) can also be brought into a random walk order scheme with normal errors.

Subject matter considerations are likely to govern the anticipated level of smoothness in the regression effects. For example, Beck (1983) argues that the $RW(2)$ scheme,

$$\beta_t = 2\beta_{t-1} - \beta_{t-2} + \omega_t \qquad \omega_t \sim N(0, V),$$

provides a more plausible smoothly changing evolution for changing regression effects in political science applications. Dangl and Halling (2007) consider dynamic linear models for asset returns, y_t, and present procedures for formal Bayes model choice between constant regression effects with $V = 0$, and differing levels of variation in β_t, via a discrete prior over a small set of covariance matrix discount factors. Cooley and Prescott (1976) argue that nonstationary regression effects in econometric models partly reflect slowly changing behavioral relationships and propose an adaptive scheme, subject to permanent and transitory changes. Thus for scalar or vector β_t,

$$\beta_t = \beta_t^p + \omega_{1t},$$

superscript p denotes the permanent component of the parameters and the permanent component follows a random walk, as in,

$$\beta_t^p = \beta_{t-1}^p + \omega_{2t}.$$

Varying regression effects are important in particular applications of dynamic generalized linear models for discrete responses (Ferreira and Gamerman, 2000; Fruhwirth-Schnatter and Fruhwirth, 2007; Gamerman, 1998). Consider y from an exponential family density,

$$p(y_t|\theta_t) \propto \frac{\exp(y_t\theta_t + b(\theta_t))}{\phi_t},$$
$$\mu_t = E(y_t|\theta_t) = b'(\theta_t),$$

where the predictors, X_t, may include past responses $\{y_{t-k}, y_{t-k}^*\}$, both observed and latent (Fahrmeir and Tutz, 2001, 345). For example, y_t^*, the latent response (e.g., utility in economic applications) when y_t is binary, may depend on previous values of both y_t^* and y_t. The link for μ_t involves random regression parameters,

$$g(\mu_t) = X_t\beta_t,$$

where the parameter vector evolves according to a linear Gaussian transition model,

$$\beta_t = G_t\beta_{t-1} + \omega_t,$$

with multivariate normal errors, $\omega_t \sim N_R(0, V_t)$, independent of lagged responses and of the initial condition, $\beta_0 \sim N_R(B_0, V_0)$.

Models for binary time series with state–space priors on the coefficients have been mentioned in several studies. Thus, Fahrmeir and Tutz (2001) consider a binary dynamic logit model involving trend and varying effects of a predictor and lagged response,

$$\text{logit}(\pi_t) = \beta_{1t} + \beta_{2t}x_t + \beta_{3t}y_{t-1},$$
$$\beta_t \sim N_3\left(\beta_{t-1}, V\right),$$

while Gamerman (1998) considers nonstationary random walk priors in a marketing application with binomial data, where $\text{logit}(\pi_t) = \beta_{1t} + \beta_{2t}x_t$, and x_t is a measure of cumulative advertising expenditure. Similarly, Fruhwirth-Schnatter and Fruhwirth (2007) consider time series of binary observations with probabilities π_t, where some predictors, X_t^r, have time varying effects, β_t, while some X_t^f have time constant parameters α, so that π_t is modeled as $\text{logit}(\pi_t) = X_t^f\alpha + X_t^r\beta_t$. They point out that existing MCMC approaches rely on Metropolis–Hastings proposal densities in possibly high-dimensional parameter space, and instead use an auxiliary mixture sampler in the augmented data model,

$$U_{1t} = X_t^f\alpha + X_t^r\beta_t + \varepsilon_t,$$

where the latent utility of the alternative option, U_{0t}, and the error, ε_t, follow type 1 extreme value distributions. The error distribution is approximated as a mixture of normals, and the model reduces to a linear Gaussian state–space scheme with heteroscedastic errors.

Example 5.8. Epileptic Seizures This example considers correlated error schemes for a count response, specifically data from a clinical trial into the effect of intravenous gamma-globulin on suppression of epileptic seizures (Wang et al., 1996). Daily seizure counts are recorded for a single patient for a period of 140 days, where the first 27 days are a baseline period without treatment, and the remaining 113 days are the treatment period. Predictors are x_{1t} = treatment, x_{2t} = days treated, and an interaction $x_{3t} = x_{1t}x_{2t}$ between days treated and treatment. A simple Poisson regression is applied initially, and a predictive p test based on the DW statistic applied. In fact, this does not appear to be significant, having a value 0.76. Monte Carlo estimates of CPO statistics do, however, indicate model failures (Figure 5.3), with the log pseudo ML standing at -592. A stationary $AR(1)$ error model as in Equation 5.11 reduces the log(psML) to -372, with ρ having mean 0.23 (0.06, 0.40).

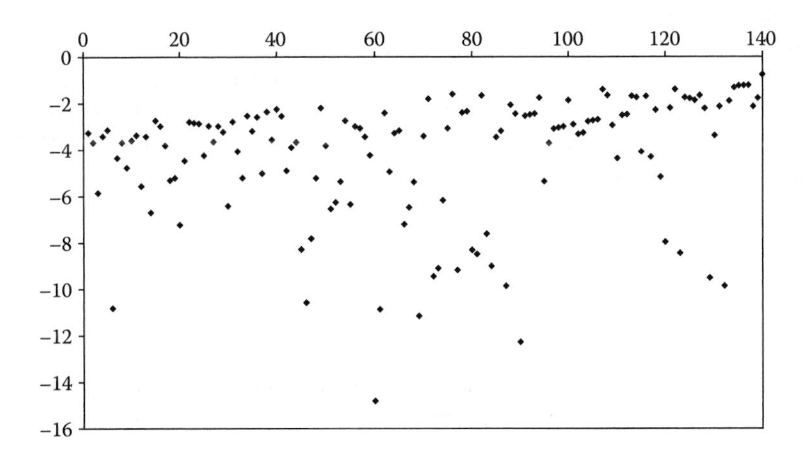

FIGURE 5.3
Log CPO plot.

Following Wang and Puterman (1999b), the dependence structure is also modeled using an unobservable finite-state Markov chain with $K = 2$ latent states. Conditional on state $S_t = j$, the Poisson mean for the seizure count on day t is represented as

$$\mu_t = \exp(\beta_{0j} + \beta_{1j}x_{1t} + \beta_{2j}x_{12} + \beta_{3j}x_{3t}) \qquad j = 1, 2.$$

An identifiability (ordered parameter) constraint is applied to the intercepts, though classical estimation makes clear that the two regimes have markedly different treatment effects, β_{1j}, and a constraint could be applied to them instead. The last 8000 iterations of a two chain run of 10,000 iterations show a further improved log(psML) of −359. State 1 has a much higher positive treatment effect, and a more negative interaction effect. If a subject is in that state on day t, the probability, π_{11}, of remaining there next day is 0.75, with probability, $\pi_{12} = 0.25$, of moving to state 2. If a subject currently occupies state 2, the respective probabilities are 0.62 and 0.38.

Example 5.9. Mortality and Environment This example illustrates time-varying regression effects, again for a count data application. It follows Smith et al. (2000) and Chiogna and Gaetan (2002) in analyzing the relationship between deaths, meteorological variables, and air pollution in Birmingham, Alabama, between August 3, 1985 and December 31, 1988 ($T = 1247$ observations). Related recent papers include Lee and Shaddick (2005) and Erbas and Hyndman (2005). The dataset contains two death series; here the over 65 deaths series is considered. In particular, Schwartz (1993) found a significant effect of PM10 on mortality for these data, whereas Smith et al. (2000) argued that results may depend on which meteorological variables are adjusted for, and how (or whether) different lagged values of PM10 are combined into a single exposure measure.

Here a time constant regression is compared with an analysis similar to that of Chiogna and Gaetan (2002) involving an $RW2$ trend on the level, and independent $RW1$ priors for time-varying coefficients on three predictors (x_1 = minimum temperature, x_2 = humidity, and x_3 = the first lag of PM10). So with $y_t \backsim Po(\mu_t)$, one has,

$$\begin{aligned} \log(\mu_t) &= \alpha_t + \beta_{1t}x_{1t} + \beta_{2t}x_{2t} + \beta_{3t}x_{3t}, \\ \alpha_t &\backsim N\left(2\alpha_{t-1} - \alpha_{t-2}, \sigma_\alpha^2\right), \\ \beta_{jt} &\backsim N(\beta_{j,t-1}, W_j), \end{aligned}$$

with $1/\sigma_\alpha^2$ assigned a $Ga(1, 1)$ prior, and the coefficient variances modeled in terms of ratios to the trend variance σ_α^2, namely,

$$\begin{aligned} W_j &= q_j\sigma_\alpha^2, \\ q_j &\backsim U(0, 10). \end{aligned}$$

As a predictive check, one step ahead predictions, $y_t^* \backsim Po(\mu_{t-1})$, are made and posterior probabilities, $Q_t = Pr(y_t^* \leq y_t|y)$, obtained. Low or high values for Q_t indicate failures of fit and/or prediction.

For the Poisson regression with constant predictor effects, the average (scaled) Poisson deviance from the last 9000 of a two chain run of 10,000 iterations is 1406, so there appears to be relatively little overdispersion in relation to the 1247 observations. Of the three regression coefficients, only β_1 has a 95% posterior interval that excludes zero, namely, −0.013 to −0.034. For the second model, the identification of time-varying predictor effects is likely to be improved using the representation,

$$\beta_{jt} = \mu_{\beta j} + b_{jt},$$
$$b_{jt} \backsim N(b_{j,t-1}, W_j),$$

where the three sets of effects, b_{jt}, are centered at each MCMC iteration to have mean zero[10].

Inferences are based on the second half of a two chain run of 20,000 iterations, reflecting relatively slow convergence in q_1. Figures 5.4 and 5.5 plot the time-varying coefficients, β_{2t} and β_{3t}. Significant effects of PM10 are limited to a central period, similar to the findings of Chiogna and Gaetan (2002). Only 22 of the 1247 points have $Q_t < 0.025$ or $Q_t > 0.975$, so judging by its one step ahead predictions, the model seems to reproduce the data satisfactorily.

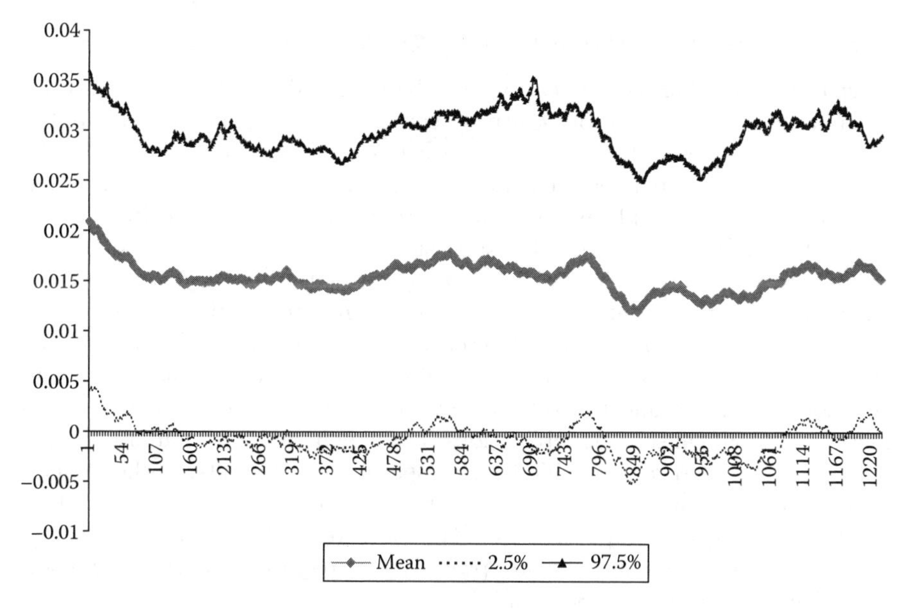

FIGURE 5.4
Time varying humidity coefficient.

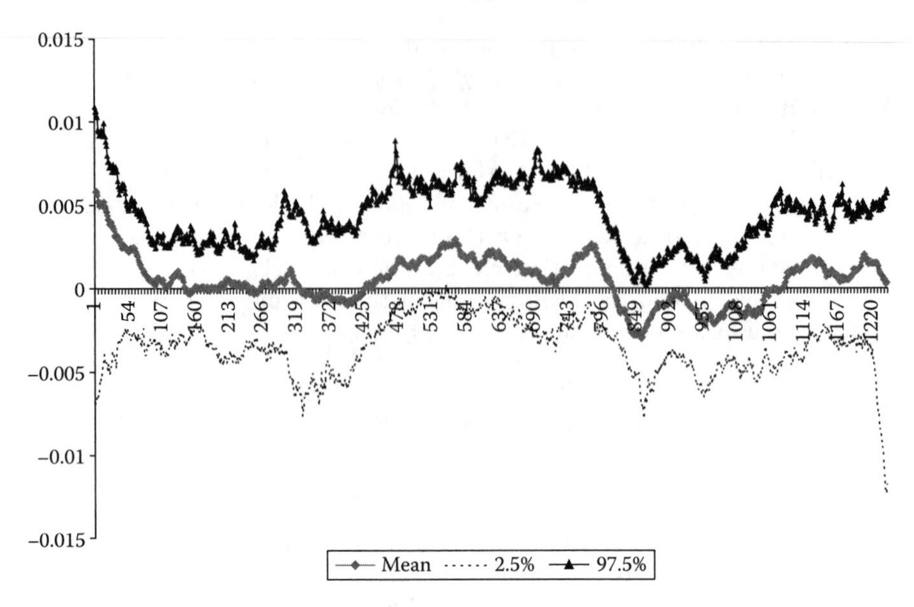

FIGURE 5.5
Time varying PM10 coefficient.

5.8 Spatial Correlation in Regression Residuals

In economic and health applications involving area data, the outcomes from neighboring areas tend to be positively correlated. If available predictors fail to account completely for variations in the outcome and the regression model treats units as independent, residual spatial correlation can bias regression parameter estimates and cause standard errors to be understated (Boyd et al., 2005). Cliff and Ord (1981, 197) suggest that possible nonlinear effects of predictors may account for residual spatial correlation, while Anselin (1988, Chapter 8) mentions that spatially correlated errors may reflect spatial correlation among predictors, spatial heterogeneity in functional form, and spatial correlation in the response when a spatially lagged response is not present in the model. Franzese and Hays (2008) review real world mechanisms in econometric and political science applications that may underlie spatial interdependence. With regard to a specific form of spatial heterogeneity, Fotheringham et al. (2002, 113) argue that spatial correlation in regression residuals often results from incorrectly applying a global model (one with homogenous regression coefficients) to a nonstationary process (that requires a varying regression coefficient approach)—see Section 5.9.

Cliff and Ord (1981, 199–206) consider test statistics (modified versions of the Moran I and Geary c statistics) to measure spatial correlation in regression residuals. In a Bayesian MCMC estimation setting, one could calculate these for observed and replicate data and apply the Gelman et al. (1996) posterior predictive check test. For example, Moran's I statistic for regression residuals, $e = y - X\beta$, from a normal linear regression involving n areas is

$$I = \frac{e'We/S_0}{e'e/n},$$

where $S_0 = \sum_i \sum_j w_{ij}$ and $W = [w_{ij}]$ is a chosen form of spatial interaction. Franzese and Hays (2008) mention Lagrange multiplier tests for linear regression residuals that could also be incorporated in a posterior predictive procedure.

Alternatively, Congdon et al. (2007) apply a test suggested by Fotheringham et al. (2002, 106) that measures spatial correlation via linear regression of appropriately defined residuals, e_i, on the spatial lag $e_i^* = \sum_j w_{ij} e_j / \sum_j w_{ij}$. The regression is simply,

$$e_i = \rho_0 + \rho_1 e_i^* + u_i,$$

where u_i are taken as unstructured. This is done at each MCMC iteration to provide a posterior mean and 95% intervals on the spatial correlation index ρ_1. If the 95% interval excludes zero then spatial correlation is present. A posterior predictive p test can also be applied with this criterion.

5.8.1 Spatial lag and spatial error models

Standard ways to deal with spatially correlated errors are to include a spatially lagged response as a predictor or to explicitly incorporate spatial effects in the residual specification. Correcting for spatial correlation in this way may affect the significance, and even the direction of predictor effects, as compared to a model with nonspatial error structure (Kuhn, 2007). Including a lagged response defines the spatial autoregressive or spatial lag model (Anselin, 1988), which for y metric has the form,

$$y_i = \rho \sum_j c_{ij} y_j + X_i \beta + u_i,$$

where $-1 \leq \rho \leq 1$, the $u_i \sim N(0, \sigma^2)$ are white noise errors, and $c_{ij} = w_{ij} / \sum_j w_{ij}$ are a row-standardized transform of the original symmetric spatial interactions, w_{ij}, with $\sum_j c_{ij} = 1$. If w_{ij} are binary and based on adjacency, this is equivalent to including the average of the neighboring responses as an extra predictor. One may express the response as spatially filtered, namely,

$$y_i^* = y_i - \rho \sum_j c_{ij} y_j,$$

with the spatial autoregressive model then taking the indirect form,

$$y_i^* = X_i \beta + u_i.$$

This model has been proposed for discrete responses and is often modeled by introducing augmented data, with a widely applied approach being known as the spatial probit (Franzese and Hays, 2007; Holloway et al., 2002). So for y binary, z_i is a latent metric variable that is positive when $y = 1$ and negative when $y = 0$. Then the usual model form is

$$z_i = \rho \sum_j c_{ij} z_j + X_i \beta + u_i,$$

with i.i.d. errors $u_i \sim N(0, 1)$. In econometric applications, this might amount to expecting individuals located at similar points in space to exhibit similar choice behavior (Smith and Lesage, 2004), or to expecting interaction between neighboring areas in the underlying propensity of the event (Coughlin et al., 2004). In matrix terms,

$$z = \rho C z + X\beta + u,$$

and solving for z gives,

$$z = (I - \rho C)^{-1} X\beta + u^*,$$

where $u^* = (I - \rho C)^{-1} u$. The disturbances are now correlated (McMillen, 1992) with $u^* \sim N(0, \Omega)$, where $\Omega = (I - \rho C)^{-1}[(I - \rho C)^{-1}]'$.

An alternative solution to spatially correlated residuals—especially if there is no strong evidence for spatial lag effects—is to include spatial structure in the errors. The rationale is that the effects of unknown predictors spill over across adjacent areas, causing spatially correlated errors. One option is the simultaneous autoregressive scheme (SAR) (Richardson et al., 1992),

$$y_i = X_i \beta + \varepsilon_i,$$
$$\varepsilon_i = \rho \sum_j c_{ij} \varepsilon_j + u_i,$$

with $u_i \sim N(0, \Sigma_u)$, and a maximum possible value of 1 for ρ since the spatial weights are standardized. The lower prior limit for ρ is typically taken as 0 since negative values are implausible.

Writing the equation for the autoregressive error vector $\varepsilon = (\varepsilon_1, \ldots, \varepsilon_n)$ as $\varepsilon = (I - \rho C)^{-1} u$, the covariance matrix for ε is

$$(I - \rho C)^{-1} \Sigma_u (I - \rho C')^{-1},$$

and with $D = I - \rho C$, the joint prior for ε is obtained as

$$(\varepsilon_1, \ldots, \varepsilon_n) \backsim N_n(0, D^{-1} \Sigma_u (D')^{-1}).$$

Assuming $\Sigma_u = \sigma^2 I$, the likelihood is

$$L\left(\alpha, \rho, \sigma^2 | y\right) = \frac{1}{2\pi\sigma^n} \left|D'D\right|^{0.5} \exp\left[-\frac{1}{2\sigma^2}\left[(y - X\beta)' D' D (y - X\beta)\right]\right]$$

An indirect procedure for estimating the SAR error model involves spatial differencing, whereby,

$$y = \rho C y + X\beta - \rho C X\beta + u,$$

or taking $y^* = y - \rho C y, X^* = X - \rho C X$,

$$y^* = X^*\beta + u,$$

where the errors are unstructured.

By contrast, conditional autoregressive errors (Besag, 1974) specify ε_i conditional on remaining effects $\varepsilon_{[i]}$. One option takes unstandardized spatial interactions with,

$$E(\varepsilon_i|\varepsilon_{[i]}) = \lambda \sum_{j \neq i} w_{ij}\varepsilon_j,$$

$$\text{Var}(\varepsilon_i|\varepsilon_{[i]}) = \sigma^2,$$

with joint covariance $\sigma^2(I - \lambda W)^{-1}$. In this case (Bell and Broemeling, 2000, 959), λ is constrained by the eigenvalues E_i of W, namely, $\lambda \in [1/E_{\min}, 1/E_{\max}]$. The conditional variances may differ between subjects with $M = \text{diag}(\sigma_i^2)$ and the covariance is then $(I - \lambda W)^{-1}M$ (Lichstein et al., 2002).

If predictor effects are written $\eta_i = X_i\beta$, this formulation may be restated in terms of an "own area" regression effect, and a filtered effect of neighboring regression residuals (Bell and Broemeling, 2000; Mollie and Richardson, 1991). Thus for y metric,

$$y_i \sim N(\eta_i + \lambda \sum_{j \neq i} w_{ij}(y_j - \eta_j), \sigma^2).$$

When y is Poisson, with means $\nu_i = E_i\rho_i$, where E_i are expected events and ρ_i are relative risks, one may similarly (Bell and Broemeling, 2000, 966) assume $r_i = \log(\rho_i)$ are normal with[11],

$$r_i \sim N\left(\eta_i + \lambda \sum_{j \neq i} w_{ij}(r_j - \eta_j), \sigma^2\right).$$

The other conditional autoregressive option takes standardized spatial interactions,

$$E(\varepsilon_i|\varepsilon_{[i]}) = \kappa \sum_{j \neq i} c_{ij}\varepsilon_j,$$

$$\text{Var}(\varepsilon_i|\varepsilon_{[i]}) = \sigma^2 \Big/ \sum_{j \neq i} w_{ij},$$

with joint covariance for the ε_i then being $\sigma^2(D - \kappa W)^{-1}$, where D is diagonal with elements $d_i = \sum_{j \neq i} w_{ij}$ (Sun et al., 1999, 342). Equivalently, in the case $w_{ij} = 1$ for neighbors and $w_{ij} = 0$ otherwise, the diagonal terms of the precision matrix are τd_i, where $\tau = 1/\sigma^2$ (Kruijer et al., 2007), while off-diagonal terms equal $-\tau\kappa$ when i and j are neighbors and 0 otherwise. In the case when $\kappa = 1$, one obtains the $CAR(1)$ prior of Besag et al. (1991) with the joint covariance matrix no longer positive definite. This approach generalizes to the convolution model with distinct spatial and white noise errors. So with regression mean, μ_i, and link, g, one has,

$$g(\mu_i) = X_i\beta + \varepsilon_i + u_i,$$

though the separate error terms are subject to possibly weak identifiability, as considered in Chapter 4.

5.9 Spatially Varying Regression Effects: Geographically Weighted Linear Regression and Bayesian Spatially Varying Coefficient Models

Further to the discussion in Fotheringham et al. (2002), linear and general linear spatial models typically assume that model parameters are homogenous over the study region without any local variation. The assumption of constant parameter values over space may often be unrealistic, and schemes that admit spatial variation or spatial nonstationarity in regression parameters may both improve fit and also account for spatially correlated residuals (e.g., Leung et al., 2000; Osborne et al., 2007).

A widely applied method (Brunsdon et al., 1998), especially for metric data, is geographically weighted linear regression (GWR). This method consists in reusing the data n times, such that the ith regression regards the ith area as the origin. With R predictors, the coefficients $\beta_{1i}, \ldots, \beta_{Ri}$ for the ith regression are derived using interaction weights $w_{ik} > 0$ between the ith area and other areas. These weightings in concert with an overall precision parameter, ϕ, define the precision parameters ϕw_{ik} for area k in a normal likelihood. So for the ith regression (centered on area i),

$$\begin{aligned} y_k &\sim N(\mu_{ik}, 1/\tau_{ik}) \qquad k = 1, \ldots, n, \\ \tau_{ik} &= \phi w_{ik}, \\ \mu_{ik} &= \beta_{1i}x_{1k} + \cdots + \beta_{Ri}x_{Rk}. \end{aligned}$$

The spatial interaction function, w_{ik}, might involve an exponential, bi-square, spherical, or Gaussian kernel function of distances d_{ik} between areas. Under weighted least squares, the estimator for $\beta_i = (\beta_{1i}, \ldots, \beta_{Ri})$ is

$$\beta_i = (X'W_iX)^{-1}X'W_iy,$$

where W_i is an $n \times n$ diagonal matrix with entries w_{ik} (Assuncao, 2003). A Gaussian decay in distance with positive bandwidth η would specify,

$$w_{ik} = \exp\left(-d_{ik}^2/2\eta^2\right) \qquad \eta > 0, \tag{5.12}$$

so that for small distances between area i and k, the kernel w_{ik} is close to 1 and nearer observations have higher precisions. A cross-validation GWR approach omits the kth response, y_k, from the kth geographically weighted regression (though w_{ik} still take account of the known geographic location of area k) with the prediction of y_k using the remaining $n-1$ areas (Farber and Paez, 2007).

Lesage (2004) notes that local linear estimates based on a distance-weighted subsample of data may suffer from weak identification as the effective number of observations used to produce estimates for some points in space may be small. This problem can be solved under a Bayesian approach by incorporating prior information. For example, a prior may be set on η in Equation 5.12, taking account of the maximum observed inter-point or inter-area distance, and one may robustify against outlying areas (Lesage, 1999) by taking a scale mixture (heavy tailed) approach. Thus, for ν degrees of freedom in a Student t density, one has,

$$\tau_{ik} = \phi w_{ik} \kappa_{ik},$$
$$\kappa_{ik} \sim Ga(0.5\nu, 0.5\nu).$$

Lesage (2004) and Lesage and Kelley Pace (2009) reframe the GWR scheme to allow spatially nonconstant variance scaling parameters, ν_i, subject to an exchangeable chi-square prior density, namely,

$$\nu_i \sim \chi^2(r),$$

with r a hyperparameter. Lesage (2004) also redefines w_{ik} as normalized distance-based weights (with $w_{ii} = 0$), such as,

$$w_{ik} = \exp(-d_{ik}/\eta) \Big/ \sum_{k=1}^{n} \exp(-d_{ik}/\eta),$$

so the terms in the row-vector $(w_{i1}, \ldots, w_{in})$ sum to unity. Then with y being the $n \times 1$ response vector and,

$$W_i y = W_i X \beta_i + \varepsilon_i,$$

the smoothing of regression effects across space is represented as,

$$\beta_i = (w_{i1} \otimes I_R, \ldots, w_{in} \otimes I_R) \begin{pmatrix} \beta_1 \\ \cdots \\ \cdots \\ \cdots \\ \beta_n \end{pmatrix} + u_i. \tag{5.13}$$

With $V_i = \text{diag}(v_1, \ldots, v_n)$, the error terms have priors,

$$\varepsilon_i \sim N(0, \sigma^2 V_i),$$
$$u_i \sim N(0, \sigma^2 \delta^2 (X' W_i^2 X)^{-1}),$$

with the specification on u_i being a form of Zellner g-prior, in which δ^2 governs adherence to the smoothing specification (Equation 5.13).

5.9.1 Bayesian spatially varying coefficient models

An alternative to the GWR approach is provided by spatially varying coefficient (SVC) models (Assuncao, 2003; Gamerman et al., 2003; Gelfand et al., 2003; Wheeler and Calder, 2006, 2007). For a continuous space perspective, let $Y(s)$ be the $n \times 1$ response vector for locations, $s = (s_1, \ldots, s_n)$, and β be a $nR \times 1$ stacked vector of spatially varying regression coefficients. Then the normal SVC model is (Gelfand et al., 2003, Section 3),

$$Y(s) \sim N(X(s)'\beta(s), \sigma^2 I),$$

where X' is a $n \times nR$ block diagonal matrix of predictors. The prior for β is

$$\beta(s) \sim N(1_{n \times 1} \otimes \mu_\beta, V_\beta),$$

where $\mu_\beta = (\mu_{\beta_1}, \ldots, \mu_{\beta_R})'$ contains the mean regression effects, and V_β is the $nR \times nR$ covariance matrix.

The latter can be expressed as,

$$V_\beta = C(\eta) \otimes \Lambda,$$

where Λ is a $R \times R$ matrix containing covariances between regression coefficients at any particular location, and $C(\eta) = [c(s_i - s_j; \eta)]$ is a $n \times n$ correlation matrix representing spatial interaction between locations or areas, with η denoting parameters governing such interaction. Hence, each of the R regression coefficients is assumed to follow the same spatial interaction structure. For example, under exponential spatial interaction (Wheeler and Calder, 2007),

$$c(s_i - s_j; \eta) = \exp(-d_{ij}/\eta),$$

where η is a positive parameter.

Setting $\beta = 1_{n \times 1} \otimes \mu_\beta + b$, where $b \sim N(0, C(\eta) \otimes \Lambda)$, the model may also be written,

$$Y = X'\mu_\beta + X'b + \varepsilon,$$

where $\varepsilon \sim N(0, \sigma^2 I)$. Integrating over b, the observed data likelihood (Gelfand et al., 2003, 390) is

$$\begin{aligned} p(y|\mu_\beta, \sigma^2, \Lambda, C(\eta)) \propto\ & [X(C(\eta) \otimes \Lambda)X' + \sigma^2 I]^{-0.5} \\ & \times \exp\{-0.5(y - X(1_{n \times 1} \otimes \mu_\beta))' \\ & \times [X(C(\eta) \otimes \Lambda)X' + \sigma^2 I]^{-1}(y - X(1_{n \times 1} \otimes \mu_\beta))\}. \end{aligned}$$

For discrete areas, distance-based kernel schemes for spatial interaction have less substantive basis. Let $i = 1, \ldots, n$ denote a set of such areas, and $\beta_i = (\beta_{1i}, \ldots, \beta_{Ri})$ be spatially varying regression effects in the linear predictor,

$$\eta_i = \sum_{r=1}^{R} \beta_{ri} x_{ri},$$

of a general linear model with mean $\mu_i = E(y_i)$, and link $g(\mu_i) = \eta_i$. With $\beta = (\beta_1, \ldots, \beta_n)$, one possible spatially structured scheme is a pairwise difference prior (Assuncao, 2003),

$$p(\beta|\Phi) \propto |\Phi|^{n/2} \exp\left\{ -0.5 \sum_i \sum_j w_{ij} (\beta_i - \beta_j)' \Phi (\beta_i - \beta_j) \right\}. \qquad (5.14)$$

with $R \times R$ precision matrix Φ, and with spatial interactions, w_{ij}, usually binary ($w_{ij} = 1$ when areas i and j are adjacent and zero otherwise).

When y_i is metric with $\mu_i = \eta_i$, precision $\tau = 1/\sigma^2$ and likelihood,

$$p(y|\beta, \tau) = (2\pi)^{-n/2} \tau \exp\left\{ -0.5\tau \sum_{i=1}^{n} (y_i - x_i\beta_i)^2 \right\},$$

one may, following Gamerman et al. (2003), scale the covariance in Equation 5.14 by τ, namely,

$$p(\beta|\Phi, \tau) \propto \tau^{nR/2} |\Phi|^{n/2} \exp\left\{ -0.5\tau \sum_i \sum_j w_{ij} (\beta_i - \beta_j)' \Phi (\beta_i - \beta_j) \right\}. \qquad (5.15)$$

Assume gamma and Wishart priors for τ and Φ, namely, $\tau \sim Ga(\nu_\tau/2, S_\tau \nu_\tau/2)$, $\Phi \sim W(\nu_\Phi/2, S_\Phi \nu_\Phi/2)$, then with normal data, y, conjugate full conditional densities can be sampled in a Gibbs scheme (Assuncao, 2003, 9; Gamerman et al., 2003, 517–18). With $\beta_{[i]} = (\beta_1, \ldots, \beta_{i-1}, \beta_{i+1}, \ldots, \beta_n)$, the posterior conditional for β_i is normal with,

$$p(\beta_i|\beta_{[i]}, \tau, \Phi) = N(H_i(x_i y_i + w_{i+} \Phi \bar{\beta}_{[i]}, \sigma^2 H_i),$$
$$H_i = [x_i x_i' + w_{i+} \Phi]^{-1},$$
$$\bar{\beta}_{[i]} = \sum_{j \neq i} w_{ij} \beta_j / w_{i+}.$$

The covariance matrix for β in Equation 5.14 is

$$K^{-1} \otimes \Phi^{-1},$$

and for Equation 5.15 is

$$\sigma^2 K^{-1} \otimes \Phi^{-1},$$

where K has elements,

$$k_{ii} = w_{i+} = \sum_{j \neq i} w_{ij},$$
$$k_{ij} = -w_{ij} \qquad i \neq j.$$

Hence priors (Equation 5.14) and (Equation 5.15) are improper because the elements in each row of K add to zero. Assuncao (2003, 460) notes that propriety can be obtained by a constraint such as $\sum_i \beta_i = A$, where A is any known R vector. This consideration leads to a practical strategy representing β_i as $\beta_i = \mu_\beta + b_i$, where b_i follow the prior in Equations 5.14 or 5.15, but are zero centered at each MCMC iteration. So b_i are effectively zero mean spatially structured deviations from the mean regression effect, $\mu_\beta = (\mu_{\beta_1}, \ldots, \mu_{\beta_R})$.

Example 5.10. Bladder Cancer in US State Economic Areas Following Wheeler and Tiefelsdorf (2005), this example considers white male bladder cancer mortality rates over 1970–1994 in 508 State Economic Areas (SEA) of the United States. The response is the age standardized mortality rate per 100,000 person-years and the predictors are x_1 = log population density; and x_2 = white male lung cancer mortality rate. Population density may proxy urban environmental risk factors, since other evidence shows bladder cancer rates increase with urban density (Ayotte et al., 2006). Lung cancer mortality rates proxy smoking, which is also a positive risk factor for bladder cancer: the attributable risk of smoking for lung cancer exceeds 80% and the attributable risk of smoking for bladder cancer exceeds 55% (Mehnert et al., 1992). The neighborhood scheme here is based on adjacency between the SEAs; this is defined according to a distance threshold of 150 km between SEA population centroids (determined using US county populations for 1982). For the 30 SEAs that have no neighbors on this basis, a single neighbor is defined as the closest SEA.

A global (i.e., stationary coefficient) model is first fitted with only the usual white noise error term. This has a DIC of 1529 ($d_e = 4$), with significant effects for the first predictor only. The posterior means (95% intervals) for β_1 and β_2 are 0.386 (0.329, 0.451) and −4.5E-6 (−0.0074, 0.0053), with the white noise variance having a mean of 1.18. The spatial residual regression coefficient (SRRC) discussed above is positive, with 95% interval (0.104, 0.115), indicating spatially correlated residuals.

One possible remedy for this is to include a specific spatially structured residual in the regression. Accordingly, a second model includes an additive $ICAR(1)$ spatial error, s_i, in addition to the usual white noise error of a normal linear regression, namely,

$$y_i = \beta_0 + \beta_1 x_{1i} + \beta_2 x_{2i} + s_i + u_i,$$

with $Ga(0.5, 0.0005)$ priors assigned to the precisions of s_i and u_i. The last 4500 of a two chain 5000 iterations run give a greatly improved DIC of 1179

($d_e = 202$) with significant effects now for both predictors, but a reduced coefficient for β_1. The posterior means (95% intervals) for β_1 and β_2 are 0.177 (0.122, 0.234) and 0.029 (0.020, 0.037), so both coefficients have entirely positive 95% credible intervals. Such a change in predictor effects demonstrates the importance of correct error specification for inferences about measured predictors. The variance of the unstructured and spatial errors are 0.40 and 1.35.

An alternative possible solution to spatially correlated residuals is to consider spatial nonstationarity in predictor effects. Here,

$$y_i = \beta_0 + \beta_{1i}x_{1i} + \beta_{2i}x_{2i} + u_i,$$

where $\beta_{ki} = \mu_{\beta_k} + b_{ki}$, and b_{ki} follow independent $CAR(1)$ priors. The b_{ki} are centered to average zero at each MCMC iteration. The last 4500 of a two chain 5000 iteration run give a DIC of 1093 ($d_e = 235$), again with significant mean effects, μ_β, for both predictors: posterior means (95% intervals) for μ_{β_1} and μ_{β_2} are 0.237 (0.167, 0.313) and 0.031 (0.022, 0.039), so that recognizing heterogeneity has also enhanced the central predictor effects. Although Wheeler and Tiefelsdorf (2005, 169) found problems with implausible negative effects when using a classical GWR approach, the Bayesian SVC approach reveals only 20 SEAs with posterior probabilities $Pr(\beta_{1i} > 0|y)$ under 0.5, and only seven SEAs with posterior probabilities $Pr(\beta_{2i} > 0|y)$ under 0.5.

Example 5.11. Discrete Poisson Mixture with Spatial Errors: Alcohol-Related Hospitalizations This example considers a discrete Poisson regression mixture for male alcohol-related hospitalizations in 354 English local authorities (data from http://www.nwph.net/nwpho/default.aspx). These are related to two composite indicators of area social structure. The first is a social fragmentation score (x_1), a standardized score derived in turn from z scores on four 2001 Census percentage indicators, namely, one person households, private rented households, recent migrants, and unmarried adults. The second is a deprivation score (x_2), also a standardized score and derived from a total of z scores for three Census percentage indicators, namely, percentages of adult workers in classes 4 and 5 (low skill manual), unemployment among the economically active, and social renting households.

One might expect spatially correlated residuals if there are major unknown risk factors. However, an initial analysis is applied to a Poisson regression mixture,

$$p(y_i|x_{1i}, \ldots, x_{Ri}) = \sum_{k=1}^{K} p_k Po\left(E_i \exp\left(\beta_{0k} + \sum_{r=1}^{R} x_{ri}\beta_{rk}\right)\right),$$

where E_i are expected events and $R = 2$. The number of mixture components K was taken to be 3, as earlier unconstrained analysis shows a stable three group mixture, with clearly separated intercepts, with means and 95% intervals −0.219 (−0.227, −0.211), 0.019 (0.011, 0.028), and 0.268 (0.254, 0.282); these can be interpreted straightforwardly as low-, medium-, and high-risk areas.

A subsequent analysis with intercepts subject to an order constraint shows the same posterior intercepts with the groups having respective probabilities (0.481, 0.384, 0.135). There are apparently stronger effects of deprivation ($\beta_{21}, \beta_{22}, \beta_{23}$) for the three groups, namely, 0.105 (0.098, 0.113), 0.105 (0.098, 0.112), 0.093 (0.084, 0.102) than of fragmentation, namely, 0.019 (0.010, 0.027), 0.027 (0.019, 0.034), and 0.009 (−4.76E-4, 0.018). However, lack of fit is apparent in a scaled deviance of 1245 and spatial correlation in the errors shows in a significant SRRC, with mean 0.39 and 95% interval (0.28, 0.49).

The analysis is repeated but including an $ICAR(1)$ spatial error (Besag et al., 1991), namely,

$$p(y_i|x_{1i}, \ldots, x_{Ri}) = \sum_{k=1}^{K} p_k Po\left(E_i \exp\left(\beta_{0k} + \sum_{r=1}^{R} x_{ri}\beta_{rk} + s_i\right)\right),$$

where s_i is centered at each MCMC iteration to have mean zero[12]. This option reduces the scaled deviance to 357, with the SRRC statistic now having a mean −0.004 close to zero with 95% interval −0.30 to 0.27. Control for spatially correlated residuals has the effect of enhancing the fragmentation effects to significance. The posterior mean fragmentation effects and the 95% intervals for ($\beta_{11}, \beta_{12}, \beta_{13}$) are now 0.096 (0.059, 0.152), 0.026 (−0.003, 0.061), and 0.100 (0.029, 0.212). Effects of deprivation are also enhanced, demonstrating as for the previous example, the importance of correct error specification for inferences about measured predictors.

Appendix: Computational Notes

1. The BayesX code in Example 5.1 is

```
dataset d
d.infile, maxobs=5000 using data\crab.txt
bayesreg b
b.regress y = x, family=nbinomial predict using d.
```

The WinBUGS code for the NBk model in this application is

```
model { for (i in 1:173) {y[i] ~dpois(mu[i]); mu[i] ~dgamma(a[i],b[i])
a[i] <- pow(xi[i],1-k)/phi; b[i] <- pow(xi[i],-k)/phi
log(xi[i]) <- beta[1]+beta[2]*(x[i]-mean(x[]))}
beta[1] ~dnorm(0,0.001); beta[2] ~dnorm(0,1)
k1~dgamma(1,1); k <- k1-1; phi~dgamma(1,1)}
```

2. The code for the overdispersed multinomial model in Example 5.2 is

```
model { for (i in 1:n) { log(phi[i,J]) <- 0; N[i] <- sum(y[i,]);
# likelihood and replicate data
y[i,1:J]~dmulti(pi[i,1:J],N[i]); yrep[i,1:J]~dmulti(pi[i,1:J],N[i])
```

```
# random choice effects
alp[i,1:JM] ~dmnorm(A[],D.inv[,]);
for (j in 1:J) {pi[i,j] <- phi[i,j]/sum(phi[i,]); yh[i,j] <- N[i]*pi[i,j]
# chi square calculations
cr[i,j] <- pow(yrep[i,j]-yh[i,j],2)/yh[i,j]; c[i,j] <- pow(y[i,j]-yh[i,j],2)/yh[i,j]}
for (j in 1:J-1) {log(phi[i,j]) <- alp[i,j]+bet[j,1]*(x1[i,j]-mean(x1[,j]))+
bet[j,2]*(x2[i,j]-mean(x2[,j]))+bet[j,3]*(x3[i]-mean(x3[]))}}
# posterior predictive check
  PPC <- step(sum(cr[,])-sum(c[,]))
# priors
for (j in 1:J-1) {A[j]~dflat()
for (k in 1:3) {bet[j,k]~dflat()}
for (k in 1:J-1) {Wsc[j,k] <- equals(j,k)}}
D.inv[1:JM,1:JM]~dwish(Wsc[,],JM)}
```

3. A WinBUGS code for the Scott (2003) approach to logit regression augmented data sampling with a single regressor is

```
model { for (i in 1:n) {Y[i] <- y[i]+1; z[i,1]~ dexp(1) I(A1[i,Y[i]],B1[i,Y[i]])
z[i,2] ~ dexp(lam1[i]) I(A2[i,Y[i]],B2[i,Y[i]])
A1[i,1] <- 0; B1[i,1] <- z[i,2]; A1[i,2] <- z[i,2]; B1[i,2] <- 10000
A2[i,1] <- z[i,1]; B2[i,1] <- 10000; A2[i,2] <- 0; B2[i,2] <- z[i,1]
log(lam1[i]) <- b[1]+b[2]*x[i]}
for (j in 1:2) {b[j] ~dflat()}}
```

4. In Example 5.3, the probit model with skewed error has WinBUGS code,

```
model{for (i in 1:n) {m[i] <-b[1]+b[2]*DVRT[i]+
b[3]*sex[i]+b[4]*fathocc[i] +delta.s*V[i]
fathocc[i] ~dnorm(mocc,tocc); V[i] ~dnorm(0,1) I(0,)
ystar[i] ~dnorm(m[i],prec) I(lo[lvcert[i]+1],up[lvcert[i]+1]) }
for (j in 1:4) {b[j] ~dnorm(0,0.01)}
mocc ~dnorm(30,0.01); tocc ~dgamma(1,0.01)
k.delta ~dbern(p.delta); p.delta ~dbeta(1,1)
delta ~dunif(-1,1); delta.s <- k.delta*delta; prec <- 1/(1-pow(delta.s,2))}
```

The code for an augmented data logit regression using the approximation of Kinney and Dunson (2007) is

```
model {for (i in 1:n) {m[i] <-b[1]+b[2]*DVRT[i]+b[3]*sex[i]+b[4]*fathocc[i]
fathocc[i]~dnorm(mocc,tocc); prec[i] <- lam[i]*0.419; lam[i]~dgamma(3.65,
3.65)
ystar[i]~dnorm(m[i],prec[i]) I(lo[lvcert[i]+1],up[lvcert[i]+1]) }
for (j in 1:4) {b[j]~dnorm(0,0.01)}
mocc~dnorm(30,0.01); tocc~dgamma(1,0.01)}
```

5. Code for a conventional ordinal probit likelihood approach in Example 5.4 is as follows, with predictive checks based on replicate data,

```
model { for (t in 1:T) { for (j in 1:J-1) {gam[t,j] <- phi(a[j] - mu[t,j])
```

```
mu[t,j] <- b[1]*x[t,1]+b[2]*x[t,2]+b[3]*x[t,3]+b[4]*x[t,4]}
pi[t,1] <- gam[t,1]; pi[t,J] <- 1-gam[t,J-1]
for (j in 2:J-1) { pi[t,j] <- gam[t,j] - gam[t,j-1]}
y[t]~dcat(pi[t,1:J]); yrep[t] ~dcat(pi[t,1:J]);
match.pred[t] <- equals(y[t],yrep[t]);
a[1] ~dnorm(0,1) I(,a[2]); a[2] ~dnorm(0,1) I(a[1],);
b[1:4]~dmnorm(b0[ ], B0[ , ]); Year.pred <- sum(match.pred[])/T}
```

The code for the latent data approach (with replicates zstar.rep now being of latent data) in Example 5.4 is

```
model { for (t in 1:T) { for (j in 1:J-1) { z[t,j] <- step(j-y[t])
zstar[t,j]~dnorm(nu[t,j],1) I(A[t,j],B[t,j])
A[t,j] <- -10*equals(z[t,j],0); B[t,j] <- 10*equals(z[t,j],1)
nu[t,j] <- a[j] - mu[t,j]; mu[t,j] <- b[1]*x[t,1]+b[2]*x[t,2]+b[3]*x[t,3]+b[4]
*x[t,4]
zstar.rep[t,j]~dnorm(nu[t,j],1); zrep[t,j] <- step(zstar.rep[t,j])}
yrep[t] <- equals(zrep[t,1],1)*equals(zrep[t,2],1)
+2*equals(zrep[t,1],0)*equals(zrep[t,2],1)+3*equals(zrep[t,1],0)*equals
(zrep[t,2],0)
match.pred[t] <- equals(yrep[t],y[t])}
a[1]~dnorm(0,1) I(,a[2]); a[2]~dnorm(0,1) I(a[1],);
b[1:4]~dmnorm(b0[ ], B0[ , ]); Year.pred <- sum(match.pred[1:T])/T}
```

6. An illustrative code for a Gaussian process with two predictors is

```
model { Tau[1:n,1:n] <- inverse(Cov[,])
y[1:n] ~dmnorm(mu[1:n],Tau[,])
for (i in 1:n){mu[i] <- b[1]+b[2]*x1[i]+b[3]*x2[i]
for (j in 1:n) {Cov[i,j] <- sig2*(K[i,j]+gam*equals(i,j))
K[i,j] <- exp(-Disx[i,j])
Disx[i,j] <- pow(x1[i]-x1[j],2)/d[1]+pow(x2[i]-x2[j],2)/d[2]}}
d ~dexp(1); gam ~dunif(0,10); sig ~dunif(0,10);
sig2 <- sig*sig; sigdiag <- sqrt(sig2+gam)
for (j in 1:3) {b[j] ~dnorm(0,0.1)}}
```

with predictor-specific range parameters, but power coefficients set to 2.

7. The variance model based on a power transform of the linear predictor (Example 5.5) is coded as

```
model {for (i in 1:113) {y[i]~dnorm(eta[i],tau[i])
ynew[i]~dnorm(eta[i],tau[i]); predF[i] <- pow(y[i]-ynew[i],2)
s[i] <- sig*pow(1+abs(eta[i]),lam)
tau[i] <- 1/(s[i]*s[i]); eta[i] <- beta[1]+beta[2]*x[i]}
sig~dunif(0,250); lam~dunif(-2,2); C.F <- sum(predF[])
for (j in 1:2) {beta[j]~dnorm(0,0.001)}}
```

8. The code for the clustered outlier model (Example 5.6) is

```
model {for (i in 1:50) {G[i] ~dcat(p[i,1:3]); y[i]~dnorm(mu[i,G[i]],tau[G[i]]);
```

```
mu[i,1] <- b1[1]+b1[2]*x[i]; mu[i,2] <- b2[1]+b2[2]*x[i]; mu[i,3] <- mu[i,2]
for (j in 1:3) {post.group.prob[j,i] <- equals(G[i],j)}
h[i] <- kap*pow(x[i]-nu,2); p[i,1] <- exp(-h[i]); p[i,2] <- (1-pi)*(1-exp
(-h[i]));
p[i,3] <- pi*(1-exp(-h[i]))}
for (j in 1:2) {tau[j] ~dgamma(1,0.001)} tau[3] <- tau[2]/D; D <- 10
pi~dbeta(1,9); kap~dgamma(1,1); nu~dunif(0.04,8.09)
for (j in 1:2) {b1[j]~dnorm(0,0.001); b2[j]~dnorm(0,0.001)}}
```

9. ZIP regression (Example 5.7) involves either a nonstandard likelihood and the Barry (2006) proposal, or a mixed Poisson obtained by using subject-specific binary variables $w[i]$. In the following, the former option is commented out:

```
model { for (i in 1:797)
# approach 1
# z[i] <- 1; Lz[i] <- -1/p[i]; Uz[i] <- 1/p[i]; z[i]~dunif(Lz[i],Uz[i]);
# approach 2
y[i] ~ dpois(mu.w[i]); mu.w[i] <- (1-w[i])*mu[i]; w[i] ~ dbern(omeg)
# likelihood
p[i] <- D0[i]*p.eq.0[i]+(1-D0[i])*p.gt.0[i]; H[i] <- 1/p[i]
D0[i] <- equals(y[i],0);
p.eq.0[i] <- omeg+(1-omeg)*exp(-mu[i])
p.gt.0[i] <- (1-omeg)*p.pois[i]
log(p.pois[i]) <- -mu[i]+y[i]*log(mu[i])-logfact(y[i])
# regression
log(mu[i]) <- b1 + b2[sex[i]+1]+b3[ethnic[i]]+b4[school[i]]+
b5*log(base[i]+0.5)}
# priors
b1~dnorm(0,0.001); b5~dnorm(0,0.001); omeg~dbeta(1,1)
b2[1] <- 0; b2[2]~dnorm(0,0.001)
b3[1] <- 0; for (j in 2:3) {b3[j]~dnorm(0,0.001)}
b4[3] <- 0; for (j in 1:2) {b4[j]~dnorm(0,0.001)}
for (j in 4:6) {b4[j]~dnorm(0,0.001)}}
```

10. The code for Example 5.9 is

```
model { for (t in 1:T) {y[t]~dpois(mu[t])
LL[t] <- -mu[t]+y[t]*log(mu[t])-logfact(y[t]); G[t] <- 1/exp(LL[t])
Dv[t] <- y[t]*log(y[t]/mu[t])-(y[t]-mu[t])
beta1[t] <- beta[1]+b1[t]; beta2[t] <- beta[2]+b3[t]; beta3[t] <- beta[3]+b3[t]
log(mu[t]) <- alp[t] + (beta[1]+b1[t])*(x1[t]-mean(x1[]))+
(beta[2]+b2[t])*(x2[t]-mean(x2[]))+(beta[3]+b3[t])*(x3[t]-mean(x3[]))}
# predictions and fit
for (t in 2:T) {ystar[t]~dpois(mu[t-1]); Q[t] <- step(y[t]-ystar[t])}
Fit[1] <- -2*sum(LL[]); Fit[2] <- 2*sum(Dv[])
# RW1 priors on beta coeffs
```

```
w[1] <- 1;    adj[1] <- 2;    n[1] <- 1
w[(T-2)*2 + 2] <- 1;    adj[(T-2)*2 + 2] <- T-1;    n[T] <- 1
for (t in 2:T-1) {w[2+(t-2)*2] <- 1; adj[2+(t-2)*2] <- t-1
w[3+(t-2)*2] <- 1;adj[3+(t-2)*2] <- t+1; n[t] <- 2}
b1[1:T] ~car.normal(adj[],w[],n[],inv.W[1])
b2[1:T] ~car.normal(adj[],w[],n[],inv.W[2])
b3[1:T] ~car.normal(adj[],w[],n[],inv.W[3])
for (j in 1:3) {q[j]~dunif(0,10); inv.W[j] <- tau.alp/q[j]; beta[j]~dnorm
(0,0.001)}
# RW2 on intercepts alp[t]
for (t in 3:T){m.alp[t] <- 2*alp[t-1]-alp[t-2];alp[t]~dnorm(m.alp[t],tau.alp)}
alp1~dnorm(3,0.01); alp2~dnorm(3,0.01);alp[1] <- alp1; alp[2] <- alp2;
tau.alp~dgamma(1,1)}
```

11. A code for this approach, oriented to the Scottish lip cancer data, with w_{ij} taken as binary on the basis of adjacency, is

```
model { for (i in 1 : NN ) { We[i] <- e[adj[i]] }
lam ~dunif(-0.326,0.175); inv.sig2 ~dgamma(1,0.001)
for (j in 1:2) {b[j] ~dnorm(0,0.001)}
for (i in 1 : N) { y[i] ~dpois(nu[i]); nu[i] <- E[i]*rho[i]
log(rho[i]) <- r[i]; r[i] ~dnorm(r.bar[i], inv.sig2);
e[i] <- r[i]-eta[i]; eta[i] <- b[1]+b[2]*x[i]
r.bar[i] <- eta[i]+lam*sum(We[cum[i] + 1:cum[i + 1] ])}}
```

where N=56 and NN=264.

12. The code for the second model in Example 5.11, including elements to derive CPO statistics, is

```
model {for (i in 1:n) {y[i]~dpois(mu[i,G[i]]); G[i]~dcat(p[1:K])
L[i] <- sum(Lcomp[i,1:K]); H[i] <- 1/L[i]; logL[i] <- log(L[i])
Dv[i] <- y[i]*log(y[i]/mu[i,G[i]])-(y[i]-mu[i,G[i]])
for (k in 1:K) {mu[i,k] <- E[i]*exp(beta1[k]+
beta2[k]*x1[i]+beta3[k]*x2[i]+s[i])
Lcomp[i,k] <- p[k]*exp(LL[i,k])
LL[i,k] <- -mu[i,k]+y[i]*log(mu[i,k])-logfact(y[i])}}
# Priors
p[1:K]~ddirch(w[1:K]); tau~dgamma(1,0.001)
for (k in 1:K) {w[k] <- 1; beta2[k] ~dnorm(0,0.001); beta3[k]
~dnorm(0,0.001)}
s[1:n]~car.normal(map[], wt[], num[], tau)
# tneigh is the total of neighbors in adjacency map
for (i in 1 : tneigh) { wt[i] <- 1}
# ordered intercept constraint
beta1[1] ~dnorm(0,0.001)
for (j in 1:2) {delbeta[j]~dgamma(1,0.001); beta1[j+1] <- beta1[j]+delbeta[j]}
Fit[1] <- -2*sum(logL[]); Fit[2] <- 2*sum(Dv[])
# Correlation in residuals
```

```
SSRC <- sum(dt[])/sum(db[]);
for (i in 1 : n) {num[i] <- cum[i+1]-cum[i];
e[i] <- (y[i]-mu[i,G[i]])/sqrt(mu[i,G[i]]); estar[i] <- mean(We[cum[i]+1:cum
[i+1]])
de[i] <- e[i]-mean(e[]); d.estar[i] <- estar[i]-mean(estar[])
dt[i] <- de[i]*d.estar[i]; db[i] <-pow(d.estar[i],2)}
for (i in 1 : tneigh) {We[i] <- e[map[i]]}}
```

6

Bayesian Multilevel Models

6.1 Introduction

The rationale for applying multilevel models to hierarchical data is well established (Skrondal and Rabe-Hesketh, 2004; Snijders and Bosker, 1999). When lower level units are nested within one or more higher level strata, conventional single level regression analysis is not appropriate since observations are no longer independent: pupils in the same schools, or households in the same communities, tend to be more similar to one another than pupils in different schools or households in different communities. Such dependency means standard errors are downwardly biased if the nesting is ignored, and spurious inferences regarding predictor effects may be made (Hox, 2002).

In multilevel analysis, predictors may be defined at any level and the interest focuses on adjusting predictor effects for the simultaneous operation of contextual and individual variability in the outcome. This may be important in health applications, for example, if the impact of individual level risk factors varies according to geographic context (Congdon and Lloyd, 2010). Another major goal is variance partitioning (Goldstein et al., 2002); for example, what proportion of area variations in mortality is due to the characteristics of those areas (what is sometimes termed contextual variation), and how much is due to the characteristics of the individuals who live in these areas (termed compositional variation) (Subramanian et al., 2003).

One may also be interested in estimates for geographic areas or institutions that include both individual and area information; for example, the multilevel model for county radon estimates discussed by Gelman (2006a). Gelman (2006a) notes that compared to estimates involving no pooling or complete pooling, inferences from multilevel models are more reasonable. Complete pooling leads to identical estimates for all units, while a no-pooling model (no borrowing strength) overfits the data, giving implausibly high or low estimates for particular units and low precisions for such estimates.

As well as predictor effects at any level, a multilevel model is likely to involve random effects defined over the clusters at higher level(s), and possibly correlation between different cluster effects. As in Chapter 3, one seeks to pool strength in inferences about clusters when the number of observations for each cluster might be quite small. While exchangeable cluster effects dominate the multilevel literature, there may well be instances where random cluster effects are better regarded as nonexchangeable, as recognized in the general design general linear mixed model of Zhao et al. (2006). For example, it is possible that the significance level of cluster (i.e., contextual) effects is overstated in area multilevel applications that disregard spatial correlation between clusters (Chaix et al., 2005).

Application of multilevel models from a Bayesian perspective exemplifies many of the issues referred to in earlier chapters regarding specification of priors on variance components such as to ensure empirical identifiability and posterior propriety. Devices such as hierarchical centering may reduce correlation in the joint posterior and increase Markov Chain Monte Carlo (MCMC) effective sample sizes, and are relevant to both nested and crossed random effects (Browne, 2004; Gelfand et al., 1995). On the other hand, the Bayesian approach has benefits in ensuring that uncertainty in variance components is fully reflected in the posterior inferences on other parameters, such as cluster effects; this is especially an issue when the number of level 2 units is small and the likelihood function of level 2 variance parameters may be asymmmetric (Selzter et al., 1996, 2002). As compared to commonly applied frequentist approaches (e.g., penalized quasi-likelihood), obtaining variance estimates of predicted random effects is computationally straightforward in MCMC approaches. The remaining sections of the chapter consider the normal linear multilevel model (Section 6.2), general linear and conjugate models for multilevel discrete data (Section 6.3), crossed factor and multiple member random effect models (Section 6.4), and robust multilevel models (Section 6.5).

6.2 The Normal Linear Mixed Model for Hierarchical Data

A multilevel model typically assumes observations to be independent conditional on fixed regression and random effects defined at one or more levels in the data hierarchy. The prototype two level model for a continuous response, y_{ij}, with repetitions $j = 1, \ldots, n_i$ (e.g., pupils, patients, households) in clusters $i = 1, \ldots, m$ (e.g., schools, hospitals, communities) tackles a similar scenario to that considered in Chapter 3, but assumes individual observations to be available rather than cluster averages. Consider observation level attribute vectors, x_{ij}, of dimension p, and z_{ij} of dimension q, typically a subvector of x_{ij} with $q \leq p$ (Chen and Dunson, 2003).

Then a widely used form of the normal linear mixed model for nested data (e.g., Snijders and Berkhof, 2002) specifies,

$$y_{ij} = x_{ij}\beta + z_{ij}b_i + u_{ij}, \tag{6.1a}$$

with b_i and u_{ij} denoting random cluster effects and observation level random effects, respectively. The intercept, $x_{1ij} = 1$, with parameter β_1 is included in x_{ij}. With $N = \sum_{i=1}^{m} n_i$ total observations, the nested form of the model is

$$y = X\beta + Zb + u, \tag{6.1b}$$

where y is $N \times 1$, $X \equiv \begin{bmatrix} X_1 \\ \cdots \\ X_m \end{bmatrix}$ is $N \times p$, with $X_i = (x_{i1}, \ldots, x_{in_i})'$ of dimension $n_i \times p$, and where the $N \times mq$ matrix Z is block diagonal with m diagonal blocks, $Z_i = (z_{i1}, \ldots, z_{in_i})'$ of dimension $n_i \times q$ (Gamerman, 1997, 61; Zhao et al., 2006, 3). Assuming Z_i is a subset of X_i, β is a $(p \times 1)$ vector of population parameters and $b_i = (b_{1i}, \ldots, b_{qi})'$ is a $q \times 1$ vector of zero mean cluster-specific deviations around those population parameters, with b_i assumed random.

While random effects models offer a way to borrow strength (e.g., when level 2 cluster sizes, n_i, are relatively small), fixed effect models, especially for varying intercepts are, however, advocated in longitudinal applications, especially in econometrics. Fixed effects for parameter collections are sometimes used in cross-sectional multilevel applications (Snijders and Berkhof, 2002). The choice between the two depends on the purpose of the statistical inference and how far the level 2 units can be regarded as a sample from a policy-relevant population (Draper, 1995). If the sampled clusters are representative of (exchangeable with) a wider population, then a random coefficient model is, in principle, appropriate (Hsiao, 1996). If statistical inference is confined to the particular unique set of level 2 units included in a data set, then a fixed effects model may be more appropriate.

The conjugate linear normal model with random cluster effects assumes multivariate normality for these effects and for the observation level errors. Assuming z_{ij} are a subvector of x_{ij}, the cluster effects have zero mean, so that,

$$(b_{1i}, \ldots, b_{qi})' \backsim N_q(0, \Sigma_b).$$

The total impact of x_{rij} in cluster i is then obtained by cumulating over fixed and random components as $\beta_r + b_{ri}$.

Assume the unstructured level 1 errors, $u_i = (u_{i1}, \ldots, u_{in_i})'$, have prior $u_i \sim N_{n_i}(0, H_i)$, where H_i represents the within-cluster dispersion matrix. The stacked form of the linear mixed model at cluster level, namely, $y_i = X_i\beta + Z_i b_i + u_i$, may then be expressed in joint likelihood form as,

$$\begin{bmatrix} y_i \\ b_i \end{bmatrix} \sim N_{n_i+q}\left(\begin{bmatrix} X_i\beta & Z_i\Sigma_b Z_i' + H_i & Z_i\Sigma_b \\ 0 & \Sigma_b Z_i' & \Sigma_b \end{bmatrix}\right),$$

or in marginal form as,

$$y_i \sim N_{n_i}(X_i\beta, Z_i\Sigma_b Z_i' + H_i).$$

The level 1 errors are typically assumed independent given cluster effects and regression terms, often with $H_i = \sigma^2 I$ for all clusters.

The conjugate model then takes inverse gamma and inverse Wishart priors for σ^2 and Σ_b, respectively (or gamma and Wishart priors on σ^{-2} and Σ_b^{-1}), and common practice is to adopt just proper priors, e.g., $\sigma^2 \sim IG(\varepsilon, \varepsilon)$, where ε is small. Recent research shows that such priors can lead to effectively improper posteriors and also that inferences are sensitive to the choice of hyperparameters (Natarajan and McCulloch, 1998). Alternatives for the level 1 variance include uniform or half t priors on σ (Gelman, 2006a), while hierarchical models for Σ_b are considered by Daniels and Kass (1999) and Daniels and Zhao (2003).

Following Gamerman (1997, 62), one may sometimes also include randomly varying predictor effects at observation level,

$$y_{ij} = x_{ij}\beta + z_{ij}b_i + w_{ij}u_{ij},$$

which is one way of specifying what is known as complex level 1 variation or heteroscedasticity related to level 1 attributes (Browne et al., 2002). This means that variances depend on subject level predictors (when subjects j are nested in clusters i), or in panel data applications that variances are changing over time (when times t are nested in subjects i). For categorical w_{ij}, one may equivalently specify complex variation in terms of category-specific variances. Thus, Goldstein (2005) considers school exam data, y_{ij} (pupils j nested in schools i), with a single predictor gender, x_{ij} (=1 for boy, 0 for girl). Then level 1 heteroscedasticity can be represented as,

$$y_{ij} = \beta_1 + \beta_2 x_{ij} + x_{ij}u_{1ij} + (1 - x_{ij})u_{0ij},$$

where $u_{0ij} \sim N(0, \sigma_0^2)$ is the prior for girl observation level errors, and $u_{1ij} \sim N(0, \sigma_1^2)$ is the prior for boy observation errors. Equivalently, setting $w_{ij} = x_{ij}$,

$$y_{ij} \sim N(\beta_1 + \beta_2 x_{ij}, \sigma^2_{w_{ij}}).$$

It can be seen that random variation over clusters or at level 1 in Equation 6.1 raises questions of empirical identification (see Chapter 1) as the fixed regression effects are confounded with the mean of the associated cluster random effect. Suppose $x_{ij} = (x_{fij}, x_{hij})$ and $\beta = (\beta_f, \beta_h)$, where x_{fij} of dimension $p - q$ contains predictors where no variation in clusters is posited, while x_{hij} contains predictors (usually including the constant term) that have a randomly varying effect over clusters. Under hierarchical centering of the cluster effects, which has been argued to improve MCMC convergence (Gelfand et al., 1995), varying cluster effects, γ_{ri}, are centered on β_r so that the rth varying predictor effect is $\gamma_{ri} = \beta_{hr} + b_{ri}$ in cluster i. The parameterization $(\beta, b_i) = ([\beta_f, \beta_h], b_i)$ with zero mean b_i, is replaced by the parameterization (β_f, γ_i), where $\gamma_i = \beta_h + b_i$. Then,

$$\begin{aligned} y_{ij} &= x_{ij}\beta + z_{ij}\gamma_i + u_{ij} (= x_{fij}\beta_f + x_{hij}\gamma_i + u_{ij}) \\ (\gamma_{1i}, \ldots, \gamma_{qi}) &\backsim N_q(\beta_h, \Sigma_\gamma), \end{aligned} \tag{6.2}$$

where the vectors z_{ij} and x_{ij} are now distinct, with x_{ij} now containing only x_{fij}, while $z_{ij} = x_{hij}$.

6.2.1 The Lindley–Smith model format

An alternative fully hierarchical presentation of the normal linear multilevel model (e.g., Candel and Winkens, 2003; Selzter, 1993) is based on the scheme of Lindley and Smith (1972). It is assumed that all the effects of level 1 predictor (e.g., pupil characteristics in a two-level educational attainment application) vary randomly over clusters, with their variability explained by cluster predictors, $W_i = (w_{1i}, \ldots, w_{ri})'$ (e.g., school level attributes).

The two level scheme is

$$\begin{aligned} y_i &= Z_i\beta_i + u_i, \\ \beta_i &= \kappa W_i + b_i \quad i = 1, \ldots, m, \end{aligned} \tag{6.3}$$

where $y_i = (y_{i1}, \ldots, y_{in_i})'$ is $n_i \times 1$, κ is $q \times r$, Z_i is $n_i \times q$, β_i is a $q \times 1$ vector of random cluster regression parameters, and the errors $u_i = (u_{i1}, \ldots, u_{in_i})'$ have prior $u_{ij} \sim N(0, \sigma^2)$. The level 2 regression for β_i involves a fixed effect parameter matrix, κ, and errors $b_i = (b_{1i}, \ldots, b_{qi})'$ with mean zero and dispersion matrix, T_b. Substituting the second equation in Equation 6.3 into the first yields the model,

$$y_i = Z_i\kappa W_i + Z_i b_i + u_i.$$

To constrain the effect of one or more level 1 predictors to have an identical effect across all clusters, the model may be reformulated as the mixed model (Equation 6.2).

In Equation 6.3, one may assume flat (uniform) priors for κ, and gamma and Wishart priors for σ^{-2} and T_b, namely, $1/\sigma^2 \sim Ga(a_u, b_u)$, $T_b \sim W(S_e, \nu_e)$. Also define $r_{ij} = y_{ij} - Z_{ij}\beta_i$, $\hat{\beta}_i = (Z_i'Z_i)^{-1}Z_i y_i$, $\tilde{V}_i = (\sigma^{-2}Z_i'Z_i + T_b)^{-1}$, $V_i = \sigma^2 Z_i'Z_i$, $\Lambda_i = (V_i^{-1} + T_b)V_i^{-1}$, $U_i = (\beta_i - \kappa W_i)$, and $G = [\sum W_i'T_b W_i]^{-1}$. Then the full conditionals for Gibbs sampling are

$$\begin{aligned} 1/\sigma^2 &\sim Ga\left(0.5(a_u + m), 0.5\left(b_u + \sum_{i=1}^{m}\sum_{j=1}^{n_i} r_{ij}^2\right)\right), \\ \beta_i &\sim N_q\big(\Lambda_i\hat{\beta}_i + (I - \Lambda_i)\kappa W_i, \tilde{V}_i\big), \\ T_b &\sim W\left(S_e + \sum_{i=1}^{m} U_i U_i', m + \nu_e\right), \\ \kappa &\sim N_r\left(G\sum_{i=1}^{m} W_i T_b \beta_i, G\right). \end{aligned}$$

Example 6.1. Exam Scores Hierarchical centering and level 1 complex variation in a mixed rather than fully hierarchical model are both illustrated in an educational example concerning 4059 pupils in 65 London schools (the "clusters" in this application). The response is a (normalized) exam score at age 16, with predictors being a standardized London Reading Test (LRT) score (obtained at age 11) and gender. The analysis assumes cluster level variation in intercepts and in the slope of the LRT score, with the effect of gender not varying by cluster. Cluster variability on the impacts of level 1 predictors is not related to school attributes. Therefore, this is a mixed model, which in hierarchical form, as in Equation 6.2, is

$$y_{ij} = x_{ij}\beta_f + z_{ij}\gamma_i + u_{ij},$$
$$(\gamma_{1i}, \gamma_{2i}) \backsim N_q(\beta_h, \Sigma_\gamma),$$

where x_{ij} = (gend) excludes an intercept, and $z_{ij} = (1, \text{LRT})$. A Wishart prior with identity scale matrix and 2 degrees of freedom is assumed for the cluster precision matrix, Σ_γ^{-1}, and a $Ga(1, 0.001)$ prior for the observation level precision σ^{-2}. Two analyzes are run, one with σ^2 constant over subjects, and a second with σ^2 differing by gender. A pseudo marginal likelihood is used to assess fit[1], with the posterior predictive density $p(y_{\text{rep}}|y)$ used for model checking.

A two chain run of 10,000 iterations[1] for the model with σ^2 constant provides early convergence, with psML = −4596.7, and shows a satisfactory check against the data: 95.2% of the 4059 observations are contained within 95% intervals of the replicates $y_{ij,\text{rep}}$. Random variation over schools is confirmed by intercept and slope standard deviations (the square roots of the diagonal terms in Σ_γ), that are bounded away from zero (MacNab et al., 2005).

The means and 95% intervals for $\sigma_{1\gamma}$ and $\sigma_{2\gamma}$ are 0.33 (0.27, 0.40) and 0.21 (0.17, 0.25), while the corresponding estimates for the observational standard deviation are 0.74 (0.725, 0.757). The intercepts and slopes are positively correlated so that schools with better exam results also show a stronger intake effect. Introducing complex level 1 variation, namely, gender-specific variances, raises the psML only slightly, to −4594.9.

6.3 Discrete Responses: General Linear Mixed Model, Conjugate, and Augmented Data Models

While conjugate multilevel structures can be developed for discrete responses such as counts or proportions (see Section 6.3.2), a more flexible approach is based on the general linear mixed model (GLMM), which extends the linear normal formulation to discrete outcomes. Thus consider univariate observations, y_{ij}, with repetitions j nested in clusters i, that conditional on cluster effects, b_i, follow an exponential family density,

$$f(y_{ij}|b_i) \propto \exp\left\{\frac{y_{ij}\theta_{ij} - d(\theta_{ij})}{\phi_{ij}} + c(y_{ij}, \phi_{ij})\right\},$$

where θ_{ij} is the canonical parameter, and ϕ_{ij} is usually a known scale parameter. Additionally, $E(y_{ij}|\theta_{ij}) = d'(\theta_{ij})$ and $\text{Var}(y_{ij}|\theta_{ij}, \phi_{ij}) = d''(\theta_{ij})\phi_{ij}$.

For example, under the Poisson, $d(u) = \exp(u)$ and for binomial data, $d(u) = \log(1+e^u)$. Taking the regression terms as $\eta_{ij} = g(\theta_{ij})$, where g is a link function, and denoting $\eta_i = (\eta_{i1}, \ldots, \eta_{in_i})'$, the observation level model (including a level 2 regression on cluster attributes) is

$$\eta_{ij} = x_{ij}\beta + z_{ij}b_i,$$
$$b_i = \kappa W_i + e,$$

where β and b_i are of dimension p and q, respectively.

For Poisson data it is also common to include a residual term, $u_i = (u_{i1}, \ldots, u_{in_i})$, to model overdispersion, so that

$$\eta_{ij} = x_{ij}\beta + z_{ij}b_i + u_{ij}. \tag{6.4}$$

Following Gamerman (1997, 62), one may have observation level predictors, g_{ij}, of dimension s with randomly varying effects at observation level. In this case, u_{ij} becomes a random vector, with

$$\eta_i = x_{ij}\beta + z_{ij}b_i + g_{ij}u_{ij}.$$

Assume priors $\beta \sim N_p(a, R)$, $b_i \sim N_q(0, \Sigma_b)$, and $u_{ij} \sim N_r(0, \Sigma_u)$, with inverse Wishart priors $\Sigma_b \sim IW(\nu_b, S_b)$ and $\Sigma_u \sim IW(\nu_u, S_u)$. Then the full posterior conditional for each b_i vector is

$$p(b_i|b_{[i]}, \beta, u, \Sigma_b, \Sigma_u) \propto \exp\left\{-0.5b_i'\Sigma_b^{-1}b_i + \sum_{j=1}^{n_i}\frac{y_{ij}\theta_{ij} - d(\theta_{ij})}{\phi_{ij}}\right\},$$

while the full conditional for each u_{ij} vector is

$$p(u_{ij}|u_{[ij]}, b, \beta, u, \Sigma_b, \Sigma_u) \propto \exp\left\{-0.5u_{ij}'\Sigma_u^{-1}u_{ij} + \frac{y_{ij}\theta_{ij} - d(\theta_{ij})}{\phi_{ij}}\right\}.$$

Additionally, the covariance matrices have inverse Wishart full conditionals, namely,

$$\Sigma_b \sim IW\left(\nu_b + m, S_b + \sum_{i=1}^{m} b_i b_i'\right),$$
$$\Sigma_u \sim IW\left(\nu_u + \sum_{i=1}^{m} n_i, S_u + \sum_{i,j} u_{ij}u_{ij}'\right).$$

The GLMM approach extends to multilevel multinomial observations in a choice setting (e.g., brand, political party),

$$(d_{ij1}, \ldots, d_{ijK}) \sim \text{Mult}(1, [p_{ij1}, \ldots, p_{ijK}]),$$

with probability, π_{ijk}, that option k is chosen by subject j in cluster i, namely, that $y_{ij} = k$ (or $d_{ijk} = 1$) where options are unordered. A particular choice

($k \in 1, \ldots, K$) for subject j in cluster i results from comparing the latent utilities of all options $(\eta_{ij1}, \ldots, \eta_{ijK})$, with,

$$p_{ijk} = Pr(y_{ij} = k) = Pr(\eta_{ijk} > \eta_{ijm}), \quad m \neq k,$$

where η_{ijk} include systematic effects and random errors, ε_{ijk}. Suppose the errors follow a Gumbel (extreme value type I) density, namely, $P(\varepsilon) = \exp(-\varepsilon - \exp(-\varepsilon))$, then since differences between Gumbel errors follow a standard logistic distribution, the choice probabilities reduce to the multinomial logit (Hedeker, 2003, 1439). Predictors in the systematic term may be defined at option-subject or at option level, but consider subject level predictors x_{ij} and z_{ij} (e.g., voter age) of respective dimensions p and q that may vary according to cluster i. Then with the final category as a reference, fixed effect parameters and random effects are specific to choices k, with $K-1$ sets of random effects, b_{ik}, each of dimension q,

$$Pr(y_{ij} = k) = \frac{\exp(\alpha_k + x_{ij}\beta_k + z_{ij}b_{ik})}{1 + \sum_{h=1}^{K-1}\exp(\alpha_h + x_{ij}\beta_h + z_{ij}b_{ih})}$$

$$Pr(y_{ij} = K) = \frac{1}{1 + \sum_{h=1}^{K-1}\exp(\alpha_h + x_{ij}\beta_h + z_{ij}b_{ih})} \qquad k = 1, \ldots, K-1.$$

The $b_i = (b_{i1}, \ldots, b_{i,K-1})$ are zero mean effects, typically assumed multivariate normal.

6.3.1 Augmented data multilevel models

Another option for multilevel binary and multinomial responses is to introduce augmented metric data, y^*_{ij}, with sampling constrained according to the observed y_{ij}, and apply the linear mixed model to y^*_{ij}. The data augmentation density depends on the assumed link. Thus, a logit link for two level binary data implies truncated standard logistic sampling to generate the augmented data, namely,

$$y^*_{ij} \sim \text{Logistic}(\eta_{ij}, 1)I(A_{ij}, B_{ij}),$$

where $A_{ij} = -\infty$ or 0, and $B_{ijk} = 0$ or ∞, according as $y_{ij} = 0$ or 1. As mentioned in Chapter 5, data augmentation leads to simpler MCMC sampling and improved residual tests. In a multilevel setting, it may further assist in assessing variance partitioning. Consider a regression with a random level 2 intercept,

$$\eta_{ij} = x_{ij}\beta + b_i,$$

where $b_i \sim N(0, \sigma^2_b)$. Since the variance of the standard logistic is $\pi^2/3$, the intraclass correlation at level 2 may be obtained as $\sigma^2_b/(\sigma^2_b + \pi^2/3)$, and monitored over MCMC iterations. Moreover, if the composite fixed effect term, $x_{ij}\beta$, is monitored and its posterior variance, $\tilde{\sigma}_F^2$, obtained, one may obtain a proportion of variance explained as $\tilde{\sigma}_F^2/[\tilde{\sigma}_F^2+\tilde{\sigma}_b^2+\pi^2/3]$, where $\tilde{\sigma}_b^2$ is the posterior mean of σ_b^2.

For multilevel ordinal outcomes with K levels, the observation,

$$(d_{ij1}, \ldots, d_{ijK}) \sim \text{Mult}(1, [p_{ij1}, \ldots, p_{ijK}]),$$

provides information about an underlying metric variable, y_{ij}^*, defined by cutpoints such that,

$$y_{ij} = k \quad (d_{ijk} = 1) \quad \text{if } \kappa_{k-1} < y_{ij}^* \leq \kappa_k \text{ for } k = 1, \ldots, K,$$

where $\kappa_0 = -\infty$ and $\kappa_K = \infty$, and by a corresponding hierarchical linear regression in the latent variables,

$$y_{ij}^* = x_{ij}\beta + z_{ij}b_i + \varepsilon_{ij}.$$

If x_{ij} excludes an intercept, there are $K-1$ unknown cutpoints $(\kappa_1, \ldots, \kappa_{K-1})$, with $y_{ij} = 1$ if $y_{ij}^* \leq \kappa_1$, $y_{ij} = 2$ if $\kappa_1 < y_{ij}^* \leq \kappa_2$, etc., and $y_{ij} = K$ if $y_{ij}^* > \kappa_{K-1}$. A standard logistic density for ε_{ij} with mean 0, variance $\pi^2/3$, and distribution function $F(\varepsilon) = (\exp(\varepsilon)/1 + \exp(\varepsilon))$ leads to a logit link for the cumulative probabilities,

$$\Delta_{ijk} = \sum_{m=1}^{k} p_{ijm} = Pr(y_{ij}^* \leq \kappa_k) = Pr(y_{ij} \leq k), \quad k = 1, \ldots, K-1,$$

with $p_{ijK} = 1 - \sum_{m=1}^{K-1} p_{ijm}$.

Taking $\varepsilon_{ij} \sim N(0,1)$ corresponds to a probit link for Δ_{ijk}. For ε logistic, the hierarchical regression is expressed as follows,

$$\begin{aligned}\Delta_{ijk} &= Pr(x_{ij}\beta + z_{ij}b_i + \varepsilon_{ij} \leq \kappa_k) \\ &= Pr(\varepsilon_{ij} \leq \kappa_k - x_{ij}\beta - z_{ij}b_i) \\ &= \frac{\exp(\kappa_k - x_{ij}\beta - z_{ij}b_i)}{1 + \exp(\kappa_k - x_{ij}\beta - z_{ij}b_i)},\end{aligned}$$

that is,

$$\text{logit}(\Delta_{ijk}) = \kappa_k - x_{ij}\beta - z_{ij}b_i.$$

6.3.2 Conjugate cluster effects

An alternative to GLLMs for count and binomial data, is provided by conjugate random effects at different levels. For example, for two level Poisson data, $y_{ij} \sim Po(\mu_{ij})$, a conjugate hierarchical model is a nested gamma mixture model, as in Lee and Nelder (2000), namely,

$$\mu_{ij} = \exp(x_{ij}\beta)\rho_i\rho_{ij},$$

where ρ_i and ρ_{ij} are both gamma distributed with mean 1. Daniels and Gatsonis (1999) also consider hierarchical conjugate priors for overdispersed data following densities from the one-parameter exponential family (e.g., Poisson, binomial) and assume both cluster-specific regression effects and scale parameters. For binomial data, $y_{ij} \sim Bin(T_{ij}, p_{ij})$, Daniels and Gatsonis (1999, 32) assume p_{ij} to be beta with means π_{ij}, and cluster-specific scale parameters δ_i. With a logit link to predictors, and level 2 regression involving cluster level predictors, W_i, one has,

$$\begin{aligned} p_{ij} &\sim Be(\pi_{ij}\delta_i, (1-\pi_{ij})\delta_i), \\ \text{logit}(\pi_{ij}) &= x_{ij}\beta + z_{ij}b_i, \\ b_i &= \kappa W_i + e_i. \end{aligned}$$

To provide robustness (e.g., to outlier clusters), the e_i are taken as Student t distributed (see Section 6.5). The prior on δ_i has the form,

$$p(\delta_i) \propto \frac{\delta_i}{(h_i + \delta_i)^2},$$

where $h_i = \min_{j \in 1,\ldots,n_i}(T_{ij})$. A Metropolis–Hastings step is needed for $\log(\delta_i)$.

For Poisson data, one has $y_{ij} \sim Po(o_{ij}\theta_{ij})$, where o_{ij} is an offset for the expected response, and $\theta_{ij} \sim Ga(\mu_{ij}\delta_i, \delta_i)$. The regression model then involves a log link for the μ_{ij},

$$\log(\mu_{ij}) = x_{ij}\beta + z_{ij}b_i.$$

More specialized models apply for particular data structures. For example, Van Duijn and Jansen (1995) suggest a model for repeated counts (e.g., tests $j = 1, \ldots, n_i$ within students $i = 1, \ldots, m$) with Poisson means,

$$\mu_{ij} = \nu_i \delta_{ij},$$

and gamma distributed student ability effects, $\nu_i \sim Ga(a_1, a_2)$, where a_1 and a_2 are additional parameters, and δ_{ij} represent subject-specific difficulty parameters for tests j, with identifiability constraint $\sum_j \delta_{ij} = 1$, and prior,

$$(\delta_{i1}, \ldots, \delta_{ij}) \sim Dir(\xi_1, \ldots, \xi_J),$$

where ξ_j are also unknowns. If the subjects fall into known (or possibly unknown) groups, $k = 1, \ldots, K$, with allocation indicators, $S_i \in (1, \ldots, K)$, then a more general model specifies,

$$(\nu_i | S_i = k) \sim Ga(a_{1k}, a_{2k}).$$

A conjugate structure for stratified area health counts is considered by Dean and McNab (2001). Thus for micro areas, $j = 1, \ldots, n_i$, nested within larger areas, $i = 1, \ldots, m$, let μ be an average event rate across all m areas and T_{ij} be populations at risk. Assume first, cluster level overdispersion represented by effects ρ_i, so that $y_{ij} \backsim Po(\mu T_{ij}\rho_i)$, where ρ_i have mean 1 and let the mean and variance of y_{i+} be $T_{i+}\mu$ and $T_{i+}\mu(1+\sigma_\rho^2)$. Under gamma mixing,

$$\rho_i \backsim Ga\left(\frac{T_{i+}\mu}{\sigma_\rho^2}, \frac{T_{i+}\mu}{\sigma_\rho^2}\right),$$

with variance $\sigma_\rho^2/(T_{i+}\mu)$. The interpretation is that ρ_i represents the average relative risk over the T_{i+} individuals in area i. An extended model considers both micro area and cluster level overdispersion with micro area effects, ρ_{2ij}, which given $\rho_{1i} \backsim Ga(T_{i+}\mu/\sigma_1^2, T_{i+}\mu/\sigma_1^2)$, are themselves also gamma with prior,

$$\rho_{2ij} \backsim Po\left(\frac{T_{ij}\rho_{1i}\mu}{\sigma_2^2}, \frac{T_{ij}\rho_{1i}\mu}{\sigma_2^2}\right).$$

Then $E(y_{ij}|\rho_{1i}) = \rho_{1i}T_{ij}\mu$ and $V(y_{ij}|\rho_{1i}) = T_{ij}\mu\rho_{1i}(1+\sigma_1^2)$, while marginally $E(y_{ij}) = T_{ij}\mu$ and

$$V(y_{ij}) = T_{ij}\mu\left[(1+\sigma_2^2) + \frac{T_{ij}}{T_{i+}}\sigma_1^2\right].$$

Example 6.2. Modern Prenatal Care This example illustrates multilevel GLMM and augmented data methods applied to three level binary data, namely, the use of modern prenatal care among Guatemalan women using some form of prenatal care. The analysis is based on 2449 births (born in a five-year period before the survey) to 1558 mothers living in one of $m = 161$ communities. The predictor variables are at all levels, namely, communities i at level 3, mothers ij at level 2, and pregnancy episodes ijk at level 1, and denoted X_i, X_{ij}, and X_{ijk}, respectively. The predictors include at level 3: proportion of community population indigenous and distance to the nearest clinic; at level 2: mother's ethnicity, mother's education, husband's education, husband's occupation, and presence of a modern toilet; and at level 1: (existing) child's age, mother's age, and birth order.

Then with $y_{ijk} \sim Bern(p_{ijk})$, the GLMM model used by Rodriguez and Goldman (2001) specifies random intercept variation at both levels 2 and 3, which is according to both mother and community. Using a noncentered parameterization and logit link, one has,

$$\text{logit}(p_{ijk}) = \alpha + b_{i3} + b_{ij2} + X_i\beta_3 + X_{ij}\beta_2 + X_{ijk}\beta_1,$$

where b_{i3} are normal community level effects, $b_{i3} \sim N(0, \sigma_3^2)$, and $b_{ij2} \sim N(0, \sigma_2^2)$ are mother level effects[2]. In a WinBUGS application, convergence is achieved by iteration 2500 in a two chain run of 5000 iterations. The posterior summary based on the second half of the run (Table 6.1) shows similar results to those obtained by Rodriguez and Goldman (2001). A BayesX application

TABLE 6.1
Modern pregnancy advice.

Fixed effects	**Mean**	**2.5%**	**97.5%**
Intercept	5.46	1.60	10.82
Pregnancy level			
Child aged 3–4 years	−1.35	−2.18	−0.64
Mother aged >25 years	1.31	0.09	2.57
Birth order 2–3	−1.04	−2.20	0.02
Birth order 4–6	−0.60	−2.07	0.89
Birth order >7	−1.40	−3.59	0.63
Mother level			
Indigenous, no Spanish	−8.32	−13.37	−4.03
Indigenous Spanish	−4.29	−8.54	−1.44
Mother's education primary	2.49	0.76	4.61
Mother's education secondary	5.52	1.22	10.73
Husband's education primary	0.89	−0.96	2.85
Husband's education secondary	4.51	0.66	8.41
Husband's education missing	−0.18	−3.16	2.70
Husband professional etc.	−0.47	−5.08	3.94
Husband agricultural self-employed	−2.58	−6.50	0.94
Husband agricultural employee	−3.66	−8.20	−0.16
Husband skilled service	−1.17	−5.00	2.57
Modern toilet in households	2.70	−0.06	5.64
Television not watched daily	2.26	−1.38	6.35
Television watched daily	2.16	−0.10	4.59
Community level			
Proportion indigenous, 1981	−6.09	−11.22	−0.73
Distance to nearest clinic	−0.07	−0.13	−0.02
Random effects (standard deviations)			
Family	10.22	7.40	13.23
Community	5.55	3.87	7.46

Reference categories for fixed effects: child aged 0–2 years, mother aged under 25, birth order 1, Ladino, mother no education, husband no education, husband not working or unskilled occupation, no modern toilet in the household, and no television in the household.

provides similar results for fixed effects, but mother and community variances have higher posterior means (117 and 34, compared to 104 and 31 from the WinBUGS analysis). Alternative model options that might be considered are (a) community variation in level 2 and level 1 predictor effects, (b) maternal variation in level 1 predictor effects, and (c) random effect priors allowing non-normality.

Note that the same results are obtained (and left as an exercise) by considering the observed 0 or 1 as arising from an underlying continuous variable,

y^*_{ijk}, so that 1 is observed when a certain threshold is exceeded, otherwise 0 is observed. A logit link means the sampling mechanism,

$$y^*_{ijk} \sim \text{Logistic}(\eta_{ijk}, 1)I(A_{ijk}, B_{ijk}),$$

where $A_{ijk} = -\infty$ or 0 (and $B_{ijk} = 0$ or ∞) according as $y_{ijk} = 0$ or 1. It follows that the intraclass correlation at level 2 may be obtained as $\sigma_2^2/(\sigma_2^2 + \pi^2/3)$, because the variance of the standard logistic is $\pi^2/3$.

Example 6.3. Ordinal Three Level Model Hedeker et al. (1994) undertake a two level analysis of a data subset from the Television School and Family Smoking Prevention and Cessation Project (Flay et al., 1987). Schools included in the project were randomized to one of four categories, defined by the presence or absence of a TV intervention (TV), and by the presence or absence of a social-resistance classroom curriculum (CC). One outcome measure from the project was a tobacco and health knowledge (THK) score obtained as the total correct answers to seven items on tobacco and health knowledge. The response variable is the postintervention THK score collapsed into $K=4$ ordinal categories, with predictors being a pre-intervention THK score, the binary intervention variables TV and CC, and a CC by TV interaction.

The analysis is three level, with schools $i = 1, \ldots, m$ at level 3 (with $m = 28$), and classrooms j within schools at level 2. With responses $y_{ijk} \in (1, \ldots, H)$, where $H = 4$, for subjects $k = 1, \ldots, n_{ij}$ in each class, consider a logit model for the cumulative probabilities,

$$\Delta_{ijkh} = Pr(y_{ijk} \leq h),$$

with two sets of higher level errors, namely, level 3 random errors u_{3i} with variance σ_3^2, and level 2 class errors u_{2ij} with variance σ_2^2 (pertaining to effects of classrooms within schools). Then,

$$\begin{aligned} \text{logit}(\Delta_{ijkh}) &= \kappa_h - \mu_{ijk}, \\ \mu_{ijk} &= x_{ijk}\beta + u_{3i} + u_{2ij} \quad h = 1, \ldots, H-1, \end{aligned}$$

with $N(0, 1000)$ priors on fixed effects and $U(0, 1000)$ priors on the random effect standard deviations[3].

A two chain run of 5000 iterations with 500 for convergence gives posterior means for the cutpoints $(\kappa_1, \kappa_2, \kappa_3)$ of −0.09, 1.2, and 2.4, with a significant coefficient of 0.89 on the curriculum intervention, but no significant effects for TV or the interaction term. The posterior means for σ_2 and σ_3 are 0.41 and 0.28 with densities bounded away from zero. By contrast, a maximum likelihood analysis using numerical quadrature reported by Rabe-Hesketh et al. (2004) finds an insignificant school variance. The analysis can also be carried out using the latent data approach, which may be useful for obtaining intraclass correlations or for model checking. This produces larger estimates of σ_2 and σ_3, namely, 0.60 and 0.38, but similar fixed predictor effects.

6.4 Crossed and Multiple Membership Random Effects

Crossed random effects at level 2 and above occur when classifications in a model are not completely nested. For a two level example, let i denote the main level 2 nesting classification and h_{ij} denote a crossed nesting. Raudenbush (1993), Browne et al. (2001, Section 3.2), and Snijders and Bosker (1999) mention educational examples, namely, pupils classified by primary school, i, and by secondary school, h_{ij}, or pupils classified by school, i, and neighborhood, h_{ij}. In the latter situation, a school can draw pupils from multiple neighborhoods, and residents in a neighborhood can choose between multiple schools for their children. By extension, if pupils are classified by primary school, secondary school, and neighborhood, then two crossed nestings (denoted say as h_{1ij} and h_{2ij}) will be involved. An important issue is the relationship between the crossed classifications since they may not be independent, and introducing extra crossed factors will typically reduce the variance explained by the main level 2 nesting.

A straightforward extension of the normal linear mixed model to account for a single extra crossed factor is to add varying intercepts and slopes according to that extra factor. Assuming h_{ij} varies between 1 and H, then adapting Equation 6.1 format for z_{ij} of dimension q,

$$y_{ij} = x_{ij}\beta + z_{ij}(b_i + c_{h_{ij}}) + u_{ij},$$

where

$$(b_{1i}, \ldots, b_{qi}) \backsim N_q(0, \Sigma_b) \qquad i = 1, \ldots, m,$$
$$(c_{h1}, \ldots c_{hq}) \backsim N_q(0, \Sigma_c) \qquad h = 1, \ldots, H.$$

Alternatively, variation over the extra crossed factor may be applied to a different predictor than those subject to random variation over the main level 2 classification. Often the additional random effects would be confined to intercept variation over the extra crossed factor, so that with $q = 1$ and $z_{i1} = 1$ also, one has,

$$y_{ij} = x_{ij}\beta + b_i + c_{h_{ij}} + u_{ij}.$$

In these situations the random effects are confounded and empirical identification may be impeded. Selection between random effects may well be needed (Browne et al., 2001).

Another possible source of variation in crossed models is defined in cells formed by cross-classification of two or more higher level factors. For example, N patients living in a particular administrative health district may be classified into subpopulations, s, based on intersections of their primary care general practitioner, $i_1 = 1, \ldots, m_1$, and small area of residence, $i_2 = 1, \ldots, m_2$ (Congdon and Best, 2000). Often there may be no subjects in certain combinations of higher level factors. So define total nonempty cells as S_n, equal to or less than the total, $S = m_1 m_2$, of all possible combinations, with different values, $s = 1, \ldots, S_n$, defined by cross-hatched factor identifiers $[i_1, i_2]$. Let $r = 1, \ldots, N$ denote a single string subject level identifier. Subjects will be classified by subpopulation, $s_r \in \{1, \ldots, S_n\}$, by higher level factor 1 classification indicator, h_{1r} (general practitioner), higher level factor 2 classification indicator, h_{2r} (small area of residence), and so on. Random intercept variation in a metric response over the two factors and the cells then takes the form,

$$y_r = x_r \beta + z_r(\alpha_{1,h_{1r}} + \alpha_{2,h_{1r}} + \eta_{s_r}) + u_r, \tag{6.5}$$

where α_{1i_1}, α_{2,i_2}, η_s, and u_r are random effects.

Multiple membership schemes are a generic weighting scheme applicable to cross-classified data (Browne et al., 2001), and may be illustrated by the case where subjects at level 1 may belong to more than one level 2 unit. Suppose a pupil's entire primary school career is of interest, then there may be moves between schools. Multiple affiliations then need to be taken account of in terms of school impacts on attainment. Another example is analysis of neighborhood health effects to take account of changes in residence (Subramanian, 2004). Suppose there are m level 2 units (clusters such as schools) that are included in the analysis, and that subjects $j = 1, \ldots, J$ (not taken to be nested within schools) have K_j level 2 affiliations with weights $\{w_{j1}, w_{j2}, \ldots, w_{jK_j}\}$, where $\sum_{k=1}^{K_j} w_{jk} = 1$. The weights would in many situations be taken as known (e.g., based on number of terms spent by a pupil in different schools). Then for pupil level predictors, z_j, of dimension q not varying over affiliations, the normal linear mixed model becomes,

$$y_j = x_j \beta + z_j \sum_{k=1}^{K_j} w_{jk} b_k + u_{ij},$$

where $(b_{1i}, \ldots, b_{qi}) \backsim N_q(0, \Sigma_b), i = 1, \ldots, m$. If the pupil predictors vary over affiliations, then,

$$y_j = x_j \beta + \sum_{k=1}^{K_j} w_{jk} z_{jk} b_k + u_j.$$

Multiple member schemes extend to data frames that are structured spatially or temporally rather than nested. A particular kind of multiple member prior can be applied to spatially configured count responses, y_i, subject to random intercept variation. Thus, let $y_i \sim Po(o_i \mu_i)$, where o_i are expected events, and where μ_i measure the Poisson intensity relative to expected levels (in spatial health applications, μ_i are termed relative risks). Then the impact of neighboring areas can be represented by random effects, b_k, while own area effects are represented by effects, u_i, in a model,

$$\log(\mu_i) = x_i \beta + \sum_{k=1}^{K_i} w_{ik} b_k + u_i,$$

where w_{ik} are row standardized with $\sum_{k=1}^{K_i} w_{ik} = 1$. They might be based on binary spatial interactions, c_{ik} ($c_{ik} = 1$ if areas i and k are contiguous, $c_{ik} = 0$ otherwise), or based on distances, d_{ik}, between area centers, such as $c_{ik} = \exp(-\eta d_{ik})$ where η is positive; then,

$$w_{ik} = c_{ik} \Big/ \sum_{k=1}^{K_i} c_{ik}.$$

Example 6.4. Educational Attainment and Neighborhood Garner and Raudenbush (1991) consider attainment data from a Scottish education authority with $R = 2310$ pupils classified by crossed factors (neighborhood and secondary school). Some neighborhoods send pupils to multiple schools, while some schools draw pupils from multiple neighborhoods. They consider effects of the "place variable," namely, neighborhood social deprivation, in reducing educational attainment after controlling for the impacts on attainment of pupil level attributes (pupil aptitude and family background). Their data involves $m_1 = 524$ neighborhoods and $m_2 = 17$ schools with child-specific predictors being gender ($1 = \text{M}, 0 = \text{F}$), and a verbal reasoning quotient (VRQ) and a reading test score (RTS) both obtained when the child was at primary school; parent-specific predictors are the father's status (metric) and three binary indicators (whether the father was educated beyond age 15, whether the mother was educated beyond 15, and whether the father was unemployed).

Two models are considered. In the first there are random effects for both neighborhoods and schools, with varying neighborhood effects linked to social deprivation. Let h_{1r} and h_{2r} denote neighborhood and school for pupils, $r = 1, \ldots, R$. Uniform $U(0, 1000)$ priors are adopted on neighborhood, school, and pupil random effect standard deviations in the model,

$$\begin{aligned}
y_r &= x_r\beta + \alpha_{1,h_{1r}} + \alpha_{2,h_{2r}} + u_r, \\
\alpha_{1i_1} &\sim N(\beta_1 + \gamma Dep_{i_1}, \sigma_1^2) \qquad i_1 = 1, \ldots, m_1, \\
\alpha_{2i_2} &\sim N(0, \sigma_2^2) \qquad i_2 = 1, \ldots, m_2, \\
U_r &\sim N(0, \sigma_3^2),
\end{aligned}$$

with x_r excluding a constant term[4]. The second half of a two chain run of 5000 iterations shows pupil variation, σ_3^2, to be more substantial than school variation, σ_2^2, and shows a negative deprivation effect γ (with mean -0.156 and 95% interval from -0.206 to -0.107). Area effects act to significantly reduce school variation. Predictor effects are close to those cited by Raudenbush (1993, 335).

A second model allows the deprivation effect to vary by school—expressing varying effectiveness in schools countering catchment area effects (also known as contextual value added effects). There are now four random variances, with

$$y_r = x_r\beta + \alpha_{1,h_{1r}} + \alpha_{2,h_{2r}} + \gamma_{h_{2r}} Dep_{h_{1r}} + u_r,$$
$$\alpha_{1i_1} \sim N(\beta_1, \sigma_1^2), \quad \alpha_{2i_2} \sim N(0, \sigma_2^2), \quad \gamma_{i_2} \sim N(\Gamma, \sigma_3^2), \quad u_r \sim N(0, \sigma_4^2).$$

In fact, this model has a very similar psML (around -2398) to the first model. The school deprivation effects, γ_{i_2}, on attainment have a mean -0.16, and vary from -0.178 $(-0.31, -0.09)$ for school 11 to -0.132 $(-0.20, -0.02)$ for school 9. Their standard deviation is 0.037.

6.5 Robust Multilevel Models

Under normality assumptions regarding errors at different levels, extreme data points can influence estimates of fixed effect and variance component parameters and reduce the precision of estimates (e.g., widen the width of credible intervals). Sensitivity of the level 2 fixed effect estimates, κ, to alternative assumptions regarding the prior on level 2 effects, b_i, in Equation 6.3 is the focus of Selzter (1993). Estimates of level 2 random cluster effects may also be sensitive when level 2 normality of effects is assumed (Selzter et al., 1996, 137). For a two level model, outliers may occur both in level 2 cluster effects and in level 1 within-cluster errors (Langford and Lewis, 1998; Pinheiro et al., 2001), and the two sources may be confounded. For example, a discordant school effect might be due to a systematic effect across all pupils or because a few pupils in the school are responsible for the discrepancy. Selzter et al. (2002) investigate how level 1 outliers affect estimation of fixed effect regression parameters and inferences regarding level 2 cluster effects (e.g., treatment contrasts for individual clusters) in two level models for continuous outcomes.

Robust alternatives to the normal linear mixed model based on the t density have been proposed by Pinheiro et al. (2001) and Staudenmayer et al. (2009), and shown to outperform Gaussian error assumptions when outliers are present in multilevel data. Daniels and Gatsonis (1999, 31) assume multivariate t random effects at level 2 by default in a GLMM, while Selzter et al. (1996) present Gibbs sampling steps for the linear mixed model case where a multivariate t with a single degrees of freedom parameter is assumed for level 2 random effects. Seltzer et al. (2002) adopt Student t priors at both levels and apply a $U(0,1)$ prior to sample from a discrete grid of values on the degrees of freedom parameter. Thus, for an equally spaced grid of potential values $\{2.1, 2.2, 2.3, \ldots, 49.9\}$ with equal prior probabilities, the cumulative probability, $Pr(\nu = 2.1) + Pr(\nu = 2.2) + \cdots$, is calculated for each point and the $U(0,1)$ draw determines which is sampled.

Following the Pinheiro et al. (2001) scheme, assume a gamma-normal hierarchical representation with $s_i \sim Ga(0.5\nu, 0.5\nu)$, and also that $e_i \sim N_q(0, I)$. Then for continuous responses, $y_i = (y_{i1}, \ldots, y_{in_i})'$, a level 2 assumption of t distributed random effects, $b_i = (b_{1i}, \ldots, b_{qi})'$, with dispersion Σ_b leads to,

$$y_i = X_i\beta + Z_ib_i + u_i,$$
$$b_i = \kappa W_i + \Sigma_b^{0.5} e_i/\sqrt{s_i}, \quad i = 1, \ldots, m.$$

For outlier clusters with low s_i, the overall dispersion, Σ_b/s_i^2, is inflated, but the fixed effect, κ, will be less distorted than under normal level 2 errors.

The degrees of freedom parameters of the level 2 multivariate t prior may be taken to vary between clusters, namely,

$$\begin{bmatrix} y_i \\ b_i \end{bmatrix} \sim t_{n_i+q}\left(\begin{matrix} X_i\beta \\ 0 \end{matrix}, \begin{bmatrix} Z_i\Sigma_b Z_i' + \Lambda_i & Z_i\Sigma_b \\ \Sigma_b Z_i' & \Sigma_b \end{bmatrix}, \nu_i\right).$$

or under a gamma-normal hierarchical representation,

$$\begin{bmatrix} y_i \\ b_i \end{bmatrix} \sim N_{n_i+q}\left(\begin{matrix} X_i\beta \\ 0 \end{matrix}, \frac{1}{s_i}\begin{bmatrix} Z_i\Sigma_b Z_i' + \Lambda_i & Z_i\Sigma_b \\ \Sigma_b Z_i' & \Sigma_b \end{bmatrix}\right),$$

where $s_i \sim Ga(0.5\nu_i, 0.5\nu_i)$. The s_i can then be used for identifying cluster outliers. An alternative to assuming cluster-specific degrees of freedom is to take $\nu_i = \nu_{g_i}$, according to a known or possibly unknown grouping variable, $g_i \in (1, \ldots, G)$, applicable to clusters, for example, type of school in an educational application. Then $\{\nu_1, \ldots, \nu_G\}$ may be extra parameters. The level 1 random prior may be specified as $u_i \sim t_{n_i}(0, \Lambda_i, \nu_{h_i})$, where h_i is another grouping indicator that may be the same as g_i.

Discrete mixtures of random effects are also possible for outlier accommodation, modeling non-normality or other asymmetry in random effects. Latent mixtures of regression effects may also be present: Muthén and Asparouhov (2009) show how latent regression classes may be misrepresented as random cluster variation. To detect outlier random effects, Daniels and Gatsonis (1999, 36) adapt the approach of Albert and Chib (1997) in their models for hierarchical conjugate priors for discrete data. For nested binomial data, $y_{ij} \sim Bin(n_{ij}, p_{ij})$, a mechanism to detect level 1 outliers may be specified with p_{ij} drawn for a two group mixture of beta densities both with means π_{ij}. For the main group, the dispersion parameters are δ_i, while for the outlier group they are deflated as δ_i/K, where $K \gg 1$. Then,

$$p_{ij} \sim (1-\lambda)Be(\pi_{ij}\delta_i, (1-\pi_{ij})\delta_i) + \lambda Be\left(\pi_{ij}\frac{\delta_i}{K}, (1-\pi_{ij})\frac{\delta_i}{K}\right).$$

If the outlier probability λ is preset to a low value (e.g., $\lambda = 0.05$), then K might be taken as an extra parameter. Weiss et al. (1999) suggest a similarly motivated prior for mixtures of normal random effects at levels 1 and 2 in Equations 6.1 and 6.3, namely,

$$b_i \sim (1-\lambda_b)N_q(0, \Sigma_b) + \lambda_b N_q(0, K_b\Sigma_b),$$
$$u_{ij} \sim (1-\lambda_u)N(0, \sigma_u^2) + \lambda_u N(0, K_u\sigma_u^2).$$

An alternative mixture prior to reduce the impact of parametric assumptions is the mixture of Dirichlet process approach (Guha, 2008; Kleinman and Ibrahim, 1998). Thus, a conventional first stage likelihood,

$$y_i \sim N(X_i\beta + Z_ib_i, \sigma^2),$$

may be combined with a semiparametric approach for $b_i = (b_{1i}, \ldots, b_{qi})'$, typically with a multivariate normal base G_0 as in,

$$\begin{aligned} b_i &\backsim G, \\ G &\backsim DP(\alpha, G_0), \\ G_0 &= N_q(0, D), \\ D^{-1} &\sim Wish(d_0, R_0). \end{aligned}$$

Gibbs sampling for D^{-1} is modified for clustering among the sampled b_i (Kleinman and Ibrahim, 1998, 94).

Example 6.5. Police Stops and Ethnicity Gelman and Hill (2006) present multilevel Poisson regression analysis of counts, y_{ij}, of "stop and frisk" over a 15-month period in 1998–1999. For each of $m = 75$ New York police precincts, counts are disaggregated both by ethnic group (1 = black, 2 = hispanic, 3 = white), and crime type ($1 = \text{violent}, 2 = \text{weapons}, 3 = \text{property}, 4 = \text{drug}$), so that there are $j = 1, \ldots, n_i$ observations, with $n_i = 12$, for each precinct i. These 12 categories are called classes here. As an offset o_{ij}, Gelman and Hill suggest arrests by precinct, ethnicity, and type in 1997 (multiplied by 15/12); in fact, $o_{ij} + 1$ is used instead since some arrest counts are zero.

Here, an initial GLMM with log link estimates fixed ethnicity effects and random errors at both precinct and precinct-class level, the latter introduced to account for overdispersion, while the former measures overall crime levels in a precinct. So, $y_{ij} \sim Po(\mu_{ij}[o_{ij} + 1])$ with,

$$\begin{aligned} \log(\mu_{ij}) &= \beta_{\text{eth}_{ij}} + b_i + u_{ij}, \\ u_{ij} &\sim N(0, \sigma_u^2), \quad b_i \sim N(0, \sigma_b^2). \end{aligned}$$

With $y_{ij,\text{rep}}$ denoting replicate data sampled from the model, predictive checks involve the underprediction criterion, $Pr(y_{ij} \geq y_{ij,\text{rep}}|y)$ (Daniels and Gatsonis, 1999), and the log(CPO), with overall fit measured by the expected predictive deviance (EPD) and log(psML). The significance of individual b_i is assessed using the probabilities, $Pr(b_i > 0|y)$.

With $U(0, 100)$ priors on the random standard deviations, the last 7000 of a two chain run of 10,000 iterations[5] produce a mean scaled deviance of 898, so extravariation is accounted for, while the maximum and minimum underprediction probabilities are 0.92 and 0.25. The ethnic coefficients have 95% intervals $(-0.87, -0.56), (-0.80, -0.55)$, and $(-1.26, -0.97)$, so whites have lower chances of being subject to "stop and frisk." Despite the presence of precinct-cell errors (which might reduce the need for separate precinct effects), a relatively high number (29) of the precinct effects, b_i, are significant in the sense that the probabilities, $Pr(b_i > 0|y)$, exceed 0.95 or are under 0.05.

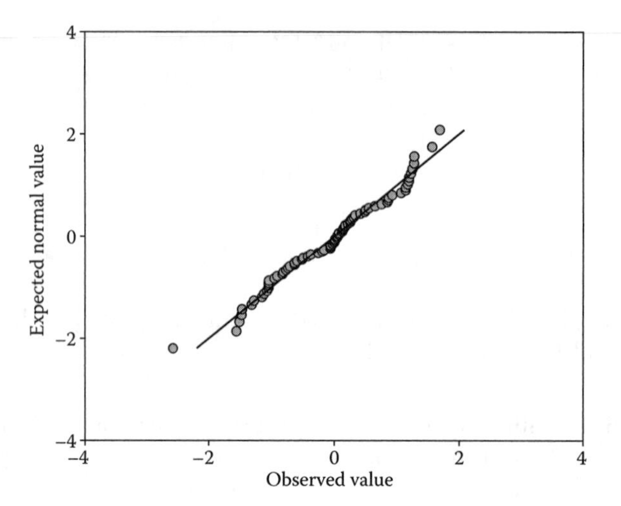

FIGURE 6.1
Normal Q–Q plot of $b[i]$.

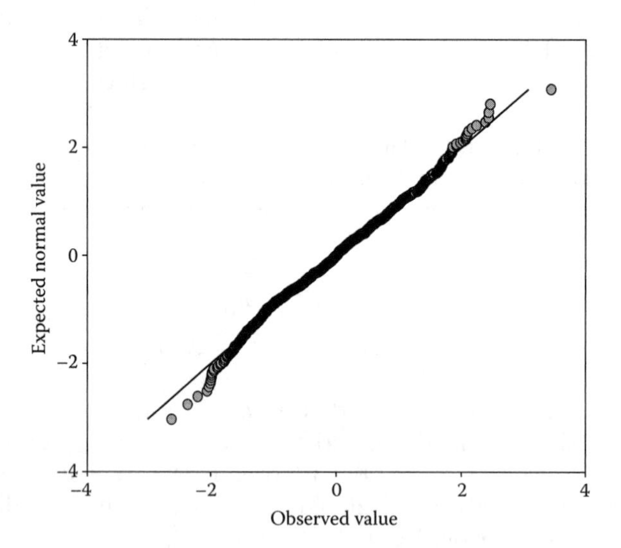

FIGURE 6.2
Normal Q–Q plot of $u[i, j]$.

Of interest in terms of the robustness of the model assumptions are the characteristics of the posterior estimates of b_i and u_{ij}. A normal Q–Q plot for the posterior means of the standardized $z_i^b = b_i/\sigma_b$ from the first model suggests no major departure from normality (Figure 6.1), and the same applies for $z_{ij}^u = u_{ij}/\sigma_u$ (Figure 6.2). However, for illustrative purposes, a discrete mixture prior is adopted for the u_{ij}, namely,

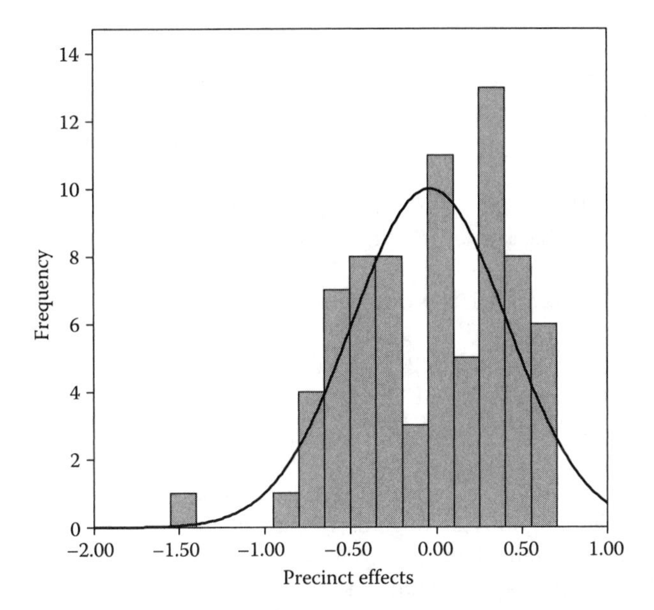

FIGURE 6.3
Random effects (Posterior means) under MDP prior.

$$u_{ij} \sim \lambda N(0,\sigma_u^2) + (1-\lambda)N(0,\sigma_u^2 K),$$
$$K = K' + 1; \quad K' \sim U(0,9); \quad \lambda \sim Be(19,1),$$

where the prior on λ favors a low outlier probability $(1-\lambda)$, and the prior on the inflation factor K specifies values between 1 and 10. This model has delayed convergence, but the second half of a two chain run of 50,000 iterations provides an estimated λ of 0.98, with $K = 3.6$. The ethnic coefficients are slightly enhanced with 95% intervals $(-1.03, -0.68)$, $(-0.88, -0.57)$, and $(-1.40, -1.10)$. The psML is -3873 compared to -3923 for the first model considered (an estimate also based on the second half of a run of 50,000 iterations). Figure 6.3 plots precinct effects under this model.

Appendix: Computational Notes

1. In Example 6.1, the code uses a stacked data form with $h \equiv [ij]$ and $h = 1, \ldots, N$, where $N = \sum_{i=1}^{m} n_i$. Both options for the observational variance are included, so one or other must be commented out:

```
model { for (i in 1:4059) {
# constant observational variance (option 1)
y[i]~dnorm(mu[i],tau); yrep[i]~dnorm(mu[i],tau)
```

```
LL[i] <- -0.5*tau*(y[i]-mu[i])*(y[i]-mu[i])+0.5*log(tau/6.28);
mu[i] <- gam[sch[i],1]+gam[sch[i],2]*LRT[i]+beta[3]*gend[i]}
# complex observational variance (option 2)
# y[i]~dnorm(mu[i],tau.cx[gend[i]+1]); yrep[i]~dnorm(mu[i],tau.cx[gend[i]
+1])
# LL[i] <- -0.5*tau.cx[gend[i]+1]*(y[i]-mu[i])*(y[i]-
# mu[i])+0.5*log(tau.cx[gend[i]+1]/6.28)
# Other priors
for (j in 1:65) {gam[j,1:2] ~dmnorm(beta[1:2],T.gam[,])
for (k in 1:2) {gamr[k,j] <- gam[j,k]}}
T.gam[1:2,1:2]~dwish(Q[,],2); tau~dgamma(1,0.001);
# inverse LKD node for CPO estimation
for (i in 1:4059) {G[i] <- 1/exp(LL[i])}
for (j in 1:2) {tau.cx[j]~dgamma(1,0.001); sig.cx[j] <- 1/sqrt(tau.cx[j])}
Sigma.gam[1:2,1:2] <- inverse(T.gam[,]); sigobs <- sqrt(1/tau)
r.gam <- Sigma.gam[1,2]/(sigclus[1]*sigclus[2]);
for (j in 1:2) {sigclus[j] <- sqrt(Sigma.gam[j,j])
for (k in 1:2) {Q[j,k] <- equals(j,k)}}
for (j in 1:3) {beta[j]~dnorm(0,0.001)}}
```

2. The GLMM logit code in Example 6.2 adopts Uniform $U(0,100)$ priors on the random effect standard deviations and normal priors with variance 1000 on the fixed effects. The code uses a stacked data form ith $h \equiv [ijk]$, namely,

```
model { for (h in 1:2449) {y[h]~dbern(p[h])
logit(p[h]) <- alph+b3[comm[h]]+b2[mom[h]]
+beta1[1]*kid3p[h]+beta1[2]*mom25p[h] +beta1[3]*ord23[h]+beta1[4]*
ord46[h]
+beta1[5]*ord7p[h]+beta2[1]*indNS[h]+beta2[2]*indS[h]
+beta2[3]*momPr[h]+beta2[4]*momSc[h]+beta2[5]*husPr[h]+beta2[6]*
husSc[h]
+beta2[7]*husDK[h]+beta2[8]*Prof[h]+beta2[9]*AgrS[h]+beta2[10]*
AgrE[h]
+beta2[11]*Sk[h]+beta2[12]*toi[h]
+beta2[13]*tvND[h]+beta2[14]*tvD[h] +beta3[1]*Ind81[h]+beta3[2]*
ssDist[h]}
alph ~dnorm(0,0.001)
for (m in 2:3) {sig[m] ~dunif(0,100); tau[m] <- 1/pow(sig[m],2)}
for (i in 1:161) {b3[i]~dnorm(0,tau[3])}
for (j in 1:1558) {b2[j]~dnorm(0,tau[2])}
for (a in 1:5) {beta1[a] ~dnorm(0,0.001)}
for (a in 1:14) {beta2[a] ~dnorm(0,0.001)}
for (a in 1:2) {beta3[a]~dnorm(0,0.001)}}.
```

The BayesX code for this example is

```
dataset d
d.infile, maxobs=5000 using data\prenatal.txt
bayesreg b
b.regress y = mom(random)+comm(random)+kid3p+mom25p+ord23+
ord46+ord7p+indNS+indS+momPr+momSc+husPr+husSc+husDK+Prof+
AgrS+AgrE+Sk+toi+tvND+tvD+Ind81+ssDist, family=binomial predict
using d
```

3. The code for the ordinal logit analysis in Example 6.3 again assumes stacked data with composite indicator, $h \equiv [ijk]$. The code is

```
model { for (h in 1:1600) { thk[h] ~dcat(p[h,])
for (j in 1:H-1) {logit(gam[h,j]) <- kap[j] - mu[h] }
p[h,1] <- gam[h,1]; p[h,H] <- 1-gam[h,H-1]
for (r in 2:H-1) { p[h,r] <- gam[h,r] - gam[h,r-1] }
mu[h] <- beta[1]*prethk[h]+beta[2]*cc[h]+beta[3]*tv[h]
+beta[4]*cctv[h]+u3[sch[h]]+u2[sch[h],cls[h]]}
for (i in 1:m) {u3[i] ~dnorm(0,tau[3]);for (j in 1:n[i]) {u2[i,j]~dnorm(0,
tau[2])}}
for (j in 2:3) {sig[j] ~dunif(0,1000); sig2[j]<-sig[j]*sig[j]; tau[j]<-1/sig2[j]}
for (j in 1:4) {beta[j] ~dnorm(0,0.001)}
# Cut points on latent scale
kap[1] ~dnorm(0, 0.01) I(,kap[2]); kap[2] ~dnorm(0, 0.01) I(kap[1],kap[3]);
kap[3] ~dnorm(0, 0.01) I(kap[2],)}
```

The code for the data augmented approach is as above, except that the first few lines are replaced by

```
model { for (h in 1:1600) { for (j in 1:H-1) {z[h,j] <- step(j-thk[h])
A[h,j] <- -10*equals(z[h,j],0); B[h,j] <- 10*equals(z[h,j],1)
ystar[h,j]~dlogis(nu[h,j],1) I(A[h,j],B[h,j])
nu[h,j] <- kap[j] - mu[h]}
mu[h] <- ...}
```

4. The code for the first crossed factor analysis in Example 6.4 is

```
model { for (r in 1:R) {y[r]~dnorm(mu[r],tau[3])
# h1 is neighborhood, h2 is school
mu[r] <- alph1[h1[r]] +alph2[h2[r]]+beta[2]*VRQ[r]+beta[3]*RTS[r]
+beta[4]*FSTAT[r]+beta[5]*FUNEM[r]
+beta[6]*FED[r]+beta[7]*MED[r]+beta[8]*MALE[r]
LL[r] <- 0.5*log(tau[3]/6.28)-0.5*tau[3]*pow(y[r]-mu[r],2); H[r] <- 1/
exp(LL[r])}
for (i1 in 1:m1) {alph1[i1]~dnorm(mu.alph[i1],tau[1])
                  mu.alph[i1] <- beta[1]+gam*DEP[i1]}
for (i2 in 1:m2) {alph2[i2]~dnorm(0,tau[2])}
```

for (j in 1:3) {sig[j]~dunif(0,1000); tau[j] <- 1/(sig[j]*sig[j])}
for (j in 1:8) {beta[j]~dflat()} gam ~dflat()}

The analysis with area deprivation impacts varying by school has code

```
model { for (r in 1:R) {y[r] ~dnorm(mu[r],tau[4])
mu[r] <- beta[1]+alph1[h1[r]]+gam[h2[r]]*DEP[h1[r]] +alph2[h2[r]]+
beta[2]*VRQ[r]
+beta[3]*RTS[r]+beta[4]*FSTAT[r]+beta[5]*FUNEM[r]+
beta[6]*FED[r]+beta[7]*MED[r]+beta[8]*MALE[r]
LL[r] <- 0.5*log(tau[4]/6.28)-0.5*tau[4]*pow(y[r]-mu[r],2); H[r] <- 1/
exp(LL[r])}
for (i1 in 1:m1) {alph1[i1] ~dnorm(0,tau[1])}
for (i2 in 1:m2) {alph2[i2] ~dnorm(0,tau[2]); gam[i2] ~dnorm(Gam,tau[3])}
for (j in 1:4) {sig[j] ~dunif(0,1000); tau[j] <- 1/(sig[j]*sig[j])}
for (j in 1:8) {beta[j] ~dflat()} Gam ~dflat()}
```

5. The code in Example 6.5 is as follows with the log(CPO) estimated using the negative of the logged mean of $H[h]$:

```
model { for (h in 1:900) {y[h] ~dpois(mu[h]); ynew[h] ~dpois(mu[h])
mu[h] <- (15/12)*(o[h]+1)*nu[h]; u[h] ~dnorm(0,tau.u); st.u[h] <-
u[h]*sqrt(tau.u)
# eth groups: Black, Hispanic, White
log(nu[h]) <- beta[eth[h]]+b[precinct[h]]+u[h]
# fit criteria
under[h] <- step(y[h]-ynew[h])+0.5*equals(y[h],ynew[h])
yp[h] <- y[h]+0.5; ypp[h] <- ynew[h]+0.5; mh[h] <- mu[h]+0.5
EPD[h] <- 2*(yp[h]*log(yp[h]/ypp[h])-(y[h]-ynew[h]))
Dev[h] <- 2*(yp[h]*log(yp[h]/mh[h])-(y[h]-mu[h]))
LL[h] <- -mu[h]+y[h]*log(mu[h])-logfact(y[h]);
H[h] <- 1/exp(LL[h])}
# cluster effects
for (j in 1:75) {b[j] ~dnorm(0,tau.b); st.b[j] <- b[j]*sqrt(tau.b)
                 step.b[j] <- step(b[j]-mean(b[]))}
sig.b ~dunif(0,100); sig.u ~dunif(0,100)
tau.u <- 1/(sig.u*sig.u); tau.b <- 1/(sig.b*sig.b)
F[1] <- sum(EPD[]); F[2] <- sum(Dev[])
for (k in 1:3) {beta[k] ~dflat()}}
```

For the mixture model, one has amended code elements

```
model { for (h in 1:900) {... u[h] ~dnorm(0,tau.u[G[h]])
G[h] ~dcat(lam[1:2]); st.u[h] <- u[h]*sqrt(tau.u[G[h]])...}
........
lam[1] ~dbeta(19,1); lam[2] <- 1-lam[1];
tau.u[2] <- tau.u[1]/(1+Kd); Kd ~dunif(0,9)}
```

7

Multivariate Priors, with a Focus on Factor and Structural Equation Models

7.1 Introduction

A range of multivariate techniques are available for modeling multivariate collections of metric, binary, or count data, or for modeling multivariate random effects or regression residuals. These include data reduction (reduced dimension) methods, such as factor and principal component analysis (e.g., Hayashi and Arav, 2006; Lopes and West, 2004), discriminant analysis (e.g., Brown et al., 1999; Rigby, 1997), data mining, as well as direct (full dimension) modeling of the joint density of the observations or regression residuals (e.g., Chib and Winkelmann, 2001). Structured multivariate effects in the analysis of spatial or time-configured data raise additional issues, such as representing intervariable correlation within units as well as nonexchangeability between units (Song et al., 2005).

Bayesian applications in the area of factor and structural equation modeling have grown considerably in recent years; for overviews, see Ando (2009), Lee (2007), and Palomo et al. (2007), while for frequentist perspectives see Hayashi et al. (2008) and Sanchez et al. (2005). The rationale for introducing latent variables lies in parsimonious representation of the covariance structure of multivariate collections of data, while also revealing underlying clustering of, or associations between, the variables, ideally one with substantive relevance and interpretability. The latent variables are typically unobservable constructs (e.g., authoritarianism, population morbidity, or a common trend shared between several time series) that can only be imperfectly measured by observed indicators. The latent variables may be continuous, as in factor analysis (Fokoue, 2004; Lopes and West, 2004) or categorical, as in latent class analysis (Berkhof et al., 2003). The original variables might themselves also be discrete or continuous. For example, item response models typically involve multiple binary observed items and a single latent continuous ability score (Albert and Ghosh, 2000; Bazan et al., 2006).

The extraction of information from multivariate observed indicators to derive a smaller set of latent variables defines a measurement model, as in confirmatory and explanatory factor analysis (Bartholomew, 1987; Skrondal and Rabe-Hesketh, 2007). The subsequent use of the latent constructs in describing causal relationships or associations leads into structural equation modeling (Lee, 2007). Both types of model have been developed especially in areas such as psychology, marketing, educational testing, and sociology, where underlying constructs are not possible to measure directly. Newer areas of development include environmental modeling (Malaeb et al., 2000; Nikolov et al., 2007), biomass models (Arhonditsis et al., 2006), and time series and spatial data analysis using common factor approaches.

The observed variables in a measurement model may be variously known as items (e.g., in psychometric tests), indicators, or manifest variables. Canonical assumptions are that (a) conditional on the constructs, the observed indicators are independent, in which case the constructs explain the observed correlations between the indicators; and (b) that the construct scores are independent over subjects. As Bollen (2002) points out, the local independence property in (a) is not an intrinsic feature of structural equation models, while recent developments in spatial and time series factor and structural equation models (Congdon et al., 2007; Hogan and Tchernis, 2004) exemplify how construct scores may be dependent over space or time.

This chapter presents a selective review of multivariate techniques, namely,

1. factor modeling via continuous latent constructs, as applied in normal linear and general linear model contexts (Sections 7.2, 7.3, and 7.4);
2. models for multivariate discrete area (lattice) data, including spatial factor models (Sections 7.6 and 7.7); and
3. models for multivariate time series (Section 7.8), with a focus on dynamic linear and general linear models.

A Bayesian approach is arguably of benefit in such multivariate applications. Many classical applications of factor and structural equation methods assume multivariate normality of the indicators and/or residuals with estimation based on minimizing a discrepancy between the observed and predicted covariance matrix—under multivariate normality, the covariance matrix is sufficient for describing the correlations between observed indicators (Hox and Bechger, 1998; Sanchez et al., 2005). Considerations of robustness to outliers and departures from normality, and the ease with which parameter restrictions may be imposed and predictions made for new cases, may point to a Bayes approach that retains the full observation set as input (Lee, 2007). Section 7.5 discusses robust approaches to factor models and structural equation models (SEMs).

The fully Bayes method has further potential advantages over maximum likelihood or empirical Bayes estimates in terms of fully taking account of uncertainty in variance component estimates, for describing the densities of the parameters of structural equation models without making asymptotic approximations (Aitkin and Aitkin, 2005), and in estimating models conventionally considered unidentifiable, for instance regressions with all predictors measured with error so that instrumental variable estimation is infeasible (Scheines et al., 1999). Maximum likelihood standard errors calculated assuming asymptotically normal estimators may be distorted for small sample sizes, whereas in the Bayesian approach, Markov Chain Monte Carlo (MCMC) samples are taken from the true posterior regardless of sample size, and so standard errors calculated from MCMC output are reliable regardless of sample size, or when there are other sources of non-normality.

7.2 The Normal Linear SEM and Factor Models

Following Joreskog (1973) and further classical presentations (e.g., Bollen, 1989; Hoyle, 1995), several Bayesian treatments of normal linear or general linear structural equation models have appeared recently (e.g., Nikolov et al., 2007; Palomo et al., 2007; Song et al., 2006). Consider observed multivariate metric indicators, y and x, and continuous endogenous and exogenous construct vectors, denoted F and H, respectively. For subjects $i = 1, \ldots, n$, the measurement model components of a normal linear SEM are

$$y_i = \alpha_y + \Lambda_y F_i + u_i,$$
$$x_i = \alpha_x + \Lambda_x H_i + e_i,$$

where $y_i = (y_{1i}, \ldots, y_{P_y i})'$ is a $P_y \times 1$ vector of indicators describing or measuring an endogenous construct vector, $F_i = (F_{1i}, \ldots, F_{Q_y i})'$, of dimension Q_y less than or equal to P_y; and x_i is a $P_x \times 1$ vector of indicators measuring an exogenous construct vector, H_i, of dimension $Q_x \leq P_x$. The individual factor variables, F_{qi}, may be independent of each other or intercorrelated, and similarly for H_{qi}. The matrices, Λ_y and Λ_x, are of dimension $P_y \times Q_y$ and $P_x \times Q_x$ and contain loading parameters describing how observed indicators are related to the latent constructs. The $\{F, H\}$ are sometimes known as common factors, while the errors $\{u, e\}$ are sometimes called unique factors (Skrondal and Rabe-Hesketh, 2007). The errors are assumed independent of the common factors.

A structural model may describe (a) interrelations between F_{qi} (namely, reciprocal flows between endogenous variables such as social authoritarianism and religiosity) and (b) effects of exogenous constructs, H_{qi}, on the endogenous ones (e.g., effects of socioeconomic status on authoritarianism or religiosity). These effects are represented by the equation system,

$$F_i = BF_i + CH_i + w_i,$$

where an intercept is typically not identified, and B is a $Q_y \times Q_y$ matrix with zero diagonal elements and off-diagonal parameters describing relations between endogenous constructs. The matrix C is $Q_y \times Q_x$ with parameters describing the impact of exogenous on endogenous constructs. The structural model may also contain further observed variables as responses or predictors.

Many multivariate reduction applications involve just a measurement model (i.e., a simple factor analysis), and so distinction between different types of observed indicator and factor is not needed. Then a normal linear factor model is

$$y_i = \alpha + \Lambda F_i + u_i, \tag{7.1}$$

where $y_i = (y_{1i}, \ldots, y_{Pi})'$ is $P \times 1$, and $F_i = (F_{1i}, \ldots, F_{Qi})'$ is of dimension $Q \leq P$, and normality of errors and factors is often assumed with $(u_{1i}, \ldots, u_{Pi})' \sim N_P(0, \Sigma)$ and $F_i \sim N_Q(0, \Phi)$. Identifying assumptions on Λ and Φ are considered below. Under a local independence assumption, the residuals $(u_{1i}, \ldots, u_{Pi})'$ are typically taken to be independent over cases i and variables, so that $\Sigma = \text{diag}(\sigma_1^2, \sigma_2^2, \ldots, \sigma_P^2)I$, and $u_{pi} \sim N(0, \sigma_p^2)$. This assumption can equivalently be stated as that the outcome variables are conditionally independent given the latent variables (Skrondal and Rabe-Hesketh, 2007).

Uncertainty in the number of factors in the normal linear factor model, or choice between models involving different numbers of factors, may be tackled using parameter expansion combined with a Bayes factor approximation (Ghosh and Dunson, 2008), by RJMCMC methods (Lopes and West, 2004) or by marginal likelihood approximation using path sampling. The latter approach may be extended to full SEMs (Lee and Song, 2008). The parameter expansion method may also improve MCMC performance (Ghosh, 2008), and involves reference model (Equation 7.1), but with standardized factors, and a lower triangular structure for Λ (see Section 7.3) including the diagonals constraint, $\lambda_{qq} > 0$. Thus, the reference model is

$$y_i = \alpha + \Lambda F_i + u_i \quad F_i \sim N_Q(0, I),$$

and the expanded model is

$$y_i = \alpha + \Lambda^* F_i^* + u_i \quad F_i^* \sim N_Q(0, \Psi),$$

where $\Psi = \text{diag}(\psi_q)$, and the loadings Λ^* are not subject to the diagonals constraint. Priors on parameters in the expanded model induce priors on (Λ, F, Σ) in the reference model, via,

$$\lambda_{pq} = S(\lambda_{qq})\lambda_{pq}^* \psi_q^{0.5}; \quad F_{qi} = S(\lambda_{qq})F_{qi}^*/\psi_q^{0.5},$$

and a Bayes factor is derived using the integrated likelihood, $y \sim N(\alpha, \Lambda\Lambda' + \Sigma)$. The sign function, $S(x) = -1$ if $x < 0$ and $S(x) = 1$ if $x \geq 0$, is used to ensure the diagonals constraint in Λ.

7.2.1 Forms of model

If all loadings, λ_{pq}, in the $P \times Q$ matrix Λ are free parameters (apart from those subject to identification constraints, as discussed below), this structure is known as an exploratory factor analysis. By contrast, in a confirmatory factor analysis or measurement model some of the loadings take preset values (usually zero) on the basis of substantive theory. A particular form of confirmatory model is known as simple structure, such that each observed variable, y_{pi}, loads on only one of the constructs, F_{qi}. For example, Fleishman and Lawrence (2003) apply a simple structure model to ordinal items from the SF12 questionnaire, assuming that each item reflects either a physical or mental health construct.

A multiple indicator-multiple cause (MIMIC) model extends confirmatory models by incorporating the effects of exogenous observed variables on latent factors (Joreskog and Goldberger, 1975). MIMIC models for normal outcomes consist of (a) measurement equations,

$$y_i = \alpha + \Lambda F_i + \varphi X_i + u_i,$$

relating multiple indicator variables, y_i, to latent constructs, F_i, and possibly also to known influences, X_i; and (b) structural equations. In the latter, the latent variables, F_i, are related both to one another and to observed exogenous variables, Z_i, which are viewed as "causes" of the factors, namely,

$$F_i = BF_i + CZ_i + w_i,$$

where Z_i excludes a constant term, and the coefficient matrix B allows reciprocal effects between latent factors. A MIMIC model with a single latent construct, as applied, for instance, in analyzes of the size of underground economies (Wang et al., 2006; Frey and Weck-Hannemann, 1984), would typically take the form,

$$y_{pi} = \alpha_p + \lambda_p F_i + \varphi_p X_i + u_{pi},$$
$$F_i = \gamma Z_i + w_i.$$

As noted by Bruesch (2005), the correlation structure in a MIMIC model may need substantive support, as it typically assumes that (i) the indicators y are conditionally independent of the causes Z, given the latent construct(s) F; and (ii) that the indicators $y_1, \ldots, y_P$ are mutually independent given F. This amounts to saying that all connections that indicator variables y have with the causal variables Z, and with one another, are transmitted through the latent variable(s). Ghosh (2008, Section 4.7) demonstrates the application of parameter expansion in a MIMIC model.

7.2.2 Marginal and complete data likelihoods, and Markov Chain Monte Carlo sampling

From Equation 7.1, the conditional likelihood of the normal linear factor model is $p(y_i|F_i, \alpha, \Lambda, \Phi, \Sigma) = N(\alpha + \Lambda F_i, \Sigma)$, with conditional covariance matrix, $V(y_i|F_i, \Sigma) = \Sigma$, and hence $\{\text{cov}(y_{ji}, y_{mi}) = 0, m \neq j\}$ if Σ is diagonal. The marginal likelihood obtained by integrating out the factor scores in the normal linear factor model (Fokoue, 2004; Lee and Shi, 2000, 724) is $p(y_i|\alpha, \Lambda, \Phi, \Sigma) = N(\alpha, \Lambda\Phi\Lambda' + \Sigma)$. The joint likelihood of y_i and F_i, obtained by multiplying the marginal density of F, $F_i \sim N_Q(0, \Phi)$, and the conditional density of y_i given F_i, is

$$\begin{bmatrix} y_i \\ F_i \end{bmatrix} \sim N_{P+Q}\left(\begin{bmatrix} \alpha \\ 0 \end{bmatrix}, \begin{bmatrix} \Lambda\Phi\Lambda' + \Sigma & \Lambda\Phi \\ \Phi\Lambda' & \Phi \end{bmatrix}\right).$$

When the factors are standardized (Bartholomew et al., 2002, 150; Lopes and West, 2004, 44), the marginal variance of y_p is, accordingly, $\lambda_{p1}^2 + \cdots + \lambda_{pQ}^2 + \sigma_p^2$ and the marginal covariance of y_p and y_m is $\lambda_{p1}\lambda_{m1} + \lambda_{p2}\lambda_{m2} + \cdots + \lambda_{pQ}\lambda_{mQ}$. The contribution, $\lambda_{p1}^2 + \cdots + \lambda_{pQ}^2$, of the common factors to explaining the marginal variability in y_p is known as the communality, while that part due to the residual error, σ_p^2, is called the unique variance or uniqueness.

The marginal likelihood structure for cov(y) as $\Lambda\Phi\Lambda' + \Sigma$ does not lead to any simple form for the posterior distributions of the unknowns, though it can be used in RJMCMC approaches to estimation and factor model selection (Lopes and West, 2004). In Gibbs sampling estimation of linear Bayesian factor and SEM models, it is simplest to approach estimation of the parameters $(F, \alpha, \Lambda, \Phi, \Sigma)$ indirectly through the conditional likelihood or complete data model (Aitkin and Aitkin, 2005; Fokoue, 2004), with the F scores regarded as missing data rather than integrated out (Lee and Shi, 2000). Setting $\theta = (\alpha, \Lambda, \Phi, \Sigma)$, the posterior density is then,

$$p(\theta, F|y) \propto L(\theta|y, F)p(\theta).$$

While MCMC sampling is typically used with the conditional likelihood, the marginal covariance, $\Lambda\Phi\Lambda' + \Sigma$, may be useful in posterior checking of model assumptions (e.g., conditional independence between y variables given the factor scores). For example, Lee and Shi (2000) suggest a posterior check using $D(y, \theta) = \sum y_i'(\Lambda\Phi\Lambda' + \Sigma)^{-1}y_i$. Following Gelman et al. (1996), replicate data $y_{\text{rep,i}}$ are sampled from the predictive distribution $p(y_{\text{rep}}|y, \theta)$ and $D(y, \theta)$ compared to $D(y_{\text{rep}}, \theta)$.

From a set of MCMC samples, one seeks the marginal posterior density, $p(\theta|y)$, of the hyperparameters, and the predictive distribution, $p(F|y)$, of the factor scores. Estimation at iteration $t + 1$ proceeds by switching between (a) sampling $\theta^{(t+1)}$ from the posterior conditional, $p(\theta|y, F^{(t)})$, for θ conditional on y and sampled F scores, and (b) updating $F^{(t+1)}$ from the conditional density, $p(F|y, \theta^{(t+1)})$. The latter corresponds to the imputation step in data augmentation (Tanner, 1996).

A range of inference issues may occur, subject to identifiability being fully considered (Section 7.3). The choice between different numbers of factors may be made via BIC or DIC criteria, or by formal Bayes factor estimates. Lee (2007) uses path sampling to estimate formal Bayes model choice criteria, and RJMCMC methods may also be used in the normal linear factor model (Lopes and West, 2004). The patterns of significant loadings and subject factor scores raise questions of substantive theory, depending on the application area. As noted by Aitkin and Aitkin (2005), one can assess the significance of parameter or factor score contrasts on the basis of the MCMC sample, such as pairwise difference or ratio comparisons of scores on the kth factor for subjects i_1 and i_2, $F_{i_1k} - F_{i_2k}$ and F_{i_1k}/F_{i_2k}. Compared to classical analysis, the posterior density characteristics of the factor scores (and of factor contrasts) are routinely obtained.

To illustrate MCMC complete data sampling, assume Σ is diagonal in the conjugate normal model (Equation 7.1) with priors $\sigma_{pp}^{-1} \sim \text{Ga}(\alpha_{0p}, \beta_{0p})$, that the precision matrix for F has a Wishart prior $\Phi^{-1} \sim W(R_0, \rho_0)$, and that the prior for Λ follows the form proposed by Press and Shigemasu (1989). Specifically, with Λ_p as the pth row of Λ,

$$\Lambda_p \sim N_Q(\Lambda_{0p}, \sigma_{pp} H_{0p}),$$

where the $Q \times Q$ matrix, H_{0p}, is positive definite. Often, simple assumptions such as $H_{0p} = I_Q$ are made (Lee and Shi, 2000, 729). Letting y_p' be the pth row of y, and denoting $\Omega_p = (H_{0p}^{-1} + F'F)^{-1}$, and $\eta_p = \Omega_p(H_{0p}^{-1}\Lambda_{0p} + Fy_p)$, the posterior conditional for the unique variances is (Lee and Shi, 2000, 725)

$$\sigma_{pp}^{-1} \sim \text{Ga}(\alpha_{0p} + n/2, \beta_{0p} + 0.5[y_p'y_p - \eta_p'\Omega_p^{-1}\eta_p + \Lambda_{0p}'H_{0p}^{-1}\Lambda_{0p}]).$$

The conditional for Λ_p is a Q-variate normal with mean, η_p, and covariance, $\sigma_{pp}\Omega_p$, and the conditional for Φ^{-1} is Wishart with scale matrix, $FF' + R_0$, and degrees of freedom, $n + \rho_0$. Finally, the conditional $p(F_i|y, \theta)$ for the factor scores for subject i is a Q-variate normal with mean $[\Phi^{-1} + \Lambda'\Sigma^{-1}\Lambda]^{-1}\Lambda'\Sigma^{-1}y_i$ and covariance $[\Phi^{-1} + \Lambda'\Sigma^{-1}\Lambda]^{-1}$.

7.3 Identifiability and Priors on Loadings

Under the model (Equation 7.1), the marginal covariance of y is $V = \Lambda\Phi\Lambda' + \Sigma$. It can be seen that the contribution, $\Lambda\Phi\Lambda'$, of the factor scores to explaining variation in y may be achieved by an infinite number of pairs (Λ, Φ), and constraints must be imposed to ensure a unique location and scale for the factor scores (Wedel et al., 2003, 358–9). One way of providing identifiability (the scaling constraint) is to define the factors to be in standardized form, with zero means and variances of 1 (Bentler and Weeks, 1980). Under the alternative anchoring constraint (Skrondal and Rabe-Hesketh, 2004), one among the set of loadings $\{\lambda_{pq}, p = 1, \ldots, P\}$ on each construct is preset for identification. The factors are still required to have zero means (providing unique location), but may have unknown variances.

For the measurement model to be identifiable, the number of unknown parameters in $\theta = (\Sigma, \Phi, \Lambda)$ must be less than the number, $P(P+1)/2$, of distinct elements in the residual variance–covariance matrix V of y. For example, in the standardized factor case, and with $\Phi = I$ excluding correlations, one has

$$V = \Lambda\Lambda' + \Sigma,$$

with $PQ + P$ parameters on the right-hand side under a local independence assumption (Σ taken as diagonal). For $P(P+1)/2 \geq PQ+P$ to apply requires that $P \geq 2Q+1$ (Geweke and Zhou, 1996). In confirmatory models, elements of Λ may be preset to zero, alleviating requirements such that Σ be diagonal or that Φ exclude covariances/correlations.

When $Q > 1$, additional identifying constraints must be set to avoid rotation invariance. Otherwise, there is no unique solution because any orthogonal transformation of Λ leaves the likelihood unchanged (Everitt, 1984, 16). Thus for $F^* = H'F$ and $\Lambda^* = \Lambda H$, where $HH' = I$,

$$y = 1\alpha + \Lambda F + u = 1\alpha + (\Lambda H)(H'F) + u = 1\alpha + \Lambda^* F^* + u,$$

where $\text{cov}(F^*) = H'\text{cov}(F)H = \text{cov}(F)$. The exception is the simple structure case (each observed variable loading on only one factor) when rotational identifiability is not an issue (Liu et al., 2005, 550; Wedel et al., 2003, 358).

In other cases, identification may be achieved by fixing enough λ_{pq} to ensure a unique solution; in the case, $Q = 2$, setting any $\lambda_{p2} = 0$ would be sufficient. Provided the variables are ordered in such a way as to ensure substantive justification, a widely adopted option is to assume Λ to be lower triangular, as in Geweke and Zhou (1996), Quinn (2004), Ghosh and Dunson (2008), and Lopes and West (2004), namely,

$$\Lambda = \begin{bmatrix} \lambda_{11} & 0 & 0 & \cdots & 0 & 0 \\ \lambda_{21} & \lambda_{22} & 0 & \cdots & 0 & 0 \\ \lambda_{31} & \lambda_{32} & \lambda_{33} & \cdots & 0 & 0 \\ \vdots & \vdots & \vdots & \ddots & \vdots & \vdots \\ \lambda_{Q-1,1} & \lambda_{Q-1,2} & \lambda_{Q-1,3} & \cdots & \lambda_{Q-1,Q-1} & 0 \\ \lambda_{Q1} & \lambda_{Q2} & \lambda_{Q3} & \cdots & \lambda_{Q,Q-1} & \lambda_{QQ} \\ \vdots & \vdots & \vdots & \ddots & \vdots & \vdots \\ \lambda_{P1} & \lambda_{P2} & \lambda_{P3} & \cdots & \lambda_{P,Q-1} & \lambda_{PQ} \end{bmatrix}.$$

To avoid potential labeling issues, this structure can be combined with the diagonals constraint,

$$\lambda_{qq} > 0,$$

and if λ_{qq} are unknowns under a fixed factor scale with $\Phi = I$, one might take,

$$\lambda_{qq} \sim N(0, \delta_{qq})I(0,).$$

Without such a constraint, the loadings and scores on a particular factor may flip over during MCMC iterations—see Geweke and Zhou (1996, 566) for discussion of this issue in multivariate factor time series. For $Q \geq 1$, setting one loading for each construct to be fixed (usually at 1.0) under an anchoring constraint, usually ensures remaining loadings conform to a consistent interpretation and direction of the factor.

An illustration is in item analysis of educational tests (with $Q = 1$) where one generally seeks a univariate positive ability measure, F_i (Albert and Ghosh, 2000; Sinharay, 2004, 465). Without suitable constraints on one or more of the loadings, λ_p (e.g., gamma or lognormal priors), a negative ability construct may be obtained. To completely avoid the possibility of label switching, a positivity constraint may be applied to all loadings; for example, Sahu (2002) applies half normal priors with informative variance in an item analysis, i.e.,

$$\lambda_q \backsim N(0, \delta_q)I(0,),$$

where $\delta_q = 0.5$ or 1.

While not necessarily a device to ensure consistent labeling but instead to seek relatively simple structure (a Bayesian version of varimax rotation), one may follow Fokoue (2004) and take the loading variances as unknown gamma variables, namely,

$$\lambda_{pq} \sim N(0, \delta_{pq}),$$
$$\delta_{pq} \sim \mathrm{Ga}(\nu, \nu),$$

with ν small.

7.3.1 An illustration of identifiability issues

To exemplify identifiability constraints, consider a spatial example involving English local authorities, and suppose six observed indicators $\{y_1, \ldots, y_6\}$ are taken to measure two latent area constructs, F_1 and F_2, deprivation and fragmentation. Thus, several studies have shown that area material deprivation (i.e., meaning economic hardship represented by observed variables such as high unemployment, and low car and home ownership) tends to be associated with higher psychiatric morbidity and suicide mortality (Gunnell et al., 1995). So also does social fragmentation, meaning relatively weak community ties associated with observed indices such as one person households, high population turnover, and many adults outside married relationships (Evans et al., 2004). Indicators $\{y_1, y_2, y_3\}$ of deprivation are provided by square roots (a normalizing transform) of the UK 2001 Census rates of renting from social landlords, and of unemployment among the economically active, together with the square root of the 2001 rate of households claiming income support. Indicators $\{y_4, y_5, y_6\}$ of social fragmentation are provided by square roots of Census rates of one person households, migration in the year 2000–2001, and people over 15 not married.

A confirmatory factor model (with simple structure) is assumed with $\{y_1, \ldots, y_3\}$ loading only on a deprivation score, F_1, and with $\{y_4, \ldots, y_6\}$

loading only on a fragmentation score, F_2. Let D_{pi} be the denominator (e.g., total population) used to define the transformed Census index, y_{pi}. Then the measurement model has the form,

$$\begin{aligned}
y_{1i} &= \alpha_1 + \lambda_{11}F_{1i} + u_{1i},\\
y_{2i} &= \alpha_2 + \lambda_{21}F_{1i} + u_{2i},\\
y_{3i} &= \alpha_3 + \lambda_{31}F_{1i} + u_{3i},\\
y_{4i} &= \alpha_4 + \lambda_{42}F_{2i} + u_{4i},\\
y_{5i} &= \alpha_5 + \lambda_{52}F_{2i} + u_{5i},\\
y_{6i} &= \alpha_6 + \lambda_{62}F_{2i} + u_{6i},
\end{aligned}$$

where u_{ji} are mutually uncorrelated, with $u_{pi} \backsim N(0, \tau_p/D_{pi})$ (Hogan and Tchernis, 2004, 316).

Since F_1 and F_2 have arbitrary location and scale, one way of providing identifiability (the variance scaling or standardization constraint) is to define them to be in standard form with zero means and variances of 1 (while still possibly allowing nonzero correlations between factors, which is possible under this confirmatory model). Under the alternative anchoring constraint (Skrondal and Rabe-Hesketh, 2007), one loading on each construct is preset for identification, e.g., $\lambda_{11} = \lambda_{42} = 1$. The F_{qi} may be assumed independent of one another, although correlation over areas i may still be incorporated via two separate univariate CAR priors (Besag et al., 1991). Alternatively, correlation both between factors and over areas may be assumed, so that $\{F_{1i}, F_{2i}\}$ follow a bivariate CAR prior (see Section 7.6). Under an anchoring constraint, the within-area factor covariance matrix would then contain three unknowns $\{\phi_{11}, \phi_{22}, \rho\}$,

$$\Phi = \begin{pmatrix} \phi_{11} & \rho\sqrt{\phi_{11}\phi_{22}} \\ \rho\sqrt{\phi_{11}\phi_{22}} & \phi_{22} \end{pmatrix},$$

whereas under a standardization constraint the diagonal elements in Φ are set to 1, and only ρ would be unknown.

Adopting an anchoring constraint has utility in helping to prevent "relabeling" of the construct scores, F_{ki}, during MCMC sampling. Since the indicators $\{y_1, \ldots, y_3\}$ in this example are positive measures of material deprivation, setting $\lambda_{11} = 1$ is consistent with the construct F_{1i} being a positive deprivation measure. If, however, one adopted the fixed scale constraint with $\phi_{pp} = 1$ and all λ_{pq} free, it would be necessary, in order to prevent label switching, to set a prior on one or possibly more loadings constraining positivity, e.g.,

$$\begin{aligned}
\lambda_{p1} &\sim N(1, 1)I(0,),\\
\lambda_{p2} &\sim N(1, 1)I(0,),
\end{aligned}$$

for one or more p.

TABLE 7.1
Maximum likelihood factor solution.

	Factor loadings		
Variable	**1**	**2**	**Uniqueness**
Murder/manslaughter	−0.07	0.65	0.57
Rape	0.46	0.67	0.34
Robbery	0.32	0.54	0.61
Assault	0.34	0.91	0.06
Burglary	0.76	0.19	0.39
Larceny	1.00	0.00	0.00
Autotheft	0.31	0.31	0.81

Example 7.1. American City Crime: Exploratory Factor Model Everitt (1984) presents crime rates per 100,000 population in 1970 for 16 American cities. There are $P = 7$ different types of crime namely, (a) murder/manslaughter, (b) rape, (c) robbery, (d) assault, (e) burglary, (f) larceny, and (g) autotheft. A maximum likelihood factor analysis of the data in Stata leads to $Q = 2$ factors with loadings in Table 7.1. The first factor is interpretable as representing nonviolent crime contrasts, and the second represents violent crime. The Stata analysis does not provide estimation intervals for the loadings.

An exploratory factor model[1] is then applied under a Bayesian approach, with the first model adopting standardized and uncorrelated factors. On the basis of the maximum likelihood analysis, the priors on selected loadings are informative with λ_{61} (the larceny loading on factor 1) constrained to be positive with the intention of ensuring a nonviolent crime score. Similarly, λ_{42} (the assault loading on factor 2) is constrained to be positive with the intention of ensuring scores on the second factor measure violent crime, and with λ_{62} set to zero to avoid rotation invariance. Otherwise, the loadings have $N(0, 1)$ priors. A second model applies an anchoring constraint with $\lambda_{61} = \lambda_{42} = 1$ and $\lambda_{62} = 0$, and has unknown variances for the factor scores, which are still assumed uncorrelated. Uniform $U(0, 1000)$ priors are assumed for the unique standard deviations and the factor standard deviations in model 2.

Results from a two chain run of 20,000 iterations (with early convergence in loading parameters and unknown standard deviations) show the solutions for both models are imprecise (Table 7.2). In the second model, λ_{22} is more precisely defined. On the other hand, the imprecision may be a valid reflection of a relatively complex model being applied to a small dataset. More precise posterior densities for the loadings are obtained using the Press-Shigemasu prior—see also Lee (2007, 83–7). Thus,

$$(\lambda_{p1}, \lambda_{p2}) \sim N_Q([\Lambda_{0p}], \sigma_{pp} H_{0p}),$$

where the 2×2 matrix, H_{0p}, is assumed to be an identity matrix, and $\Lambda_{0p} = (0, 0)$. This prior is combined with an anchoring constraint to form model 3.

TABLE 7.2
Posterior summary: Crime EFA.

	Model 1			Model 2			Model 3		
	Mean	**2.5%**	**97.5%**	**Mean**	**2.5%**	**97.5%**	**Mean**	**2.5%**	**97.5%**
λ_{11}	0.04	−0.77	0.88	0.00	−1.19	1.06	−0.17	−0.93	0.59
λ_{12}	0.61	−0.27	1.33	0.72	−0.15	1.71	0.67	0.12	1.26
λ_{21}	0.49	−0.43	1.22	0.58	−0.59	1.64	0.33	−0.31	0.96
λ_{22}	0.63	−0.23	1.27	0.76	−0.04	1.61	0.70	0.24	1.17
λ_{31}	0.33	−0.47	1.09	0.33	−0.87	1.40	0.20	−0.60	0.93
λ_{32}	0.50	−0.35	1.22	0.58	−0.39	1.45	0.57	0.03	1.14
λ_{41}	0.40	−0.50	1.23	0.42	−0.79	1.54	0.14	−0.33	0.72
λ_{42}	0.84	0.12	1.43	1.00			1.00		
λ_{51}	0.77	−0.17	1.45	0.87	−0.36	1.84	0.73	−0.04	1.38
λ_{52}	0.17	−0.49	0.81	0.21	−0.71	1.02	0.24	−0.27	0.77
λ_{61}	0.86	0.10	1.52	1.00			1.00		
λ_{62}	0.00			0.00			0.00		
λ_{71}	0.33	−0.41	1.08	0.37	−0.69	1.41	0.29	−0.54	1.11
λ_{72}	0.29	−0.50	1.03	0.32	−0.61	1.19	0.33	−0.28	0.95
σ_1	0.84	0.32	1.35	0.85	0.40	1.35	0.82	0.54	1.23
σ_2	0.63	0.13	1.10	0.63	0.17	1.06	0.66	0.43	1.01
σ_3	0.84	0.30	1.35	0.86	0.50	1.36	0.83	0.56	1.24
σ_4	0.43	0.02	0.94	0.42	0.02	0.91	0.33	0.02	0.77
σ_5	0.68	0.13	1.22	0.69	0.15	1.24	0.70	0.43	1.11
σ_6	0.61	0.06	1.26	0.65	0.06	1.27	0.50	0.03	1.14
σ_7	0.98	0.58	1.51	0.98	0.63	1.49	0.93	0.64	1.38
ϕ_1				0.68	0.04	2.00	0.92	0.06	2.36
ϕ_2				0.80	0.10	2.12	1.05	0.31	2.43

7.4 Multivariate Exponential Family Outcomes and General Linear Factor Models

The normal linear factor and structural equation models considered above generalize straightforwardly to general linear factor and SEM models for non-normal data from the exponential family density: namely, binomial, Poisson, and multinomial or ordinal data. Consider multivariate observations, $y_i = (y_{1i}, \ldots, y_{Pi})'$, that conditional on factor scores $F_i = (F_{1i}, \ldots, F_{Qi})'$ follow an exponential family density, namely,

$$p(y_{pi}|F_i) \propto \exp\left\{\frac{y_{pi}\theta_{pi} - b(\theta_{pi})}{\phi_{pi}} + c(y_{pi}, \phi_{pi})\right\},$$

where θ_{pi} is the canonical parameter, with ϕ_{pi} typically taken as known scale parameters. Denoting regression terms as $\eta_{pi} = g(\theta_{pi})$, where g is a link function, and $\eta_i = (\eta_{1i}, \ldots, \eta_{Pi})'$, intercept $\alpha = (\alpha_1, \ldots, \alpha_P)'$, and $P \times Q$ loading matrix Λ, the regression term without extra-variation is

$$\eta_i = \alpha + \Lambda F_i, \tag{7.2}$$

while allowing extravariation,

$$\eta_i = \alpha + \Lambda F_i + u_i, \tag{7.3}$$

with $u_i = (u_{1i}, \ldots, u_{Pi})'$, where u_{pi} are independent of each other under conditional independence. The errors u (if present) and factor scores F are also independent.

Normality of errors and factors is often assumed with $(u_{1i}, \ldots, u_{Pi})' \sim N_P(0, \Sigma)$, where Σ is diagonal, and $F_i \sim N_Q(0, \Phi)$, where Φ may be non-diagonal according to the form of model (e.g., exploratory or confirmatory) assumed. Compared to the normal data-normal factor model, the marginal densities of y are no longer simply derived, but involve integration over F, namely,

$$p(y_i|\theta, \psi) = \int \prod_{p=1}^{P} p(y_{pi}|F_i, \theta)p(F_i|\psi)\,dF_i,$$

where ψ are hyperparameters defining the density of F_i. The usual conditional independence assumptions are made. For example, for a P-variate categorical response (K_p categories for the pth response), the conditional probability that subject i with factor scores $F_i = (F_{1i}, \ldots, F_{Qi})'$ exhibits a particular set of responses is the product of separate categorical likelihoods,

$$\begin{aligned} &Pr(y_{1i} = k_1, y_{2i} = k_2, \ldots, y_{Pi} = k_P|F_i) \\ &\quad = Pr(y_{1i} = k_1|F_i)Pr(y_{2i} = k_2|F_i), \ldots, Pr(y_{Pi} = k_P|F_i). \end{aligned}$$

For factor reduction of binary, multinomial, or ordinal data, there may be benefit (e.g., in simplified MCMC sampling algorithms) in considering latent variables posited to underlie the observed discrete responses. The missing data then consists not only of factor scores, but of the latent scale data, y^*_{pi}, that underlie the observed data, y_{pi}. Thus for y_{pi} binary, and $y_{pi} = 1$ if $y^*_{pi} > 0$ and $y_{pi} = 0$ otherwise, one might take $y^*_i = (y^*_{1i}, \ldots, y^*_{Pi})$ to be normal or logistic, with the diagonal terms in the unique covariance matrix Σ set (usually to 1) for identifiability. For instance, a normal model taking the underlying responses to be conditionally independent given the factors, would be

$$y^*_{pi} \sim N(\alpha_p + \Lambda_p F_i, 1)I(A_{pi}, B_{pi}),$$

where Λ_p is the pth row of Λ, and the truncation ranges are determined by the observed y_{pi}.

7.4.1 Multivariate Poisson data

Factor models with $Q < P$ may be more parsimonious than full dimension error models for multivariate exponential family data. However, multivariate reduction may not always be preferred in terms of fit, so parsimony may sometimes be at the expense of predictions that reproduce the data satisfactorily. Chib and Winkelmann (2001) illustrate how multivariate count data may not always be suitable for reduction using latent factors. In their full-dimension model, $y_{pi} \sim Po(\mu_{pi})$, with outcome-specific predictors, $x_{pi} = (x_{1pi}, x_{2pi}, \ldots, x_{Rpi})'$, and

$$\mu_{pi} = \exp(\beta_p x_{pi} + u_{pi}),$$
$$(u_{1i}, \ldots, u_{pi}) \sim N_P(0, D).$$

The y_{pi} are conditionally independent given the correlated errors, $u_i = (u_{1i}, \ldots, u_{Pi})'$. Defining $v_{pi} = \exp(u_{pi})$, one has, equivalently, $y_{pi} \sim Po(\lambda_{pi} v_{pi})$ with

$$\lambda_{pi} = \exp(\beta_p x_{pi}),$$
$$(v_{1i}, \ldots, v_{pi}) \sim LN_P(\mu_v, \Sigma_v).$$

That is, v_{pi} are multivariate lognormal with mean vector $\mu_v = \exp(0.5\, \text{diag}(D))$, and covariance $\Sigma_v = \text{diag}(\mu_v)[\exp(D) - 11']\text{diag}(\mu_v)$. Other ways to generate correlated count data include the overlapping sums technique (Madsen and Dalthorp, 2007). Thus consider independent Poisson variables, Z_{12}, Z_1, and Z_2, with means θ_{12}, θ_1, and θ_2; then $y_1 = Z_1 + Z_{12}$ and $y_2 = Z_2 + Z_{12}$ are correlated with marginal means $\theta_1 + \theta_{12}$ and $\theta_2 + \theta_{12}$ and covariance θ_{12}. The mean and covariance of the corresponding joint Poisson density for three variables is provided by Karlis and Meligkotsidou (2005, 257).

Factor models for count data typically may include both normal factor scores and residuals, u_{pi}, taken as uncorrelated if the usual conditional independence assumption is made. Thus,

$$y_{pi} \sim Po(\mu_{pi}),$$
$$\mu_{pi} = \exp(\beta_p x_{pi} + \Lambda_p F_i + u_{pi}),$$

where Λ_p is $1 \times Q$, $F_i = (F_{1i}, \ldots, F_{Qi})'$ and under a standardized factor constraint, $F_i \sim N_Q(0, R_F)$, where R_F is a correlation matrix, with possibly unknown off-diagonal terms subject to identifiability. Alternatively, Wedel et al. (2003) consider gamma distributed factors in an identity link model as well as normal F scores combined with a log link. Gamma factors would have mean 1 to avoid location invariance, and taking their cumulative impact to be multiplicative, one could have

$$\mu_{pi} = \exp(\beta_p x_{pi} + u_{pi})\left(F_{1i}^{\omega_{p1}} F_{2i}^{\omega_{p2}}, \ldots, F_{Qi}^{\omega_{pQ}}\right),$$

with comparable identification restrictions on the loadings, $\Omega_p = (\omega_{p1}, \ldots, \omega_{pQ})$, to those in the normal linear factor model (see also Dunson and Herring, 2005). The constraints would differ according to whether the variance of the F scores were unknown, as in,

$$F_{qi} \sim \text{Ga}(\varphi_q, \varphi_q),$$

with φ_q to be estimated, or whether the variance of F is preset, as in $F_{qi} \sim \text{Ga}(1, 1)$.

An alternative to outcome-specific residuals, u_{pi}, in the above models is a common residual factor, especially when F_i are derived as part of a broader structural model involving further observed indicators. For example, consider count observations, y_{pi} $(p = 1, \ldots, P)$, on clinical outcomes for a set of hospitals, while also available are metric measures, x_{ri} $(r = 1, \ldots, R)$, of resource inputs, efficiency, etc. The latter variables are relevant to defining a multivariate latent "care quality" construct, F_i, in a MIMIC framework that assists in explaining the clinical outcomes, but this construct may not explain all the covariation among (or overdispersion in) the y variables and correlated residuals and/or common residual factors are needed. The cause equation defining the latent construct might take the form (for standardized x),

$$F_i = \beta x_i + w_i,$$

while the errors in the Poisson likelihood measurement equations for $y_{pi} \sim Po(\mu_{pi})$ are correlated over outcomes under a common factor model,

$$\log(\mu_{pi}) = \alpha_p + \Lambda_p F_i + \kappa_p u_i.$$

Assuming u_i is univariate, one of the loadings, κ_p, is preset if the variance, σ_u^2, of the common residual scores, u_i, is unknown. Spatial applications of common factors are exemplified by Wang and Wall (2003) and Congdon (2008b).

7.4.2 Multivariate binary data and item response models

As for counts, models for P-variate binary outcomes, $y_{pi} \sim Bern(\pi_{pi})$, may retain the observed binary data likelihood and represent joint or residual correlations by additive full-dimension multivariate effects, u_{pi}, for example,

$$\text{logit}(\pi_{pi}) = \beta_p x_{pi} + u_{pi},$$
$$(u_{1i}, \ldots, u_{pi}) \sim N_P(0, D).$$

By contrast, multivariate probit or logit models may also follow from an augmented data perspective in which unobserved metric variables, $y_i^* = (y_{1i}^*, y_{2i}^*, \ldots, y_{Pi}^*)$, result in the observed binary vector (Chen and Dey, 1998, 2000).

Thus, Bayesian estimation of the multivariate probit involves augmenting the data with the latent normal variables obtained by truncated multivariate normal sampling. Thus, with $\eta_{pi} = \beta_p x_{pi}$ and $\eta_i = (\eta_{pi}, \ldots, \eta_{Pi})$,

$$(y_{1i}^*, y_{2i}^*, \ldots, y_{Pi}^*) \sim N_P(\eta_i, \Sigma) I(A_i, B_i), \tag{7.4}$$

with the observations generated according to,

$$y_{pi} = I\{y_{pi}^* > 0\}.$$

The lower and upper sampling limits in the vectors $A_i = (A_{1i}, \ldots, A_{Pi})$ and $B_i = (B_{1i}, \ldots, B_{Pi})$ depend on the observations: sampling of the constituent y^*_{pi} is confined to values above zero when $y_{pi} = 1$, and to zero or negative values when $y_{pi} = 0$. Scale mixtures of multivariate normal densities for y^*_{pi} are also possible leading to multivariate Student t, which for particular degrees of freedom approximates a multivariate logit link (Chen and Dey, 1998). A multivariate logit regression may be achieved directly with suitable mixing strategies (Chen and Dey, 2000; O'Brien and Dunson, 2004).

The covariance matrix Σ in Equation 7.4 is not identified, and when the predictor effects vary by response only the correlation matrix can be identified (Rossi et al., 2005). The identification criteria for the multivariate probit differ from those of the multinomial probit where identification is obtained by setting one of the diagonal variance elements, σ_{pp} (e.g., the first), to 1 (McCulloch et al., 2000). While it is possible to sample the correlation matrix directly (Barnard et al., 2000; Chib and Greenberg, 1998), one may also (Edwards and Allenby, 2003; McCulloch and Rossi, 1994) sample the Σ matrix or its inverse from an unrestricted prior, and then scale both the fixed effects and the covariance matrix to their identified forms, namely,

$$\beta^*_p = \beta_p/\sqrt{\sigma_{pp}},$$

and the correlation matrix,

$$R = \Delta\Sigma\Delta,$$

where $\Delta = \text{diag}(\sqrt{\sigma_{pp}})$.

Factor models for multiple binary data most typically have the general linear mixed form,

$$\begin{aligned} y_{pi} &\sim Bern(\pi_{pi}), \\ g(\pi_{pi}) &= \beta_p x_{pi} + \Lambda_p F_i, \end{aligned}$$

where g is the link, and $F_i = (F_{1i}, \ldots, F_{Qi})'$. As for the normal linear factor model, a common assumption for the density of F is normal with known scale. If, additionally, factors are independent, then $F_{qi} \sim N(0, 1)$, $q = 1, \ldots, Q$. If instead the assumption, $F_{qi} \sim \text{logist}(0, 1)$ is made, with loadings κ_{pq} in,

$$\eta_{pi} = \alpha_p + \kappa_{p1} F_{1i} + \cdots + \kappa_{pQ} F_{Qi},$$

then $\kappa_{pq} \approx (\sqrt{3}/\pi)\lambda_{pq}$, since the variance of a standard logistic is $\pi^2/3$ (Bartholomew, 1987). Another possibility involves F scores linked (e.g., by probit or logit transforms) to uniform scores z. For example, the scheme,

$$\begin{aligned} g(\pi_{pi}) &= \alpha_p + \lambda_{p1} F_{1i} + \cdots + \lambda_{pQ} F_{Qi}, \\ F_{qi} &= \text{logit}(z_{qi}), \\ z_{qi} &\sim U(0, 1), \end{aligned}$$

corresponds to F_{qi} being logistic.

A widely applied paradigm in educational and psychometric evaluation (Albert, 1992; Fox and Glas, 2005; Rupp et al., 2004) is based on item response theory (IRT). Typically, the observation vector, y_i, consists of P binary items measuring ability, with 1 denoting a correct answer and 0 an incorrect answer, and a model seeks a single latent ability factor score, F_i. Under conditional independence the joint success probability given F_i is

$$\begin{aligned} &Pr(y_{1i} = 1, y_{2i} = 1, \ldots, y_{pi} = 1|F_i) \\ &\quad = Pr(y_{1i} = 1|F_i)Pr(y_{2i} = 1|F_i) \cdots Pr(y_{Pi} = 1|F_i). \end{aligned}$$

If the Bernoulli likelihood, $y_{pi} \sim Bern(\pi_{pi})$, is retained, one has a factor type model

$$g(\pi_{pi}) = \eta_{pi} = \alpha_p + \lambda_p F_i.$$

The intercepts α_p can be interpreted as measures of difficulty of item p, while λ_p measures an item's power to discriminate ability between subjects.

Fox and Glas (2005) describe Bayesian model choice analysis for IRT models allowing for differential item functioning (DIF)—when an item is not appropriate for measuring ability because the knowledge needed for a correct answer is culturally specific (Swanson et al., 2002). Thus, let $x_i = 0$ for a reference population and $x_i = 1$ for a focal group (e.g., disadvantaged or minority group); then DIF is indicated if the extended model,

$$\begin{aligned} Pr(y_{pi} = 1|F_i) &= \Phi(\eta_{pi}), \\ \eta_{pi} &= \alpha_p + \lambda_p F_i + x_i(\gamma_p + \delta_p F_i), \end{aligned}$$

has better fit than the standard model without group differentiation (Swaminathan and Rogers, 1990).

7.4.3 Latent scale binary models

As an alternative to binary likelihood modeling in IRT and binary SEM applications, the latent scale method may be applied with the appropriate underlying density defined by the link g. Thus, for a probit link with $g^{-1} = \Phi$, the latent metric scale, y^*, is normal, such that $y_{pi} = 1$ corresponds to the imputation scheme,

$$y^*_{pi} \backsim N(\eta_{pi}, 1)\ I(0,)$$

and $y_{pi} = 0$ corresponds to $y^*_{pi} \backsim N(\eta_{pi}, 1)\ \ I(0,)$. For a logit link, $g^{-1}(u) = L(u) = (e^u/1 + e^u)$ and sampling of y^* is from a standard logistic. Sahu (2002) considers an extra data imputation to provide three-parameter IRT models. The three-parameter probit IRT model specifies,

$$\pi_{pi} = c_p + (1 - c_p)\Phi(\alpha_p + \lambda_p F_i),$$

while the three-parameter logistic IRT is

$$\pi_{pi} = c_p + (1 - c_p)L(\alpha_p + \lambda_p F_i),$$

where c_p is interpretable as a guessing parameter.

Lee and Song (2003) adopt a latent scale approach to a structural equation model for multiple binary observations. Their model specifies,

$$y_i^* = \alpha + \Lambda F_i + u_i,$$

where the latent constructs, F_i, are partitioned into endogenous and exogenous vector components, $F_i = (F_{1i}, F_{2i})$, of dimension Q_1 and Q_2, respectively, with structural model,

$$F_{1i} = BF_{1i} + \Gamma F_{2i} + w_i.$$

For identification, $u_i \sim N_P(0, I)$, while $F_{1i} \sim N_{Q_1}(0, \Phi_1), F_{2i} \sim N_{Q_2}(0, \Phi_2), w_i \sim N_{Q_1}(0, \Sigma_w)$, and each row of Λ follows a separate normal prior. The observed binary data, y, is augmented with latent data $\{y^*, F\}$ to provide complete data $\{y, y^*, F\}$. Setting $\theta = (\alpha, \Lambda, \Phi_1, \Phi_2, \Sigma_\delta, B, \Gamma)$, the updating sequence involves sampling from conditionals $p(\theta^{(t+1)}|F^{(t)}, y^{*(t)})$, $p(F^{(t+1)}|\theta^{(t+1)}, y^{*(t)})$, and $p(y^{*(t+1)}|F^{(t+1)}, \theta^{(t+1)})$.

Dunson and Herring (2005) consider instead the case where the underlying y_{pi}^* (e.g., tumor counts) are Poisson, or overdispersed Poisson, and the observations, y_{pi} (e.g., whether tumors are present), are binary. Thus,

$$y_{pi}^* \sim Po(\exp(x_{pi}\beta)\Lambda_p \xi_i)I(A_{pi}, B_{pi}),$$

where $\xi_i = (\xi_{1i}, \ldots, \xi_{Qi})$ are gamma distributed latent constructs, and the loadings, Λ_p, are also gamma distributed. The sampling limits are ($A_{pi} = 0$, $B_{pi} = 0$) when $y_{pi} = 0$, and ($A_{pi} = 1, B_{pi} = \infty$) when $y_{pi} = 1$.

7.4.4 Categorical data

For unordered polytomous indicators, y_{pi}, with M_p categories ($p = 1, \ldots, P$), intercept and loading parameters are typically specific to the category of each item, with one category (e.g., the final one) as reference. Assume a multiple logit link (Bartholomew, 1987), with multinomial parameter, $\pi_{pi} = (\pi_{pi1}, \ldots, \pi_{piM_p})$, for subject i and indicator p. Then while factors are common across categories, loadings are specific to indicator p and category h of each indicator,

$$\begin{aligned}
y_{pi} &\sim \text{categoric}(\pi_{pi1}, \pi_{pi2}, \ldots, \pi_{piM_p}),\\
\pi_{pih} &= \varphi_{pih} \bigg/ \sum_{m=1}^{M_p} \varphi_{pim} \qquad h = 1, \ldots, M_p,\\
\log(\varphi_{pih}) &= \alpha_{ph} + \lambda_{ph1} F_{1i} + \cdots + \lambda_{phQ} F_{Qi} \qquad h = 1, \ldots, M_{p-1},\\
\varphi_{piM_p} &= 1,
\end{aligned}$$

with the usual constraints on Λ and/or F to avoid scale and rotational invariance.

Both Bartholomew et al. (2002) and Moustaki (2000) mention the item response function approach to estimating factor models for ordinal data (see Example 7.3). Thus, let $\pi_{ji}^{(m)}$ denote the probability of responding correctly to category $m \in (1, \ldots, M_j)$ of item j and denote the cumulative probability of a response to item j in category m or lower as

$$\Delta_{ji}^{(m)} = \pi_{ji}^{(1)} + \pi_{ji}^{(2)} + \cdots + \pi_{ji}^{(m)}.$$

Then with cut points, δ_{jm}, an ordinal factor model may be estimated via $M_j - 1$ binary regressions,

$$g(\Delta_{ji}^{(m)}) = \delta_{jm} - \sum_{k=1}^{Q} \lambda_{jk} F_{ik},$$

where the negative sign preceding the factor impact term indicates that as F scores increase, the responses to item j are more likely to fall at the upper end of the scale.

For ordinal data, the latent variable approach is also appealing. For example, suppose observations y consist of $P = P_1 + P_2$ variables, the first P_1 of which are continuous, and the subsequent P_2 are ordinal containing $M_1, M_2, \ldots, M_{P_2}$ categories (e.g., Lee and Shi, 2001; Lee and Song, 2004; Lee and Tang, 2006). To model correlation among these variables or introduce regression effects, one may define latent metric variables, y_{pi}^*, for $p > P_1$ with $M_p - 1$ unknown cut points δ_{pm} (and preset ones, $\delta_{p0} = 0$, $\delta_{pM_p} = \infty$) such that,

$$y_{pi} = m \iff \delta_{p,m-1} \leq y_{pi}^* \leq \delta_{pm} (m = 1, \ldots, M_p),$$
$$0 \leq \delta_{p1} \leq \cdots \leq \delta_{p,M_p-1} \leq \infty.$$

For identification, the submatrix of $\text{cov}(y)$ relating to the last P_2 outcomes is a correlation matrix. Instead of a P-dimensional multivariate model for the joint responses $\{y_{1i}, \ldots, y_{P_1 i}, y_{P_1+1,i}^*, \ldots, y_{Pi}^*\}$, a common factor model proposes that the correlation structure result from a smaller set ($Q < P$) of metric factors, $F_i \sim N_Q(0, \Phi)$, as in,

$$y_{pi} = \alpha_p + \lambda_{p1} F_{1i} + \cdots + \lambda_{pQ} F_{Qi} + u_{pi} \qquad p = 1, \ldots, P_1,$$
$$y_{pi}^* = \alpha_p + \lambda_{p1} F_{1i} + \cdots + \lambda_{pQ} F_{Qi} + u_{pi} \qquad p = P_1 + 1, \ldots, P,$$

where $\text{cov}(y) = \Lambda\Phi\Lambda + \Sigma$, and under conditional independence, the covariance matrix Σ for errors, u_i, is diagonal (Shi and Lee, 2000) with $\text{var}(u_{pi}) = 1$ for $p > P_1$. One possible framework (Moustaki, 2000) assumes that the latent variables, F_{ki}, are standard normal and uncorrelated, and the $P_2 \times P_2$ covariance submatrix of $\text{cov}(y)$ corresponding to the y_{pi}^* is a matrix with elements,

$$r_{ab} = \sum_{m=1}^{Q} \lambda_{am} \lambda_{bm}.$$

To avoid rotational invariance, there are then only $PQ - Q(Q-1)/2$ independent factor loadings.

Example 7.2. Greek Crime Totals This example considers count data and compares a common factor model to a full dimension covariance structure. The data relate to $P = 4$ counts of crimes (rapes, arsons, manslaughter, smuggling of antiquities) in 49 Greek prefectures, as used in Karlis and Meligkotsidou (2005)*. The counts are assumed to be Poisson with offset being prefecture populations, o_i (in millions). Predictors are unemployment rate (z_1), a binary indicator (z_2) for whether the prefecture is at the Greek borders, gross domestic product (GDP) per capita in euros (z_3), and a binary indicator (z_4) for whether the prefecture has at least one large city (>150,000 inhabitants).

Event rates are low in relation to populations at risk, so a reasonable sampling model takes $y_{pi} \sim Po(o_i \rho_{pi})$, with ρ_{pi} being crime rates per million. The full dimension model specifies,

$$\rho_{pi} = \exp(z_i \beta_p + u_{pi}) \quad p = 1, \ldots, P,$$
$$(u_{1i}, \ldots, u_{Pi}) \sim N_P(0, D),$$

with prior $D^{-1} \sim W(PI, P)$. A 25,000 iteration two chain analysis is run in OpenBUGS, with inferences based on the second half of the sequence. The effective parameter count is estimated at 99.5, and the mean scaled deviance at 203, comparing closely to the number of observations, namely, 196. The DIC is therefore 302.5. Adequate performance is also shown by the fact that only 9 of the 196 observations have mixed predictive p values under 0.05 or over 0.95 (Marshall and Spiegelhalter, 2007). Most predictor effects are insignificant: the only significant effects are of unemployment on the rape crime rate (with the relevant β coefficient having mean 0.064), and of the border and GDP variables on manslaughter rates. Most correlations, $r_{jm} = \text{corr}(u_{ji}, u_{mi})$, in the regression residuals have credible intervals straddling zero, though r_{13} has posterior mean 0.47 with 95% interval $(-0.05, 0.80)$.

To illustrate a factor analytic approach to these data, the four predictors are taken to be causes of a single underlying crime construct, F_i, in a MIMIC analysis. The Poisson regressions form a measurement model in which crime levels are indicators of F_i. A further common factor, u_i, is included in the model for the crime types to account for residual variation in the crime data. So,

$$\rho_{pi} = \exp(\alpha_p + \lambda_p F_i + \kappa_p u_i),$$
$$F_i = b_1 z_{1i} + b_2 z_{2i} + b_3 z_{3i} + b_4 z_{4i} + w_i,$$
$$w_i \sim N(0, 1/\tau_w); \quad u_i \sim N(0, 1/\tau_u).$$

Anchoring constraints are used to define the scale of the factor scores, F_i and u_i. So $\lambda_1 = 1$ and $\kappa_2 = 1$, with the latter setting corresponding to a belief that arson is relatively distinct from the other variables in its spatial patterning.

---

*Data kindly provided by Dimitris Karlis.

Inferences are based on the second half of a 30,000 run of two chains. The posterior means (sd) of the unknown λ_p coefficients ($p = 2, 3, 4$) are, respectively, −0.32 (0.77), 0.49 (0.29), 0.87 (0.42). These loadings tend to confirm F as a positive crime construct with positive loadings on all crime variables except arson. The posterior mean F scores range from −1.28 to 1.44, with high F scores in prefectures with above average violent crime (such as prefecture 13), or where high violent crime is combined with smuggling. By contrast, low F scores occur in prefectures with little crime (prefecture 40), or in areas where arson is unduly elevated (e.g., prefecture 48). The z_i are relatively weak predictors of F_i, though the GDP coefficient (b_3) has a mainly positive 95% interval (−0.02, 0.34). The average scaled deviance of this model (274) indicates some residual dispersion, though the complexity is lower than model 1 at 37, so the DIC is not that much higher than model 1. Model checks are adequate: 11 of the 196 observations have mixed predictive p values under 0.05 or over 0.95.

The fact that this particular data reduction method did not yield a better fit may be taken to illustrate caveats to discrete data factor reduction, as also illustrated by Chib and Winkelmann (2001). They undertake a Poisson regression analysis of six health use outcomes, and conclude that "a flexible model with a full set of correlated latent effects is needed to adequately describe the correlation structure [in the regression residuals]."

Example 7.3. Political and Economic Risk: Factor Analysis for Mixed Ordinal and Continuous Responses Quinn (2004) considers a mix of continuous, binary, and ordinal indicators of a single latent construct, namely, political–economic risk in 62 countries. The first indicator (independence of judiciary) is binary, set to 1 if the judiciary is judged to be independent. The next indicator is continuous, the log of the black-market premium in each country where the premium is the black-market exchange rate. The third measure of political–economic risk is ordinal (with $M_3 = 6$ levels), measuring the lack of expropriation risk. The fourth indicator, also ordinal with $M_4 = 6$ levels, measures lack of corruption. The final indicator is productivity, measured by logged values of GDP per worker.

The latent construct has a preset variance of 1 for identification (according to a standardized factor rather than anchoring constraint), and the discrete response variables are analyzed using metric data augmentation[2]. In particular, the model for the ordinal outcomes, y_3 and y_4, has the form,

$$y_{pi} = m \text{ if } y^*_{pi} \in (\delta_{p(m-1)}, \delta_{pm}),$$
$$y^*_{pi} = \lambda_p F_i + \varepsilon_{pi},$$
$$\varepsilon_{pi} \sim N(0, 1),$$

where $\delta_{p0} = -\infty$ and $\delta_{pM} = \infty$. This is implemented using $M_3 - 1 = 5$ and $M_4 - 1 = 5$ binary regressions as discussed in Section 7.4.4. To ensure consistent labelling of the factor, the loading λ_3 of y_3 (lack of expropriation

TABLE 7.3
Political and economic risk factor model.

	Mean	St. devn.	MC st. error	2.5%	97.5%
α_1	−0.05	0.37	0.01	−0.80	0.70
κ_{31}	−3.85	0.50	0.02	−4.79	−2.88
κ_{32}	−2.60	0.40	0.02	−3.42	−1.88
κ_{33}	−1.75	0.34	0.01	−2.42	−1.10
κ_{34}	0.06	0.30	0.01	−0.53	0.64
κ_{35}	1.95	0.42	0.02	1.17	2.80
κ_{41}	−3.20	0.45	0.02	−4.11	−2.35
κ_{42}	−1.72	0.34	0.01	−2.41	−1.08
κ_{43}	0.04	0.31	0.01	−0.58	0.65
κ_{44}	1.48	0.39	0.02	0.77	2.29
κ_{45}	2.99	0.52	0.03	1.98	4.01
λ_1	−2.98	0.81	0.04	−4.68	−1.59
λ_2	0.77	0.12	0.00	0.56	1.02
λ_3	−2.21	0.32	0.02	−2.84	−1.60
λ_4	−2.34	0.38	0.02	−3.14	−1.63
λ_5	−0.72	0.12	0.00	−0.97	−0.50

threat) on the common factor F is constrained to be negative. A two chain run of 10,000 iterations shows early convergence, with parameter estimates close to those of Quinn (2004) (see Table 7.3).

Example 7.4. Maths Tests: Item Response Model This example compares item analysis (IRT) models to a full dimensional multivariate probit for binary item data from Tanner (1996) concerning maths aptitude; there are $n = 39$ students and $P = 6$ items. For the IRT models, a probit link is adopted and two- and three-parameter models compared. The augmented data two-parameter model[3] for subjects i and item p is

$$y^*_{pi} \backsim N(\alpha_p + \lambda_p F_i, 1)\ I(A_{pi}, B_{pi}),$$

where $F_i \sim N(0,1)$ and all λ_p are unknowns. Following Sahu (2002) and others, the λ_p are assigned $N(0,1)$ priors truncated to positive values. The sampling limits (A_{pi}, B_{pi}) depend on the observed y_{pi}. The model is checked by assessing whether replicates, $y_{\text{rep},pi}$, sampled from the model are concordant with actual values, y_{pi}, though one may also compare actual and predicted totals falling into particular item response patterns (Sahu, 2002). The three-parameter probit IRT model including difficulty parameters specifies $y_{pi} \sim Bern(\pi_{pi})$ with,

$$\pi_{pi} = c_p + (1 - c_p)\Phi(\alpha_p + \lambda_p F_i).$$

This model is estimated from the binary data likelihood with truncated normal priors assumed for λ_p and $Beta(1,3)$ priors for c_p.

Convergence on both models is obtained early in two chain runs of 5000 iterations with inferences based on the last 4500. The two-parameter model has log(psML) = −161.8, and posterior mean percentages of predictive concordance for the six items (52.6, 54.1, 60.8, 62.4, 65.2, 61.6). The lowest CPO (of 0.12) is for subject 28 and item 6. The three-parameter model provides log(psML) = −159, but predictive concordance is not improved for all items standing at (53.1, 55.6, 60.1, 60.1, 62.4, 60). The guess parameters, c_p, for the first two items both have posterior means of 0.275, but for the other items are below their prior means of 0.25.

The multivariate probit, as in Equation 7.4, is applied with unrestricted Wishart sampling of Σ^{-1}, so that parameters need to be scaled by the sampled standard deviations in Σ. The prior degrees of freedom and scale matrix in the Wishart are set at P and PI, respectively, so that $E(\Sigma) = I$ under the prior. Initially, the item means in the MVN model are simply taken as unknown constants, but predictive concordancies are higher when item means are extended to take a factor form, with $\alpha_p + \lambda_p F_i$ where $F_i \sim N(0, 1)$. This amounts to a factor analysis without assuming conditional independence since,

$$(y_{1i}^*, y_{2i}^*, \ldots, y_{pi}^*) \sim N_p(\alpha_p + \lambda_p F_i, R)\; I(A_i, B_i),$$

with R a correlation matrix with unknown off-diagonal terms. This extension improves predictions: mean percent concordancies from the last 4000 of a two chain run of 5000 iterations are $(54.4, 54.8, 59.5, 62.3, 61.7, 57.7)$ though the identified loadings, $\lambda_j^* = \lambda_j/\sqrt{\sigma_{jj}}$, are estimated less precisely than under the IRT models. None of the correlations in R has a 95% interval entirely confined to positive or negative values, so setting $R = I$ might be considered.

7.5 Robust Options in Multivariate and Factor Analysis

Real world data structures may exhibit segmentation and unobserved heterogeneity such that a single multivariate density or multivariate reduction model is inappropriate. Techniques are also required to ensure inferences in multivariate models are robust against outlier observations or departures from normality. A flexible method for dealing with heterogeneity, and one adapted to elucidating subpopulations with substantive meaning (e.g., in the case of market segmentation), involves the use of discrete mixtures (Jedidi et al., 1997; Arminger et al., 1999; Lubke and Muthen, 2005).

7.5.1 Discrete mixture multivariate models

Discrete mixtures are familiar as a technique in standard multivariate analysis, where the full covariance parameterization of the error structure is retained but extended to differ between K latent populations. While conceptually appealing, discrete mixture models in practice may encounter difficulties with label switching and assessing the optimal number of components. Such problems may be reduced under a nonparametric approach via Dirichlet process and other priors (Kottas et al., 2005).

To illustrate discrete parametric mixtures, let y_i be a $P \times 1$ vector of observations for subject i on P continuous variables. Let there be K mixture components with π_k denoting the prior probability of belonging to the kth component or latent class. In a discrete parametric mixture model, it is assumed that the density of y_i, $p(y_i|\theta)$, is a mixture of K multivariate densities, $p_k(y_i|\theta_k)$, with $\theta = (\theta_1, \ldots, \theta_K)$, namely,

$$p(y_i|\theta) = \sum_{k=1}^{K} \pi_k p_k(y_i|\theta_k).$$

Commonly used mixtures for multivariate continuous data, residuals, or random effects take each component density to be multivariate normal (Muller et al., 1996) with parameters $\theta_k = \{\mu_k, \Sigma_k\}$, or multivariate Student t with distinct degrees of freedom parameters, $\theta_k = \{\mu_k, \Sigma_k, \nu_k\}$ (Lin et al., 2004).

Bayesian estimation is facilitated by considering latent group allocation indicators, $S_i = k$ where $k \in (1, \ldots, K)$, and with consequent focus on the complete data likelihood. In the case of a multivariate normal mixture,

$$p(y_i|S_i = k) = N(\mu_{ik}, \Sigma_k),$$

where μ_{ik} may be modeled via regressions, $\mu_{ik} = X_i\beta_k$, on known predictors X_i, while probabilities, π_{ik}, that subjects belong to particular subpopulations may also be predicted (e.g., in a multinomial logit regression) by further known attributes, Z_i. Rossi et al. (2005, Section 5.5) consider the application of such models in analyzing market segmentation.

An unrestricted multivariate normal or student t mixture involves distinct covariance matrices for each component and for large P and/or large K this may imply a heavily parameterized model. Hence, simplifying assumptions such as $\Sigma_k = \Sigma$ may be suitable. Banfield and Raftery (1993) propose a parsimonious reparameterization of the component-specific covariance matrix via an eigenvalue decomposition,

$$\Sigma_k = \lambda_k E_k A_k E_k',$$

where A_k is a diagonal matrix scaled such that $|A_k| = 1$ and with elements proportional to the eigenvalues of Σ_k, E_k is a matrix of eigenvectors, and $\lambda_k = |\Sigma_k|^{1/p}$.

7.5.2 Discrete mixtures in SEM and factor models

Parsimony as well as enhanced substantive knowledge are the goals of discrete mixture factor and structural equation models. Discrete mixture factor analysis and SEM models have similar features to multigroup methods where the population exhibits a known, rather than latent, segmentation based typically on demographic and psychometric variables (Jedidi et al., 1997; Lee and Song, 2003; Zhu and Lee, 2001). Under a discrete mixture factor analysis model for normal metric data one has, conditioning on membership $S_i = k$ of the kth component ($k = 1, \ldots, K$),

$$y_i = \alpha_k + \Lambda_k F_{ki} + u_{ki}, \tag{7.5}$$

where the factor scores for subject i in the kth group, $F_{ki} = (F_{k1i}, \ldots, F_{kQi})'$, are of dimension $Q \leq P$. With normality assumptions for each component,

$$(u_{k1i}, \ldots, u_{kPi})' \sim N_P(0, \Sigma_k),$$

and

$$(F_{k1i}, \ldots, F_{kQi})' \sim N_Q(0, \Phi_k),$$

where the form of Φ_k is subject to the identifying restrictions imposed on Λ_k, and vice versa. Under conditional independence, Σ_k is diagonal with elements $\{\sigma^2_{kj}, j = 1, \ldots, P\}$. The unconditional likelihood for the kth subpopulation is

$$p_k(y_i|\theta_k) = N(\alpha_k, \Lambda_k \Phi_k \Lambda'_k + \Sigma_k),$$

where $\theta_k = (\alpha_k, \Lambda_k, \Phi_k, \Sigma_k)$, and the entire observed data likelihood is

$$p(y_i|\theta) = \sum_{k=1}^{K} \pi_k N(\alpha_k, \Lambda_k \Phi_k \Lambda'_k + \Sigma_k),$$

where $\theta = (\theta_1, \ldots, \theta_K)$. Restricted covariance structures can be obtained (Lee, 2007, 322; Zhu and Lee, 2001, 139) by making some parameters or covariance matrices invariant across subpopulations.

Bayesian estimation is facilitated by considering latent multinomial allocation indicators, $S_i = k$ where $k \in (1, \ldots, K)$, with prior probabilities, $Pr(S_i = k) = \pi_k$, and the complete or augmented data likelihood,

$$p(y_i|S_i, \theta) = N(\alpha_k + \Lambda_k F_{ki}, \Sigma_k).$$

As mentioned in Chapter 3, an additional issue involved in Bayesian discrete mixture estimation via MCMC methods is that of unique labeling. In this regard, Lee and Song (2003) suggest an identifying constraint via ordering $\alpha_1 < \alpha_2 < \cdots < \alpha_K$ of component locations. For purposes of MCMC sampling, the conditional posterior for the allocation indicators, S_i, is

$$p(S_i = k|y_i, \theta) = \frac{\pi_k p_k(y_i|\theta_k)}{p(y_i|\theta)},$$

and so multinomial simulation is straightforward. It is simplest (Zhu and Lee, 2001) to apply a conjugate prior structure where the π vector follows a symmetric Dirichlet density with prior vector $(d, d, \ldots, d)$. The conditional update for the subgroup probabilities π is then Dirichlet with elements $d+n_k$, where n_k is the number allocated to the kth subpopulation at a particular MCMC iteration.

As well as facilitating estimation, assessing the fit of different models and the appropriateness of their assumptions (e.g., normality or otherwise of construct scores and measurement errors) is arguably more straightforward using augmented data. For example, predictive discrepancies using the complete likelihood are mentioned by Zhu and Lee (2001, 141). For replicate data, y_i^{rep}, sampled from the model, one possible discrepancy criterion is

$$D = \sum_{k=1}^{K} \sum_{S_i=k} (y_i^{\text{rep}} - \alpha_k - \Lambda_k F_{ki})' \Sigma_k^{-1} (y_i^{\text{rep}} - \alpha_k - \Lambda_k F_{ki}).$$

Since D is chi-square with nP degrees of freedom, one may use the predictive check at each MCMC iteration,

$$Pr(\chi^2_{nP} \geq D^{(t)}),$$

and assess the posterior probability over a long MCMC sampling run.

A discrete mixture structural equation model extends Equation 7.5 to both endogenous and exogenous manifest indicators, and to specify a structural model also specific to each subpopulation or component. Consider a $P_1 \times 1$ vector, y_i, of endogenous indicators, and a $P_2 \times 1$ vector, x_i, of exogenous indicators. Conditional on membership of the kth subpopulation, one has measurement models,

$$y_i = \alpha_{1k} + \Lambda_{1k} F_{ki} + u_{ki},$$
$$x_i = \alpha_{2k} + \Lambda_{2k} H_{ki} + w_{ki},$$

where u_k and w_k are normal with mean zero and covariances Σ_{1k} and Σ_{2k}. The endogenous indicators measure an endogenous construct vector $F_{ki} = (F_{k1i}, \ldots, F_{kQ_1 i})'$, of dimension $Q_1 < P_1$, and the exogenous indicators measure the exogenous construct vector, H_{ki}, of dimension Q_2, assumed to be $N_{Q_2}(0, \varphi_k)$. The structural model specifies,

$$F_{ki} = a_k + B_k F_{ki} + C_k H_{ki} + e_{ki},$$

where $e_{ki} \sim N_{Q_1}(0, \psi_k)$, B_k is $Q_1 \times Q_1$ with a diagonal of zeroes, and C_k is $Q_1 \times Q_2$ containing regression effects of exogenous on endogenous constructs in the kth subpopulation. Letting $B_{0k} = I - B_k$, the covariance matrix of (F_{ki}, H_{ki}) is given (Lee, 2007, 322) by,

$$\Omega_k = \begin{bmatrix} B_{0k}^{-1}(C_k \varphi_k C_k' + \psi_k)(B_{0k}^{-1})' & B_{0k}^{-1} C_k \varphi_k \\ \varphi_k C_k' (B_{0k}^{-1})' & \varphi_k \end{bmatrix},$$

and the observed data likelihood in the kth subpopulation has covariance $\Lambda_k \Omega_k \Lambda_k + \Sigma_k$, where $\Sigma_k = \text{diag}(\Sigma_{1k}, \Sigma_{2k})$ and $\Lambda_k = \text{diag}(\Lambda_{1k}, \Lambda_{2k})$.

7.5.3 Robust density assumptions in factor models

Discrete mixture estimation can be demanding with questions of the appropriate number of components, K, and labeling issues. An alternative to discrete mixtures, for instance in relation to robust random error or factor specification, involves the use of heavy-tailed or skew densities (Ando, 2009; Yuan et al., 2004). Consider the linear factor reduction model (Equation 7.1), namely, $y_i = \alpha + \Lambda F_i + u_i$. Instead of conventional normality assumptions for residuals $u_i = (u_{1i}, \ldots, u_{Pi})'$ or factor scores $F_i = (F_{1i}, F_{2i}, \ldots, F_{Qi})'$, one might use options that are robust to measurement or construct outliers. For example, a Student t model with ν_{1p} degrees of freedom for the measurement model regressions for y_{pi} is obtainable via scale mixing, with,

$$y_{pi} \sim N(\alpha_p + \Lambda_p F_i, \sigma_p^2/\zeta_{pi}),$$
$$\zeta_{pi} \sim \text{Ga}(0.5\nu_{1p}, 0.5\nu_{1p}).$$

To identify possible outliers, one may monitor the lowest weights, ζ_{pi}. Assume also standardized and uncorrelated factor scores, but following a Student t rather than normal density. Then the corresponding heavy-tailed construct score model is

$$F_{qi} \sim N(0, 1/\omega_{qi}),$$
$$\omega_{qi} \sim \text{Gamma}(0.5\nu_{2q}, 0.5\nu_{2q}).$$

Skewness in outcomes or factor scores may also be present. Following Azzalini (1985), let f and g be symmetric probability density functions, with G being the cumulative distribution function (cdf) associated with g. Then for location parameter μ and scale parameter σ, the density,

$$\frac{2}{\sigma} f\left(\frac{x-\mu}{\sigma}\right) G\left(\kappa\frac{x-\mu}{\sigma}\right),$$

is a skew pdf for any κ. If $f = \phi$ and $G = \Phi$, one obtains the skew-normal distribution. Positive (negative) values of κ indicate positive (negative) skewness, while $\kappa = 0$ provides the normal density. Bazan et al. (2006) consider application of the skew-normal density in item analysis. For binary items, $p = 1, \ldots, P$, and with $\delta_p = (\kappa_p/\sqrt{(1+\kappa_p^2)})$, they define a skew probit IRT model involving a common factor, F_i, and item-specific effects, V_{pi}, to allow for skew errors. So,

$$y_{pi}^* \backsim N(\alpha_p + \lambda_p F_i + \delta_p V_{pi}, 1 - \delta_p^2)\ I(A_{pi}, B_{pi}),$$
$$F_i \sim N(0,1); \quad V_{pi} \sim HN(0,1),$$

with sampling limits $\{A_{pi}, B_{pi}\}$ defined according to the observed binary responses. This parameterization necessitates priors for δ_p in the interval $[-1, 1]$.

Example 7.5. Market Segmentation: Discrete Mixture SEM This example follows the scenario provided by Jedidi et al. (1997, Figure 1).

The unobserved univariate affect, F, for a new food product depends on two latent product dimensions, sweetness (H_1) and richness (H_2). There are $K = 2$ unobserved consumer segments, of equal size $\pi_k = 0.5$, the first being "pleasure seeking" (with latent group indicator $S_i = 1$) and the other "health conscious" ($S_i = 2$). Sweetness and richness have positive impacts on affect for the first group, but negative impacts on affect for the second group. Thus,

$$F_{S_i,i} = \gamma_{S_i,1} H_{1i} + \gamma_{S_i,2} H_{2i} + e_{S_i,i},$$

where $e_{ki} \sim N(0, 0.5)$, and there are positive impacts, $\gamma_{11} = \gamma_{12} = 0.5$, for the first consumer group, but $\gamma_{21} = \gamma_{22} = -0.5$ for the second. The response construct, namely, affect, and the exogenous constructs, sweetness and richness, are each measured by two observed metric variables. Thus,

$$\begin{aligned} x_{1i} &= \alpha_{x1} + \lambda_{x11} H_{1i} + \lambda_{x12} H_{2i} + w_{1i}, \\ x_{2i} &= \alpha_{x2} + \lambda_{x21} H_{1i} + \lambda_{x22} H_{2i} + w_{2i}, \\ x_{3i} &= \alpha_{x3} + \lambda_{x31} H_{1i} + \lambda_{x32} H_{2i} + w_{3i}, \\ x_{4i} &= \alpha_{x4} + \lambda_{x41} H_{1i} + \lambda_{x42} H_{2i} + w_{4i}, \end{aligned}$$

where $w_{pi} \sim N(0, 0.5)$, and the intercepts are $\alpha_{x1} = \alpha_{x2} = \alpha_{x3} = \alpha_{x4} = 0$. Nonzero loadings are $\lambda_{x11} = \lambda_{x21} = \lambda_{x32} = \lambda_{x42} = 1$, and zero loadings are $\lambda_{x12} = \lambda_{x22} = \lambda_{x31} = \lambda_{x41} = 0$. The exogenous constructs (H_{1i}, H_{2i}) follow a bivariate normal density with mean zero and covariance terms, $\varphi_{11} = 1$, $\varphi_{22} = 1, \varphi_{12} = \varphi_{21} = 0.5$. Also,

$$\begin{aligned} y_{1i} &= \alpha_{y1} + \lambda_{y1} F_i + u_{1i}, \\ y_{2i} &= \alpha_{y2} + \lambda_{y2} F_i + u_{2i}, \end{aligned}$$

where $\alpha_{y1} = \alpha_{y2} = 0; \lambda_{y11} = \lambda_{y21} = 1$, and $u_{pi} \sim N(0, 0.5)$.

The observed data generated under this scheme are the y and x variables. First of all, a single population SEM is applied to the sampled data with structural model,

$$F_i = \gamma_1 H_{1i} + \gamma_2 H_{2i} + e_i,$$

and $\text{var}(e_i) = 1$ for identification. The H_{qi} are in standard form, taken to be bivariate normal with correlation matrix containing an unknown φ_{12}, with both γ coefficients then being unknowns that are assigned $N(0, 10)$ priors. The λ_x coefficients are taken to define a confirmatory model, and so only $\{\lambda_{x11}, \lambda_{x21}, \lambda_{x32}, \lambda_{x42}\}$ are taken as unknown (i.e., to have possibly nonzero values).

This simple structural model aggregates over two distinct subpopulations (of equal size) and as might be expected, the γ coefficients are estimated as effectively zero (with 95% credible intervals straddling zero) from the last 20,000 of a two chain run of 25,000 iterations. Otherwise, the original parameter structure is reproduced; for example, the posterior means for $\{\lambda_{y1}, \lambda_{y2}\}$ are $\{1.37, 0.98\}$, and for $\{\lambda_{x11}, \lambda_{x21}, \lambda_{x32}, \lambda_{x42}\}$ are $\{1.05, 1, 1.23, 0.82\}$, while for $\text{corr}(H_1, H_2)$ the posterior mean is 0.40.

For the two population mixture SEM[4], an identifying constraint is imposed (to produce unique labeling for H scores), so for $S_i = k$,

$$F_{ki} = \gamma_{k1}H_{1i} + \gamma_{k2}H_{2i} + e_{ki},$$
$$\gamma_{1k} \sim N(0,10)I(0,), \gamma_{2k} \sim N(0,10)I(,0).$$

A Dirichlet prior with unity weights is assumed for the prior subpopulation proportions. With the same run lengths as for the single population SEM, the estimated structural parameters clearly distinguish the two populations. Thus, the posterior means parameters $\{\gamma_{11}, \gamma_{12}\}$ in the pleasure-seeking population are $\{0.84, 0.59\}$ with entirely positive credible intervals, while for the health-conscious population the parameters $\{\gamma_{21}, \gamma_{22}\}$ have means $\{-0.76, -0.73\}$. The subpopulation probabilities are estimated at 0.52 and 0.48, with the correlation φ_{12} between sweetness and richness estimated at 0.46.

Example 7.6. Greek Crimes by Prefecture: Nonparametric Prior for Random Effects The analysis of the Greek crime data in Example 7.2 assumed normally distributed errors, u_{pi}, in the log link model for the crime rates, ρ_{pi}, As noted by Knorr-Held and Rasser (2000), a fully parametric specification of the random effects distribution may result in oversmoothing, and in masking local discontinuities especially when the true distribution is characterized by a finite number of locations. Here a truncated Dirichlet process prior is adopted to model the density of the residuals, u_{pi}, with potential values $\{u^*_{pk}, p = 1, \ldots, P\}$ from K clusters centered on the multivariate normal, $G_0 = N_P(0, D)$, where $P = 4$. D^{-1} has a Wishart prior with identity scale matrix and P degrees of freedom.

Thus the infinite DP representation is approximated by one truncated at $K \leq n$ components, with appropriate values, u_{pi}, for prefecture i chosen according to an allocation indicator, $S_i \in (1, \ldots, K)$. The probabilities, π_k, of allocation to clusters $\{1, \ldots, K\}$ are determined by $K - 1$ beta distributed random variables, $V_k \backsim Beta(1, \kappa)$, with unknown concentration parameter κ and $V_K = 1$ to ensure the random weights, π_k, sum to 1 (Ishwaran and James, 2001; Sethuraman, 1994). Then, $\pi_1 = V_1$ and

$$\pi_k = (1 - V_1)(1 - V_2)\cdots(1 - V_{k-1})V_k \quad k > 1.$$

Following Ishwaran and Zarepour (2000, 377), the gamma prior for κ, namely, $\kappa \sim \text{Ga}(\nu_1, \nu_2)$, has relatively large ν_1 and ν_2, with ν_2 set larger than ν_1. Such a setting discourages small and large values for κ. Here, $\nu_2 = 4$ and $\nu_1 = 2$. The maximum possible clusters is set at $K = 20$.

A two chain run of 7500 iterations[5] shows early convergence and replicates Example 7.2 in showing mostly nonsignificant predictor effects. There are, however, significant positive effects of unemployment on rape, and of urban center on manslaughter, and a significant negative effect of GDP levels on manslaughter. The posterior mean deviance $\overline{Dev}$ is 231 with 81 effective parameters, giving a DIC of 312. This is obtained by comparing $\overline{Dev}$ with

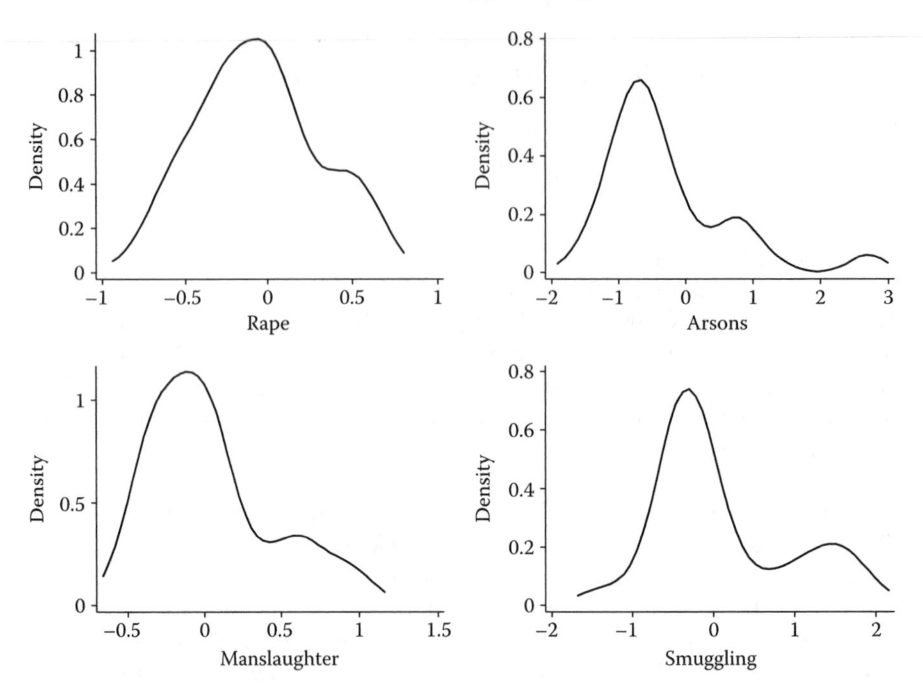

FIGURE 7.1
Residual plots, DPP mixture model.

the deviance at the posterior means of the Poisson means μ_{pi}. The posterior mean for κ is 1.32 (from the second half of the MCMC iterations) with the average number of nonempty clusters, K^*, being 8.45. Kernel smoothed plots of the posterior mean residuals demonstrate greater apparent non-normality for the last three outcomes (see Figure 7.1, which uses a Gaussian kernel with optimal half width).

Example 7.7. Maths Aptitude: Skew Probit A skew probit link is adopted for the data from Tanner (1996) following the approach of Bazan et al. (2006). So latent metric data are sampled according to,

$$y^*_{pi} \backsim N(\alpha_p + \lambda_p F_i + \delta_p V_{pi}, 1 - \delta_p^2),$$

with $F_i \sim N(0, 1)$ and λ_p all being unknowns. A $U(-1, 1)$ prior is adopted on the δ_p parameters, and,

$$\lambda_p \sim N(1, 0.5)\ I(0,),$$

providing an identifying constraint (cf. Sahu, 2002).

A two chain run of 10,000 iterations (with convergence at under 1,000) shows none of the δ_p (and hence κ_p) parameters to be significantly positive or negative. Despite the apparent absence of skew, this model has log(psML)= −149, and posterior mean percentages of predictive concordance for the six items raised to (57.3, 58.5, 67.3, 66.6, 70.6, 67.0). The gain in performance over the analysis in Example 7.4 may be related to the introduction of subject-item level parameters, V_{pi}.

7.6 Multivariate Spatial Priors for Discrete Area Frameworks

Consider multivariate spatial responses $(y_{1i}, \ldots, y_{Pi})'$ of dimension P from an exponential family density observed over n discrete areas (e.g., administrative regions). Conditional on spatial effects, $s_i = (s_{1i}, \ldots, s_{Pi})'$, of the same dimension, and predictors, $x_i = (x_{1i}, \ldots, x_{Ri})'$, one then has,

$$p(y_{pi}|s_{pi}, x_i) \propto \exp\left\{\frac{y_{pi}\theta_{pi} - b(\theta_{pi})}{\phi_{pi}} + c(y_{pi}, \phi_{pi})\right\},$$

where θ_{pi} is the canonical parameter, and ϕ_{pi} a known scale. Denoting regression terms as $\eta_{pi} = g(\theta_{pi})$ with link g, such a term is likely to include a random effect, s_{pi}, to measure spatially configured but unmeasured predictors. So one has at a minimum the representation,

$$\eta_{pi} = \alpha_p + \beta_p x_i + s_{pi},$$

where the spatial effects for area i, $s_i = (s_{1i}, \ldots, s_{Pi})'$, follow a multivariate spatial prior. For certain definitions of spatial effects, it may be appropriate to also include unstructured (i.e., exchangeable over areas) multivariate effects, in line with a multivariate form of the Besag et al. (1991) convolution prior. Thus, the full dimension analogue to the convolution prior is

$$\eta_{pi} = \alpha_p + \beta_p x_i + s_{pi} + u_{pi}, \tag{7.6}$$

where u_{pi} also follow a multivariate prior. Other possibilities, following Chapter 5, include regression effects, β_{pi}, which vary spatially as well as over response variables.

Conditions for a valid multivariate spatial prior, specifically a multivariate Gaussian Markov random field (MGMRF), are discussed by Rue and Held (2005, Section 2.2) and Banerjee et al. (2004, Section 7.4). Thus, denote the nP length vector over all areas as $s = (s_1, \ldots, s_n)'$, and denote the mean vector, possibly including regression effects, as $\mu_i = (\mu_{1i}, \ldots, \mu_{Pi})$, with $\mu = (\mu_1, \ldots, \mu_n)'$. Also denote the matrix describing observed spatial interactions in the region by $W = [w_{ij}]$, with $w_{ij} = w_{ji}$, and set $D = \text{diag}(d_1, \ldots, d_n)$, where $d_i = \sum_{j \neq i} w_{ij}$. If $w_{ij} = 1$ when areas i and j are contiguous, and zero otherwise, then d_i is the number of neighbors for area i. The neighborhood for area i is often denoted ∂_i, and if area j is a neighbor of area i, then the neighbor relation (under binary interaction) is denoted $j \sim i$.

The joint density for a normal MGMRF for P spatial effects and with $nP \times nP$ precision matrix Q may be expressed,

$$p(s|Q) = \left(\frac{1}{2\pi}\right)^{nP/2} |Q|^{0.5} \exp[(s-\mu)'Q(s-\mu)]$$
$$= \left(\frac{1}{2\pi}\right)^{nP/2} |Q|^{0.5} \sum_{ij} \exp[(s_i-\mu_i)'Q_{ij}(s_j-\mu_j)].$$

Q is block diagonal with $P \times P$ submatrix elements Q_{ij} that are nonzero (zero) if area j is (is not) a neighbor of area i. Retaining the possibility of a regression model in the means μ_i (Rue and Held, 2005), the corresponding full conditional density is

$$s_i|s_{[i]} \sim N\left(\mu_i - Q_{ii}^{-1}\sum_{j\neq i} Q_{ij}(s_j-\mu_j), \frac{1}{Q_{ii}}\right),$$

with conditional precision matrices,

$$\text{prec}(s_i|s_{[i]}) = Q_{ii} = \Delta_i.$$

Equivalently, define $P \times P$ matrices, $B_{ij} = -Q_{ij}/Q_{ii}$, with $B_{ii} = 0$, and $\Delta_i = Q_{ii}$. Then,

$$E(s_i|s_{[i]}) = \mu_i + \sum_{j\neq i} B_{ij}(s_j-\mu_j), \tag{7.7}$$
$$\text{prec}(s_i|s_{[i]}) = \Delta_i.$$

Most commonly the μ_i are set to zero.

Under the parameterization in Equation 7.7, the joint density has mean μ and precision matrix $Q = \Delta(I-B)$, where Δ is block diagonal with blocks Δ_i, and the $nP \times nP$ matrix B is block diagonal with (i,j)th block B_{ij}. The requirements for a valid joint density to exist (e.g., if specification is starting from a prior involving the full conditionals) are that (Sain and Cressie, 2007; Rue and Held, 2005, 31),

$$\Delta_i B_{ij} = \Delta_j B_{ji}.$$

For example, setting $B_{ij} = [w_{ij}/d_i]I_{P\times P}$, and $\Delta_i = d_i\zeta$ (where ζ is a $P \times P$ within area precision matrix) will ensure a valid joint density.

A number of multivariate priors that incorporate spatial dependence between areas have been proposed. The generalization of the intrinsic univariate CAR to a multivariate setting, is denoted as the multivariate CAR or MCAR prior (Mardia, 1988; Jin et al., 2005, Equation 6; Song et al., 2005, 254). This takes the vector of multivariate area effects, s, as multivariate normal with mean consisting of a vector of zeroes of length nP, and with $nP \times nP$ precision matrix, $Q = (D - \alpha W) \otimes \zeta$, namely,

$$p(s|\zeta, \alpha) = \left(\frac{1}{2\pi}\right)^{nP/2} |D - \alpha W|^{P/2} |\zeta|^{n/2} \exp\left[-\frac{1}{2} s' Q s\right], \tag{7.8}$$

where $\alpha \in (0, 1)$ is a propriety parameter. The $P \times P$ positive definite symmetric matrix, ζ^{-1}, describes covariation between the outcomes, and $D - \alpha W$ is the precision matrix for the spatial effects. The latter matrix can also be written as $D(I - \alpha B)$ where $B = D^{-1}W$. Let the effects be arranged by variable rather than subject, so that $S_1 = (s_{11}, s_{12}, \ldots, s_{1n})'$, $S_2 = (s_{21}, s_{22}, \ldots, s_{2n})'$, etc., then for $P = 2$, the joint prior is

$$\begin{pmatrix} S_1 \\ S_2 \end{pmatrix} \sim N\left(\begin{pmatrix} 0 \\ 0 \end{pmatrix}, \begin{bmatrix} \zeta_{11}(D - \alpha W) & \zeta_{12}(D - \alpha W) \\ \zeta_{12}(D - \alpha W) & \zeta_{22}(D - \alpha W) \end{bmatrix}^{-1}\right),$$

where each submatrix, $\zeta_{pq}(D - \alpha W)$, is of dimension $n \times n$.

The conditional prior under Equation 7.8 for s_i given the remaining effects, $s_{[i]} = (s_1, \ldots, s_{i-1}, s_{i+1}, \ldots, s_n)$, is multivariate normal with means $E(s_i|s_{[i]}) = (M_{1i}, \ldots, M_{Pi})$, where,

$$E(s_{pi}|s_{[i]}) = M_{pi} = \alpha \sum_{j \neq i} w_{ij} s_{pj} \Big/ \sum_{j \neq i} w_{ij},$$

and with precisions,

$$\text{prec}(s_i|s_{[i]}) = d_i \zeta.$$

If w_{ij} are set to 1 for neighboring areas and to 0 otherwise, then $M_{pi} = \alpha \sum_{j \in \partial_i} s_{pj}/d_i$ are locality averages of the spatial effect for the pth response. Setting $\alpha = 1$ provides the multivariate version of the intrinsic CAR prior of Besag et al. (1991); such intrinsic GMRFs (for spatial and nonspatial priors) are considered by Rue and Held (2005, Chapter 3).

MacNab (2007) discusses a multivariate extension of the prior of Leroux et al. (1999), which allows the data to determine the appropriate mix between spatial or exchangeable dependence. Hence, if a mix of local and global smoothing is regarded as a reasonable prior structure, this may be achieved with a single set of random effects, r_{pi}, rather than the two sets $\{s_{pi}, u_{pi}\}$ present in the multivariate extension (Equation 7.6) of the convolution prior. Thus, with $r_i = (r_{1i}, \ldots, r_{Pi})'$, parameter $\kappa \in (0, 1)$, and spatial interactions $W = [w_{ij}]$,

$$E(r_i|r_{[i]}) = [M_{1i}, \ldots, M_{pi}] = \kappa \sum_{j \neq i} w_{ij} I_P r_j \Big/ \left[1 - \kappa + \kappa \sum_{j \neq i} w_{ij}\right],$$

$$\text{prec}(r_i|r_{[i]}) = \left[1 - \kappa + \kappa \sum_{j \neq i} w_{ij}\right] \zeta,$$

where, as above, ζ is of dimension $P \times P$. Thus,

$$B_{ij} = \left[\frac{\kappa w_{ij}}{[1-\kappa+\kappa\sum_{j\neq i} w_{ij}]}\right] I_{P\times P},$$

$$\Delta_i = \left[1-\kappa+\kappa\sum_{j\neq i} w_{ij}\right]\zeta,$$

and $\Delta_i B_{ij} = \Delta_j B_{ji}$ holds. When w_{ij} are binary adjacency indicators, the conditional expectations become,

$$E(r_{pi}|r_{[i]}) = M_{pi} = \frac{\kappa\sum_{j\in\partial_i} r_{pj}}{[1-\kappa+\kappa d_i]}.$$

Define,

$$H = \text{diag}\left(1-\kappa+\kappa\sum_{j\neq 1} w_{1j},\ldots,1-\kappa+\kappa\sum_{j\neq n} w_{nj}\right) = (1-\kappa)I_n + \kappa D.$$

Then the joint density is multivariate normal with mean vector 0 and $nP \times nP$ precision matrix $(H - \kappa W)\otimes\zeta$.

Jin et al. (2005) propose a generalized MCAR (GMCAR) model whereby the joint distribution for a multivariate spatial effect is obtained by specifying a sequence of conditional and marginal models. Let effects be arranged by variable rather than subject. Then for a bivariate spatial effect with $P(S_1, S_2) = p(S_1|S_2)P(S_2)$, where $S_1 = (s_{11}, s_{12}, \ldots, s_{1n})'$ and $S_2 = (s_{21}, s_{22}, \ldots, s_{2n})'$, one has,

$$\begin{pmatrix} S_1 \\ S_2 \end{pmatrix} \sim N\left(\begin{pmatrix} 0 \\ 0 \end{pmatrix}, \begin{bmatrix} \Sigma_{11} & \Sigma_{12} \\ \Sigma_{12} & \Sigma_{22} \end{bmatrix}\right),$$

where $E(S_1|S_2) = \Sigma_{12}\Sigma_{22}^{-1}S_2$, and $\text{var}(S_1|S_2) = \Sigma_{11.2} = \Sigma_{11} - \Sigma_{12}\Sigma_{22}^{-1}\Sigma'_{12}$. Hence with $G = \Sigma_{12}\Sigma_{22}^{-1}$, one has equivalently,

$$\begin{pmatrix} S_1 \\ S_2 \end{pmatrix} \sim N\left(\begin{pmatrix} 0 \\ 0 \end{pmatrix}, \begin{bmatrix} \Sigma_{11.2} + G\Sigma_{22}G' & G\Sigma_{22} \\ (G\Sigma_{22})' & \Sigma_{22} \end{bmatrix}\right).$$

To specify the joint distribution of S_1 and S_2, it is therefore necessary to specify the matrices $\Sigma_{11.2}$, Σ_{22}, and G.

Jin et al. take $\Sigma_{11.2}^{-1} = \tau_1[D - \alpha_1 W]$, $\Sigma_{22}^{-1} = \tau_2[D - \alpha_2 W]$, and $G = \gamma_0 I + \gamma_1 W$. The marginal joint prior for the second set of effects is then,

$$S_2 \sim N(0, \tau_2^{-1}[D - \alpha_2 W]^{-1}),$$

and the conditional prior for the first set of effects is

$$S_1|S_2 \sim N(GS_2, \tau_1^{-1}[D - \alpha_1 W]^{-1}).$$

As above, the $0 < \alpha_j < 1$ are propriety parameters, and the γ_0 parameter links different variable-same area effects, namely, regresses s_{1i} on s_{2i}, while γ_1 links s_{1i} with other variable-other area effects $\{s_{2j}, j \neq i\}$. This approach is possibly more suitable for small P, as $P!$ conditional density sequences are possible, and may give different inferences or fits—though Jin et al. (2005, 957) demonstrate how initial regression analysis may lead one to prefer one sequence to another[6].

7.7 Spatial Factor Models

When high correlations are evident in ζ^{-1}, common spatial factor models may be more parsimonious (Congdon et al., 2007; Liu et al., 2005; Tzala and Best, 2007). Standard presentations of the normal and general linear factor models assume factor scores are independent over subjects, though in fact they might be spatially or temporally structured. So for P outcomes, Q factors, and for $P > Q$ in a spatial application, the F scores of dimension Q may be correlated over both variables and areas. Then for Poisson or binomial responses with mean $\mu_{pi} = g^{-1}(\eta_{pi})$ for the pth dependent variable and area i, one might have a regression term,

$$\eta_{pi} = \alpha_p + \beta_p x_i + \Lambda_p F_i,$$

where the vector Λ_p is of dimension Q, and the factor score variables $F_i = (F_{1i}, \ldots, F_{Qi})'$ are spatially dependent over areas i, as well as mutually intercorrelated. For example, a MCAR prior would specify the joint pairwise difference density for the factor scores,

$$p(F \mid \Sigma_F) \propto \mid \Sigma_F \mid^{-n/2} \exp\left[-0.5 \sum_{i,j} w_{ij}(F_i - F_j)' \Sigma_F^{-1} (F_i - F_j) \right].$$

As in other factor models, constraints are required to deal with location, scale, and rotational indeterminacy. In the multivariate CAR model for $F_i = (F_{1i}, \ldots, F_{Qi})'$, the location is fixed in practice by centering each of the Q sets of spatial factor scores at each MCMC iteration. Scale may be determined by fixing the Q variances of the F_{qi} scores at 1, or by fixing one of the loadings $(\lambda_{1q}, \ldots, \lambda_{Pq})$ linking the P manifest indicators to the qth factor. Additional loadings would need to be fixed to avoid rotational indeterminacy, typically $\lambda_{pq} = 0$ for $q > p$. For example, if $Q = 2$, and the variances of the F scores are free parameters, then the two loadings, λ_{qq}, may be set to 1 to define the scale, while rotational invariance is avoided by setting $\lambda_{12} = 0$.

A simultaneous errors approach to a spatial SEM is provided by Oud and Folmer (2008). Let $W = [w_{ij}]$ denote a row standardized spatial interaction matrix (Anselin and Hudak, 1992, 514). The structural model interrelating the factor scores has a MIMIC form involving known exogenous variables, X_i, and the spatially lagged transform (weighted average of neighboring values) of a Q dimensional factor score vector F_i,

$$F_i = \rho W F_i + \gamma X_i + \zeta_i,$$

where ζ_i is an error vector with preset variances, and ρ has maximum value 1 and can be taken to have minimum 0.

Example 7.8. Psychiatric Hospitalizations in England This example contrasts a full dimension covariance model for multivariate spatial outcomes with a common spatial factor approach. It considers gender-specific counts of hospitalizations for schizophrenia and bipolar disorder for 354 English local authorities over 2002–2003 to 2004–2005 for patients aged 15 to 64. The $P = 4$ outcomes are counts y_{pi} of male schizophrenia, female schizophrenia, male bipolar disorder, and female bipolar disorder; offsets are expected hospitaliations, E_{pi}, based on England-wide hospitalization rates specific to gender and five-year age bands (15–19, 20–24, etc.).

The first model applied is the multivariate generalization of the Leroux et al. (1999) conditional autoregressive prior (LMCAR) under a Poisson likelihood, and with binary adjacencies, w_{ij}. Then the first regression involves a constant term and spatially configured effects $(r_{1i}, \ldots, r_{Pi})$ that are of the same dimension as the response vector,

$$y_{pi} \sim Po(E_{pi} e^{\eta_{pi}}),$$
$$\eta_{pi} = \alpha_p + r_{pi}.$$

The conditional mean of r_{pi} is

$$E(r_{pi}|r_{[i]}) = M_{pi} = \frac{\kappa \sum_{k \in \partial_i} r_{pk}}{[1 - \kappa + \kappa d_i]},$$

where κ is between 0 and 1. A Wishart prior for the conditional precision matrix, ζ, with prior mean covariance I, is assumed[7].

Early convergence is attained in a two chain run of 5000 iterations, with the last 4000 showing a mean scaled deviance of 1474 (compared to $4 \times 354 = 1416$ observations), with $d_e = 1124$, and a posterior mean (sd) for κ of 0.66 (0.09) Setting $\Phi = \zeta^{-1}$, mean correlations, $r_{jk} = \Phi_{jk}/(\Phi_{jj}\Phi_{kk})^{0.5}$, between the spatial effects vary from 0.66 (between y_2 and y_3) to 0.87 (between the two bipolar disorder outcomes y_3 and y_4). The correlation between the two schizophrenia outcomes is also high, namely, 0.86.

Only one observation is not contained within the 95% intervals of replicate data sampled from the model, suggesting that a less heavily parameterized model could be found still giving a satisfactory fit. A common factor model is therefore applied with,

$$\eta_{pi} = \alpha_p + \lambda_p F_i,$$

where F_i follows a univariate (LCAR) prior. Thus for $\kappa \in (0, 1)$, conditional precision τ_F, and with $F_{[i]} = (F_1, \ldots, F_{i-1}, F_{i+1}, \ldots, F_n)$ and binary spatial interactions $W = [w_{ij}]$,

$$E(F_i|F_{[i]}) = \frac{\kappa \sum_{j \in \partial_i} F_j}{[1 - \kappa + \kappa d_i]},$$

and

$$\text{prec}(F_i|F_{[i]}) = [1 - \kappa + \kappa d_i]\tau_F.$$

With $\tau_F = \sigma_F^{-2}$ taken as unknown, with prior $\sigma_F \sim U(0, 1000)$, one of the loadings λ_j must be fixed for identification, and accordingly $\lambda_1 = 1$. This model has a posterior mean scaled deviance of 8125 and only 69% coverage of the observations by 95% intervals of replicate data sampled from the model.

To improve fit, a two factor analysis is applied. The revised model is confirmatory in the sense that certain loadings are set to zero. The full dimension error model showed high correlations between male and female outcomes with the same diagnosis, and the confirmatory model framework accordingly involves two diagnosis-specific factors, F_{1i} for schizophrenia and F_{2i} for bipolar disorder. These are, however, assumed to be correlated under a bivariate LMCAR prior. The loadings $\{\lambda_{31}, \lambda_{41}, \lambda_{12}, \lambda_{22}\}$ are set to zero in line with a confirmatory analysis.

To further ensure adequate predictive fit, outcome-specific unstructured effects are included, but with a selection mechanism to avoid excess parameterization. Thus, exchangeable errors $u_{pi} \sim N(0, 1/\tau_p)$ are added only when a binary area specific indicator δ_i is 1, with the prior probability $\pi_\delta = Pr(\delta_i = 1)$ set low at 0.05. The selection mechanism corresponds to a prior expectation that unusual psychiatric referral patterns over outcomes tend to occur in particular areas, perhaps due to distinctive care provision in such areas.

Then, $y_{pi} \sim Po(E_{pi}e^{\eta_{pi}})$ with,

$$\eta_{1i} = \alpha_1 + \lambda_{11}F_{1i} + \delta_i u_{1i},$$
$$\eta_{2i} = \alpha_2 + \lambda_{21}F_{1i} + \delta_i u_{2i},$$
$$\eta_{3i} = \alpha_3 + \lambda_{32}F_{2i} + \delta_i u_{3i},$$
$$\eta_{4i} = \alpha_4 + \lambda_{42}F_{2i} + \delta_i u_{4i}.$$

The prior variances of the two factors are taken as unknown, so an anchoring constraint needs to be used for identification, with loadings λ_{11} and λ_{32} set to 1. $N(0, 1)$ priors are assumed for the two unknown loadings. A two chain run (5000 iterations, with the last 4000 for inference) converges early and provides precise estimates for the loadings, with mean (sd) for λ_{21} of 1.05 (0.02) and for λ_{42} of 0.96 (0.02).

There is correlation of 0.77 between the two factors, with the κ parameter in the spatial prior estimated at 0.66 (with 95% interval from 0.47 to 0.91). Predictive fit seems adequate with only 4 out of 1416 observations having probabilities of underprediction (Marshall and Spiegelhalter, 2007) below 0.05 or above 0.95. In fact, all four are overpredicted; that is, they have posterior probabilities of underprediction below 0.05. However, despite apparent predictive adequacy, the posterior mean scaled deviance stands at 2026, in excess of the number of observations.

7.8 Multivariate Time Series

Multivariate time series can occur in several ways. One example is where the same measurement process (e.g., repeated environmental readings) is carried out at several locations and where high correlation between the series is expected. Another situation occurs with financial data, such as exchange rates or stock returns, where high correlations observed raise questions such as whether there are feedbacks between different series, or whether common factors (e.g., market risk) affect all series. Classical approaches using autoregressive moving average models rest on assumptions that series are stationary, typically after transformation or differencing. In econometrics, time series are said to be integrated of order d or $I(d)$, when differencing to order d of the original series is needed for stationarity. Such series are cointegrated if some linear combination of the series has a lower order of integration than the individual series (Phillips and Durlauf, 1986). A particular case of cointegration is when two series, y_t and x_t, are both $I(1)$, but there is a parameter α such that $u_t = y_t - \alpha x_t$ is stationary (integrated of order zero).

Much classical multivariate time series analysis is based on extending the ARMA model to vector responses (Tiao and Tsay, 1989). For observation vector, $y_t = (y_{1t}, \ldots, y_{Pt})'$, the vector $ARMA(r, s)$ model has the form,

$$y_t = \mu + \Phi_1 y_{t-1} + \cdots + \Phi_r y_{t-r} + u_t - \Theta_1 u_{t-1}, \ldots, -\Theta_s u_{t-s},$$

where the coefficient matrices are all of order $P \times P$, and u_t denotes P-variate white noise, with $E(u_t) = 0$, and

$$E(u_t u'_{t-k}) = 0 \qquad k \neq 0;$$
$$E(u_t u'_{t-k}) = \Sigma \qquad k = 0.$$

For the vector autoregressive or VAR model obtained by omitting moving average terms, stationarity requires that the roots of the characteristic equation,

$$\det(I - \Phi_1 z + \cdots + \Phi_r z^r) = 0,$$

lie outside the unit circle. Bayesian analyzes of the VAR model are extensive, and include treatments of cointegration (Koop et al., 2006), model selection and averaging (Andersson and Karlsson, 2007), and informative and restricted priors (Litterman, 1986; Sims and Zha, 1998). A library of MATLAB programs is available for Bayesian VAR estimation at http://home.earthlink.net/˜tzha02/ProgramCode/programCode.html.

7.8.1 Multivariate dynamic linear models

The structural model approach is widely applied in Bayesian time series studies (e.g., Carter and Kohn, 1994; West and Harrison, 1997; Durbin and Koopman, 2000) and focuses on underlying components of multiple series without requiring initial differencing. The multivariate normal dynamic linear model (DLM) specifies,

$$\begin{aligned} y_t &= F_t\theta_t + e_t, & e_t &\sim N(0, V_t), & t &= 1, \ldots, T, \\ \theta_{t+1} &= G_t\theta_t + R_t u_t, & u_t &\sim N(0, W_t), \end{aligned}$$

where y_t is a $P \times 1$ observation vector, and θ_t is a $Q \times 1$ latent state vector following a Markov process. The disturbance vectors, e_t and u_t, are normally distributed, and uncorrelated with each other and over time. The initializing prior for the state vector is typically assumed to be normal with mean m_1 and covariance matrix C_1, $\theta_1 \sim N(m_1, C_1)$. The system matrices F_t, G_t, V_t, W_t, and R_t may be assumed to be known, in which case simple updating, forecasting, and filtering densities can be derived—see West and Harrison (1997, 582). In more realistic settings where the covariances V_t and W_t are unknown, time-invariant assumptions, such as $V_t = \Sigma_e$ and $W_t = \Sigma_u$, are one possible parameterization. A simple case occurs (Koopman and Durbin, 2000) when V_t is diagonal, the assumption being that the observations are independent conditional on the latent states.

Common model forms include the local level (LL) model with measurement and transition equations,

$$\begin{aligned} y_t &= \theta_t + e_t, & e_t &\sim N(0, \Sigma_e), \\ \theta_{t+1} &= \theta_t + u_t, & u_t &\sim N(0, \Sigma_u), & t &= 1, \ldots, T, \end{aligned}$$

where y_t is a $P \times 1$ metric observation, θ_t also has dimension P, and Σ_e and Σ_u are of dimension $P \times P$. A local linear trend (LLT) includes a mechanism for trend in the underlying levels, as in,

$$\begin{aligned} y_t &= \theta_t + e_t, & e_t &\sim N(0, \Sigma_e), \\ \theta_{t+1} &= \theta_t + \delta_t + u_t, & u_t &\sim N(0, \Sigma_u), \\ \delta_{t+1} &= \delta_t + w_t, & w_t &\sim N(0, \Sigma_w), & t &= 1, \ldots, T. \end{aligned}$$

For example, Proietti (2006) applies a LL model to measuring core inflation, while Moauro and Savio (2005) apply a LLT approach to temporal disaggregation of multiple economic series. Multivariate signal models may be applied to measure latent risk, as in the accident rate and credit card use examples of Bijlevel et al. (2005). This approach involves time series or panel data on exposure totals (x_t or x_{it}), outcomes (y_t or y_{it}), and what may be generically termed "losses" (z_t or z_{it}). A simple bivariate case with x_t=vehicle registrations and y_t=motor accidents would lead to a model,

$$\log(x_t) = \theta_t^{(E)} + e_t^{(x)},$$
$$\log(y_t) = \theta_t^{(E)} + \theta_t^{(R)} + e_t^{(y)},$$

where the components of $\theta_t = (\theta_t^{(E)}, \theta_t^{(R)})$ represent underlying log exposure and log risk, which evolve according to a bivariate LLT,

$$\theta_{t+1} = \theta_t + \delta_t + u_t, \qquad u_t \sim N(0, \Sigma_u),$$
$$\delta_{t+1} = \delta_t + w_t, \qquad w_t \sim N(0, \Sigma_w).$$

A simplifying "homogenous" model (Harvey, 1989, Chapter 8) for the covariance matrices is obtained for the LL model by setting,

$$\Sigma_u = q\Sigma_e,$$

where q is an unknown signal-to-noise ratio, and for the LLT model by setting,

$$\Sigma_u = q_1\Sigma_e,$$
$$\Sigma_w = q_2\Sigma_e.$$

Generalizations to include trend, seasonal, and cyclical effects can be made in which each sort of effect is independent of the other and each follows its own multivariate evolution prior (Durbin and Koopman, 2001, 44). These assumptions lead to what is termed a seemingly unrelated time series equations (SUTSE) model (Harvey and Koopman, 1997; Harvey and Shephard, 1993), since the individual series are connected only via the correlated disturbances in the measurement and transition equations. More complex matrix normal priors (West and Harrison, 1997, 597) result from assuming interdependence between different types of parameter.

A multivariate model with level, seasonal, and cyclical effects would specify,

$$y_t = \theta_t + \gamma_t + \psi_t + e_t, \qquad e_t \sim N(0, \Sigma_e),$$
$$\theta_{t+1} = \theta_t + u_t, \qquad u_t \sim N(0, \Sigma_u) \qquad t = 1, \ldots, T,$$

where the seasonal components for the pth variable (with s seasons) evolve according to,

$$\gamma_{pt} = \gamma_{p,t-1} + \gamma_{p,t-2} + \cdots + \gamma_{p,t-s+1} + \omega_{pt},$$

with

$$(\omega_{1t}, \omega_{2t}, \ldots, \omega_{Pt}) \sim N(0, \Sigma_\omega).$$

Following Harvey and Koopman (1997), the cyclical effects, ψ_t, may be assumed "similar," namely, to have the same damping factor, ρ, and frequency,

$0 \leq \lambda \leq \pi$, across variables. The period is then $2\pi/\lambda$ with the full prior being,

$$\psi_t = (\psi_{1t}, \psi_{2t}, \ldots, \psi_{Pt}) \sim N(m_\psi, \Sigma_\eta),$$

with additional shadow period effects,

$$\psi_t^* = (\psi_{1t}^*, \psi_{2t}^*, \ldots, \psi_{Pt}^*) \sim N(m_{\psi^*}, \Sigma_{\eta^*}),$$

where means m_{ψ_p} and $m_{\psi_p^*}$ for the pth variable are obtained according to,

$$\begin{bmatrix} \psi_{pt} \\ \psi_{pt}^* \end{bmatrix} = \rho \begin{bmatrix} \cos(\lambda) & \sin(\lambda) \\ -\sin(\lambda) & \cos(\lambda) \end{bmatrix} \begin{bmatrix} \psi_{p,t-1} \\ \psi_{p,t-1}^* \end{bmatrix} + \begin{bmatrix} \eta_{pt} \\ \eta_{pt}^* \end{bmatrix}.$$

It may be noted that multivariate DLMs occur in the analysis of univariate data, for example for categorical and ordinal outcomes. Thus, Cargnoni et al. (1997) propose a model for time series of a multinomial outcome with M categories in which,

$$\begin{aligned}
(y_{1t}, y_{2t}, \ldots, y_{Mt}) &\sim \text{Mult}(n_t, [\pi_{1t}, \pi_{2t}, \ldots, \pi_{Mt}]), \\
\pi_{mt} &= \exp(\eta_{mt}) \Big/ \sum_{h=1}^{M} \exp(\eta_{ht}), \\
\eta_{mt} &= \alpha_{mt} + \beta_m x_t \qquad m = 1, \ldots, M-1, \\
\eta_{Mt} &= 0,
\end{aligned}$$

where the time-varying category intercepts, $\alpha_t = (\alpha_{1t}, \ldots, \alpha_{M-1,t})$, follow a multivariate normal random walk prior,

$$\alpha_t \sim N_{M-1}(\alpha_{t-1}, \Sigma_\alpha).$$

West and Harrison (1997, 586) assume a normal approximation to the multinomial in which,

$$y_{mt} = n_t \pi_{mt} + e_{mt},$$

and the vector of errors, $e_t = (e_{1t}, \ldots, e_{Mt})$, has a covariance matrix,

$$\begin{aligned}
V_{mmt} &= n_t \pi_{mt}(1 - \pi_{mt}) \\
V_{mkt} &= -n_t \pi_{mt} \pi_{kt} \qquad k \neq m.
\end{aligned}$$

7.8.2 Dynamic factor analysis

Time series factor models become sensible for large P, as they result in less heavy parameterization of covariance between series, and may provide insights into latent structure, as well as more efficient inferences and forecasts (Durbin and Koopman, 2001). Typically, the covariance structure between series is attributed to the common factors only, with observation errors assumed independent (Jungbacker et al., 2009). There are a number of application areas, and Bayesian approaches have been important. Prado and West (1997) consider the case of a single latent series, F_t, underlying multiple series, $y_t = (y_{1t}, \ldots, y_{Pt})$, of electroencephalogram readings, and discuss TVAR autoregressive models for the latent F_t involving time-varying coefficients that follow random walk priors. Thus with first order random walk priors in $h = 1, \ldots, r$ autoregressive parameters one has,

$$y_{pt} \sim N(\alpha_p + \lambda_p F_t, \tau_y),$$
$$F_t \sim N\left(\sum_{h=1}^{r} \phi_{ht} F_{t-h}, \tau_F\right),$$
$$\phi_{ht} \sim N(\phi_{h,t-1}, \tau_h).$$

Another example occurs in econometric modeling of asset returns, where the number of assets may exceed the length of the time series and factor models for returns are a clear option (Zivot and Wang, 2006). Factor models are also one approach to multivariate volatility (changing variances)—see Section 7.8.3.

A relatively simple approach for reducing a metric P vector y_t to a Q vector F_t involves a dynamic linear model for the factor score vector. Thus, a local linear factor or factor trend model would propose a measurement model linking indicators and factors,

$$y_t = \alpha + \Lambda F_t + e_t, \qquad e_t \sim N(0, \Sigma_e) \qquad t = 1, \ldots, T,$$

with the transition equation specifying a random walk in the factors, namely,

$$F_{t+1} = F_t + u_t, \qquad u_t \sim N(0, \Sigma_u),$$

where F_t is a $Q \times 1$ multivariate latent construct, with $Q < P$, and Λ is of dimension $P \times Q$. If the series, e_t and F_t, are uncorrelated, then the marginal mean and covariance of y_t are α and $\Lambda \Sigma_u \Lambda + \Sigma_e$, respectively. To avoid location invariance in the F scores, devices such as centering at each iteration or setting initial factor scores to known values can be used (see Example 7.10).

The loadings matrix Λ and/or the factor score covariance matrix Σ_u are parameterized to ensure identification and avoid various forms of invariance. If all elements in Σ_e are taken as unknown (i.e., off-diagonal as well as diagonal terms) and Σ_u is also unknown, then Harvey and Koopman (1997) mention the form,

$$\Lambda = \begin{pmatrix} I_Q \\ \Lambda^* \end{pmatrix},$$

with Λ^* of dimension $(P - Q) \times Q$ containing unknown loadings. If Σ_e is diagonal (that is, just residual variances are assumed unknown), and Σ_u is also diagonal but contains unknown factor variances, then one may set $\lambda_{pp} = 1$ and $\lambda_{pq} = 0$ for $q > p$. This is the anchoring constraint of Skrondal and

Rabe-Hesketh (2004), with the latter constraint used to avoid rotation invariance (Geweke and Zhou, 1996, 565–6).

If Σ_e contains just residual variances, and Σ_u is diagonal with known factor variances (typically of 1), then constraints on λ_{pq} to ensure scale identification are not needed, but the rotational constraint, $\lambda_{pq} = 0$, for $q > p$ still applies. However, Geweke and Zhou (1996) suggest $\lambda_{pp} > 0$ as an identification device in this case, to ensure a unique labeling of factors.

7.8.3 Multivariate stochastic volatility

Many multivariate series (e.g., share prices, exchange rates, asset returns) may be subject to volatility clustering, with the clustering correlated over different series. For example, Yu and Meyer (2006) mention that financial decision making needs to take correlations into account when market volatilities move together across multiple assets. Harvey et al. (1994) suggested the first multivariate stochastic volatility (MSV) model, involving metric series $(y_{1t}, y_{2t}, \ldots, y_{Pt})'$ either mean centered or in transformed form (e.g., logs of successive share prices) with effectively zero means.

Thus, for centered or appropriately transformed prices or returns, y_t, one possible MSV model is

$$\begin{aligned} y_t &= H_t e_t, \\ H_t &= \text{diag}(\exp(h_{1t}/2),) \exp(h_{2t}/2), \ldots, \exp(h_{Pt}/2)), \end{aligned}$$

with $h_t = (h_{1t}, \ldots, h_{Pt})'$, a vector of unobserved log variances (or volatilities), typically evolving according to a stationary VAR scheme, either,

$$h_{t+1} = \mu + \phi(h_t - \mu) + u_t,$$

or

$$h_t = \mu + \phi(h_{t-1} - \mu) + u_t.$$

The errors in the price series and in the volatilities are assumed to be multivariate normal or multivariate t, with the MVN assumption expressed,

$$\begin{pmatrix} e_t \\ u_t \end{pmatrix} \sim N\left[\begin{pmatrix} 0 \\ 0 \end{pmatrix} \begin{pmatrix} R_e & 0 \\ 0 & \Sigma_u \end{pmatrix}\right],$$

where R_e is a positive definite correlation matrix with a diagonal of ones, and Σ_u is a $P \times P$ covariance matrix for volatility shocks. Taking R_e to be nondiagonal means shocks in prices may be correlated, while taking Σ_u to be nondiagonal allows volatility shocks to be correlated (Yu and Meyer, 2006, 365–66). The simplest form for the coefficient vector in the VAR autoregression for h_t is $\phi = \text{diag}(\phi_{11}, \phi_{22}, \ldots, \phi_{PP})$, and this may be combined with a stationarity assumption by adopting the prior $\phi_{pp} \sim U(-1, 1)$ (Chan et al., 2006).

The autoregression can be extended to full vector autoregressive form at the expense of heavier parameterization (Asai et al., 2006).

Yu and Meyer (2006) consider various extensions using a bivariate series as an example. Thus, taking,

$$\phi = \begin{pmatrix} \phi_{11} & \phi_{12} \\ \phi_{21} & \phi_{22} \end{pmatrix},$$

to be nondiagonal amounts to allowing bilateral Granger causality in volatility between the two series. Allowing for R_e to evolve through time according to

$$R_{et} = \begin{pmatrix} 1 & \rho_t \\ \rho_t & 1 \end{pmatrix},$$

means that not only log volatilities, h_t, but also correlation coefficients are time varying. Specifically with,

$$\rho_t = \frac{\exp(g_t) - 1}{\exp(g_t) + 1},$$

an additional autoregression can be set, with,

$$g_{t+1} = \mu_g + \phi_g(g_t - \mu_g) + w_t.$$

Factor analytic models may include a correlated volatility feature (Chan et al., 2006; Pitt and Shephard, 1999). As an example, for two series $\{y_{pt}, p = 1, 2\}$ and a univariate factor, F_t, one might have,

$$y_{1t} = \lambda_1 F_t + e_{1t},$$
$$y_{2t} = \lambda_2 F_t + e_{2t},$$

with evolving variances for F_t and e_{pt}. The stochastic variance prior for the residuals, e_{pt}, may include autoregressive dependence, since a factor structure may be sufficient to account for the nondiagonal elements of the residual variance matrix of the outcomes, but not sufficient to explain all the marginal persistence in volatility (Pitt and Shephard, 1999, 551). Thus one might have $F_t \sim N(0, e^{h_{1t}})$, $e_{1t} \sim N(0, e^{h_{2t}})$, and $e_{2t} \sim N(0, e^{h_{3t}})$, with first order autoregressive dependence in the log variances, h_{pt},

$$h_{pt} = \phi_p h_{k,t-1} + u_{pt} \qquad t = 2, \ldots, T,$$

with possibly unknown initial conditions, h_{p1}. For identification, one may set one or other of the λ_p parameters to 1 (an anchoring constraint). Alternatively, a standardized factor constraint might be implemented by setting the scale of the factors at one time point, for instance by taking $F_1 \sim N(0, 1)$, that is $h_{11} = 0$.

Extensions to provide greater adaptivity in the modeling of stochastic variances can be combined with factor reduction. Chib et al. (2005) propose an MSV factor model that permits both series-specific jumps at each time, and Student t innovations with unknown degrees of freedom. For bivariate data and a univariate factor, this model has the form,

$$y_{1t} = \lambda_1 F_t + \delta_{1t} q_{1t} + \varepsilon_{1t},$$
$$y_{2t} = \lambda_2 F_t + \delta_{2t} q_{2t} + \varepsilon_{2t},$$

where $q_{pt} = 1$ with probability π_p, and ε_{pt} follow independent Student t densities with unknown degrees of freedom, ν_p. In hierarchical form,

$$\varepsilon_{pt} = \lambda_{pt} e_{pt}, \qquad \lambda_{pt} \sim \text{Ga}\left(\frac{\nu_p}{2}, \frac{\nu_p}{2}\right), \qquad [e_{1t}, e_{2t}] \sim N(0, V_t),$$

where V_t is diagonal with elements $\exp(h_{pt})$, with evolution scheme,

$$h_{p,t+1} - \mu_p = \phi_p(h_{pt} - \mu_p) + \sigma_p u_t,$$
$$u_t \sim N(0, 1).$$

The variables, $\zeta_{pt} = \log(1 + \delta_{pt})$, are assumed to be $N(-0.5\xi_p^2, \xi_p^2)$ where ξ_p are additional unknowns. The more general form for $y_t = (y_{1t}, \ldots, y_{Pt})'$ and $F_t = (F_{1t}, \ldots, F_{Qt})'$, $Q \leq P$ is

$$y_t = \Lambda F_t + \Delta_t q_t + \varepsilon_t,$$

with identification constraints $\lambda_{pp} = 1$ and $\lambda_{pq} = 0$ for $q > p$. These constraints set a scale and prevent rotation invariance. The covariance matrix for F_t is diagonal with evolution scheme as for the log diagonal elements of V_t.

Example 7.9. Minks and Muskrats: Multivariate Dynamic Linear Models Harvey and Koopman (1997) consider a bivariate series, namely, numbers of skins of minks and muskrats traded annually (1848–1909) by the Hudson Bay Company. There is a prey-predator relationship between the $P = 2$ species leading to interlinked cycles. A model is fitted including trends and similar cycles, so that,

$$\begin{aligned}
y_t &= \theta_t + \psi_t + e_t, & e_t &\sim N(0, \Sigma_e),\\
\theta_{t+1} &= \theta_t + \beta_t + u_t, & u_t &\sim N(0, \Sigma_u),\\
\beta_{t+1} &= \beta_t + w_t, & w_t &\sim N(0, \Sigma_w),\\
\psi_t &= (\psi_{1t}, \psi_{2t}) \sim N(m_\psi, \Sigma_\eta), & &\\
\psi_t^* &= (\psi_{1t}^*, \psi_{2t}^*) \sim N(m_{\psi^*}, \Sigma_{\eta^*}) & t &= 1, \ldots, T,
\end{aligned}$$

where the cyclical effects for the two species have the same damping factor, $\rho \sim U(0, 1)$, and frequency, λ, and the nondiagonal covariance matrices are of order $P \times P$. Since the series contains 64 points, an informative assumption is made that the period is between 4.2 and 21, namely, that $\lambda \sim U(0.3, 1.5)$. Taking a simple uniform prior on λ between 0 and π was associated with implausibly low λ. Covariances are linked using the homogeneity assumption, namely, $\Sigma_u = q_u \Sigma_e$, $\Sigma_w = q_w \Sigma_e$, $\Sigma_\eta = q_\eta \Sigma_e$, and $\Sigma_{\eta^*} = q_{\eta^*} \Sigma_e$, with the signal-to-noise ratios $\{q_u, q_w, q_\eta, q_{\eta^*}\}$ all assumed to follow $Ga(1, 1)$ priors. For

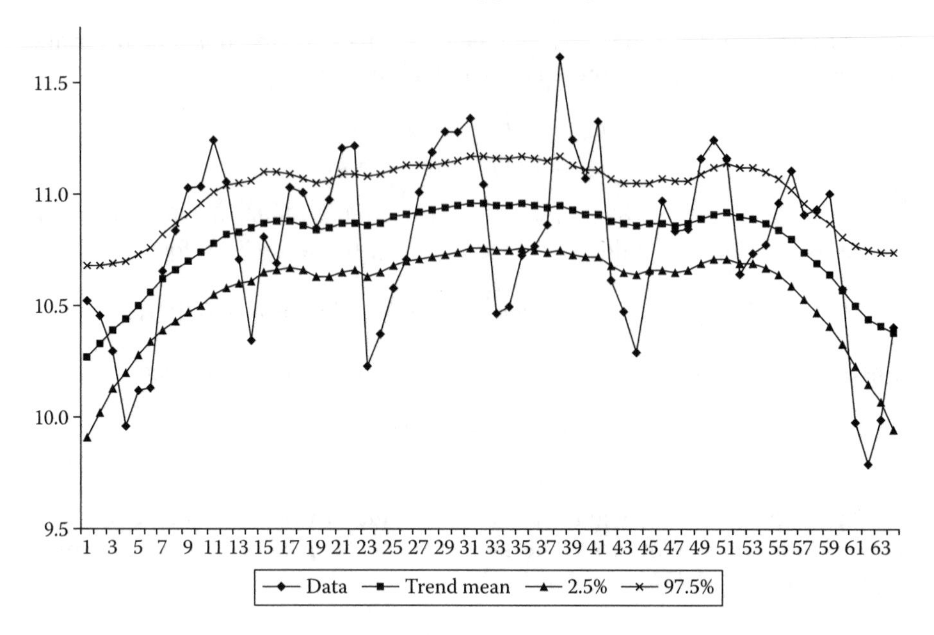

FIGURE 7.2
Data and trend, minks.

Σ_e^{-1}, a Wishart prior assumes five degrees of freedom and a prior covariance matrix based on the observed variances and covariance[8].

With inferences from the second half of a two chain run of 10,000 iterations, one finds the cycles have a mean period of 10.7 years, though this parameter has a skew posterior density with lower and upper 2.5% points of 9.3 and 20. Figures 7.2 and 7.3 show trends in the mink and muskrat series together with the original data, while Figure 7.4 plots the two cycles. The posterior means for q_u, q_w, q_η, and q_{η^*} are (0.088, 0.0036, 0.049, 0.13).

The interlinking of the two series (and its predator-prey nature) also shows in a $VAR(1)$ model with,

$$\begin{aligned}
y_{t1} &= \gamma_1 + \alpha_{11}y_{1,t-1} + \alpha_{12}y_{2,t-1} + u_{1t}, \\
y_{t2} &= \gamma_2 + \alpha_{21}y_{1,t-1} + \alpha_{22}y_{2,t-1} + u_{2t}, \\
u_t &\sim N(0, \Sigma_u) \qquad\qquad t = 2, \ldots, T,
\end{aligned}$$

with y_{11} and y_{12} taken as known, and where Σ_u is nondiagonal. The estimated α coefficient matrix from the second half of a two chain run of 5000 iterations is

$$\begin{pmatrix} 0.72 & 0.18 \\ -0.45 & 0.92 \end{pmatrix},$$

though residuals between the two series are positively correlated after accounting for the lag 1 effect of one series on the other.

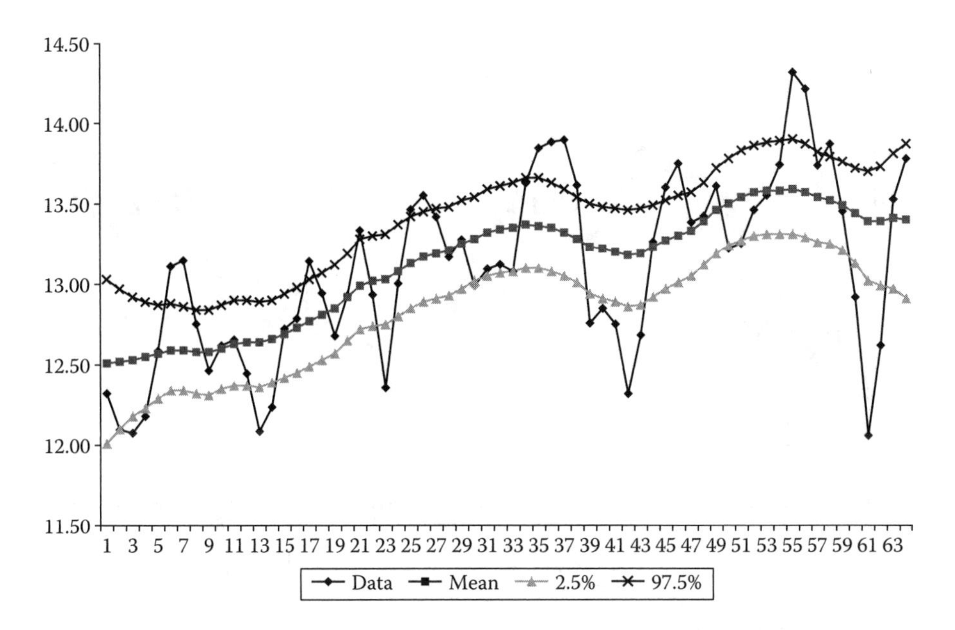

FIGURE 7.3
Data and trend, muskrats.

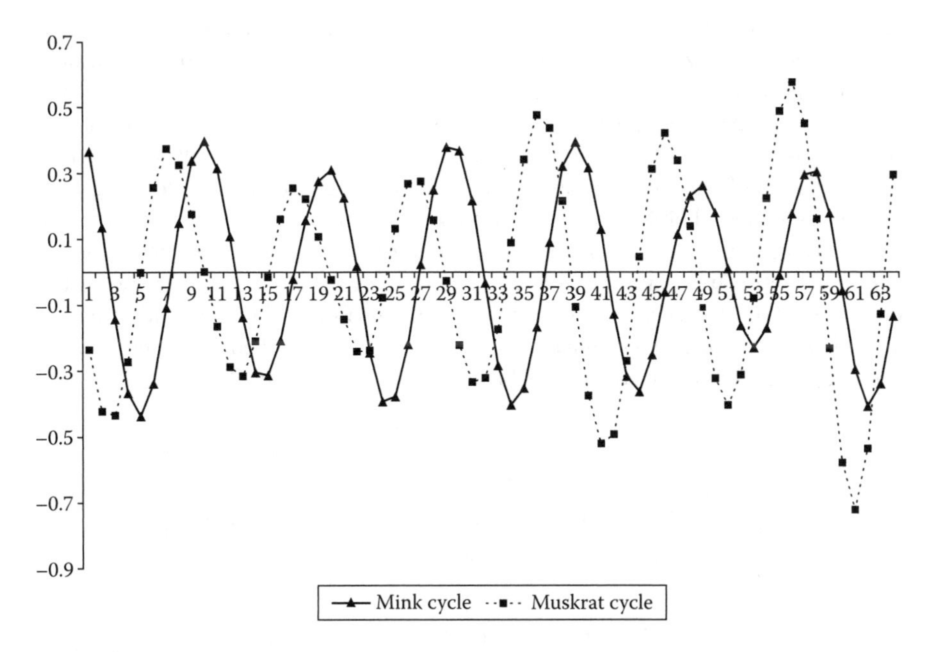

FIGURE 7.4
Estimated cycles.

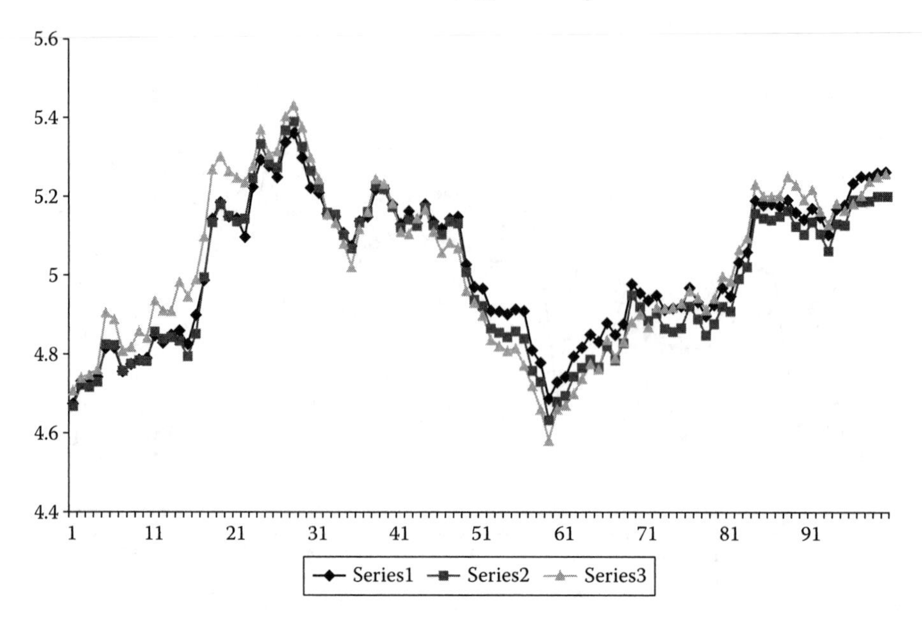

FIGURE 7.5
Flour price series.

Example 7.10. Common Factor Model for Flour Prices Tiao and Tsay (1989) analyze a trivariate series formed by the logarithms of indices of monthly flour prices in Buffalo, Minneapolis, and Kansas City between August 1972 and November 1980. The data are plotted in Figure 7.5, which shows that the series are closely related, and a common factor model is indicated. The variance of the factor scores, F_t, is taken as unknown, so a loading constraint is needed. The factor scores are assumed to follow a random walk with $F_1 = 0$ to identify the level of the scores. The residuals after accounting for the common factor are assumed multivariate normal, with a Wishart prior on the precision matrix with three degrees of freedom and a diagonal scale matrix. The elements of the scale matrix are taken as the observed variances, V_p, of the three series, leading to a data-based prior. Thus with $P = 3$,

$$\begin{aligned}
y_{pt} &= \alpha_p + \lambda_p F_t + u_{pt},\\
(u_{1t}, u_{2t}, u_{3t}) &\sim N_P(0, \Sigma_u);\\
\Sigma_u^{-1} &\sim W(PS, P); \quad S = \text{diag}(V_1, V_2, \ldots, V_P);\\
\lambda_1 &= 1; \quad \lambda_k \sim N(1, 1), k = 2, 3;\\
F_t &\sim N(F_{t-1}, \sigma_F^2) \qquad t = 2, \ldots, T;\\
F_1 &= 0; \sigma_F \sim U(0, 10).
\end{aligned}$$

An alternative model adopts a locally adaptive prior for the factor scores, allowing for changing variance through time (Lang et al., 2002). Thus,

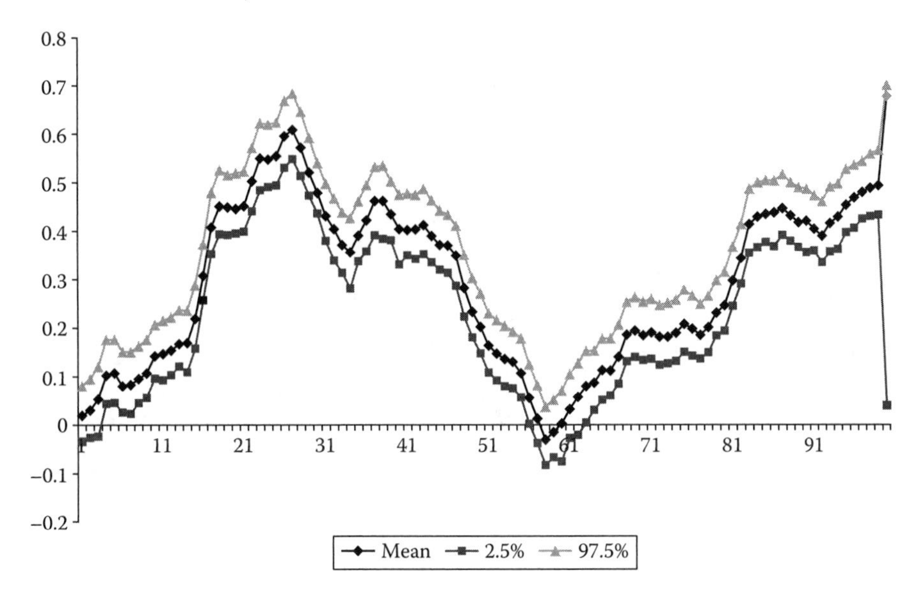

FIGURE 7.6
Common factor score, model 1.

$$
\begin{aligned}
&F_t \sim N(F_{t-1}, \exp(h_t)) \qquad t = 2, \ldots, T;\\
&h_t \sim N(h_{t-1}, \tau_h^{-1}) \qquad t = 2, \ldots, T;\\
&\tau_h \sim \text{Ga}(1, 1); \quad h_1 \sim N(0, 1); \quad F_1 = 0.
\end{aligned}
$$

Following Migon and Moreira (2004), fit is assessed using the predictive approach of Gelfand and Ghosh (1998), based on a goodness of fit term, $G = \sum_p \sum_t (y_{\text{rep},pt} - y_{pt})^2$, and a penalty term, $H = \sum_p \sum_t \text{var}(y_{\text{rep},pt})$. Figure 7.6 shows the estimated factor scores through time under the constant variance model. The posterior mean for σ_F^2 is 0.0018, with posterior mean for G of 1.416 and with $H = 1.107$. The nonconstant variance model has very similar fit criteria, namely, a posterior mean for G of 1.423 with $H = 1.106$. There seems little to choose between the models though the plot of the evolving log variances (Figure 7.7) suggests a reduction in volatility in the second half of the observation period.

Example 7.11. Multivariate Stochastic Volatility model for FTSE and S&P fluctuations during 2006–2007 This example follows the FTSE 100 and S&P 500 stock indices $\{r_{pt}, t = 1, \ldots, 253; p = 1, 2\}$ over 253 trading days (October 27, 2006–October 19, 2007, as recorded by uk.finance.yahoo.com) that include two periods of market turbulence, the second being associated with US subprime mortgage lending. Where a trading day is present in one index but not the other, the gap is filled by taking an average of the preceding and subsequent days. Figure 7.8 plots the series

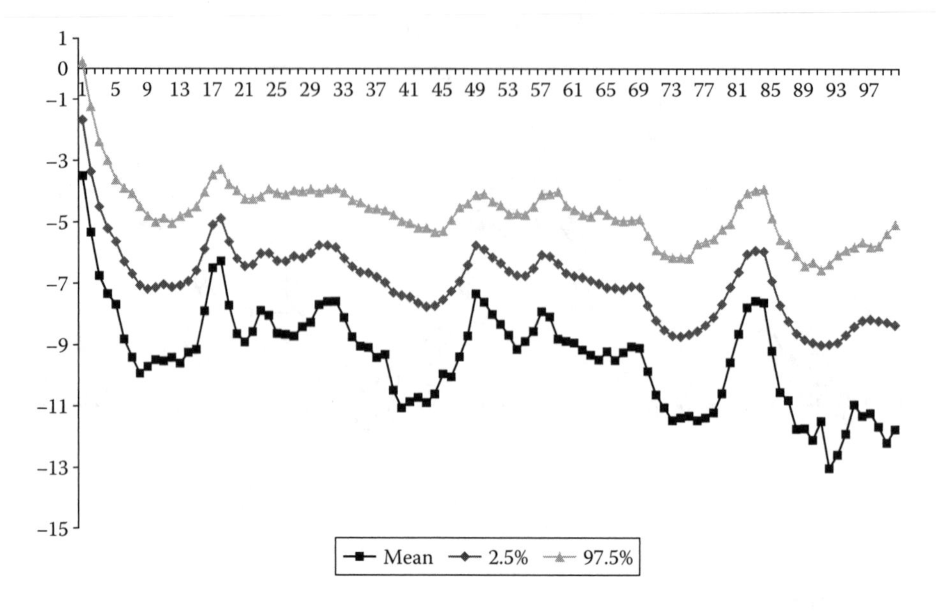

FIGURE 7.7
Nonconstant log factor variance.

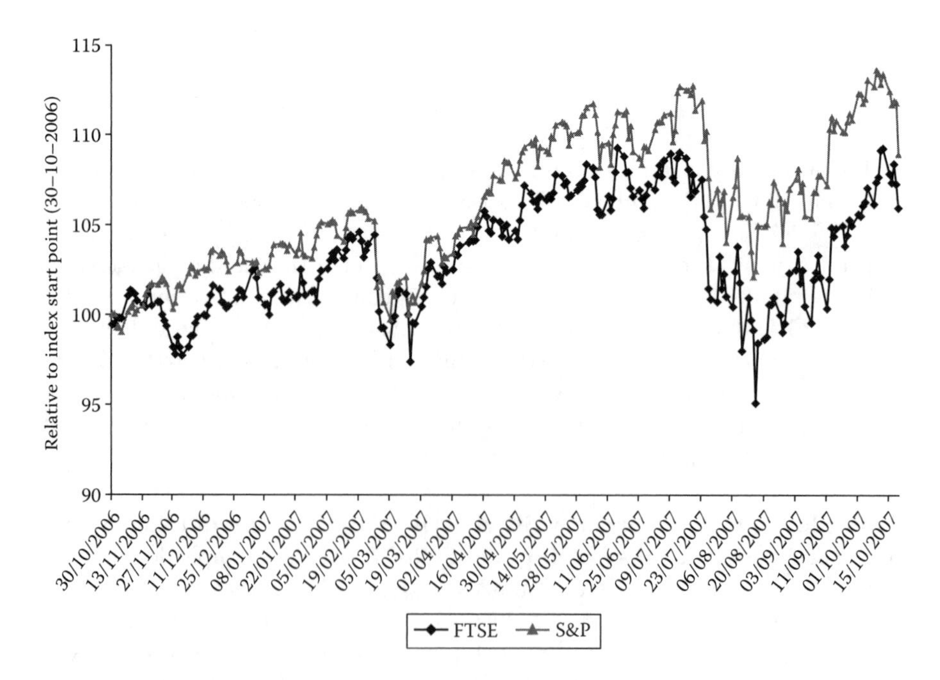

FIGURE 7.8
US and GB share indices 30th October 2006–19th October 2007.

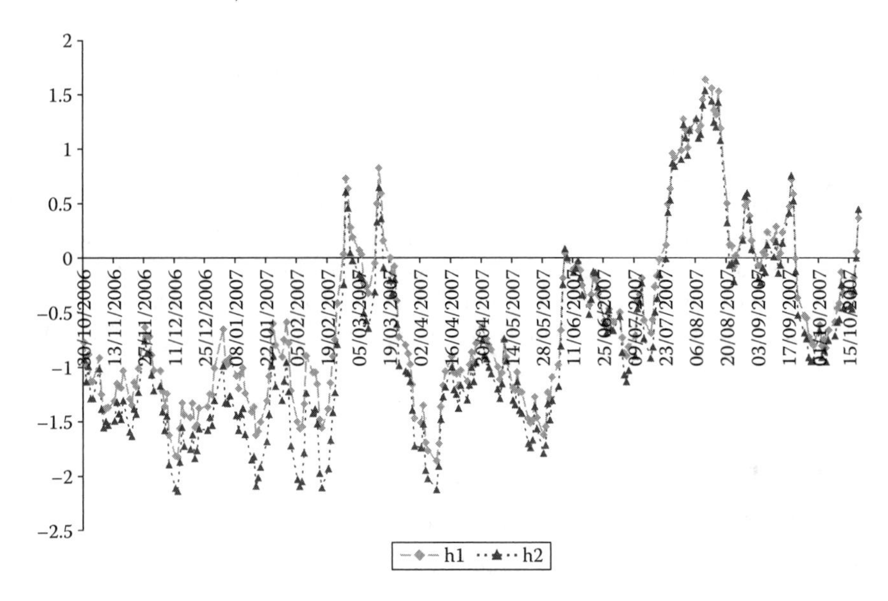

FIGURE 7.9
Log volatility plot.

relative to their start points, namely, in the form $100r_{pt}/r_{p1}$. The data to be analyzed are obtained as $y_{pt} = 100r_{p,t}/r_{p,t-1} - 100$.

The model allows for correlated shocks in the stock exchange variables and correlated log variances, so that,

$$\begin{aligned} y_t &= H_t e_t, \\ H_t &= \text{diag}[\exp(h_{1t}/2),)\exp(h_{2t}/2)], \end{aligned}$$

with unobserved log volatilities $h_t = (h_{1t}, h_{2t})'$ evolving via a $VAR(1)$ model,

$$\begin{aligned} h_{t+1} &= \mu + \text{diag}(\phi_{11}, \phi_{22})(h_t - \mu) + u_t, \\ h_1 &= \mu + u_1. \end{aligned}$$

The errors in the price series and volatilities equations are multivariate normal,

$$\begin{pmatrix} e_t \\ u_t \end{pmatrix} \sim N\left[\begin{pmatrix} 0 \\ 0 \end{pmatrix} \begin{pmatrix} R_e & 0 \\ 0 & \Sigma_u \end{pmatrix}\right].$$

where R_e is a nondiagonal correlation matrix, and Σ_u is also nondiagonal, allowing volatility shocks to be correlated (Yu and Meyer, 2006, 365–66).

The WinBUGS code[9] follows Yu and Meyer (2006) in sampling standard normal h^*_{1t} and h^*_{2t} and applying the standard deviations $(\sigma_{u1}, \sigma_{u2})$ and correlation ρ_u of the u_{1t} and u_{2t} series to the standard normal log volatilities. Thus,

$$\begin{aligned} h_{1t} &= \sigma_{u1} h^*_{1t}, \\ h_{2t} &= \sigma_{u2}\rho_u h^*_{1t} + \sigma_{u2}(1 - \rho_u^2)^{0.5} h^*_{2t}. \end{aligned}$$

This raises an identification issue for the correlation ρ_u, since $\rho_u h^*_{1t} = (-\rho_u)(-h^*_{1t})$, which is resolved by assuming ρ_u to be $U(0,1)$ rather than $U(-1,1)$. The correlation ρ_e in R_e is taken as $U(-1,1)$. The stationary autocorrelation parameters are obtained as $\phi_{pp} = 2\phi^*_{pp} - 1$, where $\phi^*_{pp} \sim Beta(19,1)$. The diagonal terms in Σ_u^{-1} are taken to be Ga(1, 1).

A two chain run of 10,000 iterations, converges after around 1500, with posterior means for ρ_e and ρ_u of 0.54 and 0.91, and with the autoregressive coefficients in the AR1 log volatility equations having means 0.88 and 0.81. Figure 7.9 plots the resulting log volatility series (posterior means of h_{1t} and h_{2t}) with the two periods of market turbulence apparent. The DIC is 1056 ($d_e = 59.4$).

Appendix: Computational Notes

1. In Example 7.1, the code for models 1 and 2 is

```
model {for (j in 1:p) {tau[j] <- 1/(s[j]*s[j]);s[j]~dunif(0,1000)}
for (j in 1:5){lam[j,1] ~dnorm(0,1)}
lam[6,1] ~dnorm(0,1) I(0,); lam[7,1] ~dnorm(0,1)
for (j in 1:3) {lam[j,2] ~dnorm(0,1)}
lam[4,2] ~dnorm(0,1) I(0,); lam[5,2] ~dnorm(0,1)
lam[6,2] <- 0; lam[7,2] ~dnorm(0,1)
for (i in 1:n) {F[1,i] ~dnorm(0,1); F[2,i] ~dnorm(0,1);
for (j in 1:p){yst[i,j] <- (y[i,j]-mean(y[,j]))/sd(y[,j]); yst[i,j]~dnorm(mu[i,j],
tau[j]);
mu[i,j] <- lam[j,1]*F[1,i] + lam[j,2]*F[2,i] }}}
```

and

```
model {for (j in 1:p) {tau[j] <- 1/(s[j]*s[j]);s[j]~dunif(0,1000)}
for (j in 1:5){lam[j,1] ~dnorm(0,1)} lam[6,1] <- 1; lam[7,1] ~dnorm(0,1)
for (j in 1:3){lam[j,2] ~dnorm(0,1)} lam[4,2] <- 1; lam[5,2] ~dnorm(0,1);
lam[6,2] <- 0; lam[7,2] ~dnorm(0,1)
for (i in 1:n) {F[1,i] ~dnorm(0,inv.phi[1]); F[2,i] ~dnorm(0,inv.phi[2]);
for (j in 1:p){ yst[i,j] <- (y[i,j]-mean(y[,j]))/sd(y[,j]); yst[i,j] ~dnorm(mu[i,j],
tau[j]);
mu[i,j] <- lam[j,1]*F[1,i] + lam[j,2]*F[2,i] }}
for (k in 1:2) {phi[k] ~dunif(0,1000); inv.phi[k] < - 1/phi[k]}}
```

2. The code relating to the first indicator, y_1, in Example 7.3 is

```
for (i in 1:n) {F[i]~dnorm(0,1); yst1[i]~dnorm(mu1[i],1) I(A1[i],B1[i])
A1[i] <- -10*equals(y1[i],0); B1[i] <- 10*equals(y1[i],1)
mu1[i] <- alph1+lam[1]*F[i]}
```

The code relating to the third indicator uses $M_3 - 1$ probit regressions:

```
for (i in 1:n) {for (j in 1:M3-1) { z3[i,j] <- step(j-y3[i])
A3[i,j] <- -10*equals(z3[i,j],0); B3[i,j] <- 10*equals(z3[i,j],1)
```

```
yst3[i,j] ~dnorm(nu3[i,j],1) I(A3[i,j],B3[i,j]); nu3[i,j] <- kap3[j] - mu3[i]}
mu3[i] <- lam[3]*F[i]}}
```

3. The code for the two parameter probit IRT model in Example 7.4, including elements to calculate CPOs, and hence obtain a pseudo marginal likelihood, is

```
model {for (i in 1:39) {F[i] ~dnorm(0,1)
for (j in 1:6) {A[i,j] <- -10*equals(y[i,j],0);B[i,j] <- 10*equals(y[i,j],1)
pi[i,j] <- phi(mu[i,j]); mu[i,j] <- lam[j]*F[i]-alpha[j];
yst[i,j]~dnorm(mu[i,j],1) I(A[i,j],B[i,j]);
# sample new data without truncation
ystnew[i,j]~dnorm(mu[i,j],1)
ynew[i,j] <- step(ystnew[i,j]); conc[i,j] <- equals(y[i,j],ynew[i,j])
LL[i,j] <- y[i,j]*log(pi[i,j])+(1-y[i,j])*log(1-pi[i,j]); H[i,j] <- 1/exp(LL[i,j]) }}
for (j in 1:6) {lam[j] ~dnorm(0,1) I(0,); alpha[j] ~dnorm(0,0.001)
conc.item[j] <- sum(conc[,j])/39 }}.
```

The code for the full multivariate probit in this example is

```
model {for (i in 1:39) {F[i] ~dnorm(0,1);
ystnew[i,1:6]~dmnorm(m[i,1:6],T[1:6,1:6])
yst[i,1:6] ~dmnorm(m[i,1:6],T[1:6,1:6]) I(A[i,1:6],B[i,1:6])
for (j in 1:6){conc[i,j] <- equals(y[i,j],ynew[i,j])
m[i,j] <- alp[j]+lam[j]*F[i]; ynew[i,j] <- step(ystnew[i,j])
A[i,j]<- -10*equals(y[i,j],0);B[i,j]<- 10*equals(y[i,j],1)}}
# Priors
T[1:6,1:6] ~dwish(Sc[,],6); Sigma[1:6,1:6] <- inverse(T[,])
for (j in 1:6) {alp[j] ~dnorm(0,0.001);lam[j] ~dnorm(0,1) I(0,)
conc.item[j] <- sum(conc[,j])/39
# scale loadings
s[j] <- sqrt(Sigma[j,j]); lamst[j] <- lam[j]/s[j];
for (k in 1:6) {Sc[j,k] <- 6*equals(j,k)}}
for (j in 2:6) {for (k in 1:j-1){r[j,k] <- Sigma[j,k]/(s[j]*s[k])}}}.
```

4. The code for the market segment model (Example 7.5) is

```
model { for (i in 1:500) {S[i] ~dcat(p[1:2])
for (j in 1:2){ y[i,j] ~dnorm(mu.y[i,j],tauy[j])
mu.y[i,j] <- alph.y[j]+lamy[j]*F[i]}
F[i] ~dnorm(mu.F[i],1)
mu.F[i] <- gam[S[i],1]*H[i,1]+gam[S[i],2]*H[i,2]}
p[1:2] ~ddirch(alph[1:2])
rho ~dunif(-1,1); tau.H <- 1/(1-rho*rho)
for (j in 1:2) {tauy[j] ~dgamma(1,0.001); alph.y[j]~dnorm(0,0.01)}
for (j in 1:4) {taux[j] ~dgamma(1,0.001); alph.x[j] ~dnorm(0,0.01)}
for (k in 1:2) {gam[1,k] ~dnorm(0,0.1) I(0,); gam[2,k] ~dnorm(0,0.1) I(,0)}
for (j in 1:2) {lamx[1,j] ~dnorm(0,0.1);
lamx[1,j+2] <- 0
lamx[2,j] <- 0; alph[j] <- 1;
```

```
lamx[2,j+2] ~dnorm(0,0.1)
lamy[j] ~dnorm(0,0.1)}
for (i in 1:500) {H[i,1] ~dnorm(0,1); m.H[i] <- rho*H[i,1]
                  H[i,2] ~dnorm(m.H[i],tau.H)
for (j in 1:4){x[i,j] ~dnorm(mu.x[i,j],taux[j])
mu.x[i,j] <- alph.x[j]+lamx[1,j]*H[i,1]+lamx[2,j]*H[i,2]}}}
```

5. The code relevant to the Dirichlet process mixture model on the residuals in Example 7.6 is

```
model { for (i in 1:n) {S[i] ~dcat(pi[1:K])
for (p in 1:P) {y[i,p] ~dpois(mu[i,p]);
log(mu[i,p]) <- log(pop[i])+beta[p,1]+beta[p,2]*unem[i]
+beta[p,3]*border[i]+beta[p,4]*GDP[i]+beta[p,5]*urb[i]+u[i,p];
u[i,p] <- ustar[S[i],p]}
for (k in 1:K) {d[i,k] <- equals(S[i],k)}}
# Priors
for (p in 1:P) { for (r in 1:R) {beta[p,r] ~dflat()}}
inv.D[1:P,1:P] ~dwish(Sc[,],P)
# truncated Dirichlet process
kap ~dgamma(2,4) I(0.1,); V[K] <- 1; pi[1] <- V[1]
for (k in 1:K-1){ V[k] ~dbeta(1,kap)}
for (j in 2:K) { pi[j] <- V[j]*(1-V[j-1])*pi[j-1]/V[j-1]}
for (j in 1:K) {ustar[j,1:P] ~dmnorm(nought[],inv.D[,])}
# total non-empty clusters
    Kstar <- sum(C[])
for (j in 1:K) {C[j] <- step(sum(d[,j])-1)}}
```

6. Consider Poisson outcomes with offsets, E_{pi}, and with a marginal/conditional sequence starting with S_1. For example, for $P = 3$, one has $P(S_1, S_2, S_3) = p(S_3|S_2, S_1)p(S_2|S_1)P(S_1)$. A WinBUGS code is then, with the entire (binary) interaction matrix W included in the data input:

```
model {for(p in 2:P) {for (q in 1:p-1) {gam0[p,q] ~dnorm(0, 0.1); gam1[p,q]
~dnorm(0, 0.1)
for (i in 1:N) {M[1,i] <- 0 ; inv.d[i] <- 1/d[i]
G0[i,p,q] <- gam0[p,q]*s[q,i]; G1[i,p,q] <- gam1[p,q]*inprod(W[i,],s[q,])}}}
for (p in 1:P) {y[i,p] ~dpois(mu[i,p]); mu[i,p]<- E[i,p]*exp(s[p,i])}}
for(p in 2:P) {M[p,i] <- sum(G0[i,p,1:p-1])+sum(G1[i,p,1:p-1])}}
for (p in 1:P){tau[p] ~dgamma(1,0.01);alph[p] ~dunif(0,0.999)
s[p,1:N] ~car.proper(M[p, ], C[], adj[], d[], inv.d[], tau[p], alph[p])}}
```

The priors on the regression effects $\{\gamma_0, \gamma_1\}$ follow the recommendations of Jin et al. (2005, 955).

7. The core code for the LMCAR model in Example 7.8 is

```
model { for (i in 1 : N) {r[i,1:P]~dmnorm(M[i,1:P],Zeta.r[i,,])
for (p in 1:P) {mu[i,p] <- nu[i,p]*E[i,p]; log(nu[i,p]) <- alph[p]+r[i,p]
```

```
    y[i,p]~dpois(mu[i,p]);
    M[i,p] <- (kap/(1-kap+kap*d[i]))*sum(radj[cum[i]+1 : cum[i+1], p ])
    for (k in 1:P) {Zeta.r[i,j,k] <- Zeta[j,k]*(1-kap+kap*d[i])}}}
    Zeta[1:P,1:P] ~dwish(Sc[,],P); kap ~dunif(0,1); Inv.Zeta[1:P,1:P]
<- inverse(Zeta[,])
    for (j in 1:P) {alph[j] ~dnorm(0,0.01)
    for (k in 1:P) {Sc[j,k] <- P*equals(j,k)}}
    for (i in 2:P) {for (j in 1:i-1){r.Zeta[i,j] <- Inv.Zeta[i,j]/sqrt(Inv.Zeta[i,i]
*Inv.Zeta[j,j])}}
    for (i in 1 : NN) { for (j in 1:P){radj[i,j] <- r[adj[i],j]}}}
```

8. The code in Example 7.9 is

```
    model { for ( t in 1:T){ for (p in 1:P) {beta.var[p,t] <- beta[t,p];
    # trends and cycles by variable p
    th.var[p,t] <- th[t,p]; psi.var[p,t] <- psi1[t,p]}
    y[t,1] <- logmink[t]; y[t,2] <- logmuskrat[t]; y[t,1:P] ~dmnorm(nu[t,1:P],
Precy[,])
    for (p in 1:P) {nu[t,p] <- th[t,p]+psi1[t,p]}}
    for (t in 2:T) {th[t,1:P] ~dmnorm(M.th[t,1:P],Precth[,])
    beta[t,1:P] ~dmnorm(beta[t-1,1:P],Precbeta[,])
    psi1[t,1:P] ~dmnorm(M.psi1[t,1:P],Precps1[,])
    psi2[t,1:P] ~dmnorm(M.psi2[t,1:P],Precps2[,])
    for (p in 1:P) {M.th[t,p] <- th[t-1,p]+beta[t,p]
    M.psi1[t,p] <- rho*cos(lam)*psi1[t-1,p]+rho*sin(lam)*psi2[t-1,p]
    M.psi2[t,p] <- -rho*sin(lam)*psi1[t-1,p]+rho*cos(lam)*psi2[t-1,p]}}
    for (p in 1:P){beta1[p] ~dnorm(0,0.01);th1[p] ~dnorm(0,0.01)
    psi11[p] ~dnorm(0,0.01); psi21[p] ~dnorm(0,0.01)
    psi1[1,p] <- psi11[p]; psi2[1,p] <- psi21[p]
    beta[1,p] <- beta1[p]; th[1,p] <- th1[p]}
    rho ~dunif(0,1); lam ~dunif(0.3,1.5); Period <- 2*pi/lam
    # prior precision matrices
    Precy[1:P,1:P] ~dwish(Scy[1:P,1:P],nu.y);
    for (a in 1:4) {q[a]~dgamma(1,1)}
    for (p1 in 1:P){ for (p2 in 1:P) {Scy[p1,p2] <- Sigy[p1,p2]*nu.y
    Precth[p1,p2] <- Precy[p1,p2]/q[1]; Precbeta[p1,p2] <- Precy[p1,p2]/q[2]
    Precps1[p1,p2] <- Precy[p1,p2]/q[3]; Precps2[p1,p2] <- Precy[p1,p2]/q[4]}}}
```

9. The code in Example 7.11, including direct calculation of the precision matrix for y, is

```
    model { for (t in 1:T) {D[t] <- exp(h[1,t]+h[2,t])*(1-rho.e*rho.e);
    y.Prec[t,1,1] <- exp(h[2,t])/D[t]; y.Prec[t,2,2] < - exp(h[1,t])/D[t];
    y.Prec[t,1,2] <- -rho.e*exp(0.5*h[1,t]+0.5*h[2,t])/D[t]; y.Prec[t,2,1]
<- y.Prec[t,1,2];
    y[t,1:2] ~dmnorm(nought[1:2],y.Prec[t,,])}
    # log volatility VAR
```

```
for (p in 1:P) { h.st[1,p] ~dnorm(mu[p],1)
for (t in 2:T) { h.st[t,p] ~dnorm(h.mu[t,p],1)
h.mu[t,p] <- mu[p] + ph[p]*(h.st[t-1,p]-mu[p]) }}
for (t in 1:T) {h[1,t] <- sig.u[1]*h.st[t,1];
h[2,t] <- sig.u[2]*rho.u*h.st[t,1]+sig.u[2]*sqrt(1-rho.u*rho.u)*h.st[t,2] }
# priors
for (p in 1:P) {inv.sig2.u[p] ~dgamma(1,1); sig.u[p] <- 1/sqrt(inv.sig2.u[p])
phstar[p] ~dbeta(19,1); ph[p] <- 2*phstar[p] -1; mu[p] ~dnorm(0,1)}
rho.e ~dunif(-1,1); rho.u ~dunif(0,1)}
```

8

Hierarchical Models for Panel Data

8.1 Introduction

Panel or longitudinal datasets occur when continuous or discrete observations y_{it} on a set of subjects or units $i = 1, \ldots, n$ are repeated over a number of measuring occasions $t = 1, \ldots, T_i$ possibly differing between subjects (with $N = \sum T_i$). There are many contexts for such data to occur, and many variations in study design and data format as well as variations in how subjects or units are defined. For instance in economic and marketing applications (Rossi et al., 2005), the panel is typically at the level of an individual consumer, household or firm, and relates to questions such as economic participation or brand choice (for households or consumers), or investments and outputs at firm level. In actuarial applications (Antonio and Beirlant, 2007; Frees et al., 2001) the "subjects" may consist of groups of policyholders (risk classes) with responses being insurance claim counts. In clinical trials, patients are randomly assigned to treatments and measures taken during follow-up to compare treatment effects.

As well as regular panel surveys in which repeat measurements on all subjects are contemporaneous, continuous measurement of variables, causing unbalanced longitudinal data (Daniels and Hogan, 2008), may be the most suitable method of identifying treatment effects or assessing social change. In the latter situation, the times $\{a_{it}, t = 1, \ldots, T_i\}$ at which events are recorded will differ between subjects. Furthermore, measuring occasions may be over more than one time scale. An example is analysis of mortality or cancer incidence by calendar time and age at onset or death, leading to a further implicit cohort scale defined by the difference between age and time (Berzuini and Clayton, 1994; Lagazio et al., 2003; Schmid and Held, 2007); see Section 8.7.

There are also a variety of approaches to analyzing panel data, such as the generalized estimating equation method proposed by Liang and Zeger (1986), as well as general linear mixed models (Molenberghs and Verbeke, 2006), and conditional log-linear models for categorical outcomes (Gilula and Haberman, 1994). One may further distinguish between random effects or conditional models on the one hand, and marginal or population-averaged approaches on the other (Heagerty and Zeger, 2000; Lee and Nelder, 2004), with parameters obtained under the former not necessarily having the same interpretation as under a marginal model (Neuhaus et al., 1991). The focus of this chapter is on conditionally specified hierarchical and random effect models, and on

Markov Chain Monte Carlo (MCMC) estimation via conditional likelihood with random effects as part of the parameter set (e.g., Chib and Carlin, 1999; Daniels and Hogan, 2008); see Section 8.2.

Panel data offer major advantages over cross-sectional designs in the analysis of causal interrelationships between variables, including developmental and growth processes and clinical studies (Davies, 1994; Finkel, 1995; Menard, 2002; Ruspini, 1999). The accumulation of information over both times and subjects increases the power of statistical methods to identify treatment effects or values added (Lockwood et al., 2003), and permits the estimation of parameters (e.g., permanent random effects or "frailties" for subjects i) that are not identifiable from cross-sectional analysis or from repeated cross-sections on different subjects.

On the other hand, analysis of panel data may be problematic if the panel sequences are subject to missing observations; see Section 8.8. Missingness may involve intermittently missing values of responses (or predictors) or total loss to observation after a certain point. The latter is variously known as attrition (Schafer and Graham, 2002) or dropout (Hogan et al., 2004). A particular question relevant to estimation of the main structural model is whether such permanent exit is random (independent of the response that would otherwise have been observed) or informatively related to the missing response (Goodman and Blum, 1996).

Bayesian estimation via repeated sampling from posterior densities facilitates hierarchical modeling of panel data, whether of permanent subject effects (often simply called "random effects"), correlated or unstructured observation level errors, time varying regressor effects or common factors in multivariate panel data. As noted by Davidian and Giltinan (2004), the random effects are treated as parameters in Bayesian MCMC estimation, and are ordinarily not integrated out as often done in frequentist approaches. The integrated or marginal likelihood estimation approach may become infeasible or unreliable in complex varying coefficient models (Tutz and Kauermann, 2003), and different parameter estimates may be obtained according to the maximization methods used (Molenberghs and Verbeke, 2004). Bayesian modeling perspectives are also important in the application of latent metric augmentation to categorical panel outcomes (binary, multinomial) (Chib and Carlin, 1999), and for dealing with missing data, especially attrition of subjects (Little and Rubin, 2002). By contrast, quasi-likelihood methods for categorical panel outcomes may be biassed, especially for binary panel data (Lesaffre and Spiessens, 2001).

8.2 General Linear Mixed Models for Panel Data

Following Tutz and Kauermann (2003), the first stage of general linear mixed model may be characterized by two components, the distributional and structural assumptions. Thus, conditional on predictors and random effects it is assumed that data y_{it} for subjects i and times t are distributed according to the exponential family,

$$p(y_{it}|\theta_{it}, \phi) = \exp\left(\frac{y_{it}\theta_{it} - a(\theta_{it})}{\phi} + C(y_{it}, \phi)\right),$$

where θ_{it} denotes the natural parameter (e.g., see Chib and Carlin, 1999; Gamerman, 1997; Gamerman, 1998; Natarajan and Kass, 2000; Tsai and Hsiao, 2008). The structural assumption governs the forms assumed for the conditional means $E(y_{it}) = \mu_{it} = a'(\theta_{it})$, with link $g(\mu_{it}) = \eta_{it}$ to a regression term η_{it}, and for the variances $V_{it} = \phi\text{Var}(\mu_{it})$. This involves questions such as whether the conditional mean is linear or nonlinear in predictors and random effects, and at what levels random effects are present.

Thus there are likely to be enduring differences between subjects that may be represented by permanent (time-invariant) random effects. To represent excess dispersion in discrete outcomes there may also need to be observation-level random effects. Fixed effects are also often used for representing subject level heterogeneity, especially in econometrics (Frees, 2004). This is equivalent to generating dummy variables for each subject and works best for relatively few subjects and more time periods, as there is no pooling strength in fixed effects models and the parameter count increases with the number of subjects n. In this chapter the focus is on random subject effects that are exchangeable over subjects, i.e., unstructured random effects, though if the units are spatially configured (say) then a structured prior for the permanent effects can be used. It is relevant to note that Marshall and Spiegelhalter (2007) assess prior-likelihood conflict in hierarchical models by p-tests that involve replicates sampled from hierarchical as against fixed effects priors for subject effects.

Conjugate priors may be suitable to handle unstructured random variation at subject or observation level for exponential family responses, especially if random variation does not involve predictors. For example, suppose y_{it} are overdispersed counts, then one possible scheme (Lee and Nelder, 2000, 593) involves multiplicative permanent effects h_i and observation level random effects h_{it}, so that the structural assumption takes the form:

$$E(y_{it}|h_i, h_{it}) = \exp(X_{it}\beta)h_i h_{it}.$$

If the $1 \times P$ predictor vector $X_{it} = (x_{1it}, \ldots, x_{Pit})$ includes an intercept, then both h_i and h_{it} have mean 1 for identification and would be gamma variables, $h_i \sim \text{Ga}(\varphi_1, \varphi_1)$, $h_{it} \sim \text{Ga}(\varphi_2, \varphi_2)$, with unknown hyperparameters.

However, more flexible models, possibly involving subject-specific regression effects as well as varying intercepts, involve a vector of subject level effects $b_i = (b_{1i}, \ldots, b_{Qi})'$ in a general linear mixed model format. These are typically taken to be normal with covariance D, and mean $B = (B_1, \ldots, B_Q)$ which is zero or nonzero depending on how the predictors are defined (see Section 8.2.1). As considered above (Chapter 6), mixed models extend linear regression by including both random and fixed effects in the model for the mean. So with link g, the structural assumption specifies

$$g[E(y_{it}|b_i.u_{it})] = \eta_{it} = X_{it}\beta + Z_{it}b_i + u_{it}, \quad (8.1a)$$

where $Z_{it} = (z_{1it}, \ldots, z_{Qit})$ is $1 \times Q$. In a typical analysis b_i and $u_i = (u_{i1}, \ldots, u_{iT_i})'$ are assumed independent, and both also taken to be normal, at least initially. Variations on the independence assumption might be to take the cluster effects to be spatially correlated, or the observation errors to be correlated through time. In some applications (e.g., Poisson data without overdispersion) the observation level errors u_{it} may not be present.

For a normal linear mixed model where responses are assumed independent given fixed effects β, and normal random effects b_i with covariance matrix D, one has

$$y_{it} = X_{it}\beta + Z_{it}b_i + u_{it},$$

with $u_i = (u_{i1,}, \ldots, u_{iT_i})'$ unstructured normal with mean zero, $T_i \times T_i$ covariance matrix $\Sigma_i = \sigma^2 I$, and conditional expectations,

$$E(y_{it}|b_i) = \eta_{it} = X_{it}\beta + Z_{it}b_i.$$

Equivalently therefore $y_{it}|b_i \backsim N(X_{it}\beta + Z_{it}b_i, \sigma^2)$. The normal linear mixed model may be achieved with latent rather than actual data, when the observed data are binary or categorical (e.g., Chib, 2008; Chib and Jeliazkov, 2006; Czado, 2000). Thus, for binary y_{it},

$$\begin{aligned} y_{it}|b_i &\backsim N(X_{it}\beta + Z_{it}b_i, 1) \quad I(0,) \text{ if } y_{it} = 1, \\ y_{it}|b_i &\backsim N(X_{it}\beta + Z_{it}b_i, 1) \quad I(,0) \text{ if } y_{it} = 0, \end{aligned}$$

with unstructured residuals under conditional independence having known variance $\sigma^2 = 1$, but with D still unknown.

Stacked over times the conditional mean in Equation 8.1a is expressed as

$$\eta_i = X_i\beta + Z_ib_i + u_i, \quad (8.1b)$$

where η_i is $T_i \times 1$, X_i is $T_i \times P$, and Z_i is $T_i \times Q$, while the normal mixed model is

$$y_i = X_i\beta + Z_ib_i + u_i.$$

For the linear normal mixed model, the marginal model (with b_i integrated out) is obtainable analytically as

$$y_i \backsim N(X_{it}\beta, Z_iDZ_i' + \sigma^2 I),$$

which is a feature not present for the broader class of general linear mixed models (Molenberghs and Verbeke, 2006).

Given the fixed effect regression and the full set of subject permanent effects $b = (b_1, \ldots, b_n)$, the assumption that repeated observations on the same subject (or more generally within units or clusters) are conditionally independent (Kleinman and Ibrahim, 1998; Tutz and Kauermann, 2003), means that the conditional likelihood factors as

$$p(y|b, \beta, \sigma^2) = \prod_{i=1}^{n} p(y_i|b_i, \beta, \sigma^2),$$

where $p(y_i|b_i, \beta, \sigma^2) = \prod_{t=1}^{T_i} p(y_{it}|b_i, \beta, \sigma^2)$. Similarly the joint density,

$$p(y, b|\beta, D, B, \sigma^2) = p(y|b, \beta, \sigma^2)p(b|B, D)$$

factors into subject specific elements $\left(\prod_{t=1}^{T_i} p(y_{it}|b_i, \beta, \sigma^2)\right)p(b_i|B, D)$.

If model checking reveals that a conditional independence assumption does not provide an adequate fit to the actual correlation structure, then the model is incomplete and requires elaboration. For example, such checking may show that in fact the regression errors are correlated through time, and the white noise assumption for the observation level errors u_{it} has to be reconsidered, or lagged effects in the response included in the predictor sets X_{it} or Z_{it} (Frees, 2004, 279)—see Sections 8.3 and 8.5.

One may also generalize Equation 8.1a to include predictors H_{it} of dimension $1 \times R$ with effects varying by time, as well as by subject. Denoting time varying regression coefficients as $c_t = (c_{1t}, \ldots, c_{Rt})'$, the structural assumption becomes

$$g[E(y_{it}|b_i, c_t.u_{it}] = X_{it}\beta + Z_{it}b_i + H_{it}c_t + u_{it}. \tag{8.2}$$

A particular example (with $Q = R = 1$) is the two-way error component model (Baltagi, 2003; Hjellvik and Tjøstheim, 1999) with

$$g[E(y_{it}|b_i, c_t.u_{it}] = X_{it}\beta + b_i + c_t + u_{it},$$

where the c_t represent effects over time influencing all the series. Tutz and Kauermann (2003) consider a time varying regression form of Equation 8.1a, namely

$$\eta_{it} = X_{it}\beta_t + Z_{it}b_i + u_{it}$$

with time acting as an "effect modifier." For example, they mention that assuming a constant effect of treatment variables in clinical trials may often be unrealistic.

8.2.1 Centered or noncentered priors

The parameterization adopted in the prior for b_i depends on whether X_{it} and Z_{it} are specified to be overlapping or distinct, and also on MCMC convergence considerations. In many presentations of the GLMM in panel analysis, the Z_{it} are assumed to be a subset of X_{it}, in which case the b_i are typically taken to be zero mean random effects usually following a standard density (e.g., multivariate normal) (Section 8.2.2). If the X_{it} and Z_{it} are nonoverlapping, then the b_{qi} may be taken to have nonzero means equal to the central (fixed effect) regression parameters B_q for z_{qit} (Chib and Carlin, 1999, 20), namely

$$g[E(y_{it}|b_i.u_{it})] = X_{it}\beta + Z_{it}b_i + u_{it},$$
$$b_i \sim N(B, D).$$

Such hierarchical centering may assist precise identification and MCMC convergence. If no predictors have fixed effect coefficients, one has what is sometimes termed a random coefficient regression, namely

$$g[E(y_{it}|b_i.u_{it})] = Z_{it}b_i + u_{it},$$

where b_{qi} have nonzero means B_q (e.g., Daniels and Hogan, 2008, 22; Yang and Chen, 1995).

Papaspiliopoulos et al. (2003) compare MCMC convergence for centered, noncentered, and partially noncentered hierarchical model parameterizations, and mention that hierarchical centering may be less effective when the latent effects b_i are relatively weakly identified. Consider the normal linear mixed model in the form:

$$y_i = X_i\beta + Z_ib_i + (\sigma^2 I_{T_i})e_i$$
$$b_i = B + D^{0.5}v_i,$$

where e_i and v_i, of dimension T_i and Q, respectively, are standard normal variables.

Then the noncentered parameterization (NCP) and partially noncentered parameterizations (PNCP) are, respectively,

$$\widetilde{b}_i = b_i - B$$

and

$$\widetilde{b}_i^w = b_i - W_iB,$$

where $W_i = U_iD^{-1}, U_i = (1/\sigma^2)Z_iZ_i + D^{-1}$, and

$$\widetilde{b}_i \sim N(0, D),$$
$$\widetilde{b}_i^w \sim N(B - W_iB, D).$$

The proportion of B subtracted from b_i under the PNCP form (that has favorable MCMC convergence properties) is observation specific. The longitudinal model under the NCP and PNCP becomes

$$y_i = X_i\beta + Z_iB + Z_i\widetilde{b}_i + (\sigma^2 I_{T_i})e_i,$$
$$y_i = X_i\beta + Z_iW_iB + Z_i\widetilde{b}_i^w + (\sigma^2 I_{T_i})e_i.$$

The NCP form has potential use in random effects selection (see Section 8.2.3).

8.2.2 Priors on permanent random effects

The most commonly adopted prior for the random subject or cluster effects $b_i = (b_{1i}, \ldots, b_{Qi})$ is an unstructured multivariate normal

$$(b_{1i}, \ldots, b_{Qi}) \backsim N_Q(B, D), \tag{8.3}$$

where the means $B = (B_1, \ldots, B_Q)$ are either zeroes or unknown fixed effects, and $D = [d_{rs}]$ represents covariation within subjects between the rth and sth random effects b_{ri} and b_{si}. If the Z_{it} are a subset of the X_{it}, then the means B_q will be zero. For robustness against non-normality or outliers, other forms of mixture, including scale mixtures of normals or discrete mixtures of random effects, may be assumed for subject effects (Section 8.6). For spatially configured units a prior for $(b_{1i}, \ldots, b_{Qi})$ including correlation over areas is likely to be relevant. For doubly nested data (e.g., observations y_{ijt} within subjects i within clusters j), the second stage parameters are likely to be cluster specific and possibly also randomly varying, as in

$$(b_{1ij}, \ldots, b_{Qij}) \backsim N_Q(B_j, D_j),$$
$$(B_{1j}, \ldots, B_{Qj}) \backsim N_Q(M_B, C_B).$$

In many applications the Z_{it} will be of relatively small dimension, confined to the intercept or simple time functions. For example, if $Q = 1$ and $Z_{it} = 1$, one has the normal linear form:

$$y_{it} = b_i + X_{it}\beta + u_{it}, \tag{8.4}$$

where b_i represent permanent subject effects, namely enduring differences between subjects due to unmeasured attributes. If X_{it} excludes (or includes) an intercept, then the b_i will be normal with mean B (or zero) and variance D.

In growth curve and educational attainment applications the Z_{it} typically include transforms of time or age, and the mean level for an individual changes with time or age (e.g., linearly or quadratically) with growth rate coefficients specific to each subject. For example, under a linear growth model with $Q = 2$, each subject has their own linear growth rate (Darby and Fearn, 1979; Weiss, 2005)

$$y_{it} = b_{1i} + b_{2i}t + X_{it}\beta + u_{it},$$

where D_{12} measures the correlation between intercepts and slopes. Assuming X_{it} omits an intercept and linear time term, one may take

$$(b_{1i}, b_{2i}) \sim N([B_1, B_2], D).$$

In particular, Weiss (2005, 254) refers to random intercept and slope (RIAS) models in which

$$y_{it} = b_{1i} + b_{2i}t + u_{it},$$

so that an individual's observed performance will differ from his/her mean level at a particular time or age by a random term u_{it}. Another option is to replace known time functions by an unknown time varying function, δ_t, as in

$$y_{it} = b_{1i} + b_{2i}\delta_t + X_{it}\beta + u_{it},$$

with δ_t subject to identifying constraints (e.g., $\delta_1 = 1$), or with the variance of b_i preset. The δ_t are interpretable as factor loadings by time. Related models are considered by Lee and Carter (1992) and Tutz and Reithinger (2007).

To illustrate MCMC sampling, consider the random coefficient normal linear model, namely

$$y_{it} = Z_{it}b_i + u_{it}$$

with $u_{it} \sim N(0, \sigma^2)$, and $(b_{1i}, b_{2i}, \ldots, b_{Qi}) \sim N_Q([B_1, B_2, \ldots, B_Q], D)$. Let $\tau = 1/\sigma^2$ and assume, following Wakefield et al. (1994), that $\tau \sim \text{Ga}(\nu_0/2, \tau_0\nu_0/2)$. Also assume a multivariate normal prior for the second-stage population means B, and a Wishart prior for D^{-1}, namely

$$\begin{aligned} B &\sim N(B_0, C), \\ D^{-1} &\sim W([\rho R]^{-1}, \rho). \end{aligned}$$

Setting $N = \sum_{i=1}^{n} T_i, E_i^{-1} = \tau Z_i' Z_i + D^{-1}, V^{-1} = nD^{-1} + C^{-1}$, and $\bar{b} = \sum_{i=1}^{n} b_i/n$, then Gibbs sampling involves the posterior conditionals

$$\begin{aligned} b_i &\sim N\left(E_i[\tau Z_i' y + D^{-1}B], E_i\right), \quad i = 1, \ldots, n, \\ B &\sim N\left(V\left[nD^{-1}\bar{b} + C^{-1}B_0\right], V\right), \\ D^{-1} &\sim W\left(\left[\sum_{i=1}^{n}(b_i - B)(b_i - B)' + \rho R\right]^{-1}, n + \rho\right), \\ \tau &\sim \text{Ga}\left(\frac{\nu_0 + N}{2}, \frac{1}{2}\left[\sum_{i=1}^{n}(y_i - Z_i b_i)'(y_i - Z_i b_i) + \nu_0\tau_0\right]\right). \end{aligned}$$

When predictors are otherwise available that might explain heterogeneity between subjects (e.g., treatment allocations), regression priors may be used for the permanent random effects b_{qi} (Chib, 2008, 481). Regression priors are illustrated by growth curve applications in which randomly varying effects of time variables are related to fixed subject attributes in a higher stage regression (Candel and Winkens, 2003). Thus, Muthen et al. (2002) consider a model with varying intercepts b_{1i}, varying linear growth effects b_{2i}, and varying quadratic growth effects b_{3i}. Subjects are observed at differentially spaced time points $\{a_{i1}, a_{i2}, \ldots, a_{iT_i}\}$. So

$$y_{it} = b_{1i} + b_{2i}a_{it} + b_{3i}a_{it}^2 + u_{it},$$

where random growth coefficients are related to an intervention variable Tr_i according to

$$b_{1i} = B_1 + e_{1i},$$
$$b_{2i} = B_2 + \delta_2 Tr_i + e_{2i}, \quad (8.5)$$
$$b_{3i} = B_3 + \delta_3 Tr_i + e_{3i},$$

where $(e_{1i}, \ldots, e_{Qi}) \backsim N_Q(0, D)$. Treatment is randomized so the permanent (equivalently baseline) effects b_{1i} are taken to be independent of the intervention Tr_i.

8.2.3 Priors for random covariance matrix and random effect selection

Inferences may be sensitive to the form of prior adopted for D, and the amount of information it contains. Improper or overly diffuse priors on D or other variance hyperparameters may be associated with actual or effectively improper posterior densities. For example, the Jeffrey's rule prior, namely

$$p(D) \propto \det(D)^{-(Q+1)/2}$$

may lead to an improper joint posterior for D and β under certain conditions (Natarajan and Kass, 2000). The conjugate model for $Q > 1$ involves a Wishart prior for D^{-1}, $D^{-1} \sim \text{Wish}(S, \nu)$, or

$$p(D^{-1}|S, \nu) \propto |S|^{\nu/2}|D^{-1}|^{0.5(\nu-Q-1)} \exp(-0.5\text{tr}(SD^{-1}),$$

where $E(D^{-1}) = \nu S^{-1}$ and $E(D) = S/(\nu - Q - 1)$.

Setting the elements in S may be difficult in the absence of substantive information. Greater flexibility, including random effects selection, may be gained with matrix decomposition alternatives to the Wishart (e.g., Frühwirth-Schnatter and Tüchler, 2008; Hedeker, 2003; Kinney and Dunson, 2007; Tutz and Kauermann, 2003), or with adaptations of the uniform shrinkage prior (Natarajan and Kass, 2000; Tsai and Hsiao, 2008).

Consider the Cholesky decomposition $D = CC'$ where C is a lower triangular matrix, with $D_{pq} = \sum_{r=1}^{Q} c_{pr}c_{qr}$ and variances obtained as

$$D_{qq} = \sum_{r=1}^{Q} c_{qr}^2.$$

Then if Z_{it} is a subset of X_{it}, Equation 8.1a may be expressed,

$$\eta_{it} = X_{it}\beta + Z_{it}C\zeta_i + u_{it},$$

where $(\zeta_{1i}, \ldots, \zeta_{Qi}) \backsim N_Q(0, I)$. For example with $Q = 2$,

$$(z_{1it}, z_{2it}) \begin{pmatrix} c_{11} & 0 \\ c_{21} & c_{22} \end{pmatrix} \begin{pmatrix} \zeta_{1i} \\ \zeta_{2i} \end{pmatrix} = \zeta_{1i}(z_{1it}c_{11} + z_{2it}c_{21}) + \zeta_{2i}z_{2it}c_{22}.$$

Instead of a Wishart prior on D^{-1}, priors are then adopted for each element of C. To ensure D is positive definite, the diagonal terms c_{11} and c_{22} need to be assigned positive priors, while the prior c_{21} is unconstrained. For $Q = 3$, one has

$$(z_{1it}, z_{2it}, z_{3it}) \begin{pmatrix} c_{11} & 0 & 0 \\ c_{21} & c_{22} & 0 \\ c_{31} & c_{32} & c_{33} \end{pmatrix} \begin{pmatrix} \zeta_{1i} \\ \zeta_{2i} \\ \zeta_{3i} \end{pmatrix}$$
$$= \zeta_{1i}(z_{1it}c_{11} + z_{2it}c_{21} + z_{3it}c_{31}) + \zeta_{2i}(z_{2it}c_{22} + z_{3it}c_{32}) + \zeta_{3i}z_{3it}c_{33},$$

with three positive unknowns c_{qq} and three unconstrained lower diagonal unknowns.

Cai and Dunson (2006) and Chen and Dunson (2003) use the Cholesky decomposition

$$D = \Lambda\Omega\Omega'\Lambda,$$

where $\Lambda = \text{diag}(\lambda_1, \ldots, \lambda_Q)$ and Ω is lower triangular,

$$\Omega = \begin{pmatrix} 1 & 0 & \cdots & 0 \\ \omega_{21} & 1 & \cdots & 0 \\ \cdots & \cdots & \ddots & 0 \\ \omega_{Q1} & \omega_{Q2} & \cdots & 1 \end{pmatrix}.$$

Hence C in $D = CC'$ can be written

$$C = \begin{pmatrix} \lambda_1 & 0 & \cdots & 0 \\ \omega_{21}\lambda_2 & \lambda_2 & \cdots & 0 \\ \cdots & \cdots & \ddots & 0 \\ \omega_{Q1}\lambda_Q & \omega_{Q2}\lambda_Q & \cdots & \lambda_Q \end{pmatrix}.$$

Then positive priors (e.g., log-normal, gamma) are taken for the elements of Λ, while normal $N(0, V_\Omega)$ priors may be assumed for the unconstrained elements of Ω. Chen and Dunson (2003, 865) mention that taking $V_\Omega = 0.5$ leads to relatively diffuse priors on the correlations between the random effects b_{qi}. Retention or otherwise of the terms in Λ is determined by binary indicators $\gamma_{qq} \sim \text{Bern}(\pi_{qq})$, where π_{qq} may be preset or extra unknowns. Retention or otherwise of the unknown terms in Ω is determined both by binary indicators $\gamma_{qr} \sim \text{Bern}(\pi_{qr})$, and also by whether λ_q and λ_r are retained; if either of $\{\lambda_q, \lambda_r\}$ is omitted, then ω_{qr} necessarily is.

If Z_{it} is not a subset of X_{it}, Frühwirth-Schnatter and Tüchler (2008) adopt the noncentered parameterization:

$$\eta_{it} = X_{it}\beta + Z_{it}B + Z_{it}C\zeta_i + u_{it},$$

where $(\zeta_{1i}, \ldots, \zeta_{Qi}) \backsim N_Q(0, I)$. As above, diagonal terms c_{qq} need to be assigned positive priors, while priors for c_{qr} $(q > r)$ are unconstrained. Selection of which c_{qq} and c_{qr} terms to retain may be based on binary indicators $\{\gamma_{qq} \sim \text{Bern}(\pi_{qq}), \pi_{qq} \backsim \text{Be}(a_{qq}, b_{qq})\}, \{\gamma_{qr} \sim \text{Bern}(\pi_{qr}), \pi_{qr} \backsim \text{Be}(a_{qr}, b_{qr}), q > r\}$ where $a_{qq} = b_{qq} = a_{qr} = b_{qr} = 1$ is a default option. In effect the model involves composite terms:

$$G_{qq} = c_{qq}\gamma_{qq},$$
$$G_{qr} = c_{qr}\gamma_{qr}, \quad q > r$$

so that for $Q = 2$:

$$\eta_{it} = X_{it}\beta + \zeta_{1i}(z_{1it}G_{11} + z_{2it}G_{21}) + \zeta_{2i}z_{2it}G_{22} + u_{it}.$$

The posterior estimate for D would be based on MCMC monitoring of $D_{qr} = \sum_{s=1}^{Q} G_{qs}G_{rs}$.

A regression approach to covariance estimation for panel data is proposed by Daniels and Pourahmadi (2002); Pourahmadi (1999, 2000), and Pourahmadi and Daniels (2001). For the essence of the method, consider normal metric data y_{it} for subjects $i = 1, \ldots, n$ with individual specific covariance matrices Σ_i of dimension $T \times T$. The model $y_{it} \sim N(\mu_{it}, \Sigma_i)$ may be re-expressed (for the purposes of decomposing Σ_i) as an antedependence model:

$$y_{it} - \mu_{it} = \sum_{j=1}^{t-1} \phi_{itj}(y_{ij} - \mu_{ij}) + u_{it},$$

where the errors $u_{it} \sim N(0, h_{it})$ are uncorrelated. Denote,

$$H_i = \text{diag}(h_{i1}, \ldots, h_{iT}),$$

together with the lower triangular matrix,

$$F_i = \begin{bmatrix} 1 & & & & \\ -\phi_{i21} & 1 & & & \\ -\phi_{i31} & -\phi_{i32} & 1 & & \\ \cdots & \cdots & \cdots & \cdots & \\ -\phi_{iT1} & -\phi_{iT2} & \cdots & -\phi_{iT,T-1} & 1 \end{bmatrix}.$$

One then has the decomposition,

$$\text{Var}(u_i) = H_i = F_i\Sigma_i F_i'.$$

The parameters ϕ_{itj} and h_{it} may be referred to respectively as the generalized autoregressive parameters and the innovation variances of Σ_i (Daniels and Pourahmadi, 2002).

A parsimonious covariance model, especially for large T, may then be achieved by using predictors z_{it} and w_{itj} in the regressions,

$$\log(h_{it}) = z_{it}\gamma,$$
$$\phi_{itj} = w_{itj}\lambda.$$

Often one might take $\Sigma_i = \Sigma$, in which case

$$y_{it} - \mu_{it} = \sum \phi_{tj}(y_{ij} - \mu_{ij}) + u_{it}$$

with $u_{it} \sim N(0, h_t)$, $H = \text{diag}(h_1, \ldots, h_T)$, and

$$F = \begin{bmatrix} 1 & & & & \\ -\phi_{21} & 1 & & & \\ -\phi_{31} & -\phi_{32} & 1 & & \\ \cdots & \cdots & \cdots & \cdots & \\ -\phi_{T1} & -\phi_{T2} & \cdots & -\phi_{T,T-1} & 1 \end{bmatrix},$$

with $\text{Var}(u_i) = H = F\Sigma F'$. The covariates used for covariance model become $\{z_t, w_{tj}\}$ where the w_{tj} might simply be powers in $(t-j)$ as illustrated by Cepeda and Gamerman (2004), and the z_t are simply powers of t. A possible drawback to using polynomial functions of time is the multicollinearity that may be encountered, and Bayesian regression selection may then be applied. One may also consider autoregressive or random walk priors in h_t and modeling ϕ_{tj} as a collection of unstructured random effects under a shrinkage prior strategy (Daniels and Pourahmadi, 2002, 558).

8.2.4 Priors for multiple sources of error variation

Estimation of variance components and convergence of MCMC samplers for longitudinal data may also be sensitive to the assumed prior interlinkages (or not) between multiple sources of random variation. Consider a random intercept model

$$y_{it} = b_i + X_{it}\beta + u_{it},$$

with unknown variances $D = \text{Var}(b_i)$ and $\sigma^2 = \text{Var}(u_{it})$. The conjugate approach with the advantage of simple posterior conditionals involves separate gamma priors on D^{-1} and $\tau = \sigma^{-2}$. These could be informative (e.g., downweighted results from a maximum likelihood fit), but are often taken to be diffuse with small scale and shape parameters, leading to potentially delayed convergence of Gibbs sampling methods since sampling is from an almost improper posterior (Natarajan and McCulloch, 1998). These problems may increase if an autocorrelated error term is added to the white noise error as in

$$y_{it} = b_i + X_{it}\beta + u_{it} + \varepsilon_{it},$$
$$\varepsilon_{it} = \rho\varepsilon_{i,t-1} + v_{it},$$

where $v_{it} \backsim N(0, \sigma_v^2)$, and there are three variances.

An alternative is to allow for potential interdependence between variance components via adaptations of the uniform shrinkage prior (Natarajan and Kass, 2000). The uniform shrinkage principle extends to beta priors on the relative shares for two variances and to Dirichlet priors on the relative shares for three or more variance components. So one might set a prior on one or other of D or σ^2, but then specify a variance partitioning rule such as $\kappa = D/(D+\sigma^2) \sim U(0,1)$ to obtain the other. A related strategy might take $\kappa = D/(D+\sigma^2) \sim \text{Be}(a_\kappa, b_\kappa)$, where a_κ and b_κ are preset or hyperparameters. Lee and Hwang (2000) use uniform shrinkage priors in a multilevel panel context when repetitions $t = 1, \ldots, T_{ij}$ are for subjects i nested within clusters j, and where there is an autocorrelated error ε_{ijt} as well as a white noise error u_{ijt}. An extension to multivariate b_i of the uniform shrinkage prior proposed by Natarajan and Kass (2000) takes the form:

$$p(D) \propto \det\left[I_Q + \left\{\frac{1}{n}\sum_{i=1}^{n} Z_i' W_i Z_i\right\} D\right],$$

where W_i is diagonal of dimension T_i with elements $1/V_{it}[\partial\eta_{it}/\partial\mu_{it}]^2$, where $V_{it} = \phi\text{Var}(\mu_{it})$ and $g(\mu_{it}) = \eta_{it}$.

Example 8.1. Growth Model Simulation To illustrate random effects selection in a panel setting, observations y_{it} $(i = 1, \ldots, 500; t = 1, \ldots, 5)$ are generated according to

$$\begin{aligned} y_{it} &= b_{1i} + b_{2i}t + b_{3i}x_{it} + u_{it}, \\ x_{it} &\sim U(-1,1), \\ (b_{1i}, b_{2i}, b_{3i}) &\sim N(B, D), \\ u_{it} &\sim N(0, 1/\tau); \tau = 0.5, \\ D^{-1} &= \begin{pmatrix} 1 & 0 & 0 \\ 0 & 100 & 0 \\ 0 & 0 & 10{,}000 \end{pmatrix}, \\ B &= (5, 0.5, 0.5), \end{aligned}$$

with the random effects b_{qi} having respective standard deviations $\{1, 0.1, 0.01\}$.

The parameters are then re-estimated under a "standard model" with conjugate prior for the precision matrix $D^{-1} \backsim W(I,3)$, and with $\tau \sim \text{Ga}(1, 0.001)$ and $(B_q \sim N(0,100), q = 1,3)$. A more diffuse prior on D^{-1} (e.g., a scale matrix with diagonal terms 0.001) is avoided as it delays convergence on the parameters in D. The last 4000 of a 5000 iteration two chain run provides estimated means (sd) for $\sigma_{bq} = D_{qq}^{0.5}$ of 0.99 (0.09), 0.206 (0.02), and 0.401 (0.063). All three standard deviations σ_{bq} suggest significant variation (at odds with the original simulation parameters), and additionally an off-diagonal term in D, namely D_{12}, is significant, in the sense of a posterior 95% interval confined to entirely negative values.

The posterior results for the standard deviations σ_{bq} from the standard model are used to set priors on the diagonal Cholesky terms c_{qq} in a random effects selection model (Fruhwirth-Schnatter and Tuchler, 2008). Inferences from such selection may be sensitive to the prior used on the Cholesky elements, and a full analysis would consider several choices of prior. Then as mentioned in Section 8.2.3, with composite terms $G_{qr} = c_{qr}\gamma_{qr}$ and $Z_{it} = (1, t, x_{it})$, the linear predictor is

$$\begin{aligned} y_{it} &= Z_{it}B + Z_{it}G\zeta_i + u_{it} \\ &= B_1 + B_2 t + B_3 x_{it} + \zeta_{1i}(z_{1it}G_{11} + z_{2it}G_{21} + z_{3it}G_{31}) \\ &\quad + \zeta_{2i}(z_{2it}G_{22} + z_{3it}G_{32}) + \zeta_{3i}z_{3it}G_{33} + u_{it}. \end{aligned}$$

For the selection indicators, $\gamma_{qr} \sim \text{Bern}(\pi_{qr})$, where π_{qr} are assigned uniform priors, while the off-diagonal Cholesky terms c_{qr} $(q > r)$ are assigned $N(0, 1)$ priors. The posterior means and standard deviations on σ_{bq} from the standard model analysis can be summarized by gamma priors Ga(122, 124), Ga(100, 489), and Ga(40, 100). These are used to set gamma priors on c_{qq} in the selection model, but with precision downweighted a 100 times, namely Ga(1.22, 1.24), Ga(1, 4.9), and Ga(0.4, 1); heavier downweighting (e.g., a thousandfold) is avoided as it may lead to over-diffuse priors.

The last 4000 of a 5000 iteration two chain run using random effect selection[1] with $D = CC'$ provides estimated means (medians) for $\sigma_{bq} = D_{qq}^{0.5}$ of 0.95 (0.95), 0.012 (0), and 0.092 (0.023). The posterior densities for the standard deviations σ_{b2} and σ_{b3} both have spikes at zero, while no off-diagonal terms in D are significant. While not completely reproducing the original simulation parameters, the selection approach has provided more accurate estimates of the original σ_{bq}, namely $\{1, 0.1, 0.01\}$.

Random effect selection is also undertaken with $D = \Lambda\Omega\Omega'\Lambda$, namely $C = \Lambda\Omega$. Priors on λ_q are the same as for c_{qq} under the decomposition $D = CC'$, while $N(0, 0.5)$ priors are used for the elements of Ω, and Bernoulli priors with preset probability 0.5 for γ_{qr}. The last 4000 of a 5000 iteration two chain run provides estimated means (medians) for $\sigma_{bq} = D_{qq}^{0.5}$ of 0.95 (0.95), 0.012 (0), and 0.041 (0). The posterior densities for the standard deviations σ_{b2} and σ_{b3} again both have spikes at zero, and no off-diagonal terms in D are significant.

It may be noted that either of the Cholesky decomposition approaches could be applied to estimate the random effects covariance parameters, but without selection being applied (in effect $\gamma_{qr} = 1$); this is left as an exercise. For example, with Ga(1, 1) priors on λ_q and $N(0, 0.5)$ priors on the ω_{qr}, the Chen and Dunson (2003) method gives posterior means (and 95% intervals) for σ_{b2} and σ_{b3} of 0.04 (0.002, 0.12), and 0.18 (0.01, 0.43), with σ_{b3} less inflated (as compared to the true value) than under the conjugate Wishart prior.

Example 8.2. Joint Regression for Mean and Covariance This example follows Cepeda and Gamerman (2004) in applying the method of Pourahmadi (1999) to $T = 24$ monthly height readings y_{it} for $n = 6$ students. The antedependence re-expression of the model $y_{it} \sim N(\mu_t, \Sigma)$ is

$$y_{it} - \mu_t = \sum \phi_{tj}(y_{ij} - \mu_j) + u_{it}$$

with $u_{it} \sim N(0, h_t)$. Taking $H = \text{diag}(h_1, \ldots, h_T)$ and

$$F = \begin{bmatrix} 1 & & & & \\ -\phi_{21} & 1 & & & \\ -\phi_{31} & -\phi_{31} & 1 & & \\ \cdots & \cdots & \cdots & \cdots & \\ -\phi_{T1} & -\phi_{T1} & \cdots & -\phi_{T,T-1} & 1 \end{bmatrix}$$

provides the covariance decomposition $\text{Var}(u_i) = H = F\Sigma F'$. The covariates $\{w_{tj}, z_t\}$ used for the covariance regression model are powers of $t - j$ for w_{tj}, and powers of t for z_t. Then the model takes

$$\begin{aligned} \mu_t &= \beta_1 + \beta_2 t + \beta_3 t^2 + \beta_4 t^3, \\ \log(h_t) &= \gamma_1 + \gamma_2 t + \gamma_3 t^2, \\ \phi_{tj} &= \lambda_1 + \lambda_2(t-j) + \lambda_3(t-j)^2, \end{aligned}$$

with the model for ϕ_{tj} here extending only to a quadratic term in $(t-j)$ rather than a quartic as in Cepeda and Gamerman (2004).

The last 8000 iterations from a two chain run of 10,000 iterations[2] provide posterior mean and standard deviation estimates for the parameters as follows: $\beta_1 = 94.25(0.37)$, $\beta_2 = 0.82(0.11)$, $\beta_3 = -0.021(0.011)$, $\beta_4 = 3.9E-4(3.3E-4)$, $\gamma_1 = 0.36(0.38)$, $\gamma_2 = -0.194(0.077)$, $\gamma_3 = 0.0087(0.0032)$, $\lambda_1 = 0.37(0.038)$, $\lambda_2 = -0.06(0.01)$, and $\lambda_3 = 0.0021(5.2E-4)$. Predictions from the model reproduce the observations satisfactorily, with 139 of the 144 (i.e., 96.5%) data points contained within 95% credible intervals of replicate data.

8.3 Temporal Correlation and Autocorrelated Residuals

Correlation between regression errors at different times is obtained as a by-product of other random effect schemes, not only from explicit time series priors. The random intercept model in Equation 8.4, applied widely in econometrics and clinical studies, illustrates how subject level random effects induce temporal correlation. It is important to control for such heterogeneity in order to avoid spurious "state dependence," namely dependence of the current outcome or probability on past outcomes or occurrences (Chib and Jeliakov, 2006). Thus for metric data, suppose

$$y_{it} = b_i + X_{it}\beta + u_{it},$$

where X_{it} includes an intercept, $b_i \sim N(0, D)$, and $u_{it} \sim N(0, \sigma^2)$. On the assumption that u_{it} and b_i are independent, the correlation between $\omega_{it} = u_{it} + b_i$ and $\omega_{is} = u_{is} + b_i$ at periods t and s is

$$\kappa = \text{cov}(\omega_{it}, \omega_{is})/\text{Var}(\omega_{it}) = D/(D + \sigma^2),$$

sometimes called the intra-class correlation. The random intercept model leads to the "compound symmetry" form for the intra-subject covariance matrix Σ_i (Weiss, 2005, 246–50), with diagonal terms $\Sigma_{itt} = \sigma^2$, and off-diagonal terms $\Sigma_{ist} = \sigma^2\kappa, s \neq t$. Equivalently

$$\Sigma_i = \sigma^2[(1-\kappa)I + \kappa J],$$

where J is an $T_i \times T_i$ matrix of ones.

A factor analytic form of the random intercept model (Weiss, 2005, 269) includes period-specific loadings D_t,

$$y_{it} = D_t^{0.5} b_i + X_{it}\beta + u_{it},$$

with $b_i \sim N(0,1)$ having a known variance to provide identification, so that

$$\kappa_t = D_t/(D_t + \sigma^2),$$

and the correlation between $\omega_{it} = D_t^{0.5} b_i + u_{it}$ at times t and s is $(\kappa_t \kappa_s)^{0.5}$. The corresponding RIAS model has loadings $D_{1t}^{0.5}$ and $D_{2t}^{0.5}$ and standard normal random effects b_{1i} and b_{2i}, so that

$$y_{it} = D_{1t}^{0.5} b_{1i} + D_{2t}^{0.5} b_{2i} t + X_{it}\beta + u_{it}.$$

For discrete data, the temporal correlation under random intercept and RIAS models may be confined to positive values only. Thus for Poisson counts y_{it}, with $\log[E(y_{it}|b_i)] = \log(\mu_{it}) = X_{it}\beta + b_i$, and $\gamma_{it} = \exp(X_{it}\beta)$, one has under conditional independence that

$$\begin{aligned}\text{cov}(y_{it}, y_{is}) &= E[\text{cov}(y_{it}, y_{is}|b_i)] + \text{cov}[E(y_{it}|b_i), E(y_{is}|b_i)] \\ &= \text{cov}([e^{b_i}\gamma_{it}], [e^{b_i}\gamma_{is}]) = \gamma_{it}\gamma_{is}\text{var}(e^{b_i}),\end{aligned}$$

while

$$\begin{aligned}\text{var}(y_{it}) &= E[\text{var}(y_{it}|b_i)] + \text{var}[E(y_{it}|b_i)] \\ &= E[e^{b_i}\gamma_{it}] + \text{var}[e^{b_i}\gamma_{it}] = \mu_{it} + \gamma_{it}^2\text{var}(e^{b_i}),\end{aligned}$$

with correlation then necessarily positive.

8.3.1 Explicit temporal schemes for errors

When residuals from Equations 8.1 and 8.2 show temporal correlation, autocorrelated residuals may be used instead of, or in addition to, the white noise errors u_{it}. Let these take generic form:

$$(\varepsilon_{i1}, \ldots, \varepsilon_{iT_i}) \backsim N(0, \Sigma_i),$$

where Σ_i is a unit level covariance matrix of dimension $T_i \times T_i$. Commonly adopted schemes for such residuals include low-order random walks (e.g., first order or $RW1$ priors), or low-order stationary schemes (typically $AR1$ or $MA1$). For example, Xu et al. (2007), Oh and Lim (2001), and Ibrahim et al. (2000) adopt stationary $AR1$ errors in models for panel count data, with $y_{it} \sim \text{Po}(\mu_{it})$,

$$\log(\mu_{it}) = X_{it}\beta + \varepsilon_{it},$$
$$\varepsilon_{it} = \rho\varepsilon_{i,t-1} + \upsilon_{it},$$

where $\upsilon_{it} \backsim N(0, \sigma_\upsilon^2)$ are unstructured and $|\rho| < 1$. For metric data, a stationary $AR1$ error scheme with

$$y_{it} = X_{it}\beta + \varepsilon_{it},$$
$$\varepsilon_{it} = \rho\varepsilon_{i,t-1} + \upsilon_{it},$$

where $|\rho| < 1$ and $\upsilon_{it} \backsim N(0, \sigma_\upsilon^2)$, leads to error covariance matrix Σ_i with elements

$$\Sigma_{ist} = \text{var}(\varepsilon_{it})\rho^{|s-t|} = \frac{\sigma_\upsilon^2}{1-\rho^2}\rho^{|s-t|},$$

with (e.g., Witkovsky, 1996),

$$\Sigma_i = \frac{\sigma_\upsilon^2}{1-\rho^2}\begin{pmatrix} 1 & \rho & \rho^2 & \cdots & \rho^{T_i-1} \\ \rho & 1 & \rho & \rho^2 & \cdots \\ \cdots & \cdots & \cdots & \cdots & \cdots \\ \cdots & \rho^2 & \rho & 1 & \rho \\ \rho^{T_i-1} & \cdots & \rho^2 & \rho & 1 \end{pmatrix}.$$

Assuming homogenous parameters across subjects so that $\Sigma_i = \Sigma$, and that subjects are independent, the full population covariance matrix is

$$\Phi = \frac{\sigma_\upsilon^2}{1-\rho^2} I_n \otimes \Sigma,$$

where I_n is an identity matrix of order n. With $e_{it} = y_{it} - X_{it}\beta$, the marginal likelihood for parameters $\chi = (\rho, \sigma_\upsilon^2, \beta)$ is then of the form $L(\chi|y) = \text{const} - 0.5\log|\Phi| + e'\Phi^{-1}e$.

A stationary first-order moving average or $MA1$ scheme (Baltagi and Li, 1995), with

$$y_{it} = X_{it}\beta + u_{it} + \theta u_{i,t-1},$$

and with $|\theta| < 1$, leads to a particular form of a Toeplitz covariance matrix (Weiss, 2005, 267). Thus, set $\varphi^2 = \text{var}(u_{it} + \theta u_{i,t-1}) = \sigma^2(1+\theta^2)$ and $\gamma = \theta/(1+\theta^2)$, then

$$\Sigma_i = \varphi^2\begin{pmatrix} 1 & \gamma & 0 & 0 & \cdots \\ \gamma & 1 & \gamma & 0 & \cdots \\ \cdots & \cdots & \cdots & \cdots & \cdots \\ \cdots & 0 & \gamma & 1 & \gamma \\ \cdots & \cdots & 0 & \gamma & 1 \end{pmatrix} = \sigma^2\begin{pmatrix} 1+\theta^2 & \theta & 0 & 0 & \cdots \\ \theta & 1+\theta^2 & \theta & 0 & \cdots \\ \cdots & \cdots & \cdots & \cdots & \cdots \\ \cdots & 0 & \theta & 1+\theta^2 & \theta \\ \cdots & \cdots & 0 & \theta & 1+\theta^2 \end{pmatrix}.$$

Stationary or random walk models for errors can be extended in various ways. Thus for unequally spaced data at points $\{a_{i1}, a_{i2}, \ldots, a_{iT}\}$, the $AR1$ model becomes

$$\varepsilon_{it} = \rho^{|a_{it}-a_{i,t-1}|}\varepsilon_{i,t-1} + \upsilon_{it},$$

with covariance between errors given by (Baltagi and Wu, 1999),

$$\text{cov}(\varepsilon_{it}, \varepsilon_{is}) = \rho^{|a_{it}-a_{is}|}\sigma_\upsilon^2.$$

Another option when T_i is relatively large are subject varying autocorrelation parameters, with Ryu et al. (2007) proposing independently distributed $\rho_i \sim U(-1, 1)$. Autoregressive and random walk priors are also applicable for subject level predictor effects β_{it} which are changing over time. Thus, an autoregressive prior could be centered on an average subject effect B_i, as in

$$\beta_{it} = B_i + \rho_\beta(\beta_{i,t-1} - B_i) + e_{it},$$

where B_i is itself random and the e_{it} are unstructured (see Example 8.3).

The use of autocorrelated or random walk effects raises issues about how to specify the initial conditions (initial random effects) such as ε_{i1} under an $AR1$ or $RW1$ prior on ε_{it}, and $\{\varepsilon_{i1}, \varepsilon_{i2}\}$ under an $AR2$ or $RW2$ prior. For stationary autoregressive errors, such as the $AR1$ prior

$$\varepsilon_{it} = \rho\varepsilon_{i,t-1} + \upsilon_{it},$$

the variances of ε_{it} and υ_{it} are analytically linked, so that the initial conditions are necessarily specified as part of the prior. So for stationary $AR1$ dependence in ε_{it} and equally spaced data, one has

$$\varepsilon_{i1} = \upsilon_{i1}/(1 - \rho^2)^{0.5},$$

and

$$\text{var}(\varepsilon_{i1}) = \sigma_\upsilon^2/(1 - \rho^2),$$

and the joint distribution of the ε_{it} is obtained (Xu et al., 2007, 418) as

$$p(\varepsilon_{i1}) \prod_{t=2} p(\varepsilon_{it}|\varepsilon_{i,t-1}),$$

where

$$p(\varepsilon_{it}|\varepsilon_{i,t-1}) = \frac{1}{\sigma_\upsilon(2\pi)^{0.5}} \exp(-0.5[\varepsilon_{it} - \rho\varepsilon_{i,t-1}]^2/\sigma_\upsilon^2).$$

In nonstationary and random walk models without the constraint $|\rho| < 1$, the initial conditions are usually specified by diffuse fixed effect priors, though Chib and Jeliakov (2006) interlink the variance of the initial conditions with that of the main sequence of effects to provide a proper joint prior on $\{\varepsilon_{i1}, \ldots, \varepsilon_{iT_i}\}$ under a $RW2$ prior. One may also link initial conditions ε_{i1} and subject heterogeneity, as in

$$b_i \sim N(\psi\varepsilon_{i1}, \sigma_b^2),$$

where ψ can be positive or negative (Chamberlain and Hirano, 1999). This amounts to assuming a bivariate density for b_i and ε_{i1}.

Regressions involving low-order autoregressive errors may sometimes be transformable to models with unstructured errors by differencing the regression means. Thus, the difference $\varepsilon_{it} - \varepsilon_{i,t-1}$ between a random walk error ε_{it} and its predecessor is white noise, as is the difference $\varepsilon_{it} - \rho\varepsilon_{i,t-1}$ under an $AR1$ error assumption (Lee and Nelder, 2001). So for y continuous the specification

$$\begin{aligned} y_{it} &= X_{it}\beta + \varepsilon_{it}, \\ \varepsilon_{it} &= \rho\varepsilon_{i,t-1} + \upsilon_{it}, \quad t = 2, \ldots, T, \end{aligned}$$

where υ_{it} is unstructured is equivalent to

$$y_{it} = \rho y_{i,t-1} + X_{it}\beta - \rho X_{i,t-1}\beta + \upsilon_{it}, \quad t = 2, \ldots, T. \tag{8.6}$$

Example 8.3. Capital Asset Pricing Model This example considers a metric outcome for what may be termed a medium n, large T scenario (not typical of many panel datasets). Following the theme of the capital asset pricing model from financial economics, Frees (2004, 302) considers links between the performance of (i.e., returns from) a particular security and market performance in general. The particular application is to $n = 90$ insurance firms observed over $T = 60$ months (January 1995–December 1999). The response y_{it} is the security return for firm i in excess of the risk-free rate and the predictor x_t is the market return in excess of the risk-free rate.

To allow for varying impacts of x_t on y_{it}, a baseline model (Model 1) proposed by Frees (2004) is the RIAS specification:

$$y_{it} = b_{1i} + b_{2i}x_t + u_{it},$$

with unstructured errors $u_{it} \sim N(0, 1/\tau_u)$. The coefficients b_{2i} measure how far the return of security i is attributable to market factors. A bivariate normal prior is assumed for $\{b_{1i}, b_{2i}\}$, with mean (B_1, B_2) and covariance D. A Wishart $W(4I, 4)$ prior for D^{-1} is assumed, with the prior mean for the covariance matrix D then being the identity matrix.

The last 4000 of a 5000 iteration two chain run provide a significant effect for x_t with B_2 having posterior mean (95% credible interval) of 0.72 $(0.63, 0.81)$. To assess whether first-order autoregressive dependence might be present, define realized residuals $e_{it} = y_{it} - b_{1i} - b_{2i}x_t$. Then a firm-specific statistic proposed by Baltagi and Li (1995) can be obtained, namely

$$T^{0.5}(\tilde{\rho}_{2i} - \tilde{\rho}_{1i}^2)/(1 - \tilde{\rho}_{2i}),$$

where

$$\tilde{\rho}_{1i} = \sum_{t=2}^{T} e_{it}e_{i,t-1} \bigg/ \sum_{t=1}^{T} e_{it}{}^2,$$
$$\tilde{\rho}_{2i} = \sum_{t=2}^{T} e_{it}e_{i,t-2} \bigg/ \sum_{t=1}^{T} e_{it}{}^2.$$

This statistic tends to $N(0,1)$ for large T, and over half the 90 firms have significantly negative or positive values on this basis. Another evaluation involves a posterior predictive check based on an average of Durbin–Watson (DW) statistics taken over all 90 firms. Thus, at each iteration r, a DW statistic is derived for each firm, namely

$$DW_i^{(r)} = \sum_{t=2}^{T} \left(e_{it}^{(r)} - e_{i,t-1}^{(r)}\right)^2 \bigg/ \sum_{t=1}^{T} \left(e_{it}^{(r)}\right)^2.$$

A summary statistic for autocorrelation is then the average over firms $\overline{DW}^{(r)} = \sum_i DW_i^{(r)}/n$. This can be obtained for both actual data $\overline{DW}_{\text{obs}}^{(r)}$, and for replicate data $\overline{DW}_{\text{new}}^{(r)}$, and the procedure of Gelman et al. (1996) applied. The resulting posterior probability $Pr(\overline{DW}_{\text{obs}} \geq \overline{DW}_{\text{new}}|y)$ is 1, indicating inadequate fit.

Accordingly a revised model[3] (Model 2) includes a stationary $AR1$ error, so that

$$y_{it} = b_{1i} + b_{2i}x_t + \varepsilon_{it},$$
$$\varepsilon_{it} = \rho_\varepsilon \varepsilon_{i,t-1} + \upsilon_{it},$$

and $\rho_\varepsilon \sim U(-1,1)$. A 5000 iteration two chain run (with the last 4000 for inference) gives significant ρ_ε, B_2, and diag(D) estimates, with respective means and 95% intervals −0.09 (−0.12, −0.06), 0.74 (0.64, 0.84), 0.63 (0.29, 1.20), and 0.17 (0.12, 0.25). The densities for D_{11} and D_{22} are bounded away from zero. This model lowers the DIC by 27 compared to model 1 (from 39,721 to 39,694), albeit with complexity d_e rising from 69 to 89.

Frees (2004) suggests an extension to unit-time specific predictor effects β_{it}. Thus Model 3 specifies,

$$y_{it} = b_{1i} + \beta_{it}x_t + \varepsilon_{it},$$
$$\varepsilon_{it} = \rho_\varepsilon \varepsilon_{i,t-1} + \upsilon_{it},$$

where a stationary $AR1$ prior is assumed for the evolving market effects β_{it}:

$$\beta_{it} = b_{2i} + \rho_\beta(\beta_{i,t-1} - b_{2i}) + e_{it}.$$

Priors are as above on $\{b_{1i}, b_{2i}\}, D$, and ρ_ε, and additionally $\rho_\beta \sim U(-1,1)$, $e_{it} \sim N(0, \sigma_\beta^2)$, $\beta_{i1} \sim N(0, \sigma_{\beta 1}^2)$, $1/\sigma_{\beta 1}^2 \sim \text{Ga}(1, 0.001)$, and $1/\sigma_\beta^2 \sim \text{Ga}(1, 0.001)$.

This model shows a predominantly negative density for ρ_β, but not significantly negative, with 95% interval $(-0.38, 0.05)$, and a similar estimate for ρ_ε to that obtained under Model 2. The DIC falls to 39,566, despite the complexity index d_e rising to 630. With a diffuse Ga(1,0.001) prior on $1/\sigma_\beta^2$, the estimates for β_{it} have relatively low precision and a prior stressing smoothly varying parameter change may be preferred.

8.4 Panel Categorical Choice Data

Repeated categorical data involving ordered or unordered options or choices $k = 1, \ldots, K$ by subjects $i = 1, \ldots, n$ for repetitions $t = 1, \ldots, T_i$ are often found in brand choice, labor market and political science applications (Niedermeier and Von Eye, 1999; Pettitt et al., 2006; Rossi et al., 2005). These may be expressed via binary indicators $d_{ikt} = 1$ if category or choice k applies ($d_{ikt} = 0$ for remaining categories) or by categorical responses $y_{it} \in (1, \ldots, K)$. Clinical and pharmaceutical applications commonly involve ordinal rating scales (e.g., Agresti and Natarajan, 2001; Qiu et al., 2002; Tan et al., 1999; Zayeri et al., 2005). Particular issues raised by such data include the possibility that permanent subject effects vary between choices (or more generally between categories), and that predictor effects may vary over one or more of choices, as well as over subjects or times. If lagged effects of the dependent variable are included (Section 8.5), these may include both own category and cross-category lags, leading to categorical transition models (Fokianos and Kedem, 2003).

Chintagunta et al. (2001) consider repeated brand choice data and allows subject heterogeneity in relation to attributes of the choices (e.g., variable consumer responsiveness to brand prices), as well as randomly varying subject-choice intercepts b_{ik}. A Bayesian perspective, including optimal MCMC sampling schemes, on consumer heterogeneity in multinomial panel data for purchase choices is provided by Rossi et al. (2005, Chap. 5). For identifiability, choice or category-specific parameters must be set to a fixed value (usually zero) in a reference category. For example, the probability that a consumer chooses brand k in period t might be modeled using a multinomial logit (MNL) regression,

$$\pi_{ikt} = Pr(y_{it} = k) = \phi_{ikt} \Big/ \sum_{k=1}^{K} \phi_{ikt},$$

$$\log(\phi_{ikt}) = \beta_{0k} + b_{ik} + P_{kt}\beta_k + A_{kt}\gamma, \quad k = 1, \ldots, K-1,$$

$$\log(\phi_{iKt}) = A_{Kt}\gamma,$$

where β_{0k} are intercept terms, P_{kt} and A_{kt} are brand-time specific characteristics (e.g., price and advertising spend) varying in whether associated regression parameters are choice specific, and b_{ik} are random consumer-brand taste effects. These are typically taken as multivariate normal of dimension $K-1$, with $b_{iK}=0$ for identifiability (Malchow-Moller and Svarer, 2003).

Consumer variation in response to prices or attributes would involve making the β_k and γ coefficients specific to each consumer and defining hyperparameters for the densities of β_{ki} and γ_i. For P_{kt} of dimension R, Rossi et al. (2005, 136) propose a conjugate normal hierarchical prior structure for subject effects $\beta_i=(\beta_{1i},\ldots,\beta_{Ri})$ with mean $Z_i\Delta$, where Z_i are consumer attributes, and with variance V_β of dimension $\dim(\beta_i)=(R-1)R$. V_β is assigned an inverse Wishart prior having with expectation I and $\dim(\beta_i)+$ three degrees of freedom. They demonstrate the improved MCMC convergence for β_i obtained by using a random walk Metropolis with increments that have covariance $s^2(H_i+(V_\beta^{(r)})^{-1})^{-1}$, where H_i is the Hessian of a composite likelihood based on multiplying the MNL subject-specific likelihood by the pooled (all subject) likelihood raised to power $\rho_i=(T_i/cN)$, and $c>1$ and $s=2.93/\text{sqrt}[\dim(\beta_i)]$ are tuning constants.

Hedeker (2003) considers categorical longitudinal data with subject level predictors only, namely X_{it} and Z_{it} of dimension P and Q, and category-specific fixed regression effects, namely

$$\log(\phi_{ikt})=\beta_{0k}+X_{it}\beta_k+Z_{it}b_{ik},$$
$$\log(\phi_{iKt})=0,$$

where $Z_{it}b_{ik}=z_{1it}b_{ik1}+z_{2it}b_{ik2}+\cdots+z_{Qit}b_{ikQ}$. Assuming the X_{it} and Z_{it} are nonoverlapping, one may adopt Q independent sets of subject-category effects each of dimension $K-1$, one for each predictor z_{qit},

$$(b_{i1q},\ldots,b_{i,K-1,q})\sim N_{K-1}(B_q,D_q).$$

Alternatively the covariance matrix of the random effects may be of dimension $(K-1)Q$ with the b_{ik} correlated over both categories and predictors. Hedeker (2003) in fact considers the case where Z_{it} is a subset of X_{it}, so the b_{ikq} are zero mean random effects, and takes the covariance matrices to be choice-specific D_k of dimension Q, so that

$$(b_{ik1},\ldots,b_{ikQ})\sim N_Q(0,D_k).$$

This permits a latent variable interpretation based on a Cholesky decomposition of D_k and standardized random effects ζ_{ik}, namely

$$\log(\phi_{ikt})=\beta_{0k}+X_{it}\beta_k+Z_{it}C_k\zeta_{ik},$$

where $C_kC_k'=D_k$.

Examples of repeated ordinal observations are provided by labor market perception data (Spiess, 2006), changing attitudes to divorce (Berrington et al., 2005), and repeated ordinal scores in horticultural research (Parsons et al., 2006). Suppose responses y_{it} have K ordered categories, with corresponding latent responses y^*_{it} specified by thresholds, possibly time varying κ_{kt}, or subject varying κ_{ik}. For time varying thresholds

$$y_{it} = k \quad \text{if} \quad \kappa_{k-1,t} < y^*_{it} \leq \kappa_{kt},$$

with predictor effects also possibly varying over (at least one of) categories, subjects or times. For example, Spiess (2006) considers predictor effects varying over times, as in

$$y^*_{it} = X_{it}\beta_t + \varepsilon_{ikt},$$

where $P(\varepsilon_{ikt})$ is usually a normal or logistic distribution; these distributions are very similar though the logistic places more probability in the tails (Hedeker, 2003). So

$$Pr(y^*_{it} \leq \kappa_{kt}) = Pr(X_{it}\beta_t + \varepsilon_{ikt} \leq \kappa_{kt}) = Pr(\varepsilon_{ikt} \leq \kappa_{kt} - X_{it}\beta_t).$$

According to the form for $P(\varepsilon_{ikt})$, one has

$$Pr(y^*_{it} \leq \kappa_{kt}) = \Phi(\kappa_{kt} - X_{it}\beta_t),$$

or

$$Pr(y^*_{it} \leq \kappa_{kt}) = 1/(1 + \exp(-[\kappa_{kt} - X_{it}\beta_t])).$$

Let $\gamma_{ikt} = Pr(y^*_{it} \leq \kappa_{kt})$, then

$$Pr(y_{it} = k) = Pr(\kappa_{k-1,t} < y^*_{it} \leq \kappa_{kt}) = \gamma_{ikt} - \gamma_{i,k-1,t}.$$

An equivalent specification of this model involves sets of $K-1$ binary variables for each subject–time pairing, namely $g_{ikt} = 1$ if $y_{it} \leq k$, and $g_{ikt} = 0$ otherwise. Then for ε logistic,

$$Pr(y_{it} \leq k)) = Pr(g_{ikt} = 1) = 1/(1 + \exp(-\kappa_{kt} + X_{it}\beta_t).$$

Example 8.4. Yoghurt Purchases Data on yoghurt brand choice from Chen and Kuo (2001) and Frees (2004), exemplify household heterogeneity in consumer behavior, and panel analysis of unordered choices data, as considered by authors such as Chintagunta et al. (2001). The yoghurt choice data relate to repeated purchases by $i = 1, \ldots, n$ households ($n = 100$) between $K = 4$ brands, with widely varying numbers of repetitions T_i for each household (between 4 and 185). The total of observations is $N = \sum_{i=1}^n T_i = 2412$. Known influences on brand choice are brand and time specific, namely features A_{kt} (=1 if the brand k was subject to an advertising feature at the time t of purchase, = 0 otherwise), and shelf price P_{kt}. Of potential interest is the benefit of considering (a) random choice variation due to subject and/or choice heterogeneity, and (b) consumer heterogeneity in response to known features.

A baseline fixed effects model[4] (Model 1) has the form:

$$\pi_{ikt} = Pr(d_{ikt} = 1) = \phi_{ikt} \Big/ \sum_{k=1}^{K} \phi_{ikt},$$

$$\log(\phi_{ikt}) = \beta_{0k} + A_{kt}\gamma_1 + P_{kt}\gamma_2, \quad k = 1, \ldots, K-1,$$
$$\log(\phi_{iKt}) = A_{Kt}\gamma_1 + P_{Jt}\gamma_2.$$

A random intercepts model (Model 2) allows for heterogeneity at subject-choice level but homogenous impact of brand attributes, and has the form:

$$\log(\phi_{ikt}) = A_{kt}\gamma_1 + b_{ik} + P_{kt}\gamma_2, \quad k = 1, \ldots, K-1,$$
$$\log(\phi_{iKt}) = A_{Kt}\gamma_1 + P_{Kt}\gamma_2,$$
$$(b_{i1}, \ldots, b_{i,K-1}) \sim N(B, D),$$

where the vector B denotes the average category intercepts $(\beta_{01}, \ldots, \beta_{0,K-1})$.

Inferences are based on the last 4000 of a two chain run of 5000 iterations in Model 1, and on the second half of a two chain run of 10,000 iterations in Model 2. The fixed effect model has DIC of 5324 ($d_e = 5.1$) and log(psML) of -2662, whereas the random heterogeneity model has a DIC of 2540 ($d_e = 251$) and log(psML)$= -1330$. Estimates of the covariance matrix D under Model 2 show brand 2 choice to be negatively related to choice of both brand 1 and brand 3, while the covariance parameter between brands 1 and 3 straddles zero. While Model 2 yields a pronounced gain in fit, it has not controlled for consumer variation in price or advertising responsiveness. Such variation might be modeled by additional random effects over subjects (i.e., making the γ_1 and γ_2 coefficients household specific) without necessarily also assuming random variation over subject–brand combinations; this is left as an exercise.

Example 8.5. National Institute of Mental Health Schizophrenic Collaborative Study: Ordinal Symptom Score This study exemplifies panel modeling of repeated ordinal outcomes, and involves an evaluation of four drug treatments to alleviate symptoms in schizophrenia subjects: chloropromazine, fluphenazine, thioridazine, and a placebo. Similar effects were obtained for the three anti-psychotic drugs, and so here the treatment is reduced to a binary comparison of any drug vs. the placebo (cf. Hedeker and Gibbons, 1994). Symptom severity scores y_{it} are observed for $n = 324$ subjects on three occasions after the first reading (at week 0), which is coincident with treatment commencing, namely at weeks 1, 3, and 6. The score is ordinal with $K = 7$ levels, namely 1 = normal, 2 = borderline, 3 = mildly ill, 4 = moderately ill, 5 = markedly ill, 6 = severely ill, and 7 = extremely ill. Random intercepts are assumed, together with random slopes on a time variable Z_t obtained as the square root of weeks. Fixed effect predictors X_{it} are the treatment effect, a treatment by time interaction and the patient's sex.

In clinical applications it may be useful to have continuous measures of different types of patient morbidity or "caseness," for example, a measures of any symptom level as against normal status, or of markedly ill or worse symptoms vs. less severe symptoms. Rather than a single latent metric frailty, this indicates the utility of several latent scales in some settings, as defined by the $K-1$ binary variables g_{ikt} implied by each ordinal subject–time pairing y_{it}. So the analysis here[5] involves augmentation with latent metric data w^*_{ikt} defined by truncated normal or logit sampling according as $g_{ikt}=1$ (if $y_{it} \leq k$) and $g_{ikt}=0$ (if $y_{it} > k$). Then for a probit link,

$$w^*_{ikt} \sim N(\kappa_k - \mu_{it}, 1)I(0,), \quad \text{when } g_{ikt}=1,$$
$$w^*_{ikt} \sim N(\kappa_k - \mu_{it}, 1)I(,0), \quad \text{when } g_{ikt}=0,$$

where

$$\mu_{it} = b_{1i} + b_{2i}Z_t + \beta_1 \text{Tr}_i + \beta_2 Z_t \text{Tr}_i + \beta_3 \text{Gend}_i,$$
$$(b_{1i}, b_{2i}) \sim N_2(B, D)$$

and $B=(B_1,B_2)$ contains an overall intercept and time slope. With B_1 as an unknown, identification of the $K-1=6$ thresholds requires setting κ_1 to zero, and estimating the remaining five threshold parameters subject to monotonicity constraints. Here $\kappa_k = \kappa_{k-1} + \delta_k$, where $\delta_k \sim \text{Ga}(1,1)$.

Table 8.1 summarizes posterior densities obtained from the second half of a two chain run of 20,000 iterations, with s_{b_1} and s_{b_2} equivalent to $D_{11}^{0.5}$ and $D_{22}^{0.5}$, and ρ_b equivalent to $D_{12}/(D_{11}^{0.5} D_{22}^{0.5})$. The main treatment effect β_1 is not significant, but there is a steeper decline in morbidity for treated subjects—shown by the significantly negative coefficient β_2. Wide variation

TABLE 8.1
Posterior parameter summary.

	Mean	St. devn.	Monte Carlo SE	2.5%	97.5%
β_1	0.085	0.232	0.01621	−0.449	0.524
β_2	−0.735	0.100	0.00686	−0.920	−0.553
β_3	0.133	0.131	0.00864	−0.126	0.375
η_1	4.996	0.230	0.01524	4.500	5.500
η_2	−0.570	0.091	0.00542	−0.736	−0.393
κ_2	1.305	0.120	0.00718	1.068	1.546
κ_3	2.355	0.124	0.00745	2.121	2.609
κ_4	3.569	0.129	0.00785	3.328	3.828
κ_5	5.163	0.142	0.00870	4.894	5.441
κ_6	7.412	0.181	0.01034	7.065	7.762
S_{b1}	1.353	0.089	0.00456	1.178	1.531
S_{b2}	0.817	0.049	0.00233	0.725	0.917
ρ_b	−0.433	0.064	0.00270	−0.551	−0.299

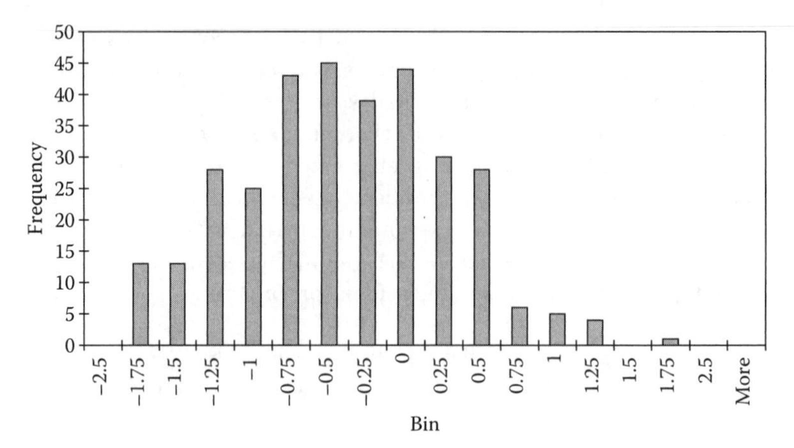

FIGURE 8.1
Histogram plot, posterior means of time slopes.

in time paths is apparent in the density of s_{b_2}, and in a plot (Figure 8.1) of the posterior means of b_{2i}. Diagnostic tests such as Q–Q plots and Jarque-Bera tests support normality of the permanent effects (b_{1i}, b_{2i}). The averages of the posterior mean b_{2i} over treated and placebo subjects are very similar, but an alternative approach (left as an exercise) is to relate variation in the permanent effects directly to treatment, as in

$$\begin{aligned}\mu_{it} &= b_{1i} + b_{2i}Z_t + \beta_1 \text{Gend}_i,\\ (b_{1i}, b_{2i}) &\sim N_2([B_{1i}, B_{2i}], D),\\ B_{qi} &= \gamma_{0q} + \gamma_{1q}\text{Tr}_i.\end{aligned}$$

8.5 Observation-Driven Autocorrelation: Dynamic Panel Models

Differences in behavior or event proneness between individuals (e.g., in econometric or health applications) may operate through an autoregression in the observations, latent or observed. Panel models including lagged observations are often termed dynamic panel models, whereas static panel models do not include lagged response values (e.g., Nerlove, 2002). A canonical dynamic model for metric data involves lagged values of the dependent variable with the overall error combining a time-invariant individual effect and observation level random noise (Bond, 2002).

Thus with a first-order lag in the response, one has

$$y_{it} = \phi y_{i,t-1} + X_{it}\beta + b_i + u_{it}, \quad t = 2, \ldots, T,$$

where the $u_{it} \sim N(0, \sigma^2)$ are independent of each other, and under standard assumptions (Ahn and Schmidt, 1995) are also uncorrelated with the initial observations y_{i1} and with permanent subject effects b_i. If X_{it} contains a constant term, then the b_i have mean zero and $b_i \sim N(0, D)$. Allowing for subject level variation in a Q length vector of predictors Z_{it} as well as for first-order lagged response leads to

$$y_{it} = \phi y_{i,t-1} + X_{it}\beta + Z_{it}b_i + u_{it}.$$

Assuming a stationary process with $|\phi| < 1$, one possible model for y_{i1} is

$$y_{i1} = \frac{Z_{it}b_i}{1-\phi} + \frac{X_{it}\beta}{1-\phi} + u_{i1}$$

with $u_{i1} \sim N(0, \sigma^2/(1-\phi^2))$ (Gourieroux et al., 2006). A simplifying approach, more feasible for large T, is to condition on the first observation in a model involving a first-order lag in y, so that y_1 is nonstochastic (Bauwens et al., 1999, 135; Hjellvik and Tjøstheim, 1999). Geweke and Keane (2000) and Lancaster (2002) consider Bayesian approaches to the dynamic linear panel model, in which the model for period 1 is not necessarily linked to those for subsequent periods in a way consistent with stationarity.

Maximum likelihood analysis of dynamic panel models is subject to an initial condition problem if in fact there is correlation between the permanent subject effects b_i and the initial observations (Hsiao, 1996). In case of such correlation, possible options are a joint random prior (e.g., bivariate normal) involving b_i and u_{i1} (Dorsett, 1999), or a prior for b_i that is conditional on y_{i1}, such as (Hirano, 2002; Wooldridge, 2005)

$$b_i | y_{i1} \sim N(\varphi y_{i1}, \sigma_1^2).$$

Dynamic linear models may be extended in several ways, to include ARMA(p, q) error schemes, effects of time functions, or random variation over subjects or times in the impacts of lagged predictors. For example, Ulrick (2007) uses a dynamic earnings model with $AR1$ autocorrelated errors,

$$y_{it} = b_i + \phi y_{i,t-1} + X_{it}\beta + W_i\gamma + \varepsilon_{it},$$
$$\varepsilon_{it} = \rho\varepsilon_{i,t-1} + \upsilon_{it},$$

where W_i are fixed human capital attributes. Such a specification follows related earnings modeling work by authors such as Geweke and Keane (2000), Lillard and Willis (1978), and MacCurdy (1982). The earnings regression equation may be extended (Galler, 2001) to a RIAS form:

$$y_{it} = b_{1i} + b_{2i}t + \phi y_{i,t-1} + X_{it}\beta + W_i\gamma + \varepsilon_{it},$$

where the random effects b_{1i} and b_{2i} allow subject-specific variation in wage level and wage growth. Taking the time function to be an unknown function of t, δ_t, lead to the autoregressive latent trait (ALT) models considered by Bollen and Curran (2004, 2006). Allowing for time varying coefficients on lagged responses $y_{i,t-1}$ as well as random subject intercepts and growth rates one might then have

$$y_{it} = b_{1i} + b_{2i}\delta_t + X_{it}\beta + \phi_t y_{i,t-1} + u_{it},$$

with δ_t subject to identifying constraints, such as $\delta_1 = 1$.

8.5.1 Dynamic panel models for discrete data

For discrete data, a range of dynamic panel approaches have been proposed, varying according to the type of response (e.g., count or binary) and the initial conditions method. For example, Wooldridge (2005) considers models for binary and count data using a conditional prior method relating b_i and y_{i1}, while Pettitt et al. (2006) follow the Heckman (1981) strategy, whereby the model for the initial period does not include permanent effects or a lagged response effect.

For counts y_{it} taken as Poisson, $y_{it} \sim Po(\mu_{it})$, problems with taking a linear impact of the first lag outcome $y_{i,t-1}$, as in

$$\mu_{it} = \exp(X_{it}\beta + \phi y_{i,t-1} + b_i), \quad t = 2, \ldots, T$$

are mentioned by Fahrmeir and Tutz (2001, 244). This option for modeling lag response impacts defines the Markov property scheme studied by Fotouhi (2005) and Fotouhi (2007), under which the initial observation is modeled as

$$\mu_{i1} = \exp(X_{i1}\widetilde{\beta} + c_i),$$

where X_{i1} includes any relevant predictors for the first period, and the subject effects b_i and c_i follow a bivariate normal with correlation ρ.

Alternatively the impact of a lagged count response may be modeled by a log or other transform $g(y)$, with extra preset or unknown parameters in case the lagged y is zero. Thus if $g(y) = \log(y + c)$ where $c = 1$ (say), one has

$$\mu_{it} = \exp(X_{it}\beta + \phi g(y_{i,t-1}) + b_i), \quad t = 2, \ldots, T$$

and following Wooldridge (2005), one might assume

$$b_i | y_{i1} \sim N(\varphi y_{i1}, \sigma_1^2).$$

By contrast, an extension of the conditional linear autoregressive process to panel data (Grunwald et al., 2000) would lead to Poisson means

$$\mu_{it} = \phi y_{i,t-1} + \exp(X_{it}\beta + b_i),$$

while a particular form of the full autoregressive conditional Poisson specification of Jung et al. (2006) specifies

$$\mu_{it} = \phi y_{i,t-1} + \eta\mu_{i,t-1} + \exp(X_{it}\beta + b_i).$$

By contrast to count regression, regression for binary responses $y_{it} \sim \text{Bern}(\pi_{it})$ may straightforwardly include lags in observed outcomes $y_{i,t-s}$ leading to Markov chain models of various orders (Azzalini, 1994; Frees, 2004; Hamerle and Ronning, 1995). In an early paper, Korn and Whittemore (1979) consider a logit model,

$$\text{logit}(\pi_{it}) = \alpha_0 + \alpha_1 y_{i,t-1} + X_{it}\beta$$

with first-order Markov dependence. They find a highly significant α_1, and so strong evidence against an assumption that successive responses are independent. Erkanli et al. (2001) consider higher-order Markov logistic regression combined with random heterogeneity,

$$\text{logit}(\pi_{it}) = \alpha_0 + \sum_{s=1}^{S_{\max}} \alpha_k y_{i,t-s} + X_{it}\beta + b_i,$$

where $b_i \sim N(0, D)$, and $S_{\max}$ is determined by a coefficient selection process. Alternatively fixed predictor effects β, and parameters (B, D) for random effects b_i, may vary according to the previous value of the binary response. So for one lag dependence, regression parameters and random effect hyperparameters $\{\beta_s, B_s, D_s\}$ may be specific to previous response $\{y_{i,t-1} = s, s = 0$ or $1\}$ (Islam and Chowdhury, 2006).

Such alternatives extend in principle to multinomial outcomes $y_{it} \in (1, \ldots, K)$ or equivalently $d_{ikt} = 1$ if category k applies (or is chosen) and $d_{itk} = 0$ otherwise. So

$$(d_{it1}, \ldots, d_{itK}) \sim \text{Mult}(n_{it}, [\pi_{it1}, \ldots, \pi_{itK}]),$$

where $n_{it} = 1$. Use of lags is complicated by the possible influence of cross-category lags as well as own-category lags. Pettitt et al. (2006) consider a Bayesian hierarchical multinomial model for changes in employment status (a trichotomy), with one period lags in status as predictors. Thus, with employment status 1 as the reference (and so $\phi_{i1t} = 1$), one has for $t > 1$

$$\pi_{ikt} = Pr(y_{it} = k) = \phi_{ikt} \Big/ \sum_{k=1}^{K} \phi_{ikt},$$

$$\log(\phi_{ikt}) = b_{ik} + \beta_k X_{it} + \gamma_{k1} I(y_{it} = 2) + \gamma_{k2} I(y_{it} = 3), \quad k = 2, \ldots, K,$$

where b_{ik} are category-specific random effects. For the initial period, they adopt a static multinominal login model which does not include lag effects or b_{ik} (Pettitt et al., 2006, 102), and has distinct fixed regression effects, namely

$$\log(\phi_{ik1}) = \delta_k X_{i1}, \quad k = 2, \ldots, K.$$

This follows from a linear approximation to the reduced form obtained when lagged response variables are replaced by their specifications under the dynamic model for periods preceding $t = 1$.

Dynamic modeling approaches may also be applied using latent metric responses associated with binary or ordinal observations. Suppose observations y_{it} are binary such that the latent continuous response $y^*_{i,t} > 0$ if and only if $y_{it} = 1$, and $y^*_{i,t} \leq 0$ if $y_{it} = 0$. Then one might specify,

$$y^*_{i,t} = X_{it}\beta + \phi_1 y_{i,t-1} + \phi_2 y^*_{i,t-1} + u_{it} \qquad I(0,), \quad \text{for } y_{it} = 1,$$
$$y^*_{i,t} = X_{it}\beta + \phi_1 y_{i,t-1} + \phi_2 y^*_{i,t-1} + u_{it} \qquad I(,0), \quad \text{for } y_{it} = 0,$$

with standard normal white noise errors,

$$u_{it} \sim N(0,1),$$

and lag one dependence on both previous events and latent utilities. Chib and Jeliazkov (2006) propose a scheme with lags only in observed binary outcomes but with both heterogeneity and autocorrelated errors. Thus they take,

$$y^*_{i,t} = X_{it}\beta + Z_{it}b_i + \sum_{k=1}^{K} \phi_k y_{i,t-k} + \varepsilon_{it} \qquad I(0,), \quad \text{for } y_{it} = 1,$$
$$y^*_{i,t} = X_{it}\beta + Z_{it}b_i + \sum_{k=1}^{K} \phi_k y_{i,t-k} + \varepsilon_{it} \qquad I(,0), \quad \text{for } y_{it} = 0,$$

so distinguishing between alternative sources of intertemporal dependence. The first such source is state dependence on lagged responses $y_{i,t-k}$; the second is serial correlation in the errors, namely

$$\varepsilon_{it} = \rho_1 \varepsilon_{i,t-1} + \cdots + \rho_S \varepsilon_{i,t-S} + \upsilon_{it},$$

where $\upsilon_{it} \sim N(0,1)$; and third is intraclass correlation arising because of heterogeneity. In this way one may avoid spurious state dependence in which previous responses proxy unobserved variation (Heckman, 1981).

Example 8.6. National Longitudinal Study of Youth: Lagged Earnings Model This example considers a metric outcome, namely earnings data from the US National Longitudinal Survey (NLS) relating to young women aged between 14 and 26 in 1968, and either already in the labor market in 1968, or entering the labor market during the period 1968–1988. In this period there were 15 measuring occasions, namely each year during 1968–1988 except 1974, 1976, 1979, 1981, 1984, and 1986. There are 4711 subjects varying considerably in their observed histories; many subjects are subject to attrition or intermittent observation. The analysis here is based on a 10% sample of the 4164 subjects who have at least two measurements on yearly log earnings, where earnings figures for each subject are divided by calendar year averages to correct for inflation. In this way the earnings profile of a subject observed over 1968–1975 (say) can be compared with that for a subject observed over 1978–1985. An alternative might be to have fixed or random effects for each calendar year to model population trends in average income (Hjellvik and Tjoshteim, 1999).

Although not all years were subject to survey updates, the analysis here takes a subject's entire observation span (obtained by comparing initial and last observation year) to define that subject's total times T_i. Any intervening years without observations are treated as missing data, whether this is due to intermittent missingness or the absence of a NLS update in particular years. Thus, the first subject is observed on 12 occasions (in the studies in 1970, 1971, 1972, 1973, 1975, 1977, 1978, 1980, 1983, 1985, 1987, and 1988), but that subject's total times T_i is set at 19, with the intervening years without observations (e.g., 1974, 1976, etc.) treated as missing data. Missingness is taken to be at random, not depending on the possibly missing response value.

With y_{it} denoting (inflation corrected) log earnings, the initial regression model includes subject effects b_i, and fixed binary attributes $\{W_{1i}, W_{2i}, W_{3i}\}$, with W_{1i} for college graduate (= 1, 0 otherwise), W_{2i} for white ethnicity (= 1, 0 for other ethnicities), and an interaction $W_{3i} = W_{1i}W_{2i}$. So for $i = 1, \ldots, n$ where $n = 416$,

$$y_{it} = \beta_1 + b_i + W_i\gamma + u_{it}, \quad t = 1, \ldots, T_i,$$

$b_i \sim N(0, D)$, and $u_{it} \sim N(0, \sigma^2)$. Uniform $U(0, 10)$ priors are assumed for σ and $D^{0.5}$, and $N(0, 1000)$ priors are assumed for the fixed effects $\{\beta_1, \gamma_1, \gamma_2, \gamma_3\}$.

The last 9000 of a two chain run of 10,000 gives a DIC of -1424 with $d_e = 354$. The estimated γ coefficients show significantly higher earnings for college graduates, and a positive but not significant white ethnicity effect, with posterior means (sd) for γ_1 and γ_2 of 0.32 (0.05) and 0.033 (0.022). The effect of the interaction term is significantly negative, with mean (sd) of -0.11 (0.055), suggesting a greater positive impact of college education on earnings for nonwhite subjects. The posterior mean for the standard deviation of the b_i is 0.18, so that a subject for whom b_i is one standard deviation above the average would have earnings about 20%, namely $100 \times \exp(0.18)$, above average, given observed personal characteristics W_i. Taking $\hat{u}_{it} = y_{it} - \beta_1 - b_i - W_i\gamma$, there is evidence of autocorrelated errors, with the 95% interval for the statistic $r_u = \Sigma_{i=1}^{n}\Sigma_{t=2}^{T_i}\hat{u}_{it}\hat{u}_{i,t-1}/\Sigma_{i=1}^{n}\Sigma_{t=2}^{T_i}\hat{u}_{it}^2$ being $(0.06, 0.12)$.

To improve fit a second dynamic model is nonstationary and similar to Geweke and Keane (2000), in that there is a distinct model for the first period for each subject, and a one period lag effect in earnings in subsequent periods, with this effect not constrained to stationarity. Permanent subject effects are also included in the model for periods $t = 2, \ldots, T_i$ so that

$$\begin{aligned} y_{it} &= b_i + W_i\gamma + \phi y_{i,t-1} + u_{it}, \quad t = 2, \ldots, T_i, \\ y_{i1} &= W_i\gamma_1 + u_{i1}, \end{aligned}$$

where a $N(0, 1)$ prior on ϕ is adopted, and the errors $u_{it} \sim N(0, \sigma^2)$ and $u_{i1} \sim N(0, \sigma_1^2)$ are independent.[6] The last 9000 of a two chain run of 10,000 show a 95% interval for ϕ of (0.55, 0.63), along with considerably reduced autocorrelation, with 95% interval for r_u now from -0.058 to 0.000. Fit is improved with DIC now lower at -1541 with $d_e = 582$. The γ coefficients are reduced in absolute size, but the college effect γ_1 remains significant, with 95% interval $(0.09, 0.18)$. The posterior mean for the standard deviation of the b_i is also reduced to 0.074.

Example 8.7. Epileptic Seizure Data: Lagged Count Model These data from Thall and Vail (1990) have been widely analyzed, albeit not necessarily including lagged response effects. They relate to counts of epileptic seizures in four successive two-week periods. An anti-epileptic drug treatment (progabide) was applied for some of the $n = 59$ patients, with others receiving a placebo. A pre-treatment eight-week baseline seizure count was also obtained, and may be treated either as exogenous, or as an endogenous initial condition (Fotouhi, 2007). Here the baseline count is included in the outcome profile, so that $T = 5$ with the baseline seizure count denoted as y_{i1}. The analysis here follows Lindsey (1993) in terms of including a lagged response as one predictor. The predictor set for periods $t = 2, \ldots, T$ is a subset of that in Table 12 of Fotouhi (2007), as initial runs showed most effects in the full set to be nonsignificant. The predictors for $t \geq 2$ are then age, treatment, and lagged seizure count, while the predictor set $X_{1it} = X_{1i}$ consists of age at baseline only. A bivariate normal model correlates the permanent effects b_i for periods $t \geq 2$ with the errors c_i in the model for y_{i1}.

So with X_{it} including an intercept and age, Model 1 takes

$$y_{it} \backsim Po(\mu_{it}), t = 1, 5$$

with the means modeled as

$$\begin{aligned}
\mu_{it} &= \exp(X_{it}\beta + \phi y_{i,t-1} + b_i), \qquad t = 2, \ldots, T, \\
\mu_{i1} &= \exp(X_{i1}\tilde{\beta} + c_i), \\
(b_i, c_i) &\sim N_2(0, D),
\end{aligned}$$

where $D^{-1} \sim W(I, 2)$, and fixed effects are assigned $N(0,10)$ priors. The last 9000 of a two chain run of 10,000 iterations[7] show that in fact this model leaves excess dispersion: the mean scaled deviance of 496 is well in excess of the number, $5 \times 59 = 295$, of observations. This issue is returned to in Example 8.10 below. The model for $t \geq 2$ shows a 95% credible interval for treatment $(-0.61, 0.02)$ mostly confined to negative values, while the 95% interval for lagged seizures is $(0.0001, 0.0042)$. The correlation between b_i and c_i is 0.81.

Fotouhi (2007) demonstrates that inferences for these data may be affected by assuming the first observation as exogenous or endogenous. Accordingly a second model takes

$$y_{it} \backsim Po(\mu_{it}), \quad t = 2, 5$$

TABLE 8.2
Epilepsy count data, with and without model for initial observations.

	Mean	**St. devn.**	**MC St. Error**	**2.5%**	**97.5%**
With Period 1 Model (N = 295), Permanent Effects Bivariate Normal					
Scaled Deviance	495.7	16.0	0.2	465.6	528.8
Const	1.74	0.15	0.0073	1.45	2.027
Age	−0.007	0.02	0.0011	−0.048	0.036
Trt	−0.29	0.16	0.0064	−0.61	0.02
Lagcount	0.0022	0.0011	0.00002	0.0001	0.0042
Corr(b_i, c_i)	0.81	0.05	0.0006	0.69	0.90
Without Period 1 Model (N = 236), Permanent Effects Univariate Normal					
Scaled Deviance	438.3	12.5	0.1	415.9	464.6
Const	1.74	0.18	0.0097	1.392	2.091
Age	−0.008	0.02	0.0009	−0.047	0.030
Trt	−0.30	0.25	0.0139	−0.80	0.20
Lagcount	0.0025	0.0010	0.00002	0.0004	0.0045
With Period 1 Model (N = 295), Permanent Effects Bivariate Skew-Normal					
Scaled Deviance	494.2	16.0	0.2	464.8	527.4
Const	1.62	0.26	0.02	1.12	2.14
Age	−0.008	0.020	0.001	−0.047	0.033
Trt	−0.29	0.16	0.006	−0.62	0.02
Lagcount	0.0021	0.0011	0.00002	0.0000	0.0042
Corr(b_i, c_i)	0.83	0.05	0.001	0.70	0.91
Skew parameter c_i	−0.027	0.256	0.015	−0.517	0.454
Skew parameter b_i	0.126	0.294	0.018	−0.453	0.673

and conditions on y_{i1}, so that b_i is now a univariate permanent effect. Thus, the means are

$$\mu_{it} = \exp(X_{it}\beta + \phi y_{i,t-1} + b_i), \qquad t = 2, \ldots, T,$$
$$b_i \sim N(0, D),$$

where the precision parameter of the b_i is assigned a gamma prior, $D^{-1} \sim \text{Ga}(1, 0.001)$. The main impact of this changed specification is a lessening of the treatment effect (Table 8.2). The mean scaled deviance of 438, now for $4 \times 59 = 236$ data points, again shows excess dispersion. This indicates that a complete model for these data requires observation level errors.

A third analysis (Model 3) replicates Model 1 except that it investigates whether taking (b_i, c_i) as multivariate skew normal affects fit or inferences (see Section 8.6), as unobserved morbidity may not be symmetric. Thus,

$$\mu_{it} = \exp(X_{it}\beta + \phi y_{i,t-1} + b_i + \delta_2 W_{2i}), \quad t = 2, \ldots, T,$$
$$\mu_{i1} = \exp(X_{i1}\tilde{\beta} + c_i + \delta_1 W_{1i}),$$
$$(b_i, c_i) \sim N_2(0, D),$$

where the $\{W_{1i}, W_{2i}\}$ are independently half normal HN$(0, 1)$ a priori, and the skew parameters have $\delta_k \sim N(0, 10)$ priors. Fit is not improved under this model (i.e., excess dispersion remains), and the skew parameters are not significant. Similar conclusions are reached using a multivariate skew t (Model 4) for patient heterogeneity—namely that excess dispersion remains. As expected the lowest scale factors ξ_i, in the notation of Equations 8.8a through c below, are for the exceptionally ill subjects 18 and 49, namely $\xi_{18} = 0.84$ and $\xi_{49} = 0.71$ (cf. Fotouhi, 2007, 601). A $U(0.01, 0.5)$ prior is adopted on the inverse of the multivariate t degrees of freedom parameter ν. The posterior mean for ν is 24, not strongly indicating heavy tailed permanent intercepts.

8.6 Robust Panel Models: Heteroscedasticity, Generalized Error Densities, and Discrete Mixtures

Preceding sections consider the normal linear mixed model for continuous longitudinal outcomes $y_i = (y_{i1}, \ldots, y_{iT_i})'$ assuming normal errors at both levels, and constant variances (or dispersion matrices) across subjects and observations. Thus, assuming Z_{it} is a subset of X_{it}, one has

$$y_{it} = X_{it}\beta + Z_{it}b_i + u_{it},$$
$$(b_{1i}, b_{2i}, \ldots, b_{Qi}) \sim N_Q(0, D),$$

and

$$(u_{i1}, \ldots, u_{iT_i}) \sim N(0, \sigma^2 I_{T_i}).$$

The general linear mixed model for y possibly being a nonmetric response from the exponential family may not include observation level residuals, and then takes the form:

$$g[E(y_{it}|b_i)] = X_{it}\beta + Z_{it}b_i,$$

with $(b_{1i}, b_{2i}, \ldots, b_{Qi}) \sim N_Q(0, D)$, where again normality of errors and constant dispersion D are default assumptions.

Violation of standard assumptions regarding the forms of error density, or of homoscedasticity, are likely to affect inferences. For example, Neuhaus et al. (1992) show that mis-specification of the density for random intercepts in a mixed logistic regression with $\pi_{it} = E(y_{it}|b_i)$ and

$$\text{logit}(\pi_{it}) = b_i + X_{it}\beta$$

results in inconsistent estimation of the β coefficients. Among principles that may provide a robust approach to departures from such standard assumptions is that of embedding the model in a more general framework (Ma et al., 2004; Rice, 2005), with conventional assumptions (e.g., normality and homoscedasticity of errors) as special cases of a broader model.

Following Chapter 5, assumptions of homoscedasticity at level 1 (repeated observations within subjects) or at level 2 (heterogeneity between subjects) may be modified to allow more general variance functions varying over subjects, times, or both, including dependence of the variance on subject or observation level attributes. For example, studies by Baltagi and Griffin (1988) and Roy (2002) are concerned with heteroscedasticity in the permanent random effects component of panel models, with variance linked to predictors in a positive function. For varying intercepts b_i as in Equation 8.4, one might relate the subject-specific variances D_i to predictor values averaged over time, $\overline{X_i}$, as in

$$D_i = \alpha^2(1 + \varphi\overline{X_i})^2,$$

where terms in the scalar or vector φ are positive. Li and Stengos (1994) consider heteroscedasticity at observation level, so that for $y_{it} = X_{it}\beta + Z_{it}b_i + u_{it}$ one might take

$$u_{it} \backsim N(0, \sigma_{it}^2),$$
$$\sigma_{it}^2 = \alpha^2(1 + \varphi X_{it})^2.$$

Wakefield et al. (1994) in a nonlinear pharmacokinetic panel analysis with positive structural effects η_{it} specify a Bayesian heteroscedastic model at observation level. Thus, $y_{it} \sim N(\eta_{it}, \eta_{it}^{\omega}/\tau)$, where ω is an unknown power and τ is an overall precision parameter, and $\omega = 0$ corresponds to homoscedasticity.

Similarly, more general error densities allowing for skewness, heavy tails or other non-normal features may be adopted, with the standard assumptions embedded within them. Alternatives to assuming multivariate normal subject effects may include heavy tail Student t heterogeneity (Chib, 2008; Lin and Lee, 2006), skew normal and skew-t densities, and skew-elliptical densities (Ma et al., 2004). Thus, the normal linear mixed model can be embedded within a wider class of scale mixture normal densities, with the subject or observation level scale parameters measuring outlier status (Chib, 2008; Wakefield et al., 1994). Thus, the model of (8.1b), with normal cluster effects b_i and normal residuals u_{it}, is a special case of a scale mixture model with

$$y_{it} = X_{it}\beta + Z_{it}b_i + u_{it},$$
$$u_{it} \backsim N\left(0, \frac{1}{\lambda_i}\Sigma_i\right),$$
$$b_i \backsim N\left(B, \frac{1}{\xi_i}D\right),$$
$$\lambda_i \backsim G_\lambda, \xi_i \backsim G_\xi.$$

A widely applied option takes the densities $\{G_\lambda, G_\xi\}$ to be gamma densities with equal scale and shape $\nu_\lambda/2$ and $\nu_\xi/2$, respectively, leading to t density random effects with $\{\nu_\lambda,\ \nu_\xi\}$ degrees of freedom. This provides resistance to atypical data at both observation and cluster levels.

For possibly skew residual or subject effects, transformation is one option: Candel and Winkens (2003) consider transformed exponential errors for modeling heterogeneity, obtained by sampling $c_{qi} \sim E(\eta_q)$, and centering the sampled c_{qi} giving $b_{qi} = c_{qi} - \overline{c_q}$, and providing errors that may be strongly skewed to the right rather than symmetric. Ghosh et al. (2007) consider bivariate skew-normal errors at both subject and observation level in a linear panel model for metric responses, while Jara et al. (2008) allow both subject random effects and observation level errors to follow a multivariate skew-t distribution. Thus, for a mixed panel model for a metric response of dimension T_i:

$$y_i = X_i\beta + Z_ib_i + u_i \tag{8.7}$$

suppose y_i follows the multivariate skew-t density (Sahu et al., 2003). So

$$y_i|\beta, b_i, \sigma^2, R_i, \Delta_i \sim ST_\nu(X_i\beta + Z_ib_i, \sigma^2 R_i, \Delta_i),$$

where ν is the degrees of freedom, R_i is a $T_i \times T_i$ matrix and $\Delta_i = \text{diag}(\delta_{1i}, \ldots, \delta_{T_i,i})$ contains skewness parameters relevant to the observation level residuals that may in principle be specific to individuals and times. The density of the entire observation set $y = (y_1, \ldots, y_n)$ conditional on collections of b_i, R_i and Δ_i is (Jara et al., 2008)

$$\begin{aligned} &p(y|\beta, b, \sigma^2, R, \Delta) \\ &\quad \propto \prod_{i=1}^{n} 2^{T_i} t_{T_i,\nu}(y_i|X_i\beta + Z_ib_i, \sigma^2 R_i + \Delta_i^2) \times \int_0^\infty t_{T_i,\nu}(w_i|\mu_w, \Sigma_w)dw_i, \end{aligned}$$

where $t_{m,\nu}(x|\mu_x, \Sigma_x)$ denotes a multivariate t density of dimension m. When R_i reduces to an identity matrix I_{T_i} and the subject–time skewness parameters δ_{it} to a global parameter δ, namely $\delta_{it} = \delta$, the conditional expectation and variances for each subject are

$$E(y_i|\beta, b_i, \sigma^2, \delta) = X_i\beta + Z_ib_i + (\nu/\pi)^{0.5}\frac{\Gamma[(\nu-1)/2]}{\Gamma(\nu/2)}\delta 1,$$

$$\text{Var}(y_i|\beta, b_i, \sigma^2, \delta) = \frac{\nu}{\nu-2}(\sigma^2 + \delta^2)I_{T_i} + (\nu/\pi)\left[\frac{\Gamma[(\nu-1)/2]}{\Gamma(\nu/2)}\right]^2 \delta^2 I_{T_i}.$$

Under the reductions $R_i = I_{T_i}, \delta_{it} = \delta$, the conditional density may be described by a mixture of normal distributions by conditioning on positive variables $w_i = (w_{1i}, \ldots, w_{T_ii})$ obtained by truncated sampling from a multivariate normal with identity covariance matrix of dimension T_i and subject-specific scalings $\lambda_i \sim \text{Ga}(\nu/2, \nu/2)$, so that

$$y_i|\beta, b_i, \sigma^2, w_i, \lambda_i, \delta) \sim N_{T_i}\left(X_i\beta + Z_ib_i + \delta w_i, \frac{\sigma^2}{\lambda_i}I\right).$$

$$w_i \sim N_{T_i}\left(0, \frac{1}{\lambda_i}I\right) \quad I(0,).$$

In the (usual) case when $X_i\beta + Z_ib_i$ contains an intercept, then for identifiability reasons the elements in the vector w_i may be centered (subsequent to truncated sampling) (Jara et al., 2008). Thus, at each iteration the average of the w_{it} can be obtained, and then the centered variables $W_{it} = w_{it} - \bar{w}_i$, so that

$$y_{it} \sim N\left(X_{it}\beta + Z_{it}b_i + \delta W_{it}, \frac{\sigma^2}{\lambda_i}\right).$$

Additionally in the model Equation 8.7, the permanent random effects b_i may also be taken as skew multivariate t. Assuming the Z predictors are a subset of the X predictors, one then has

$$b_i|D, \Gamma_i \sim ST_{\nu_b}(0, D, \Gamma_i), \tag{8.8a}$$

where D is $Q\times Q$, ν_b is the degrees of freedom, and $\Gamma_i = \text{diag}(\gamma_{1i}, \ldots, \gamma_{Qi})$ contains skewness parameters relevant to the permanent effects. Assuming common skew parameters $\Gamma_i = \Gamma = \text{diag}(\gamma_1, \ldots, \gamma_Q)$, and conditional on a Q vector of positive variables, $h_i = (h_{1i}, \ldots, h_{Qi})$, with

$$\begin{aligned} h_i &\sim N_Q\left(0, \frac{1}{\xi_i}I\right) \quad I(h_i > 0), \\ \xi_i &\sim \text{Ga}\left(\frac{\nu_b}{2}, \frac{\nu_b}{2}\right), \end{aligned} \tag{8.8b}$$

the random effects are mixtures of normals, namely

$$b_i \sim N_Q\left(\Gamma h_i, \frac{1}{\xi_i}D\right). \tag{8.8c}$$

For improved identification, the h_i can be centered around their means (at each MCMC iteration), namely $H_{qi} = h_{qi} - \overline{h_{q.}}$, so that

$$b_i \sim N_Q\left(\Gamma H_i, \frac{1}{\xi_i}D\right).$$

8.6.1 Robust panel data models: discrete mixture models

Another way of reducing the impact of arbitrarily selecting a particular parametric form for random variation in b_i and/or u_{it} is by using discrete mixtures of random effects priors—for example, see Butler and Louis (1992), Verbeke and Lesaffre (1996), Sharples (1990), Weiss et al. (1999). A discrete mixture prior may be more flexible in dealing with unusual cases, skewness, and multiple modes. The possibly conflicting criteria required in the case of a prior on b_i are considered by Muller and Rosner (1997): namely that the prior should be flexible to allow for heterogeneity in the population, though on the other hand, unusual cases should not have an undue predictive influence.

An often suitable approach would involve two group normal mixture priors with the groups typically being a main group and outlier group (Weiss et al., 1999, 1563). Such schemes apply both for heterogeneity b_i:

$$b_i \backsim \pi_b N_Q(0, D) + (1 - \pi_b) N_Q(0, \varphi_b^2 D),$$

and for unstructured observation level u_{it}:

$$u_{it} \backsim \pi_u N(0, \sigma^2) + (1 - \pi_u) N(0, \varphi_u^2 \sigma^2),$$

where the factors $\{\varphi_b > 1, \varphi_u > 1\}$ are used for variance inflation for the outlier group. The prior probabilities of being in the main population are set high (e.g., $\pi_b = \pi_u = 0.95$), and variance inflation factors are typically large, e.g., $\varphi_b = \varphi_u = 5$ or 10. Provided one or other of the parameter sets $\{\pi_b, \pi_u\}$ or $\{\varphi_b, \varphi_u\}$ is assumed known (i.e., is assigned preset values), the other set may be taken as unknowns.

Another option are "switching" or shift priors whereby one group has zero effects, but a minority group has nonzero effects. These may be used for unstructured errors introduced to reflect overdispersion in count or binomial data. For example, for $y_{it} \sim Po(\mu_{it})$, one may have

$$\log(\mu_{it}) = X_{it}\beta + Z_{it}b_i + \sigma k_{it} u_{it},$$

where σ is a scale factor, $k_{it} \sim \text{Bern}(\pi_u)$, $u_{it} \sim N(0, 1)$, such that observation level effects are zero when $k_{it} = 0$. One may preset π_u low, say $\pi_u = 0.05$. For a longitudinal series with level c_t subject to possible shifts, and X_{it} not containing an intercept. one may similarly propose in an adaptation of Gerlach et al. (2000) that

$$\begin{aligned} y_{it} &= c_t + X_{it}\beta + Z_{it}b_i + \sigma u_{it}, \\ c_t &= c_{t-1} + k_t \sigma w_t, \end{aligned}$$

where $u_{it} \sim N(0, 1)$, $w_t \sim N(0, 1)$, and $k_t \sim \text{Bern}(\pi_c)$ with π_c low.

A different emphasis is when there is a substantive rationale for assuming subject level effects b_{qi} follow a discrete prior at subject level. The hyperparameters governing the subject effects $\{b_{1i}, b_{2i}, \ldots, b_{Qi}\}$ then become specific for the latent category. Thus, in a growth curve model for modeling changes in aggression ratings, Muthen et al. (2002) assume that a small number of latent trajectories characterize growth in aggression. For subject i, let the latent category be denoted $k_i \in (1, \ldots, K)$. Then conditional on $k_i = k$, Equation 8.5 would become

$$\begin{aligned} b_{1i} &= B_{1k} + e_{1i}, \\ b_{2i} &= B_{2k} + \delta_{2k} Tr_i + e_{2i}, \\ b_{3i} &= B_{3k} + \delta_{3k} Tr_i + e_{3i}, \end{aligned}$$

where $(e_{1i}, e_{2i}, e_{3i}) \backsim N_3(0, D_k)$. Observation level dispersion parameters may also differ according to latent group.

Flexible discrete mixture models are also obtained under Dirichlet process and related semi-parametric priors, as considered for repeated binary data by Quintana et al. (2008), and for panel count data by Kleinman and Ibrahim (1998). Averaging over different number of mixture components K is possible under discrete parametric mixture models using the reversible jump MCMC algorithm—see Ho and Hu (2008) for an application to the linear mixed model. In the nonparametric mixture approach, the number of clusters is an outcome of other parameters such as the DP mass parameter κ. Under the truncated Dirichlet process (Ohlssen et al., 2007), one may set a maximum K_m possible clusters, with the realized number at each iteration being $K \leq K_m$. The posterior density of K will indicate whether the assumed maximum K_m is sufficient.

Hirano (2000, 2002) discusses nonparametric alternatives regarding white noise observation errors u_{it} in panel data, while Kleinman and Ibrahim (1998) and Muller and Rosner (1997) consider mixed Dirichlet process (MDP) modeling of permanent effects b_i. Under the MDP option, one has b_i following a density G which is itself unknown, centered on a specified base density G_0 with precision κ. For example, with a base density $G_0 = N_Q(B, D)$, one has

$$\begin{aligned} g[E(y_{it}|b_i)] &= \eta_{it} = X_{it}\beta + Z_{it}b_i + u_{it}, \\ u_{it} &\sim N(0, \sigma^2), \\ b_i &\sim G, \\ G &\sim \text{DP}(\kappa G_0), \\ G_0 &= N_q(B, D), \end{aligned}$$

where priors on β, B, D, σ^2 are typically as considered above. This is the conjugate MDP prior for the normal linear mixed model which tends to the conventional hierarchical prior as $\kappa \to \infty$.

The model considered by Hirano (2002) is also conjugate, and based on a dynamic model,

$$y_{it} = b_i + \rho y_{i,t-1} + u_{it},$$

where the b_i are zero mean effects that are modeled parametrically, and $u_{it} = y_{it} - b_i - \rho y_{i,t-1}$ may have nonzero means. One has for $\theta_{it} = \{\mu_{it}, \sigma^2_{it}\}$,

$$\begin{aligned} u_{it} &\sim N\left(\mu_{it}, \sigma^2_{it}\right), \\ \theta_{it} &\sim G, \\ G &\sim \text{DP}(\kappa G_0), \end{aligned}$$

where G_0 specifies,

$$G_0(\mu, \sigma^2) : \frac{1}{\sigma^2} \sim \frac{\chi^2(s)}{sL}; \mu \sim N(m, b\sigma^2).$$

where s, L, m and b are preset. As discussed in Chapter 3, κ may be preset or taken as an unknown. Thus Kleinman and Ibrahim (1998, 2592) consider defaults such as $\kappa = 1.5$ and $\kappa = 100$, while Hirano (2002) takes $\kappa \sim \text{Ga}(2, 0.5)$.

Example 8.8. A Pharmacokinetic Application To exemplify heteroscedastic panel analysis, this example considers pharmacokinetic panel data relating to plasma concentrations y_{it} of the drug Cadralazine in $n = 10$ cardiac failure patients at various times $t = 1, \ldots, T_i$ (in hours) following administration of a single dosage of $G = 30$ mg. Wakefield et al. (1994) propose a one compartment nonlinear model for these data with mean concentration at time t expressed as

$$\eta_{it} = (G/\alpha_i)\exp(-\beta_i t/\alpha_i),$$

where $\alpha_i > 0$ and $\beta_i > 0$ are, respectively, the volume of distribution and clearance parameters for each subject. A hierarchical model is proposed with second stage consisting of a multivariate normal or multivariate t for the transformed subject effects $(b_{1i}, b_{2i}) = \{\log(\alpha_i), \log(\beta_i)\}$. For the first stage model, one option is a log-normal since y is positive, or a truncated normal, with y_{it} constrained to be positive. Under the latter, a heteroscedastic power model, with a single precision parameter τ, leads to a variance η_{it}^{ω}/τ, and the first stage model is

$$y_{it} \sim N(\eta_{it}, \eta_{it}^{\omega}/\tau) \quad I(0,).$$

Another option for the first stage model involves a normal scale mixture, namely

$$y_{it} \sim N(\eta_{it}, \eta_{it}^{\omega}/[\lambda_i\tau]) \quad I(0,),$$
$$\lambda_i \sim \text{Ga}(0.5\nu, 0.5\nu).$$

Here these options are compared under the priors $\tau \sim \text{Ga}(1, 0.001)$, and $1/\nu \sim U(0.01, 0.5)$. A lognormal LN(0, 1) prior was assumed here for ω, rather than $\omega \sim U(0, 5)$ as in Wakefield et al. (1994), as the latter led to convergence problems. At the second stage, a bivariate normal for (b_{1i}, b_{2i}) is assumed with

$$(b_{1i}, b_{2i}) \sim N_2([B_1, B_2], D),$$
$$B \sim N(B_0, C),$$
$$D^{-1} \sim W([\rho R]^{-1}, \rho),$$

with ρ, R, B, and C as in Wakefield et al. (1994).

Inferences are based on the second halves of 100 thousand iteration runs in WinBUGS14 (with two chains); the scale mixture model with variances $\eta_{it}^{\omega}/[\lambda_i\tau]$ has a lower log(psML), namely 48.4 vs. 74.3, than the model with variances η_{it}^{ω}/τ. This may reflect the relatively heavy parameterization in the scale mixture approach, even though the average ν under the scale mixture is 7.6. The power ω is estimated at 0.92 under the better performing model, whereas a log-normal model would imply $\omega \simeq 2$. Out-of-sample predictions of concentrations are important and these are made at a duration 32 hours.

For subject 2, whose plasma concentrations remain relatively high compared to other subjects, the mean prediction is 0.109.

An alternative robustification[8] involves a scale mixture at the second stage (i.e., in the prior for random permanent effects) with

$$\begin{aligned} y_{it} &\sim N(\eta_{it}, \eta_{it}^{\omega}/\tau) \quad I(0,), \\ (b_{1i}, b_{2i}) &\sim N_2([\mu_1, \mu_2], D/\xi_i), \\ \xi_i &\sim \text{Ga}(0.5\nu, 0.5\nu). \end{aligned}$$

This provides a log(psML) of 53.6, with ω having mean 0.88. The lowest weight ($\xi_2 = 0.66$) is for subject 2, whose mean prediction at 32 hours becomes 0.115.

Example 8.9. Skewed Cholesterol Data This example relates to longitudinal data on cholesterol levels collected during the Framingham heart study for $n = 200$ randomly selected subjects, as considered by Zhang and Davidian (2001). Relevant subject attributes are sex (1 = M, 0 = F) and age at baseline. Several studies have re-considered the linear mixed model used by those authors, namely

$$y_{it} = X_{it}\beta + Z_{it}b_i + u_{it} = \beta_1 + \beta_2\text{Sex}_i + \beta_3\text{Age}_i + \beta_4 a_{it} + b_{1i} + b_{2i}a_{it} + u_{it}, \quad (8.9)$$

where y_{it} is cholesterol level divided by 100, and a_{it} is $(\text{time} - 5)/10$, where time is years from baseline. Total periods T_i differ between subjects, varying from 1 to 6.

Here two models are considered to reflect positive skew apparent from plots of the outcome. One may, for example, consider multivariate normal or t skew in the permanent effects (b_{1i}, b_{2i}). For skew bivariate normal permanent effects one has (Model 1)

$$b_i | D, \Gamma \sim SN(0, D, \Gamma),$$

where D is 2×2, and $\Gamma = \text{diag}(\gamma_1, \gamma_2)$. Equivalently, conditional on the positive standard normal effects

$$h_i \sim N_Q(0, I) \quad I(0,),$$

(with $Q = 2$), the random intercepts and slopes in Equation 8.9 are obtained as

$$b_i \sim N_Q(\Gamma h_i, D).$$

However, an alternative perspective (Model 2) is provided[9] by allowing changing skew through time. This involves a T_i vector of period-specific skewness parameters $\delta_i = (\delta_1, \ldots, \delta_{T_i})$, that is $\delta_{it} = \delta_t$, in a multivariate skew normal scheme for the observation level errors. Hence

$$\begin{aligned} u_{it} | \sigma^2, w_{it}, \delta_t &\sim N_{T_i}\left(\delta_t w_{it}, \sigma^2\right), \\ w_{it} &\sim N(0, 1) \quad I(0,). \end{aligned}$$

While centered positive variables h_i and w_{it} may be preferred for identification, this slowed MCMC analysis considerably and uncentered effects are used for illustration.

The second half of a two chain run of 7500 iterations for Model 1 shows significant skewness in subject intercepts b_{1i}, but not in the time slopes, with the respective γ parameters having 95% intervals (0.40, 0.61) and (−0.26, 0.22). The log(psML) obtained by monitoring inverse likelihoods is 7.7. A slightly lower log(psML) of 6 is obtained for a run of the same length for Model 2. The δ_t coefficients show greater skewness at earlier follow-up times with δ_1, δ_2, and δ_3 having positive 95% credible intervals, respectively (0.10, 0.30), (0.07, 0.26), and (0.01, 0.22), but later δ_t parameters having credible intervals straddling zero.

Example 8.10. Robust GLMM for Epilepsy Data This example considers forms of robust modeling for the seizure data discussed in Example 8.7. For example, Yau and Kuk (2002) consider sensitivity of fixed effects parameter estimates to the specification of the random effects components at both subject level and observation level, and seek inferences that are robust to outliers at both levels. They consider data for the 59 patients over the last $T = 4$ visits by conditioning on the initial observation, though they include it as a baseline measure of severity. The five predictors (X_1 to X_5) that they use are: log of baseline seizure (Base), treatment (Tr), treatment interaction with baseline, log of patient age, and a binary variable equal to 1 for the final visit, V4. Yau and Kuk consider first two normal random effects models, one with such effects at unit level only, namely

$$\log(\mu_{it}) = \alpha + X_{it}\beta + b_i$$

with $b_i \sim N(0, D)$, and the other with random variation at both unit and observation levels, namely

$$\log(\mu_{it}) = \alpha + X_{it}\beta + b_i + u_{it}.$$

The first model applied here (Model 1) has subject level effects only, and assumes a uniform prior on $D^{0.5}$ and flat priors on fixed effects; convergence is obtained before iteration 1000 in a two chain run of 10,000 iterations. This model shows predictive coverage limitations, since only 91% of the data are contained within 95% intervals of replicates $y_{\text{rep},it}$ sampled from the model. There is also excess heterogeneity with posterior mean scaled deviance of 432 compared to 236 observations. Standardized subject effects (namely posterior means of b_i divided by their posterior standard deviations) include three values over 5 (patients 25, 56, and 35 with standardized effects 5.47, 6.26, and 6.62). Examination of the observation level CPO estimates suggests poor fits to some points, for example, the third visit of patient 25, whose series is $\{18, 24, 76, 25\}$.

A second model (Model 2) has random variation at both subject and subject–time levels, with hierarchically centered random effect priors (cf. Roberts and Sahu, 2001), namely

$$u_{it} \sim N\left(b_i, \sigma_u^2\right),$$
$$b_i \sim N(\alpha, D),$$
$$\mu_{it} = u_{it} + X_{it}\beta,$$

with a $U(0,100)$ prior for $D^{0.5}$. Additionally a uniform shrinkage prior (Natarajan and Kass, 2000) is adopted in relation to the other variance component $\text{var}(u_{it}) = \sigma_u^2$, with $\phi = D/(D + \sigma_u^2) \sim U(0,1)$. Convergence in a two chain run of 10,000 iterations is obtained before iteration 2500, and iterations 2501–10,000 give posterior means for σ_u^2 and D of 0.135 and 0.255, close to the restricted maximum likelihood (REML) estimates of Yau and Kuk (2002). The coefficient on V4 loses its significance, but predictive coverage is now over 99% and the posterior mean of the scaled deviance is now 248. The pseudo marginal likelihood for this model is −605, compared to −671 for the simpler model with subject level random intercepts only. The standardized subject level effects for patients 25, 56, and 35 are less extreme, namely 3.38, 3.74, and 4.01, but still suggest outlier status.

Yau and Kuk (2002) also mention models where the heterogeneity is modeled using a discrete mixture of intercepts, namely

$$\log(\mu_{it}) = \alpha_{k_i} + x_{it}\beta,$$

where the latent categorical allocation $k_i \in (1, \ldots, K)$ is multinomial with probabilities $(\pi_1, \ldots, \pi_K)$ following a diffuse Dirichlet prior. The intercepts $\alpha_1, \ldots, \alpha_K$ are subject to an order constraint. For illustrative purposes this approach is applied with $K = 4$ (as Model 3), but is clearly inferior to a two level random effect approach. The posterior mean deviance is 449.

Model 4 retains observation errors u_{it} as in Model 2, but with a selection mechanism for these effects.[10] This model adapts to the scenario where many patients may exhibit a stable differential over the visits (modeled by a level 2 effect b_i) with only a subset of patients exhibiting erratic trajectories that require a random effect for each visit. Thus binary indicators $\delta_i \sim \text{Bern}(\pi_\delta)$ are introduced for each subject in a model where

$$\log(\mu_{it}) = \alpha + x_{it}\beta + b_i + \delta_i u_{it},$$

with a $U(0,100)$ prior for $D^{0.5}$, and $\phi = D/(D + \sigma_u^2) \sim U(0,1)$. One may set π_δ to be an unknown or preset its value. Here a value $\pi_\delta = 0.05$ is adopted initially, so that the posterior values $Pr(\delta_i = 1|y)$ can provide clear contrasts to the prior values $Pr(\delta_i = 1) = \pi_\delta$. A two chain run of 10,000 iterations converges early and from the last 7500 iterations it emerges that 10 patients have values for $Pr(\delta_i = 1|y)$ that exceed 0.1, and four patients, namely 25,

56, 10, and 16, have values that exceed 0.9. This model produces a mean scaled deviance of 291 with predictive coverage of 98.3%. However, its psML is slightly worse than Model 2, namely −608. Setting π_δ at 0.10 reduces the psML to −605, comparable to Model 2.

Finally, Model 5 reverts to a random effect only model, but allows for possibly non-normality via a mixed Dirichlet process. The random effects are bivariate, with nonzero means, one for the intercept (α in above models) and one for a linear slope on visit. Thus

$$\log(\mu_{it}) = X_i\beta + Z_{it}b_i,$$

where $X_i = (\text{Base}, \text{Tr}, \text{Base} * \text{Tr}, \text{Age})$, with predictor variables defined as in Kleinman and Ibrahim (1998, 2592), and with $Z_{it} = (1, \text{Visit})$, where the visit times are centered weeks/10. The patient random effects have prior,

$$\begin{aligned}(b_{1i}, b_{2i}) &\sim G,\\ G &\sim \text{DP}(\kappa G_0),\\ G_0 &= N_Q(B, D),\\ \kappa &\sim \text{Ga}(2, 4),\\ (B_1, B_2) &\sim N_2(0, 1000I),\\ D^{-1} &\sim W(R, \rho),\\ R &= \text{diag}(20), \quad \rho = 10,\end{aligned}$$

so that $E(D^{-1}) = \text{diag}(0.5)$, as in Kleinman and Ibrahim (1998). The maximum number of possible clusters is set at $K_m = 20$.

A two chain run of 10,000 iterations converges after 5000. Conditional on the particular choice made for the prior on κ, one obtains a mean scaled deviance of 392, better than Model 1 but still leaving excess variability. The posterior density for the number of clusters has 0.025 and 0.975 percentiles at 6 and 13, with mean 8.9, while κ has posterior mean 1.36. Histograms of the mean $\{b_{1i}, b_{2i}\}$ with superimposed normal curves suggest excess kurtosis rather than skewness (Figures 8.2 and 8.3). The treatment effect under this model is significant, with 95% interval (−0.66, −0.03).

8.7 Multilevel, Multivariate, and Multiple Time Scale Longitudinal Data

Applications involving longitudinal data often involve contextual nesting of subjects, multiple responses or multiple time scales. Consider first data y_{ijt} for repetitions $t = 1, \ldots, T_{ij}$ for subjects $i = 1, \ldots, n_j$ nested within clusters $j = 1, \ldots, J$. The general linear mixed model now assumes that conditional on predictors and random effects, the data are distributed independently according to the exponential family,

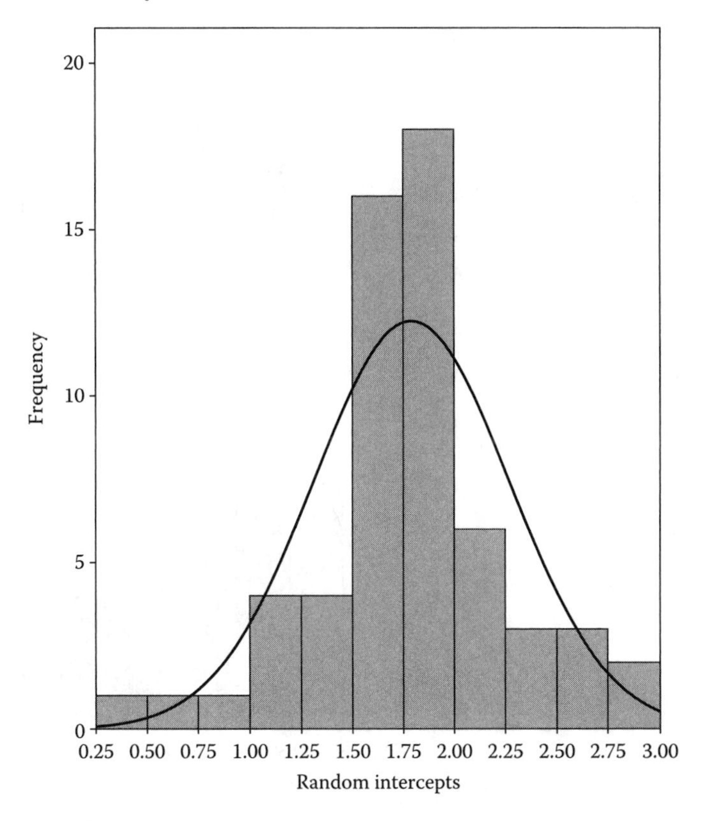

FIGURE 8.2
Random intercepts (seizure data).

$$p(y_{ijt}|\theta_{ijt},\phi) = \exp\left(\frac{y_{ijt}\theta_{ijt} - a(\theta_{ijt})}{\phi} + C(y_{ijt},\phi)\right)$$

with conditional means $E(y_{ijt}) = \mu_{ijt} = a'(\theta_{ijt})$, and link $g(\mu_{ijt}) = \eta_{ijt}$ to regression terms η_{ijt}. The structural model may specify permanent random effects $\{d_j, b_{ij}\}$ for both clusters and subjects within clusters. Fixed effect regression parameters may now be cluster specific, namely

$$g[E(y_{ijt}|\beta_j, c_i, b_{ij})] = X_{ijt}\beta_j + W_{ijt}d_j + Z_{ijt}b_{ij}.$$

Taking β_j as fixed effects is appropriate when the categorization $j = 1, \ldots, J$ refers to a small number of treatment groups or demographic categories, as in the much analyzed data from Pothoff and Roy (1964), or the data from Oman et al. (1999) where four groups are formed by crossing treatment by gender—see Example 8.11.

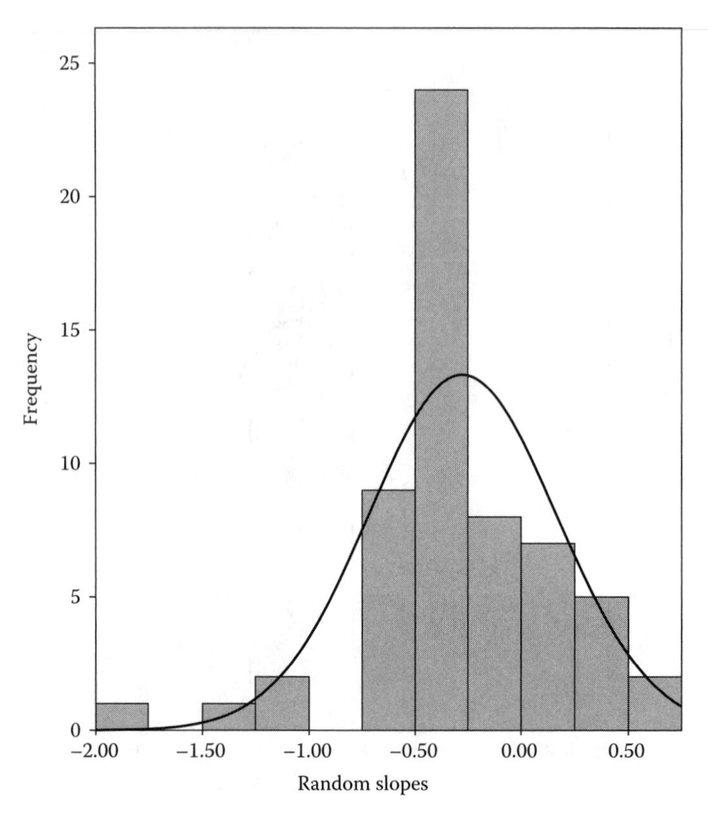

FIGURE 8.3
Random slopes (seizure data).

Where predictor effects vary randomly over time as in Equation 8.2, one may now include time and cluster-specific effects c_{jt}, so that

$$g\left[E\left(y_{ijt}|\beta_j, c_i, b_{ij}\right)\right] = X_{ijt}\beta_j + W_{ijt}d_j + H_{ijt}c_{jt} + Z_{ijt}b_{ij}.$$

Autocorrelated errors may also be required to model temporal dependencies, so that the unexplained variance may be due to a number of sources. For example, Lee and Hwang (2000) consider a normal mixed effects model, applicable in growth curve applications with multiple groups of subjects $j = 1, \ldots, J$, with

$$y_{ijt} = X_{ijt}\beta_j + b_{ij} + \varepsilon_{ijt} + u_{ijt},$$

where

$$\varepsilon_{ijt} = \rho\varepsilon_{ij,t-1} + v_{ijt}$$

and the variances of b_{ij}, u_{ijt}, and v_{ijt} are subject to uniform shrinkage priors (see Section 8.2.3).

For nested panel data inferences (e.g., on growth patterns) may be improved by borrowing strength over clusters. Similarly with panel data on multiple outcomes, y_{mit} for subjects $i = 1, \ldots, n$, outcomes $m = 1, \ldots, M$, and repetitions $t = 1, \ldots, T$, inferences on particular outcomes may be strengthened by incorporating correlations between outcomes. An example might be for panel data on correlated but relatively rare spatially configured health events, such as cancer types. Multiple outcome panel data are common in clinical and educational applications, and the effectiveness of interventions may be judged in terms of multiple (usually) correlated outcomes rather than by a single criterion (Dunson, 2007). In environmental applications multiple outcomes with related aetiology are likely to be correlated (e.g., Jorgensen et al., 1999; Liu and Hedeker, 2006).

With metric or discrete data y_{mit} for multiple outcomes, the general linear mixed model with time homogenous, time varying and subject varying predictor effects becomes

$$g\left[E\left(y_{mit}|\beta_m, b_{mi}, c_{mt}\right)\right] = \eta_{mit} = X_{it}\beta_m + Z_{it}b_{mi} + H_{it}c_{mt},$$

where X_{it}, Z_{it}, and H_{it} are of length P, Q, and R. For example, Agresti (1997) considers multivariate repeated binary responses,

$$y_{mit} \sim \text{Bern}(\pi_{mit}),$$

and a prediction model based on outcome–subject and outcome–time effects, namely

$$\text{logit}(\pi_{mit}) = \beta_m + b_{mi} + c_{mt}.$$

Given multivariate random outcome–subject effects b_{mi}, and fixed effects c_{mt} subject to an identifying corner constraint such as $c_{m1} = 0$ or $c_{mT} = 0$, the y_{mit} are assumed conditionally independent.

The corresponding normal linear mixed model for multivariate metric responses is

$$y_{mit}|\beta_m, b_{mi}, c_{mt}, \sigma_m^2 = X_{it}\beta_m + Z_{it}b_{mi} + H_{it}c_{mt} + u_{mit},$$

where the residuals u_{mit} are typically unstructured normal with variances σ_m^2 specific to outcome m. The M sets of permanent effect priors b_{mi}, each of dimension Q, may be correlated between predictors q within outcomes m, or between outcomes m within variables q, or most generally over both variables q and outcomes m. The same applies to the outcome–time effects c_{mt}, which may be random, and incorporate short-range temporal dependence. For example, time varying intercepts $c_t = (c_{1t}, \ldots, c_{Mt})$ in the case $R = 1$ (and $H_{it} = 1$) could follow autoregressive or random walk priors correlated over outcomes, as in

$$(c_{1t}, \ldots, c_{Mt}) \sim N_M(c_{t-1}, \Sigma_c).$$

Suppose T_{mi} responses are observed on outcome m for subject i, with $S_i = \sum_m T_{mi}$. In vector form the multivariate normal longitudinal model is then

$$Y_i = X_i\beta + Z_ib_i + H_ic_t + u_i,$$

where Y_i is of length S_i. For example, Beckett et al. (2004) consider a bivariate response ($M = 2$), with Z_{itm} of dimension $Q = 2$, and Z_i of dimension $(S_i \times 4) = (S_i \times QM)$. Heterogeneity involves a covariance matrix of dimension QM with outcome-specific random intercepts and random time slopes, and parameters of interest include covariance between person-specific initial levels, D_{13}, and covariance between person-specific slopes, D_{24}.

Conjugate structures (e.g., Poisson–gamma, beta–binomial) may also be used instead of the GLMM approach for discrete multivariate panel outcomes. For example, overdispersed count data y_{mit} may be assumed Poisson with

$$y_{mit} \sim Po(\mu_{mit}\theta_{im}\xi_{mit}),$$

where $\mu_{mit} = \exp(X_{mit}\beta_m)$, and

$$\theta_{im} \sim \text{Ga}(a_m, a_m),$$

represent subject–outcome permanent random effects. The ξ_{mit} represent observation level effects that are unstructured, or autoregressive, as in

$$\xi_{mit} \sim \text{Ga}(b_m\xi_{mi,t-1}, b_m)$$

with variance parameters b_m (e.g., Jorgensen et al., 1999).

8.7.1 Latent trait longitudinal models

As M increases the full dimensional approach becomes cumbersome, and factor analytic or latent trait approaches may pool information just as effectively and more parsimoniously (e.g., Dunson, 2003, 2006, 2007; Jorgensen et al., 1999; Roy and Lin, 2000). Panel data on multiple outcomes raise the possibility of shared random effects across outcomes, instead of outcome-specific effects. For example, in spatio-temporal health applications, it is common to have correlated count responses y_{mit} such as different types of cancer or psychiatric illness. Observed risk factors for such outcomes may be limited or incomplete. Common unobserved area-time risks may be summarized in effects r_{it}, with loadings λ_m linking the common factor scores to each outcome. These may be taken as unstructured (Tzala and Best, 2008), or assumed to be spatially and/or temporally correlated. For identifiability, one may either set $\text{var}(r_{it}) = 1$, in which case the loadings are free parameters, or set one of the loadings to a fixed value, such as $\lambda_1 = 1$, in which case $\text{var}(r_{it})$ is an unknown. Time and area common effects, r_{it}, may be combined with common area effects b_i with loadings γ_m, and common time effects c_t with loadings κ_m, and the same type of identifying rules. Then $y_{mit} \sim Po(\mu_{mit})$, with fixed regression effects that might vary over outcomes, as in an extension of Equation 8.2,

$$\log(\mu_{mit}) = X_{it}\beta_m + \lambda_m r_{it} + \gamma_m b_i + \kappa_m c_t.$$

Additionally unstructured effects u_{mit} may be included to represent remaining overdispersion.

In item analysis and psychometric longitudinal applications, a measurement model might involve both constant and time-varying common factors. Thus, for M items or tests carried out on T occasions, responses may be determined by M item-specific factors b_{mi} and by T time-specific factors r_{it} (Eid, 1996; Marsh and Grayson, 1994). The impact of these is governed by time and outcome specific loadings γ_t and λ_m respectively, so that

$$y_{mit} = \alpha_m + \gamma_t b_{mi} + \lambda_m r_{it} + u_{mit}.$$

Structural equation models for longitudinal data typically involve both response indicators y_{mit} of dimension P_y which measure latent outcomes η_{qit} of dimension $Q_y < P_y$, and exogenous predictors x_{kit} of dimension P_x which measure latent causal influences ξ_{qit} of dimension $Q_x < P_x$ (Dunson, 2007). For example, for $Q_y = Q_x = 1$, ξ_{it} might be a time varying stress severity scale related to short-term stressors $\{x_{kit}, k = 1, \ldots, P_x\}$, and η_{it} might be a time varying latent depression scale related to mood scale measures $\{y_{mit}, m = 1, \ldots, P_y\}$. Then the measurement model is

$$\begin{aligned} y_{mit} &= \alpha_{1m} + \lambda_{1m}\eta_{it} + u_{1mit}, \quad m = 1, \ldots, P_y, \\ x_{kit} &= \alpha_{2k} + \lambda_{2k}\xi_{it} + u_{2kit}, \quad k = 1, \ldots, P_x, \end{aligned}$$

while the structural model might include a linear effect, possibly time varying, of ξ_{it} on η_{it}. Additional aspects of the structural model might be lagged effects in each factor score on its current value, as in

$$\begin{aligned} \eta_{it} &= \rho_1\eta_{i,t-1} + \beta_t\xi_{it} + W_{it}\gamma_1 + \delta_{1it}, \\ \xi_{it} &= \rho_2\xi_{i,t-1} + W_{it}\gamma_2 + \delta_{2it}, \end{aligned}$$

where $\{\delta_{hit} \sim N(0,1), h = 1,2\}$ have known scale for identifiability, and the W_{it} are additional known predictors relevant to the factors, but there are no intercepts (Daniels and Normand, 2006).

A simple common factor model may be applied when there are alternative measuring scales, typically a gold standard measure, and one or more measures of the same quantity but less expensive to obtain. Consider a situation where bivariate metric data $\{y_{1ijt}, y_{2ijt}\}$ are obtained for subjects i within clusters j, where y_{1ijt} denotes repetitions on the standard measure, and y_{2ijt} denotes repetitions on the proxy measure. The goal is to assess the reliability of the proxy measure. One may postulate a shared permanent effect b_{ij} between the two outcomes as well as a unique permanent effect c_{ij} for the proxy measure. In the absence of intercepts for the y_1 model one has

$$\begin{aligned} y_{1ijt} &= b_{ij} + u_{1ijt}, \\ y_{2ijt} &= \alpha_j + \lambda_j b_{ij} + c_{ij} + u_{2ijt}, \end{aligned}$$

where the b_{ij} have nonzero cluster means B_j, but the c_{ij} are zero mean effects, namely

$$b_{ij} \sim N(B_j, D_{1j}), c_{ij} \sim N(0, D_{2j}).$$

The residuals are distributed as $u_{mijt} \sim N(0, 1/\tau_{mj})$. The hypothesis that $\{\alpha_j = 0, \lambda_j = 1\}$ corresponds to y_1 and y_2 being identically calibrated in group j (Oman et al., 1999, 43), that is, they both measure the same quantity on the same scale (see Example 8.11).

8.7.2 Multiple scale panel data

Aggregate health and demographic event data are often available as totals y_{ixt} for multiple time scales, for example by age group $x = 1, \ldots, X$, as well as by period $t = 1, \ldots, T$, and possibly also by area or actuarial risk group $i = 1, \ldots, n$. A further cohort dimension $c = 1, \ldots, C$ is implicit in biological age–time data via the relation $c = t - x + X$, and there have been extensive developments in Bayesian age–period–cohort (APC) and area APC models (AAPC) models (Baker and Bray, 2005; Bray, 2002; Lagazio et al., 2003; Schmid and Held, 2004). For rare event totals y_{ixt} in relation to large populations N_{ixt}, and assuming $y_{ixt} \sim Po(N_{ixt}\mu_{ixt})$, a baseline model age–period (AP) model might assume independence of age and period dimensions, with $\mu_{ixt} = \exp(\eta_{ix})\exp(\theta_{it})$, or equivalently

$$\log(\mu_{ixt}) = \kappa + \eta_{ix} + \theta_{it},$$

where structured (e.g., random walk or autoregressive) priors might be adopted for age–area effects η_{ix} and area–time effects θ_{it}, and the intercept κ is identified according to possible constraints on the random effects.

Thus, Clayton and Schifflers (1987) consider data of the form y_{xt} (i.e., without further stratification), with means μ_{xt} where

$$\log(\mu_{xt}) = \eta_x + \theta_t,$$

with both sets of effects assumed to be random, though fixed effects may be used when X or T is small. In the absence of an overall intercept in this model, one or other series (say η_x) sets the level, and identifiability may be gained by centering the remaining series θ_t at zero (possibly repeatedly at each MCMC iteration), or by setting one parameter in the remaining series to a fixed value, e.g., $\theta_1 = 0$. If the model includes an overall intercept κ, then centering both sets of effects, namely $\sum_x \eta_x = \sum_t \theta_t = 0$, provides a way of ensuring identifiability. An APC model including a mean and structured age, period and cohort effects is

$$\log(\mu_{xt}) = \kappa + \eta_x + \theta_t + \gamma_c$$

and identifiability requires either that the three sets of effects be centered, or that edge constraints such as $\eta_1 = \theta_1 = \gamma_1 = 0$ are used to avoid confounding of the three series. Additionally the relation $c = X - x + t$ means an extra constraint is needed for full identification, for example by taking $\gamma_1 = \gamma_2 = 0$ (Clayton and Schifflers, 1987).

The convolution prior of Besag et al. (1991) may be generalized by adopting structured and unstructured effects for each time scale, as well as for areas (Knorr-Held, 2000). Hence an APC model would become

$$\log(\mu_{xt}) = \kappa + \eta_x + \theta_t + \gamma_c + u_{1x} + u_{2t} + u_{3c},$$

where u_{1x}, u_{2t}, and u_{3c} are unstructured zero mean random effects, while $\{\eta_x, \theta_t, \gamma_c\}$ follow structured (i.e., random walk or other autoregressive) form. For area–age–period data, $y_{ixt} \sim Po(N_{xit}\mu_{xit})$, this approach leads to

$$\log(\mu_{ixt}) = \kappa + \eta_x + \theta_t + \gamma_c + s_i + u_{1x} + u_{2t} + u_{3c} + u_{4i},$$

where s_i follows a structured spatial autoregressive prior, but the u_{4i} are unstructured zero mean random effects.

In the preceding models, the dimensions are independent and multiplicative in the risk scale (additive in the log risk scale). In particular, Hoem (1987) discusses multiplicative models for data y_{ix} observed over ages and other strata such as areas. In practice, interactions between one or more of the different time scales, or between the time scales and the units (e.g., areas or actuarial risk groups), are likely. There may also be overdispersion, necessitating more complex parameterizations. Interactions ψ_{xc} between age and cohort are relevant if the age slope is changing between cohorts (e.g., cancer deaths at younger ages are less common in recent cohorts), while in mortality forecasting, age–time interactions ψ_{xt} are of interest since different age groups may be subject to different mortality improvements (Lee and Carter, 1992; Pedroza, 2006). In area APC models, area–cohort and area–time interactions might be relevant (Lagazio et al., 2003), while in area life table models (Congdon, 2006a), age–area interactions may be investigated, since deprived areas may have relatively high "premature" mortality or illness (premature mortality is sometimes defined by death before age 75).

In area–time models, one may extend the RIAS principle, and assume area-specific random variation for both the level and a time covariate. This amounts to taking the interaction ψ_{it} as a linear trend model, with neighboring areas having similar trend parameters, as in Bernardinelli et al. (1995). Thus with $y_{it} \sim Po(N_{it}\mu_{it})$,

$$\log(\mu_{it}) = \kappa + \omega_{1i} + \omega_{2i}(t - \bar{t}),$$

where ω_{1i} and ω_{2i} are spatially correlated over areas. One may further adopt a bivariate spatial (e.g., bivariate CAR) prior for $\{\omega_{1i}, \omega_{2i}\}$, allowing level and trend parameters to be correlated. Additionally, a convolution form may be adopted both for level and trend, so that

$$\log(\mu_{it}) = \kappa + \omega_{1i} + u_{1i} + (\omega_{2i} + u_{2i})(t - \bar{t}),$$

where u_{1i} and u_{2i} are unstructured random effects. Equivalently letting $c_{ji} = \omega_{ji} + u_{ji}$ one has

$$\log(\mu_{it}) = \kappa + c_{1i} + c_{2i}(t - \bar{t}).$$

A variation is to introduce an overall nonlinear trend via parameters δ_t, along with time-specific spatial and unstructured effects $\{\omega_{it}, u_{it}\}$, and stationary $AR1$ dependence in the total lagged spatial effect $c_{it} = \omega_{it} + u_{it}$ (Martinez-Beneito et al., 2008). Thus for $t > 2$,

$$\log(\mu_{it}) = \kappa + \delta_t + c_{it} + \rho c_{i,t-1},$$

with $\rho \in (-1, 1)$, while for $t = 1$,

$$\log(\mu_{it}) = \kappa + \delta_1 + \frac{c_{i1}}{(1-\rho^2)^{0.5}}.$$

This is equivalent to assuming $\log(\mu_{it}) = \kappa + \delta_t + \rho^{t-1}(1-\rho^2)^{-0.5}c_{i1} + \sum_{k=2}^{t} \rho^{t-k} c_{ik}$, where the last term is zero when $t = 1$.

In area–age–time models, area–age–time interactions ψ_{ixt} may be parsimoniously modeled by separate linear trends for each age and area, namely

$$\psi_{ixt} = (\omega_{1x} + \omega_{2i})(t - \bar{t})$$

as in Sun et al. (2000), where the random coefficients ω_{1x} and ω_{2i} may be structured over ages and areas, respectively. Sun et al. (2000) actually assume a spatial CAR(ρ) prior with mean zero for the ω_{2i} (Section 4.8.2), but take the ω_{1x} to be unrelated fixed effects. The full model of Sun et al. (2000) also includes unstructured age–area–time effects, u_{ixt}, so that

$$\log(\mu_{ixt}) = \kappa + s_i + \eta_x + (\omega_{1x} + \omega_{2i})(t - \bar{t}) + u_{ixt}.$$

Alternatively the time function in ψ_{ixt} may be unknown, as in

$$\psi_{ixt} = (\omega_{1x} + \omega_{2i})\delta_t,$$

where positive loadings ω_{1x} and ω_{2i} specify which ages are most sensitive to trend effects δ_t. For identification, the δ_t are centered at zero or have a corner constraint such as $\delta_1 = 0$, and the loadings ω_{1x} and ω_{2i} may be centered at 1, constrained to sum to 1, or have a minimum of 1 (e.g., Osborn, 1975). So for declining mortality, represented by δ_t following (say) a first-order random walk, larger ω_{1x} and ω_{2i} indicate which age groups and areas contribute most to the mortality decline. Lee and Carter (1992) apply the age–time product model $\psi_{xt} = \omega_x\delta_t$ in mortality forecasting, with identification obtained by ensuring δ_t sum to zero and that the ω_x sum to 1.

Interaction priors may also be based on a Kronecker product of the structure matrices for the relevant dimensions (Clayton, 1996; Knorr-Held, 2000), where a structure matrix is a constituent part of the precision (inverse covariance) matrix. For example, if the structure matrix of separate area and age effects are denoted K_s and K_x, then $K_{sx} = K_s \otimes K_x$ defines the structure matrix for the joint prior for ψ_{ix}, and conditional priors on ψ_{ix} can be obtained from K_{sx}. Thus, an $RW1$ prior in age has a structure matrix with off-diagonal elements $K_{x[ab]} = -1$ if ages a and b are adjacent, and $K_{x[ab]} = 0$

otherwise. Diagonal elements are 1 if $a = b = 1$ or $a = b = X$, and equal 2 for other diagonal terms. An $RW2$ prior for age has structure matrix:

$$K_x = \begin{bmatrix} 1 & -2 & 1 & & & & & & & \\ -2 & 5 & -4 & & & & & & & \\ 1 & -4 & 6 & -4 & 1 & & & & & \\ & 1 & -4 & 6 & -4 & 1 & & & & \\ & & \cdot & \cdot & \cdot & \cdot & & & & \\ & & & 1 & -4 & 6 & -4 & 1 & & \\ & & & & 1 & -4 & 6 & -4 & 1 \\ & & & & & 1 & -4 & 5 & -2 \\ & & & & & & 1 & -2 & 1 \end{bmatrix}.$$

The CAR(1) prior for spatially structured errors $s = (s_1, \ldots, s_n)$ based on adjacency of areas is multivariate normal with precision matrix $\tau_s K_s$, where τ_s is an overall precision parameter, and off-diagonal terms $K_{s[ij]} = -1$ if areas i and j are neighbors, and $K_{s[ij]} = 0$ for nonadjacent areas. The diagonal terms in K_s are L_i where L_i is the cardinality of area i (its total number of neighbors). Then an area–age interaction effect ψ_{ix} formed by crossing an $RW1$ age prior with a CAR(1) spatial effect has joint precision

$$\frac{1}{\sigma^2_\psi} K_s \otimes K_x,$$

and full prior conditionals with variances σ^2_ψ / L_i when $x = 1$ or $x = X$, and $\sigma^2_\psi/(2L_i)$ otherwise. With ∂_i denoting the neighborhood of area i, the prior conditional means Ψ_{ix} for ψ_{ix} are

$$\begin{aligned} \Psi_{i1} &= \psi_{i2} + \sum_{j \in \partial_i} \psi_{j1}/L_i - \sum_{j \in \partial_i} \psi_{j2}/L_i, \\ \Psi_{ix} &= 0.5(\psi_{i,x-1} + \psi_{i,x+1}) + \sum_{j \in \partial_i} \psi_{jx}/L_i \\ &\quad - \sum_{j \in \partial_i} (\psi_{j,x+1} + \psi_{j,x-1})/(2L_i), \quad 1 < x < X, \\ \Psi_{iX} &= \psi_{i,X-1} + \sum_{j \in \partial_i} \psi_{jX}/L_i - \sum_{j \in \partial_i} \psi_{j,X-1}/L_i. \end{aligned}$$

For identification, the ψ_{ix} should be doubly centered at each iteration (over areas for a given age x, and over ages for a given area i).

Example 8.11. Alternative Measures of Creatinine Clearance Oman et al. (1999) compare a standard measure of creatinine clearance (MCC) with a proxy measure ECC. MCC is obtained as the ratio of the amount of creatinine (CR24) excreted in the urine over 24 hours, divided by serum creatinine (SERUMCR) concentration and by the number of minutes in the period, namely

$$\text{MCC} = \text{CR24}/(\text{SERUMCR} \times 60 \times 24).$$

ECC is obtained from patient age and weight WT as

$$\text{ECC} = (140 - \text{Age}) * \text{WT}/(\text{SERUMCR} \times 60 \times 24),$$

with a further scaling by 0.85 for women only. There are four patient groups formed by crossing gender with whether third-space body fluids were present on at least one visit. The $J = 4$ groups are then (1 = female, no fluids; 2 = female, fluids; 3 = male, no fluids; 4 = male, fluids), with group sizes $n = (51, 12, 41, 9)$ and total visits within groups $N = (211, 42, 148, 36)$.

The repeated responses for patients $i = 1, \ldots, n_j$ within groups j are $y_{1ijt} = \log(\text{MCC}_{ijt})$ and $y_{2ijt} = \log(\text{ECC}_{ijt})$, with the common factor model described in Section 8.7.1 then being applied, namely

$$\begin{aligned}
y_{1ijt} &= b_{ij} + u_{1ijt}, \\
y_{2ijt} &= \alpha_j + \lambda_j b_{ij} + c_{ij} + u_{2ijt}, \\
b_{ij} &\sim N(B_j, D_{1j}), \quad c_{ij} \sim N(0, D_{2j}), \quad u_{mijt} \sim N(0, 1/\tau_{mj}).
\end{aligned}$$

Gamma priors with index and shape parameters of unity are assumed for the precisions $\{1/D_{mj}, \tau_{mj}\}$, and $N(0,1000)$ priors for the fixed effects $\{\alpha_j, \lambda_j, B_j\}$.

A two chain run of 50,000 iterations[11] converges after around 15,000 iterations, with the the second half of the run providing posterior mean estimates (with standard deviations) for λ_j of 0.62 (0.14), 0.82 (0.81), 0.85 (0.14), and 0.91 (0.91). In fact, only the third (male-no fluid) group has a 95% interval for λ both straddling 1 and confined to positive values. The α coefficients straddle zero for all groups except the first. So identical calibration only seems to hold for group 3.

The lack of precision for the λ coefficients for groups 2 and 4 may reflect the small samples and the heavily parameterized model being applied to them. It may in fact be reasonable to apply more informative priors such as constraining the λ_j to be positive, or adopting priors (with downweighted precision) based on the frequentist results of Oman et al. (1999).

Example 8.12. Mortality Change, with Area and Age Dimensions
Consider deaths and population data $\{y_{ixt}, P_{ixt}\}$ for areas $i = 1, \ldots, n$, ages $x = 1, \ldots, X$, times $t = 1, \ldots, T$. One may assume Poisson sampling $y_{ixt} \sim Po(P_{ixt}\mu_{ixt})$ with a log link for the mortality rates μ_{ixt} (e.g., Hoem, 1987; Sun et al., 2000). The application here involves annual male deaths over the period 1999–2006 (T = eight years), in $n = 32$ London boroughs (in alphabetic order, with Hackney including the City of London), and $X = 19$ age bands (ages under 1, 1–4, 5–9, 10–14, …, 80–84, and 85+). Populations P_{ixt} are from the UK Office of National Statistics mid-year population estimates. This example considers age–area interactions in level and trend and how these can be modeled parsimoniously.

Thus one might adopt a linear trend model (Model 1) with independent age and area impacts (η_x and s_i) on the mortality level, and parallel effects (ρ_{1x} and ρ_{2i}) on the trend also. Thus

$$\log(\mu_{ixt}) = \kappa + \eta_x + s_i + (\rho_{1x} + \rho_{2i})(t - \bar{t}) + u_{ixt},$$

where the intercept κ is assigned a normal $N(0, 1000)$ prior, the area effects s_i follow a spatial CAR(1) prior, and age effects η_x following a normal first-order random walk. The associated conditional precisions (τ_s, τ_η) are assigned gamma Ga(1, 0.01) priors. The ρ_{1x} and ρ_{2i} linear trend coefficients are taken to be unstructured normal random effects with zero means, and with precisions $\tau_{\rho 1}$ and $\tau_{\rho 2}$ that are assigned gamma Ga(1, 0.01) priors. Model 1 allows for miscellaneous departures (e.g., including age–area interactions in level and trend) from a linear trend by adding unstructured Normal errors $u_{ixt} \sim N(0, 1/\tau_u)$ for each observation, where $\tau_u \sim \text{Ga}(1, 0.01)$.

By contrast, Model 2 allows for nonlinear trend and for explicit age–area interactions in both level and trend. Thus

$$\log(\mu_{ixt}) = \kappa + \eta_x + \lambda_x s_i + \omega_{ix}\delta_t,$$

where priors on κ, s_i, η_x, and (τ_s, τ_η) are as for Model 1. To ensure identification, the ω_{ix} are positive random effects with mean 1, $\omega_{ix} \sim \text{Ga}(\alpha_\omega, \alpha_\omega)$ with $\alpha_\omega \sim \text{Ga}(1, 0.01)$. Similarly, the δ_t are assumed to be fixed effects with identifying corner constraint $\delta_1 = 0$, and $\delta_t \sim N(0, 1000), t = 2, \ldots, T$. The λ_x parameters are gamma with mean 1, namely $\lambda_x \sim \text{Ga}(\alpha_\lambda, \alpha_\lambda)$ with $\alpha_\lambda \sim \text{Ga}(1, 0.01)$.

The way the age–area interaction in the level of mortality is expressed is that ages with elevated λ_x (significantly above 1) are those that are most sensitive to (in other words, are most relevant to defining) the spatial effects s_i. For example, high mortality areas (those with s_i significantly above zero) may particularly demonstrate excess mortality (elevated λ_x) at younger and middle ages. Similarly high posterior ω_{ix} identify age–area combinations that strongly contribute to the trend expressed in the δ_t. In the case where interactions in age and area are absent for both level and trend (i.e., the λ_x and ω_{ix} parameters are effectively all equal to 1), Model 2 reduces to an independent dimension model,

$$\log(\mu_{ixt}) = \kappa + \eta_x + s_i + \delta_t.$$

Whether independence (multiplicativity of area, age, and time in the risk scale) is supported or not would be ascertainable from the posterior 95% intervals for λ_x and ω_{ix}. If such intervals all straddle 1, then both the λ_x (for all x) and the ω_{ix} (for all x, i combinations) are effectively 1.

Comparisons of model fit use firstly the DIC, based on monitoring the scaled Poisson deviance,

$$D = \sum_i \sum_x \sum_t y_{ixt} \log(y_{ixt}/P_{ixt}\mu_{ixt}) - (y_{ixt} - P_{ixt}\mu_{ixt}),$$

with the complexity estimate d_e obtained by comparing the average deviance $\bar{D}$ over MCMC chains with the deviance $D(\bar{\mu})$ at the posterior means of μ_{ixt}. The second criterion is the pseudo marginal likelihood (psML), derived using Monte Carlo estimates of conditional predictive ordinates $p(y_{ixt}|y_{[ixt]})$, namely the density for y_{ixt} when the model is estimated using data $y_{[ixt]}$ for remaining areas, groups, ages, and times (Ibrahim et al., 2001). Model checking involves sampling predictions $y_{\text{rep},ixt}$ from the posterior predictive density $p(y_{\text{rep}}|y)$ which are compared with the actual observations (Gelfand, 1996). Concordance with the data may be represented by the probabilities $Pr(y_{ixt,\text{rep}} \leq y_{ixt}|y)$, with extreme values (e.g., under 0.05 or over 0.95) indicating cases where the model doesn't fit the data well. These probabilities are estimated in practice by counting MCMC iterations r where the constraint $y^{(r)}_{ixt,\text{rep}} \leq y_{ixt}$ holds, which for discrete death totals is based on the probability (Marshall and Spiegelhalter, 2003)

$$Pr(y_{\text{rep},ixt} < y_{ixt}|y) + 0.5Pr(y_{\text{rep},ixt} = y_{ixt}|y).$$

Inferences for both models are based on the second halves of two chain runs of 5000 iterations. Table 8.3 shows that the heavy parameterization involved in Model 1 (an extra nXT parameters to account for model discrepancies) results a high d_e, and the DIC in fact prefers Model 2. The log(psML) also prefers Model 2. The close fit achieved by Model 1 (albeit with a heavy parameterization) does, however, have the benefit that predictive discrepancies are low.

Figure 8.4 shows variation between age bands in their impact on the spatial effects; the posterior mean for α_λ is 3.1, implying a variance in the λ_x of around 0.3. The ages between 35–39 and 65–69 are those most associated with area contrasts in the level of mortality. Table 8.4 shows the age–area combinations with the highest and lowest ω_{ix}—those most and least associated with the trend demonstrated in the Δ_t. The highest ω_{ix} (significantly exceeding 1) associated with the mortality decline are concentrated in central London areas (such as Kensington and Chelsea, Westminster) and some inner boroughs (e.g., Wandsworth, Hammersmith) where falls in mortality over 1999–2006 were largest. For example, the observed SMR for males in Kensington and Chelsea (obtained as $100\sum\sum y_{it}/\sum\sum E_{it}$ with expected deaths E_{it} based on England and Wales death rates in 2006) fell sharply: from 104 in 1999 to 94 (2000), 86 (2001), 86 (2002), 82 (2003), 72 (2004), 66 (2005), and 61 (2006).

TABLE 8.3
Model fit criteria, Example 8.12.

	Deviance at $\bar{\mu}$	Mean deviance	d_e	DIC	log(psML)	% of observations y_{ixt} over or underpredicted*
Model 1	3887	5234	1347	6581	−14,500	4.0
Model 2	5440	5784	344	6128	−14,115	8.7

*Observations where $Pr(y_{\text{rep}} = y|y) + 0.5\ Pr(y_{\text{rep}} = y|y)$ exceeds 0.95 or underceeds 0.05.

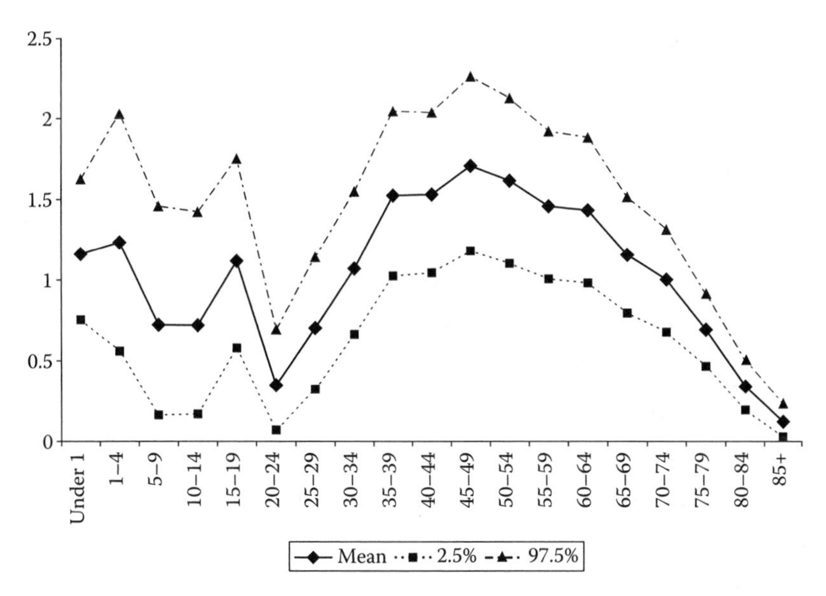

FIGURE 8.4
Age weights on spatial level effects.

8.8 Missing Data in Panel Models

Attrition and intermittently missing data are frequently found in panel data, and disregarding the process underlying such missingness may lead to biased and inefficient estimates, though different mechanisms may apply for attrition as opposed to intermittent missingness (Ma et al., 2005). In particular, missingness may be nonignorable, meaning that the probability of a missing observation or of permanent drop-out is associated with the value or values of the variable that would otherwise have been observed (Troxel et al., 1998). Thus in clinical trials, patients may drop out because of adverse treatment effects, or because they don't feel the treatment is of benefit, leading to biased estimates of treatment effects unless the missing data mechanism is allowed for.

Missingness generates an additional form of binary (or sometimes categorical) data R, depending on whether responses Y and/or predictor variables X are missing. Li et al. (2007) obtain a categorical (trinomial) missing data indicator by distinguishing between intermittent and permanent missingness. Similarly, if missing data is entirely due to attrition, it may be summarized in a single multinomial indicator $R_i = j$ if an individual drops out between the $(j-1)$th and jth measurement (Fitzmaurice et al., 2004, Section 14.4; Hedeker and Gibbons, 2006, 290). In pattern mixture models for attrition, the dropout pattern may be summarized in various ways, most simply via a binary variable contrasting completers as against dropouts, regardless of when

TABLE 8.4
Area–age weights, highest and lowest posterior means.

Borough	**Age**	**Mean**	**2.5%**	**97.5%**
Highest Weights				
Wandsworth	30–34	4.36	2.78	6.10
Kensington-Chelsea	85+	3.37	2.81	3.96
Wandsworth	25–29	3.16	1.61	4.88
Westminster	80–84	2.75	2.20	3.33
Lewisham	30–34	2.56	1.15	4.14
Kensington-Chelsea	80–84	2.50	1.90	3.14
Kensington-Chelsea	75–79	2.46	1.80	3.17
Westminster	85+	2.38	1.92	2.90
Newham	85+	2.27	1.76	2.81
Brent	70–74	2.21	1.68	2.78
Hammersmith	75–79	2.20	1.59	2.86
Westminster	75–79	2.11	1.57	2.68
Lowest Weights				
Lewisham	80–84	0.40	0.11	0.77
Brent	50–54	0.38	0.08	0.85
Kingston upon Thames	65–69	0.37	0.07	0.83
Tower-Hamlets	60–64	0.35	0.07	0.80
Barnet	85+	0.35	0.12	0.61
Barking-Dagenham	70–74	0.33	0.07	0.69
Redbridge	85+	0.32	0.09	0.61
Hounslow	85+	0.30	0.07	0.61
Havering	80–84	0.27	0.07	0.55
Barking-Dagenham	80–84	0.22	0.04	0.48
Greenwich	85+	0.20	0.04	0.45
Havering	85+	0.19	0.04	0.41
Waltham-Forest	80–84	0.19	0.04	0.43
Waltham-Forest	85+	0.14	0.03	0.33

the dropout occurred (Hedeker and Gibbons, 1997). Finally, for longitudinal datasets with continuous measurement at differing observation times a_{it}, one may record actual drop out times U_i (Hogan et al., 2004), and use these in the model for the observed Y.

However, initially consider binary indicators $R_{it} = 1$ when a response variable Y_{it} is missing, whether intermittent and permanent, and $R_{it} = 0$ for when the response is observed. Further, let $Y = (Y_{\text{obs}}, Y_{\text{mis}})$ denote the observed and unobserved response data. The totality (R, Y) is sometimes known as the complete or full data (Daniels and Hogan, 2008, 89; Ibrahim et al., 1999).

How one deals with missing data depends on the generating mechanism assumed. Two broad missing data schemes (the selection approach and the pattern mixture approach) involve a different conditioning for the joint density $P(Y, R|\theta_Y, \theta_R)$ of the responses and the missingness indicators. The pattern mixture model (Little, 1993) starts with a model for the missing data $P(R|\theta_R)$, and models Y conditional on R, namely $P(Y|R,\theta_Y)$. When dropout times are discrete, the model for $P(R|\theta_R)$ is often not specified (Hogan et al., 2004), or when missingness is expressed in various dropout patterns, $P(R|\theta_R)$ may be specified simply by the relevant multinomial probabilities of different dropout options (Curran et al., 2002, 13). By contrast, the selection model (Diggle and Kenward, 1994; Heckman, 1976) starts with the data likelihood $P(Y|\theta_Y)$, and models missingness conditional on the responses, $P(R|Y,\theta_R)$, so that $P(Y,R|\theta_Y,\theta_R) = P(R|Y,\theta_R)P(Y|\theta_Y)$.

A classification of missingness mechanisms is set out by Little and Rubin (2002), and framed in terms of the selection approach, though is applicable also to pattern mixture analysis. They distinguish between

(a) missingness completely at random (abbreviated as MCAR), when the probability $Pr(R=1)$ of a missing response is independent of both observed and missing data $Y = (Y_{\text{obs}}, Y_{\text{mis}})$, namely $P(R|Y) = P(R)$;

(b) missingness at random (MAR), when missingness is independent of the unobserved data Y_{mis}, but may depend on observed data Y_{obs}, such as when the chance that $R_{it} = 1$ depends on preceding observations $y_{i,t-s}$; in this case one has the simplification $P(R|Y) = P(R|Y_{\text{obs}})$;

(c) missingness not at random (MNAR), when the probabilities of missingness depend on unobserved missing responses, namely $P(R|Y) = P(R|Y_{\text{obs}}, Y_{\text{mis}})$. Since the data are partly missing, and R now depends on the complete outcome data $(Y_{\text{obs}}, Y_{\text{mis}})$, the selection model factors the joint distribution into a complete outcome model and a missing-data mechanism given the partially unobserved complete outcomes (Troxel et al., 2004).

An additional distinction is made between ignorable and nonignorable missingness. Assume a MAR mechanism and that the missing-data model is independent of the response data parameters θ_Y. Then the missing-data process is ignorable in the sense that that a model for missingness is not needed in order to make valid inferences from the main Y-likelihood (Rubin, 1976; Fichman and Cummings, 2003). However, for nonignorable missingness both the R-likelihood and Y-likelihood must be modeled.

As an illustration, drop-out at time t is classed as being at random if $Pr(R_t = 1|Y) = Pr(R_t = 1|Y_1, \ldots, Y_{t-1})$, namely when the missingness probability is related to lagged observed responses. However, if the probability of missingness at time t is related to the current outcome Y_t, possibly missing, so that $Pr(R_t = 1|Y) = Pr(R_t = 1|Y_1, \ldots, Y_t)$ then missingness is nonrandom or informative (Diggle and Kenward, 1994). In practice, informative missingness is assessed empirically, and would require a significant effect of Y_{it} on $\pi_{it} = Pr(R_{it} = 1)$ in a binary regression also involving other influences on missingness, with the regression taken over subjects i and repetitions T_i. For dropouts one takes $T_i = T_{i,\text{obs}} + 1$ where $T_{i,\text{obs}}$ is the last interval where data on subject i was obtained (Roy and Lin, 2002). Since MNAR missingness can never be excluded as a generating mechanism, a sensitivity analysis under different mechanisms may be considered (Kenward, 1998). This means estimating the model under a "range of assumptions about the nonignorability parameters and assessing the impact of these parameters on key inferences" (Ma et al., 2005).

A common set of predictors X_{it} may be relevant to modeling both the data Y_{it} and missingness indicators R_{it}, or different predictors W_{it} may be used in the R model. King (2001) accordingly presents a statement of the MCAR–MAR–MNAR alternatives as above, but replacing Y by $D = (Y, X)$, namely predictor and outcome data combined, and where $D = (D_{\text{obs}}, D_{\text{mis}})$ denotes the subdivision of the data according to observation status. For example, the MCAR assumption then requires $P(R|D) = P(R)$, while missingness at random requires $P(R|D) = P(R|D_{\text{obs}})$.

An alternative less stringent definition of MCAR missingness is used by Little (1995), in which missingness is independent of Y, whether observed or not, but may depend on fully observed covariates X (Curran et al., 2002, 12; Daniels and Hogan, 2008, 92). Such covariates might for instance include time, as missingness rates often increase at later stages of panels (Hedeker and Gibbons, 2006, 281). So given X_{obs}, R is independent of both Y_{obs} and Y_{mis}, leading to what is sometimes termed covariate dependent MCAR missingness.

8.8.1 Forms of missingness regression (selection approach)

A logit or probit regression is the most common approach to predicting $\pi_{it} = Pr(R_{it} = 1)$, and to assessing ignorability and MCAR assumptions. For example, π_{it} might at a minimum be a function of immediately preceding and current Y values, namely (Curran et al., 2002; Mazumdar et al., 2007)

$$\text{logit}(\pi_{it}) = \gamma_1 + \gamma_2 y_{it} + \gamma_3 y_{i,t-1},$$

with a significant γ_2 indicating nonignorable missingness. Refinements, especially in problems with intermittently missing data, include transition probability approaches (Li et al., 2007) with the model for

$$\pi_{i01t} = Pr(R_{it} = 1|R_{i,t-1} = 0)$$

having distinct parameters from that for

$$\pi_{i11t} = Pr(R_{it} = 1|R_{i,t-1} = 1).$$

If missingness is restricted to dropout only (i.e., there is no intermittent missingness), then one may use a logit or clog–log link for the probability that $R_i = j|R_i \geq j$, where $R_i = j$ if a subject drops out between the $(j-1)$th and jth measurement.

Choice of additional predictors in the missingness model is an area of potential sensitivity in terms of whether the coefficient on the current Y value is found to be significant. Hedeker and Gibbons (2006) use logit or clog–log link models to assess whether a covariate dependent MCAR assumption applies for a given dataset. They relate $Pr(R_i = j|R_i \geq j)$ to observed covariates X_{obs} such as time and treatment, as well as to the history $h(y_{it})$ of observed Y values, and to interactions between X_{obs} and $h(y)$. For example, $h(y_{ij})$ might be the average of all y_{it} between periods 1 and j. Then to test for covariate-dependent MCAR, one might use a logit regression for $Pr(R_i = j|R_i \geq j)$ that includes main effects $\{t, \text{Tr}, h(y)\}$, as well as interactions between $h(y)$ and t, between $h(y)$ and Tr, and between $h(y)$, Tr and t jointly.

The missingness model may have a role not only as part of a likelihood analysis allowing nonrandom missing data or testing for different types of missingness, but as a method for imputing missing data. Thus a "propensity score" analysis may be based on categorizing the regression terms η_{it} in

$$\text{logit}(\pi_{it}) = \eta_{it}$$

into quantile groups (e.g., quartiles) (Rosenbaum and Rubin, 1983). Within subjects located within particular quantiles of η_{it}, some subjects will exit but some remain. Sampling of the missing y_{it} for exiting subjects may be based on sampling with replacement from the known y_{it} values of stayers in the same quantile—this is sometimes called the approximate Bayesian bootstrap method (Lavori et al., 1995; Rubin and Schenker, 1986). In multiple imputation, this imputation process would be repeated several times to provide multiple filled-in datasets.

8.8.2 Common factor models

Latent variables may be introduced to explain both the Y and R data. Thus a latent data perspective on the selection model might consider bivariate data (Y, Z) where Y_{it} is observed if the latent data Z_{it} is positive (Copas and Li, 1997). Furthermore let X_{it} be predictor data potentially relevant to explaining both Y and Z and define bivariate standard normal errors $(\varepsilon_{1i}, \varepsilon_{2i})$ with correlation ρ. Assume a linear regression for Y with

$$y_{it} = X_{it}\beta + \sigma_1\varepsilon_{1i},$$

and a missingness model

$$Z_{it} = X_{it}\gamma + \varepsilon_{2i}.$$

Then if $\rho \neq 0$ the missing data are informative or nonignorable, whereas $\rho = 0$ corresponds to missingness at random.

A similar principle involves low dimension random effects F, also known as common factors, that are shared between outcome and missingness models; similar shared frailty models are used for models with outcome-dependent follow-up (Ryu et al., 2007). As often in factor models, the outcome data and missingness patterns may be viewed as conditionally independent given the common factors (Albert et al., 2002; Roy and Lin, 2002; Song and Belin, 2004; Ten Have et al., 1998). Equivalently, it is assumed that "all information about the missing data in the observed response is accounted for through the shared random effects" (Albert and Follmann, 2007). In fact, Li et al. (2007) and Yang and Shoptaw (2005) distinguish such models as an alternative to selection and pattern mixture methods, since under conditional independence one may represent the (R, Y, F) joint density as

$$\begin{aligned} P(R_i, Y_i, F_i|\theta_R, \theta_Y, \theta_F) &= P(R_i, Y_i|\theta_Y, \theta_R, F_i)P(F_i|\theta_F) \\ &= P(R_i|\theta_R, F_i)P(Y_{\text{obs},i}, Y_{\text{mis},i}|\theta_Y, F_i)P(F_i|\theta_F). \end{aligned}$$

Integrating out the F_i, one has

$$P(R_i, Y_i|\theta_Y, \theta_R) = \int P(R_i|\theta_R, F_i)P(Y_{\text{obs},i}, Y_{\text{mis},i}|\theta_Y, F_i)P(F_i|\theta_F)dF_i.$$

Other assumptions are possible, as under the "conditional linear model" (Daniels and Hogan, 2008, 112), with the conditioning sequence,

$$P(R_i, Y_i, F_i|\theta_R, \theta_Y, \theta_F) = P(Y_i|\theta_Y, F_i, R_i)P(F_i|\theta_F, R_i)P(R_i|\theta_R).$$

One form of common effect that may be used to model informative missingness is based on shared heterogeneity (e.g., Chib, 2008, 507; Li et al., 2007). An example is a general linear mixed model with permanent subject random effects $b_i = (b_{1i}, \ldots, b_{Qi})$;

$$g[E(y_{it}|b_i)] = X_{it}\beta + Z_{it}b_i,$$

where the model for $Pr(R_{it} = 1)$ also conditions on the b_i, and possibly on separate predictors W_{it}, and on the history of responses $H_{it} = \{y_{i1}, \ldots, y_{it}\}$. Consider the case $Q = 1$ with $z_{it} = 1$, and suppose predictors W_{it} are relevant to dropout (e.g., baseline health status in a clinical trial). Then a common factor model adapted to predicting $\pi_{it} = Pr(R_{it} = 1|W_{it}, H_{it})$ might take the form:

$$\begin{aligned} g[E(y_{it}|b_i)] &= X_{it}\beta + b_i, \\ \text{logit}(\pi_{it}) &= W_{it}\gamma + \lambda b_i + y_{it}\delta_1 + y_{i,t-1}\delta_2, \end{aligned} \tag{8.10}$$

where b_i are zero mean random effects, and the predictors $\{X_{it}, W_{it}\}$ both include an intercept. For example, Li et al. (2007) consider Poisson data with $y_{it} \sim Po(\lambda_{it})$,

$$\log(\lambda_{it}) = X_{it}\beta + b_i,$$

and with binary indicators for missingness, and a lagged outcome scheme adapted to counts, one would obtain

$$\text{logit}(\pi_{it}) = W_{it}\gamma + \lambda b_i + \log(y_{it} + 1)\delta_1 + \log(y_{i,t-1} + 1)\delta_2.$$

In fact the model of Li et al. (2007) distinguishes between intermittently missing data and permanent attrition via a multinomial rather than binary regression, and uses a transition probability missingness model.

A model with shared latent effects exemplified by Equation 8.10 imposes possibly restrictive assumptions on the correlations among repeated responses for a given subject. Conditional on the time-invariant shared effects b_i, observations on a subject are uncorrelated (Albert and Follmann, 2007). An alternative is a shared autoregressive process, as in

$$
\begin{aligned}
g[E(y_{it}|F_{it})] &= X_{it}\beta + F_{it}, \\
F_{it} &= \rho F_{i,t-1} + u_{it}, \\
\text{logit}(\pi_{it}) &= W_{it}\gamma + \lambda F_{it} + Y_{it}\delta_1 + Y_{i,t-1}\delta_2,
\end{aligned}
$$

where the u_{it} are white noise and $\rho \in (-1, 1)$.

For multivariate responses $\{y_{mit}, m = 1, \ldots, M\}$, one might propose common factors to model both correlation between the observed responses, and the probabilities of missing response, especially attrition affecting all outcomes (Lin et al., 2004). Thus, consider a single time varying factor F_{it}, and loadings $\{\lambda_m, \kappa\}$ in the Y and R likelihoods, and let H_{it} denote a subset of the history of the observed X and Y variables up to time t. Then for outcomes $m = 1, \ldots, M$, one might have

$$g[E(y_{mit}|X_{it}, F_{it})] = X_{it}\beta_m + \lambda_m F_{it}$$

while the drop out probability $R_{it} \sim \text{Bern}(\pi_{it})$ is modeled as

$$\text{logit}(\pi_{it}) = W_{it}\gamma + \varphi H_{i,t-1} + \kappa F_{it}$$

for $t = 1, \ldots, T_i$, where for dropouts $T_i = T_{i,\text{obs}} + 1$ and $T_{i,\text{obs}}$ is the last interval where data was observed. Furthermore, the factor scores may depend on known predictors $\{U_{it}, Z_{it}\}$ and zero mean random permanent effects b_i, as in

$$F_{it} = U_{it}\eta + Z_{it}b_i + \upsilon_{it}$$

with $\upsilon_{it} \sim N(0, 1)$ if all loadings $\{\kappa, \lambda_m\}$ are unknowns, and with U_{it} omitting an intercept for identifiability (Roy and Lin, 2002, 42). The missingness model is nonignorable by virtue of dependence of π_{it} on F_{it}, which represents possibly missing y_{mit} (Roy and Lin, 2002, 43).

8.8.3 Missing predictor data

Often panel data will have missingness on covariates as well as on the response, so that binary or categorical indicators R_X are defined according as covariates have missing values or not. With $R = (R_Y, R_X)$, the joint density under a selection approach has the form:

$$p(Y, X, R_Y, R_X|\eta, \beta, \theta) = p_R(R_X, R_Y|Y, X, \eta)p_Y(Y|X, \beta)p_X(X|\theta),$$

where p_X now models the likelihood of the predictors. If R_Y is conditional on all the components of R_X one has

$$\begin{aligned} &p(Y, X, R_Y, R_X|\eta, \beta, \theta) \\ &\quad = p(R_Y|R_X, Y, X, \eta_Y)p(R_X|Y, X, \eta_X)p_Y(Y|X, \beta)p_X(X|\theta). \end{aligned}$$

Alternatively R_Y may be modeled jointly with the R_X, though complexity increases as the number of predictors subject to missingness rises, giving rise to different possible conditional sequences for R_Y and the components of R_X.

Suppose a subset of q predictors have missing values, with $R_{ji} = 1$ if X_{ji} is missing, and $R_{ji} = 0$ otherwise. If Y is fully observed, a selection approach specifies

$$p(Y, X, R_X|\eta, \beta, \theta) = p(R_X|Y, X, \eta)p_Y(Y|X, \beta)p_X(X|\theta),$$

where $p(R_X)$ is a multinomial with 2^q cells. To define p_X, one needs to specify the joint distribution of $X_{i,\text{mis}} = \{X_{1i}, \ldots, X_{qi}\}$. Suppose the incompletely observed covariates $X_{\text{mis}} = (X_1, \ldots, X_q)$ are both continuous $X_{\text{mis},C} = \{X_1, \ldots, X_r\}$ and categorical $X_{\text{mis},D} = \{X_{r+1}, \ldots, X_q\}$, with fully observed covariates denoted $X_{\text{obs}} = \{X_{q+1}, \ldots, X_p\}$. Ibrahim et al. (1999) propose the joint density of X_{mis} be specified as a series of conditional distributions, namely

$$\begin{aligned} &p(X_1, \ldots, X_q|\theta\} \\ &\quad = p_q(X_q|X_{q-1}, \ldots, X_1, \theta_q, X_{\text{obs}}\} \cdots p_2(X_2|X_1, \theta_2, X_{\text{obs}})p_1(X_1|\theta_1, X_{\text{obs}}) \end{aligned}$$

though there may be sensitivity to the which of the $q!$ conditioning sequences is adopted. The completely observed predictors may be used in predicting the missing covariates. For continous predictors the form of density (e.g., gamma, normal) can be adapted to whether only positive values are observed. Ibrahim et al. (1999, 180) suggest one-dimensional or joint distributions for the continuous predictors in the lower stages (p_1, p_2, etc.), with the higher stages being models for categorical predictors that are based on the imputed continuous covariates (e.g., logistic regression models).

Another general scheme for specifying the jont density of X_{mis} adopts a different strategy by first representing the joint density of categorical predictors. This is the general location model (e.g., Cho and Schenker, 1999)

$$\begin{aligned} p(X_{\text{mis}}|\theta, X_{\text{obs}}) &= p(X_{\text{mis},C}, X_{\text{mis},D}|\theta_C, \theta_D, X_{\text{obs}}) \\ &= p(X_{\text{mis},C}|X_{\text{mis},D}, \theta_C, X_{\text{obs}})p(X_{\text{mis},D}|\theta_D, X_{\text{obs}}), \end{aligned}$$

typically involving a multivariate normal or multivariate Student t distribution for the continuous predictors, conditional on a given combination of values of the categorical covariates. For example, means and covariances for the multivariate normal model could be specific to each combination of the categorical predictors. The first stage of the joint density for predicting missing categorical covariates $p(X_{\text{mis},D}|\theta_D, X_{\text{obs}})$ would be a multinomial distribution, or possibly loglinear regression, over discrete outcomes, missing and observed.

Possible approaches for modeling the covariate missingness indicators $p(R_{ji}|Y_i, X_i, \eta)$ under a selection approach include a joint log-linear model with $X_i = (X_{i,\text{mis}}, X_{i,\text{obs}})$ as predictors, or equivalently a multinomial model with all possible classifications of nonresponse as categories (Schafer, 1997, Chapter 9). For example, if $X_{i,\text{mis}}$ contains two variables subject to missingness, then there are four possible combinations of values of R_{1i} and R_{2i} for each subject. The joint density of missingness indicators can be expressed (Ibrahim et al., 1999) as a series of conditional distributions, namely

$$p(R_{1i}, \ldots, R_{qi}|\eta, X_i, Y_i\}$$
$$= p(R_{qi}|R_{q-1,i}, \ldots R_{1i}, \eta_q, X_i, Y_i) \cdots p(R_{2i}|R_{1i}, \eta_2, X_i, Y_i) p(R_{1i}|\eta_1, X_i, Y_i),$$

which in practice implies a series of binary regressions. For assessing nonrandomness in covariate missingness, one allows $Pr(R_{2i} = 1|R_{1i}, \eta_2, X_i, Y_i)$ to depend on predictors X_i that may be subject to missing values, as well as on earlier R_{ji} in the conditional sequence.

In practice, a multivariate density for a set of continuous variables might be represented indirectly by a series of regressions, and missing values for binary or categorical data items modeled or imputed via regressions on other predictors—see Austin and Escobar (2005) for an illustration of such methods. Such procedures are related to multivariate imputation procedures for covariates and possibly responses also (Allison, 2000; Schafer, 1997). Consider the case where Z and X are predictors, with Z subject to missingness. Schafer (1997) proposes random regression imputation by initially regressing Z on X and Y, but using only cases with Z observed, and from this regression forms point estimates $\hat{Z}$ for cases with missing data. Let $\hat{\sigma}$ be the square root of the mean square error from the observed data regression, then for subjects with missing Z, one obtains imputations $\tilde{Z} = \hat{Z} + \hat{\sigma}U$ where U is a draw from a standard normal. For cases with observed Z, one sets $\tilde{Z} = Z$. One then carries out a filled-in data regression of Y on X and $\tilde{Z}$ for all subjects. The Z-imputation and filled-in data regressions may be repeated M times. Such a procedure is, however, not proper in the sense of Rubin (1987).

8.8.4 Pattern mixture models

Pattern mixture models may have a benefit in avoiding intricate modeling of the missingness indicators. For regular panel data (repeat measures at fixed intervals for all subjects) subject to missingness only through attrition, a pattern mixture analysis (Little, 1995) might simply involve differentiating regression effects in the Y-model according to discrete drop out times $U_i \in (2, \ldots, T-1)$, as well as completers with $U_i = T$. Thus "the missing-data patterns can be used as grouping variables in the [Y regression] analysis" (Hedeker and Gibbons, 1997). If there are h_m subjects in the M different missingness patterns, with associated proportions $\phi_m = h_m/n$, then the "marginal" or composite parameter (e.g., the regression impact of a predictor x_p) is obtained as a weighted average of the pattern-specific parameters β_{pm}, namely $\beta_p = \sum_{m=1}^{M} \phi_m \beta_{pm}$ (Curran et al., 2002). A Bayesian analysis might involve repeated multinomial sampling of the ϕ_m at each MCMC iteration, and monitoring the composite parameters $\beta_p^{(r)} = \sum_{m=1}^{M} \phi_m^{(r)} \beta_{pm}^{(r)}$. Often the preliminary model for missingness $P(R|\theta_R)$ would be confined to such multinomial sampling.

For example, in a clinical application, separate intercepts, growth coefficients, and treatment effects would be estimated (in the Y-model) according to dropout category; the variance or covariance parameters for random effects may also be differentiated. In an initial analysis, droput category might just be binary, differentiating between completers and dropouts, regardless of the interval when the dropout occurred. Thus, set $G_i = 1$ for dropouts and $G_i = 2$ for completers, and consider a regression model for fixed interval (balanced) panel data y_{it} with intercept, time, treatment (Tr), and time–treatment interaction (e.g., Hedeker and Gibbons, 1997; Mazumdar et al., 2007). Then a grouped regression with varying intercepts could take the form:

$$
\begin{aligned}
y_{it} &= \beta_{1,G_i} + \beta_{2,G_i} t + \beta_{3,G_i} \mathrm{Tr}_i + \beta_{4,G_i}(t.\mathrm{Tr}_i) + b_{1i} + b_{2i}t + e_{it},\\
b_i &\sim N(0, D_{G_i}); e_{it} \sim N(0, \sigma^2_{G_i}).
\end{aligned}
$$

Curran et al. (2002, 13) allow for an additional autocorrelated error ε_{it} with pattern-specific covariance matrix R_{G_i}.

The conditional linear model (Hogan et al., 2004; Paddock, 2007; Wu and Bailey, 1989) is a version of the pattern mixture model that may be applied to continuously recorded longitudinal data (rather than fixed interval panel data). The impact of missingness on Y involves functions $\beta_j(U_i)$ of possibly continuous dropout times U_i though this reduces to a grouping approach for fixed intervals; that is, the $\beta_j(U_i)$ become step functions (Hogan et al., 2004, 856). At their most simple such functions are linear in U, but polynomial functions or nonparametric models (e.g., splines) can be used. In the preceding example, one might have

$$
\begin{aligned}
y_{it} &= \beta_1(U_i) + \beta_2(U_i)t + \beta_3(U_i)\mathrm{Tr}_i + \beta_4(U_i)(t.\mathrm{Tr}_i) + b_{1i} + b_{2i}t + e_{it},\\
b_i &\sim N(0, D); e_{it} \sim N(0, \sigma^2),\\
\beta_j(U_i) &= \alpha_{j0} + \alpha_{j1}U_i, \quad j = 1, \ldots, 4,
\end{aligned}
$$

and a test for missingness at random is whether the α_{j1} are zero. Paddock (2007) applies a Bayesian regression selection approach to coefficients in models involving quadratic effects of U_i.

Example 8.13. Cocaine Use and Desipramine These data are used to compare some of the models for missing data described above, including common factor and pattern mixture approaches. They are from a trial of the antidepressant desipramine in cocaine-dependent patients with depressive comorbidity, and relate to fixed interval panel data on 106 patients, with 52 in the treatment arm, and the remainder given a placebo (Ma et al., 2005). The responses y_{it} are average dollars per day spent on cocaine use. Only 47 patients completed the full 12 weeks of observation. Let $T_i^* = 12$ for completers, while for drop-outs let T_i^* denote the week subsequent to the last week $T_i^* - 1$ when an observation is obtained. So $T_i^* = 7$ if a subject is observed for the first six weeks, but is missing for all the last six weeks.

A plot of the average responses for the two arms (including the baseline) shows that the treatment group begins with a higher average baseline spending level, and reduces its cocaine spending more. The Y-model involves predictors $X = \{1, \text{Tr}, t, \text{Tr}.t, B\}$ where B = baseline cocaine spending. So

$$y_{it} = \beta_1 + \beta_2 \text{Tr}_i + \beta_3(t - \bar{t}) + \beta_4 \text{Tr}_i(t - \bar{t}) + \beta_5 B_i + u_{it},$$

where $u_{it} \sim N(0, 1/\tau_u)$. Assessment of desipramine efficacy focuses especially on the coefficient for treatment–time interaction, Tr.t. $N(0, 100)$ priors are assumed on the first three predictors, but for numeric stability a more informative $N(0, 0.1)$ prior is assumed for the impact of baseline spend (as large predictor values are observed). Assessing whether missingness is informative or not is initially based on a selection approach, with $R_{it} \sim \text{Bern}(\pi_{it}), t = 1, \ldots, T_i^*$, and

$$\text{logit}(\pi_{it}) = \gamma_1 + \gamma_2 y_{it} + \gamma_3 y_{i,t-1},$$

using both y_{it} and $y_{i,t-1}$ as predictors (Mazumdar et al., 2007).

Posterior estimates (from iterations 1001–10,000 of a two chain run) show a significant positive effect on π_{it} of the lagged outcome $y_{i,t-1}$, but also an impact (of borderline significance) of the possibly unobserved current outcome y_{it}, with 95% interval $\{-0.015, 0.0005\}$. The coefficient for the treatment–time interaction in the Y-model is negative, but with an inconclusive 95% credible interval $\{-2.4, 0.6)$. The DIC is 9395 ($d_e = 11$) with 8890 for the Y-model and 505 for the R-model.

An alternative model involves a common factor F_i that depends on B_i = baseline spend; the mean for F_i omits an intercept for identifiability. The missingess model now involves a lagged response and the common factor, while the Y likelihood no longer involves baseline spending.[12] Thus

$$\begin{aligned}
F_i &\sim N(\eta B_i, 1), \\
y_{it} &= \beta_1 + \beta_2 \text{Tr}_i + \beta_3(t - \bar{t}) + \beta_4 \text{Tr}_i(t - \bar{t}) + \lambda F_i + u_{it}, \\
R_{it} &\sim \text{Bern}(\pi_{it}), \quad t = 1, \ldots, T_i^*, \\
\text{logit}(\pi_{it}) &= \gamma_1 + \gamma_2 y_{i,t-1} + \kappa F_i,
\end{aligned}$$

where a $N(1, 1)$ prior is adopted for κ and the prior on λ is constrained to positive values.

Posterior estimates (from iterations 1001–10,000 of a two chain run) show the coefficient for treatment–time to be still inconclusive (see Table 8.5), but

TABLE 8.5
Cocaine use, common factor model, posterior summary.

	Mean	St. devn.	Monte Carlo SE	2.5%	97.5%
β_1	25.1	2.2	0.048	20.8	29.4
β_2	2.868	2.865	0.048	−2.718	8.499
β_3	−0.255	0.502	0.008	−1.220	0.745
β_4	−1.077	0.690	0.010	−2.436	0.273
η	0.964	0.247	0.006	0.488	1.470
γ_1	−2.243	0.155	0.002	−2.555	−1.940
γ_2	−0.005	0.004	0.000	−0.014	0.002
κ	0.326	0.143	0.003	0.064	0.631
λ	9.3	0.8	0.012	7.7	10.8

with a 95% interval more clearly focused on negative values. The common factor is a positive function of baseline spend and its impact on π_{it} is positive; so the chance of a missing value increases with the common factor. The DIC is reduced to 9213 despite a higher complexity count of 48, with 8683 for the Y-model and 529 for the R-model. While the generating mechanism for missing data is always unknown, this model describes both the original observations and missing data indicators more effectively.

Finally, a pattern mixture analysis is applied, distinguishing simply between noncompleters ($G_i = 1$) and completers ($G_i = 2$). The assumed model is

$$y_{it} = \beta_{1,G_i} + \beta_{2,G_i}\mathrm{Tr}_i + \beta_{3,G_i}t + \beta_{4,G_i}\mathrm{Tr}_i(t - \bar{t}) + \beta_{5,G_i}B_i + b_i + e_{it},$$
$$b_i \sim N(0, D_{G_i}); e_{it} \sim N\left(0, \sigma^2_{G_i}\right).$$

The precisions $1/D_j$ and $1/\sigma_j^2$ are assumed to follow independent gamma priors with shape 1 and scale 1.

Convergence is attained by iteration 10,000 in a two chain run of 20,000 iterations, with the last 10,000 showing the dropouts to have higher cocaine spending (β_{11} and β_{12} have respective means 29.2 and 16.8) together with a significantly positive baseline effect, β_{51}, for dropouts, with 95% interval $(0.09, 0.29)$. Completers do not have a significant baseline effect, and their time-treatment effect, namely -1.31 with 95% interval $(-2.8, 0.14)$ is more precisely estimated than that for droputs, namely -1.64 $(-4.0, 0.8)$. The pooled estimates for β_2 and β_4 (pooling over dropout patterns) show an insignificant main treatment effect, but β_4 is virtually significant, with mean and 95% interval -1.52 $(-3.1, 0.07)$. The DIC for this model is 8576, with $d_e = 77.1$.

Example 8.14. Shared Effect Missingness Model for IMPS (Inpatient Multidimensional Psychiatric Scale) Data Hedeker and Gibbons (2006, 297–302) consider a shared latent effect model for these data, relating to psychiatric morbidity; in particular, item 79 of the IMPS scale is a positive measure of morbidity with values ranging from 0 (normal) to 7 (extremely ill). The analysis here follows Hedeker and Gibbons (2006) in treating the outcomes as metric, but adopts a simpler shared effects model. The data involve $n = 437$ patients with up to $T_i = 5$ repeat measurements not necessarily at the same times (in weeks) after the baseline at 0 weeks; most patients have four measurements. Follow up is terminated after six weeks, with most patients only measured at weeks $a_{i1} = 0$, $a_{i2} = 1$, $a_{i3} = 3$, and $a_{i4} = 6$; completers are those terminating at six weeks, with all sequences ending in earlier weeks considered as dropouts. So in a similar way to that used for discrete time hazards in Chapter 9 (Section 9.5), one may define event indicators $\{w_{ij} = 0, j = 1, \ldots, T_i\}$ for completer subjects whose last week is 6, and $\{w_{ij} = 0, j = 1, \ldots, T_i - 1; w_{iT_i} = 1\}$ for subjects whose last observation is before six weeks.

Let $\text{Drug}_i = 1$ for the treatment group subjects, with $\text{Drug}_i = 0$ otherwise. Also let $S_{it} = a_{it}^{0.5}$ be the square root of the number of weeks at which the tth observation of patient i is obtained. The model for the morbidity outcome then has the form:

$$y_{it} = \beta_1 + \beta_2\text{Drug}_i + \beta_3 S_{it} + \beta_4 S_{it}\text{Drug}_i + b_{1i} + b_{2i}S_{it} + u_{it},$$

with priors

$$\begin{aligned}(b_{1i}, b_{2i}) &\sim N([0,0], D); D^{-1} \sim \text{Wish}(I, 2),\\ u_{it} &\sim N(0, 1/\tau_u); \tau_u \sim \text{Ga}(1, 0.001).\end{aligned}$$

The missing data model[13] is a complementary log–log regression, sharing the random intercept b_{1i} and with an interaction between treatment and the shared effect. Thus

$$\log(-\log[1 - Pr(w_{ij} = 1 | T_i \geq j)]) = \gamma_{1j} + \gamma_2\text{Drug}_i + \alpha_1 b_{1i} + \alpha_2 b_{1i}\text{Drug}_i,$$

with $\{\gamma_{1j} \sim N(0, 1000), j = 1, \max(T_i)\}$, $\gamma_2 \sim N(0, 1000)$, and $\{\alpha_k \sim N(1, 1), k = 1, 2\}$. Nonignorable missingness corresponds to any of the α_k coefficients being distinct from zero (Hedeker and Gibbons, 2006, 298).

A two chain run of 5000 iterations shows early convergence, and posterior means on β coefficients (from the last 2500 iterations) similar to those reported by Hedeker and Gibbons (2006). In particular β_4 has mean (sd) of -0.65 (0.08) consistent with a greater reduction in morbidity for the treatment group. Dropout is lower for treated patients with γ_2 having mean (sd) of -0.75 (0.22). Both the α coefficients have 95% credible intervals excluding zero, and so can be said to be "significant": α_1 has mean and 95% interval 0.86 (0.22, 1.5) indicating that (in general) more ill patients are likely to dropout, while α_2 has mean and 95% interval -1.4 ($-2.1, -0.7$), showing that for those being treated, the more ill are in fact less likely to dropout.

Appendix: Computational Notes

1. For the random effects selection Example 8.1, the code is

```
model { for (i in 1:500) { for (t in 1:5) {
y[i,t] ~dnorm(mu[i,t],tau); z1[i,t] <-1; z2[i,t] <- t; z3[i,t] <- x[i,t]
mu[i,t] <- B[1]+B[2]*t+B[3]*x[i,t]
+zeta[i,1]*(z1[i,t]*G[1,1]+z2[i,t]*G[2,1]+z3[i,t]*G[3,1])
+zeta[i,2]*(z2[i,t]*G[2,2]+z3[i,t]*G[3,2])+zeta[i,3]*(z3[i,t]*G[3,3])}
for (k in 1:3) {zeta[i,k]~dnorm(0,1)}}
G[1,2] <- 0; G[1,3] <- 0; G[2,3] <- 0
for (p in 1:Q) { sigb[p] <- sqrt(D[p,p])
for (q in 1:Q) {D[p,q] <- sum(G.s[p,q,])
for (r in 1:Q) {G.s[p,q,r] <- G[p,r]*G[q,r]}}}
tau~dgamma(1,0.001); sig2 <- 1/tau
for (k in 1:Q) {B[k]~dnorm(0,0.01); c[k,k]~dgamma(a[k],b[k]);
G[k,k] <- c[k,k]*gam[k,k]; gam[k,k]~dbern(pi[k,k]); pi[k,k]~dbeta(1,1)
for (m in 1:k-1){c[k,m]~dnorm(0,1);G[k,m] <- c[k,m]*gam[k,m]
gam[k,m]~dbern(pi[k,m]);pi[k,m]~dbeta(1,1)}}}
```

The code for the Chen–Dunson method replaces the last four lines by

```
for (k in 1:Q) {B[k]~dnorm(0,0.01); lam[k]~dgamma(a[k],b[k]);
gam[k,k]~dbern(0.5); omeg[k,k] <- 1; G[k,k] <- lam[k]*gam[k,k]
for (m in k+1:Q){omeg[k,m]<- 0; gam[k,m]<- 0; G[k,m] <- 0}
for (m in 1:k-1) {omeg[k,m]~dnorm(0,2); gam[k,m]~dbern(0.5)
G[k,m] <- lam[k]*omeg[k,m]*gam[k,k]*gam[m,m]*gam[k,m]}}}.
```

2. The central code for the regression approach to covariance matrix specification (Example 8.2) is

```
model { for (i in 1:n) { for (t in 1:T) {y[i,t] ~dnorm(nu[i,t],inv.d[i,t]);
log(d[i,t]) <- gam[1]+gam[2]*t+gam[3]*t*t; inv.d[i,t] <- 1/d[i,t]
mu[i,t] <- beta[1]+beta[2]*t+beta[3]*t*t+beta[4]*t*t*t}
nu[i,1] <- mu[i,1]
for (t in 2:T) {nu[i,t] <- mu[i,t]+sum(F[i,t,1:t-1])
for (j in 1:t-1) {ph[i,t,j] <- lam[1]+lam[2]*(t-j)+lam[3]*(t-j)*(t-j)
F[i,t,j] <- ph[i,t,j]*(y[i,j]-mu[i,j])}}}
for (j in 1:3) {lam[j] ~dnorm(0,0.001); gam[j] ~dnorm(0,0.001)}
for (j in 1:4) {beta[j] ~dnorm(0,0.001)}}
```

3. The response data input in this example is in stacked form with 5400 rows, with each row containing firm identifier, time identifier and the return y_{it}. Differencing the regression means as in Equation 8.6, and taking $1/\sigma_v^2 \sim G(1, 0.001)$, leads to code for Model 2 as follows:

```
model { for (i in 1:5400) { y[id[i],time[i]] <- Y[i]}
for (i in 1:n) { b[i,1:2] ~dmnorm(B[1:2],D.inv[1:2,1:2])
```

```
for (t in 1:T) {mu[i,t] <- b[i,1]+b[i,2]*x[t]}
              y[i,1] ~dnorm(mu[i,1],tau1);
for (t in 2:T) {y[i,t] ~dnorm(nu[i,t],tau)
nu[i,t] <- mu[i,t]-rho*mu[i,t-1]+rho*y[i,t-1]}}
# Priors
for (j in 1:2) {B[j] ~dnorm(0,0.001);
for (k in 1:2) {Wsc[j,k] <- 4*equals(j,k);
C[j,k] <- D[j,k]/sqrt(D[j,j]*D[k,k])}}
D.inv[1:2,1:2] ~dwish(Wsc[,],4); D[1:2,1:2] <- inverse(D.inv[,])
tau ~dgamma(1,0.001); rho ~dunif(-1,1); tau1<- (1-rho*rho)*tau}
```

4. Assume stacked household-repetition (i.e., purchase level) data with $N = 2412$ observations, and binary indicators $\{D_{i1}, D_{i2}, \ldots, D_{iK}\}$ for each observation with $D_{ik} = 1$ if the kth brand is chosen. The data for each observation contains the J feature and price attributes relevant to each purchase. Then a model with fixed effects only and the final category as the reference, and with KM $= K - 1$ is coded as

```
model {for (i in 1:N) {for (k in 1:K) {pi[i,k] <- ph[i,k] / sum(ph[i,])
d[hh[i],rep[i],1:K] ~dmulti( pi[i,1:K] , 1 )
d[hh[i],rep[i],k] <- D[i,k]; LL[i,k] <- d[hh[i],rep[i],k]*log(pi[i,k]);
log(ph[i,k]) <- alph[k] + del[1]*feature[i,k]+ del[2]*price[i,k]}}
# priors
alph[K] <- 0; for (k in 1:KM) {alph[k] ~dnorm(0,Pr)}
for (k in 1:2) {del[k] ~dnorm(0,Pr)}}
```

where Pr is a known (low) precision, e.g., $Pr = 0.0001$. Introducing subject specific-random effects (which are also necessarily choice specific) leads to the code

```
model {for (i in 1:n) {alph[i,K] <- 0;
alph[i,1:KM] ~dmnorm(mu.alph[],invD[1:KM,1:KM])}
for (i in 1:N) {for (k in 1:K) {pi[i,k] <- ph[i,k] / sum(ph[i,])
d[hh[i],rep[i],1:K] ~dmulti( pi[i,1:K] , 1 ); d[hh[i],rep[i],k] <- D[i,k];
LL[i,k] <- d[hh[i],rep[i],k]*log(pi[i,k]);
log(ph[i,k]) <- alph[hh[i],k] + del[1]*feature[i,k]+ del[2]*price[i,k]}}
# priors
for (k in 1:2) {del[k] ~dnorm(0,Pr)}
invD[1:KM,1:KM] ~dwish(ScD[,],KM)
for (k1 in 1:KM) {mu.alph[k1] ~dnorm(0,Pr)
for (k2 in 1:KM) {ScD[k1,k2] <- equals(k1,k2)}}}
```

5. The code in Example 8.5 is

```
model { for (i in 1:N) {y[id[i],time[i]] <- Rank[i]
Z[id[i],time[i]] <- sqrt(Week[i])}
P.slope.tr <- step(av.slope.tr[1]-av.slope.tr[2])
for (j in 1:2) {av.slope.tr[j] <- sum(slope.tr[j,1:n])/tot.tr[j]}
```

```
for (i in 1:n) {b[i,1:2] ~dmnorm(eta[1:2],InvD[,])
slope.tr[1,i] <- b[i,2]*equals(Tr[i],1); slope.tr[2,i] <- b[i,2]*equals(Tr[i],0);
for (t in 1:T) { mu[i,t] <- b[i,1]+b[i,2]*Z[i,t]+beta[1]*Tr[i]+ beta[2]*Tr[i]
*Z[i,t]+beta[3]*Gend[i]
for (j in 1:C-1) { d[i,t,j] <- step(j-y[i,t]); nu[i,t,j] <- kap[j]-mu[i,t]
# truncated sampling for latent scale
                wstar[i,t,j] ~dnorm(nu[i,t,j],1) I(A[i,t,j],B[i,t,j])
A[i,t,j] <- -10*equals(d[i,t,j],0); B[i,t,j] <- 10*equals(d[i,t,j],1)}}}
# thresholds for latent variable
kap[1] <- 0
for (k in 2:CM){ kap[k] <- kap[k-1]+del[k]; del[k] ~dgamma(1,1)}
# permanent effects prec'n matrix
InvD[1:2,1:2] ~dwish(Q.b[,],2); D[1:2,1:2] <- inverse(InvD[,])
for (i in 1:2) { eta[i] ~dnorm(0,0.01); sig[i] <- sqrt(D[i,i])
for (j in 1:2) {corr.b[i,j] <- D[i,j] / (sig[i]*sig[j])}}
Q.b[1,1] <- 1; Q.b[2,2] <- 1; Q.b[2,1] <- 0; Q.b[1,2] <- 0
# covariate effects
for (j in 1:3) {beta[j] ~dnorm(0,0.01)}}
```

6. The code for the second analysis in Example 8.6 is

```
model {for (i in 1:2849) {y[id[i],time[i]] <- log.earnings[i]}
for (i in 1:416) {b[i] ~dnorm(beta[1],invD)
P1[i] <- sum(p1[i,2:T[i]]); P2[i] <- sum(p2[i,2:T[i]])
w1[i] <- college[i]; w2[i]<- equals(eth[i],1)
y[i,1] ~dnorm(mu[i,1],tau1);
# regression for t=1
mu[i,1] <- beta1.1+gam1[1]*w1[i]+gam1[2]*w2[i]+gam1[3]*w1[i]*w2[i]
for (t in 1:T[i]) { e[i,t] <- y[i,t]-mu[i,t]; p2[i,t] <- pow(e[i,t],2)
LL[i,t] <- 0.5*log(tau/6.28)-0.5*tau*pow(y[i,t]-mu[i,t],2)
H[i,t] <- 1/exp(LL[i,t])}
for (t in 2:T[i]) {y[i,t] ~dnorm(mu[i,t],tau);
# elements for testing autocorrelation
p1[i,t] <- e[i,t]*e[i,t-1]
# regression for t>1
mu[i,t] <- ph*y[i,t-1]+gam[1]*w1[i]+gam[2]*w2[i]+gam[3]*w1[i]*w2[i]+b[i]}}
ph ~dnorm(0,1); sig ~dunif(0,10); sig1~dunif(0,10);
tau <- 1/(sig*sig); tau1 <- 1/(sig1*sig1);
sqrtD ~dunif(0,10); invD <- 1/(sqrtD*sqrtD)
beta[1] ~dnorm(0,0.001);beta1.1 ~dnorm(0,0.001);
for (j in 1:3) {gam[j] ~dnorm(0,0.001); gam1[j] ~dnorm(0,0.001)}
ARerr <- sum(P1[])/sum(P2[]); test.ARerr <- step(ARerr)}
```

7. The code for the first analysis in Example 8.7 is

```
model {for (i in 1:n) {for (t in 1:T) {y[i,t] ~dpois(mu[i,t])
ynew[i,t] ~dpois(mu[i,t])
```

```
LL[i,t] <- -mu[i,t]+y[i,t]*log(mu[i,t])-logfact(y[i,t]); G[i,t] <- 1/exp(LL[i,t])
dv[i,t] <- y[i,t]*log((y[i,t]+0.5)/(mu[i,t]+0.5)) -(y[i,t]-mu[i,t])}}
for (i in 1:n) {b[i,1:2] ~dmnorm(zero[1:2],invD[1:2,1:2])
Age[i] <- age[i]-mean(age[]))
log(mu[i,1]) <- beta1[1]+beta1[2]*Age[i]+b[i,1]
for (t in 2:T) {log(mu[i,t]) <- beta[1]+beta[2]*Age[i]+beta[3]*Trt[i]+ph
*y[i,t-1]+b[i,2]}}
for (j in 1:3) {beta[j] ~dnorm(0,0.1)}
for (j in 1:2) {beta1[j] ~dnorm(0,0.1)}
Fit[1] <- 2*sum(dv[,]); Fit[2] <- -2*sum(LL[,])
invD[1:2,1:2] ~dwish(Q[,],2); D[1:2,1:2] <- inverse(invD[,])
sigb <- sqrt(D[2,2]); rh <- D[1,2]/sqrt(D[1,1]*D[2,2]); ph ~dnorm(0,0.1)}
```

8. The code for the second analysis in Example 8.8 is

```
model { for (i in 1:N) { y[i] ~dnorm(mu[i], tau.pow[i]) I(0,)
LL[i] <- 0.5*(log(tau.pow[i]/6.28)-tau.pow[i]*pow(y[i]-mu[i],2))
tau.pow[i] <- tau/pow(mu[i],kap)
mu[i] <- dose[i] *exp(-beta[subj[i]] * time[i]/alpha[subj[i]])/alpha[subj[i]]}
for (i in 1:n) {mu.new[i] <- 30 *exp(- beta[i] * 32/alpha[i])/alpha[i]
tau.pow.new[i] <- tau/pow(mu.new[i],kap)
lamb[i] ~dgamma(nub.2,nub.2)
ynew[i] ~dnorm(mu.new[i], tau.pow.new[i]) I(0,)
alpha[i] <- exp(b[i,1]) ; beta[i] <- exp(b[i,2]) ;
for (k1 in 1:2) { for (k2 in 1:2) {
InvDsc[i,k1,k2] <- lamb[i]*Inv.D[k1,k2]}}
b[i,1:2] ~dmnorm(mu.b[1:2], InvDsc[i,1:2, 1:2]) }
mu.b[1:2] ~dmnorm(eta[1:2], Inv.C[1:2, 1:2])
Inv.D[1:2, 1:2] ~dwish(R[1:2, 1:2], 2)
VOLpop <- exp(eta[1]) ; CLRpop <- exp(eta[2]) ;
D[1:2,1:2] <- inverse(Inv.D[1:2, 1:2])
nub.2 <- nub/2; nub <- 1/inv.nub; inv.nub ~dunif(0.01,0.5)
kap ~dlnorm(0,1); tau ~dgamma(1,0.001)}
```

9. The code for the second analysis in Example 8.9 is

```
model { for (i in 1:N) {y[Subj[i],rep[i]] <- chol[i]/100
years[Subj[i],rep[i]] <- yr[i]}
for (i in 1:n) {b[i,1:2] ~dmnorm(nought[1:2],invD[1:2,1:2])
age[i] <- ag[i]-mean(ag[])
tLL[i] <- sum(LL[i,1:T[i]])
for (t in 1:T[i]) {y[i,t] ~dnorm(mu[i,t],tau)
LL[i,t] <- 0.5*log(tau/6.28)-0.5*tau*pow(y[i,t]-mu[i,t],2)
G[i,t] <- 1/exp(LL[i,t])
a[i,t] <- (years[i,t]-5)/10; w[i,t] ~dnorm(0,1) I(0,)
mu[i,t] <- beta[1]+beta[2]*sx[i]+beta[3]*age[i]
+beta[4]*a[i,t]+b[i,1]+b[i,2]*a[i,t] +del[t]*w[i,t]}}
```

```
for (j in 1:4) {beta[j] ~dnorm(0,0.01)}
for (j in 1:6) {del[j] ~dnorm(0,0.01)}
sig ~dunif(0,10); s2 <- sig*sig; tau <- 1/s2
TLL <- sum(tLL[])
D[1:2,1:2] <- inverse(invD[,]); invD[1:2,1:2] ~dwish(ScD[,],2)}
```

10. The code for the fourth analysis in Example 8.10 is

```
model {for (i in 1:n) {bh[i]~dnorm(beta[1],invD);
b[i] <- bh[i]-beta[1];
# selection indicators for random effects u[i,t]
delu[i] ~dbern(0.1)
ln.base[i] <- log(y0[i]/4); ln.age[i] <- log(age[i])
Base[i] <- ln.base[i]-mean(ln.base[]); Age[i] <- ln.age[i]-mean(ln.age[])
for (t in 1:T) {y[i,t] ~dpois(mu[i,t]);ynew[i,t] ~dpois(mu[i,t])
u[i,t] ~dnorm(0,tau2u)
LL[i,t] <- -mu[i,t]+y[i,t]*log(mu[i,t])-logfact(y[i,t]); H[i,t] <- 1/exp(LL[i,t])
dev[i,t] <- y[i,t]*log((y[i,t]+0.5)/(mu[i,t]+0.5)) -(y[i,t]-mu[i,t]);
# regression
log(mu[i,t]) <- bh[i]+beta[2]*Base[i]+beta[3]*Trt[i]+
beta[4]*Base[i]*Trt[i]+beta[5]*Age[i]+beta[6]*V4[t]+delu[i]*u[i,t]}}
for (j in 1:6) {beta[j] ~dflat()}
Fit[1] <- 2*sum(dev[,]); Fit[2] <- -2*sum(LL[,])
sqrtD ~dunif(0,100); D <- sqrtD*sqrtD; invD <- 1/D;
sig2u <- (D-v*D)/v; tau2u <- 1/sig2u; v ~dunif(0,1)}
```

The MDP code (fifth analysis) is

```
model {for (i in 1 : n) { ln.base[i] <- log(y0[i]/4); ln.age[i] <- log(age[i])
Base[i] <- ln.base[i]-mean(ln.base[]); Age[i] <- ln.age[i]-mean(ln.age[])
for( t in 1 : T) { y[i, t] ~dpois(mu[i, t])
log(mu[i, t]) <- b[1,i]+beta[1]*Base[i]
+beta[2]*Trt[i]+beta[3]*Base[i]*Trt[i]+beta[4]*Age[i]+b[2,i]*Visit[t]
LL[i,t] <- -mu[i,t]+y[i,t]*log(mu[i,t])-logfact(y[i,t]);
dev[i,t] <- y[i,t]*log((y[i,t]+0.5)/(mu[i,t]+0.5)) -(y[i,t]-mu[i,t])}}
Fit[1] <- 2*sum(dev[,]); Fit[2] <- -2*sum(LL[,])
# select cluster
for (i in 1:n) { S[i] ~dcat(p[1:Km])
for (k in 1:Km) {Sr[i,k] <- equals(k,S[i])}
# realized heterogeneity
for (j in 1:2) {b[j,i] <- bstar[S[i],j]}}
alpha ~dgamma(2,4); V[Km] <- 1; p[1] <- V[1]
for (k in 1:Km-1){V[k] ~dbeta(1,alpha)}
for (k in 2:Km) { p[k] <- V[k]*(1-V[k-1])*p[k-1]/V[k-1]}
# Base Density
for (k in 1:Km) { bstar[k,1:2] ~dmnorm(B[], invD[,])
# total non-empty clusters
clusn[k] <- sum(Sr[,k]); nonempty[k] <- step(clusn[k]-1)}
```

```
K <- sum(nonempty[])
# priors:
invD[1:2,1:2] ~dwish(R[ , ], 10); D[1:2,1:2] <- inverse(invD[,])
for (j in 1:2) {B[j] ~dnorm(0,0.001)}
for (j in 1:4) {beta[j] ~dnorm(0,0.001)}}
```

11. The code in Example 8.11 is

```
model { for (i in 1:437) {y[i] ~dnorm(mu2[Group[i],PatGroup[i],Visit[i]],
tauy[Group[i]])
x[i] ~dnorm(mu1[Group[i],PatGroup[i],Visit[i]],taux[Group[i]])}
# Models for each of J=4 clusters
for (i in 1:n[1]) { b[1,i] ~dnorm(B[1],invD1[1]);c[1,i] ~dnorm(0,invD2[1])
for (t in 1:nvis1[i]) {
mu1[1,i,t] <- b[1,i]; mu2[1,i,t] <- alph[1]+lam[1]*b[1,i]+c[1,i]}}
for (i in 1:n[2]) { b[2,i] ~dnorm(B[2],invD1[2]);c[2,i] ~dnorm(0,invD2[2])
for (t in 1:nvis2[i]) {
mu1[2,i,t] <- b[2,i]; mu2[2,i,t] <- alph[2]+lam[2]*b[2,i]+c[2,i]}}
for (i in 1:n[3]) {b[3,i] ~dnorm(B[3],invD1[3]);c[3,i] ~dnorm(0,invD2[3])
for (t in 1:nvis3[i]) {
mu1[3,i,t] <- b[3,i]; mu2[3,i,t] <- alph[3]+lam[3]*b[3,i]+c[3,i]}}
for (i in 1:n[4]) {b[4,i] ~dnorm(B[4],invD1[4]);c[4,i] ~dnorm(0,invD2[4])
for (t in 1:nvis4[i]) {
mu1[4,i,t] <- b[4,i]; mu2[4,i,t] <- alph[4]+lam[4]*b[4,i]+c[4,i]}}
# Priors
for (j in 1:J) {tauy[j] ~dgamma(1,1); taux[j] ~dgamma(1,1); invD2[j]
~dgamma(1,1);
invD1[j] ~dgamma(1,1); B[j] ~dnorm(0,0.001); alph[j] ~dnorm(0,0.001);
lam[j] ~dnorm(0,0.001)}}
```

12. For the cocaine-use data the second missingness (common factor) model has code

```
model { for (i in 1:106) {F[i]~dnorm(mu.F[i],1); mu.F[i] <- eta*basey[i]/100
for (t in 1:Tstar[i]) {y[i,t]~dnorm(mu[i,t],tau) I(0,)
mu[i,t] <- beta[1]+beta[2]*Trt[i]+beta[3]*(t-6)+beta[4]*Trt[i]*(t-6)+lam*F[i]}
for (t in 2:Tstar[i]) {R[i,t]~dbern(p[i,t])
logit(p[i,t]) <- gam[1]+gam[2]*y[i,t-1]+kap*F[i]}}
tau~dgamma(1,0.001); lam~dnorm(1,1) I(0,);kap~dnorm(1,1)
eta~dnorm(0,0.01); beta[1]~dnorm(20,0.001);
for (j in 2:4) {beta[j]~dnorm(0,0.01)}
for (j in 1:2) {gam[j]~dnorm(0,1)}}
```

13. The code for the model in Example 8.14 is

```
model {# response data in stacked form
for (i in 1:N) {y[i]~dnorm(mu[subj[i],rep[i]],tau); S[subj[i],rep[i]] <-
sqrt(week[i])}
```

```
for (i in 1:n) {b[i,1:2]~dmnorm(nought[],D.inv[,])
# model means for nested data form
for (t in 1:T[i]) {mu[i,t] <- beta[1]+beta[2]*drug[i]+beta[3]*S[i,t]
                    +beta[4]*drug[i]*S[i,t]+b[i,1]+b[i,2]*S[i,t]
w[i,t]~dbern(p[i,t])
# equivalent cloglog forms
p[i,t] <- 1-exp(-exp(eta[i,t]))
# cloglog(p[i,t]) <- eta[i,t]
eta[i,t] <- gam1[t]+gam2*drug[i]+alph[1]*b[i,1]+alph[2]*b[i,1]*drug[i]}}
# set up missingness indicators
for (i in 1:n) {w[i,T[i]] <- 1-step(Lastweek[i]-6)
for (t in 1:T[i]-1) {w[i,t] <- 0}}
# Priors
for (j in 1:4) {beta[j]~dnorm(0,0.001); gam1[j]~dnorm(0,0.001)}
gam1[5] <- gam1[4]; gam2~dnorm(0,0.001)
for (j in 1:2) { alph[j]~dnorm(1,1)}
tau~dgamma(1,0.001); D.inv[1:2,1:2]~dwish(Sc[,],2)}
```

9

Survival and Event History Models

9.1 Introduction

In many applications in the health and social sciences the response of interest is duration to a certain event, such as age at first maternity, survival time after diagnosis, or times spent in different jobs or places of residence. In clinical applications the interest is typically in representing and comparing the distribution of times to an event among different patient groups (e.g., treatment vs. control groups), whereas in social science applications the interest may focus on the impacts of demographic or socioeconomic attributes on human behaviors. Typically durations or event times are not observed for all subjects, either because not all subjects are followed up, or because for some events the event may never occur (e.g., age at first marriage). So some times are missing or censored, and the missingness mechanism is generally assumed to be at random. The most common form is right-censoring, when the event has not occurred by the end of the observation period; the unknown failure time exceeds the subject's survival time c when observation ceased. A failure time is left censored at c if its unobserved actual value is less than c (e.g., a population census may record limiting illness status by current age, but not the age when it commenced). A failure time is interval censored if it is known only that it lies in the interval (c_1, c_2).

Distributions of durations or survival times are equivalently described by hazard rates, also known as failure rates, exit rates, or forces of mortality according to the application. The modeling of the hazard rate through time may be undertaken parametrically. Alternatively, one may adopt nonparametric methods, such as assuming piecewise constancy in the rates within subintervals of the observation span (Ibrahim et al., 2001). Pooling strength through correlated priors is then relevant as rates in successive intervals tend to be similar. Imposing smoothness conditions on the baseline hazard also provides stable estimators when observations are sparse at particular durations (Omori, 2003).

Variations in failure rates between subjects or other units may be explained to a large degree by observed characteristics, the impact of which may also vary over intervals or time. However, unobserved random variations between subjects are present in many applications and may be modeled by introducing subject level frailty (see Section 9.4). Additionally duration times may be hierarchically stratified (e.g., patient survival by hospital or by area of residence), or they may be differentiated by types of possible exit, as in competing risk analysis (see Sections 9.6 and 9.7). One may also consider multivariate survival outcomes, as in multiple component failure (Damien and Muller, 1998) or in familial survival studies (Viswanathan and Manatunga, 2001). In such situations, shared frailty models may account for correlated unobserved variation over different strata or causes of exit.

9.2 Survival Analysis in Continous Time

Let T denote a survival time. The distribution function of T, providing the probability of exit before time $T = t$, is then

$$F(t) = Pr(T \leq t),$$

while the probability of surviving beyond t is $S(t) = 1 - F(t) = Pr(T > t)$. Note that one has $S(\infty) = 0$. So the density of T can be expressed as

$$f(t) = \frac{dF(t)}{dt} = -\frac{dS(t)}{dt}.$$

The chance of an event occurring in a short interval $(t, t{+}dt)$, given survival to t, is

$$Pr(t < T \leq t + dt | T > t) = \frac{Pr(t < T \leq t + dt)}{Pr(T > t)} = \frac{F(t + dt) - F(t)}{S(t)}.$$

The hazard function $h(t)$ is the instantaneous event rate, obtained as $dt \to 0$ in the ratio of the preceding probability to the length of the interval dt. That is

$$h(t) = \lim_{dt \to 0} \frac{F(t + dt) - F(t)}{dt} \frac{1}{S(t)} = \lim_{dt \to 0} \frac{S(t) - S(t + dt)}{dt} \frac{1}{S(t)} = \frac{f(t)}{S(t)}.$$

Since $-f(t)$ is the derivative of $S(t)$, one obtains that $h(t) = (-S'(t)/S(t))$, and so

$$h(t) = \frac{-d \log S(t)}{dt}. \tag{9.1}$$

On integrating both sides in Equation 9.1, one obtains the cumulative hazard rate

$$\begin{aligned} H(t) &= \int_0^t h(u) du = \int_0^t \left[\frac{-d \log S(u)}{du} \right] du \\ &= -\int_0^{-\log S(t)} d \log S(u) = -\log S(t) \end{aligned}$$

and so

$$S(t) = \exp[-H(t)] = \exp\left[-\int_0^t h(u)du\right].$$

If predictors Z_i are available their impact is most simply modeled using a proportional hazards form (e.g., Cox, 1972; Kiefer, 1988; Li, 2007):

$$h(t|Z) = h_0(t)\exp(Z_i\beta),$$

where $h_0(t)$ is known as the baseline hazard and the regression impact is constant across time. Letting $\eta_i = Z_i\beta$, the associated survivor function is

$$\begin{aligned} S(t|Z_i) &= \exp\left[-\int_0^t h(u|Z_i)du\right] = \exp[-H_0(t)e^{\eta_i}] = [S_0(t)]^{\exp(\eta_i)} \\ &= \exp\{-\exp[\eta_i + \log H_0(t)]\}, \end{aligned}$$

where $H_0(t)$ is the integrated baseline hazard. The proportional hazard assumption is often restrictive, though Yin and Ibrahim (2006) show the proportional hazard model (PHM) may be nested in a broader class of transformation hazard models, with parameter $0 \leq \gamma \leq 1$ and

$$h(t|Z) = [h_0(t)^\gamma + \exp(Z_i\beta)]^{1/\gamma}$$

which reduces to the proportional model when $\gamma = 0$ and to an additive model when $\gamma = 1$.

Consider an absorbing (nonrepeatable) type of exit, and let $d_i = 1$ for an observed exit and $d_i = 0$ for a censored time. Assuming censoring is noninformative, the likelihood contribution for subject i is

$$f(t_i|Z_i) = h(t_i|Z_i)S(t_i|Z_i)$$

if $d_i = 1$, and $S(t_i|Z_i)$ if $d_i = 0$. The likelihood contribution may be expressed in equivalent forms as

$$h(t_i|Z_i)^{d_i}S_i(t_i|Z_i) = f(t_i|Z_i)^{d_i}S_i(t_i|Z_i)^{1-d_i}.$$

For a PHM the likelihood contribution also may be written (Aitkin and Clayton, 1980; Orbe and Nunez-Anton, 2006) as

$$\begin{aligned} &[h_0(t_i)\exp\{Z_i\beta - H_0(t_i)\exp(Z_i\beta)\}]^{d_i}[\exp\{-H_0(t_i)\exp(Z_i\beta)\}]^{1-d_i} \\ &= \left(\mu_i^{d_i}e^{-\mu_i}\right)\left(\frac{h_0(t_i)}{H_0(t_i)}\right)^{d_i}, \qquad (9.2) \end{aligned}$$

where

$$\mu_i = H_0(t_i)\exp(Z_i\beta) \qquad (9.3)$$

and the second bracketed term in Equation 9.2 depends only on the baseline hazard and is independent of β. The first term in Equation 9.2 is the kernel of a Poisson likelihood for the event status indicators $d_i \sim Po(\mu_i)$. From Equation 9.3, the corresponding log-linear model is

$$\log(\mu_i) = \log(H_0(t_i)) + Z_i\beta, \tag{9.4}$$

where $\log(H_0(t_i))$ is an offset using the observed time, whether censored or uncensored.

9.2.1 Counting process functions

For repeated events in continuous time, especially with successive durations not necessarily independent, it is advantageous to use additional functions. The count of failures $N(t)$ occurring over $(0, t]$ for a given individual or component system defines a counting process satisfying $N(s) \leq N(t)$ for $s < t$. For a nonrepeatable event the counting process may still be useful (e.g., in modeling time varying predictors), and one may denote $N(t) = I(T \leq t)$, namely by an indicator of whether the event has occurred by t.

For all event types considered over sufficiently small intervals, the counting process increments $dN(t) = N(t) - N(t-)$ are either 1 or 0, where $N(t-)$ denotes $\lim_{\delta \downarrow 0} N(t - \delta)$ (Manda et al., 2005). Let $A(t-)$ denote the antecedent history of the event sequence up to but not including t. Then conditional on $A(t-)$, the probability that $dN(t) = 1$ can be written in terms of an intensity process $\lambda(t)$, namely,

$$Pr\{N(t + \delta) - N(t-) = 1 | A(t-)\} \simeq \lambda(t)\delta.$$

Equivalently

$$Pr\{dN(t) = 1 | A(t-)\} \simeq d\Lambda(t),$$

where $\Lambda(t) = \int_0^t \lambda(u)\, du$ is the integrated intensity, with $\Lambda(t) = E(N(t))$.

The intensity is equal to the hazard while the subject or system is still under observation, that is still at risk, but is zero when the event has happened (when the event is nonrepeatable), or when a sequence of (repeatable) events has finished. An example of the latter might be when a repairable system subject to repeated breakdowns is finally decommissioned—see Watson et al. (2002) for a counting process analysis of failure times of water pipes. Let $Y(t) = I(T \geq t)$ denote the at risk indicator then

$$\lambda(t) = Y(t)h(t).$$

This representation of the intensity function generalizes to include predictors and random effects (or frailties). So for proportional hazards, predictor effects would be included via

$$\lambda(t_i | Z_i) = Y(t_i)h_0(t_i)\exp(Z_i\beta).$$

One may then compare observed and predicted counts via the Martingale residual at t, defined as

$$M_i(t) = N(t_i) - \Lambda_0(t_i|Z_i) = N(t_i) - \int_0^{t_i} Y_i(u)\exp(Z_i\beta)dH_0(u).$$

The total residual $M_i = M_i(\infty)$ for a subject with observation time t_i is obtainable for a nonrepeatable event as

$$M_i = d_i - \Lambda_0(t_i|Z_i).$$

Deviance residuals r_i are obtained as

$$r_i = \text{sgn}(M_i)\sqrt{2\left[M_i - N_i(\infty)\log\left(\frac{N_i(\infty) - M_i}{N_i(\infty)}\right)\right]}.$$

9.2.2 Parametric hazards

The hazard rate $h(t)$ is called duration dependent if its value changes over t; under negative duration dependence (often observed in job or residential careers), $h(t)$ decreases with time. In practice, plots of survivor proportions are often jagged with respect to time, and semi- or nonparametric methods for representing the hazard function reflect this. However, parametric lifetime models are also often applied to test whether certain basic features of duration dependence are supported by the data.

The simplest parametric model is the exponential model, under which the leaving rate is constant, defining a stationary process with hazard

$$h(t) = \lambda,$$

survival function $S(t|\lambda) = \exp(-\lambda t)$, and density

$$f(t|\lambda) = \lambda\exp(-\lambda t).$$

With covariates and assuming proportionality $h(t|Z_i) = \lambda e^{Z_i\beta}$. Equivalently under the Poisson likelihood approach of Aitkin and Clayton (1980), one has $d_i \sim Po(\mu_i)$ and from Equation 9.4, a log-linear model

$$\log(\mu_i) = \log(\lambda t_i) + Z_i\beta,$$

since $H_0(t) = \lambda t$. Absorbing λ into the regression term one has

$$\log(\mu_i) = \log(t_i) + Z_i\beta.$$

This Poisson likelihood device can be used in piecewise exponential models as considered below.

Another commonly used parametric form is the Weibull with scale parameter λ and shape κ, namely

$$h(t|\lambda,\kappa) = \lambda\kappa t^{\kappa-1},$$

so that $S(t|\lambda,\kappa) = \exp[-\lambda t^{\kappa}]$, and $f(t|\lambda,\kappa) = \lambda\kappa t^{\kappa-1}\exp[-\lambda t^{\kappa}]$. The Weibull hazard rate is monotonic, with positive duration dependence if $\kappa > 1$ (and a 95% credible interval excludes 1), and negative dependence if $\kappa < 1$.

Since $\log(S(t|\lambda,\kappa)) = -\lambda t^{\kappa}$, one has $\log(-\log(S(t|\lambda,\kappa))) = \log\lambda + \kappa\log t$. Therefore a plot of $\log(-\log(S(t|\lambda,\kappa)))$ against $\log(t)$ should be approximately linear when a Weibull is appropriate. An initial assessment can be made using a Kaplan–Meier estimate of $S(t)$ in the R package. Assume a dataset named survdat containing variables time (corresponding to t_i) and status (corresponding to d_i). Then the procedure is

```
library(survival)
KMinputs <- Surv(survdat$time,survdat$status)
KM <- survfit(KMinputs)
plot(log(KM$time), log(-log(KM$surv)), type="S").
```

However, many processes exhibit peaks in exit rates; for example, the rate may at first increase but after reaching a peak tail off again (Gore et al., 1984; Shao and Zhou, 2004). Parametric models accommodating such a pattern include the log-logistic model and the sickle model (Bennett, 1983; Brüderl and Diekmann, 1995; Diekmann and Mitter, 1983). The log-logistic density has hazard

$$h(t) = \lambda\kappa t^{\kappa-1}[1 + \lambda t^{\kappa}]^{-1},$$

and survivor function

$$S(t) = [1 + \lambda t^{\kappa}]^{-1},$$

where all parameters are positive, and the scale parameter λ can be adapted to model the impact of predictors; see Li (1999) for a Bayesian application to Chapter 11 bankruptcies. An alternative common parameterization (Florens et al., 1995) sets $\lambda = \nu^{\kappa}$, so that

$$h(t) = \frac{\nu^{\kappa}\kappa t^{\kappa-1}}{[1 + (\nu t)^{\kappa}]}. \tag{9.5}$$

The sickle model has corresponding functions

$$\begin{aligned} h(t) &= ct\, e^{-t/\lambda} \\ S(t) &= \exp[-\lambda c\{\lambda - (t+\lambda)e^{-t/\lambda}\}] \end{aligned}$$

with both c and λ positive. The sickle model has a permanent survival probability or "cure rate" (Chen et al., 1999) in that $S(\infty) > 0$ (see Section 9.2.3). In general one may define a cure rate $r = 1 - \pi$ as the limit as $t \to \infty$ of the survivor function, namely (Tsodikov et al., 2003)

$$r = \lim_{t\to\infty} S(t) = \exp\left[-\int_0^{\infty} h(u)du\right]$$

with π denoting the proportion of susceptibles.

9.2.3 Accelerated hazards

In contrast to the proportional hazard model with $h(t_i|Z_i) = h_0(t_i)\exp(Z_i\beta)$ in an accelerated failure time (AFT) model the explanatory variates are assumed to act multiplicatively on time (Wei, 1992). So with $B_i = \exp(Z_i\beta)$, one has

$$h(t_i|Z_i) = h_0(t_iB_i)B_i,$$
$$S(t_i|Z_i) = [S_0(tB_i)]^{B_i}$$

and the effect of the predictors Z_i on survival time is more direct, acting to accelerate or decelerate the time to failure. To illustrate this in the case of a treatment comparison, assume Z_i excludes an intercept and that the baseline hazard includes a scale parameter to model the mean hazard (e.g., the parameter λ in exponential and Weibull models). Also assume a single predictor such as $z_i = 1$ for a new treatment and $z_i = 0$ for control. Then with $B_i = e^{\beta z_i} = e^{\beta}(= \phi)$ for a treated subject, one has a hazard $\phi h_0(\phi t_i)$ and survivor function $S(\phi t_i)$ for such a subject, but a hazard $h_0(t_i)$ and survivor function $S(t_i)$ for a control subject. So the lifetime under the new treatment is ϕ times the lifetime under the control regime.

More inclusive schemes are possible. For example, defining $G_i = \exp(Z_i\gamma)$, one has

$$h(t_i|Z_i) = h_0(t_iG_i)B_i,$$

which includes the AFT and PHM forms as special cases (Chen and Jewell, 2001). For example, for the log-logistic this would imply

$$h(t_i|Z_i) = B_i\lambda\kappa(t_iG_i)^{\kappa-1}[1 + B_i\lambda(tG_i)^{\kappa}]^{-1}.$$

Apart from avoiding the assumption of proportional hazards, the AFT approach has the advantage of a direct regression form which may be useful in modeling nonlinear effects of predictors (Orbe and Nunez-Anton, 2006). Let Z_i be of dimension p and T_i denote the completed failure time which for censored subjects is unobserved. Then $T_i = t_i$ when $d_i = 1$ but $T_i > t_i$ when $d_i = 0$, so truncated sampling with the censored time as the lower limit is necessary.[1] The regression formulation is then

$$\log(t_i) = Z_i\gamma + \sigma u_i,$$

where σ is a scale parameter, and the errors are defined by the survivor function, namely

$$S(t_i) = Pr(T_i > t_i) = Pr(\log(T_i) > \log(t_i))$$
$$= Pr\left(u_i > \frac{\log(t_i) - \gamma_0 - \gamma_1 z_{1i} - \cdots - \gamma_p z_{pi}}{\sigma}\right).$$

A positive γ_j coefficient means that z_j leads to longer survival or length of stay.

Taking u to be standard normal with variance 1 corresponds to a log-normal density for failure times t_i, under which

$$S(t_i) = 1 - \Phi\left(\frac{\log(t_i) - \gamma Z_i}{\sigma}\right).$$

Taking u to be standard logistic with density $p(u) = e^u/(1 + e^u)^2$ corresponds to a log-logistic failure time density with

$$S(t_i) = \left[1 + \exp\left\{\frac{\log(t_i) - \gamma Z_i}{\sigma}\right\}\right]^{-1}$$

with σ corresponding to the inverse of the shape parameter κ. Finally, consider a Weibull density for failure times with hazard $h(t_i|Z_i) = \lambda\kappa t_i^{\kappa-1}\exp(\beta Z_i)$ where Z_i excludes a constant term. Taking u to follow a standard extreme value density, namely $p(u) = \exp(u - e^u)$, the AFT regression takes the form (Keiding et al., 1997):

$$\log(t_i) = -\frac{\log\lambda}{\kappa} - z_{1i}\frac{\beta_1}{\kappa}\cdots - z_{pi}\frac{\beta_p}{\kappa} + \frac{u_i}{\kappa}.$$

Example 9.1. Nursing Home Stays Morris et al. (1994) analyze lengths of stay t_i for $n = 1601$ nursing home patients, with stay usually terminated by death. There are 322 censored lengths of stay. Predictor effects are assessed via a proportional Weibull hazard,

$$h(t|\lambda_i, \kappa) = \lambda_i\kappa t_i^{\kappa-1},$$
$$\lambda_i = \exp(Z_i\beta).$$

This is equivalent to accelerated hazards regression for logged length of stay with error u_i

$$\log(t_i) = \gamma Z + \sigma u_i,$$

where $\gamma = -\beta\sigma$ and $\sigma = 1/\kappa$. The γ coefficients[2] express influences on length of stay (i.e., survival) while the β coefficients express influences on mortality.

Parameter estimates (Table 9.1) for the treatment and attribute variables replicate those of Morris et al. (1994), with the age covariate included in the regression in the form age/100. The estimates are based on the second half of a two chain run of 10,000 iterations. Health status is measured in terms of dependency in activities of daily living; with health = 2 if there are four or fewer activities with dependence, health = 3 for five dependencies, health = 4 for six dependencies, and health = 5 if there were special medical conditions requiring extra care. It can be seen that higher ADL dependency is associated with earlier mortality and lower stays. The age effect is not significant, but its negative sign is possibly misleading and may reflect varying frailty (selection effects)—see Section 9.4. The κ coefficient shows negative

TABLE 9.1
Nursing home stays.

	WinBUGS			BayesX		
	Mean	**2.5%**	**97.5%**	**Mean**	**2.5%**	**97.5%**
Influences on Mortality						
Intercept	−3.22	−3.85	−2.58	−6.93	−7.27	−6.63
Age	−0.44	−1.18	0.27			
Treatment	−0.13	−0.23	−0.02	−0.06	−0.17	0.05
Male	0.35	0.22	0.48	0.34	0.21	0.47
Married	0.16	0.00	0.31	0.16	0.01	0.31
ADL Status 3	−0.03	−0.18	0.12	−0.03	−0.19	0.11
ADL Status 4	0.23	0.07	0.38	0.22	0.06	0.37
ADL Status 5	0.53	0.33	0.73	0.52	0.32	0.72
κ	0.61	0.58	0.64			
Influences on Length of Stay						
Intercept	5.25	4.24	6.26			
Age	0.72	−0.45	1.92			
Treatment	0.21	0.03	0.38			
Male	−0.57	−0.79	−0.36			
Married	−0.25	−0.51	0.00			
ADL Status 3	0.05	−0.20	0.30			
ADL Status 4	−0.37	−0.62	−0.12			
ADL Status 5	−0.87	−1.19	−0.54			
σ	1.64	1.56	1.71			

duration dependence. The DIC for this model (based on the unstandardized deviance) is 16,462 ($d_e = 9$).

To illustrate use of BayesX for right-censored survival data a piecewise exponential (PE) model (Section 9.3) is also applied. This application additionally models the impact of age using a B-spline with second-order random walk penalty on the parameters (see Chapter 10). Due to the different baseline hazard the intercept changes, but otherwise influences on mortality are similar, except that the treatment variable is no longer significant. The changing impact of age on the log hazard is shown in Figure 9.1, and shows that the 95% interval straddles zero for all ages. The changing baseline parameters are shown in Figure 9.2, and confirm negative duration dependence. The negative duration effect may be diminished if frailty were allowed for, and including frailty in the BayesX model is left as an exercise; this involves adding an extra patient identifier column, patno, into the data input, and adding a term for patno(random) in the model. The DIC falls to 16,362 ($d_e = 17$) under the PE model without frailty, and to 15,984 for a PE model with frailty (under the default inverse gamma parameter settings).

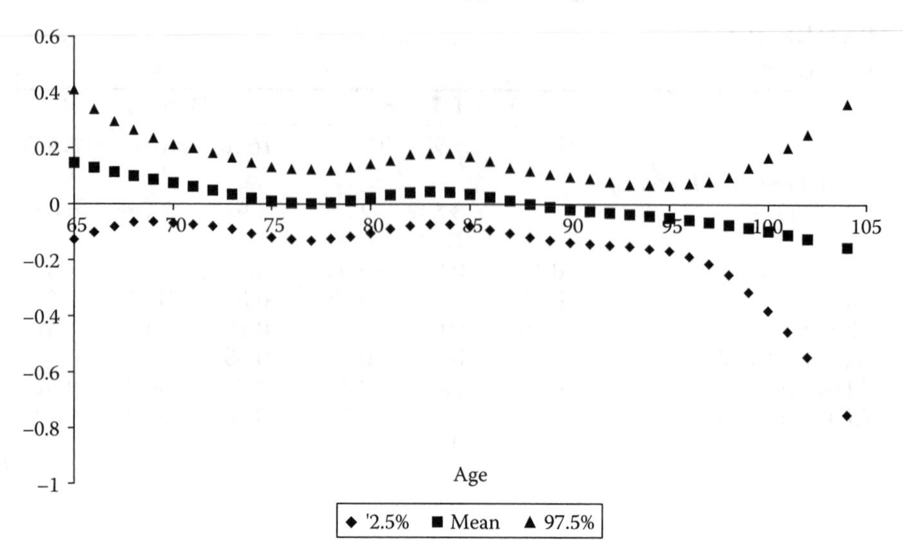

FIGURE 9.1
Age effect, Nursing home stays.

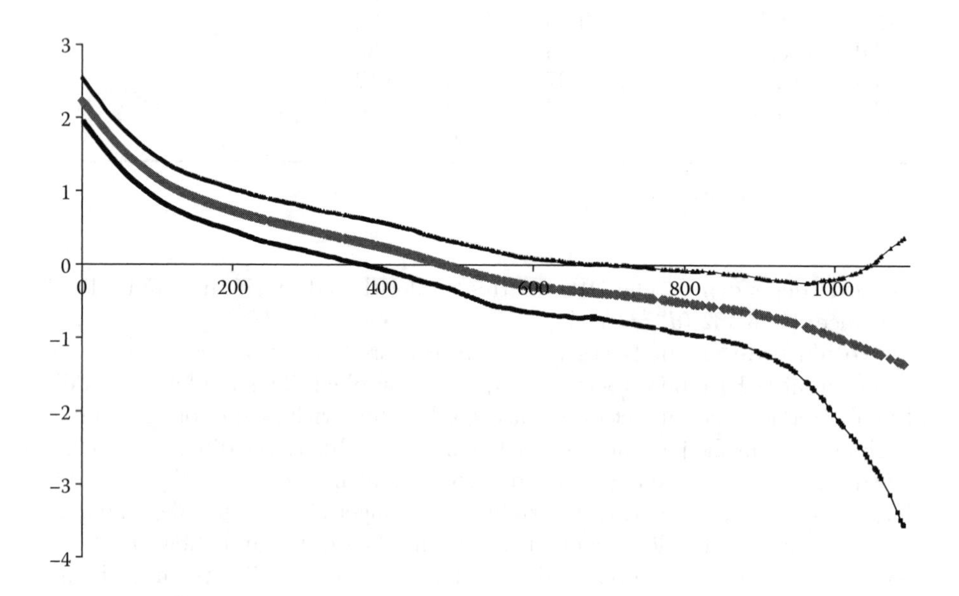

FIGURE 9.2
Piecewise log hazard, Nursing home stays.

9.3 Semi-Parametric Hazards

In the proportional hazards model

$$h(t|Z) = h_0(t)\exp(Z\beta)$$

it may be difficult to choose a parametric form for the baseline hazard $h_0(t)$, and semi-parametric or nonparametric approaches are often preferable. These have benefits in avoiding possible mis-specification of parametric hazard forms, and in facilitating other aspects of hazard regression, such as time varying predictor effects (Gamerman, 1991). Such aproaches have been applied to the cumulative hazard, and implemented in counting process models (Clayton, 1991). However, they may also be specified for the baseline hazard h_0 itself (e.g., Gamerman, 1991; Sinha and Dey, 1997) and typically use only information about the intervals in which exit times occur.

Consider a partition of the response time scale into J intervals $(a_0, a_1], \ldots, (a_{J-1}, a_J]$, where a_J equals or exceeds the largest observed time, censored or uncensored (Ibrahim et al., 2001, 106). The partition scheme can be based on distinct values in the profile of observed times $\{t_1, \ldots, t_n\}$, whether censored or not, or by siting knots a_j at selected points in the range $(t_{\min}, t_{\max})$. Yin and Ibrahim (2006, 173) propose that the partitioning should ensure an approximately equal number of failures in each of the J intervals, with each interval containing at least one failure. Among alternatives are knots sited at $(j - 1/J)$th quantiles of observed times (Gustafson et al., 2000), or evenly spaced along the range of the observed t values. As the number of intervals J tends to infinity, a truly nonparametric model is obtained but not likely to be empirically well identified (Lopes et al., 2007).

Different approaches may be based on the asumption that the baseline hazard is constant within each interval. Thus Ibrahim et al. (1999) and Ibrahim et al. (2001, 55) consider discrete approximation to the gamma process of Dykstra and Laud (1981). This involves a prior on the increments

$$\Delta_j = h_0(a_j) - h_0(a_{j-1}), \quad j = 1, \ldots, J$$

in the baseline hazard, and use of the approximate survival function

$$S(t|Z_i) = \exp\left[-B_i \int_0^t h_0(u)du\right] \backsimeq \exp\left[-B_i \sum_{j=1}^{J} \Delta_j (t - a_{j-1})_+\right],$$

where $B_i = e^{Z_i\beta}$, and $(u)_+ = u$ if $u > 0$ and is zero otherwise. The probability of exit in interval j is then

$$\begin{aligned} q_j &= S(a_{j-1}) - S(a_j) \\ &\backsimeq \left[\exp\left\{-B_i \sum_{m=1}^{j-1} \Delta_m (a_{j-1} - a_{m-1})\right\}\right]\left[1 - \exp\left\{-B_i (a_j - a_{j-1}) \sum_{m=1}^{j} \Delta_m\right\}\right]. \end{aligned}$$

Piecewise exponential priors (Brezger et al., 2008, 23; Gamerman, 1991; Ibrahim et al., 2001, 106; Sinha et al., 1999) specify a baseline parameter λ_j for each interval, possibly combined with interval specific regression parameters β_j, so that

$$h(t_i \in (a_{j-1}, a_j]|Z_i) = \lambda_j \exp(Z_i\beta_j),$$

where Z_i excludes an intercept. Let $B_{ij} = \exp(Z_i\beta_j)$. For a subject surviving beyond the jth interval (with $t_i > a_j$) the likelihood contribution during interval j is

$$\exp(-\lambda_j(a_j - a_{j-1})B_{ij}).$$

For a subject with $a_{j-1} < t_i \leq a_j$, either failing ($w_{ij} = 1$) in interval j, or censored but nevertheless exiting ($w_{ij} = 0$) in the jth interval, the likelihood contribution is

$$[\lambda_j B_{ij}]^{w_{ij}} \exp[-\lambda_k(t_i - a_{j-1})B_{ij})].$$

So a Poisson likelihood approach may be applied[3] as in Equations 9.2 through 9.4, with responses defined by the event type in each interval and with offsets defined according to whether the subject survives the interval.

The successive baseline parameters λ_j are likely to be correlated, but also possibly to show erratic fluctuations or be imprecisely estimated if treated as fixed effects. Hence a smoothing prior is indicated. One might assume a parametric model (e.g., polynomial in j) but allowing for additional random variation. Thus Albert and Chib (2001) and Omori (2003) assume a polynomial for $\alpha_j = \log(\lambda_j)$,

$$\alpha_j = \psi_0 + \psi_1(j-1) + \psi_2(j-1)^2 + u_j,$$

where $u_j \sim N(0, 1/\tau_u)$. Pooling strength under hierarchical autocorrelated priors linking successive λ_j or α_j is also widely applied. These are known as correlated prior processes or Martingale prior processes for the baseline hazard. Possibilities are first or second-order random walks in the α_j, possibly adjusted to reflect unequal width $\delta_j = (a_j - a_{j-1})$ of the intervals. Thus one might take a first-order random walk,

$$\alpha_j \sim N(\alpha_{j-1}, \sigma_\alpha^2\delta_j),$$

with α_1 a separate fixed effect, and with $\tau_\alpha = 1/\sigma_\alpha^2$ following a gamma or uniform prior. Alternatively, as in Gustafson et al. (2003), one may take $w_j = 0.5(a_j + a_{j+1})$, $\zeta_j = w_j - w_{j-1}$, and

$$\alpha_j \sim N\left(\alpha_{j-1} + (\alpha_{j-1} - \alpha_{j-2})\frac{\zeta_j}{\zeta_{j-1}}, \sigma_\alpha^2(\zeta_j/\bar{\zeta})^2\right).$$

Since setting particular partitions of the time scale involves an element of arbitrariness, Sahu and Dey (2004) apply RJMCMC techniques in which J is an additional unknown; they specify a sparse precision matrix formulation for the joint prior for the $(\alpha_1, \ldots, \alpha_J)$ under an RW1 prior for given J.

Since random walk priors of degree r set a mean level not on the α_j themselves, but on differences of order r (e.g., an RW1 prior specifies a zero mean for $\alpha_j - \alpha_{j-1}$), identifiability may require that a separate regression intercept is omitted or that the α_j are centered to sum to zero at each MCMC iteration, by the operation $\alpha'_j = \alpha_j - \bar{\alpha}$. Alternatives are to set any value, say the hth, to zero (by the operation $\alpha'_j = \alpha_j - \alpha_h$ at each iteration), or set the first effect α_1 to zero (Sahu and Dey, 2004).

A gamma prior in the baseline hazard rates is also possible (Arjas and Gasbarra, 1994), namely

$$\lambda_j \sim \text{Ga}(b, b/\lambda_{j-1}),$$

where λ_1 is a separate positive effect, and larger b values lead to smoother sequences of λ_j. The same identifiability issues obtain as for $\alpha_j = \log(\lambda_j)$ and, if a regression intercept is used, devices such as normalization of the λ_j (to value 1) at each iteration may be applied.

Piecewise priors may also be used to model nonconstant predictor effects, though typically values of time varying regression coefficients β_j in successive intervals are expected to be close (Sinha et al., 1999). Sargent (1997) considers alternative gamma priors for the precision $\tau_\beta = 1/\sigma^2_\beta$ of regression coefficients assumed to evolve according to a first-order random walk with normal errors. Prior knowledge in his application (the Veterans Administration lung cancer trial) suggests that values of time varying coefficients on successive days would differ by at most 0.001. Taking this as the standard deviation of the normal distribution, the prior mean precision for the gamma is 10^6. This corresponds to quartiles (0.0027, 0.0038, 0.0059) for σ_β. An alternative prior adopted by Sargent has mean precision 10^5. Posterior inferences for the mean precision were different under the alternative priors, but not those on the estimated β_j. Fahrmeir and Knorr-Held (1997, 432) suggest gamma $\text{Ga}(1, b)$ priors on precision parameters τ_α on varying log baseline rates, or precisions τ_β on varying predictor effects. Sensitivity is gauged with b taking alternative values (e.g., $b = 0.05$ and $b = 0.0005$), since b determines how close to zero the variances are allowed to be a priori.

9.3.1 Cumulative hazard specifications

Semi-parametric approaches may also be applied to the cumulative baseline hazard H_0 (Kalbfleisch, 1978). Consider a counting process approach with data $(N_i(t), Y_i(t), Z_i(t))$ and independent priors on $\beta(t)$ and H_0. For an individual i exiting or censored before t, so that $Y_i(s) = 0$ for $s > t$, one may apply a Poisson likelihood with binary responses $dN_i(t)$ and means $Y_i(t)\exp(Z_i(t)\beta)dH_0(t)$. An independent gamma increments prior for dH_0 may be adopted (assuming a constant baseline hazard in each interval), namely

$$dH_0(t) \sim \text{Ga}(c[dH^*(t)], c),$$

where $dH^*(t)$ is a prior estimate of the hazard rate per unit time. Other possibilities include normal priors on $\log(dH_0)$.

Let $J+1$ intervals $(s_0, s_1], \ldots, (s_J, s_{J+1}]$ be defined by the J distinct failure times in a dataset, with s_1 equal to the minimum observed failure time, and s_{J+1} exceeding the largest failure time s_J (Sargent, 1997, 16). The likelihood for individual i exiting or censored before s_j, so that $Y_{ij} = 0$ for $t > s_j$, reduces to a discretized form of Poisson likelihood[4] over all possible intervals j with binary resposes dN_{ij} and means $Y_{ij} \exp(Z_{ij}\beta_j)dH_{0j}$. This model may be adapted to allow for unobserved covariates or other sources of heterogeneity ('frailty') as considered in Section 9.4. It also allows for autoregressive dependencies between intervals.

Example 9.2. Veterans Lung Cancer Trial To illustrate the implementation of semi-parametric hazards and the opportunity they offer to model time varying regression effects, consider data from the Veteran's Administration lung cancer trial. In this trial, $n = 137$ male subjects with advanced inoperable lung cancer were randomized to either a standard or a test chemotherapy, with the end point being time to death in days. Only 13 of the 137 survival times are censored. Most analyzes find the treatment to be insignificant and consider the remaining predictors, namely:

celltype (1 = squamous, 2 = smallcell, 3 = adeno, 4 = large);

the Karnofsky score or KS, based on a patient's ability to perform common tasks (with values 0–100, where a score of 100 signifies normal physical abilities with no evidence of disease);

prior therapy or PT, (0 = no, 1 = yes);

an interaction between KS and PT.

Sargent (1997) considers a counting process version of the Cox model for these data (see Section 9.3.1) with time varying effects on the KS predictor, and finds its relevance to be less at higher durations.

Here a piecewise exponential model

$$h(t_i \in (a_{j-1}, a_j]|Z_i) = \lambda_j \exp(Z_i\beta_j)$$

is adopted, with the partitioning of the time scale involving 51 equally spaced points between 0 and the maximum survival time of 999 days. The first model[5] assumes a constant Karnofsky score effect, but time varying (log) baseline hazard, with $\alpha_j \sim N(\alpha_{j-1}, \sigma_\alpha^2)$, a $U(0, 10^8)$ prior on σ_α^2, and $\alpha_1 \sim N(0, 1000)$. With 20,000 (and 5000 burn-in) iterations in a two chain run, the posterior density of σ_α is found to be bounded away from zero, with median 0.14. There is a slight upward shift to higher mortality between 500 and 1000 days in the trial (Figure 9.3). The DIC is 993 ($d_e = 11$). The coefficient on the Karnofsky score has 95% interval $(-0.32, -0.12)$, while the interaction term has coefficient with 95% interval $(-0.54, -0.10)$.

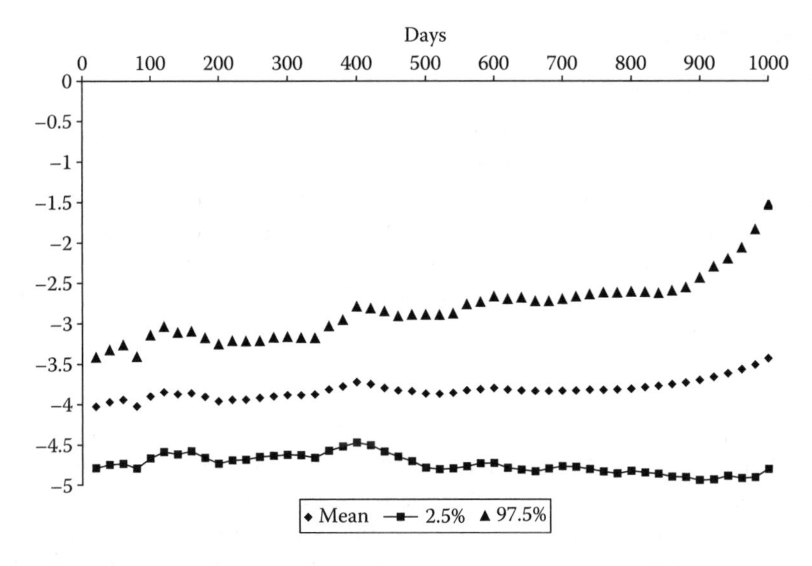

FIGURE 9.3
Baseline log hazard.

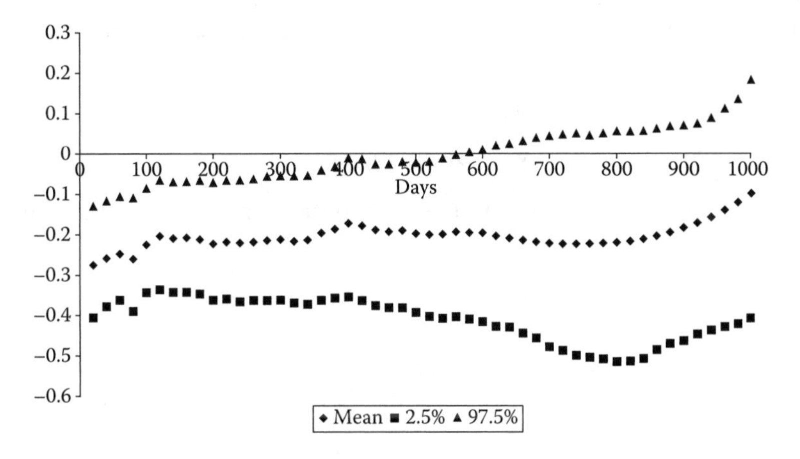

FIGURE 9.4
Varing effect of Karnofsky score.

A second model instead takes the Karnofsky score to have a time varying coefficient. The DIC is slightly smaller at 991.5 ($d_e = 13$) and there is a slight upward trend in the coefficient, with an insignificant effect becoming apparent at higher days (Figure 9.4). Topics for further investigation might be the sensitivity of the form of time variation in the KS effect to the partitioning scheme of the durations, or to unobserved heterogeneity between subjects.

9.4 Including Frailty

Subjects with a given profile of attributes are still likely to show variations in survival times due to unobserved factors. Such factors mean that subjects have different frailties and the most frail will exit before others (Aalen, 1988), so that survivors are subject to a selection effect. Inferences from survival analysis may be incorrect if unobserved heterogeneity is ignored (Lancaster, 1990). The canonical form for introducing unobserved differences between observation is via a multiplicative frailty, γ_i, distributed independently of Z_i and t_i, with

$$h(t_i|Z_i, \gamma_i) = \gamma_i h_0(t_i) \exp(Z_i\beta)$$

leading to mixed proportional hazard or MPH models (Abbring and van den Berg, 2007; Mosler, 2003; Van den Berg, 2001). Except for the case of positive stable frailty distributions, the MPH model is inconsistent with the usual Cox proportional hazard formulation (Henderson and Oman, 1999).

A typical assumption for the distribution $p(\gamma_i)$ of multiplicative frailties is that they are gamma distributed (Perperoglou et al., 2006), typically $\gamma_i \sim \text{Ga}(k, k)$ where k is unknown. So the frailties have mean 1 and variance $1/k$ to ensure identification when Z_i includes an intercept. Another possibility is to include the regression effect $\exp(Z_i\beta)$ in the specification of the frailty density. So, for example,

$$h(t_i|Z_i, \gamma_i) = \gamma_i h_0(t_i)$$
$$\gamma_i \backsim \text{Ga}(k \exp[Z_i\beta], k).$$

Sohn et al. (2007) assume Weibull distributed survival times, with density form

$$f(t_i) = \frac{a}{\gamma_i} t_i^{\alpha-1} \exp(t_i^{\alpha}/\gamma_i)$$

and then take γ_i to be inverse gamma. With the form $x \backsim \text{IG}(a, b)$ corresponding to $f(x) = [b^a/\Gamma(a)]x^{-(\alpha+1)} \exp[-b/x]$ the frailty density is then

$$\gamma_i \backsim \text{IG}(\alpha + 1, \alpha \exp[Z_i\beta]).$$

Other positive parametric densities can be used to represent frailty, such as the log-normal (Gustafson, 1997). An advantage of gamma frailty combined with Weibull hazard is that joint and marginal survival functions can be obtained analytically. An alternative is to assume the γ_i have a positive stable distribution (Hougaard, 2000), in which case the proportional hazards property is preserved after the γ_i are integrated out (Aalen and Hjort, 2002). Ravishanker and Dey (2000) consider a finite mixture of stable densities for robust frailty modeling.

Estimates resulting from the mixed proportional hazard model are often sensitive to the functional form of the heterogeneity distribution, and may be biased if the functional form of the distribution is mis-specified (Baker and

Melino, 2000; Keiding et al., 1997). Heckman and Singer (1984) report sensitivity of regression estimates according to different parametric distributions of frailty. They propose discrete mixture models with finite support at a small number K of points, so that

$$h(t_i|Z_i, \gamma_i) = \gamma_{G_i} h_0(t_i) \exp(Z_i\beta),$$

where G_i is a multinomial indicator with K categories. Sahu and Dey (2004) compare gamma, stable and skewed log-t frailty models and show how the gamma assumption may attenuate covariate effects as compared to the other forms.

Despite such sensitivity it is important to consider possible heterogeneity. One can show (Lancaster, 1990) that a model neglecting frailty will show spurious duration dependence, and specifically over-estimate the extent of negative duration dependence in the true baseline hazard, and under-estimate the extent of positive duration dependence. This is a consequence of selection since in the presence of negative duration dependence, subjects with high values of γ exit faster, so survivors at a given survival time are increasingly biased toward relatively low γ values and lower hazard rates. These features can be illustrated with the MPH assumption and particular parametric hazards. Conditional on a particular value of γ_i, the survivor function is

$$S(t_i|Z_i, \gamma_i) = \exp\left[-\gamma_i \exp(Z_i\beta) \int_0^{t_i} h_0(u)du\right],$$

or in terms of the cumulative hazard $H_0(t_i)$,

$$S(t_i|Z_i, \gamma_i) = \exp[-\gamma_i \exp(Z_i\beta) H_0(t_i)].$$

The unconditional survival function (integrating out the frailties) is therefore

$$\begin{aligned} S(t_i|Z_i) &= \int_0^{\infty} S(t_i|Z_i, \gamma_i) p(\gamma_i) d\gamma_i \\ &= \int_0^{\infty} p(\gamma_i) \exp[-\gamma_i H_0(t_i) e^{Z_i\beta}] d\gamma_i. \end{aligned}$$

For γ_i following a gamma density, $\gamma_i \sim \text{Ga}(a, b)$, the unconditional survivor function is (Vaupel et al., 1979)

$$S(t_i|Z_i) = b^a \left[b + H_0(t_i) e^{Z_i\beta}\right]^{-a}$$

which for $a = b = k$ (with Z_i including a constant) reduces to

$$S(t_i|Z_i) = \left[1 + k^{-1} H_0(t_i) e^{Z_i\beta}\right]^{-k}.$$

Consider exponentially distributed times so that $h_0(t_i) = 1$ and $H_0(t_i) = t_i$. Then

$$\begin{aligned} S(t_i|Z_i) &= [1 + k^{-1} e^{Z_i\beta} t_i]^{-k}, \\ f(t_i|Z_i) &= e^{Z_i\beta}[1 + k^{-1} e^{Z_i\beta} t_i]^{-k-1}, \\ h(t_i|Z_i) &= e^{Z_i\beta}[1 + k^{-1} e^{Z_i\beta} t_i]^{-1}. \end{aligned}$$

For a frailty variance $1/k > 0$, the hazard rate is a decreasing function of t, an example of spurious duration dependence. If frailty is present but ignored not only will duration effects be mis-stated, but covariate effects will be underestimated (Hougaard et al., 1994; Pickles and Crouchley, 1995). Lancaster (1990) confirmed this analytically for uncensored Weibull survival data.

More general forms of subject level random variation can be achieved by a general linear mixed model form where the impact of selected predictors $w_i = (w_{1i}, \ldots, w_{ri})$ is assumed to vary over subjects, or clusters of subjects. Thus

$$h(t_i|Z_i, w_i) = h_0(t)\exp(Z_i\beta + w_i b_i).$$

When $r = 1$ and $w_i = 1$, the random effect $b_i \sim N(B, 1/\tau_b)$ is used to represent variability in frailties between subjects. If Z_i contains an intercept, the b_i are constrained to have zero mean, namely $b_i \sim N(0, 1/\tau_b)$.

The general linear mixed model form for frailty may take account of spatial locations of subjects, as in geoadditive hazard regression (Henderson et al., 2002; Kneib, 2006); so the b_i could be spatially correlated with local pooling of strength. This may make sense if individuals in neighboring locations are subject to similar environmental risks that affect survival. Heterogeneity in risks may be combined with nonparametric modeling of predictor effects, as illustrated in the BayesX analysis in Example 9.1.

In accelerated failure time models (Section 9.2.3), frailty is conveniently obtained by discrete mixture modeling of the error term. Following Roeder and Wasserman (1997) a mixture of normals provides a flexible model for estimation of densities. Suppose membership of latent subgroups is denoted by a categorical variable G_i with K options, and prior $G_i \sim \text{Mult}(1, [\pi_1, \ldots, \pi_K])$. Assuming a log-normal density for exit times, one may transform observed failure or censoring times as $r_i = \log(t_i)$, and to account for right-censoring define lower sampling limits $L_i = \log(t_i)$ if $d_i = 0$, and $L_i = 0$ if $d_i = 1$ and all failure times are at least 1. Then the discrete mixture adopts varying group intercepts and variances in the survivor function:[6]

$$S(t_i) = 1 - \sum_{k=1}^{K} \pi_k \Phi\left(\frac{\log t_i - \gamma_{0k} - \gamma_1 z_{1i} - \gamma_2 z_{2i} - \cdots}{\sigma_k}\right).$$

Mixed Dirichlet process and Polya Tree priors for the errors u in an AFT regression are used by Kuo and Mallick (1997) and Walker and Mallick (1999).

9.4.1 Cure rate models

A particular form of heterogeneity may arise when permanent survival from an event is possible. Demographic examples are provided by age at first marriage or age at first maternity. The issue is then to identify latent subpopulations in the censored group, namely to distinguish a permanent survival subgroup from a subgroup still liable or susceptible to experience the event but exhibiting extended survival. Not allowing for permanent survival when it can occur will distort the failure time parameter estimates for the true susceptible population. Herring and Ibrahim (2002) point out—in the context of cancer survival—that improved treatment means that a substantial proportion of patients may now be cured, whereas traditional survival analysis, including the Cox (1972) regression model, assume that no patients are cured but that all remain at risk of death or relapse. Similarly in the context of component reliability, Sinha et al. (2003) consider the case where if a unit is free of manufacturing faults, it will never fail in its technological lifetime under usual stress levels.

The most common approach to modeling events with a permanent survival fraction or cure rate assumes the total survival rate is a binary mixture (Ibrahim et al., 2001). One subpopulation has $S_c(t) = 1$ with probability $(1 - \pi)$, and the other (the noncured or suceptible subpopulation) follows a conventional survival pattern in which $S_n(t) \rightarrow 0$ as $t \rightarrow \infty$. So the overall survivor function is

$$S^*(t) = (1 - \pi) + \pi S_n(t),$$

and the overall distribution function (Bruderl and Diekmann, 1995) is

$$F^*(t) = \pi F_n(t).$$

Ibrahim et al. (2001, 157) point out that if covariate effects are modeled via binary regression for π_i then the proportional hazard property no longer obtains.

Let R_i be a partially unobserved binary indicator with $R_i = 1$ if a subject is susceptible. Banerjee and Carlin (2004) and Schmidt and Witte (1989) follow the standard cure rate model and take R_i to be Bernoulli with $Pr(R_i = 1) = \pi_i$ being a propensity to experience the event (e.g., propensity to relapse). For simplicity, omit the subscript n in the survivor function for susceptibles. Then for subjects observed to fail, namely with $d_i = 1$, it necessarily follows that $R_i = 1$, and so the likelihood contribution from such cases is

$$Pr(R_i = 1)f(t_i) = \pi_i f(t_i).$$

Censored subjects may be either susceptibles or nonsusceptibles with likelihood contribution

$$Pr(R_i = 0) + Pr(R_i = 1)Pr(T > t_i) = (1 - \pi_i) + \pi_i S(t_i).$$

The total likelihood contribution[7] is then

$$[\pi_i f(t_i)]^{d_i}[(1 - \pi_i) + \pi_i S(t_i)]^{1-d_i},$$

which reduces to the usual form $f(t_i)^{d_i} S(t_i)^{1-d_i}$ when $R_i = 1$ for all subjects, and so $\pi_i = 1$ (i.e., there is no permanent survivor fraction (PSF)). Any form of binary regression (e.g., logit) may be used for predicting π_i (Schmidt and Witte, 1989). Banerjee and Carlin (2004) carry out a Bayesian analysis with individual level regression in the scale parameter of the failure distribution $f(t)$, but without a regression for the susceptible probability. However, their observations are hierarchical (spatially configured) response times t_{ij} (subjects i within areas j), and they allow spatial variability in the propensities so that $\pi_{ij} = \pi_j$; see also Cooner et al. (2006).

Chen et al. (1999) describe an alternative structure in which there is a latent count of risks C_i, taken to be Poisson with mean θ (for example, tumor cells remaining after treatment that have varying potentials to cause relapse), and unobserved times $U_{i1}, \ldots, U_{iC_i}$ associated with each of these risks. The U_{ir} are assumed to follow the same failure distribution $F(t) = 1 - S(t)$. An observed failure time t_i is the minimum of these times. If $C_i = 0$ then a subject survives permanently from the event being modeled (e.g., a form of cancer). In this case the composite survival function is

$$S^*(t_i) = Pr(C_i = 0) + Pr(U_{i1} > t_i, \ldots, U_{iC_i} > t_i | C_i \geq 1),$$
$$= \exp(-\theta) + \sum_{k=1}^{\infty} S(t)^k \frac{\theta^k}{k!} \exp(-\theta),$$
$$= \exp(-\theta + \theta S(t)) = \exp(-\theta F(t))$$

and the composite hazard rate is

$$h^*(t_i) = \theta f(t_i).$$

An alternative derivation of this model, not tied to the notion of multiple latent risks, is that the cumulative hazard $H(t) = \int_0^t h(u)du$ tends to a finite positive limit θ as $t \to \infty$ (Tsodzikov et al., 2003). Chen et al. (1999) and Ibrahim et al. (2001, 158) mention that the survivor function of the noncured subpopulation can be written as

$$S_n(t_i) = \frac{\exp(-\theta F(t_i)) - \exp(-\theta)}{1 - \exp(-\theta)}$$

so that the composite survival function is in fact also representable as a binary mixture, namely

$$S^*(t_i) = \exp(-\theta) + (1 - \exp(-\theta))S_n(t_i).$$

Chen et al. (1999) introduce covariates into a Poisson regression model for subject-specific θ_i. Consider Weibull distributed times with $F(t_i|Z_i) = 1 - \exp[-\lambda_i t_i^{\kappa}], \lambda_i = \exp(Z_i\beta)$, and $f(t_i|\kappa, Z_i) = \lambda_i \kappa t^{\kappa-1} \exp(-\lambda_i t_i^{\kappa})$. The likelihood[8] when predictors are used to explain both θ_i and λ_i, and with d_i being event status indicators, is then

$$[h^*(t_i)]^{d_i} S^*(t_i) = [\theta_i \lambda_i \kappa t_i^{\kappa-1} \exp(-\lambda_i t_i^{\kappa})]^{d_i} \exp(-\theta_i\{1 - \exp(-\lambda_i t_i^{\kappa})\}).$$

Multiplicative frailty, as in the MPH setup above, can be introduced in cure rate models but identifiability may be weak because susceptibility responses are partially unobserved themselves. Models for frailty in multivariate cure fraction models are considered by Yin (2005). Thus for times t_{ij} observed on subjects i and events j, Yin proposes multiplicative frailty at subject level combined with Poisson regression for θ_{ij} in the cure fractions $\exp(-\theta_{ij})$. One option takes

$$S^*(t_{ij}) = \exp(-\theta_{ij}\gamma_i F(t_{ij})),$$

with hazard rates $h^*(t_{ij}) = \theta_{ij}\gamma_i f(t_{ij})$.

Example 9.3. Age at First Maternity To illustrate frailty modeling in a cure rate model, this example follows Winkelmann and Boes (2005) in analyzing ages at first maternity for 1517 women in the German General Social Survey for 2002. The subsample considered by Winkelmann and Boes involves 1371 women, comprised of (a) 1154 uncensored subjects who may have been over 40 at the time of the survey, but whose age at first maternity (AFM) was under 40 and (b) 217 women aged under 40 in 2002, but who had not yet had a child. Here we consider all 1517 women (including childless women aged over 40).

A log-logistic model with hazard $h(t) = (\lambda\kappa t^{\kappa-1}/[1+\lambda t^{\kappa}])$ and survivor function $S(t) = [1+\lambda t^{\kappa}]^{-1}$ is appropriate to the form of hazard for first maternity, typically peaking between ages 20 and 30. A standard log-logistic is here compared with a log-logistic model with a PSF, modeled according to the latent count approach (Chen et al., 1999). The PSF log-logistic model is then generalized to allow for unmeasured heterogeneity in the age at first maternity. Permanent survivorship in this case is equivalent to a woman never undergoing a maternity, and at population level is essentially equivalent to the rate of childlessness.

Regression effects are included in the scale parameter of the log-logistic hazard via $\lambda_i = \exp(Z_i\beta)$, with θ assumed constant. However, a Poisson regression for θ_i could be included. Predictors Z_i and regression effects β under the standard log-logistic are as in Table 9.2. Predictors are binary apart from number of siblings and education years. The modal age $\chi = [(\kappa-1)\exp(-Z_T\beta)]^{1/\kappa}$ reported in Table 9.2 is based on a predictor vector Z_T for a white subject with 13 years of education, and three siblings.

The standard log-logistic model gives a DIC of 8775 from the second half of a two chain run of 5000 iterations. The significant coefficients show that delayed AFM is associated with greater education, being white and immigrant status. Allowing for a permanently childless subpopulation, but without allowing for frailty, reduces the DIC to 7989 with inferences based on the final third of a two chain run of 7500 iterations. Finally, adding log-normal frailty via

$$\lambda_i = \exp(Z_i\beta + u_i),$$
$$u_i \sim N(0, 1/\tau_u)$$

reduces the DIC to 7877, albeit with a large increase in model dimension ($d_e = 290$). This analysis[9] uses starting values based on an earlier

TABLE 9.2
First maternity: log-logistic model parameters.

	Standard log-logistic			Log-logistic with childless fraction			Log-logistic with childless fraction and frailty		
Predictor	**Mean**	**2.5%**	**97.5%**	**Mean**	**2.5%**	**97.5%**	**Mean**	**2.5%**	**97.5%**
Years of education	−0.203	−0.241	−0.164	−0.275	−0.312	−0.244	−0.315	−0.365	−0.261
Number of siblings	0.019	−0.009	0.051	0.022	−0.009	0.053	0.025	−0.016	0.062
White	−0.607	−0.850	−0.380	−0.921	−1.148	−0.683	−1.086	−1.349	−0.798
Immigrant	−0.325	−0.611	−0.071	−0.577	−0.905	−0.270	−0.661	−1.054	−0.276
Low income at age 16	0.040	−0.202	0.261	0.207	−0.042	0.468	0.259	−0.006	0.539
Living in city at age 16	−0.077	−0.254	0.101	−0.017	−0.207	0.171	−0.015	−0.251	0.200
Modal age (typical individual)	33.0	32.1	34.0	31.8	31.0	32.6	30.8	30.2	31.4
Shape parameter, κ	5.0	4.7	5.3	8.8	8.2	9.4	10.3	9.9	10.9
Proportion childless				0.17	0.15	0.19	0.17	0.15	0.19
Frailty standard deviation							1.036	0.947	1.181

analysis, with inferences based on the second half of a two chain run of 7500 iterations.

Allowing for a childless subpopulation is a form of frailty in itself, and enhances (absolutely) the coefficients on significant predictor effects. Formally including frailty in the modeling of the failure density scale parameter further enhances predictor effects. The low income effect comes close to being "significant" in the log-normal frailty model, in the sense of having a 95% credible interval confined to positive values. The childless fraction (i.e., the permanent survival fraction), $\exp(-\theta)$, is estimated at around 0.17, regardless of the presence or not of frailty. A standard log-logistic model leads to a significantly later modal age than the extended models. In fact, a better representation of the age at first maternity process is provided by the generalized log-logistic of Brüderl and Diekmann (1995), as discussed in Congdon (2008c).

9.5 Discrete Time Hazard Models

In applications with interval censored times, analysis using a discrete time scale becomes appropriate, and in fact such analysis has certain benefits also for modeling time varying or lagged predictor effects (Fahrmeir and Tutz, 2001, 410). Let the time scale be grouped into J intervals $A_1 = [a_0, a_1), \ldots$, $A_J = [a_J, a_{J+1})$ with interval j being $[a_{j-1}, a_j)$, and $a_0 = 0$, $a_{J+1} = \infty$, where a_J denotes either the end of the observation period, or the largest time (censored or failed). The intervals may be of equal length $\delta_j = a_j - a_{j-1}$ but are not necessarily so. Instead of continuous observed failure times, only the discrete times $t_i \in A_j$ are observed. Equivalently let $t_i = j$ denote that a time of failure or censoring is observed within $[a_{j-1}, a_j)$.

With S_j denoting the probability of surviving to the end of interval j, the unconditional probability of failing in interval j is

$$f_j = Pr(t \in [a_{j-1}, a_j)) = S_{j-1} - S_j,$$

and the hazard function (the conditional probability of failing in interval j given survival till the start of the interval) is

$$q_j = Pr(t \in [a_{j-1}, a_j)|t \geq a_{j-1}) = Pr(t = j|t \geq j) = f_j/S_{j-1} = \frac{S_{j-1} - S_j}{S_{j-1}}.$$

Alternatively stated, q_j is the proportion of subjects at risk at the beginning of interval j who experience the event sometime during the interval. The survivor function (the probability of surviving beyond interval j) is obtained as

$$S_j = Pr(t > a_j) = \prod_{k=1}^{j}(1 - q_k) = f_{j+1} + f_{j+2} + \cdots + f_J = S_{j-1}(1 - q_j),$$

though an alternative survivor function $\tilde{S}_j = Pr(t > a_{j-1})$ may be defined as the probability of surviving to the start of interval j (Aitkin et al., 2004, 350; Fahrmeir and Tutz, 2001, 396).

Let $w_{ij} = 1$ if individual i undergoes the event during interval j and w_{ij} otherwise. The likelihood up to interval k for that individual is then (Aitkin et al., 2004, 351),

$$\begin{aligned} f_{ik}^{w_{ik}} S_{ik}^{1-w_{ik}} &= (q_{ik} S_{i,k-1})^{w_{ik}} [S_{i,k-1}(1-q_{ik})]^{1-w_{ik}} \\ &= S_{i,k-1} {q_{ik}}^{w_{ik}} (1-q_{ik})^{1-w_{ik}} \\ &= q_{ik}^{w_{ik}} (1-q_{ik})^{1-w_{ik}} \prod_{j=1}^{k-1} (1-q_{ij})^{(1-w_{ij})} \\ &= \prod_{j=1}^{k} q_{ij}^{w_{ij}} (1-q_{ij})^{(1-w_{ij})}. \end{aligned}$$

This shows that the likelihood involves binary responses $w_{ij} \sim \text{Bern}(q_{ij})$, where the q_{ij} may vary between time intervals, but are assumed constant within them. So the hazard probability becomes

$$q(j|Z_{ij}) = Pr(t = j | t \geq j, Z_{ij}) = F(\alpha_j + Z_{ij}\beta),$$

where F is a suitable distribution function, and α_j models the baseline hazard (Singer and Willetts, 1993). If the predictors include lagged event status indicators $\{w_{i,j-1}, w_{i,j-2}, \text{etc.}\}$ one is led to discrete Markov event histories (e.g., Barmby, 2002); lagged predictor effects may also be used (Fahrmeir and Tutz, 2001, 410).

A benefit of the discrete framework is that the baseline hazard can be modeled via polynomial functions of j (Efron, 1988; Mantel and Hankey, 1978), for example:

$$\alpha_j = \psi_0 + \psi_1(j-1) + \psi_2(j-1)^2 + u_j,$$

where $u_j \sim N(0, 1/\tau_u)$. Parametric time models can also be modeled straightforwardly: a Weibull model is represented in a complementary log–log link by taking the log of the time interval as a covariate (Allison, 1997). Nonparametric models for time (e.g., via splines) can also be applied, or a correlated random effect prior assumed, as in Section 9.2.2. Time varying predictor effects are straightforward to use (Muthen and Masyn, 2005), and nonproportional effects are readily modeled by including interactions between subject attributes Z_{ij} and j.

Commonly used links for the probabilities q_{ij} are the logit, probit, and complementary log–log. For example, a logit link with time varying intercepts and predictor effects (where the vector Z_{ij} excludes a constant term) would mean

$$q(j|Z_{ij}) = \frac{\exp(\alpha_j + Z_{ij}\beta_j)}{1 + \exp(\alpha_j + Z_{ij}\beta_j)}.$$

Adopting a logit link means the log-odds of the event occurring are modeled as functions of predictors and time (i.e., interval). The complementary log–log link model with

$$q(j|Z_{ij}) = 1 - \exp(-\exp(\alpha_j + Z_{ij}\beta_j)),$$

can be derived by assuming an underlying proportional hazard in continuous time, under which

$$S(t_i|Z_i) = \exp\left[-\int_0^{t_i} h(u|Z_i)du\right] = \exp\{-\exp[Z_i\beta + \log H_0(t_i)]\}.$$

Then taking $\alpha_j = \log \int_{a_{j-1}}^{a_j} h_0(t)$ leads to the complementary log–log model, with the same predictor effects as under a PH model (Fahrmeir and Tutz, 2001, 401; Kalbfleisch and Prentice, 1980).

If correlated priors (e.g., random walks) on the α_j and β_j are adopted, the setting of priors on the hyperparameters (e.g., precisions) following the same considerations as discussed above in connection with semi-parametric models for continuous time hazards (Section 9.2.2). Fahrmeir and Knorr-Held (1997) discuss alternative Hastings sampling schemes for collections of time varying coefficients $\{\alpha_j, \beta_{j1}, \ldots, \beta_{jp}\}$ in discrete hazard regression.

As for continuous time survival modeling, neglecting unobserved heterogeneity may mean that the estimated baseline hazard parameters are biased downwards, the impact of constant covariates is underestimated, or that spurious time-dependent effects for observed predictors are obtained. For improved identification, frailties may be included at subject level, rather than at subject-interval level, though bilinear schemes are possible. Thus a log-normal frailty might specify

$$q_{ij} = F(\alpha_j + Z_{ij}\beta + b_i),$$

where $b_i \sim N(0, \sigma_b^2)$. Alternatively a bilinear scheme might be used

$$q_{ij} = F(\alpha_j + Z_{ij}\beta + \delta_j b_i),$$

where one of the δ_j is set to a fixed value for identification if the variance of b_i is unknown. Muthen and Masyn (2005) use a discrete mixture approach in which $G_i \in (1, \ldots, K)$ are latent groups (e.g., developmental trajectories in educational applications). Then

$$F^{-1}(q_{ij}) = \alpha_{j,G_i} + Z_{ij}\beta_{G_i} + \delta_{j,G_i} b_i,$$

where the probability that $G_i = k$ is defined by predictors U_i in a separate multiple logit regression. The factor scores b_i may be defined by $b_i \sim N(0, 1/\tau_b)$, or by a hierarchical linear regression on the predictors U_i.

9.5.1 Life tables

Life tables are a particular way of analyzing discrete time survival data. They may be applied to situations where permanent survival or withdrawal is possible, such as marital status life tables (Schoen and Weinick, 1993), or to population mortality. The intervals in such applications refer to age or duration bands and discretization may extend beyond that present in the data, as in abridged life tables (Kostaki and Panousis, 2001). The intervals are not necessarily of equal length (Wong, 1977). For example, in one common scheme for human life tables, ages under 1 form the first interval, ages 1–4 comprise the second interval, ages 5–9 interval 3, and so on for successive five-year bands, with the final interval typically open ended, such as ages over 90. Often human life tables are estimated from population deaths data over a specified calendar period, to provide "period" life tables, based on current mortality in individuals born in different periods, as distinct from cohort life tables, based on follow-up studies of mortality in a group of individuals born in the same time period (Richards and Barry, 1998).

Following life table conventions, ages are denoted x and age intervals are denoted $[x, x+n)$, e.g., $n = 5$ if intervals are five years in length. Let t denote a random variable for the total lifetime (age of death) of an individual. Also in line with life table conventions, the probability $Pr(T > x)$ that the age of death T is x or higher (the survivor function) is denoted $l(x)$. The hazard rate—also called the force of mortality in life table applications—is then

$$h(x) = \lim_{\Delta x \to \infty} \frac{l(x) - l(x + \Delta x)}{l(x)\Delta x} = \frac{-l'(x)}{l(x)},$$

with solution

$$l(x) = l(0) \exp\left[-\int_0^x h(u)du\right].$$

With $l(0) = 1$, the density of the age at death is $f(x) = h(x)l(x)$. The probability of surviving from age x to age $x+n$, given survival to x, namely $Pr(t > x + n|t > x)$, is denoted ${}_np_x$ with

$${}_np_x = l(x+n)/l(x) = \frac{\exp[-\int_0^{x+n} h(u)du]}{\exp[-\int_0^x h(u)du]} = \exp\left[-\int_0^n h(x+u)du\right]$$

while the probability of dying before age $x+n$ conditional on reaching age x is

$${}_nq_x = 1 - {}_np_x = 1 - l(x+n)/l(x) = \frac{l(x) - l(x+n)}{l(x)}.$$

Important in linking these functions to estimable quantities is the central rate of mortality, which represents a weighted average of the force of mortality applying over the interval $[x, x+n)$. Let $P(x)$ denote the population of age x. Then the death rate for age interval $[x, x+n)$ is

$${}_nM_x = \int_x^{x+n} h(a)P(a)da \Big/ \int_x^{x+n} P(a)da.$$

Assuming linearity of $l(a)$ in the interval from x to $x + n$, this can be simplified (Namboodiri and Suchindran, 1987, 36) to

$$_nM_x = \frac{l(x) - l(x+n)}{0.5n[l(x) + l(x+n)]}.$$

Hence the survivor probability can be written

$$\frac{l(x+n)}{l(x)} = {_np_x} = \frac{1 - 0.5n(_nM_x)}{1 + 0.5n(_nM_x)}$$

giving

$$_nq_x = \frac{n(_nM_x)}{1 + 0.5n(_nM_x)}.$$

To clarify the operations involved, life tables involve hypothetical populations of initial size $l_0 = 100{,}000$ (the radix) with l_x denoting numbers still alive at age x from the initial population. The number dying between age x and $x + n$ is denoted $_nd_x = l_x - l_{x+n}$ and from above

$$_nq_x = 1 - l_{x+n}/l_x = \frac{l_x - l_{x+n}}{l_x} = \frac{_nd_x}{l_x}.$$

To develop the life table from observed deaths and populations requires an estimator for the probability $_nq_x$. Let D_x denote observed deaths for age band $[x, x+n)$ over a certain period, P_x denote observed mid-period populations at risk (or person-years) and M_x denote age specific death rates. One estimator of probability of dying in interval $[x, x + n)$ conditional on being alive at the start of the interval is then (Chiang, 1984)

$$_nq_x = \frac{n_nM_x}{1 + n(1 - {_na_x})_nM_x},$$

where $_na_x$ is the fraction of the interval lived by those dying during it. For most age groups $_na_x$ is taken as a half but for infants (ages under 1) can be taken as 0.1, and for the 1–4 age group as 0.4.

Under conventional life table methods that are usually applied to large populations, the M_x are treated as unrelated fixed effects and estimated by assuming binomial sampling $D_x \sim \text{Bin}(P_x, M_x)$ or Poisson sampling $D_x \sim Po(M_xP_x)$. In a Bayesian version of the fixed effect approach, the M_x would be assigned diffuse beta or gamma priors with known hyperparameters, e.g., $M_x \sim \text{Beta}(1, 1)$. Overdispersed versions of binomial or Poisson densities may also be used, namely hierarchical schemes for "borrowing strength" over correlated mortality rates, with a higher stage density for the M_x involving unknown hyperparameters. An example might be when age-specific deaths D_{ix} for a set of areas or hospitals $(i = 1, \ldots, I)$ are to be analyzed and populations at risk are relatively small. Then the conjugate binomial-beta approach would mean taking death rates M_{ix} to be distributed according a hierarchical model, namely

$$D_{ix} \sim \text{Bin}(P_{ix}, M_{ix}),$$
$$M_{ix} \sim \text{Beta}(a, b),$$

where $\{a, b\}$ are unknown parameters. Congdon (2009) adopts a general linear mixed model approach for data involving an additional stratifying group g in which

$$D_{ixg} \sim \text{Bin}(P_{ixg}, M_{ixg}),$$

and a logistic regression with group-specific autoregressive area and age effects has the form:

$$\text{logit}(M_{ixg}) = \alpha_g + s_{ig} + h_{xg}.$$

Other options might be to model the impact of age by a parametric function; for example, Neves and Migon (2007) use Makeham's Law, by which

$$D_x \backsim Po(M_x P_x),$$
$$M_x = \alpha + \beta\delta^x$$

and extend this to a time series model for age-specific death rates, namely

$$M_{xt} = \alpha_t + \beta_t \delta_t^x.$$

Example 9.4. Cancer Survival This example illustrates discrete survival with potential heterogeneity (frailty). It involves survival times in months for 48 participants in a cancer drug trial. Of the 48 patients, 28 receive an experimental drug treatment (drug = 1) and 20 receive a control treatment (drug = 0). The other predictor is patient age at the start of the trial, ranging from 47 to 67 years. The observed times provide the month of death or the last month the patient was known to be alive.

With a complementary log–log link, Weibull time dependence (model 1) is compared with a semi-parametric baseline hazard modeled via a first-order random walk (model 2).[10] Convergence in the latter is assisted by omitting an intercept β_0, and instead using the parameterization $\alpha_j = \beta_0 + a_j$ where a_j are random walk effects with precision $\tau_a = 1/\sigma_a^2$ centered to have mean zero at each MCMC iteration. For numeric stability age values are divided by 100.

A two chain run of 100 thousand iterations on the semi-parametric model gives a DIC of 234.3 with effective parameter count $d_e = 3.9$. The low d_e despite 39 random effect parameters suggests relatively little fluctuation about the central value of β_0 which has posterior mean -7.3. The impression is confirmed by plot of the α_j showing virtual linearity (Figure 9.5 with 80% credible intervals), and by the density plot of the standard deviation σ_a of the a_j which has a spike at zero. A Weibull model has a lower DIC (namely 230.8), but in fact the Weibull parameter has a 95% interval straddling one: its mean

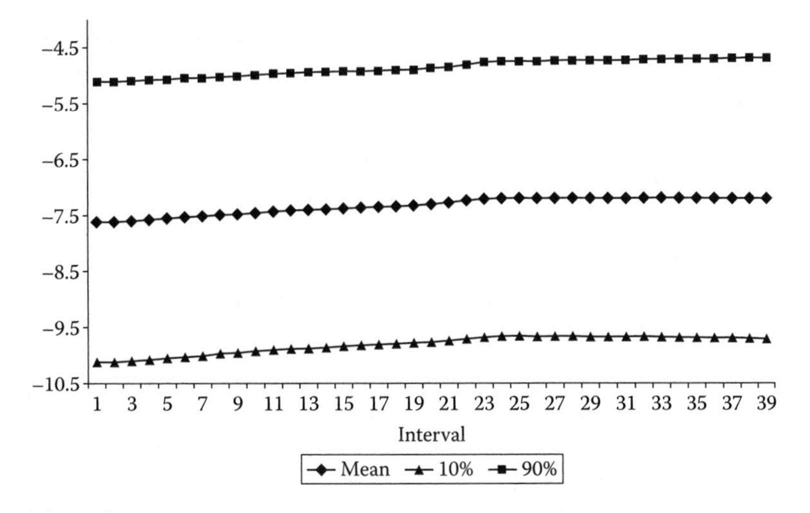

FIGURE 9.5
Varying intercept, Cancer survival data.

(95% interval) is 0.59 (0.15, 1.12) while the treatment and age effects are −2.2 (−3, −1.4) and 11.5 (3.6, 18.5). Mortality declining with time is possibly an unexpected effect, and might be due to unmodeled heterogeneity.

A lognormal frailty effect at subject level is then added to the Weibull model, so that

$$q(j|Z_{ij}) = 1 - \exp(-\exp(\beta_0 + \text{Drug}_i\beta_1 + \text{Age}_i\beta_2 + \kappa\log(j) + \sigma_b b_i)),$$

where $b_i \sim N(0, 1)$ and σ_b is assigned a uniform prior with maximum 10. Convergence in a two chain run is obtained after around 6000 iterations, and basing inferences on a subsequent 6000 leads to a κ coefficient with mean 3.3 and 95% interval from 0.8 to 7.6. There are much amplified treatment and age effects, namely −6.4 (−13, −2.4) and 33.5 (12.6, 57). These large shifts in parameter values may partly reflect the small sample, but point to the way inferences may be affected by neglecting frailty. The DIC falls to 196, with $d_e = 26$.

9.6 Dependent Survival Times: Multivariate and Nested Survival Times

Multivariate and nested survival data can occur in a number of different ways; for discussions, see Hougaard (1987) and Sinha and Ghosh (2005) for a Bayesian perspective. Examples are when each subject may experience repetitions of the same event; when subjects may experience more than one event; when times are for subjects arranged in clusters; or in competing risks situations (considered in Section 9.7). For example, bivariate survival models can be used to analyze:

- survival data on twins or other types of matched pair (Anderson et al., 1992);
- reliability data when the lifetime of one component is related to the lifetimes of other components;
- failure times of paired human organs (Sahu and Dey, 2000; Tosch and Holmes, 1980).

Examples of grouped or clustered data are provided by Gustafson (1997) as when several response times are measured for a single patient in a clinical trial, or when responses are for patients categorized according to clinic of treatment. Multivariate perspectives on more specialized survival models are exemplified by Bayesian multivariate cure rate models (Chen et al., 2002; Yin, 2005), and multivariate counting processes (Sinha and Ghosh, 2005).

The statistical model applied to such data needs to account for the intra-cluster or inter-event correlation. It may be possible to model the dependence structure directly, for example, via multivariate versions of widely adopted parametric survival models (Yashin et al., 2001). Thus Sahu and Dey (2000) consider bivariate exponential and Weibull survival models for data on times to visual impairment for paired eyes, while Damien and Muller (1998) provide a Bayesian treatment of a bivariate Gumbel model. The multivariate log-normal is another possibility, which adapts to the situation of conditional multivariate data, when durations on a second event are obtained conditional on the duration in a first event (Henderson and Prince, 2000).

The other approach is to introduce random frailty terms at the cluster level or common frailties across events. The frailty term represents common influences across clusters or events that are neglected or not observed, and responses on members of a cluster (or on correlated events) are typically assumed independent given the value of the cluster effect (or shared frailty factor). Sahu and Dey (2004, 325) describe how different frailty assumptions lead to different correlations between log survival times in a bivariate situation (under the assumption a Weibull baseline hazard). Let t_{ij} be the failure time for the jth component or outcome ($j = 1, \ldots, m_i$) of the ith subject ($i = 1, \ldots, n$). Then the hazard function assuming a common multiplicative frailty takes the form (Sahu et al., 1997; Yin and Ibrahim, 2005):

$$h(t_{ij}|Z_{ij}, \gamma_i) = \gamma_i h_0(t_{ij}) \exp(Z_{ij}\beta_j)$$

with the unit frailty effect γ_i distributed independently of Z_{ij} and t_{ij}. If γ_i is high then all hazards are raised, and so times t_{ij} tend to be low; if γ_i is low then all hazards are lowered and the t_{ij} tend to be relatively extended. In this way the common frailty induces a positive association between observed times. In the case of repeated occurrences $r = 1, \ldots, R_i$ of the same outcome to the same subject (e.g., multiple occupation shifts or repeat cardiac events), the hazard function conditional on γ_i is independent of the number r of previous occurrences (Sinha, 1993). Unconditionally, however, the hazard for the $(r+1)$th occurrence is

$$h_r(t_{ir}|Z_i) = h_0(t_{ir})\exp(Z_i\beta)(1 + r\text{Var}(\gamma_i)).$$

The same scenario applies when subjects i are nested within clusters j, with cluster effects γ_j shared between the n_j individuals in the same cluster:

$$h(t_{ij}|Z_{ij}, \gamma_j) = \gamma_j h_0(t_{ij})\exp(Z_{ij}\beta_j), \quad i = 1, \ldots, n_j; j = 1, \ldots, J.$$

If the γ_j are assumed gamma distributed $\gamma_j \sim \text{Ga}(h, h)$ with variance $1/h$ then smaller values of h signify a closer relationship between subjects in the same group and greater heterogeneity between the groups. For models including cure rates, Yin (2005) proposes mutiplicative frailty at cluster level combined with Poisson regression for θ_{ij} in the cure fractions $\exp(-\theta_{ij})$. One option takes

$$S^*(t_{ij}) = \exp(-\theta_{ij}\gamma_j F(t_{ij})),$$

with hazard rates $h^*(t_{ij}) = \theta_{ij}\gamma_j f(t_{ij})$.

Survival time data are often highly skewed and this may affect the appropriate form of frailty. Frailty models allowing for fat tails and skewness are obtained under the skew log-normal or skew log-t common frailty approach of Sahu and Dey (2004); in practice they consider only positively skewed frailty. Consider a parametric hazard (e.g., Weibull) for multiple event time data (subjects $i = 1, \ldots, n$ and events $j = 1, \ldots, m$) and with subject level scale parameters λ_{ij} for event j, namely

$$h(t_{ij}|\lambda_{ij}, \kappa) = \lambda_{ij}\kappa_j t_{ij}^{\kappa_j - 1}.$$

Then a skew log-normal frailty model implies

$$\log(\lambda_{ij}) = Z_i\beta_j + b_i + \delta u_i,$$

where $b_i \sim N(0, \sigma_b^2), \delta$ is positive, and $u_i \sim N(0,1)I(0,)$ with u_i independent of b_i. Under the skew log-t model,

$$b_i \sim t(0, \nu, \sigma_b^2),$$

where ν is a degrees of freedom parameter and $u_i \sim N(0,1)I(0,)$.

In practice, this kind of model may need informative priors for stable identification, bearing in mind that censoring reduces identifiability of complex random effect models, that b_i and v_i are to some extent overlapping in their roles and σ_b^2 and δ^2 are confounded in $\text{var}(b_i + \delta u_i) = \sigma_b^2 + \delta^2$. To illustrate relevant strategies for priors on the variance components, uncensored bivariate times $(n = 100, m = 2)$ are generated with Weibull hazards,[11] and scales

$\lambda_{ij} = \exp(\beta_{1j} + \beta_{2j}x_i + b_i + \delta u_i)$ where the x_i are standard normal, with $\beta_1 = (-5, -6), \beta_2 = (0.5, 1)$, $\kappa = (1.5, 2)$, $\delta = 0.5$, and $1/\sigma_b^2 = \tau_b = 2$, so that $\sigma_b \simeq 0.7$. A $U(0, 5)$ prior on δ is adopted in the analysis to re-estimate the parameters (cf. Sahu and Dey, 2004). The re-estimated parameters lead to considerable under-estimation of σ_b with the second half of a single chain run of 10,000 iterations leading to posterior mean of 0.054, whereas the posterior mean of δ is 1.6. Assuming instead a $U(0, 100)$ prior on $V = \delta^2 + \sigma_b^2$, and a $U(0, 1)$ prior on the ratio $\delta^2/(\delta^2 + \sigma_b^2)$, improves the estimation of σ_b with posterior mean 0.55, while the posterior mean of δ is now 1.17.

As for univariate models, considerable generality is obtained by adopting a semi-parametric hazard while allowing also for common frailty. An example involves a semi-parametric counting process including multiplicative frailty for repeated occurrences of the same event (Sinha, 1993). The semi-parametric hazard is based on $J-1$ intervals $A_j = [a_{j-1}, a_j)$ obtained by considering distinct failure times, with a_J equal to the maximum time (censored or failed). Thus for subject-occurrence index i, subject s, and interval j define an intensity function:

$$\lambda(t_{ij}|Z_i) = Y(t_{ij})b_0(t_{ij})\exp(Z_{ij}\beta)\gamma_s,$$

where $b_0(t)$ is the baseline intensity function, and the γ_s represent subject level frailty. The integrated baseline intensity $B_0(t) = \int_0^t b_0(u)du$ is assumed to follow an independent increments gamma process, namely

$$dB_0(t) \sim \text{Ga}(cdB_0^*(t), c),$$

where $B_0^*(t)$ is an assumed mean intensity. The likelihood kernel for each spell within each subject is Poisson[12] in form with response variables $dN_{ij} = 1$ or 0, and means $dB_0(t_{ij})\exp(Z_{ij}\beta)\gamma_s$.

Example 9.5. Clustered Trial of Infection Treatment This example involves two forms of nesting: repetitions of a single event within patients and multi-level nesting of patients within hopitals. The data are from Fleming and Harrington (1991) and Yau (2001), and concern a randomized trial of gamma interferon in treating infections among patients with chronic granulomatous disease. The 126 patients were nested in 13 hospitals and patients may experience more than one infection. Of 63 patients in the treatment group, 14 had at least one infection and 20 infections were recorded in all, whereas in the placebo group, 30 patients had at least one infection and there were 56 infections in all.

The $n = 201$ observations are therefore at three levels: infections at level 1, patients at level 2, and hospitals as level 3 units. Let t_{klm} be times between recurrent infections, with k denoting events within patients, l denoting patients ($l = 1, \ldots, L$), and m denoting hospitals ($m = 1, \ldots, M$). The analysis seeks to assess the effect of gamma interferon in reducing the rate of infection as well as taking account of the clustering in the data; ignoring such

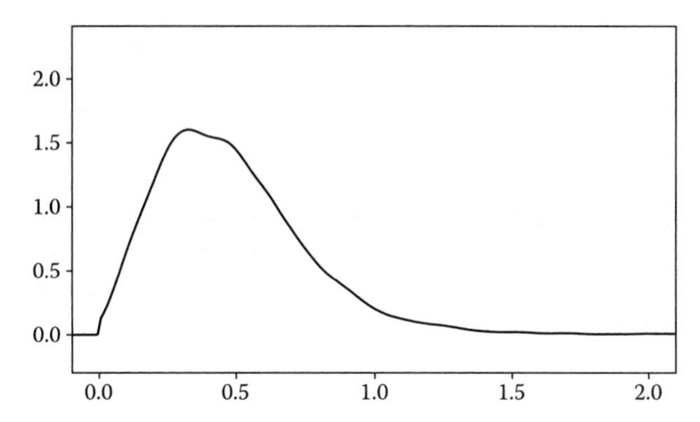

FIGURE 9.6
Posterior density of sigma.u.

clustering may affect the estimated treatment effect. A piecewise exponential baseline hazard[13] is assumed with $J = 20$ intervals $(a_{j-1}, a_j]$ based on 5th percentiles of the observed times, with a_J being the maximum time of 389. Then with a single predictor ($z_{lm} = 1$ for treated subjects, 0 otherwise)

$$h(t_{\text{klm}} \in (a_{j-1}, a_j]|z_{lm}) = \lambda_j \exp(z_{lm}\beta + e_{lm} + u_m)$$

with a gamma process prior on the λ_j, and normally distributed patient and hospital effects $e_{lm} \sim N(0, \sigma_e^2)$, and $u_m \sim N(0, \sigma_u^2)$. Since the two sources of variation are confounded, a uniform prior $V \sim U(0, 100)$ is adopted on the total variance $V = \sigma_u^2 + \sigma_e^2$, and a $U(0, 1)$ prior on the ratio $\sigma_u^2/(\sigma_u^2 + \sigma_e^2)$.

A two chain run shows convergence in the variance parameters after around 3500 iterations. Although Yau (2001) reported no significant hospital variation, here posterior means for σ_e and σ_u of 1.01 and 0.49 are obtained (using iterations 5000 to 10,000). The posterior density for σ_u (Figure 9.6) has no spike at zero but rather is bounded away from zero (cf. MacNab et al., 2004, 13). The treatment effect is estimated as -1.22 ($sd = 0.37$) compared to the estimate of -1.07 reported by Yau (2001). Centered hospital effects have posterior means varying from 0.43 (hospital 2) to -0.26 (hospital 10).

Example 9.6. Bivariate Survival The Diabetic Retinopathy Study was conducted by the National Eye Institute to assess the effect of laser photocoagulation in delaying onset of severe visual loss in patients with diabetic retinopathy. One eye of each patient was randomly selected for photocoagulation and the other was observed without treatment, with patients followed up over several years for the occurence of blindness in one or other eye. The follow-up time is in months, with 80 patients censored on both eyes at the same time, while 36 patients have onset in both eyes. Censoring is caused by dropout, death, or termination of the study. Following Huster et al. (1989), a subset of the dataset containing $n = 197$ high-risk patients is considered here, so there are $i = 1, \ldots, n$ patients and $j = 1, \ldots, m$ events with $m = 2$. The correlation between pairs of uncensored observations for patients in both treatment and control groups is 0.28, indicating possible dependence between the two times.

The Weibull might be appropriate for these data, as a plot of the transformed Kaplan–Meier survivor function namely $\log[-\log(S_{\text{KM}}(t))]$ on $\log(t)$, is approximately linear when either t_{i1} or t_{i2} are considered.[14] Alternative analyzes consider Weibull survival with and without log-normal frailty at patient level. Under the latter

$$\begin{aligned}
t_{ij} &\sim \text{Wei}(\kappa_j, \lambda_{ij}),\\
\log(\lambda_{ij}) &= \beta_0 + \beta_1 \text{Age}_i/10 + \beta_2 \text{Trt}_{ij} + \beta_3 \text{Type}_i + b_i,\\
b_i &\sim N(0, \sigma_b^2),
\end{aligned}$$

where Age is age at diagnosis and Type relates to diabetes type. A model without frailty produces significant Weibull shape effects—both shape parameters κ_j have 95% intervals below 1, suggesting a lesser chance of impairment at longer follow-ups. However, predictor effects, including treatment, are not significant. The DIC is 1684, and log(PsML), based on Monte Carlo estimates of log(CPO) statistics, is −1460. The worst fitted observation is eye 2 for patient 68. The hazard ratio $\theta = e^{-\beta_2}$ for untreated eyes averages 2.1, but has a 95% interval straddling 1.

In the fraility model, a $U(0, 10)$ prior on the standard deviation σ_b of the effects is adopted, and this model suggests the time effect was spurious (the 95% intervals for the shape parameters now straddle 1). The DIC improves to 1616 ($d_e = 88$) and the log(psML) rises to −1422, in line with significant unobserved heterogeneity. The treatment effect increases, with θ now averaging 2.4, but remains nonsignificant. A histogram and normal Q–Q plots of the posterior mean b_i show a subgroup with high negative values (see Figure 9.7)

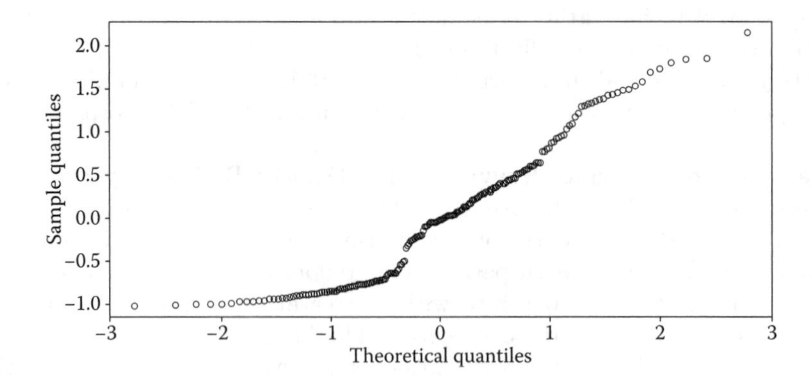

FIGURE 9.7
Normal Q–Q plot of $b[i]$.

suggesting that a discrete mixture approach (e.g., a two group normal) to frailty might be appropriate.

9.7 Competing Risks

Competing risks (CR) models involve the tracking of multiple durations corresponding to different types of exit or transition. With nonrepeatable events subjects are observed until the first exit and completion of one of the multiple durations, but for repeatable events (e.g., occupational or migration histories) event histories might include repeated transitions between different job or residential destinations. Assume that there are K possible mutually exclusive causes of exit or K possible outcomes that a subject is at risk of. Let C_i be a subject level categorical random variable with K possible levels. Under the latent failure time approach (Box-Steffensmeier and Jones, 2004; Crowder, 2001; Gelfand et al., 2000; Kozumi, 2004) with independent risks, there is a latent failure time T_{ik} corresponding to each outcome, but only the minimum time is observed when individual i exits for cause k_i, so that $t_i = \min(T_{i1}, \ldots, T_{ik})$ with $k_i = \text{argmin}(T_{i1}, \ldots, T_{ik})$. The remaining times are censored. All times are censored if an individual does not exit for any of the K possible reasons.

With these assumptions and conditioning on possibly cause-specific predictors Z_k, the hazard rate may be expressed as a sum of cause-specific hazards, $h(t|Z_k) = \sum_{k=1}^{K} h_k(t|Z_k)$, where

$$h_k(t|Z_k) = \lim_{\Delta t \to 0} \frac{Pr(t < T \leq t + \Delta t, C = k|T > t, Z_k)}{\Delta t}.$$

The survival function may be decomposed into K marginal survival functions, with

$$S(t|Z) = \prod_{k=1}^{K} S_k(t|Z_k).$$

Assuming a failure to risk C_i is observed, the contribution of the ith subject to the likelihood has the form:

$$f_{C_i}(t_i|Z_{iC_i}) \prod_{l \neq C_i}^{K} S_l(t_i|Z_{il}) = h_{C_i}(t|Z_{iC_i}) \prod_{l=1}^{K} S_l(t_i|Z_{il}),$$

while for a subject censored on all risks the contribution is $\prod_{l=1}^{K} S_l(t_i|Z_{il})$. With event indicators $d_{ik} = 1$ if $C_i = k$, and $d_{ir} = 0$ for $r \neq k$, the likelihood contribution is equivalently

$$\prod_{r=1}^{K} [f_r(t_i|Z_{ir})]^{d_{ir}} [S_r(t_i|Z_{ir})]^{1-d_{ir}}.$$

For continous survival times, one may assume parametric forms for the time effect, e.g., a Weibull hazard

$$h_k(t) = \lambda_k \kappa_k t^{\kappa_k - 1},$$

or model risk-specific semi-parametric hazard sequences that may be correlated over causes. Possible label switching problems under the latent failure approach may require parameter constraints, such as ordering the shape parameters κ_k (Gelfand et al., 2000).

Often competing risk models are applied to repeated transitions between occupational, residential, or marital states. The hazard rate then generalizes to reflect moves between the mth observed state and the $(m+1)$th state. If T_{im} denotes the time spent in the mth state, and occupancy of the mth state for subject i is denoted $C_{im} = k$, then

$$h_{kl}(t|Z_{ik}) = \lim_{\Delta t \to 0} Pr(t < T_{im} \leq t + \Delta t, C_{i,m+1} = l | T_{im} > t, C_{im} = k, Z_{ik})/\Delta t,$$

is the instantaneous risk of moving from state k to state l (with $l \neq k$), given survival in the mth state until t. Under independent risks, the overall hazard for leaving state k is then

$$h_k(t|Z_{ik}) = \sum_{l \neq k}^{K} h_{kl}(t|Z_{ik}).$$

For discrete time data the functions described in Section 9.5 similarly generalize to the competing risk case. For nonrepeated events, intervals $[a_{j-1}, a_j)$ for $j = 1, \ldots, J+1$, and $C_i \in (1, \ldots, K)$:

$$f_{jk} = Pr(t \in [a_{j-1}, a_j), C = k),$$

with risk-specific hazard function:

$$\begin{aligned} q_{jk} &= Pr(t \in [a_{j-1}, a_j), C = k | t > a_{j-1}), \\ &= f_{jk}/S_{j-1}, \end{aligned}$$

and survivor function obtained as

$$S_j = \prod_{m=1}^{j} \prod_{h=1}^{K} (1 - q_{mh}).$$

Define event indicators $d_{imh} = 1$ when risk h occurs in interval m, and 0 otherwise. Then for subject i undergoing the kth risk in the jth interval, the event indicators are $d_{ijk} = 1, \{d_{ijh} = 0, h \neq k\}$ and $d_{i1h} = d_{i2h} = \cdots d_{i,j-1,h} = 0$ for all h, with likelihood

$$q_{ijk} \left[\prod_{m=1}^{j-1} \prod_{h=1}^{K} (1 - q_{imh}) \right] = q_{ijk} S_{i,j-1}.$$

The response at each interval is multinomial, and to model the impact of predictors different links may be used such as the multiple logit, or multiple probit. Consider a multiple logit link with $K + 1$ categories (K alternate risks plus an extra category for survival, denoted by $C_i = 0$). Let the first category be for survival, and define regression coefficients β_k for the kth risk. Then identification is obtained if one category is taken as a reference category. Using the survivor category as reference, and assuming the β_r do not contain an intercept, would lead to

$$q(t \in [a_{j-1}, a_j), C_i = 0) = \frac{1}{1 + \sum_{r=1}^{K} \exp(\alpha_{jr} + Z_{ir}\beta_r)},$$

$$q(t \in [a_{j-1}, a_j), C_i = h|Z_{ih}) = \frac{\exp(\alpha_{jh} + Z_{ih}\beta_h)}{1 + \sum_{r=1}^{K} \exp(\alpha_{jr} + Z_{ir}\beta_r)}, \qquad h = 1, \ldots, K,$$

where the parameters α_{jh} describe the baseline hazard for risk. K-dimensional versions of the correlated prior processes discussed in Section 9.3 may be used for the α_{jh}, for example, multivariate normal first or second-order random walks.

9.7.1 Modeling frailty

Assuming independent risks one may introduce unobserved frailties γ_{ik} that impact on each risk but are uncorrelated across risks, such as independent gamma densities with mean 1 for each possible cause. Under proportionality, the risk-specific hazard in a continuous time CR hazard is then

$$h_k(t_i|Z_{ik}) = \gamma_{ik} h_{0k}(t_i) \exp(Z_{ik}\beta_k).$$

The assumption of independent risks may not hold in practice because particular groups of subjects may be more likely to experience subsets of the events. Just as it may be unrealistic in multinomial discrete choice situations to assume independence of irrelevant alternatives (i.e., that ratios of choice probabilities of any two alternatives are unaffected by changes in utilities of any other alternatives, or by their removal), so it may be unrealistic in survival analysis that the relative risks of two outcomes will be unaffected by the removal of a third (Gordon, 2002).

To allow for dependent competing risks, especially for multiple spell data, one may assume correlated or dependent frailties. In a generalization of the MPH scheme, Abbring and van den Berg (2003) mention that the joint distribution of $(T_{i1}, \ldots, T_{ik})$ given predictors Z_{ik} and correlated frailties $(\gamma_{i1}, \ldots, \gamma_{ik})$ factorizes into independent densities $f(T_k|Z_{ik}, \{\gamma_{i1}, \ldots, \gamma_{ik}\})$ which are fully characterized by cause-specific hazard rates

$$h(T_k|Z, \{\gamma_1, \ldots, \gamma_K\}) = \gamma_k \lambda_k(t) \exp(Z_k\beta_k).$$

Correlated frailties are also obtained by expanding the regression term to a general mixed form, as in Section 9.4, so that in a continuous time analysis,

$$h_k(t_i|b_{ik}, Z_{ik}) = \lambda_k(t_i) \exp(\beta_k Z_{ik} + b_{ik}),$$

where b_{ik} are zero mean effects that might be multivariate normal, discrete mixtures of multivariate normal, etc. Assuming a multivariate normal with covariance matrix Σ_b, dependent risks will be apparent in significant off-diagonal terms. Whether there are significant correlations in the frailty effects over different risks will depend in part on whether observed predictors successfully explain variations in event proneness. Another possibility is a common frailty model with risk-specific loadings, so that

$$b_{ik} = \lambda_k b_i,$$

where $\lambda_k > 0$ and $b_i \sim N(0,1)$ for identification.

Example 9.7. Political Careers This example consider data on career paths in U.S. House of Representative incumbents (Box-Steffensmeier and Jones, 2004, 169) and adopts a discrete time approach to yearly data. The data relate to all new incumbents during 1950–1976, and a competing risk analysis is undertaken to possibly repeated events for each incumbent. There are $K = 4$ types of career exit: these are $k = 1$ for loss in general election, $k = 2$ for loss in primary election, $k = 3$ for retirement, and $k = 4$ for moves to seek alternative office. The predictors are

Party: Republican (1), Democrat (0);

Redistrict: whether or not the incumbent's district was substantially redistricted;

Scandal: whether or not the incumbent was involved in scandal;

OpenGub: whether or not an open gubernatorial seat was available during election cycle;

OpenSen: whether or not an open U.S. Senatorial seat was available during election cycle;

Leadership: whether or not the incumbent had a leadership position in the House;

Age (divided by 10);

Prior Margin: % of votes the incumbent (or party) received in his/her previous election (divided by 10).

Baseline hazard time effects are assumed to be quadratic in time with coefficients differing across causes. Following Fukumuto (2005), the predictors Z_{ik} are cause specific: the full set of predictors is used only for loss in general election; for loss in primary election, the predictors (apart from time) are party and age; for retirement, they are party and redistrict; and for alternative office they are party, OpenGub, OpenSen, and Age.

The data are arranged so that each year's observation for each incumbent is a separate line, with the time variable incrementing until an exit occurs; there are 5429 year-person records for 997 subjects. In a year when no type of exit occurs, the categorical response C_i is 0. An independent competing risk model (model 1) with fixed effects only is applied using a multiple logit link with $K+1$ possible outcomes, including no exit as the reference. This is compared to a model (model 2) including multivariate normal frailty at subject level.[15] So for years $j = 1, \ldots, L_i$ for subject i, and with $\alpha_{jk} = \alpha_0 k + \alpha_{1k} j + \alpha_{2k} j^2$ as the baseline hazard,

$$q(t \in [a_{j-1}, a_j), C_i = 0) = \frac{1}{1 + \sum_{h=1}^{K} \exp(\alpha_{jh} + Z_{ih}\beta_h + b_{ih})},$$

$$q(t \in [a_{j-1}, a_j), C_i = k|Z_{ik}) = \frac{\exp(\alpha_{jk} + Z_{ik}\beta_k + b_{ik})}{1 + \sum_{h=1}^{K} \exp(\alpha jh + Z_{ih}\beta_h + b_{ih})},$$
$$k = 1, \ldots, K,$$

where $(b_{i1}, b_{i2}, \ldots, b_{ik}) \sim N(0, \Sigma_b)$.

A Wishart prior with identity scale matrix for Σ_b^{-1} is adopted for Σ_b. Inferences are from the second half of a two chain run of 10,000 iterations. The variances (diagonal terms in Σ_b) are estimated as 0.16 (0.12, 0.22), 0.70 (0.48, 0.90), 0.89 (0.73, 1.11), and 0.96 (0.80, 1.26). There is negative correlation between risk 1 and risks 3 and 4; risk 2 is also negatively correlated with the third and fourth risks. So there is departure from independent risks. The DIC falls from 6340 ($d_e = 25$) for model 1 to 6160 ($d_e = 236$) for model 2.

Appendix: Computational Notes

1. In WinBUGS one may transform observed failure or censoring times as $r_i = \log(t_i)$, and to account for right-censoring define lower sampling limits $L_i = \log(t_i)$ if $d_i = 0$, and $L_i = 0$ if $d_i = 1$ and all failure times are at least 1. The lower limit L_i may be set at a negative value rather than zero if some failure times are between 0 and 1. Then a log-normal AFT regression involves truncated sampling of the r_i as in the WinBUGS code

```
model { for (i in 1:n) {r[i] ~dnorm(mu[i],invsig2) I(L[i],)
mu[i] <- gam0+gam[1]*z[1,i]+··· }}
```

while a log-logistic would use a logistic rather than normal model. The log-normal or log-logistic AFT regression can be extended to allow finite mixtures of survival densities, and hence model frailty (Section 9.4).

2. The WinBUGS code for the Weibull model of nursing home mortality (Example 9.1) includes code to derive the log-likelihoods. Thus

```
model {for (i in 1 : 1601) {t[i] ~dweib(kap, mu[i])I(t.cen[i],)
L[i] <- pow(f[i],1-cens[i])*pow(S[i],cens[i]); LL[i] <- log(L[i])
S[i] <- exp(-mu[i]*pow(t[i],kap)); f[i] <- mu[i]*kap*pow(t[i],kap-1)*S[i]
```

```
log(mu[i]) <- beta[1] + beta[2]*age[i]/100 + beta[3]*trt[i] + beta[4]*gender[i]
                    + beta[5]*marstat[i] + c[hltstat[i]-1]}
kap ~dgamma(1,1); TLL <- sum(LL[])
# impacts on exit rate
for (j in 1:5) {beta[j] ~dnorm(0,0.01)}
c[1] <- 0; for (j in 2:4) {c[j] ~dnorm(0,0.01); beta[4+j] <- c[j]}
# impacts on stay length
sigma <- 1/kap; for (j in 1:8) {gam[j] <- -beta[j]/kap}}
```

The BayesX code for the nursing home stay piecewise exponential model involves recoding the categorical health status variable which has categories 2,3,4, and 5. Thus with category 2 as reference,

```
dataset d
d.infile using data\nurshome.txt
d.generate hlth3 = 0
d.replace hlth3 = 1 if hlth=3
d.generate hlth4 = 0
d.replace hlth4 = 1 if hlth=4
d.generate hlth5 = 0
d.replace hlth5 = 1 if hlth=5
bayesreg b
b.regress d = t(baseline)+trt+sex+marr+hlth3+hlth4+hlth5+age(psp-
linerw2), family=cox predict using d
```

3. For illustration, define $J+1$ knots sited evenly along $(0, t_{\max})$, where $t_{\max}$ is the maximum observed time (including both exits and censored times). Also set $d_i = 1$ for an observed exit at time t_i and $d_i = 0$ for a censored time. A generic WinBUGS code with a single predictor (omitting priors) is then

```
model { for (j in 1:J+1) {a[j] <- ranked(t[],n)*(j-1)/J}
for (i in 1:n) for (j in 1:J){ w[i,j] <- d[i]*step(t[i]-a[j])*step(a[j+1] - t[i])
# offset term
  o[i,j] <- (min(t[i],a[j+1])-a[j])*step(t[i]-a[j]);
# Poisson likelihood with offset
  w[i,j] ~dpois(mu[i,j]);    mu[i,j] <- o[i,j]*lambda[j]*exp(beta[j]*z[i,j])}}}
```

4. To represent a generic BUGS code for n subjects, suppose exit or censoring times are denoted t[i], the J distinct exit times as s[j], exit indicators as d[i], and a single time varying predictor as z[i,j]. Then relevant functions and the regression are defined as follows, with a gamma increment prior on dH_{0j},

```
model { for (j in 1:J) { beta1[j] ~dflat();beta2[j] ~dflat();
  for (i in 1:n) {Y[i,j] <- step(t[i] - s[j] + eps)
   dN[i, j] <- Y[i, j] * step(s[j + 1] - t[i] - eps) * d[i]
   dN[i, j] ~dpois(mu[i, j]); mu[i,j] <- mu1[i,j]
   mu1[i, j] <- Y[i, j] * exp(beta1[j] * z[i,j]) * dH0[j]
   mu2[i, j] <- Y[i, j] * exp(beta2[j] * z[i,j] + alpha[j]) }
   alpha[j] ~dnorm(0,tau.alph); dH0[j] ~dgamma(cstar[j], c)
```

```
    cstar[j] <- dH0star[j] * c; dH0star[j] <- r * (s[j + 1] - s[j])}
  # M[] are martingale residuals, dvres[] are deviance residuals
    for (i in 1:n) { M[i] <- sum(dN[i,]) - sum(mu[i,]);
    dvres[i] <- step(M[i])-1)*sqrt(2*(-M[i]-sum(dN[i,])*log((sum(dN[i,])-
M[i]))/M[i]))}}
```

In this code eps is a small positive constant and the hyperparameters {r,c} may be preset or taken as unknowns. The code includes total Martingale residuals M[i], namely the difference over the completed observation period between the actual number of events for a subject and the expected number. Martingale residuals are not symmetrically distributed, even when the fitted model is correct. The deviance residual dvres[] is a normalized transform of the martingale residual that is more symmetric about zero. Observations with large deviance residuals are poorly predicted by the model.

5. The WinBUGS code for the first model applied to the Veterans data (Example 9.2) is

```
  model {for (j in 1:J+1){ a[j] <- ranked(t[],n)*(j-1)/J}
  for (j in 2:J) {alph[j] ~dnorm(alph[j-1],tau.0)}
  tau.0 <- 1/sig2.0; sig.0 <- sqrt(sig2.0)
  sig2.0 ~dunif(0,100000000); alph[1] ~dnorm(0,0.001)
  for (i in 1:n) {KSr[i] <- KS[i]/10
  for (j in 1:J){w[i,j] <- d[i]*step(t[i]-a[j])*step(a[j+1] - t[i]);
  # offset term
  o[i,j] <- (min(t[i],a[j+1])-a[j])*step(t[i]-a[j]);
  log(th[i,j]) <- alph[j]+b.cell[cell[i]] + b.PT[PT[i]] + b.KS*KSr[i] +
b.int*KSr[i]*(PT[i]-1)
  # Poisson likelihood with offset
  mu[i,j] <- o[i,j]*th[i,j]; w[i,j] ~dpois(mu[i,j])}}
  # Priors:
  b.KS ~dnorm(0.0, 0.01); b.int ~dnorm(0.0, 0.001);
  b.PT[1] <- 0; b.PT[2] ~dnorm(0.0, 0.0001);
  b.cell[1] <- 0; for (k in 2:4) { b.cell[k] ~dnorm(0.0, 0.0001)}}
```

6. The basic WinBUGS code is then

```
  model { for (i in 1:n) {G[i]~dcat(pi[1:K]);
  r[i] ~dnorm(mu[i],invsig2[G[i]]) I(L[i],)
  mu[i] <- gam0[G[i]]+gam[1]*z[1,i]+...}}.
```

7. Consider log-logistic failure times as in Equation 9.5, with regression in both the susceptibility and failure time aspects of the process, and respective predictors w[1:n,1:Q] and z[1:n,1:P]. The basis of a WinBUGS code, including the uniform trick for nonstandard densities (Barry, 2006), can then be as follows:

```
  model { for (i in 1:n) {# log-likelihood
  log(L[i]) <- d[i]*log(pi[i]*f[i])+(1-d[i])*log(1-pi[i]+pi[i]*S[i])
```

```
logit(pi[i]) <- gam[1]+gam[2]*w[i,1]+..
log(nu[i]) <- beta[1]+beta[2]*z[i,1]+...
S[i] <- 1/(1+pow(nu[i]*t[i],alph))
log(f[i]) <- log(alph)+alph*log(nu[i])+(alph-1)*log(t[i])-2*log(1+pow(nu[i]
*t[i],alph))
y[i] <- 1; y[i] ~dunif(A1[i],A2[i]); A1[i] <- -1/L[i]; A2[i] <- 1/L[i]}
# priors
for (j in 1:P) {beta[j] ~dflat()}; for (j in 1:Q) {gam[j] ~dflat()}
alph ~dgamma(a.alph,b.alph)}
```

8. This model can be applied using the uniform trick (Barry, 2006) in the following WinBUGS code, with predictors w[1:n,1:Q] and z[1:n,1:P], both including a constant, a Ga(1,0.01) prior on the Weibull shape parameter, and L[i] denoting likelihoods:

```
model { for (i in 1:n) {log(th[i]) <- gam[1]+gam[2]*w[i,1]..;
log(lam[i]) <- beta[1]+beta[2]*z[i,1]+...
log(L[i]) <- d[i]*(log(th[i])+log(kap)+log(lam[i])
+(kap-1)*log(t[i])-lam[i]*pow(t[i],kap))-th[i]*(1-exp(-lam[i]*pow(t[i],kap)))
z[i] <- 1; z[i] ~dunif(G[i],H[i]); G[i] <- -1/L[i]; H[i] <- 1/L[i]}
# priors
for (j in 1:P) {beta[j] ~dflat()}; for (j in 1:Q) {gam[j] ~dflat()}
kap ~dgamma(1,0.01)}
```

9. The code for the final analysis in Example 9.3 (age at first maternity) is

```
model { for (i in 1:n) {Sstar[i] <- exp(-th*F[i]); fstar[i] <- th*f[i]*Sstar[i]
b[i] ~dnorm(0,tau.b); F[i] <- 1-1/(1+lam[i]*pow(t[i],kap))
log(f[i]) <- log(kap)+log(lam[i])+(kap-1)*log(t[i])-2*log(1+lam[i]*pow(t[i],
kap))
log(lam[i]) <- beta[1]+beta[2]*educ[i]+beta[3]*sibs[i]+beta[4]*white[i]
+beta[5]*immig[i]+beta[6]*lowinc[i]+beta[7]*city[i]+b[i]
# log-likelihood
log(L[i]) <- d[i]*log(fstar[i])+(1-d[i])*log(Sstar[i]); LL[i] <- log(L[i]);
H[i] <- 1/L[i]; G[i] <- -1/L[i]; z[i] <- 1; z[i] ~dunif(G[i],H[i])}
Dv <- -2*sum(LL[]); p.nochild <- exp(-th)
# regn function, hazard and modal age (modeT) for "typical" individual
log(lamT) <- beta[1]+beta[2]*13+beta[3]*3+beta[4]
for (age in 15:50) {log(hazT[age]) <-
log(lamT)+log(kap)+(kap-1)*log(age)-log(1+pow(age,kap)*lamT)}
modeT <- pow((kap-1)/lamT,1/kap)
# priors
th ~dgamma(1,0.01); for (j in 1:P) {beta[j] ~dflat()}
kap ~dgamma(1,0.01); sig.b <- 1/sqrt(tau.b); tau.b~dgamma(1,0.01)}
```

Taking the negative of the logs of the posterior means of H[i] in the above code provides Monte Carlo estimates of log(CPO) statistics and (on totalling over subjects) a pseudo log marginal likelihood.

10. In Example 9.4 (cancer survival), let t[i] denote time of death or censoring, and d[i] be binary according to whether death is observed. The code (including options on modeling time effects) is

```
model {for (i in 1:n) {w[i,t[i]] <- d[i]; for (j in 1:t[i]-1) {w[i,j] <- 0}
for (j in 1:t[i]) {w[i,j] ~dbern(q[i,j])
# alternative time effects (Weibull or semi-parametric)
# cloglog(q[i,j]) <- beta0+beta[1]*drug[i]+beta[2]*age[i]/100 + kap*log(j)
cloglog(q[i,j]) <- beta0+beta[1]*drug[i]+beta[2]*age[i]/100 + a[j]}}
beta0 ~dnorm(0,0.001); for (j in 1:2) {beta[j] ~dnorm(0,0.001)}
tau.a ~dgamma(1,0.001); sig.a <- 1/sqrt(tau.a); kap ~dgamma(1,0.001)
# RW1 prior with centered values
a[1:J] ~car.normal(adjage[],wage[],nage[],tau.a)
adjage[1] <- 2; adjage[(J-1)*2] <- J-1;
for (j in 2:J-1) {adjage[2+(j-2)*2] <- j-1; adjage[3+(j-2)*2] <- j+1}
for (j in 1:J) {alph[j] <- beta0+a[j]
alphtr[1,j] <- alph[j]; alphtr[2,j] <- alph[j]+beta[1]}}
```

11. The BUGS code to generate the simulated data is

```
model {for (i in 1: n) { b[i]~dnorm(0,tau.b)
x[i]~dnorm(0,1); u[i] ~dnorm(0,1)I(0,)
for (j in 1:m) {t[i,j] ~dweib(kap[j],lam[i,j]);
log(lam[i,j]) <- beta1[j]+beta2[j]*x[i]+b[i]+ delta*u[i]}}}
```

with input values $\beta_1 = (-5, -6)$, $\beta_2 = (0.5, 1)$, $\kappa = (1.5,2)$, $\delta = 0.5$, and $1/\sigma_b^2 = \tau_b = 2$, so that $\sigma_b \simeq 0.7$. The simulated values of t[] and x[] are obtained using gen inits/save state, and then the model is re-estimated using the code:

```
model {for (i in 1: n) { b[i]~dnorm(0,tau.b); u[i]~dnorm(0,1)I(0,)
for (j in 1:m) {t[i,j]~dweib(kap[j],lam[i,j])
log(lam[i,j]) <- beta1[j]+beta2[j]*x[i]+b[i]+ delta[j]*u[i]}}
for (j in 1:m) { delta[j] ~dunif(0,5); kap[j] ~dgamma(1,0.001);
beta1[j] ~dnorm(0,0.001) ; beta2[j] ~dnorm(0,0.001) }
tau.b ~dgamma(1,0.001)}
```

with initial values file list(beta1=c(-5,-5),kap=c(1,1),beta2=c(0,0),delta=1, tau.b=1).

12. An example of the computation involves repeated times to mammary tumor in rats randomly assigned to treatment and control groups (Sinha, 1993). Totals of tumors diagnosed in each rat vary between 0 and 13; so spell totals for each rat (including possibly censored final spells) range from 1 to 14. There are $n = 253$ spells in all for $K = 48$ rats, and $J = 35$ distinct times relevant to defining the intervals, with $a_J = t_{\max} = 182$. A code for such an analysis, including gamma frailty for each rat, a treatment covariate, and indicators $d[i]$ of tumor occurrence or censoring, is

```
model {for (j in 1:J) { for(i in 1:n) {# Y indicates whether case still at
risk
```

```
Y[i,j] <- step(t[i] - a[j] + eps)
dN[i, j] <- Y[i, j] * step(a[j + 1] - t[i] - eps) * d[i]
dN[i, j] ~dpois(lam[i, j])
lam[i, j] <- Y[i, j] * exp(beta * trt[i]) * dB0[j] * gam[rat[i]]}
# independent increment gamma process
dB0[j] ~dgamma(mu[j], c); mu[j] <- dB0.star[j] * c
dB0.star[j] <- M * (a[j + 1] - a[j])
# Survivorship in two groups
S.tr[j] <- pow(exp(-sum(dB0[1 : j])), exp(beta));
S.cntr[j] <- exp(-sum(dB0[1 : j]))}
c <- 1; M ~dexp(1); beta ~dnorm(0,0.001)
# frailty prior
for (k in 1:K) {gam[k] ~dgamma(h,h)}
h ~dgamma(1,0.001); vargam <- 1/h; th <- exp(beta)}
```

where eps is a small positive value to ensure at risk and counting indices are correctly defined. The gamma process includes an unknown parameter M defining the mean intensity.

13. The code in Example 9.5 (clustered trial of infection treatment) assumes data in stacked rather than nested form. Then the code, with centered hospital effects denoted u.r[], is

```
model {for (i in 1:n) { for (k in 1:J) {# risk status for subject i at interval k,
    y[i,k] <- d[i]*step(t[i] - a[k])*step(a[k+1] - t[i])
# time spent in interval k
    o[i,k] <- (min(t[i], a[k+1]) - a[k])*step(t[i] - a[k])
# piecewise exponential
    theta[i,k] <- lam[k]*exp(beta*trt[i]+e[pat[i]]+u[hos[i]])
    mu[i,k] <- o[i,k]*theta[i,k]; y[i,k] ~dpois(mu[i,k]);
# likelihood (nu used to avoid logs of zero)
nu[i,k] <- equals(mu[i,k],0) +(1-equals(mu[i,k],0))*mu[i,k]
LL[i,k] <- y[i,k]*log(nu[i,k])-mu[i,k]-logfact(y[i,k])}}
# multi-level variation: patient effects
for (j in 1:L) {e[j]~dnorm(0,tau.e); e.r[j] <- e[j]-mean(e[])}
# hospital effects
for (k in 1:M) {u[k] ~dnorm(0,tau.u); u.r[k] <- u[k]-mean(u[])}
V ~dunif(0,100); r ~dunif(0,1); sig2.u <- r*V; sig2.e <- V-sig2.u
tau.e <- 1/sig2.e; tau.u <- 1/sig2.u
sig[1] <- 1/sqrt(tau.e); sig[2] <- 1/sqrt(tau.u)
# correlated process prior on baseline hazard
for (k in 2:J) {lam[k] ~dgamma(a0, b0[k]); b0[k] <- a0/lam[k-1]}
lam[1] ~dgamma(0.1,0.1)
# treatment parameter
beta ~dnorm(0, 0.001); a0~dgamma(0.1,0.1)
# Cum Hazard and Survivorship
H0[1] <- lam[1]*a[1]; for (k in 2:J) {H0[k] <- lam[k]*(a[k]-a[k-1])}
```

```
for (j in 1:J) {S[1,j] <- pow(exp(-sum(H0[1:j])), exp(beta))
S[2,j] <- exp(-sum(H0[1:j]))}
# deviance
Dv <- -2*sum(LL[,])}
```

14. For example, after creating a file vi.csv with columns headed t1, t2, d1, and d2, the commands for the first eye survival times t_{i1} are

```
vis <- read.csv("vi.csv", header = TRUE)
vis.inp <- Surv(vis$t1,vis$d1)
KM.vis <- survfit(vis.inp)
t <- KM.vis$time
S <- KM.vis$surv
plot(log(t),log(-log(S)),type="S").
```

For the Weibull survival analysis the data are arranged in successive lines for each patient, one line for each eye. The predictors are at patient level, namely age at diagnosis of diabetes, and type of diabetes (type 1 or 2), and the eye-level variable, treatment (0=untreated eye, 1=treated eye). The WinBUGS code for the bivariate survival model allows different shape parameters for each eye, namely

```
model {for (i in 1:n) {b[i] ~dnorm(0, tau.b)
for (j in 1:m) {t[i,j] ~dweib(kap[j],lam[i,j]) I(t.cen[i,j],)
    tnew[i,j] ~dweib(kap[j],lam[i,j])
    log(lam[i,j]) <- beta[1] + beta[2] *age[i]/10 + beta[3]*trt[i,j]+beta[4]
*type[i]+b[i]
    S[i,j] <- exp(-lam[i,j]*pow(t[i,j],kap[j])); h[i,j] <- kap[j]*lam[i,j]
*pow(t[i,j],kap[j]-1)
# log-likelihood and inverse likelihood
    LL[i,j] <- d[i,j]*log(h[i,j])+log(S[i,j]); g[i,j] <- 1/exp(LL[i,j])}}
    Dv <- -2*sum(LL[,]);
# Excess risk for untreated patients
    th <- exp(-beta[3])
# uniform prior on SD(frailty)
    sig.b~dunif(0,10); tau.b <-1/(sig.b*sig.b);
    for (j in 1:2) {kap[j] ~dgamma(1,0.001)}
    for (j in 1:4) {beta[j]~dnorm(0, 0.001)}}
```

15. The code for the second model in Example 9.7 (Political Careers) with K=4 risks, KP=K+1, and P=11 predictors, is

```
model { for (i in 1:5429) {Y[i] <- C[i]+1; Y[i] ~dcat(p[i,1:KP])
dv[i] <- log(p[i,Y[i]]); D[i] <- (1+sum(ph[i,1:K])); p[i,1] <- 1/D[i]
for (k in 1:K) {p[i,k+1] <- ph[i,k]/D[i]
log(ph[i,k]) <- beta[k,1]+beta[k,2]*rep[i]+beta[k,3]*redist[i]
+beta[k,4]*scand[i]+beta[k,5]*opengov[i]
+beta[k,6]*opensen[i]+beta[k,7]*lead[i]
+beta[k,8]*age[i]/10+beta[k,9]*primrg[i]/10
```

```
+beta[k,10]*time[i]+beta[k,11]*time[i]*time[i]+b[subj[i],k]}}
for (i in 1:997) {b[i,1:4] ~dmnorm(nought[],T.b[,])}
T.b[1:4,1:4] ~dwish(Q[,],4); Sig.b[1:4,1:4] <- inverse(T.b[,])
for (k in 1:4) {nought[k] <- 0; for (m in 1:4) {Q[k,m] <- equals(k,m)}
# inc[k,m] define included predictors for particular risks
for (m in 1:P) {gam[k,m] ~dnorm(0,0.001); beta[k,m] <- inc[k,m]*gam[k,m]}}
Dev <- -2*sum(dv[])}
```

10

Hierarchical Methods for Nonlinear Regression

10.1 Introduction

Standard versions of the normal linear model and general linear models assume additive and linear predictor effects in the regression mean, and a constant variance. While linear regression effects are often suitable, nonlinear predictor effects and heteroscedasticity are common in areas as diverse as economics, hydrology (Qian et al., 2005), and epidemiology (Natario and Knorr-Held, 2003). Simple nonlinear forms such as polynomials or logarithmic transformations of predictors or responses may often be suitable, but arguably are seeking to provide a global parameterization when local flexibility is needed to reproduce observed reponse patterns (Beck and Jackman, 1998). In some applications there may be a theoretical basis for a particular form of nonlinearity, though some elements of specification will be uncertain—see Borsuk and Stow (2000) for an example on biochemical oxygen demand and Meyer and Millar (1998) for models of fishery stock.

In other situations the form of nonlinearity is unknown and to be assessed from the data—hence the term "nonparametric" since a particular form for the mean function is not assumed. In many applications, a nonlinear effect is present or suspected in only a subset of predictors leading to partially linear models or semiparametric regression models (Beck and Jackman, 1998, 600). Consider outcomes $\{y_i, i = 1, \ldots, n\}$ from an exponential density:

$$p(y_i|\theta_i, \phi) = \exp\left(\frac{y_i\theta_i - a(\theta_i)}{\phi} + c(y_i, \phi)\right)$$

with $E(y_i) = \mu_i = a'(\theta_i)$ and link $g(\mu_i) = \eta_i$ to a regression term η_i. Suppose it is intended that R metric predictors $(w_{1i}, w_{2i}, \ldots, w_{Ri})$ be modeled nonparametrically via unknown smooth functions $S(w_{ri})$,

$$g(\mu_i) = \eta_i = \alpha + X_i\beta + S_1(w_{1i}) + \cdots + S_R(w_{Ri}) + u_i,$$

with $u_i \sim N(0, \sigma^2)$. For instance, Engle et al. (1986) analyze the relationship between temperature and monthly electricity sales (y metric and u normal, and with g an identity link) for four US cities. The impact of electricity price, month (11 dummy variables), and income is modeled parametrically, but an unknown smooth function is adopted to model the impact of monthly temperature.

Residual errors u_i will be present when y_i is metric and may also be present for overdispersed discrete outcomes. While an assumption of independent errors with constant variance is standard, nonparametric regression for the regression mean may be extended to modeling heteroscedastic errors (Yau and Kohn, 2003). When the observations are taken through time or over space it may also be important to control for correlations in the u. Kohn et al. (2000) and Smith et al. (1998) consider the case when the observations y_t are arranged in time, smooth functions are used for predictor effects, and the u_t are autocorrelated; the estimate of the smooth function will be adversely affected if independent residuals are incorrectly assumed.

The two major forms of nonparametric regression involve basis functions (e.g., polynomial spline methods) and general additive methods based on smoothness priors; these are considered in Sections 10.2 and 10.5, respectively. Extending nonparametric regression to multiple predictors raises the same issues as multiple linear regression, for example, whether interactions are necessary and how the presence of smooths for other predictors alters the smooth for a given predictor—see Section 10.3. Robustness in nonparametric regression (e.g., to heteroscedastic errors) may be obtained through spatially adaptive methods which allow the level of smoothness to vary over the space of the covariates (Baladandayuthapani et al., 2005; Wood et al., 2002)—see Section 10.4. A major application area for nonparametric regression is in longitudinal settings, as discussed in Section 10.6.

10.2 Nonparametric Basis Function Models for the Regression Mean

A wide range of methods for nonparametric regression in one or more predictors typically assume linear combinations of basis functions $S_r(w_r)$ of predictors $(w_1, \ldots, w_R)$. Numerous basis functions can be used, including truncated polynomial functions, B-spline functions, radial basis functions (Yau et al., 2003), logistic functions (Hooper, 2001), trigonometric basis functions, and wavelets (Dennison et al., 2002). For exponential family responses y_i with mean μ_i and link g, a truncated polynomial spline (or piecewise polynomial spline) regression on a single predictor w_i has the form (Dennison et al., 2002, 52):

$$g(\mu_i) = \alpha + \sum_{k=1}^{K} \beta_k (w_i - \kappa_k)_+^q + u_i, \tag{10.1}$$

where $u_i \sim N(0, \sigma^2)$, q is a known positive integer, and the κ_k are knots placed within the range $[w_{\min}, w_{\max}]$ of w. In Equation 10.1, the piecewise polynomials are fitted in each interval $[\kappa_k, \kappa_{k+1})$ and preferably join smoothly at each knot (e.g., this applies for a cubic spline as it has continous first and second derivatives at each knot).

An alternative spline specification (e.g., Meyer, 2005; Tutz and Reithinger, 2007, 2877) matches the degree q of the truncated function $T(w_i) = \sum_{k=1}^{K} \beta_k(w_i - \kappa_k)_+^q$ by a standard polynomial of order q, namely $Q(w_i) = \varphi_1 w_i + \cdots + \varphi_q w_i^q$. So the total smooth is

$$S(w_i) = Q(w_i) + T(w_i)$$

and one has

$$g(\mu_i) = \alpha + \varphi_1 w_i + \cdots + \varphi_q w_i^q + \sum_{k=1}^{K} \beta_k(w_i - \kappa_k)_+^q + u_i. \tag{10.2}$$

Values $q = 1, 2,$ or 3 are most typical with $q = 1$ often being suitable for reproducing a smooth function given a large enough set of knots (Ruppert et al., 2003, 68), but also capable of reproducing abrupt changes in the underlying function (Denison et al., 2002, 52).

The knots in Equations 10.1 and 10.2 may be known, or unknown. If known, then they are typically much less than the sample size in number. They could be sited at percentile points (e.g., deciles) of w, or possibly placed more densely at points where the function is known to be rapidly changing and less densely elsewhere. Choosing too few knots can result in oversmoothing and choosing too many in overfitting—see the light detection and ranging (LIDAR) data examples discussed by Ruppert et al. (2003, 63). Coull et al. (2001, 540) suggest allocation of one knot for every four to five observations, up to a maximum of about 40 knots. Yau and Kohn (2003) suggest fitting a model with a small number of knots first and gradually increasing their number until estimates and fit stabilize. An alternative procedure known as smoothing splines places a knot at every observed distinct predictor value (Berry et al., 2002; Dias and Gamerman, 2002). The most general model-averaging approach takes both the number of knots and their sitings as unknowns, while both Biller (2000) and Denison et al. (1998) assume a large number of potential but prespecified candidate knot locations. If knots are taken to have unknown locations within $[w_{\min}, w_{\max}]$, identification may rely on order constraints such as $\kappa_k > \kappa_{k-1}$ and analysis resembles time series with multiple change points.

Assuming the β_k in Equations 10.1 through 10.3 are modeled as fixed effects, predictor coefficient selection is open as a way of achieving model parsimony, and is especially indicated under the smoothing spline method (Smith and Kohn, 1996). With a large number of preset potential knot sitings, predictor selection involves obtaining posterior probabilities $Pr(\delta_{jk} = 1|y)$

on binary indicator variables δ_{1k} $(k = 1, \dots, q)$ for retaining coefficients in the $Q(w)$ component and δ_{2k} $(k = 1, \dots, K)$ in the $T(w)$ component. One then has

$$g(\mu_i) = \alpha + \delta_{11}\varphi_1 w_i + \cdots + \delta_{1q}\varphi_q w_i^q + \sum_{k=1}^{K} \delta_{2k}\beta_k (w_i - \kappa_k)_+^q + u_i, \qquad (10.3)$$

with coefficients estimated by means of the products $\delta_{1j}\varphi_j$ and $\delta_{2k}\beta_k$.

10.2.1 Mixed model splines

By contrast to models 10.1 through 10.3 where all coefficients are fixed effects, under the mixed model spline regression, or penalized spline method, the coefficients in $Q(w_i)$ usually remain fixed effects, but the coefficients in $T(w_i)$ follow a penalizing random effects or P-spline prior (Brumback et al., 1999; Ruppert et al., 2003; Wand, 2003). In Yau and Kohn (2003) all coefficients are treated as random, e.g., both φ and b in Equation 10.5. Currie and Durban (2002) argue that under a P-spline approach the problem of choosing the number and position of the knots is largely overcome, since providing enough knots are used, the penalty function should ensure that the resulting fits are very similar. Under the P-spline approach, Equations 10.1 and 10.2 become

$$g(\mu_i) = \alpha + S(w_i) = \alpha + \sum_{k=1}^{K} b_k (w_i - \kappa_k)_+^q + u_i, \qquad (10.4)$$

$$g(\mu_i) = \alpha + \varphi_1 w_i + \cdots + \varphi_q w_i^q + \sum_{k=1}^{K} b_k (w_i - \kappa_k)_+^q + u_i, \qquad (10.5)$$

where b_k is a collection of random parameters from a common density with unknown hyperparameters.

Possible priors for the random b_k include an unstructured normal (Currie and Durban, 2002; Ruppert et al., 2003)

$$b_k \backsim N(0, \phi), \qquad (10.6)$$

which, by comparison with a fixed effects prior, imposes a restriction on the b_k when $\phi < \infty$ and tends to shrink the b_k leading to a smooth fit (Wand, 2003). A common choice for ϕ is an inverse gamma prior, or a gamma prior may be used for $\theta = 1/\phi$. A standard approach recommended by Lang and Brezger (2004) is $\phi \sim \mathrm{IG}(g, h)$ with $g = 1$ and h small (e.g., 0.001, 0.0001 or 0.00001), though Jullion and Lambert (2007) report sensitivity to the value chosen for h. Jullion and Lambert (2007) suggest a prior $\theta \sim \mathrm{Ga}(\frac{\nu}{2}, \delta\frac{\nu}{2})$ with $\nu \sim U(0, K)$ with K large (e.g., 100) and $\delta \sim \mathrm{Ga}(g_\delta, h_\delta)$ with $g_\delta = h_\delta$ taken small. They also suggest a mixture prior over alternative plausible values of h in the prior $\phi \sim \mathrm{IG}(1, h)$.

To illustrate equivalence to mixed models, define design matrices

$$W = \underset{1 \leq i \leq n}{[1, w_i, \ldots, w_i^q]},$$
$$Z = \underset{1 \leq k \leq K,\ 1 \leq i \leq n}{[(w_i - \kappa_k)_+^q]}$$

and vectors $\beta = (\alpha, \varphi_1, \ldots, \varphi_q)'$ and $b = (b_1, \ldots, b_K)'$. Then under normal error assumptions, model Equation 10.5 can be written in the mixed model form:

$$g = X\beta + Zb + u,$$
$$\begin{bmatrix} b \\ u \end{bmatrix} \sim N\left(\begin{matrix} 0 \\ 0' \end{matrix} \begin{bmatrix} \phi I & 0 \\ 0 & \sigma^2 I \end{bmatrix}\right).$$

Opsomer et al. (2008) extend this framework to encompass random area effects relevant for small area estimation in survey data modeling.

Alternatives schemes for b_k are a random walk penalty (Eilers and Marx, 1996), such as $\Delta^d b_k \backsim N(0, \phi)$. For instance, taking $d = 1$ gives

$$b_k \backsim N(b_{k-1}, \phi). \tag{10.7}$$

Another option providing monotonic smooths in applications where such smooths have a substantive rationalization (Brezger and Steiner, 2008) involves monotonically constrained b_k. Thus one may have $b_k \backsim N(0, \phi)$, but subject to

$$b_k \geq b_{k-1}, \quad k = 2, \ldots, K$$

for an increasing function $T(w_i) = \sum_{k=1}^{K} b_k (w_i - \kappa_k)_+^q$ or $b_k \leq b_{k-1}$ for a decreasing function. Brezger and Steiner argue that unconstrained estimation of spline models (with nonmonotonic b_k) may produce artifactual features (e.g., local upturns and downturns in a dose–response or economic relations at odds with substantive knowledge) due to sparse data and "overfitting" due to excess flexibility in the smooth function.

The function $S(w_i)$ resulting from a fixed effects prior on $\{\beta_k\}$ in Equations 10.1 through 10.3 may be quite rough, due to the large number of truncated polynomials being fitted, whereas the shrinkage prior under the mixed model approach tends to penalize large coefficients and lead to a smoother fit (Meyer, 2005; Ngo and Wand, 2004; Yau et al., 2003). Under an unstructured prior $b_k \sim N(0, \phi)$, and smoothing or penalty parameter λ, the mode of the posterior density of $\{\varphi, b, \phi\}$ is the same as that obtained by maximizing a penalized likelihood:

$$\text{PL} = \log[P(y|\varphi, b, \phi)] - \lambda \sum_{k=1}^{K} b_k^2,$$

where the form of λ (in terms of variance parameters) depends on whether or not there is an unstructured residual term u_i in the regression model.

For a metric outcome and $u_i \sim N(0, \sigma^2)$, one has $\lambda = \sigma^2/\phi$ (Fahrmeir and Knorr-Held, 2000). This penalized likelihood is analogous to "ridge" penalties sometimes used with correlated predictors (Eilers and Marx, 2004). For random walk priors of order d, one has (Lang and Brezger, 2004)

$$\mathrm{PL} = \log[P(y|\varphi, b, \phi)] - \lambda \sum_{k=d+1}^{K} (\Delta^d b_k)^2.$$

10.2.2 Model selection

Nonparametric regressions are often heavily parameterized and parameter redundancies are likely (Belitz and Lang, 2008; Panagiotelis and Smith, 2008; Yau et al., 2003). As a simple approach, one might consider examining a profile of penalized likelihoods for a series of values of ϕ, including the null value $\phi = 0$ corresponding to the spline term being unnecessary. Taking ϕ as an unknown is more general, and one option, following Albert and Chib (1997), is a discrete prior on alternative values of ϕ that include $\phi = 0$. Yau et al. (2003) discuss variable and model selection in nonparametric basis models using binary indicators J_r for retaining or dropping the random effect term for the rth of R predictors. So for $r = 1, \ldots, R$ predictors modeled by smooths $S_r(w_{ri}) = Q_r(w_{ri}) + T_r(w_{ri})$, where $T_r(w_{ri}) = \sum_{k=1}^{K_r} b_{rk}(w_{ri} - \kappa_{rk})_+^{q_r}$ the model including random component selection indices would be

$$\begin{aligned} g(\mu_i) &= \alpha + Q_1(w_{1i}) + \cdots + Q_R(w_{Ri}) + J_1 T_1(w_{1i}) + J_2 T_2(w_{2i}) \\ &\quad + \cdots + J_R T_R(w_{Ri}) + u_i \\ &= \alpha + \sum_{r=1}^{R} Q_r(w_{ri}) + J_1 \sum_{k=1}^{K_1} b_{1k}(w_{1i} - \kappa_{1k})_+^{q_1} \\ &\quad + J_2 \sum_{k=1}^{K_2} b_{2k}(w_{2i} - \kappa_{2k})_+^{q_2} \cdots + J_R \sum_{k=1}^{K_R} b_{Rk}(w_{Ri} - \kappa_{Rk})_+^{q_R} + u_i. \end{aligned}$$

Yau et al. (2003) mention that selection for retention ($J_r = 1$) is influenced by the degree of informativeness of the prior adopted for the variances ϕ_r of $b_r = (b_{r1}, \ldots, b_{rK_r})$. Flat priors will tend to lead to low posterior probabilities $Pr(J_r = 1|y)$ for retaining random components. They suggest initial runs with diffuse priors to develop an informative data-based prior. Suppose the Markov Chain Monte Carlo (MCMC) iterates $h = 1, \ldots, H$ from the initial run were $\phi_r^{(1)}, \phi_r^{(2)}, \ldots, \phi_r^{(H)}$. A possible data-based prior in a subsequent run is log-normal with

$$p(\phi_r|m_r, V_r) = \frac{1}{\phi_r\sqrt{2\pi n V_r}} \exp\left(-\frac{1}{2nV_r}(\log(\phi_r) - m_r)^2\right),$$

where (m_r, V_r) are the posterior median and variance of the logged variances from the initial run, $\log(\phi_r^{(1)}), \log(\phi_r^{(2)}), \ldots, \log(\phi_r^{(H)})$. The use of a data-based prior with inflation of the variance V_r by the sample size n follows the development of Shively et al. (1999) who relate the resulting posterior selection to the Bayesian Information Criterion (BIC). Related approaches include Panagiotelis and Smith (2008) who adopt hierarchical log-normal priors on ϕ_r, namely

$$\begin{aligned}\log(\phi_r) &\sim N(a_r, b_r),\\ a_r &\sim N(0, 100),\\ b_r &\sim \mathrm{IG}(101, 10100),\end{aligned}$$

independent of those for J_r.

10.2.3 Basis functions other than truncated polynomials

The fixed or random coefficient approaches can equally be applied with other basis functions to represent $T(w_i)$. Truncated polynomial basis functions span the space of degree q polyomials with knots located at $\kappa_1, \ldots, \kappa_K$ (Friedman, 1991). This property also holds for radial basis functions based on distances $r_{ik} = |w_i - \kappa_k|$, such as the polyharmonic spline $T(w_i) = \sum_{k=1}^{K} H(r_{ik})$, with

$$\begin{aligned}H(r_{ik}) &= r_{ik}^q, \quad q = 1, 3, 5, \ldots\\ H(r_{ik}) &= r_{ik}^q \log(r_{ik}), \quad q = 2, 4, 6,\end{aligned}$$

of which the thin plate spline (Kohn et al., 2001; Koop and Tole, 2004)

$$H(r_{ik}) = r_{ik}^2 \log(r_{ik})$$

is a special case. These are examples of functions which are radially symmetric around knots κ_k, such that the value of the function at w_i depends only on the distance between w_i and the knot location. They have the form $H(u) = H(|w - \kappa_k|)$, where $|v| = \sqrt{v'v}$ is the length of the vector v. Other types of radial basis include Gaussian functions (Konishi et al., 2004) with

$$H_k(w_i) = \exp\left(-\frac{|w - \kappa_k|}{2\nu\eta_k}\right),$$

where ν is the same over different knots. As for truncated splines, smoothing based on radial basis functions may include a parametric polynomial term to degree q to match the degree of the radial function. For example, with $q = 1$:

$$g(\mu_i) = \alpha + \varphi_1 w_i + \sum_{k=1}^{K} \beta_k |w_i - \kappa_k| + u_i.$$

Both radial and truncated power splines may be ill-conditioned in terms of broader regression considerations (Eilers and Marx, 2004). An alternative basis less prone to ill-conditioning is provided by B-splines, with health mapping applications exemplified by MacNab and Gustafson (2007) and Silva et al. (2008). B-splines are defined to be nonzero for at most $q+2$ interior knots for a qth degree B-spline (also called a B-spline of order $q+1$), which means the condition number of the design matrix product is relatively low (Biller, 2000; Dennison et al., 2002, 75; Eilers and Marx, 1996, 90). A B-spline of degree q consists of $q+1$ polynomial pieces of degree q and overlaps with $2q$ of its neighbors. For K knots, and so $K+1$ intervals, in the domain $[w_{\min}, w_{\max}]$ of a predictor, there will be $K^* = K+1+q$ B-spline schedules because extra knots are placed outside the domain of w to get q overlapping B-splines in each interval.

Let $B_k(w_i, q)$ be the value at w_i of the kth B-spline of degree q, with $k = 1, \ldots, K^*$. Successive B-spline values are defined by the recursion

$$B_k(w_i, 0) = I(\kappa_k \leq w_i < \kappa_{k+1})$$

$$B_k(w_i, q) = \frac{w_i - \kappa_k}{\kappa_{k+q} - \kappa_k} B_k(w_i, q-1) + \frac{\kappa_{k+q+1} - w_i}{\kappa_{k+q+1} - \kappa_{k+1}} B_{k+1}(w_i, q-1).$$

The initial terms in the recursion are simply binary indicators defining a partition of the w values. For equally spaced knots a simplified B-spline recursion applies involving differences in truncated power splines (Eilers and Marx, 2004).[1]

B-spline bases for $T(w)$ can be combined with random or fixed effects priors for the spline coefficients; for example, random b_k in an analysis with a single predictor w_i leads to

$$g(\mu_i) = \alpha + \varphi_1 w_i + \cdots + \varphi_q w_i^q + \sum_{k=1}^{K^*} b_k B_k(w_i, q) + u_i.$$

In particular, Eilers and Marx (1996) combine a B-spline basis with a penalty on dth-order differences in adjacent b_k coefficients. As mentioned above, difference penalties can be achieved by random walk priors under a Bayesian approach (e.g., a second-order random walk prior if $d = 2$).

Bayesian application of spectral (Fourier series) basis functions is discussed by Fahrmeir and Tutz (2001, chap. 5), Kitagawa and Gersch (1996), and Lenk (1999). Here the smooth may be represented by the series

$$T(w_i) = \sum_{k=1}^{\infty} b_k H_k(w_i),$$

with H_k including sine and/or cosine terms. Setting $z_i = (w_i - w_{\min})/(w_{\max} - w_{\min})$ and including only cosine terms in H_k as in Lenk (1999) gives

$$H_k(w_i) = \left(\frac{2}{w_{\max} - w_{\min}}\right)^{0.5} \cos(\pi k z_i).$$

Since a smooth T will not have high-frequency components, a natural prior on the b_k expresses decay as k increases (penalizes terms at higher k values) as in

$$b_k \backsim N(0, \phi \exp[-\delta c_k]),$$

where c_k can be taken as a known increasing function of k, and δ determines the rate of decay of the Fourier coefficients. Possibilities are $c_k = \log(k)$ with $\delta > 1$, and $c_k = k$ with $\delta > 0$. An alternative is a power function such as

$$b_k \backsim N(0, \phi\delta^k),$$

where $\delta \in (0, 1)$. Sampling of parameters is from standard densities except for the decay parameter δ, for which Lenk (1999) uses slice sampling. For practical application, the Fourier Series is truncated above at K, namely

$$T(w_i) = \sum_{k=1}^{K} b_k H_k(w_i).$$

where K can be regarded as another parameter (cf. Ruppert et al., 2003, 86).

Example 10.1. Fossil Data This example concerns a continuous outcome, namely ratios Y of strontium isotopes in fossil shells, with shell age as predictor (Ngo and Wand, 2004; Ruppert et al., 2003), and with response transformed as $y = 100{,}000Y - 70{,}700$. The initial analysis compares a linear truncated spline with a cubic B-spline ($q = 3$ in $B_k(w_i, q)$). These two models are fitted in WinBUGS using $K = 19$ knots sited at the 5th, 10th,...,95th percentiles of age. So the linear spline model is

$$y_i = \alpha + \varphi_1 \text{Age}_i + \sum_{k=1}^{19} b_k(\text{Age}_i - \kappa_k)_+ + u_i,$$

where $u_i \sim N(0, \sigma^2)$. In the truncated spline model an unstructured normal random effect prior $b_k \sim N(0, \phi)$ is assumed on the spline coefficients, while the B-spline model assumes a first-order random walk, penalizing first differences in the b_k, and with b_1 assigned a diffuse $N(0, 1000)$ prior. A linear term in age is also included. For precisions $1/\sigma^2$ and $\theta = 1/\phi$, gamma Ga(1,0.001) priors are assumed.[2]

The Bayesian version of the cubic P-spline method of Eilers and Marx (1996) as implemented in BayesX is also applied[3]—see Brezger and Lang (2006) and Lang and Brezger (2004). In this analysis, inverse gamma priors on the variance parameters have shape and scale parameters both set at 0.001. Furthermore, there are $K = 20$ equally spaced knots in the domain of age, and a second-order random walk penalty is used for the B-spline coefficients.

The last 40,000 iterations of two chain runs of 50,000 iterations in WinBUGS show similar smooths, but with fit values favoring the truncated linear spline. The respective DIC (and d_e) values are 511 (12) and 517 (16.1). Figure 10.1 shows the B-spline fit. The BayesX implementation of a B-spline smooth gives a DIC (d_e) of 510.3 (13.0), with the corresponding smooth shown in Figure 10.2.

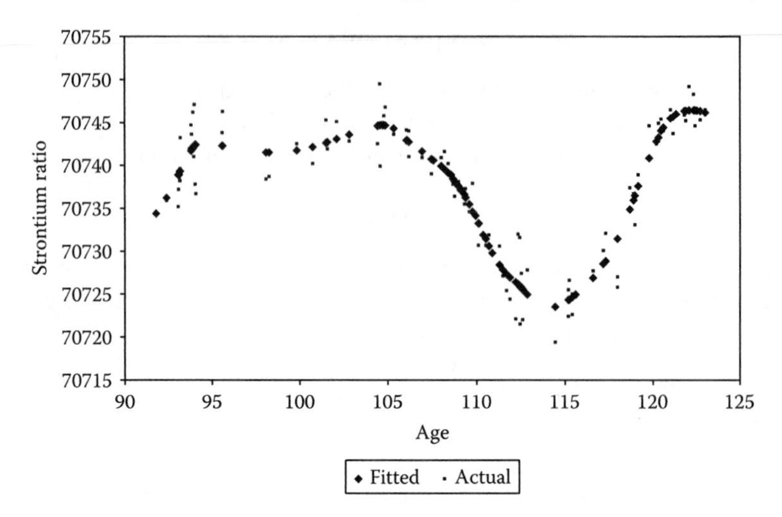

FIGURE 10.1
B-spline fit and observations.

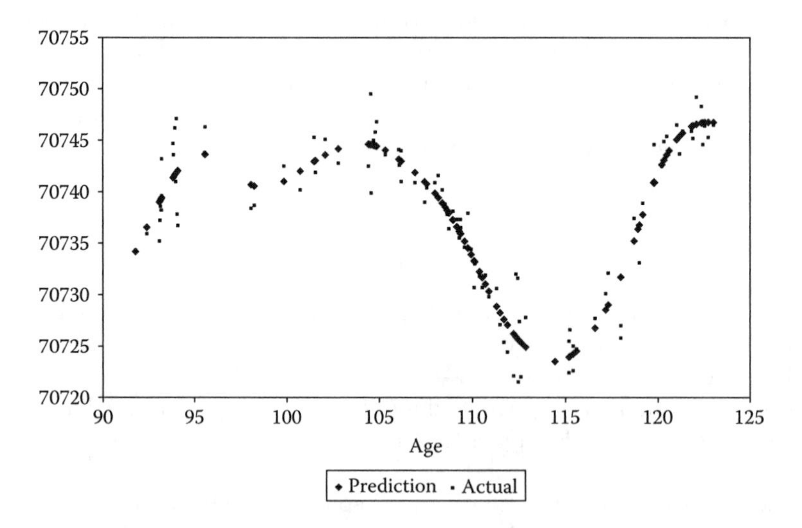

FIGURE 10.2
BayesX smooth for fossil data.

10.3 Multivariate Basis Function Regression

Generalization of Bayesian basis function methods to multiple metric predictors follows three main methodologies. The first involves tensor product-truncated polynomial bases, including the multivariate adaptive regression spline (MARS) method of Friedman (1991); the second is the generalization of radial basis methods (e.g., Yau et al., 2003); and the third is the generalization of mixed model P-splines (e.g., Durban et al., 2006). The full tensor product approach is a multiplicative generalization of Equation 10.2 or 10.5, with particular versions discussed by Brezger et al. (2005), Chen (1993), Denison et al. (2002, 104), Ruppert et al. (2003, 240), and Smith and Kohn (1997, 1524). Interactions between categorical and metric predictors in nonparametric regression are considered by Coull et al. (2001) and Ruppert et al. (2003).

Assume a predictor vector $w_i = (w_{1i}, \ldots, w_{Ri})$ of dimension R, with spline degree q for all predictors. Omitting the corresponding standard polynomial effects $Q_r(w_r)$, the tensor product generalization of Equation 10.1 for two or more metric predictors involves an analysis of variance type representation with main and various order interaction effects,

$$
\begin{aligned}
g(\mu_i) = \alpha &+ \sum_{r=1}^{R}\sum_{k=1}^{K_r} \beta_{rk}(w_{ri} - \kappa_{rk})_+^q + \sum_{r\neq s}^{R}\sum_{k=1}^{Kr}\sum_{l=1}^{K_s} \gamma_{rs,kl}(w_{ri} - \kappa_{rk})_+^q (w_{si} - \kappa_{sl})_+^q \\
&+ \sum_{r\neq s\neq t}^{R}\sum_{k=1}^{K_r}\sum_{l=1}^{K_s}\sum_{m=1}^{K_t} \delta_{rst,klm}(w_{ri} - \kappa_{rk})_+^q \\
&\times (w_{si} - \kappa_{sl})_+^q (w_{ti} - \kappa_{tm})_+^q + \cdots + u_i,
\end{aligned}
$$

where K_r is the number of knots for predictor w_r. There may be R main effects, $\binom{R}{2}$ second-order interactions, $\binom{R}{3}$ third-order interactions and so on, with the associated parameters $\{\beta, \gamma, \delta, \ldots,\}$ having dimension determined by the number of knots in K_r, $\{K_r, K_s\}$, $\{K_r, K_s, K_t\}$, etc. Higher-order interactions may be excluded even if definable in principle, as an acceptable smooth may often be obtained by restricting attention to main effects and low-order interactions; so a model with main and second-order effects only would have $R + \binom{R}{2}$ parameter sets. Gustafson (2000) considers a BWISE approximation to smooth functions involving main effects $S_1(w_1), \ldots, S_R(w_R)$, and second-order interactions only, namely

$$
S_{12}(w_1, w_2), \ldots, S_{1R}(w_1, w_R), \ldots, S_{(R-1),R}(w_{R-1}, w_R).
$$

Main effects are assumed to be either conventional linear regression effects or cubic splines. The form of the interaction depends on which form of main effect is selected for predictors w_r and w_s.

As an example, consider a tensor product of truncated polynomials with $q = 1$ and $R = 3$, so that $w_i = (w_{1i}, w_{2i}, w_{3i})$ and with $K_1 = K_2 = K_3 = 5$. Also just consider the step functions $(w - \kappa)_+$ though Dennison et al. (2002) consider sign reversed steps $(\kappa - w)_+$. Then there may be $R = 3$ main effects, $\binom{R}{2} = 3$ second-order interactions, and $\binom{R}{3} = 1$ third-order interactions. In a model confined to main effects and second-order interactions, the main effects would be terms $\sum_{k=1}^{K_1} \beta_{1k}(w_{1i} - \kappa_{1k})_+$, $\sum_{k=1}^{K_2} \beta_{2k}(w_{2i} - \kappa_{2k})_+$, and $\sum_{k=1}^{K_3} \beta_{3k}(w_{3i} - \kappa_{3k})_+$ involving 15 parameters. The second-order interactions would be terms $\sum_{k=1}^{K_1}\sum_{l=1}^{K_2}\gamma_{12,kl}(w_{1i}-\kappa_{rk})_+(w_{2i}-\kappa_{2l})_+$, $\sum_{k=1}^{K_1}\sum_{l=1}^{K_3}\gamma_{13,kl}(w_{1i}-\kappa_{1k})_+(w_{3i}-\kappa_{3l})_+$, and $\sum_{k=1}^{K_2}\sum_{l=1}^{K_3}\gamma_{23,kl}(w_{2i}-\kappa_{2k})_+(w_{3i}-\kappa_{3l})_+$ involving 75 parameters. If the coefficients $\{\beta_{rk},\gamma_{rs,kl}\}$ are assumed to be fixed effects, then predictor selection methods are relevant as in Smith and Kohn (1996) or the RJMCMC methods discussed by Denison et al. (2002, 105). If the $\{\beta_{rk},\gamma_{rs,kl}\}$ are assumed to be random effects, smoothness may be achieved by penalizing large coefficients, and parsimony achieved by selection between zero and positive variance components $\{\phi_{\beta_1},\phi_{\beta_2},\phi_{\beta_3},\phi_{\gamma_{12}},\phi_{\gamma_{13}},\phi_{\gamma_{23}}\}$.

The tensor product generalization of Equation 10.2 or 10.5 includes interactions between the terms in $T(w)$ and $Q(w)$ (Ruppert et al., 2003, 240; Smith and Kohn, 1997). Consider a two predictor situation, with K_1 knots in w_{1i} and K_2 knots in w_{2i}. For a linear spline ($q=1$), and random effect spline coefficients $\{b_{rk},d_{rsk},c_{rskm}\}$ one would have

$$\begin{aligned}g(\mu_i) &= \alpha+\varphi_1 w_{1i}+\varphi_2 w_{2i}+\varphi_3 w_{1i}w_{2i}+\sum_{k=1}^{K_1} b_{1k}(w_{1i}-\kappa_{1k})\\ &\quad+\sum_{k=1}^{K_2} b_{2k}(w_{2i}-\kappa_{2k})_+ +\sum_{k=1}^{K_2} d_{12,k}w_{1i}(w_{2i}-\kappa_{2k})\\ &\quad+\sum_{k=1}^{K_1} d_{21k}w_{2i}(w_{1i}-\kappa_{1k})_+\\ &\quad+\sum_{k=1}^{K_1}\sum_{m=1}^{K_2} c_{12km}(w_{1i}-\kappa_{1k})_+(w_{2i}-\kappa_{2m})_+ + u_i,\end{aligned}$$

where there are six variance components $(\phi_{b_1},\phi_{b_2},\phi_{d_{12}},\phi_{d_{21}},\phi_{c_{12}},\sigma^2)$. In the bivariate example of Smith and Kohn (1997, 1530), $K_1=K_2=9$ and $q=3$ leading to a (fixed effects) analysis involving 169 coefficients.

A similar scheme applies when interactions between metric and categorical predictors are considered. Thus, let $C_i\in(1,\ldots,L)$ be a categorical predictor and w_{1i} and w_{2i} be metric predictors. Suppose that only the smooth in w_2 is postulated to vary according to the level of C, and define

$$\begin{aligned}z_{il} &= 1 \quad \text{if } C_i = l\\ &= 0 \quad \text{otherwise.}\end{aligned}$$

Also consider a metric response y_i and assume that interactions between w_1 and w_2 are not present. Then with a qth degree truncated polynomial basis in both predictors, Coull et al. (2001) suggest the model

$$\begin{aligned}y_i &= \alpha+Q_1(w_{1i})+Q_2(w_{2i})+T_1(w_{1i})+T_{2,C_i}(w_{2i})+u_i\\ &= \alpha+\varphi_{11}w_{1i}+\cdots+\varphi_{1q}w_{1i}^q+\varphi_{21}w_{2i}+\cdots+\varphi_{2q}w_{2i}^q+\sum_{k=1}^{K_1} b_{1k}(w_{1i}-\kappa_{1k})_+^q\\ &\quad+\sum_{k=1}^{K_2} b_{2k}(w_{2i}-\kappa_{2k})_+^q+\sum_{l=2}^{L} z_{il}\left\{\sum_{k=1}^{K_2} c_{kl}(w_{2i}-\kappa_{2k})_+^q\right\},\end{aligned}$$

where $b_{1k} \backsim N(0, \phi_{b1}), b_{1k} \backsim N(0, \phi_{b1}), c_{kl} \backsim N(0, \phi_{cl})$, and $u_i \backsim N(0, \sigma^2)$. The amount of smoothing under $S_1 = Q_1 + T_1$ and $S_{2,C_i} = Q_2 + T_{2,C_i}$ then depends on the ratios σ^2/ϕ_{b1} and $\sigma^2/[\phi_{b2} + \phi_{cl}]$.

In a multivariate mixed model generalization of the radial basis, Yau et al. (2003) consider thin-plate functions with exponents $(2q - d)$ specified by integer combinations (q, d), where d is the dimension of the covariate vectors in the relevant interaction. So

$$H_k(z) = |z - t_k|^{(2q-d)} \log(|z - t_k|) \quad \text{for } (2q - d) \text{ even}$$
$$H_k(z) = |z - t_k|^{(2q-d)} \quad \text{for } (2q - d) \text{ odd,}$$

where z are univariate or multivariate vector predictor values, and t_k are univariate or multivariate knots. In applying such functions, Yau and Kohn (2003) argue that heavily parameterized multivariate spline models are often not likely to be well identified, and suggest additive models involving just univariate smooths in each predictor (with $d = 1$), and all possible bivariate interactions (with $d = 2$). Consider the setting $q = 2$, with predictors, w_1 and w_2, and let $z_i = (w_{1i}, w_{2i})$ denote bivariate covariate combinations, with $K_{1,2}$ bivariate centers $t_k = (t_{1k}, t_{2k})$ that might be provided by an initial cluster analysis. Also denoting distances $h_{ik} = |z_i - t_k|$, the bivariate basis for $2q - d = 2$ is of the form $h^2 \log(h)$. With linear terms in the parametric component $Q(w)$, this leads to the representation,

$$\begin{aligned} g(\mu_i) &= \alpha + S_1(w_{1i}) + S_2(w_{2i}) + S_{12}(w_{1i}, w_{2i}) \\ &= \alpha + \varphi_1 w_{1i} + \sum_{k=1}^{K_1} b_{1k} |w_{1i} - \kappa_{1k}|^3 + \varphi_2 w_{2i} \\ &\quad + \sum_{k=1}^{K_2} b_{2k} |w_{2i} - \kappa_{2k}|^3 + \sum_{k=1}^{K_{1,2}} c_k h_{ik}^2 \log(h_{ik}). \end{aligned}$$

In general with K_r knots $\{\kappa_{r1}, \ldots, \kappa_{rK_r}\}$ for predictor w_{ri}, the R main effects are

$$S_r(w_{ri}) = \varphi_r w_{ri} + \sum_{k=1}^{K_r} b_{rk} |w_{ri} - \kappa_{rk}|^3,$$

where the R sets of coefficients $\{[b_{r1}, b_{r2}, \ldots, b_{rK_r}], r = 1, \ldots, R\}$ are assumed to be random with variances $\phi_{b1}, \ldots, \phi_{bR}$. Let the $K_{r,s}$ bivariate knots (or centers) for first-order (w_r, w_s) interaction effects be denoted $t_{rs,k} = (t_{rk}, t_{sk})$. Then the interaction bases have the form:

$$T_{rs}(w_{ri}, w_{si}) = \sum_{k=1}^{K_{r,s}} c_{rs,k} |(w_{ri}, w_{si}) - (t_{sk}, t_{rk})|^2 \log(|(w_{ri}, w_{si}) - (t_{rk}, t_{sk})|).$$

The $\binom{R}{2}$ sets of coefficients $c_{rs,k}$ are also assumed to be random.

A simply applied multivariate nonparametric regression involves the adaptive logistic basis (ALB) proposed by Hooper (2001), with the ALB L_q estimator obtained by minimizing $\sum |y_i - S(w)|^q$. The ALB basis uses an M component discrete mixture of multinomial logit regressions, namely

$$g(\mu_i) = S(w_{1i}, \ldots, w_{Ri}) = \sum_{m=1}^{M} \alpha_m B_m(w_i),$$

where

$$B_m(w_i) = \exp(\gamma_m + w_i\delta_m) \Big/ \sum_{m=1}^{M} \exp(\gamma_m + w_i\delta_m),$$

with the vectors $\delta_m = (\delta_{m1}, \ldots, \delta_{mR})'$ each of dimension R and $\gamma_M = \delta_M = 0$ for identifiability. By virtue of the multinomial logit form one has $\sum_{m=1}^{M} B_m(w) = 1$ and there are $1+(R+2)(M-1)$ unknowns. As for trigonometric function bases, the complexity and smoothness of the fit are controlled by M; typically $M = 2R+1$ where R is the number of predictors (Hooper, 2001, 350). The α_m parameters (which may be assigned diffuse normal priors) are fixed effects in Hooper (2001), and are not identified unless an overall intercept α is omitted in the specification for $g(\mu_i)$. They may be interpreted as a grid of intercepts, and an ordering constraint on the α_m may be imposed to achieve unique labeling. Marginal smooths in predictor w_r may be obtained by setting $\delta_{ms} = 0$ for $s \neq r$, and monitoring

$$S(w_{ri}) = \sum_{m=1}^{M} \alpha_m B_m^r(w_i),$$

where $B_m^r(w_i) = \exp(\gamma_m + w_{ri}\delta_{mr}) / \sum_{m=1}^{M} \exp(\gamma_m + w_{ri}\delta_{mr})$.

Example 10.2. Under 18 Conceptions These data relate to conceptions (y_i) to women aged under 18 in 352 English local authorities over a three-year period 2003–2005; denominators n_i are populations of women aged 15–17 in 2004 (times 3). Binomial sampling $y_i \backsim \text{Bin}(n_i, \mu_i)$ is assumed with potential explanatory factors being area deprivation, measured by an Index of Multiple Deprivation (IMD), and the percentage of 15-year-old pupils *not* achieving 5 or more GCSE subjects at grade C or above. The acronym GCSE refers to the General Certificate of Secondary Education, and educational proficiency is set by the criterion of grade C or above. Plots of moment estimates $r_i = y_i/n_i$ against both predictors suggest some nonlinearity (Figures 10.3 and 10.4).

The first model is BayesX analysis with cubic B-splines for both main effects and a bivariate interaction in $w_1 = \text{IMD}$ and $w_2 = \text{GCSE}$, with the interaction effect involving a tensor product of two one-dimensional B-splines (Lang and Brezger, 2004). As well as the smooth functions of the predictors, an unstructured residual $u_i \sim N(0, 1/\tau_u)$ is included in a logit link regression[4]

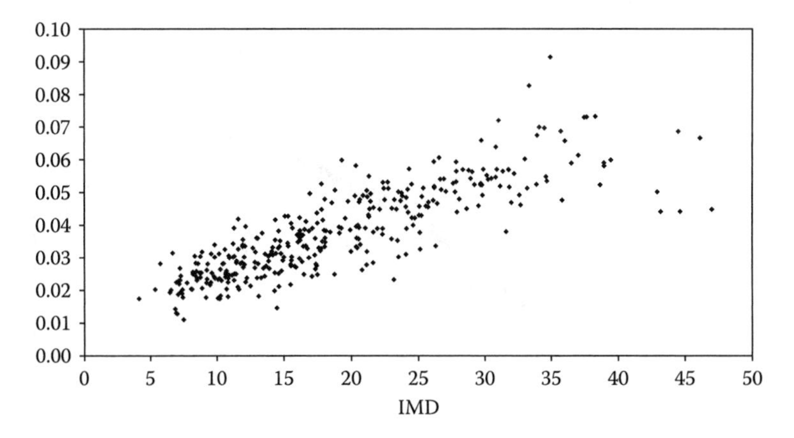

FIGURE 10.3
Conception rates against IMD scores.

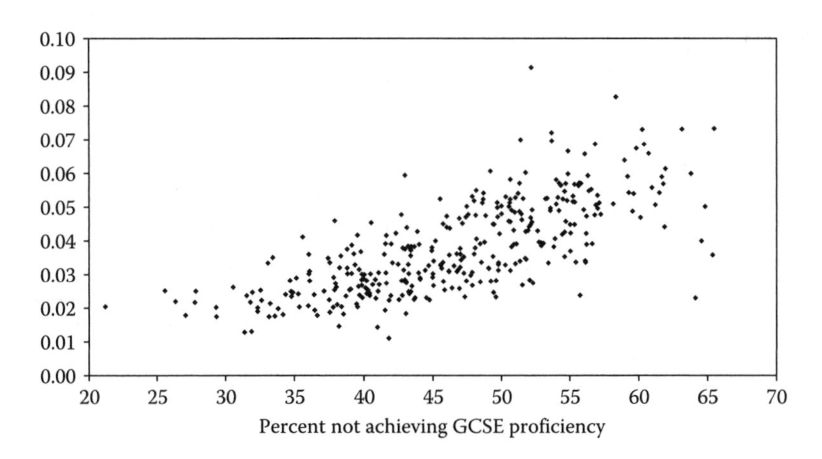

FIGURE 10.4
Conception rates and low GCSE attainment.

to account for overdispersion, where $\tau_u \sim \text{Ga}(1, 0.001)$. The regression then specifies

$$\begin{aligned}
\text{logit}(\mu_i) &= \alpha + S_1(w_{1i}) + S_2(w_{2i}) + S_{12}(w_{1i}, w_{2i}) + u_i, \\
S_r(w_{ri}) &= \sum_k^{K_r^*} b_{rk} B_k(w_{ri}, 3), \quad r = 1, 2, \\
S_{12}(w_{1i}, w_{2i}) &= \sum_k^{K_1^*} \sum_l^{K_2^*} c_{kl} B_k(w_{1i}, 3) B_l(w_{2i}, 3).
\end{aligned}$$

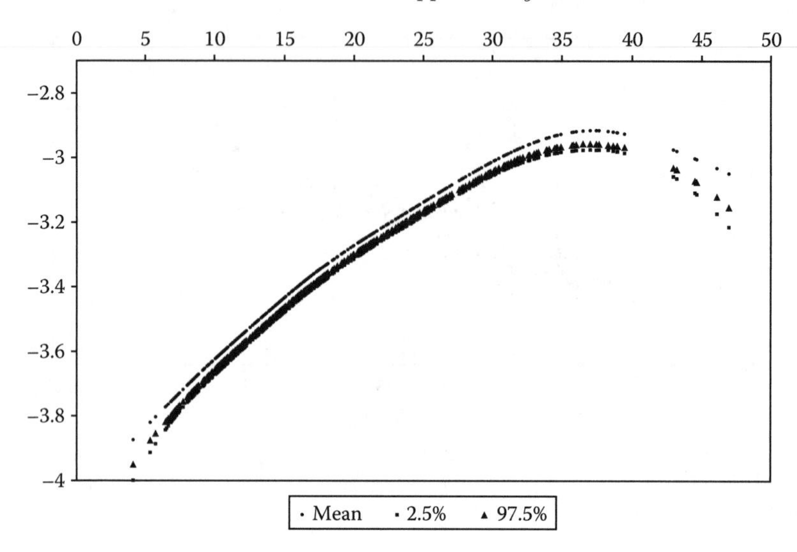

FIGURE 10.5
Logit conception probability and IMD, B-spline basis.

The main effect coefficients are subject to a second-order random walk roughness penalty, and the interaction coefficients are subject to an RW1 penalty. There are $K_1 = K_2 = 10$ equally spaced knots, where $\min(w_{ri}) = \kappa_{r0} < \kappa_{r1} \cdots < \kappa_{r,K_r+1} = \max(w_{ri})$.

The second model uses the Hooper (2001) ALB basis with $M = 10$ components, with the additional feature that the α_m parameters are random around a central parameter,

$$\alpha_m \sim N(\alpha, 1/\tau_\alpha),$$

where $\tau_\alpha \sim \mathrm{Ga}(1, 0.001)$. Then

$$\mathrm{logit}(\mu_i) = S(w_{1i}, w_{2i}) + u_i$$
$$S(w_{1i}, w_{2i}) = \sum_{m=1}^{M} \alpha_m B_m(w_i),$$
$$B_m(w_i) = \exp(\gamma_m + w_i\delta_m) \Big/ \sum_{m=1}^{M} \exp(\gamma_m + w_i\delta_m),$$

with the vectors γ_m and δ_m of dimension R, and $\gamma_M = \delta_M = 0$ for identifiability.

The first model produces a mean scaled deviance of 366 and $d_e = 293$, so the DIC is 659. Figures 10.5 and 10.6 represent the main effect smooths (these include the overall intercept α with posterior mean -3.226). A two chain run of

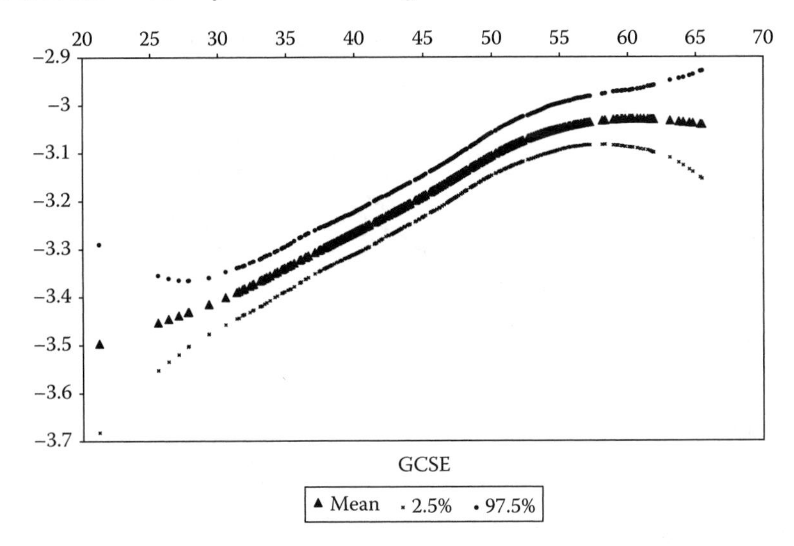

FIGURE 10.6
Logit conception probability and GCSE, B-spline basis.

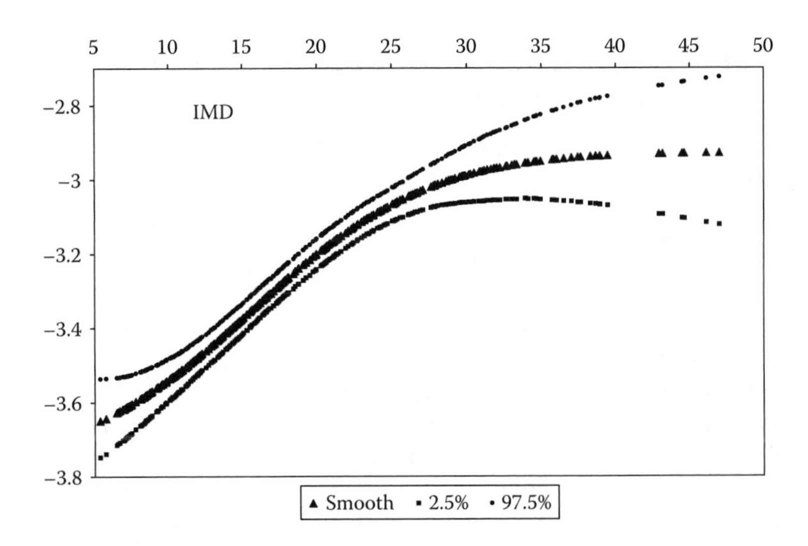

FIGURE 10.7
Logit conception probability and IMD.

5000 iterations of the ALB model (with second half for inferences) produces a mean scaled deviance of 367 and $d_e = 294.5$, so the DIC is 661.5. Figures 10.7 and 10.8 contain the marginal smooths for the two predictors under the second model, obtained as described above.

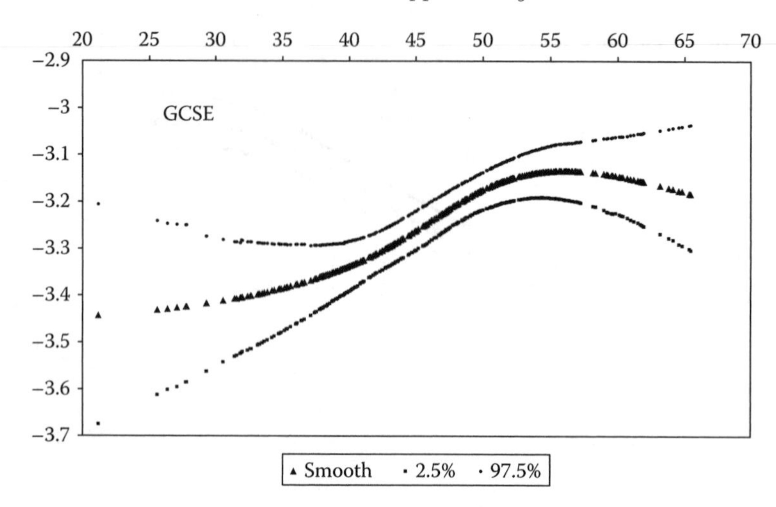

FIGURE 10.8
Logit conception probability and GCSE.

10.4 Heteroscedasticity via Adaptive Nonparametric Regression

As mentioned above, a random effects spline regression in a predictor w_i typically takes the form:

$$g(\mu_i) = \alpha + S(w_i) = \alpha + \varphi_1 w_i + \cdots + \varphi_q w_i^q + \sum_{k=1}^{K} b_k (w_i - \kappa_k)_+^q + u_i,$$

where $u_i \sim N(0, \sigma^2)$, and the spline coefficients may be taken as normal, for example, $b_k \sim N(0, \phi_b)$. This approach is spatially homogenous (in terms of the predictor space) whereas a spatially adaptive regression allows for heteroscedasticity, where such heteroscedasticity is also related to w values, or possibly to the values of other predictors (Currie and Durban, 2002). Thus with $u_i \sim N(0, \sigma_i^2)$, the variances $\sigma_i^2 = \exp(h_i)$ may be modeled by a subsidiary spline regression with M knots in the same predictor,

$$h_i = \gamma_0 + \gamma_1 w_i + \cdots + \gamma_q w_i^q + \cdots + \sum_{m=1}^{M} c_m (w_i - \psi_m)_+^q,$$

with $c_m \sim N(0, \phi_c)$. Ruppert and Carroll (2000) suggest M be taken much less than K, and apply the end knot constraints $\{\psi_1 = \kappa_1, \psi_M = \kappa_K\}$.

Jerak and Lang (2005) instead consider two alternative approaches. One is to use random walk priors in h_i, such as an RW1

$$h_i \sim N(h_{i-1}, 1/\tau_h).$$

The other involves independent local variances

$$\sigma_i^2 \sim \text{IG}\left(\frac{\nu}{2}, \frac{\nu}{2}\right)$$

which may be especially useful for functions with discontinuities. Taking $\nu = 1$ provides the Cauchy distribution. Jerak and Lang in fact consider heteroscedasticity in nonparametric regression for binary outcomes, and use the latent normal approach to representing the probit link, as in Wood and Kohn (1998).

Wood et al. (2002) suggest spatially adaptive nonparametric regression involving a discrete mixture over two or more smoothing functions, with the mixture probabilities based on multinomial logit regression involving additional covariates x_i. For y metric and M mixture components one might have

$$p(y_i|x_i, w_i) \sim \sum_{m=1}^{M} \pi_m(x_i) N(S_m(w_i, \theta_m), V_m),$$

$$\sum_{m=1}^{M} \pi_m(x_i) = 1,$$

where each smooth function $S_m(w, \theta_m)$ has its own parameter set θ_m.

Example 10.3. Elementary School Attainment The data for this example are a random sample of 400 elementary schools from California Education Department's API datafile for 2000 (http://www.cde.ca.gov/ta/ac/ap/apidatafiles.asp), which reports school academic performance (y_i) together with school characteristics such as average class size and the poverty rate among the pupil intake (Chen et al., 2003). A linear regression analysis involves regressing 2000 performance on the percent of pupils receiving free meals (FSM), percent of English language pupils (ELP), and percent of teachers with emergency credentials (EMCRED).

Let $\sigma^2 = \text{Var}(u_i)$ and assume $1/\sigma^2 \sim \text{Ga}(1, 0.001)$ in a homoscedastic linear regression

$$y_i = \alpha + \beta_1 \text{FSM}_i + \beta_2 \text{ELP}_i + \beta_3 \text{EMCRED}_i + u_i.$$

This provides a DIC of 4386.8, with $d_e = 4.9$. However, a plot of the residuals shows residual variation to decrease as fitted attainment increases (Figure 10.9). All three predictors have significant (negative) effects on attainment, but the highest ratio of posterior mean to standard deviation is for FSM, and a plot of the residual variation against FSM (Figure 10.10) suggests such variation increases with FSM.

A second model therefore specifies $y_i \sim N(\mu_i, \sigma_i^2)$:

$$\mu_i = \alpha + \beta_1 \text{FSM}_i + \beta_2 \text{ELP}_i + \beta_3 \text{EMCRED}_i + u_i,$$

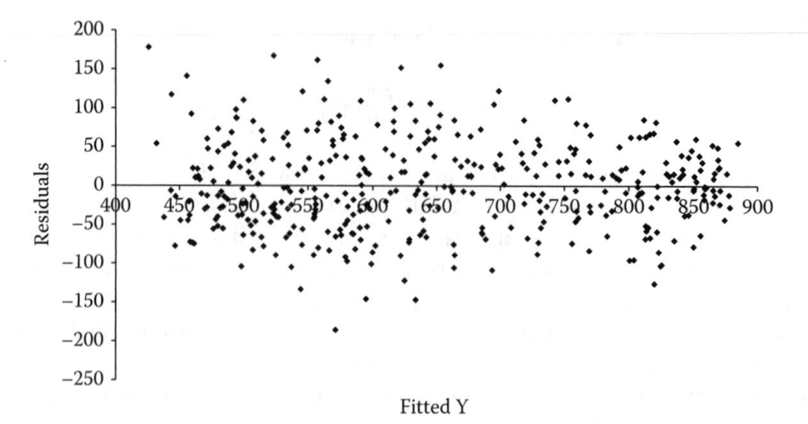

FIGURE 10.9
Fitted Y against residuals.

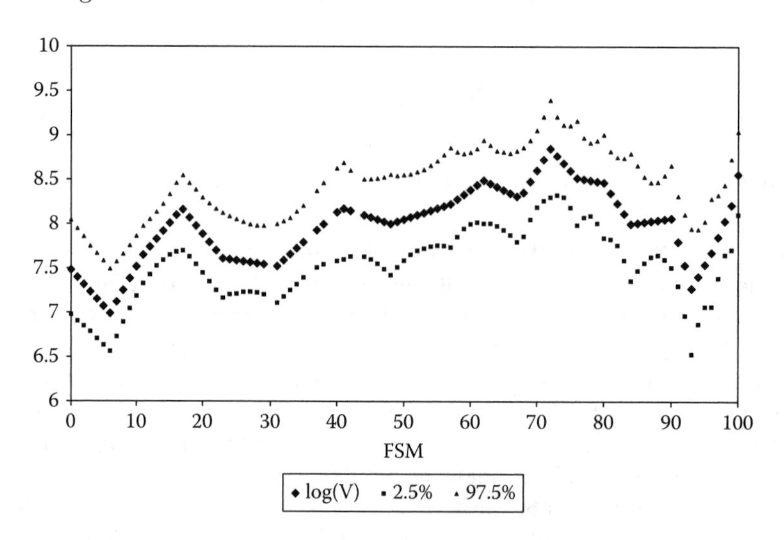

FIGURE 10.10
$\log(\sigma_i^2)$ against FSM.

where $u_i \sim N(0, \sigma_i^2)$, with σ_i^2 modeled by a spline regression

$$\log(\sigma_i^2) = \gamma_1 + \gamma_2 \text{FSM}_i + \sum_{m=1}^{M} c_m (\text{FSM}_i - \psi_m)_+.$$

The spline coefficients are random $c_m \sim N(0, \phi_c)$ with $1/\phi_c \sim \text{Ga}(1, 1)$. There are $M = 18$ knots, sited at the 5th, 10th,...,85th, and 90th percentiles of FSM. The 95th percentile of FSM is not included as it is 100%. The second half of a two chain run of 50,000 iterations gives an estimate for $\phi_c^{0.5}$ of 0.41

with 95% interval (0.29,0.60) whereas homoscedasticity would imply $\phi_c^{0.5} = 0$. There is still nonconstancy in $\log(\sigma_i^2)$ as FSM varies, though no a consistent monotonic upward or downward trend in variability as FSM increases. The DIC under the second model falls to 4372 ($d_e = 16.7$).

10.5 General Additive Methods

Consider ranked values of a single predictor $w_1, \ldots, w_n$ such that

$$w_1 < w_2 < \cdots < w_n$$

and let $S_t = S(w_t)$ be a smooth function representing the locally changing impact of w_t on $g(\mu_t)$ as it varies over its range. Thus

$$g(\mu_t) = \alpha + S(w_t) + u_t,$$

where $u_t \sim N(0, \sigma^2)$ and depending on identification procedures used the intercept α may not be present (Koop and Poirier, 2004). Appropriate priors for S_t reflect the ordering and spacing of the w values, and typically follow dynamic linear priors or other time series schemes. Normal or Student t random walks in the first, second or higher differences of S_t are one possibility (Fahrmeir and Lang, 2001; Knorr-Held, 1999). For identifiability, especially when there are smooths $S_{rt} = S(w_{rt})$ in several predictors one may adopt devices such as centering of the S_{rt}, or corner constraints (e.g., $S_{r1} = 0$). Alternatively to expedite computing speed, one may monitor identified quantities such as the centered series $S_{rt} - \bar{S}_r$without actually imposing centering constraints within the estimation. Because there is only local smoothing, inferences may also be sensitive to priors assumed for evolution variance τ^2 and other aspects of the model.

If the w values are equally spaced and distinct, then first and second-order random walk priors are just

$$S_t \sim N(S_{t-1}, \tau^2),$$
$$S_t \sim N(2S_{t-1} - S_{t-2}, \tau^2),$$

where smaller values of τ^2 result in a smoother curve. For metric or overdispersed discrete responses the parameterization $\tau^2\lambda = \sigma^2$ may be used, allowing for trade off between the residual variance and the variance of the smooth (Koop and Poirier, 2004).

In ordinary regression applications, values of the w_t are typically unequally spaced and there may be tied values. To take account of unequal spacing between successive w_t, the prior is modified such that for second and higher-order walks, the weighting on lagged values is varied according to how distant they are from the current value (Fahrmeir and Lang, 2001). In all orders of random walk, the precision of S_t is reduced the wider the gap between w_t and its preceding ordered values. Let gaps between points be denoted $\delta_2 = w_2 - w_1$, $\delta_3 = w_3 - w_2, \ldots, \delta_n = w_n - w_{n-1}$ (with $\delta_1 = 0$). Then a first-order Normal random walk becomes (for $t > 1$)

$$S_t \sim N(S_{t-1}, \delta_t \tau^2),$$

and a second order one becomes (for $t > 2$)

$$S_t \sim N([1 + \delta_t/\delta_{t-1}]S_{t-1} - [\delta_t/\delta_{t-1}]S_{t-2}, \delta_t \tau^2).$$

Separate usually fixed effect priors are assumed for the initial values (e.g., S_1 in a first-order random walk). A scheme allowing choice between RW1 and RW2 dependence for unequally spaced w is proposed by Berzuini and Larizza (1996), namely

$$s_t \sim N(M_t, \delta_t \tau^2)$$

where

$$M_t = s_{t-1}[1 + (\delta_t/\delta_{t-1})\exp(-\eta\delta_t)) - s_{t-2}[(\delta_t/\delta_{t-1})\exp(-\eta\delta_t)).$$

Larger values of $\eta > 0$, such that $\exp(-\eta\delta_t)$ tends to zero, imply an approximate RW1 prior and less smoothness.

If there are ties in the w values with only $m < n$ distinct values, denoted $\{w_j^*, j = 1, \ldots, m\}$, then the above priors would be on the differences $\delta_j = w_j^* - w_{j-1}^*$ in the ranked distinct values, and it is necessary to specify a grouping index G_t (ranging between 1 and m) for each observation $t = 1, \ldots, n$ to indicate which distinct value it takes. Assuming an RW1 prior in the smooth of the predictor effects, the regression in w_t can then be written

$$g(\mu_t) = \alpha + S(G_t) + u_t, \quad t = 1, \ldots, n,$$
$$S_j \sim N(S_{j-1}, \delta_j \tau^2), \quad j = 1, \ldots, m,$$

with $G_t \in (1, \ldots, m)$.

If there is more than one predictor then a semi-parametric model might be adopted with smooth functions $S_r(w_r)$ on a subset $r = 1, \ldots, q$ of R predictors, with the remainder modeled by assuming global linearity. So

$$g(\mu_t) = \alpha + S_1(w_{1t}) + S_2(w_{2t}) + \cdots + S_q(w_{qt}) + \beta_1 w_{q+1,t} + \cdots + \beta_{R-q} w_{R,t} + u_t.$$

If nonparametric functions are estimated for several regressors $w_{1t}, w_{2t}, \ldots, w_{qt}$, then a unique ordering across all predictors is usually infeasible and grouping indices $G_{1t}, G_{2t}, \ldots, G_{qt}$ for each of q regressors are necessary, even if the regressors have no tied values. In the case of tied values the indices range between 1 and m_1,1 and $m_2, \ldots$, 1 and m_q (rather than between 1 and n).

Another approach (Biller and Fahrmeir, 1997; Wahba, 1983; Wood and Kohn, 1998) to Bayesian general additive modeling involves the state–space version of the polynomial smoothing spline. For a spline of general order $2h-1$, $S_t = S(w_t)$ is generated by a differential equation:

$$\frac{d^h S_t}{dt^h} = \tau \frac{dW_t}{dt}$$

with W_t a Weiner process and τ^2 the evolution variance. The state vector

$$Z_t = \left(S_t, \frac{dS_t}{dt}, \frac{d^2 S_t}{dt^2}, \ldots, \frac{d^{(h-1)} S_t}{dt^{(h-1)}} \right)$$

is then of order h, evolving stochastically according to

$$Z_t = F_t Z_{t-1} + e_t, \tag{10.8}$$

where F_t is an $h \times h$ transition matrix and e_t is a multivariate error. For the cubic spline case with $h = 2$, $Z_t = (S_t, dS_t/dt)$ is bivariate and the transition matrix is

$$F_t = \begin{pmatrix} 1 & \delta_t \\ 0 & 1 \end{pmatrix},$$

where $\delta_t = w_{t+1} - w_t$. The e_t are also bivariate, for example MVN with zero mean and covariance $\tau^2 E_t$, where

$$E_t = \begin{pmatrix} \delta_t^3/3 & \delta_t^2/2 \\ \delta_t^2/2 & \delta_t \end{pmatrix}.$$

As usual there may be ties in the w values and the prior in Equation 10.8 would be on $j = 1, \ldots, m$ distinct ranked values. Each observation for $t = 1, \ldots, n$ would have a grouping index G_t with values between 1 and m.

Example 10.4. Conceptions under 18, RW2 Smooths This example considers again the data on conceptions to women aged under 18 (y_i) in 352 English local authorities. The model involves additive RW2 priors in w_1 = IMD and w_2 = GCSE. Let G_{1i} and G_{2i} indicate which of the unique IMD and GCSE values is taken by area i, where such unique values are ranked, with $m_1 = 352$ and $m_2 = 351$ unique values (there is a single tie in the GCSE values). Some of these distinct values are, however, very close to each other (a consideration relevant in a BayesX application). With $j = 1, \ldots, m_r$ denoting ranked predictor values,

$$\begin{aligned} y_i &\sim \text{Bin}(n_i, \mu_i), \\ \text{logit}(\mu_i) &= \alpha + s_1(w_{1,G_{1i}}) + s_2(w_{2,G_{2i}}), \\ s_{rj} &\sim N([1 + \delta_{rj}/\delta_{r,j-1}]s_{r,j-1} - [\delta_{rj}/\delta_{r,j-1}]s_{r,j-2}, \tau_r^2 \delta_{rj}), \\ & \quad r = 1, 2; j = 1, \ldots, m_r, \end{aligned}$$

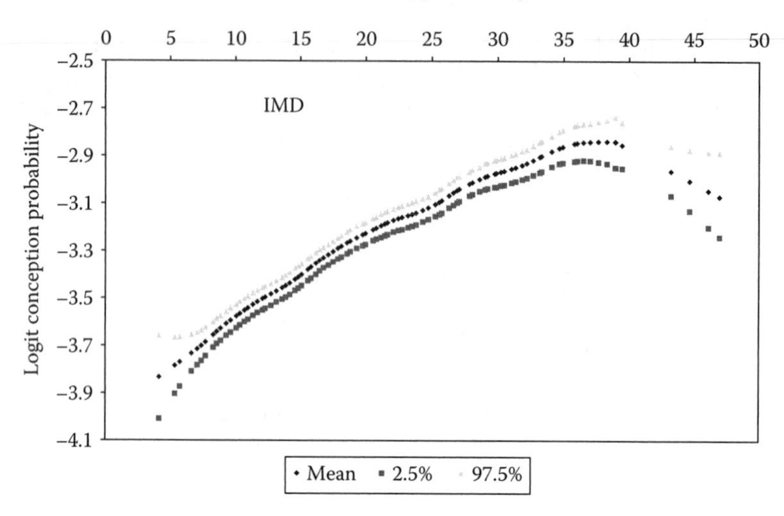

FIGURE 10.11
RW2 smooth in IMD.

where τ_r^2 is the variance for the randomly varying s_{rj}. There is excess dispersion which may be removed by a model also including an unstructured effect,

$$\text{logit}(p_i) = \alpha + s_1(w_{1,G_{1i}}) + s_2(w_{2,G_{2i}}) + u_i,$$

where $u_i \sim N(0, \sigma_u^2)$.

A BayesX analysis[5] is applied with inverse gamma priors IG(g, h) on variance parameters, and an initial setting of $\{g = 1, h = 0.001\}$ on all three variances. In a second run, the setting on h is changed to 0.0001 for the two variances τ_r^2. To avoid estimation of a large number of coefficients, BayesX performs internal grouping if a covariate has a large number of distinct values (for first- and second-order random walks), so the actual number of distinct values used will be lower than the observed m_r. The posterior means of the precisions (smoothing parameters) $\theta_r = 1/\tau_r^2$ are increased from {5192,5357} to {37178,35604} as h is lowered. Plots of the smooths under $h = 0.0001$ are based on 94 distinct IMD values and 91 distinct GCSE values; Figures 10.11 and 10.12 (which include the intercept) do not show implausible short-term fluctuations present in the smooths under $h = 0.001$. The DICs (based on the scaled deviance) are very similar between the two options, 662.9 under $h = 0.001$ and 662.6 under $h = 0.0001$.

Internal grouping of distinct covariate values can be avoided with the option "maxint," though for nonequidistant observations some grouping might be performed even if the number of distinct values is smaller than maxint. It is advisable to use a value of maxint that is larger than the number of distinct values. Setting maxint = 1000 produces an analysis based on 239 distinct values of IMD and 241 values of GCSE. With h set at 0.0001 this produces a

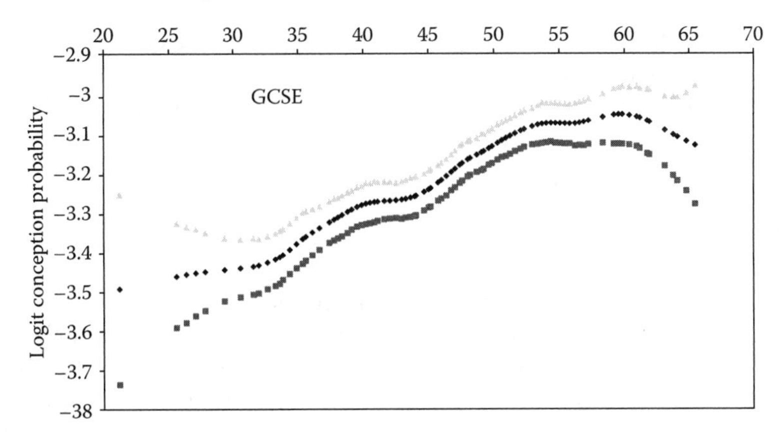

FIGURE 10.12
RW2 smooth in GCSE.

slight increase in DIC to 664.2, and reintroduces short-term fluctuations into the smooth in GCSE.

10.6 Nonparametric Regression Methods for Longitudinal Analysis

Two major applications of nonparametric regression to longitudinal datasets are to time varying regression coefficients and subject specific curves (Wu and Zhang, 2006; James et al., 2000). Time varying regression effects are a special case of the general varying coefficient model of Hastie and Tibshirani (1993), namely

$$g(\mu_i, \upsilon) = \beta_0(\upsilon_0) + w_{1i}\beta_1(\upsilon_1) + \cdots + w_{Ri}\beta_R(\upsilon_R),$$

where the effect modifiers $\upsilon = (\upsilon_1, \ldots, \upsilon_R)$ govern the effect of predictors $w = (w_1, \ldots, w_R)$. If the modifiers are all the same (e.g., time) with $\upsilon_1 = \upsilon_2 = \cdots = \upsilon_R = t$ then

$$g(\mu_{it}) = \beta_0(t) + w_{1i}\beta_1(t) + \cdots + w_{Ri}\beta_R(t)$$

and the time varying coefficient model, or dynamic general linear model (West and Harrison, 1997), is obtained. This extends to time varying predictors w_{rit}, with

$$g(\mu_{it}) = \beta_0(t) + w_{1it}\beta_1(t) + \cdots + w_{Rit}\beta_R(t).$$

Time varying intercept or regression effects $\beta_r(t)$ of unknown form can be fitted by any nonparametric method, such as regression or penalized splines, or random walks. For example, a B-spline approach would take

$$\beta_r(t) = \sum_{k=1}^{K^*} b_{rk} B_k(w_{rit}, q),$$

where b_{rk} are modeled as fixed or random effects. The fixed effects approach would typically be combined with selection of significant coefficients.

Allowing for intercepts or regression effects to vary by subject makes random effects a more sensible option. A comprehensive review of frequentist approaches to such nonparametric mixed models is provided by Wu and Zhang (2006)—see also Chapter 9 in Ruppert et al. (2003). A typical application is in growth curve analysis and involves subject specific nonparametric growth curves in time or age. For example, a growth curve model where observations at each wave included age could be modeled using a truncated spline,

$$g(\mu_{it}) = \alpha_t + c_i + S_i(\text{Age}_{it}) = \alpha_t + c_i + \sum_{k=1}^{K} b_{ik}(\text{Age}_{it} - \kappa_k)_+^q + u_{it},$$

where $u_{it} \sim N(0, \sigma^2)$, with σ^2 representing within subject variation, while $c_i \sim N(0, \sigma_c^2)$ with σ_c^2 measuring between subject heterogeneity. The subject specific spline coefficients b_{ik} are subject to a roughness penalty, such as a normal first difference penalty,

$$b_{ik} \sim N(b_{i,k-1}, 1/\theta_i),$$

with subject-specific precisions potentially modeled hierarchically. For example, one might take the $\log(\theta_i)$ to be normal with unknown variance. For applications with distinct recording times a_{it}, Wu and Zhang (2006, 205) suggest extended general linear mixed models with

$$g(\mu_{it}) = X_{it}\beta + \eta(a_{it}) + Z_{it}b_i + S_i(a_{it}) + u_{it},$$

where $\eta(a)$ is the population mean function, estimated nonparametrically, and $S_i(a)$ are subject-specific deviation functions. Silva et al. (2008) consider cubic B-spline bases to model region-wide and area-specific trends for health outcomes $y_{it} \backsim \text{Bin}(n_{it}, \pi_{it})$, namely

$$\text{logit}(\pi_{it}) = \alpha + \eta(t) + S_i(t) + d_i = \alpha + \sum_{k=1}^{K^*} b_k B_k(t, 3) + \sum_{k=1}^{K^*} c_{ik} B_k(t, 3) + d_i,$$

where d_i and c_{ik} are random area effects.

Another possible scheme for allowing variability across subjects is by random "slopes" around the population smooth functions, also sometimes denoted as random scaling of nonlinear functions (Tutz and Reithinger, 2007).

For example, consider a longitudinal (e.g., growth curve) application with a single predictor w_{it} the impact of which is modeled at population level by a smooth function $S(w_{it})$. Then one may wish to allow both for intercept (baseline) variation and for subject level variation around the average function $S(w)$. Thus

$$\begin{aligned} g(\mu_{it}) &= \alpha + b_{1i} + S(w_{it}) + b_{2i}S(w_{it}) + u_{it} \\ &= \alpha + b_{1i} + S(w_{it})(1 + b_{2i}) + u_{it}, \end{aligned}$$

where $(b_{1i}, b_{2i}) \sim N(0, D)$, and for identification $\sum_{it} S_{it} = 0$ where $S_{it} = S(w_{it})$. The smooth function $S(w_{it})$ represents the mean effect of predictor w_{it}, but this effect is stronger for subjects with $b_{2i} > 0$, and weaker for subjects with $b_{2i} < 0$. So b_{2i} acts to amplify or attenuate the nonparametric impact of the variable w_{it}. For some subjects one may even obtain large negative estimates, $b_{2i} < -1$, so that the effect of w_{it} is inverted. This model adapts to cross-sectional data where

$$g(\mu_i) = \alpha + S(w_i) + b_i S(w_i) + u_i$$

particularly in cases where the units are nonexchangeable, for example if the units were areas and b_i followed a spatial prior.

The impact of $(1 + b_{2i})$ on the unknown function $S(w_{it})$ is analogous to (subject-specific) factor loadings operating on factor scores and is subject to identifiability (label switching) issues, since $[-(1+b_{2i})][-S(w_{it})] = S(w_{it})(1+b_{2i})$. However, labeling issues (see Section 7.3) should be avoided in practice if the impact of w_{it} represented by $S(w)$ is well identified by the data. An alternative product scheme is applied by Congdon (2006b) based on the Lee and Carter (1992) mortality forecasting model. In this scheme subject-specific weights q_i that sum to 1 over all subjects operate on $S(w_{it})$, so that for $\sum_i q_i = 1$ the product scheme is $q_i S(w_{it})$. The effect of w is stronger for subjects with higher q_i, and weaker for subjects with lower q_i, with the average q_i being $1/n$.

Example 10.5. Progesterone Readings over Menstrual Cycle This example uses progesterone readings y_{it} in a study of early pregnancy loss (Brumback and Rice, 1998; Wu and Zhang, 2006). There are $i = 1, \ldots, 91$ observed cycles of length $T = 24$ days, so $N \times T = 2184$; the days are coded as $-8, -7, \ldots, 13, 14, 15$ with 0 as day of ovulation. There are $J = 2$ groups of observations, the first 69 cycles being nonceptive, the last 22 being conceptive. Two models are fitted, differing in that one involves nonparametric regression at conceptive group level, while the other involves subject level growth paths. So instead of a linear or polynomial function in the days variable, cubic B-splines are used with knots at $(-5, 0, 5, 10)$. The $K^* = 8$ basis functions are obtained from the R-splines package via the command:

```
bs(cycval,df=NULL,knots=c(-5,0,5,10),degree=3,intercept=TRUE,Boundary
.knots=range(cycval))
```

where cycval = c$(-8, -7, \ldots, 14, 15)$.

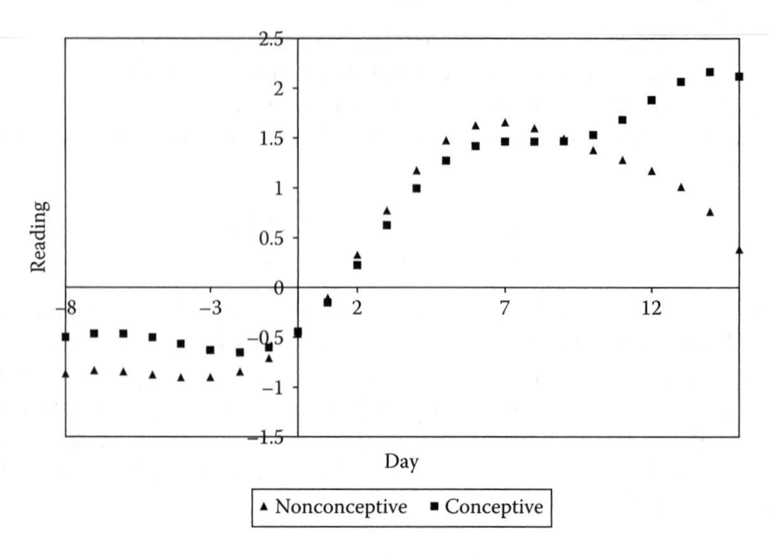

FIGURE 10.13
Smooths (posterior means) for groups.

In a baseline model, the spline coefficients are group specific random coefficients $\{b_{jk}, j = 1, 2, k = 1, K^*\}$, with group-specific precisions.[6] Let $G_i \in (1, 2)$ denote conceptive group, then $y_{it} \sim N(\mu_{it}, 1/\tau)$ with

$$\mu_{it} = \alpha_{G_i} + \sum_{k=1}^{K^*} b_{G_i k} B_k(t, 3),$$
$$b_{jk} \sim N(0, 1/\theta_j)$$

and $\theta_j \sim \text{Ga}(1, 0.001), \tau \sim \text{Ga}(1, 0.001)$. A two chain run of 5000 iterations (with the last 4000 for inference) shows a similar path between the two groups (posterior mean b_{jk}) up to the week after ovulation but distinct trends thereafter (Figure 10.13). The DIC is 5510 (d_e=16.5). A subject-specific model adds both intercept heterogeneity and subject (cycle) specific growth effects, so that

$$\mu_{it} = \alpha_{G_i} + b_{i0} + \sum_{k=1}^{K^*} b_{ik} B_k(t, 3)$$

with

$$b_{i0} \sim N(0, 1/\tau_0), \{i = 2, \ldots, n\}, \quad b_{10} = 0,$$
$$b_{i1} \sim N(0, 1/\tau_1),$$
$$b_{ik} \sim N(b_{i,k-1}, 1/\theta_{G_i}), \quad k = 2, K^*,$$
$$\tau_j \sim \text{Ga}(1, 0.001), \quad j = 0, 1; \theta_j \sim \text{Ga}(1, 0.001), j = 0, 1.$$

The corner constraint $b_{10} = 0$ is not strictly needed but aids distinct identification of the group intercepts α_j, and the subject heterogeneity effects. The DIC for this model is 2772 ($d_e = 515$). Note that instead of assuming b_{i0} and b_{i1} are independent one might assume they are correlated (see Section 8.5).

Example 10.6. Birthweight and Maternal Age Neuhaus and McCulloch (2006) consider a subset of data from a more extensive longitudinal study that involves the birthweights of babies born to $n = 878$ mothers from the state of Georgia, USA, all of whom has at least $T = 5$ babies. The analysis here is focused on the impact on birthweight y_{it} of mother's age at birth w_{it}, and the extent to which there is heterogeneity in the overall smooth $S(w_{it})$, which is based on a second-order random walk. Thus for each five birth history for mother i one may stipulate,

$$y_{it} = \beta_0 + b_{1i} + S(w_{it}) + b_{2i}S(w_{it}) + u_{it},$$

where $(b_{1i}, b_{2i}) \sim N(0, D)$, and D^{-1} follows a Wishart prior with identity scale matrix and two degrees of freedom. The random walk smooth is estimated over all (i, t) pairs using a normal RW2 prior with a single variance parameter, rather than on the basis of successive ages within each fertility sequence, which would permit distinct variance parameters for each subject.[7] The smooth involves 31 random parameters, namely for maternal ages 12–42.

A two chain run of 5000 iterations shows early convergence, with signficant heterogeneity in the b_{2i}, namely a posterior mean for var(b_2) of 1.35, and 95% interval {0.95, 1.74}. Figure 10.14 shows the varying nonparametric impact of

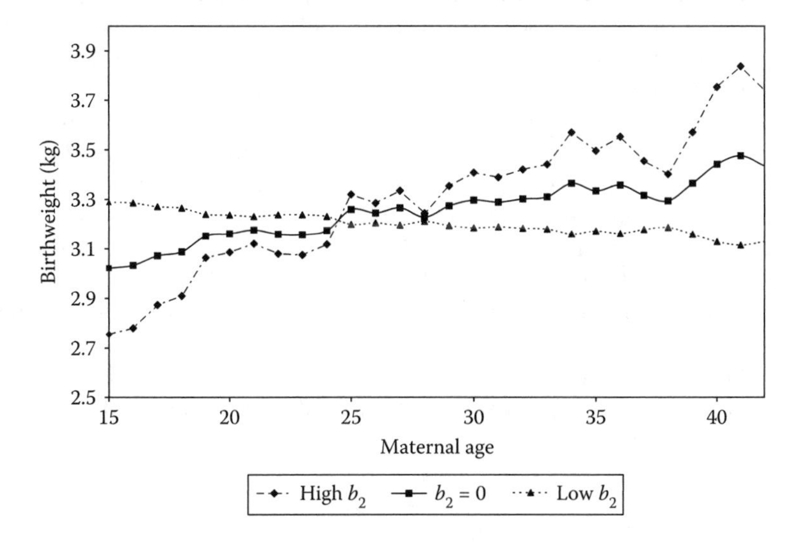

FIGURE 10.14
Heterogenity in nonlinear impact of maternal age.

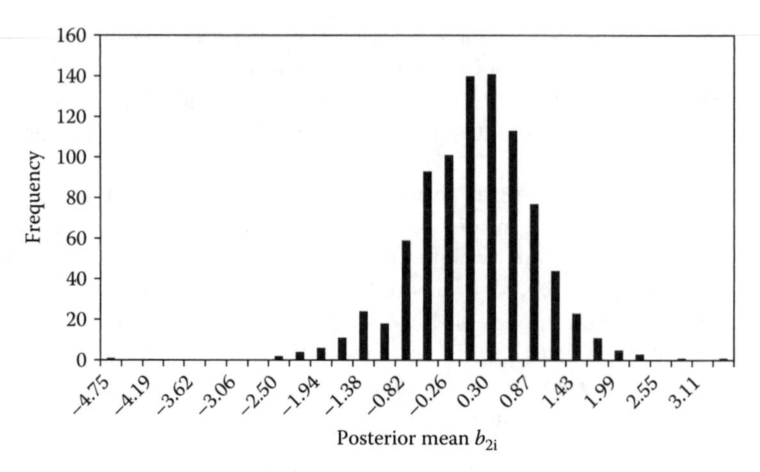

FIGURE 10.15
Histogram of mean b_2.

maternal age w_{it} according to b_{2i}, namely for subjects with $b_{2i} = sd(b_2)$, $b_{2i} = 0$ and $b_{2i} = -sd(b_2)$, where the standard deviations are those at particular MCMC iterations. A histogram plot of the posterior mean b_{2i} (Figure 10.15) indicates normality, though an extreme negative outlier of -4.75 occurs for subject 470, whose fourth and fifth infants weighed under 1 kg, whereas the first two exceeded 3 kg, in weight.

Example 10.7. Suicide Mortality and Deprivation This spatio-temporal panel example is used to illustrate both random spatial scaling of spline smooths and autocorrelated errors. The data are suicide death totals y_{it} (to all persons) over years $t = 1, \ldots, T$ (namely 1996–2006 so that $T = 11$) in 33 London boroughs. These are taken as Poisson, $y_{it} \sim \text{Po}(E_{it}\rho_{it})$, with expected deaths E_{it} based on England wide age-sex suicide rates for 2006. The smooth of interest involves an area deprivation score w_i from the 2001 Census, and constant and time varying B-spline smooths expressing how suicide relative risk varies with deprivation. $K = 9$ knots are set at the 10th, 20th, ..., 90th deciles of deprivation, and with $q = 3$ and $K^* = 13$, the $n \times K^*$ cubic B-spline function values $B_k(w_i, 3)$ are obtained using the R-splines routine.

The first model involves a constant B-spline smooth $S(w_i)$, and a bivariate CAR spatial effect (c_{1i}, c_{2i}), together with fixed effect time trend parameters Δ_t subject to a corner constraint $\Delta_1 = 0$. Hence

$$\log(\rho_{it}) = c_{1i} + (1 + c_{2i})S(w_i) + \Delta_t,$$

$$S(w_i) = \sum_{k=1}^{K^*} b_k B_k(w_i, 3),$$

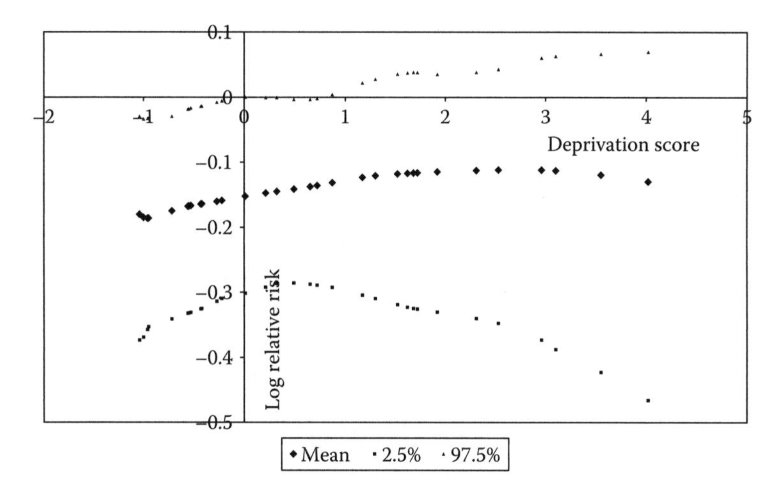

FIGURE 10.16
Suicide risk and deprivation.

where the b_k follow a first-order random walk, and to avoid confounding between the spline function and an overall intercept term, the latter is omitted. To assess autocorrelation in residuals $e_{it} = (y_{it} - \mu_{it})\mu_{it}^{-0.5}$, the statistic $r_e = \Sigma_{i=1}^{n}\Sigma_{t=2}^{T} e_{it}e_{i,t-1}/\Sigma_{i=1}^{n}\Sigma_{t=2}^{T} e_{it}^2$ is monitored. This model converges within 5000 iterations of a two chain run and iterations 5000–10,000 produce a DIC of 1960 ($d_e = 37$). Figure 10.16 reproduces the smooth $S(w)$ and it can be seen that varying deprivation is associated with a relatively small variation in relative risk, though there is a positive effect. The autocorrelation parameter has 95% interval (0.05,0.13).

A second model involves time varying splines, without random spatial scaling, so that

$$\log(\rho_{it}) = c_i + S_t(w_i) + \Delta_t,$$
$$S_t(w_i) = \sum_{k=1}^{K^*} b_{tk}B_k(w_i, 3),$$

with year-specific random walk penalty priors

$$b_{tk} \sim N(b_{t,k-1}, 1/\theta_t), k = 2, K^*,$$
$$b_{t1} \sim N(0, 1/\tau_1).$$

This model converges after 50,000 iterations of a two chain run and iterations 50,000–100,000 produce a slightly improved DIC of 1957 ($d_e = 43$). The positive deprivation effect now shows more in later years such as 2004 (Figure 10.17). The autocorrelation parameter now has a 95% interval (0.03,0.13).

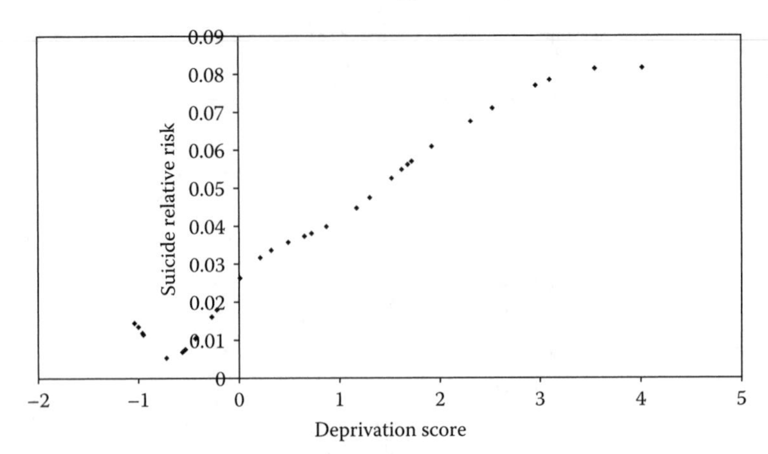

FIGURE 10.17
Time varying spline, smooth for 2004.

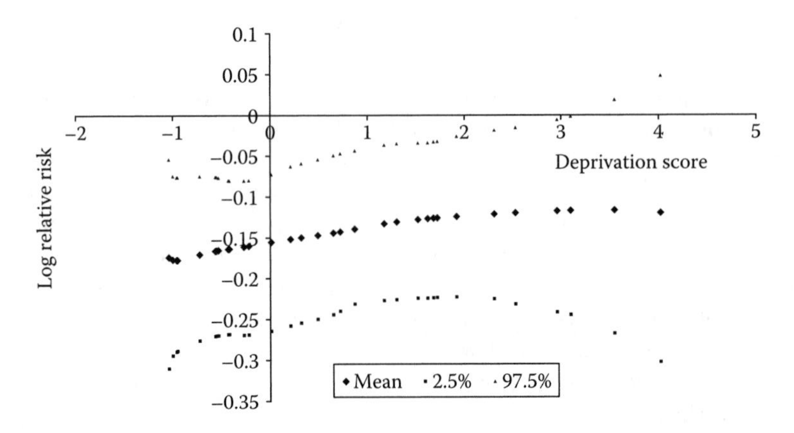

FIGURE 10.18
Suicide risk and deprivation, AR errors.

An alternative to time varying splines is to include a stationary $AR1$ error in the random spatial scaling model. So,

$$\log(\rho_{it}) = c_{1i} + (1 + c_{2i})S(w_i) + \Delta_t + \varepsilon_{it},$$
$$\varepsilon_{it} \sim N(\rho\varepsilon_{it}, \sigma^2), \quad t > 1,$$
$$\varepsilon_{i1} \sim N(0, \sigma^2(1 - \rho^2)^{-1}),$$

with $\rho \sim U(-1, 1)$. The second half of a two chain run of 50,000 iterations for this model gives a DIC of 1947 ($d_e = 62$), and produces a 95% interval for r_e that now straddles zero, namely $(-0.04, 0.10)$, with ρ estimated at 0.73 (0.39, 0.93). Compared to Figure 10.16 the mean smooth of relative risk on deprivation is, however, virtually unchanged, though the credible interval is narrower (see Figure 10.18).

Appendix: Computational Notes

1. Suppose y is observed at 15 points in time ($w_1 = 1, w_2 = 2, \ldots, w_{15} = 15$), and $K = 2$ knots are chosen at times 5 and 10. Then cubic B-spline values $\{B_k(w_i, 3), k = 1, K^*\}$ with $K^* = 6$, may be obtained by loading the splines package in R, inputting the values of $\{w_1, \ldots, w_{15}\}$ with name wvec, and then using the command

```
bs(wvec, df = NULL, knots = c(5,10), degree = 3, intercept = TRUE,
Boundary.knots = range(wvec)).
```

Eilers and Marx (1996, 100) provide MATLAB® code for obtaining B-spline values assuming equally spaced knots. B-spline schedules obtained by differencing truncated polynomials, and with the property $\sum_{k=1}^{K^*} B_k(w_i, q) = 1$, are produced by the R code (for $K1 = K + 1$ intervals) (Eilers and Marx, 2004),

```
trunc <- function(x, t, p) {(x - t) ^p * (x > t)}
B <- function(wvec, wmin, wmax, K1, q){
dw <- (wmax - wmin) / K1
knots <- seq( - q * dw, wmax + q * dw, by = dw)
P <- outer(wvec, knots, trunc, q); n <- dim(P)[2]
D <- diff(diag(n), diff = q + 1) / (gamma(q + 1) * dw ^ q)
B <- (-1) ^(q + 1) * P %*% t(D)
B}
```

2. The linear truncated spline in Example 10.1 is coded in WinBUGS as

```
model { for (i in 1:106) {y[i] <- 100000*SR[i]-70700; y[i] ~dnorm(mu[i],inv.s2)
mu[i] <- beta0+beta1*age[i]+sum(linpol[i,1:K])
for (k in 1:K) {linpol[i,k] <- b[k]*max(0,age[i]-kap[k])}}
# penalty prior
for (k in 1:K) {b[k] ~dnorm(0,th)}
beta0 ~dnorm(37,0.0001); beta1 ~dnorm(0,0.0001)
inv.s2 ~dgamma(1,0.001); th ~dgamma(1,0.001)}
```

The code for the cubic B-spline model (with the same knots) is the same except for lines 2–4 which are

```
mu[i] <- beta0+beta1*age[i]+sum(bsterms[i,1:K+q+1])
for (k in 1:K+q+1) {bsterms[i,k] <- b[k]*BS[i,k]}}
# random walk penalty
b[1] ~dnorm(0,0.001); for (k in 2:K+3) {b[k] ~dnorm(b[k-1],th)}
```

The B-spline coefficients are obtained in R by first inputting the age values, siting knots at every 5th percentile, and then using the command bs(age, df = NULL, knots = c(93.2,93.9,95.2,100.7,104.4,104.8,106.2,108.2, 109.0,109.5,110.3,111.5,112.4,113.7,115.4,118.0,119.3,120.8,122), degree = 3, intercept = TRUE,Boundary.knots = range(age)).

3. The BayesX program code for the Fossil data assumes the data file c:\data\stront.raw (ascii column format) has variable headings y and age. The code is then

```
dataset dstront
dstront.infile, maxobs=5000 using c:\data\stront.raw
bayesreg bstront
bstront.regress y = age(psplinerw2),family=gaussian predict using dstront
```

The entire code can be named stront.prg and saved in c:\program files\ BayesX. Implementing it then requires typing the command

```
usefile stront.prg
```

in the Command window of BayesX. The data for the smoothed plot can be obtained from the output file bstront_predictmean.raw.

4. The BayesX program code for the B-spline analysis of the conceptions data (with 10 knots rather than the default 20) is

```
dataset d
d.infile, maxobs=5000 using c:\data\concep.raw
bayesreg b
b.regress y = imd(psplinerw2,nrknots=10)+gcse(psplinerw2,nrknots=10)
+gcse*imd(pspline2dimrw1,nrknots=10)+areaid(random) weight femp,family
=binomial predict using d
```

The plots of the smooths are based on relevant *.res files.

The WinBUGS code for the ALB model as applied in Example 10.2 is

```
model { for (i in 1:n) {u[i] ~dnorm(0,tau.u)
y[i] ~dbin(p[i],femp[i]); logit(p[i]) <- sum(phi[i,])+u[i]
# total smooths in predictors 1 and 2
S[1,i] <- sum(S1[i,]); S[2,i] <- sum(S2[i,])
yhat[i] <- p[i]*femp[i]; yhatp[i] <- yhat[i]+con
yp[i] <- y[i]+con; dy[i] <- femp[i]-y[i]+con
dyhat[i] <- femp[i]-yhat[i]+con
dv[i] <- y[i]*log(yp[i]/yhatp[i]) +dy[i]*log(dy[i]/dyhat[i])
# total regression function
for (m in 1:M) {phi[i,m] <- alph[m]*EX[i,m]/sum(EX[i,1:M])
# components of "marginal" smooths in predictors 1 and 2
S1[i,m] <- alph[m]*EX1[i,m]/sum(EX1[i,1:M]);
S2[i,m] <- alph[m]*EX2[i,m]/sum(EX2[i,1:M])
EX[i,m] <- exp(gam[m]+del1[m]*(imd[i]-mean(imd[]))
+del2[m]*(gcse[i]-mean(gcse[])))
EX1[i,m] <- exp(gam[m]+del1[m]*(imd[i]-mean(imd[])))
EX2[i,m] <- exp(gam[m]+del2[m]*(gcse[i]-mean(gcse[])))}}
# scaled deviance
```

```
Dv <- 2*sum(dv[])
# priors
tau.u ~dgamma(1,0.001);
for (m in 1:M) {alph[m] ~dnorm(alph.m,tau.alph) }
alph.m ~dnorm(0,0.01); tau.alph ~dgamma(1,0.001)
for (m in 1:M-1) {gam[m] ~dnorm(0,0.01);
del1[m] ~dnorm(0,0.01); del2[m] ~dnorm(0,0.01)}
gam[M] <- 0; del1[M] <- 0; del2[M] <- 0}
```

5. The BayesX code for the RW2 general additive priors in the teenage conceptions analysis (Example 10.4) is

```
dataset d
d.infile, maxobs=5000 using c:\data\concep.raw
bayesreg b
b.regress y = imd(rw2,a=1,b=0.0001)+gcse(rw2,a=1,b=0.0001)+areaid
(random,a=1,b=0.001) weight femp,family=binomial predict using d
```

6. The progesterone reading data are arranged in stacked form in terms of cycles 1 to 91 and times -8 to 15. With G[1:91] containing the binary conceptive group values, the code is

```
model { for (i in 1:2184) {y[i] ~dnorm(mu[cyc[i],time[i]+9),tau)}
for (i in 1:91) { for (t in 1:T) {mu[i,t] <- mean[G[i],t)}}
for (j in 1:2) {for (t in 1:T) {mean[j,t] <- alph[j]+sum(bspl[j,t,1:Kstar])
for (k in 1:Kstar) {bspl[j,t,k] <- b[j,k]*B[t,k]}}}
for (j in 1:2) {alph[j] ~dnorm(0,0.001); th[j] ~dgamma(1,0.001)
for (k in 1:Kstar) {b[j,k] ~dnorm(0,th[j])}}
tau ~dgamma(1,0.001)}
```

The code for the subject specific model is

```
model { for (i in 1:2184) {y[i] ~dnorm(mu[cyc[i],time[i]+9),tau)}
for (i in 1:91) { for (t in 1:T) {mu[i,t] <- alph[G[i]]+
b0[i]+sum(bspl[i,t,1:Kstar])
for (k in 1:Kstar) {bspl[i,t,k] <- b[i,k]*B[t,k]}}}
for (j in 1:2) {alph[j] ~dnorm(0,0.001); th[j] ~dgamma(1,0.001)}
b0[1] <- 0; for (i in 2:91) {b0[i] ~dnorm(0,tau0)}
for (i in 1:91) { b[i,1] ~dnorm(0,tau1)
for (k in 2:Kstar) {b[i,k] ~dnorm(b[i,k-1],th[G[i]])}}
tau ~dgamma(1,0.001); tau0 ~dgamma(1,0.001);
tau1 ~dgamma(1,0.001)}
```

7. The code for the random heterogeneity maternal age effect model, including weight and adjacency information for the RW2 prior using the car.normal function is as follows:

```
model { for (i in 1:n) {b[i,1:2] ~dmnorm(nought[],Dinv[,])
for (t in 1:T) {Y[i,t] <- y[i,t]/1000; Y[i,t] ~dnorm(mu[i,t],tau)
mu[i,t] <- alpha+b[i,1]+S[i,t]+b[i,2]*S[i,t]
```

```
S[i,t] <- s[age[i,t]-11]}}
# bwt profiles by matage according to different b2
for (m in 15:42) {ageprof[1,m] <- alpha+s[m-11]*(1+sig.b[2]);
ageprof[2,m] <- alpha+s[m-11]
ageprof[3,m] <- alpha+s[m-11]*(1-sig.b[2])}
# RW2 prior in matage
s[1:X] ~car.normal(adj[], w[], nage[], taus)
w[1] <- 2; adj[1] <- 2; nage[1] <- 2
w[2] <- -1; adj[2] <- 3; nage[2] <- 3
w[3] <- 2; adj[3] <- 1; w[4] <- 4; adj[4] <- 3; w[5] <- -1; adj[5] <- 4
for (x in 3:X-2) {nage[x] <- 4;
                    w[6+(x-3)*4] <- -1; adj[6+(x-3)*4] <- x-2
                    w[7+(x-3)*4] <- 4;  adj[7+(x-3)*4] <- x-1;
                    w[8+(x-3)*4] <- 4; adj[8+(x-3)*4] <- x+1
                    w[9+(x-3)*4] <- -1;  adj[9+(x-3)*4] <- x+2}
w[(X-4)*4 + 6] <- 2; adj[(X-4)*4 + 6] <- X;
w[(X-4)*4 + 7] <- 4; adj[(X-4)*4 + 7] <- X-2
w[(X-4)*4 + 8] <- -1;  adj[(X-4)*4 + 8] <- X-3;
w[(X-4)*4 + 9] <- 2;  adj[(X-4)*4 + 9] <- X-1
w[(X-4)*4 + 10] <- -1;  adj[(X-4)*4 + 10] <- X-2;
nage[X-1] <- 3; nage[X] <- 2
# other priors
alpha ~dnorm(3,0.001); Dinv[1:2,1:2] ~dwish(ScD[,],2)
D[1:2,1:2] <- inverse(Dinv[,]); for (j in 1:2) {sig.b[j] <- sqrt(D[j,j])}
tau ~dgamma(1,0.01); taus ~dgamma(5,1)}
```

Appendix 1: Using WinBUGS and BayesX

A1.1 WinBUGS: Compiling, Initializing, and Running Programs

WinBUGS provides a versatile programming environment that is potentially adapted to a wide range of applications. The most usual approach involves specifying priors on parameters and stating the data likelihood. WinBUGS then selects conditional posterior distributions and sampling methods via an inbuilt expert system. However, for parameters obtainable using Gibbs sampling, one can also directly code the full conditionals.

Unlike other programming environments such as R, C, or Fortran, a WinBUGS program does not operate sequentially, and there is no prescribed order for programming statements of different kinds (e.g., it is not necessary that priors be specified before the likelihood). The program is first and foremost a description of the model and of the stochastic nodes that may be monitored at each Markov Chain Monte Carlo (MCMC) step. One may nevertheless perform more traditional programming tasks such as data manipulations. In the event of unexplained computational problems, it is often worth trying OpenBUGS as an alternative to WinBUGS14 and vice versa.

The following brief guide is intended for new users and applies to both programs. More comprehensive guides to developing and applying WinBUGS, as well as focused reviews, are provided by Fryback et al. (2001), Giminez et al. (2008), Lunn et al. (2009), Ntzoufras (2009), Scollnik (2002), and Woodworth (2004, Appendix B).

A1.2 WinBUGS Steps in Program Checking and Execution

A. Loading and checking the program: Open the program document (usually with an odc suffix) using “file” then “open.” Select “model” then “specification” from the main menu. Highlight the word model (or just the first few letters of the word model). Alternatively, select the entire model program code. Then select “check model.” Any syntax errors will be reported at this stage.

B. Loading the data: After loading the program, all datasets must be loaded. Two data formats are possible for datasets: what are known as s-files and ascii or formatted column files with each variable headed by a name in the first row. For s-files (starting with the word list) one loads by highlighting the whole data file or just the word list itself, or even just the first letter or two of the word list. Then select "load data." For ascii files one may select the whole file or just the first few letters in the first row. Then again select "load data." It is possible to have multiple data files, either s-files or ascii files. Note that "NA" means missing data, which implies that the model must include a mechanism to randomly generate them.

C. Setting the number of MCMC chains: The default is a single chain, so reset the number of chains if multiple chains are to be run.

D. Compile the Program. Select "compile": Errors reported here are many and various but may, for instance, include unknowns without prior distributions assigned, or variables mentioned in the data files but not in the model code.

E. Specifying initial parameter values: If compilation is successful, initial parameter values must then be assigned or randomly generated. Usually assigned (preset) values are chosen for at least a subset of the parameters, if not for the totality of parameters. The preset values are put together in what is colloquially known as an "inits" file, which also may be s-files or formatted column files. For an s-file, one may highlight the entire contents of the inits file or just the first letter or two of the word list at the start of the inits. Then select "load inits." If more than one inits file is used then repeat the procedure. An inits value is set to NA when the parameter is preset, or which has no prior itself but can be obtained from free parameters that do have priors. An example for a preset value is a corner constraint $\beta_1 = 0$ on a parameter set $\{\beta_k, k = 1, 4\}$ so that the initial values might be beta $= c(\text{NA}, 0, 0, 0)$. If there are parameters not initialized then one may press "gen inits" to generate them from the priors specified in the model code. If applied for a set of random effects (say), this can generate extreme values when the prior on the precision or precision matrix is diffuse; so an informative initial setting on the hyperparameter (e.g., a random effects precision of 10 or 100) should be used if initial values of the associated random effects are to be generated at random.

F. Running the estimation: To starting the MCMC sampling, select "model" from the main menu and then "update." An "update tool" icon appears. Often (e.g., in complex models) the default "refresh" of 100 will need to be reset to a smaller value (e.g., just 5 or 1). Also usually more than 1000 iterations are needed for a model to converge and produce sensible estimates; so resetting "updates" from 1000 to at least 5000 or 10,000 is advisable. Then select the stippled light-blue "update" icon. Only at the refresh point can the model run be stopped by selecting the now stippled update box, e.g., to list out current parameter estimates or to assess convergence. To set the model running again, select the stippled light-blue update box again.

G. Monitoring model parameters and other outputs: Either when estimation of the model is temporarily halted (as described in F), or before selecting "model" and "update," it is usually necessary to select which parameters or "nodes" are to be monitored. So select "inference" from the main menu and then "samples." In the "node" box enter the name of the parameter (or other model-related quantity) to be monitored. The word parameter is here used generically to include vectors and matrices. Then press "set". If more than one chain is running it is useful also to select the "trace" option to assess convergence visually.

H. Obtaining summaries: To obtain the current summary posterior statistics and/or density profile, select the stippled "update" button on the update icon to temporarily halt the run and select the appropriate node name and press "stats" or "density." If more than one chain is running one may also select "bgr diag" to check on convergence of that parameter or parameter set. The "history" button will also indicate the degree of mixing over chains.

I. Using inference/summary: To monitor large parameter sets (e.g., theta[1:N] where $N = 1000$) it is better to use the "inference" then "summary" option rather than the "inference/samples" option. Otherwise the memory may become overloaded. The inference/summary option however provides only summary posterior statistics, not features like density plots.

J. Running programs in batch mode using script: The above steps can be replicated in batch mode using a script file. Assume one has a program, data, and single inits file named rain_model.odc, rain_data.odc, and rain_inits.odc. These need to be placed in the WinBUGS program directory. The program is

```
model {for (i in 1:n) {y[i] ~dnorm(mu[i],tau)
mu[i] <- beta[1] + beta[2]*(x[i]-mean(x[]))}
# priors
beta[1] ~dnorm(40,0.0001); beta[2] ~dnorm(0,0.001)
tau ~dgamma(1,0.001); sig2 <- 1/tau}.
```

The data file is

```
list(n=10, y=c(41,52,18.7,55,40,29.2,51,17.6,46.6,57),
x=c(23.9,43.3,36.3,40.6,57,52.5,46.1,142,112.6,23.7)),
```

and the inits file is

```
list(beta=c(40,0),tau=1).
```

One needs to open a new file using File/New and then paste in the list of relevant "script" commands, such as

```
display('log')
check('rain_model.odc')
data('rain_data.odc')
compile(1)
inits(1,'rain_inits.odc')
thin.samples(5)
update(10,000)
set(beta)
set(sig2)
```

```
update(10,000)
stats(*)
history(*)
density(*)
autoC(*).
```

Then select Model/Script and the compilation, data and initial value loading and estimation stages will run automatically.

A1.3 Using BayesX

At the time of writing, BayesX is less general than WinBUGS in its programming syntax, and in features such as specifying alternative priors or nonstandard densities. For example, there is no equivalent to a for loop, and use of categorical predictors (with J levels) involves J-1 commands defining J-1 binary variables. There are also limits on windows functionality; for example, one cannot paste text to the command window, or select and copy text from any window.

However, for the problem types where BayesX does apply (e.g., nonparametric nonlinear regression, general linear models, spatial models with univariate CAR priors, univariate survival models), it can provide considerable computational advantages in speed and automatically provides extensive useful outputs. Calculation of the DIC and effective parameters may be included in the output, as well as files for graphical presentation of posterior densities, trace plots, etc. Reviews of BayesX include Brezger et al. (2005) and Kneib et al. (2008).

A current important limitation is that multivariate random effects, whether unstructured or spatially structured are not possible to specify. Similarly only univariate responses can be specified, not multivariate responses. Univariate autoregressive priors over area units (CAR priors) are included, but temporally autoregressive or moving average priors are not, except for random walk or P-spline priors in time. Thus, following Section 8.1.4 of the BayesX Reference Manual (version 2.00, from www.stat.uni-muenchen.de/~bayesx/manual/reference_manual.pdf) assume a Gaussian response y at quarters t, and dataset object named d containing values of y, t, and a predictor x. Then a Bayes analysis with a P-spline trend subject to second-order penalty, together with a seasonal component for the time scale t, may be specified via

b.regress $y = x + t$(psplinerw2) $+$ t(season,period$=$ 4), family=gaussian using d.

BayesX is configured in such a way that editing program files (with usual extension *.prg) is easiest in another file preparation program such as wordpad. Then one may use the batch facility in the command window of the BayesX screen, for example by typing in

usefile credit.prg

Although it is not possible to paste in text to the command window, one may restore command entries (and then edit them) using the review window. The usual working directory is c:\program files\Bayesx, so one suggestion is to save program files to a subdirectory c:\program files\Bayesx\programs relative to the usual working directory. To use the usefile command one then types in

usefile programs\credit.prg

Similarly data files with extension *.raw or *.txt may be saved to a subdirectory of c:\program files\Bayesx\data and then invoked from that subdirectory root.

To illustrate such features consider a Poisson regression using male lung cancer deaths in 32 London boroughs and standard rates provided by England annual age-specific rates 2004–2006; the standard rates are applied to area populations to provide expected deaths which are then log transformed to obtain the offset variable logE below. Assuming the subdirectory options as discussed above, a regression with spatial and unstructured random effects would then be invoked by typing in

usefile programs\london.prg

at the command prompt. The data file and model form are specified in london.prg as

```
dataset d
d.infile, maxobs=50 using data\london.txt
map m
m.infile, graph using data\london.gra
bayesreg b
b.regress y = logE(offset) + area(spatial,map=m) + area(random),
family=poisson predict using d
```

Inclusion of the word predict in the model (b.regress) command ensures the DIC and effective parameters are produced. To run and compare two models, one with and one without an unstructured error, one may use the code

```
dataset d
d.infile, maxobs=50 using data\london.txt
map m
```

m.infile, graph using data\london.gra
bayesreg b1
b1.regress y = logE(offset) + area(spatial,map=m), family=poisson predict using d
bayesreg b2
b2.regress y = logE(offset) + area(spatial,map=m) + area(random), family=poisson predict using d

The datafile london.txt includes a header line and 33 subsequent records.

area	y	logE
0	4	1.98
1	170	4.82
.....		
32	128	5.05

The areas are numbered 0 to 32 in order to coincide with the area numbering in the spatial adjacency file london.gra. This file has first line with the number of areas (here 33). There are then 3*33=99 remaining lines in the gra file. For each area, one line specifies its code, the second its number of neighbors, and third, the codes of the neighbors. Hence the file has structure

```
33
0
6
6 11 18 27 29 32
1
5
3 10 15 24 25
2
5
9 13 6 4 14
....
32
6
0 6 21 31 19 4
```

References

Aalen O (1988) Heterogeneity in survival analysis. *Statistics in Medicine*, 7, 1121–37.

Aalen O, Hjort N (2002) Frailty models that yield proportional hazards. *Statistics & Probability Letters*, 58, 335–42.

Abbring J, van den Berg G (2003) The identifiability of the mixed proportional hazards competing risks model. *Journal Royal Statistical Society: Series B*, 65, 701–10.

Abbring J, van den Berg G (2007) The unobserved heterogeneity distribution in duration analysis. *Biometrika*, 94, 87–99.

Abraham B, Ledolter J (1983) *Statistical methods for forecasting.* New York: Wiley.

Abrams K, Gillies C, Lambert P (2005) Meta-analysis of heterogeneously reported trials assessing change from baseline. *Statistics in Medicine*, 24, 3823–44.

Abrams K, Lambert P, Sanso B, Shaw S, Marteau T (2000) Meta-analysis of heterogeneously reported study results: A Bayesian approach. In D Berry, D Stangl (eds), *Meta-Analysis in Medicine and Health Policy.* Amsterdam: Marcel Dekker, 29–64.

Agresti A (1997) A model for repeated measurements of a multivariate binary response. *Journal of the American Statistical Association*, 92, 315–21.

Agresti A, Hitchcock D (2005) Bayesian inference for categorical data analysis. *Statistical Methods and Applications*, 14, 297–330.

Agresti A, Natarajan R (2001) Modeling clustered ordered categorical data: A survey. *International Statistical Review*, 69, 345–71.

Ahn S, Schmidt P (1995) Efficient estimation of models for dynamic panel data. *Journal of Econometrics*, 68, 5–27.

Aitchison J, Ho C (1989) The multivariate Poisson-log normal distribution. *Biometrika*, 76, 643–53.

Aitchison J, Shen S (1980) Logistic-normal distributions: Some properties and uses. *Biometrika*, 67, 261–72.

Aitkin M, Aitkin I (2005) Bayesian inference for factor scores. In A Maydeu-Olivares, J McArdle (eds), *Contemporary psychometrics.* Mahwah, NJ: Lawrence Erlbaum Associates.

Aitkin M, Clayton D (1980) The fitting of exponential, Weibull and extreme value distributions to complex censored survival data using GLIM. *Applied Statistics*, 29, 156–63.

Akaike H (1973) Information theory and an extension of the maximum likelihood principle. In B Petrov, F Csaki (eds), *The second international symposium on information theory*. Budapest: Akademiai Kiado, 267–81.

Alanko T, Duffy J (1996) Compound binomial distributions for modeling consumption data. *The Statistician*, 45, 269–86.

Albert J (1988) Computational methods using a Bayesian hierarchical generalized linear model. *Journal of American Statistical Association*, 83, 1037–45.

Albert J (1992) Bayesian estimation of normal ogive response curves using Gibbs sampling. *Journal of Educational Statistics*, 17, 251–69.

Albert J (1996a) A MCMC algorithm to fit a general exchangeable model. *Communications in Statistics: Simulation and Computation*, 25, 573–92.

Albert J (1996b) Bayesian selection of log-linear models. *Canadian Journal of Statistics*, 24, 327–47.

Albert J (1999) Criticism of a hierarchical model using Bayes factors. *Statistics in Medicine*, 18, 287–305.

Albert J (2007) *Bayesian Computation with R*. New York: Springer.

Albert J, Chib S (1993) Bayesian analysis of binary and polychotomous response data. *Journal of the American Statistical Association*, 88, 669–79.

Albert J, Chib S (1997) Bayesian tests and model diagnostics in conditionally independent hierarchical models. *Journal of American Statistical Association*, 92, 916–25.

Albert J, Chib S (2001) Sequential ordinal modeling with applications to survival data. *Biometrics*, 57, 829–36.

Albert J, Ghosh M (2000) Item response modeling. In D Dey, S Ghosh, B Mallick (eds), *Generalized linear models: A Bayesian perspective*. New York: Addison–Wesley, 173–93.

Albert J, Gupta A (1982) Distributions and estimation in contingency tables. *Annals of Statistics*, 10, 1261–1268.

Albert J, Pepple P (1989) A Bayesian approach to some overdispersion models. *Canadian Journal of Statistics*, 17, 333–44.

Albert P, Follmann D (2007) Random effects and latent processes approaches for analyzing binary longitudinal data with missingness: A comparison of approaches using opiate clinical trial data. *Statistical Methods in Medical Research*, 16, 417–39.

Allison P (1997) *Survival analysis using the SAS system: A practical guide*. Cary, NC: SAS Institute Inc.

Allison P (2000) Multiple imputation for missing data: A cautionary tale. *Sociological Methods and Research*, 28, 301–309.

Alqallaf F, Gustafson P (2001) On cross-validation of Bayesian models. *Canadian Journal of Statistics*, 29, 333–40.

Altaleb A, Chauveau D (2002) Bayesian analysis of the logit model and comparison of two Metropolis–Hastings strategies. *Computational Statistics & Data Analysis*, 39, 137–52.

Anderson J, Louis T, Holm N, Harvald B (1992) Time-dependent association measures for bivariate survival distributions. *Journal of American Statistical Association*, 87, 641–50.

Andersson M, Karlsson S (2007) Bayesian forecast combination for VAR models. Working Paper, Öebro University.

Ando T (2007) Bayesian predictive information criterion for the evaluation of hierarchical Bayesian and empirical Bayes models. *Biometrika*, 94, 443–58.

Ando T (2009) Bayesian factor analysis with fat-tailed factors and its exact marginal likelihood. *Journal of Multivariate Analysis*, 100, 1717–26.

Andrews R, Currim I (2003) Retention of latent segments in regression-based marketing models. *International Journal of Research in Marketing*, 20, 315–21.

Andrieu C, Moulines E (2006) On the ergodicity properties of some adaptive MCMC algorithms. *Annals of Applied Probability*, 16 (3), 1462–1505.

Angers J-F, Biswas A (2003) A Bayesian analysis of zero-inflated generalized Poisson model. *Computational Statistics & Data Analysis*, 42, 37–46.

Anselin L (1988) *Spatial econometrics: Methods and models.* Dordrecht: Kluwer Academic.

Anselin L, Bera A (1998) Spatial dependence in linear regression models, with an introduction to spatial econometrics. In A Ullah, D Giles (eds), *Handbook of applied economic statistics.* New York: Marcel Dekker, 237–90.

Anselin L, Hudak S (1992) Spatial econometrics in practice: A review of software options. *Regional Science & Urban Economics*, 22, 509–36.

Antonio K, Beirlant J (2007) Actuarial statistics with generalized linear mixed models. *Insurance: Mathematics and Economics*, 40, 58–76.

Arends L (2006) Multivariate meta-analysis: Modelling the heterogeneity. Repub/EUR Repository; http://repub.eur.nl/publications/med_hea.

Arhonditsis G, Paerl H, Valdes-Weaver L, Stow C, Steinberg J, Reckhow K (2006) Application of Bayesian structural equation modeling for examining phytoplankton dynamics in the Neuse River Estuary. *Estuarine, Coastal & Shelf Science*, 72, 63–80.

Arjas E, Gasbarra D (1994) Nonparametric Bayesian inference from right censored survival data, using the Gibbs sampler. *Statistica Sinica*, 4, 505–24.

Arminger G, Stein P, Wittenberg J (1999) Mixtures of conditional mean- and covariance-structure models. *Psychometrika*, 64, 475–94.

Asai M, McAleer M, Yu J (2006) Multivariate stochastic volatility: A review. *Econometric Reviews*, 25, 145–75.

Aslanidou H, Dey D, Sinha D (1998) Bayesian analysis of multivariate survival data using Monte Carlo methods. *Canadian Journal of Statistics*, 26, 33–48.

Assunção R (2003) Space varying coefficient models for small area data. *Environmetrics*, 13, 1–21.

Austin P, Escobar, M (2005) Bayesian Modeling of Missing Data in Clinical Research. *Computational Statistics and Data Analysis*, 49, 821–36.

Ayotte J, Baris D, Cantor K, Colt J, Robinson G, Lubin J, Karagas M, Hoover R, Fraumeni J, Silverman D (2006) Bladder cancer mortality and private well use in New England: An ecological study. *Journal of Epidemiology and Community Health*, 60, 168–72.

Azzalini A (1985) A class of distributions which includes the normal ones. *Scandinavian Journal of Statistics*, 12, 171–78.

Azzalini A (1994) Logistic regression and other discrete data models for serially correlated observations. *Statistical Methods and Applications*, 3, 169–79.

Bae S, Famoye F, Wulu J, Bartolucci A, Singh K (2005) A rich family of generalized Poisson regression models with applications. *Mathematics and Computers in Simulation*, 69, 4–11.

Baker A, Bray I (2005) Bayesian projections: What are the effects of excluding data from younger age groups? *American Journal of Epidemiology*, 162, 798–805.

Baker M, Melino A (2000) Duration dependence and nonparametric heterogeneity: A Monte Carlo study. *Journal of Econometrics*, 96, 357–93.

Baladandayuthapani V, Mallick B, Carroll R (2005) Spatially adaptive Bayesian penalized regression splines (P-splines). *Journal of Computational and Graphical Statistics*, 14, 378–94.

Balke N (1993) Detecting level shifts in time series. *Journal of Business & Economic Statistics*, 11, 81–92.

Baltagi B (2003) *Econometric Analysis of Panel Data*, 2nd edn. Chichester, Sussex: Wiley.

Baltagi B, Griffin, M (1988) A generalized error component model with heteroskedastic disturbances. *International Economic Review*, 29, 745–53.

Baltagi B, Li Q (1995) Testing AR(1) against MA(1) disturbances in an error component model. *Journal of Econometrics*, 68, 133–51.

Baltagi B, Wu P (1999) Unequally spaced panel data regressions with AR(1) disturbances. *Econometric Theory*, 15, 814–23.

Banerjee S, Carlin B (2004) Parametric spatial cure rate models for interval-censored time-to-relapse data. *Biometrics*, 60, 268–75.

Banerjee S, Carlin B, Gelfand A (2004) *Hierarchical modeling and analysis for spatial data.* Boca Raton, FL: Chapman & Hall/CRC.

Banfield J, Raftery A (1993) Model-based Gaussian and non-Gaussian clustering. *Biometrics*, 49, 803–21.

Banister J, Hill K (2004) Mortality in China 1964–2000. *Population Studies*, 58, 55–75.

Barmby T (2002) Worker absenteeism: A discrete hazard model with bivariate heterogeneity. *Labour Economics*, 9, 469–47.

Barnard J, McCulloch R, Meng X (2000) Modeling covariance matrices in terms of standard deviations and correlations, with applications to shrinkage. *Statistica Sinica*, 10, 1281–1311.

Barnett G, Kohn R, Sheather S (1996) Bayesian estimation of an autoregressive model using Markov chain Monte Carlo. *Journal of Econometrics*, 74, 237–54.

Barry R (2006) An alternative to the 'ones' trick?. *BUGS Mailing List* 08/11/2006 (http://www.jiscmail.ac.uk/lists/BUGS.html).

Bartholomew D (1987) *Latent variable models and factor analysis.* London: Charles Griffin.

Bartholomew D, Steele F, Moustaki I, Galbraith J (2002) *The analysis and interpretation of multivariate data for social scientists.* Boca Raton, FL: CRC Press.

Bartlett M (1957) A comment on D.V. Lindley's statistical paradox. *Biometrika*, 44, 533–34.

Basu S (1996) Bayesian tests for unimodality. In Proceedings Sect. on Bayesian Statistical Science. Amer. Statistical Assn., 77–82.

Batterham M (2005) Investigating heterogeneity in studies of resting energy expenditure in persons with HIV/AIDS: a meta-analysis. *American Journal of Clinical Nutrition*, 81, 702–13.

Bauwens L, Lubrano M (1998) Bayesian inference on GARCH models using the Gibbs sampler. *Econometrics Journal*, 1, C23–C46.

Bauwens L, Lubrano M, Richard J-F (1999) *Bayesian inference in dynamic econometric models.* Oxford: Oxford University Press.

Bayarri M, Berger J (1999) Quantifying Surprise in the Data and Model Verication. In J Bernardo, J Berger, A Dawid, A Smith (eds), *Bayesian statistics 6.* London: Oxford University Press, 53–82.

Bayarri M, Berger J (2000) P-values for composite null models. *Journal of the American Statistical Association*, 95, 1127–42.

Bayarri M, Castellanos M (2007) Bayesian checking of the second levels of hierarchical models. *Statistical Science*, 22, 363–67.

Bazan J, Bolfarine H, Branco M (2005) A general skew-probit link for binary response. In *Proceedings of the 9th School of Regression Models.* Sao Pedro, Brazil: Associacao Brasileira de Estatstica, 267–81.

Bazan J, Branco M, Bolfarine H (2006) A skew item response model. *Bayesian Analysis*, 1, 861–92.

Beck N (1983) Time-varying parameter regression models. *American Journal of Political Science*, 27, 557–600.

Beck N (2004) Time series. In M Lewis-Beck, A Bryman, T Futing Liao (eds), *Encyclopedia of social science research methods.* Thousand Oaks, California: Sage.

Beck N, Jackman S (1998) Beyond linearity by default: Generalized additive models. *American Journal of Political Science*, 42, 596–627.

Belitz C, Lang S (2008) Simultaneous selection of variables and smoothing parameters in structured additive regression models. *Computational Statistics & Data Analysis*, 53, 61–81.

Bell B, Broemeling L (2000) A Bayesian analysis for spatial processes with application to disease mapping. *Statistics in Medicine*, 19, 957–74.

Benjamin M, Rigby R, Stasinopoulos D (2003) Generalized autoregressive moving average models. *Journal of the American Statistical Association*, 98, 214–23.

Bennett S (1983) Log-logistic regression models for survival data. *Applied Statistics*, 32, 165–71.

Bentler P, Weeks D (1980) Linear structural equations with latent variables. *Psychometrika*, 45, 289–308.

Bera A, Jarque C (1980) Efficient tests for normality, homoscedasticity and serial independence of regression residuals. *Economics Letters*, 6, 255–59.

Berger J (1990) Robust Bayesian analysis: Sensitivity to the prior. *Journal of Statistical Planning and Inference*, 25, 303–28.

Berger J, Bernardo J (1992) On the development of reference priors. In J Bernardo, J Berger, A Dawid, A Smith (eds), *Bayesian Statistics 4*. Oxford: Clarendon Press, 35–60.

Berger J, Strawderman W, Tang D (2005) Posterior propriety and admissibility of hyperpriors in normal hierarchical models. *Annals of Statistics*, 33, 606–46.

Berke O (2004) Exploratory disease mapping: Kriging the spatial risk function from regional count data. *International Journal of Health Geographics*, 3, 18.

Berkes I, Horvath L, Ling S (2009), Estimation in nonstationary random coefficient autoregressive models. *Journal of Time Series Analysis*, 30, 395–416.

Berkhof J, van Mechelen I, Gelman A (2003) A Bayesian approach to the selection and testing of mixture models. *Statistica Sinica*, 13, 423–42.

Berkhof J, van Mechelen I, Hoijtink H (2000) Posterior predictive checks: Principles and discussion. *Computational Statistics*, 3, 337–54.

Berliner L (1996) Hierarchical Bayesian time series models. In K Hanson and R Silver (eds), *Maximum entropy and Bayesian methods*. Amsterdam: Kluwer Academic, 15–22.

Bernardinelli L, Clayton D, Montomoli C (1995) Bayesian estimates of disease maps: How important are priors? *Statistics in Medicine*, 14, 2411–31.

Bernardinelli L, Clayton D, Pascutto C, Montomoli C, Ghislandi M, Songini M (1995) Bayesian analysis of space-time variation in disease risk. *Statistics in Medicine*, 14, 2433–43.

Bernardo J, Smith A (1994) *Bayesian theory*. Chichester, Sussex: Wiley.

Berrington A, Hu Y,Ramirez-Ducoing K, Smith P (2005) *Multilevel modelling of repeated ordinal measures: An application to attitude towards divorce*. Southampton Statistical Sciences Research Institute Applications and Policy Working Paper M05/10 and ESRC Research Method Programme Working Paper No 26.

Berry S, Carroll R, Ruppert D (2002) Bayesian smoothing and regression splines for measurement error problems. *Journal of the American Statistical Association*, 97, 160–69.

Berzuini C, Clayton D (1994) Bayesian analysis of survival on multiple time scales. *Statistics in Medicine*, 13, 823–38.

Berzuini C, Larizza C (1996) A unified approach for modeling longitudinal and failure time data, with application in medical monitoring. *IEEE Transactions on Pattern Analysis and Machine Intelligence*, 18, 109–23.

Besag J (1989) Towards Bayesian image analysis. *Journal of Applied Statistics*, 16, 395–407.

Besag J (1974) Spatial interaction and the statistical analysis of lattice systems. *Journal of the Royal Statistical Society: Series B*, 36, 192–225.

Besag J, Green P (1993) Spatial statistics and Bayesian computation. *Journal of the Royal Statistical Society: Series B*, 55, 25–37.

Besag J, Green P, Higdon D, Mengersen K (1995) Bayesian computation and stochastic systems. *Statistical Science*, 10, 3–66.

Besag J, Kooperberg C (1995) On conditional and intrinsic autoregressions. *Biometrika*, 82, 733–46.

Besag J, York J, Mollie A (1991) Bayesian image restoration, with two applications in spatial statistics. *Annals of the Institute of Statistical Mathematics*, 43, 1–21.

Best N, Ickstadt K, Wolpert R (2000) Spatial Poisson regression for health and exposure data measured at disparate resolutions. *Journal of the American Statistical Association*, 95, 1076–88.

Bhattacharjee A, Jensen-Butler C (2006) Estimation of the spatial weights matrix in the spatial error model. Discussion Paper, School of Economics and Finance, University of St. Andrews.

Bijleveld F, Commandeur J, Gould P, Koopman S (2005) Model-based measurement of latent risk in time series with applications. Tinbergen Institute Discussion Paper No. 05-118/4. Available at SSRN: http://ssrn.com/abstract=873466.

Bijma F, De Munck J, Huizenga H, Heethaar R, Nehorai A (2005) Simultaneous estimation and testing of sources in multiple MEG data sets. *IEEE Transactions on Signal Processing*, 53, 3449–60.

Biller C (2000) Adaptive Bayesian regression splines in semiparametric generalized linear models. *Journal of Computational and Graphical Statistics*, 9, 122–40.

Biller C, Fahrmeir L (1997) Bayesian spline-type smoothing in generalized regression models. *Computational Statistics*, 12, 135–51.

Birkes D, Dodge Y (1993) *Alternative methods of regression.* Chichester: John Wiley.

Bivand R, Gebhardt A (2000) Implementing functions for spatial statistical analysis using the R language. *Journal of Geographical Systems*, 2, 307–17.

Bockenholt U (1999) An INAR(1) negative multinomial regression model for longitudinal count data. *Psychometrika*, 64, 53–68.

Bohning D (1999) *Computer-assisted analysis of mixtures and applications: Meta-analysis, disease mapping and others.* New York: Chapman & Hall.

Bohning D, Dietz E, Schlattmann P, Mendonca L, Kirchner U (1999b) The zero-inflated Poisson model and the decayed, missing and filled teeth index in dental epidemiology. *Journal of the Royal Statistical Society A*, 162, 195–209.

Bollen K (1989) *Structural equations with latent variables.* New York: Wiley.

Bollen K, Curran P (2004) Autoregressive latent trajectory (ALT) models: A synthesis of two traditions. *Sociological Methods & Research*, 32, 336–83.

Bollen K, Curran P (2006) *Latent curve models: A structural equation approach.* Hoboken, NJ: Wiley.

Bond S (2002) Dynamic panel data models: A guide to microdata methods and practice. *Portuguese Economic Journal*, 1, 141–62.

Borsuk M, Stow C (2000) Bayesian parameter estimation in a mixed-order model of BOD decay. *Water Research*, 34, 1830–36.

Bos C (2002) A comparison of marginal likelihood computation methods. In W Härdle, B Ronz (eds), *COMPSTAT 2002: Proceedings in computational statistics*. Heidelberg: Physica-Verlag, 111–17.

Box-Steffensmeier J, Jones B (2004) *Event history modeling*. Cambridge: Cambridge University Press.

Boyd H, Flanders W, Addiss D, Waller L (2005) Residual spatial correlation between geographically referenced observations: A Bayesian hierarchical modeling approach. *Epidemiology*, 16, 532–41.

Brown H, Prescott R (1999) *Applied mixed models in medicine*. Chichester: Wiley.

Brüderl J, Diekmann A (1995) The log-logistic rate model: Two generalizations with an application to demographic data. *Sociological Methods & Research*, 24, 158–86.

Bradlow E, Hardie B, Fader P (2002) Bayesian inference for the negative binomial distribution via polynomial expansions. *Journal of Computational and Graphical Statistics*, 11, 189–201.

Brandt P, Freeman J (2006) Advances in Bayesian time series modeling and the study of politics: Theory testing, forecasting, and policy analysis. *Political Analysis*, 14, 1–36.

Brandt P, Williams J (2007) *Modeling multiple time series*. Thousand Oaks, CA: Sage.

Breusch T (2005) Estimating the underground economy using MIMIC models. Working Paper, National University of Australia, Canberra.

Brezger A, Kneib T, Lang S (2005) BayesX: Analysing Bayesian structured additive regression models. *Journal of Statistical Software*, 14 (11), http://www.jstatsoft.org/v14/i11/paper/

Brezger A, Kneib T, Lang S (2008) *BayesX: Software for Bayesian inference in structured additive regression models, Methodology Manual*. Munich, Version 1.5.

Brezger A, Lang S (2006) Generalized structured additive regression based on Bayesian P-Splines. *Computational Statistics and Data Analysis*, 50, 967–91.

Brezger A, Steiner W (2008) Monotonic regression based on Bayesian P-splines: An application to estimating price response functions from store-level scanner data. *Journal of Business and Economics Statistics*, 26, 90–104.

Brooks S, Gelman A (1998) Alternative methods for monitoring convergence of iterative simulations. *Journal of Computational and Graphical Statistics*, 7, 434–56.

Brooks S, Roberts G (1998) Convergence assessment techniques for Markov chain Monte Carlo. *Statistics and Computing*, 8, 319–35.

Broto, C, Ruiz E (2004) Estimation methods for stochastic volatility models: A survey. *Journal of Economic Surveys*, 18, 613–49.

Brown P, Fearn T, Haque M (1999) Discrimination with many variables. *Journal of the American Statistical Association*, 94, 1320–29.

Browne W (2004) An illustration of the use of reparameterisation methods for improving MCMC efficiency in crossed random effect models. *Multilevel Modelling Newsletter*, 16, 13–25.

Browne W, Draper D (2006) A comparison of Bayesian and likelihood-based methods for fitting multilevel models. *Bayesian Analysis*, 1, 473–550.

Browne W, Draper D, Goldstein H, Rasbash J (2002) Bayesian and likelihood methods for fitting multilevel models with complex level-1 variation. *Computational Statistics and Data Analysis*, 39, 203–25.

Browne W, Goldstein H, Rasbash J (2001) Multiple membership multiple classification (MMMC) models. *Statistical Modelling*, 1, 103–24.

Browne W, Steele F, Golalizadeh M (2009) The use of simple reparameterizations to improve the efficiency of Markov chain Monte Carlo estimation for multilevel models with applications to discrete time survival models. *Journal of the Royal Statistical Society: Series A*, 172, 579–98.

Brumback B, Rice J (1998) Smoothing spline models for the analysis of nested and crossed samples of curves. *Journal of the American Statistical Association*, 93, 961–75.

Brumback B, Ruppert D, Wand M (1999) Variable selection and function estimation in additive nonparametric regression using a data-based prior: Comment. *Journal of the American Statistical Association*, 94, 794–97.

Brunsdon C, Fotheringham A, Charlton M (1998) Spatial nonstationarity and autoregressive models. *Environment and Planning A*, 30, 957–73.

Buck C, Sahu S (2000) Bayesian models for relative archaeological chronology building. *Applied Statistics*, 49, 423–44.

Buenconsejo J, Fish D, Childs J, Holford T (2008) A Bayesian hierarchical model for the estimation of two incomplete surveillance data sets. *Statistics in Medicine*, 27, 3269–85.

Bulmer M (1974) On fitting the Poisson log-normal distribution to species abundance data. *Biometrics*, 30, 101–110.

Burnham K, Anderson D (2002) *Model selection and multimodel inference: A practical information-theoretic approach.* 2nd edn. New York: Springer-Verlag.

Burr D, Doss H (2005) A Bayesian semi-parametric model for random effects meta analysis. *Journal of the American Statistical Association*, 100, 242–51.

Butler S, Louis T (1992). Random effects models with non-parametric priors. *Statistics in Medicine*, 11, 1981–2000.

Cai B, Dunson D (2006) Bayesian covariance selection in generalized linear mixed models. *Biometrics*, 62, 446–57.

Cai B, Dunson D (2008) Bayesian variable selection in generalized linear mixed models. In D. Dunson (ed.), *Random effect and latent variable model selection.* New York: Springer, 63–83.

Cai Z, Yao Q, Zhang W (2002) Smoothing for discrete-valued time series. *Journal of the Royal Statistical Society: Series B*, 357–75.

Calder C (2003) Exploring latent structure in spatial temporal processes using process convolutions. PhD thesis, Duke University.

Calder C (2007) Dynamic factor process convolution models for multivariate space–time data with application to air quality assessment. *Environmental and Ecological Statistics*, 14, 229–47.

Cameron A, Trivedi P (1998) *Regression analysis of count data*. Cambridge: Cambridge University Press.

Candel J, Winkens B (2003) Performance of empirical Bayes estimators of level-2 random parameters in multilevel analysis: A Monte Carlo study for longitudinal designs. *Journal of Educational and Behavioral Statistics*, 28, 169–94.

Cao G, West M (1996) Practical Bayesian inference using mixtures of mixtures. *Biometrics*, 52, 1334–41.

Cargnoni C, Muller P, West M (1997) Bayesian forecasting of multinomial time series through conditionally Gaussian dynamic models. *Journal of the American Statistical Association*, 92, 587–606.

Carlin B, Klugman S (1993) Hierarchical Bayesian Whittaker graduation. *Scandinavian Actuarial Journal*, 183–96.

Carlin B, Louis T (2000) *Bayes and empirical Bayes methods for data analysis*. 2nd edn. Chapman and Hall: London.

Carlin B, Polson D, Stoffer D (1992) A Monte Carlo approach to nonnormal and nonlinear state space modelling. *Journal of the American Statistical Association*, 87, 493–500.

Carlin B, Polson N (1992) Monte Carlo Bayesian methods for discrete regression models and categorical time series. In J Bernardo, J Berger, A Dawid, A Smith (eds), *Bayesian Statistics 4*. Oxford: Oxford University Press, 577–86.

Carlin J (1992) Meta-analysis for 2×2 tables: A Bayesian approach. *Statistics in Medicine*, 11, 141–58.

Carroll R, Ruppert D (1982) Robust estimation in heteroscedastic models. *Annals of Statistics*, 10, 429–44.

Carroll D, Ruppert D (1984) Power transformations when fitting theoretical models to data. *Journal of the American Statistical Association*, 79, 321–28.

Carter C, Kohn R (1994) On Gibbs sampling for state space models. *Biometrika*, 81, 541–53.

Celeux G, Forbes F, Robert C, Titterington M (2006) Deviance information criteria for missing data models. *Bayesian Analysis*, 1, 651–74.

Celeux G, Hurn M, Robert C (2000) Computational and inferential difficulties with mixture posterior distributions. *Journal of the American Statistical Association*, 95, 957–70.

Cepeda E, Gamerman D (2000) Bayesian modeling of variance heterogeneity in normal regression models. *Brazilian Journal of Probability and Statistics*, 14, 207–21.

Cepeda E, Gamerman D (2004) Bayesian modeling of joint regressions for the mean and covariance matrix. *Biometrical Journal*, 46, 430–40.

Cepeda-Benito A, Reynoso N, Erath S (2004) Meta-analysis of the efficacy of nicotine replacement therapy for smoking cessation: Differences between men and women. *Journal of Consulting and Clinical Psychology*, 72, 712–22.

Chaix B, Merlo J, Chauvin P (2005) Comparison of a spatial approach with the multilevel approach for investigating place effects on health: The example of healthcare utilisation in France. *Journal of Epidemiology and Community Health*, 59, 517–26.

Chaloner K, Brant R (1988) A Bayesian approach to outlier detection and residual analysis. *Biometrika*, 75, 651–60.

Chamberlain G, Hirano K (1999) Predictive distributions based on longitudinal earnings data. *Annales d'Economie et de Statistique*, 55, 211–42.

Chan D, Kohn R, Kirby C (2006) Multivariate stochastic volatility models with correlated errors. *Econometric Reviews*, 25, 245–74.

Chan K, Ledolter J (1995) Monte Carlo EM estimation for time series models involving counts. *Journal of the American Statistical Association*, 90, 242–52.

Chang Y, Gianola D, Heringstad B, Klemetsdal G (2006) A comparison between multivariate Slash, Student's t and probit threshold models for analysis of clinical mastitis in first lactation cows. *Journal of Animal Breeding and Genetics*, 123, 290–300.

Chatuverdi A, Kumar J (2005) Bayesian unit root test for model with maintained trend. *Statistics & Probability Letters*, 74, 109–15.

Chen C, Liu L (1993) Joint estimation of model parameters and outlier effects in time series. *Journal of the American Statistical Association*, 88, 284–97.

Chen C, So M (2006) On a threshold heteroscedastic model. *International Journal of Forecasting*, 22, 73–89.

Chen M, Ibrahim J (2000) Bayesian predictive inference for time series count data. *Biometrics*, 56, 678–85.

Chen M-H (2005) Computing marginal likelihoods from a single MCMC output. *Statistica Neerlandica*, 59, 16–29.

Chen M-H, Dey D (1998) Bayesian modeling of correlated binary responses via scale mixture of multivariate normal link functions. *Sankhya* 60A, 322–43.

Chen M-H, Dey D (2000) Bayesian analysis for correlated ordinal data models. In D Dey, S Ghosh, B Mallick (eds), *Generalized linear models: A Bayesian perspective*. New York: Marcel Dekker, 133–58.

Chen M-H, Ibrahim J (2006) The relationship between the power prior and hierarchical models. *Bayesian Analysis*, 1, 551–74.

Chen M-H, Ibrahim J, Shao Q-M (2000) Power prior distributions for generalized linear models. *Journal of Statistical Planning and Inference*, 84, 121–37.

Chen M-H, Ibrahim J, Sinha D (1999) A new Bayesian model for survival data with a surviving fraction. *Journal of the American Statistical Association*, 94, 909–19.

Chen M-H, Ibrahim J, Sinha D (2002) Bayesian inference for multivariate survival data with a cure fraction. *Journal of Multivariate Analysis*, 80, 101–26.

Chen M-H, Shao Q-M (1999) Monte Carlo estimation of Bayesian credible and HPD intervals. *Journal of Computational & Graphical Statistics*, 8, 69–92.

Chen X, Ender P, Mitchell M, Wells C (2003) Regression with Stata, from http://www.ats.ucla.edu/stat/stata/webbooks/reg/default.htm

Chen Y, Jewell N (2001) On a general class of semiparametric hazards regression models. *Biometrika*, 88, 687–702.

Chen Z (1993) Fitting multivariate regression functions by interaction spline models. *Journal of the Royal Statistical Society: Series B*, 55, 473–91.

Chen Z, Dunson D (2003) Random effects selection in linear mixed models. *Biometrics*, 59, 762–69.

Chen Z, Kuo L (2001) A note on the estimation of the multinomial logit model with random effects. *The American Statistician*, 55, 89–95.

Chi E, Reinsel G (1989) Models for longitudinal data with random effects and AR(1) errors. *Journal of the American Statistical Association*, 84, 452–59.

Chiang C (1984) *The life table and its applications*. Malabar, FL: R.E. Krieger.

Chiang J, Chib S, Narasimhan C (1999) Markov Chain Monte Carlo and models of consideration set and parameter heterogeneity. *Journal of Econometrics*, 89, 223–48.

Chib S (1993) Bayes regression with autoregressive errors: A Gibbs sampling approach. *Journal of Econometrics*, 58, 275–94.

Chib S (2008) Panel Data Modeling and Inference: A Bayesian Primer. In L Matyas, P Sevestre (eds), *The econometrics of panel data*. 3rd edn. Berlin: Springer-Verlag, 479–515.

Chib S, Carlin B (1999) On MCMC sampling in hierarchical longitudinal models. *Statistics & Computing*, 9, 17–26.

Chib S, Greenberg E (1994) Bayes inference in regression models with ARMA (p, q) errors. *Journal of Econometrics*, 64, 183–206.

Chib S, Greenberg E (1995) Understanding the Metropolis-Hastings algorithm. *The American Statistician*, 49, 327–35.

Chib S, Greenberg E (1998) Analysis of multivariate probit models. *Biometrika*, 85, 347–61.

Chib S, Jeliazkov I (2001) Marginal likelihood from the Metropolis-Hastings output. *Journal of the American Statistical Association*, 96, 270–81.

Chib S, Jeliazkov I (2006) Inference in semiparametric dynamic models for binary longitudinal data. *Journal of the American Statistical Association*, 101, 685–700.

Chib S, Nardari F, Shephard N (2002) Markov Chain Monte Carlo methods for stochastic volatility models. *Journal of Econometrics*, 108, 281–316.

Chib S, Nardari F, Shephard N (2005) Analysis of high dimensional multivariate stochastic volatility models. *Journal of Econometrics*, 134, 341–71.

Chib S, Winkelmann R (2001) Markov chain Monte Carlo analysis of correlated count data. *Journal of Business & Economic Statistics*, 19, 428–35.

Chintagunta P, Kyriazidou E, Perktold J (2001) Panel data analysis of household brand choices. *Journal of Econometrics*, 103, 111–53.

Chiogna M, Gaetan C (2002) Dynamic generalized linear models with application to environmental epidemiology. *Applied Statistics*, 51, 453–68.

Cho M, Schenker N (1999) Fitting the Log-F accelerated failure time model with incomplete covariate data. *Biometrics*, 55, 826–33.

Christensen O, Waagepetersen R (2002) Bayesian prediction of spatial count data using generalised linear mixed models. *Biometrics*, 58, 280–86.

Christiansen C, Morris C (1996) Fitting and checking a two-level Poisson model: Modeling patient mortality rates in heart transplant patients. In D Berry, D Stangl (eds), *Bayesian Biostatistics*. New York: Marcel Dekker, 467–501.

Christiansen C, Morris C (1997) Hierarchical Poisson regression modeling. *Journal of the American Statistical Association*, 92, 618–32.

Chun Y, Sumichrast R (2007) Bayesian inspection model with the negative binomial prior in the presence of inspection errors. *European Journal of Operational Research*, 182, 1188–1202.

Chung H, Loken E, Schafer J (2004) Difficulties in drawing inferences with finite-mixture models: A simple example. *The American Statistician*, 58, 152–58.

Clayton D (1991) A Monte Carlo method for Bayesian inference in frailty models. *Biometrics*, 47, 467–85.

Clayton D (1996) Generalized linear mixed models. In W Gilks, S Richardson, D Spiegelhalter (eds), *Markov Chain Monte Carlo in practice*. London: Chapman and Hall, 275–301.

Clayton D, Kaldor J (1987) Empirical Bayes estimates of age-standardised relative risks for use in disease mapping. *Biometrics*, 43, 671–82.

Clayton D, Schifflers E (1987) Models for temporal variation in cancer rates. II: Age-period-cohort models. *Statistics in Medicine*, 6, 467–810.

Cliff A, Ord J (1973) *Spatial autocorrelation*. London: Pion.

Cliff A, Ord J (1981) *Spatial processes: Models and applications*. London, Pion.

Cohen J, Nagin D, Wallstrom G, Wasserman L (1998) Hierarchical Bayesian analysis of arrest rates. *Journal of the American Statistical Association*, 93, 1260–70.

Collett D (2002) *Modelling binary data*, 2nd edn. New York, CRC Press.

Congdon P (2003) *Applied Bayesian modelling*. Chichester: Wiley.

Congdon P (2006a) A model framework for mortality and health data classified by age, area and time. *Biometrics*, 61, 269–78.

Congdon P (2006b) A model for geographical variation in health and total life expectancy. *Demographic Research*, 14, 157–78.

Congdon P (2007a) Model weights for model choice and averaging. *Statistical Methodology*, 4, 143–57.

Congdon P (2007b) Mixtures of spatial and unstructured effects for spatially discontinuous health outcomes. *Computational Statistics and Data Analysis*, 51, 3197–3212.

Congdon P (2008a) A spatially adaptive conditional autoregressive prior for area health data. *Statistical Methodology*, 5, 552–63.

Congdon P (2008b) Modelling the impact of socioeconomic structure on spatial health outcomes. *Computational Statistics and Data Analysis*, 53, 3047–56.

Congdon P (2008c) A bivariate frailty model for events with a permanent survivor fraction and non-monotonic hazards; with an application to age at first maternity, *Computational Statistics & Data Analysis*, 52, 4346–56.

Congdon P (2009) Life expectancies for small areas: a Bayesian random effects methodology. *International Statistical Review*, 77, 222–40.

Congdon P, Almog M, Curtis S, Ellerman R (2007) A spatial structural equation modelling framework for health count responses. *Statistics in Medicine*, 26 (29), 5267–84.

Congdon P, Lloyd P (2010) Estimating small area diabetes prevalence in the US using the behavioral risk factor surveillance system. *Journal of Data Science*, 8 (2), forthcoming.

Conlon E, Louis T (1999) Addressing multiple goals in evaluating region-specific risk using Bayesian methods. In A Lawson, A Bigger, D Bohning, E Lesaffre, J Viel, R Bertollini (eds), *Disease mapping and risk assessment for public health*. Chichester: John Wiley, 31–47.

Conlon E, Song N, Liu A (2007) Bayesian meta-analysis models for microarray data: A comparative study. *BMC Bioinformatics*, 8, 80 doi:10.1186/1471-2105-8-80.

Consul P (1989) *Generalized Poisson distributions*. New York: Marcel Dekker.

Cook D, Pocock S (1983) Multiple regression in geographical mortality studies, with allowance for spatially correlated errors. *Biometrics*, 39, 361–71.

Cooley T, Prescott E (1976) Estimation in the presence of stochastic parameter variation. *Econometrica*, 44, 167–84.

Cooner F, Banerjee S, McBean A (2006) Modelling geographically referenced survival data with a cure fraction. *Statistical Methods in Medical Research*, 15, 307–24.

Copas J, Li H (1997) Inference for non-random samples. *Journal of the Royal Statistical Society B*, 59, 55–95.

Coughlin C, Garrett T, Hernandez-Murillo R (2004) Spatial probit and the geographic patterns of state lotteries. No 2003-042, Working Papers from Federal Reserve Bank of St. Louis.

Coull B, Ruppert D, Wand, M (2001) Simple incorporation of interactions into additive models. *Biometrics*, 57, 539–45.

Cox D (1970) *The analysis of binary data*. London: Methuen.

Cox D (1972) Regression models and life-tables. *Journal of the Royal Statistical Society: Series B*, 34, 187–220.

Cox D (1981) Statistical analysis of time series: Some recent developments. *Scandinavian Journal of Statistics*, 8, 93–115.

Cressie N, Kapat P (2008) Some diagnostics for Markov random fields. *Journal of Computational and Graphical Statistics*, 17, 726–49.

Crowder M (1978) Beta-binomial Anova for proportions. *Applied Statistics*, 27, 34–37.

Crowder M (2001) *Classical competing risks.* Boca Raton, FL: CRC Press.

Curran D, Molenberghs G, Aaronson N, Fossa S, Sylvester R (2002) Analyzing longitudinal continuous quality of life data with dropout. *Statistical Methods in Medical Research*, 11, 5–23.

Currie I, Durban M (2002) Flexible smoothing with P-splines: A unified approach. *Statistical Modelling*, 2, 333–49.

Czado C (2000) Multivariate regression analysis of Panel data with binary outcomes applied to unemployment data. *Statistical Papers*, 41, 281–304.

Czado C, Erhardt V, Min A, Wagner S (2007) Zero-inflated generalized Poisson models with regression effects on the mean, dispersion and zero-inflation level applied to patent outsourcing rates. *Statistical Modelling*, 7, 125–53.

Damien P, Muller P (1998) A Bayesian bivariate failure time regression model. *Computational Statistics & Data Analysis*, 28, 77–85.

Dangl T, Halling M (2007) Predictive regressions with time-varying coefficients (March 15, 2007). Available at SSRN: http://ssrn.com/abstract=971712.

Daniels M (1999) A prior for the variance in hierarchical models. *Canadian Journal of Statistics*, 27, 569–80.

Daniels M, Gatsonis C (1999) Hierarchical generalized linear models in the analysis of variations in health care utilization. *Journal of the American Statistical Association*, 94, 29–42.

Daniels M, Hogan J (2008) *Missing data in longitudinal studies: Strategies for Bayesian modeling and sensitivity analysis.* Boca Raton, FL: CRC/Chapman & Hall.

Daniels M, Kass R (1999) Nonconjugate Bayesian estimation of covariance matrices and its use in hierarchical models. *Journal of the American Statistical Association*, 94, 1254–63.

Daniels M, Kass R (2001) Shrinkage estimators for covariance matrices. *Biometrics*, 57, 1174–84.

Daniels M, Normand S (2006) Longitudinal profiling of health care units based on continuous and discrete patient outcomes. *Biostatistics*, 7, 1–15.

Daniels M, Pourahmadi M (2002) Bayesian analysis of covariance matrices and dynamic models for longitudinal data. *Biometrika*, 89, 553–66.

Daniels M, Zhao Y (2003) Modelling the random effects covariance matrix in longitudinal data. *Statistics in Medicine*, 22, 1631–47.

Darby S, Fearn T (1979) The Chatham blood pressure study. An application of Bayesian growth curve models to a longitudinal study of blood pressure in children. *International Journal of Epidemiology*, 8, 15–21.

Das S, Dey D (2006) On Bayesian analysis of generalized linear models using the Jacobian technique. *The American Statistician*, 60, 264–68.

Das S, Dey D (2007) On Bayesian Analysis of Generalized Linear Models: A New Perspective. Technical Report 2007-8, Statistical and Applied Mathematical Sciences Institute, UNC (www.samsi.info).

Dauxois J-Y, Druilhet P, Pommeret D (2006) A Bayesian choice between Poisson, binomial and negative binomial models. *Test*, 15, 423–32.

Davidian M, Giltinan D (2003) Nonlinear models for repeated measures data: An overview and update. *Journal of Agricultural, Biological, and Environmental Statistics*, 8, 387–419.

Davies R (1994) From cross-sectional to longitudinal analysis. In A Dale, R Davies (eds), *Analysing social and political change: A casebook of methods*. London: Sage Publications, 20–40.

Davis R, Dunsmuir W, Streett S (2003) Observation-driven models for Poisson counts. *Biometrika*, 90, 777–90.

De Jong P, Penzer J (1998) Diagnosing shocks in time series. *Journal of the American Statistical Association*, 93, 796–806.

De Jong P, Penzer J (2004) The ARMA model in state space form. *Statistics & Probability Letters*, 70, 119–25.

De Jong P, Shephard N (1995) The simulation smoother for time series models. *Biometrika*, 82, 339–50.

de la Horra J, Rodrıguez-Bernal M (2005) Bayesian model selection: A predictive approach with losses based on distances. *Statistics & Probability Letters*, 71, 257–65.

De Oliveira V, Kedem B, Short D (1997) Bayesian prediction of transformed Gaussian random fields. *Journal of the American Statistical Association.* 92, 1422–33.

Dean C, MacNab Y (2001) Modeling of rates over a hierarchical health administrative structure. *Canadian Journal of Statistics*, 29, 405–19.

Deely J, Smith A (1998) Quantitative refinements for comparisons of institutional performance. *Journal of the Royal Statistical Society: Series A*, 161, 5–12.

Dellaportas P, Smith A (1993) Bayesian inference for generalized linear and proportional hazards models via Gibbs sampling. *Applied Statistics*, 42, 443–59.

Delucchi K, Bostrom A (2004) Methods for analysis of skewed data distributions in psychiatric clinical studies: Working with many zero values. *American Journal of Psychiatry*, 161, 1159–68.

Denison D, Mallick B, Smith A (1998) Automatic Bayesian curve fitting. *Journal of the Royal Statistical Society B*, 60, 333–50.

Dennison D, Holmes C, Mallick B, Smith A (2002) *Bayesian methods for non-linear classification and regression.* John Wiley: Chichester.

Dey D, Ghosh S, Mallick B (eds) (2000) *Generalized linear models: A Bayesian perspective.* New York: Marcel Dekker.

Dey D, Ravishanker N (2000) Bayesian approaches for overdispersion in generalized linear models. In D Dey, S Ghosh, B Mallick (eds), *Generalized linear models: A Bayesian perspective.* New York: Dekker, 73–88.

Diaconis P, Ylvisaker D (1979) Conjugate priors for exponential families. *Annals of Statistics*, 7, 269–81.

Dias R, Gamerman D (2002) A Bayesian approach to hybrid splines nonparametric regression. *Journal of Statistical Computation and Simulation*, 72, 285–98.

Diebolt N, Robert C (1994) Estimation of finite mixture distributions through Bayesian sampling. *Journal of the Royal Statistical Society: Series B*, 56, 363–75.

Diekmann A, Mitter P (1983) The "Sickle Hypothesis": A time-dependent Poisson model with applications to deviant behavior and occupational mobility. *Journal of Mathematical Sociology*, 9, 85–101.

Diggle P, Kenward M (1994) Informative dropout in longitudinal data analysis. *Journal of the Royal Statistical Society: Series C*, 43, 49–94.

Diggle P, Liang K, Zeger S (1994) *Analysis of longitudinal data.* Oxford: Clarendon Press.

Diggle P, Ribeiro P (2001) Bayesian inference in Gaussian model-based geostatistics. *Geographical and Environmental Modelling*, 6, 129–46.

Diggle P, Ribeiro P, Christensen O (2003) An introduction to model based geostatistics. In J Möller (ed.), *Spatial statistics and computational methods. Lecture Notes in Statistics*, Vol. 173. New York: Springer, 43–86.

Diggle P, Tawn J, Moyeed R (1998) Model based geostatistics. *Applied Statistics*, 47, 299–350.

Diggle P, Zeger S (1989) A non-Gaussian model for time series with pulses. *Journal of the American Statistical Association*, 84, 354–59.

Disease Mapping Collaborative Group (2000) Disease mapping models: An empirical evaluation. *Statistics in Medicine*, 19, 2217–41.

Diserud O, Engen S (2000) A general and dynamic species abundance model, embracing the lognormal and the gamma models. *The American Naturalist*, 155, 497–511.

Dorsett R (1999) An econometric analysis of smoking prevalence among lone mothers. *Journal of Health Economics*, 18, 429–41.

Draper D (1995) Assessment and propagation of model uncertainty. *Journal of the Royal Statistical Society: Series B*, 57, 45–97.

Draper D (1995) Inference and hierarchical modeling in the social sciences. *Journal of Educational and Behavioral Statistics*, 20, 115–47.

Droguett E, Groen F, Mosleh A (2006) Bayesian assessment of the variability of reliability measures. *Pesquisa Operacional*, 26, 109–127.

DuMouchel W (1996) Predictive cross-validation of Bayesian meta-analyses. In Bernardo J, Berger J, Dawid A, Smith A (eds), *Bayesian Statistics 5.* Oxford: Oxford University Press, 107–27.

DuMouchel W, Waternaux C (1992) Discussion of "Hierarchical models for combining information and for meta-analysis," by C Morris and S Normand. In J Bernardo, J Berger, A Dawid, A Smith (eds), *Bayesian Statistics*, Vol. 4. Oxford: Clarendon Press, 338–41.

Duan J, Guindani M, Gelfand A (2007) Generalized spatial Dirichlet process models. *Biometrika*, 94, 809–25.

Dunson D (2001) Commentary: Practical advantages of Bayesian analysis of epidemiologic data. *American Journal of Epidemiology*, 153, 1222–26.

Dunson D (2003) Dynamic latent trait models for multidimensional longitudinal data. *Journal of the American Statistical Association*, 98, 555–63.

Dunson D (2006) Bayesian dynamic modeling of latent trait distributions. *Biostatistics*, 7, 551–68.

Dunson D (2007) Bayesian methods for latent trait modeling of longitudinal data. *Statistical Methods in Medical Research*, 16, 399–415.

Dunson D, Herring A (2005) Bayesian latent variable models for mixed discrete outcomes. *Biostatistics*, 6, 11–25.

Dunson D, Pillai N, Park J (2007) Bayesian density regression. *Journal of the Royal Statistical Society: Series B*, 69, 163–83.

Durban M, Currie I, Eilers P (2006) Multidimensional P-spline mixed models: A unified approach to smoothing on large grids. Working Paper.

Durbin J (2000) The Foreman lecture: The state space approach to time series analysis and its potential for official statistics. *Australian & New Zealand Journal of Statistics*, 42, 1–24.

Durbin J, Koopman S (2000) Time series analysis of non-Gaussian observations based on state space models from both classical and Bayesian perspectives. *Journal of the Royal Statistical Society*, 62B, 3–56.

Durbin J, Koopman S (2001) *Time series analysis by state space methods.* Oxford: Oxford University Press.

Eaton W (1974) Mental hospitalization as a reinforcement process. *American Sociological Review*, 39, 252–60.

Edwards Y, Allenby G (2003) Multivariate analysis of multiple response data. *Journal of Marketing Research*, 40, 321–34.

Efron B (1986) Double exponential families and their use in generalized linear regression. *Journal of the American Statistical Association*, 81, 709–21.

Eid M (1996) Longitudinal confirmatory factor analysis for polytomous item responses: Model definition and model selection on the basis of stochastic measurement. *Methods of Psychological Research Online*, 1, 65–85.

Eilers P, Marx B (1996) Flexible smoothing with B-splines and penalties. *Statistical Science*, 11, 89–121.

Eilers P, Marx B (2004) Splines, knots, and penalties. Working Paper (www.stat.lsu.edu/faculty/marx/)

Eksler V (2008) Exploring spatial structure behind the road mortality of regions in Europe. *Applied Spatial Analysis and Policy*, 1, 133–50.

Engle R (1982) Autoregressive conditional heteroscedasticity with estimates of variance of United Kingdom inflation. *Econometrica*, 50, 987–1008.

Engle R, Granger C, Rice J, Weiss A (1986) Semiparametric estimates of the relation between weather and electricity sales. *Journal of the American Statistical Association*, 81, 310–20.

Epstein D, O'Halloran S (1996) Divided government and the design of administrative procedures: a formal model and empirical test. *Journal of Politics*, 58, 373–97.

Erbas B, Hyndman R (2005) Sensitivity of the estimated air pollution–respiratory admissions relationship to statistical model choice. *International Journal of Environmental Health Research*, 15, 437–48.

Erkanli A, Soyer R, Angold A (2001) Bayesian analyses of longitudinal binary data using Markov regression models of unknown order. *Statistics in Medicine*, 20, 755–70.

Escobar M, West M (1998) Computing nonparametric hierarchical models. In D Dey, P Muller, D Sinha (eds), *Practical Nonparametric and Semiparametric Bayesian Statistics.* New York: Springer-Verlag, 1–22.

Evans J, Middleton N, Gunnell D (2004) Social fragmentation, severe mental illness and suicide. *Social Psychiatry and Psychiatric Epidemiology*, 39, 165–70.

Everitt B (1984) *An introduction to latent variable models.* London: Chapman and Hall.

Everitt B, Hand D (1981) *Finite Mixture Distributions.* London: Chapman & Hall.

Everson P, Morris C (2000) Inference for multivariate normal hierarchical models. *Journal of the Royal Statistical Society: Series B*, 62, 399–412.

Fahrmeir L, Knorr Held L (1997) Dynamic discrete time duration models. *Sociological Methodology*, 27, 417–52.

Fahrmeir L, Knorr-Held L (2000) Dynamic and semiparametric models. In M Schimek (ed.), *Smoothing and regression: Approaches, computation and application.* New York: John Wiley, 513–43.

Fahrmeir L, Lang S (2001) Bayesian inference for generalized additive mixed models based on Markov random field priors. *Journal of the Royal Statistical Society: Series C*, 50, 201–20.

Fahrmeir L, Osuna L (2003) Structured count data regression. Sonderforschungsbereich 386, Discussion Paper 334, University of Munich.

Fahrmeir L, Osuna L (2006) Structured additive regression for overdispersed and zero-inflated count data. *Applied Stochastic Models in Business and Industry*, 22, 351–69.

Fahrmeir L, Tutz G (2001) *Multivariate statistical modelling based on generalized linear models*, 2nd edn. *Springer Series in Statistics.* New-York: Springer Verlag.

Farber S, Páez A (2007) A systematic investigation of cross-validation in GWR model estimation: Empirical analysis and Monte Carlo simulations. *Journal of Geographical Systems*, 9, 371–96.

Fay R, Herriot R (1979) Estimates of income for small places: An empirical Bayes application of James-Stein procedures to census data. *Journal of the American Statistical Association*, 78, 269–77.

Fernandez C, Green P (2002) Modelling spatially correlated data via mixtures: A Bayesian approach. *Journal of the Royal Statistical Society: Series B*, 64, 805–26.

Fernandez C, Ley E, Steel M (2001) Benchmark priors for Bayesian model averaging. *Journal of Econometrics*, 100, 381–427.

Fernandez C, Steel M (1998) On Bayesian modeling of fat tails and skewness. *Journal of the American Statistical Association*, 93, 359–71.

Fernandez C, Steel M (2000) Bayesian regression analysis with scale mixtures of normals. *Econometric Theory*, 16, 80–101.

Ferreira M, Gamerman D (2000) Dynamic generalized linear models. In D Dey, S Ghosh, B Mallick (eds), *Generalized linear models: A Bayesian perspective.* New York: Marcel Dekker, 57–72.

Fichman M, Cummings J (2003) Multiple imputation for missing data: Making the most of what you know. *Organizational Research Methods*, 6, 282–308.

Finkel S (1995) *Causal analysis with panel data.* Beverly Hills: Sage Publications.

Finley A, Banerjee S, Carlin B (2006) spBayes: An R package for univariate and multivariate hierarchical point-referenced spatial models. Source: blue.fr.umn.edu/spatialBayes/.

Flay B, Hansen W, Johnson C, Collins L, Dent C, Dwyer K, Grossman L, Hockstein G, Rauch J, Sobol J, Sobel D, Sussman S, Ulene A (1987) Implementation effectiveness trial of a social influences smoking prevention program using schools and television. *Health Education Research*, 2, 385–400.

Fleishman J, Lawrence W (2003) Demographic variation in SF-12 scores: True differences or differential item functioning? *Medical Care*, 41, 75–86.

Fleming T, Harrington D (1991) *Counting processes and survival analysis.* Chichester: John Wiley.

Florens J, Fougere D, Mouchart M (1995) Duration models. In L Matyas, P Sevestre (eds), *The econometrics of panel data.* Amsterdam: Kluwer, 491–534.

Fokianos K, Kedem B (2003) Regression theory for categorical time series. *Statistical Science*, 18, 357–76.

Fokoue E (2004) Stochastic determination of the intrinsic structure in Bayesian factor analysis. SAMSI Technical Report #2004-17 (http://www.samsi.info/reports/index.shtml).

Fonseca T, Ferreira M, Migon H (2008) Objective Bayesian analysis for the Student-t regression model. *Biometrika*, 95, 325–33.

Fotheringham A, Brunsdon C, Charlton, M (2002) *Geographically weighted regression: the analysis of spatially varying relationships.* Chichester, UK: Wiley.

Fotouhi A (2005) The initial conditions problem in longitudinal binary process: A simulation study. *Simulation Modelling Practice and Theory*, 13, 566–83.

Fotouhi A (2007) The initial conditions problem in longitudinal count process: A simulation study. *Simulation Modelling Practice and Theory*, 15, 589–604.

Fox A (1972) Outliers in time series. *Journal of the Royal Statistical Society: Series B*, 34, 350–63.

Fox J, Glas C (2005) Bayesian modification indices for IRT models. *Statistica Neerlandica*, 59, 95–106.

Franzese R, Hays J (2007) The spatial probit model of interdependent binary outcomes: Estimation, interpretation, and presentation. The Society for Political Methodology, Working Papers.

Franzese R, Hays J (2008) Empirical models of spatial interdependence. In J Box-Steffensmeier, H Brady, D Collier (eds), *Oxford handbook of political methodology.* Oxford University Press.

Frees E (2004) *Longitudinal and panel data.* Cambridge: Cambridge University Press.

Frees E, Young V, Luo Y (2001) Case studies using panel data models. *North American Actuarial Journal*, 5, 24–42.

Frey B, Weck-Hannemann H (1984) The hidden economy as an unobserved variable. *European Economic Review*, 26, 33–53.

Friedman J (1991) Multivariate adaptive regression splines. *Annals of Statistics*, 19, 1–67.

Friel N, Pettitt A (2008) Marginal likelihood estimation via power posteriors. *Journal of the Royal Statistical Society: Series B*, 70, 589–607.

Friesen M, MacNab Y, Marion S, Demers P, Davies H, Teschke K (2006) Mixed models and empirical Bayes estimation for retrospective exposure assessment of dust exposures in Canadian sawmills. *Annals of Occupational Hygiene*, 50, 281–88.

Fruhwirth-Schattner S (2001) Markov Chain Monte Carlo estimation of classical and dynamic switching and mixture models. *Journal of the American Statistical Association*, 96, 194–209.

Fruhwirth-Schnatter S (1994) Data augmentation and dynamic linear models. *Journal of Time Series Analysis*, 15, 183–202.

Fruhwirth-Schnatter S (1999) Bayes factors and model selection for random effect models. Working Paper, Department of Statistics, University of Business Administration and Economics, Vienna.

Fruhwirth-Schnatter S (2004) Estimating marginal likelihoods for mixture and Markov switching models using bridge-sampling techniques. *The Econometrics Journal*, 7, 143–67.

Fruhwirth-Schnatter S (2006) *Finite mixture and Markov switching models.* New York: Springer.

Fruhwirth-Schnatter S, Fruhwirth R (2007) Auxiliary mixture sampling with applications to logistic models. *Computational Statistics and Data Analysis*, 51, 3509–28.

Fruhwirth-Schnatter S, Otter T, Tuchler R (2004) Bayesian analysis of the heterogeneity model. *Journal of Business & Economics Statistics*, 22, 2–15.

Fruhwirth-Schnatter S, Tuchler R (2008) Bayesian parsimonious covariance estimation for hierarchical linear mixed models. *Statistics & Computing*, 18, 1–13.

Fryback D, Stout N, Rosenberg M (2001) An elementary introduction to Bayesian computing using WinBUGS. *International Journal of Technology Assessment in Health Care*, 17, 96–113.

Fukumoto K (2005) Survival analysis of systematically dependent competing risks: An application to the U.S. Congressional Careers, 22nd annual summer meeting of the Society for Political Methodology, Tallahassee, FL, USA.

Gabriel K (1962) Ante-dependence analysis of an ordered set of variables. *Annals of Mathematical Statistics*, 33, 201–12.

Gail M, Santner T, Brown C (1980) An analysis of comparative carcinogenesis experiments based on multiple times to tumor. *Biometrics*, 36, 255–66.

Galler H (2001) On the dynamics of individual wage rates—heterogeneity and stationarity of wage rates of west german men. In R Friedmann, L Knüppel, H Lütkepohl (eds), *Econometric studies. A Festschrift in Honour of Joachim Frohn.* Münster: LIT, 269–93.

Gamerman D (1991) Dynamic Bayesian models for survival data. *Journal of the Royal Statistical Society C*, 40, 63–79.

Gamerman D (1997) Efficient sampling from the posterior distribution in generalized linear mixed models. *Statistics and Computing*, 7, 57–68.

Gamerman D (1998) Markov chain Monte Carlo for dynamic generalized linear models. *Biometrika*, 85, 215–27.

Gamerman D, Moreira A, Rue H (2003) Space-varying regression models: Specifications and simulation. *Computational Statistics & Data Analysis*, 42, 513–33.

Gao S (2004) Combining binomial data using the logistic normal. *The Journal of Statistical Computation and Simulation*, 74, 293–306.

Garner C, Raudenbush S (1991) Neighborhood effects on educational attainment: A multilevel analysis. *Sociology of Education*, 64, 251–62.

Geisser S, Eddy W (1979) A predictive approach to model selection. *Journal of the American Statistical Association*, 74, 153–60.

Gelfand A (1996) Model determination using sampling based methods. In W Gilks, S Richardson, D Spiegelhalter (eds), Chapter 9 in *Markov Chain Monte Carlo in practice.* Boca Raton, FL: Chapman & Hall/CRC.

Gelfand A, Dey D (1994) Bayesian model choice: asymptotics and exact calculations. *Journal of the Royal Statistical Society: Series B*, 56, 501–14.

Gelfand A, Dey D, Chang H (1992) Model determination using predictive distributions with implementations via sampling-based methods. In J Bernardo et al., *Bayesian statistics 4.* Oxford: Oxford University Press, 147–68.

Gelfand A, Ghosh S (1998) Model choice: A minimum posterior predictive loss approach. *Biometrika*, 85, 1–11.

Gelfand A, Ghosh S, Christiansen C, Soumerai S, McLaughlin T (2000) Proportional hazard models: A latent competing risk approach. *Applied Statistics*, 49, 385–97.

Gelfand A, Kim H, Sirmans C, Banerjee S (2003) Spatial modelling with spatially varying coefficient models. *Journal of the American Statistical Association*, 98, 387–96.

Gelfand A, Kottas A, MacEachern S (2005b) Bayesian nonparametric spatial modeling with Dirichlet process mixing. *Journal of the American Statistical Association*, 100, 1021–35.

Gelfand A, Latimer A, Wu S, Silander J (2005a) Building statistical models to analyse species distributions. In J Clark, A Gelfand (eds), *Hierarchical modelling for the environmental sciences, statistical methods and applications.* Oxford: Oxford University Press, 33–50.

Gelfand A, Sahu S (1999) Identifiability, improper priors, and Gibbs sampling for generalized linear models. *Journal of the American Statistical Association*, 94, 247–53.

Gelfand A, Sahu S, Carlin B (1995) Efficient parameterization for normal linear mixed models. *Biometrika*, 82, 479–88.

Gelfand A, Sahu S, Carlin B (1996) Efficient parameterizations for generalised linear models. In J Bernardo, J Berger, A Dawid, A Smith (eds), *Bayesian Statistics 5*. Oxford: Clarendon Press, 165–80.

Gelfand A, Smith A (1990) Sampling-based approaches to calculating marginal densities. *Journal of the American Statistical Association*, 85, 398–409.

Gelfand A, Smith A, Lee T (1992) Bayesian analysis of constrained parameter and trucated data problems using Gibbs sampling. *Journal of the American Statistical Association*, 87, 523–32.

Gelfand A, Vlachos P (2003) On the calibration of Bayesian model choice criteria. *Journal of Statistical Planning and Inference*, 111, 223–34.

Gelman A (2006a) Prior distributions for variance parameters in hierarchical models. *Bayesian Analysis*, 1, 515–33.

Gelman A (2006b) Multilevel (hierarchical) modeling: What it can and can't do. *Technometrics*, 48, 432–35.

Gelman A, Carlin J, Stern H, Rubin D (2004) *Bayesian data analysis*. 2nd ed. Boca Raton, FL: Chapman & Hall/CRC.

Gelman A, Hill J (2006) *Data analysis using regression and multilevel/ hierarchical models*. Cambridge: Cambridge University Press.

Gelman A, Meng X (1998) Simulating normalizing constants: from importance sampling to bridge sampling to path sampling. *Statistical Science*, 13, 163–85.

Gelman A, Meng X, Stern H (1996) Posterior predictive assessment of model fitness via realized discrepancies. *Statistica Sinica*, 6, 733–807.

Gelman A, Rubin D (1996) Markov Chain Monte Carlo methods in biostatistics. *Statistical Methods in Medical Research*, 5, 339–55.

Gelman A, van Dyk D, Huang Z, Boscardin J (2008) Using redundant parameterizations to fit hierarchical models. *Journal of Computational and Graphical Statistics*, 17, 95–12.

Genton M (2004) *Skew-elliptical distributions and their applications: A journey beyond normality*, Edited Volume. Boca Raton, FL: Chapman & Hall/CRC.

George E, McCulloch R (1993) Variable selection via Gibbs sampling. *Journal of the American Statistical Association*, 88, 881–89.

George E, Makov U, Smith A (1993) Conjugate likelihood distributions. *Scandinavian Journal of Statistics*, 20, 147–56.

George E, McCulloch R (1997) Approaches for Bayesian variable selection. *Statistica Sinica*, 7, 339–73.

George E, Zhang Z (2001) Posterior propriety in some hierarchical exponential family models. In A Saleh (ed.), *Data Analysis from Statistical Foundations: Festschrift in Honor of Donald A.S. Fraser*. New York: Nova Science Publishers.

Gerlach R, Bird R, Hall A (2002) Bayesian variable selection in logistic regression: predicting company earnings direction. *Australian & New Zealand Journal of Statistics*, 44, 155–68.

Gerlach R, Carter C, Kohn R (1999) Diagnostics for time series analysis. *Journal of Time Series Analysis*, 20, 309–30.

Gerlach R, Carter C, Kohn R (2000) Efficient Bayesian inference for dynamic mixture models. *Journal of the American Statistical Association*, 95, 819–28.

Geweke J (1989) Bayesian inference in econometric models using Monte Carlo integration. *Econometrica*, 57, 1317–39.

Geweke J (1992) Evaluating the accuracy of sampling-based approaches to calculating posterior moments. In J Bernardo, J Berger, A Dawid, A Smith (eds), *Bayesian Statistics*. Vol. 4. New York: Oxford University Press, 169–93.

Geweke J (1993) Bayesian treatment of the Student's-t linear model. *Journal of Applied Econometrics*, 8, S19–S40.

Geweke J (2007) Interpretation and inference in mixture models: Simple MCMC works. *Computational Statistics and Data Analysis*, 51, 3529–50.

Geweke J, Keane M (2000) An empirical analysis of earnings dynamics among men in the PSID: 1968–1989. *Journal of Econometrics*, 96, 293–56.

Geweke J, Terui N (1993) Bayesian threshold auto-regressive models for nonlinear time series. *Journal of Time Series Analysis*, 14, 441–54.

Geweke J, Zhou G (1996) Measuring the pricing error of the arbitrage pricing theory. *Review of Financial Studies*, 9, 557–87.

Geyer C, Thompson E (1995) Annealing Markov Chain Monte Carlo with applications to ancestral inference. *Journal of the American Statistical Association*, 90, 909-20.

Ghosh J (2008) Efficient Bayesian computation and model search in linear hierarchical models. PhD thesis ISDS. Duke University.

Ghosh J, Dunson D (2008) Bayesian model selection in factor analytic models. In D Dunson (ed.), *Random effect and latent variable model selection*. New York: Springer, 151–63.

Ghosh K, Tiwari R (2007) Prediction of U.S. cancer mortality counts using semiparametric Bayesian techniques. *Journal of the American Statistical Association*, 102, 7–15.

Ghosh P, Branco M, Chakraborty H (2007) Bivariate random effect model using skew-normal distribution with application to HIV-RNA. *Statistics in Medicine*, 26, 1255–67.

Ghosh S, Kim H (2007) Semiparametric inference based on a class of zero-altered distributions. *Statistical Methodology*, 4, 371–83.

Ghosh S, Mukhopadhyay P, Lu J-C (2006) Bayesian analysis of zero-inflated regression models. *Journal of Statistical Planning and Inference*, 136, 1360–75.

Gilks W (1996) Full conditional distributions. In W Gilks, S Richardson, D Spiegelhalter (eds), *Markov Chain Monte Carlo in practice*. Chapman and Hall: London, 75–88.

Gilks W, Richardson S, Spielgelhalter D (1996) Introducing Markov Chain Monte Carlo. In W Gilks, S Richardson, D Spiegelhalter (eds), *Markov chain Monte Carlo in practice.* Chapman and Hall: London, 1–19.

Gilks W, Wang C, Yvonnet B, Coursaget P (1993) Random-effects models for longitudinal data using Gibbs sampling. *Biometrics*, 38, 963–74.

Gilks W, Wild P (1992) Adaptive rejection sampling for Gibbs sampling. *Applied Statistics*, 41, 337–48.

Gill J, Casella G (2009) Nonparametric priors for ordinal Bayesian social science models: Specification and estimation. *Journal of the American Statistical Association*, 104, 453–64.

Gilula Z, Haberman S (1994) Conditional log-linear models for analyzing categorical panel data. *Journal of the American Statistical Association*, 89, 645–56.

Giminez O, Bonner S, King R, Parker R, Brooks S, Jamieson L, Grosbois V, Morgan B, Thomas L (2008) WinBUGS for population ecologists: Bayesian modeling using Markov Chain Monte Carlo methods. In D Thomson, E Cooch, M Conroy, (eds), *Modelling demographic processes in marked populations. Environmental and ecological statistics.* New York: Springer, 883–916.

Givens G, Hoeting J (2005) *Computational statistics.* Chichester, Sussex: Wiley.

Glosten L, Jagannathan R, Runkle D (1994) On the relation between the expected value and the variance of the nominal excess return on stocks. *Journal of Finance*, 48, 1791–801.

Godolphin E, Triantafyllopoulos K (2006) Decomposition of time series models in state-space form. *Computational Statistics & Data Analysis*, 50, 2232–46.

Godsill S, Doucet A, West M (2004) Monte Carlo smoothing for nonlinear time series. *Journal of the American Statistical Association*, 99, 156–68.

Goldfelfd S, Quandt R (1975) Estimation in a disequilibrium model and the value of information. *Journal of Econometrics*, 3, 325–48.

Goldstein H (2005) Heteroscedasticity and complex variation. In B Everrit, D Howell (eds), *Encyclopedia of statistics in behavioral science.* Chichester: Wiley, Vol. 2. 790–95.

Goldstein H, Browne W, Rasbash J (2002) Partitioning variation in multilevel models. *Understanding Statistics*, 1, 223–32.

Goldstein H, Spiegelhalter D (1996) League tables and their limitations: Statistical issues in comparisons of institutional performance. *Journal of the Royal Statistical Society A*, 159, 385–443.

Gómez, E, Gómez-Villegas M, Marín J (2002) Continuous elliptical and exponential power linear dynamic models. *Journal of Multivariate Analysis*, 83, 22–36.

Goodman J, Blum T (1996) Assessing the non-random sampling effects of subject attrition in longitudinal research. *Journal of Management*, 22, 627–52.

Gopalan R, Berry D (1998) Bayesian multiple comparisons using Dirichlet process priors. *Journal of the American Statistical Association*, 93, 1130–39.

Gore S, Pocock S, Kerr G (1984) Regression models and non-proportional hazards in the analysis of breast cancer survival. *Applied Statistics*, 33, 176–95.

Gordon S (2002) Stochastic dependence in competing risks. *American Journal of Political Science*, 46, 200–17.

Gosoniu L, Vounatsou P, Sogoba N, Smith T (2006) Bayesian modelling of geostatistical malaria risk data. *Geospatial Health*, 1, 127–39.

Gotway C, Wolfinger R (2003) Spatial prediction of counts and rates. *Statistics in Medicine*, 22, 1415–32.

Gourieroux C, Phillips P, Yu J (2006) *Indirect Inference for dynamic panel models.* Cowles Foundation Discussion Paper 1550, Yale University.

Gramacy R (2007) tgp: an R package for Bayesian nonstationary, semiparametric nonlinear regression and design by reed Gaussian process models. *Journal of Statistical Software*, 19(9).

Gramacy R, Lee H (2008) Gaussian processes and limiting linear models. *Computational Statistics & Data Analysis*, 53, 123–36.

Granger C, Machina M (2006) Structural attribution of observed volatility clustering. *Journal of Econometrics*, 135, 15–29.

Green M, Medley G, Browne W (2009) Use of posterior predictive assessments to evaluate model fit in multilevel logistic regression. *Veterinary Research*, 40(4): 30.

Green P, O'Hagan A (1998) Model choice with MCMC on product spaces without using pseudo priors. Technical Report, Department of Statistics, University of Nottingham.

Green P, Richardson S (1997) On Bayesian analysis of mixtures with an unknown number of components *Journal of the Royal Statistical Society: Series B*, 59, 731–92.

Green P, Richardson S (2000) Spatially correlated allocation models for count data. Technical Report. University of Bristol (http://en.scientificcommons.org/336472).

Green P, Richardson S (2001) Modelling heterogeneity with and without the Dirichlet process. *Scandinavian Journal of Statistics*, 28, 355–75.

Green P, Richardson S (2002) Hidden Markov models and disease mapping. *Journal of the American Statistical Association*, 97, 1055–70.

Greene W (2007) Functional form and heterogeneity in models for count data. *Foundations and Trends in Econometrics*, 1, 113–218.

Greenland S (2003) Generalized conjugate priors for Bayesian analysis of risk and survival regressions. *Biometrics*, 59, 92–99.

Greenland S (2006) Smoothing observational data: A philosophy and implementation for the Health Sciences. *International Statistical Review*, 74 (1), 31–46.

Greenland S (2007) Bayesian perspectives for epidemiological research. II. Regression analysis. *International Journal of Epidemiology*, 36, 195–202.

Greenland S, Christensen R (2001) Data augmentation priors for Bayesian and semi-Bayes analyses of conditional-logistic and proportional-hazards regression. *Statistics in Medicine*, 20, 2421–28.

Greenland S, Draper, D (1998) *Exchangeability. Encyclopedia of Biostatistics*, P Armitage, T Colton (eds), London: Wiley.

Griffin J, Steel M (2006) Order-based dependent Dirichlet processes. *Journal of the American Statistical Association*, 101, 179–94.

Grunwald S (2005) *Environmental soil-landscape modeling: Geographic information technologies and pedometrics*. Boca Raton, FL: CRC Press.

Grunwald G, Hyndman R, Tedesco L, Teeedie R (2000) Non-Gaussian conditional AR(1) models. *Australian & New Zealand Journal of Statistics*, 42, 479–95.

Gschlößl S, Czado C (2006) Modelling count data with overdispersion and spatial effects. Technische Universität München, Statistical Papers. DOI 10.1007/s00362-006-0031-6.

Gschlößl S, Czado C (2008) Modelling count data with overdispersion and spatial effects. *Statistical Papers*, 49, 531–32.

Guha S (2008) Posterior simulation in the generalized linear mixed model with semiparametric random effects. *Journal of Computational and Graphical Statistics*, 17, 410–25.

Gunnell D, Peters T, Kammerling R, Brooks J (1995) Relation between parasuicide, suicide, psychiatric admissions and socio-economic deprivation. *British Medical Journal*, 311, 226–30.

Gustafson P (1996) Local sensitivity of inferences to prior marginals. *Journal of the American Statistical Association*, 91, 774–81.

Gustafson P (1996) The effect of mixing-distribution misspecification in conjugate mixture models. *Canadian Journal of Statistics*, 24, 307–18.

Gustafson P (1997) Large hierarchical Bayesian analysis of multivariate survival data. *Biometrics*, 53, 230–42.

Gustafson P (2000) Bayesian regression modelling with interactions and smooth effects. *Journal of the American Statistical Association*, 95, 795–806.

Gustafson P, Aeschliman D, Levy A (2003) A simple approach to fitting Bayesian survival models. *Lifetime Data Analysis*, 9, 5–19.

Gustafson P, Hossain S, MacNab Y (2006) Conservative priors for hierarchical models. *Canadian Journal of Statistics*, 34, 377–90.

Hadjicostas P, Berry S (1999) Improper and proper posteriors with improper priors in a Poisson-gamma hierarchical model. *Test*, 8, 147–66.

Hall D (2000) Zero-inflated Poisson and binomial regression with random effects: A case study. *Biometrics*, 56, 1030–39.

Hamerle A, Ronning G (1995) Panel analysis for qualitative variables. In G Arminger et al. (eds), *Handbook of statistical modeling for social and behavioral sciences*. New York: Plenum Press, 401–51.

Hamilton J (2009) Regime–switching models. In S Durlauf, L Blume (eds), *The New Palgrave Dictionary of Economics*. 2nd edn. Basingstoke, England: Palgrave Macmillan.

Han C, Carlin B (2001) Markov Chain Monte Carlo methods for computing Bayes factors: A comparative review. *Journal of the American Statistical Association*, 96, 1122–32.

Hanson T, Branscum A, Johnson W (2005) Nonparametric Bayesian data analysis: An introduction. In C Rao, D Dey (eds), *Handbook of Statistics 25*. Amsterdam: Elsevier, 245–78.

Hanson T, Johnson W (2002) Modeling regression error with a mixture of Polya trees. *Journal of the American Statistical Association*, 97, 1020–33.

Haran M, Hodges J, Carlin B (2003) Accelerating computation in Markov random field models for spatial data via structured MCMC. *Journal of Computational & Graphical Statistics*, 12, 249–64.

Harvey A (1989) *Forecasting, structural time series models and the Kalman filter*. Cambridge: Cambridge University Press.

Harvey A, Koopman S (1997) Multivariate structural time series models. In C Heij, H Schumacher, B Hanzon, C Praagman (eds), *Systematic dynamics in economic and financial models*. Chichester: Wiley, 269–98.

Harvey A, Ruiz E, Shepherd N (1994) Multivariate stochastic variance models. *Review of Economic Studies*, 61, 247–64.

Harvey A, Shephard N (1993) Structural time series models. In G S Maddala et al. (eds), *Handbook of Statistics*, Vol. 11. Barking: Elsevier Science.

Harvey A, Todd P (1983) Forecasting economic time series with structural and Box-Jenkins models: A case study. *Journal of Business & Economic Statistics*, 1, 299–307.

Harvey A, Trimbur T, Van Dijk H (2007) Trends and cycles in economic time series: A Bayesian approach. *Journal of Econometrics*, 140: 618–49.

Hasegawa H, Chaturvedi A, van Hoa T (2000) Bayesian unit root tests in nonnormal AR(1) models. *Journal of Time Series Analysis*, 21, 261–80.

Hastie T, Tibshirani T (1993) Varying coefficient models. *Journal of the Royal Statistical Society: Series B*, 55, 757–96.

Hastings W (1970) Monte-Carlo sampling methods using Markov Chains and their applications. *Biometrika*, 57, 97–109.

Hausman J, Wise D (1978) A conditional probit model for qualitative choice: Discrete decisions recognizing interdependence and heterogeneous preferences. *Econometrica*, 46, 403–26.

Hayashi K, Arav M (2006) Bayesian factor analysis when only a sample covariance matrix is available. *Educational and Psychological Measurement*, 66, 272–84.

Hayashi K, Bentler P, Yuan K-H (2008) Structural equation modelling. In C Rao, J Miller, D Rao (eds), *Epidemiology and medical statistics, handbook of statistics*, Chapter 13, Vol 27. Amsterdam: Elsevier, 395–428.

Heagerty P, Zeger S (2000) Marginalized multilevel models and likelihood inference. *Statistical Science*, 15, 1–26.

Heckman J (1976) The common structure of statistical models of truncation, sample selection, and limited dependent variables and a simple estimator for such models. *Annals of Economic and Social Measurement*, 5, 475–92.

Heckman J (1981) The incidental parameters problem and the problem of initial conditions in estimating a discrete time-discrete data stochastic process. In C Manski, D McFadden (eds), *Structural analysis of discrete data with econometric applications.* Cambridge: MIT Press, 179–95.

Heckman J, Singer B (1984) A method for minimizing the impact of distributional assumptions in econometric models for duration data. *Econometrica*, 52, 271–320.

Hedeker D (2003) A mixed-effects multinomial logistic regression model. *Statistics in Medicine*, 22, 1433–46.

Hedeker D, Gibbons R (1994) A random-effects ordinal regression model for multilevel analysis. *Biometrics*, 50, 933–44.

Hedeker D, Gibbons R (1997) Application of random-effects pattern-mixture models for missing data in longitudinal studies. *Psychological Methods*, 2, 64–78.

Hedeker D, Gibbons R (2006) *Longitudinal data analysis.* Hoboken, New Jersey: Wiley-Interscience.

Hedeker D, Gibbons R, Flay B (1994) Random effects regression models for clustered data: With an example from smoking research. *Journal of Consulting and Clinical Psychology*, 62, 757–65.

Henderson R, Oman P (1999) Effect of frailty on marginal regression estimates in survival analysis. *Journal of the Royal Statistical Society: Series B*, 61, 367–79.

Henderson R, Prince H (2000) Choice of conditional models in bivariate survival. *Statistics in Medicine*, 19, 563–74.

Henderson R, Shimakura S, Gorst D (2002) Modeling spatial variation in leukemia survival data. *Journal of the American Statistical Association*, 97, 965–72.

Hensher D, Greene W (2003) The mixed logit model: The state of practice. *Transportation*, 30, 133–76.

Heringstad B, Rekaya R, Gianola D, Klemetsdal G, Weigel K (2001) Bayesian analysis of liability of clinical mastitis in Norwegian cattle with a threshold model: Effects of data sampling method and model specification. *Journal of Dairy Science*, 84, 2337–46.

Herring A, Ibrahim J (2002) Maximum likelihood estimation in random effects cure rate models with nonignorable missing covariates. *Biostatistics*, 3, 387–405.

Higdon D (1998) A process-convolution approach to modelling temperatures in the North Atlantic Ocean. *Environmental and Ecological Statistics*, 5, 173–90.

Higdon D (2007) A primer on space-time modelling from a Bayesian perspective. In Finkelstadt, Held and Isham (eds), *Statistical methods for spatio-temporal systems.* Boca Raton, FL: CRC Press, 217–80.

Hildreth C, Houck J (1968) Some estimators for a linear model with random coefficients. *Journal of the American Statistical Association*, 63, 584–95.

Hirano K (2002) Semiparametric Bayesian inference in autoregressive panel data models. *Econometrica*, 70, 781–99.

Hirano K (1998) A semiparametric model for labor earnings dynamics. In D Dey, P Mueller, D Sinha (eds), *Practical nonparametric and semiparametric Bayesian statistics*. New York: Springer–Verlag, 355–67.

Hjellvik V, Tjøstheim D (1999) Modelling panels of intercorrelated autoregressive time series. *Biometrika*, 86, 573–90.

Ho R, Hu I (2008) Flexible modelling of random effects in linear mixed models- a Bayesian approach. *Computational Statistics & Data Analysis*, 52, 1347–61.

Hobert J, Casella G (1996) The effect of improper priors on Gibbs sampling in hierarchical linear mixed models. *Journal of the American Statistical Association*, 91, 1461–73.

Hodge R, Evans M, Marshall J, Quigley J, Walls L (2001) Eliciting engineering knowledge about reliability during design-lessons learnt from implementation. *Quality and Reliability Engineering International*, 17, 169–79.

Hodges J, Carlin B, Fan Q (2003) On the precision of the conditionally autoregressive prior in spatial models. *Biometrics*, 59, 317–22.

Hoem J (1987) Statistical analysis of a multiplicative model and its application to the standardization of vital rates: A review. *International Statistical Review*, 55, 119–52.

Hoff P (2003) Nonparametric modelling of hierarchically exchangeable data Technical Report 421, Department of Statistics, University of Washington.

Hogan J, Lin X, Herman B (2004) Mixtures of varying coefficient models for longitudinal data with discrete or continuous non-ignorable dropout. *Biometrics*, 60, 854–64.

Hogan J, Tchernis R (2004) Bayesian factor analysis for spatially correlated data, with application to summarizing area-level material deprivation from census data. *Journal of the American Statistical Association*, 99, 314–24.

Holloway G, Shankar B, Rahman S (2002) Bayesian spatial probit estimation: A primer with an application to HYV rice adoption. *Agricultural Economics*, 27, 383–402.

Holmes C, Held L (2006) Bayesian auxiliary variable models for binary and multinomial regression. *Bayesian Analysis*, 1, 145–68.

Hooper P (2001) Flexible regression modeling with adaptive logistic basis functions. *Canadian Journal of Statistics*, 29, 343–78.

Hougaard P (1987) Modelling multivariate survival. *Scandinavian Journal of Statistics*, 14, 291–304.

Hougaard P (2000) *Analysis of multivariate survival data*. Springer: New York.

Hougaard P, Myglegaard P, Borch-Johnsen K (1994) Heterogeneity models of disease susceptibility, with application to diabetic nephropathy. *Biometrics*, 50, 1178–88.

Howley P, Gibberd R (2003) Using hierarchical models to analyse clinical indicators: A comparison of the gamma-Poisson and beta-binomial models. *International Journal for Quality in Health Care*, 15, 319–29.

Hox J (2002) *Multilevel analysis: Techniques and applications.* Mahwah, NJ: Lawrence Erlbaum Associates.

Hox J, Bechger T (1998) An introduction to structural equation modeling. *Family Science Review*, 11, 354–73.

Hoyle R (ed.) (1995) *Structural equation modeling: Concepts, issues, and applications.* Thousand Oaks, California: Sage.

Hsiao C (1996) Random coefficient models. In L Matyas, P Sevestre (eds), *The econometrics of panel data.* Dordrecht: Kluwer, 77–99.

Hsiao C (1997) Approximate Bayes factors when a mode occurs on the boundary. *Journal of the American Statistical Association*, 92, 656–63.

Huerta G, West M (1999) Priors and component structurres in autoregressive time series. *Journal of the Royal Statistical Society*, 61B, 881–99.

Hurn M, Justel A, Robert C (2003) Estimating mixtures of regressions. *Journal of Computational and Graphical Statistics*, 12, 1–25.

Hurn M, Justel A, Robert C (2003) Estimating mixtures of regressions. *Journal of Computational and Graphical Statistics*, 12, 55–79.

Huster W, Brookmeyer R, Self S (1989) Modeling paired survival data with covariates. *Biometrics*, 45, 145–56.

Hyndman R (1996) Computing and graphing highest density regions. *American Statistician*, 50, 361–65.

Ibrahim J, Chen M-H (2000) Power prior distributions for regression models. *Statistical Science*, 15, 46–60.

Ibrahim J, Chen M-H, MacEachern S (1999) Bayesian variable selection for proportional hazards models. *The Canadian Journal of Statistics*, 27, 701–17.

Ibrahim J, Chen M-H, Ryan L (2000a) Bayesian variable selection for time series count data. *Statistica Sinica*, 10, 971–87.

Ibrahim J, Chen M-H, Sinha D (2001) *Bayesian survival analysis.* New York: Springer-Verlag.

Ibrahim J, Lipsitz S, Chen M-H (1999) Missing covariates in generalized linear models when the missing data mechanism is non-ignorable. *Journal of the Royal Statistical Society: Series B*, 61, 173–90.

Ibrahim, J, Chen M, Sinha D (2001) Criterion-based methods for Bayesian model assessment. *Statistica Sinica*, 11, 419–43.

Imai K, Lu Y, Strauss A (2010) eco: R Package for ecological inference in 2×2 Tables. *Journal of Statistical Software*, Forthcoming.

Imai K, Ying L, Strauss A (2008) Bayesian and likelihood inference for 2×2 ecological tables: An incomplete data approach. *Political Analysis*, 16, 41–69.

Ishwaran H, James L (2001) Gibbs sampling methods for stick-breaking priors. *Journal of the American Statistical Association*, 96, 161–73.

Ishwaran H, James L (2002) Approximate Dirichlet process computing in finite normal mixtures: Smoothing and prior information. *Journal of Computational and Graphical Statistics*, 11, 508–32.

Ishwaran H, James L (2003) Generalized weighted Chinese restaurant processes for species sampling mixture models. *Statistica Sinica*, 13, 1211–35.

Ishwaran H, Zarepour M (2000) Markov chain Monte Carlo in approximate Dirichlet and beta two-parameter process hierarchical models. *Biometrika*, 87, 371–90.

Ishwaran H, Zarepour M (2002) Exact and approximate sum-representations for the Dirichlet process. *Canadian Journal of Statistics*, 30, 269–83.

Islam M, Chowdhury R (2006) A higher order Markov model for analyzing covariate dependence. *Applied Mathematical Modelling*, 30, 477–88.

Jacquier E, Polson N, Rossi P (2004) Bayesian analysis of stochastic volatility models with fat-tails and correlated errors. *Journal of Econometrics*, 122, 185–212.

Jaffrézic F, Thompson R, Hill G (2003) Structured antedependence models for genetic analysis of repeated measures on multiple quantitative traits. *Genetics Research*, 82, 55–65.

Jaffrézic F, Venot E, Laloë D, Vinet A, Renand G (2004) Use of structured antedependence models for the genetic analysis of growth curves. *Journal of Animal Science*, 82, 3465–73.

James G, Hastie T, Sugar C (2000) Principal component models for sparse functional data. *Biometrika*, 87, 587–602.

Jara A (2007) Applied Bayesian non- and semi-parametric inference using DP package. *R News*, 7/3, 17–26.

Jara A, Quintana F, San Martin E (2008) Linear effects mixed models with skew-elliptical distributions: A Bayesian approach. Interuniversity Attraction Pole Report TR08010, http://www.stat.ucl.ac.be/IAP/.

Jarque C, Bera A (1980) Efficient tests for normality, homoscedasticity and serial independence of regression residuals. *Econometric Letters*, 6, 255–59.

Jasra A, Holmes C, Stephens D (2005) MCMC methods and the label switching problem in Bayesian mixture modelling. *Statistical Science*, 20, 50–67.

Jedidi K, Jagpal H, DeSarbo W (1997) Finite-mixture structural equation models for response-based segmentation and unobserved heterogeneity. *Marketing Science*, 16, 39–59.

Jerak A, Lang S (2005) Locally adaptive function estimation for binary regression models. *Biometrical Journal*, 47, 151–66.

Jin X, Carlin B, Banerjee S (2005) Generalized hierarchical multivariate CAR models for areal data. *Biometrics*, 61, 950–61.

Jiruše M, Machek J, Beneš V, Zeman P (2004) A Bayesian estimate of the risk of tick-borne diseases. *Applications of Mathematics*, 49, 389–404.

Johannes M, Polson N (2006) MCMC methods for continuous-time financial econometrics. In Y Ait-Sahalia, L Hansen (eds), *Handbook of Financial Econometrics*. Amsterdam: North Holland, 1–72.

Johnson V (2004) A Bayesian χ^2 test for goodness-of-fit. *Annals of Statistics*, 32, 2361–84.

Johnson V, Albert J (1999) *Ordinal data modeling*. New York: Springer.

Jones G, Haran M, Caffo B, Neath R (2006) Fixed-width output analysis for Markov Chain Monte Carlo. *Journal of the American Statistical Association*, 101, 1537–47.

Jonsen I, Myers R, James M (2006) Robust hierarchical state–space models reveal diel variation in travel rates of migrating leatherback turtles. *Journal of Animal Ecology*, 75, 1046–57.

Jöreskog K (1973) A general method for estimating a linear structural equation system. In A Goldberger, O Duncan (eds), New York: Seminar Press, 85–112.

Joreskog K, Goldberger A (1975) Estimation of a model with multiple indicators and multiple causes of a single latent variable. *Journal of the American Statistical Association*, 70, 631–39.

Jorgensen B, Lundbye-Christensen S, Song P, Sun L (1999) A state space model for multivariate longitudinal count data. *Biometrika*, 86, 169–81.

Jowaheer V, Sutradhar B (2002) Analysing longitudinal count data with overdispersion. *Biometrika*, 89, 389–99.

Jullion A, Lambert P (2007) Robust specification of the roughness penalty prior distribution in spatially adaptive Bayesian P-splines models. *Computational Statistics & Data Analysis*, 51, 2542–58.

Jung R, Kukuk M, Liesenfeld R (2006) Time series of count data: Modeling, estimation and diagnostics. *Computational Statistics & Data Analysis*, 51, 2350–64.

Jungbacker B, Koopman S, van der Wel M (2009) Dynamic factor models with smooth loadings for analyzing the term structure of interest rates. Tinbergen Institute Discussion Paper, TI 2009-041/4.

Kacker R, Forbes A, Kessel R, Sommer K-D (2008) Bayesian posterior predictive p-value of statistical consistency in interlaboratory evaluations. *Metrologia*, 45, 512–23.

Kahn M, Raftery A (1996) Discharge rates of Medicare stroke patients to skilled nursing facilities: Bayesian logistic regression with unobserved heterogeneity. *Journal of the American Statistical Association*, 91, 29–41.

Kalbfleisch J (1978) Non-parametric Bayesian analysis of survival time data. *Journal of the Royal Statistical Society B*, 40, 214–21.

Kalbfleisch J, Prentice R (1980) *The statistical analysis of failure time data.* New York: Wiley.

Karlis D, Meligkotsidou L (2005) Multivariate Poisson regression with covariance structure. *Statistics and Computing*, 15, 255–65.

Kashiwagi N, Yanagimoto T (1992) Smoothing serial count data through a state-space model. *Biometrics*, 48, 1187–94.

Kass R, Carlin B, Gelman A, Neal R (1998) Markov Chain Monte Carlo in practice: A round table discussion. *The American Statistician*, 52, 93–100.

Kass R, Raftery A (1995) Bayes factors. *Journal of the American Statistical Association*, 90, 773–95.

Kass R, Steffey D (1989) Approximate Bayesian inference in conditionally independent hierarchical models (parametric empirical Bayes models). *Journal of the American Statistical Association*, 84, 717–26.

Kass R, Wasserman L (1996) The selection of prior distributions by formal rules. *Journal of the American Statistical Association*, 91, 1343–70.

Kato B, Hoijtink H (2004) Testing homogeneity in a random intercept model using asymptotic, posterior predictive and plug-in p-values. *Statistica Neerlandica*, 58, 179–96.

Kedem B, Fokianos K (2002) *Regression models for time series analysis.* Chichester, UK: Wiley.

Keiding N, Andersen P, Klein J (1997) The role of frailty models and accelerated failure time models in describing heterogeneity due to omitted covariates. *Statistics in Medicine*, 16, 215–24.

Kelsall J, Wakefield J (2002) Modelling spatial variation in disease risk: A geostatistical approach. *Journal of the American Statistical Association*, 97, 692–770.

Kendall, M (1943) *The advanced theory of statistics.* London: Griffin.

Kenward M (1998) Selection models for repeated measurements with non-random dropout: An illustration of sensitivity. *Statistics in Medicine*, 17, 2723–32.

Kettl S (1991) Accounting for heteroscedasticity in the transform both sides regression model. *Journal of Applied Statistics*, 40, 261–68.

Key J, Pericchi L, Smith A (1999) Bayesian model choice: what and why? In J Bernardo, J Berger, A Dawid, A Smith (eds), *Bayesian Statistics 6.* Oxford: Oxford Science Publications, 343–70.

Kiefer N (1988) Economic duration data and hazard functions. *Journal of Economic Literature*, 26, 646–79.

Kim H, Sun D, Tsutakawa R (2002) Lognormal vs. gamma: Extra variations. *Biometrical Journal*, 44, 305–23.

Kim S, Shephard N, Chib S (1998) Stochastic volatility: Likelihood inference and comparison with ARCH models. *The Review of Economic Studies*, 65, 361–93.

King G (1997) *A solution to the ecological inference problem: Reconstructing individual behavior from aggregate data.* Princeton, NJ: Princeton, University Press.

King G (2001) Analyzing incomplete political science data: An alternative algorithm for multiple imputation. *American Political Science Review*, 95, 49–69.

King G, Rosen O, Tanner M (eds) (2004) *Ecological inference: New methodological strategies.* New York: Cambridge University Press.

Kinney S, Dunson D (2007) Fixed and random effects selection in linear and logistic models. *Biometrics*, 63, 690–98.

Kinney S, Dunson D (2008) Bayesian model uncertainty in mixed effects models. In D Dunson (ed.), *Random effect and latent variable model selection.* New York: Springer.

Kitagawa G, Gersch W (1985) A smoothness priors time-varying AR coefficient modeling of nonstationary covariance time series. *IEEE Transactions on Automatic Control*, 30, 48–56.

Kitagawa G, Gersch W (1996) *Smoothness priors analysis of time series, lecture notes in statistics 116.* New York: Springer-Verlag.

Kitanidis P (1997) *Introduction to geostatistics: Applications in hydrogeology.* Cambridge: Cambridge University Press.

Kleinman K, Ibrahim J (1998) A semi-parametric Bayesian approach to generalized linear mixed models. *Statistics in Medicine*, 17, 2579–96.

Kleinman K, Ibrahim J (1998) A semiparametric Bayesian approach to the random effects model. *Biometrics*, 54, 921–38.

Kneib T (2006) Mixed model-based inference in geoadditive hazard regression for interval-censored survival times. *Computational Statistics & Data Analysis*, 51, 777–92.

Kneib T, Belitz C, Brezger A, Lang S (2008) BayesX–Bayesian inference in structured additive regression. *Software Highlight in ISBA Bulletin*, 15(1): 11–13.

Knorr-Held L (1999) Conditional prior proposals in dynamic models. *Scandinavian Journal of Statistics*, 26, 129–44.

Knorr-Held L (2000) Bayesian modelling of inseparable space-time variation in disease risk. *Statistics in Medicine*, 19, 2555–67.

Knorr-Held L, Becker N (2000) Bayesian modelling of spatial heterogeneity in disease maps with application to German cancer mortality data. *Journal of the German Statistical Society*, 84, 121–40.

Knorr-Held L, Rainer E (2001) Projections of lung cancer mortality in West Germany: A case study in Bayesian prediction. *Biostatistics*, 2, 109–29.

Knorr-Held L, Rasser G (2000) Bayesian detection of clusters and discontinuities in disease maps. *Biometrics*, 56, 13–21.

Kohn R, Schimek M, Smith M (2000) Spline and kernel regression for dependent data. In M Schimek (ed.), *Smoothing and regression approaches, computation and estimation*, Chapter 6. Chichester, Sussex: John Wiley, 135–58.

Kohn R, Smith M, Chan D (2001) Nonparametric regression using linear combinations of basis functions. *Statistics and Computing*, 11, 313–22.

Konishi S, Ando T, Imoto, S (2004) Bayesian information criteria and smoothing parameter selection in radial basis function networks. *Biometrika*, 91, 27–43.

Koop G (2003) *Bayesian econometrics.* Chichester, Sussex: John Wiley.

Koop G, Poirier D (2004) Bayesian variants of some classical semiparametric regression techniques. *Journal of Econometrics*, 123, 259–82.

Koop G, Strachan R, van Dijk H, Villani M (2006) Bayesian approaches to cointegration. In K Patterson, T Mill (eds), *The Palgrave handbook of theoretical econometrics.* MacMillan.

Koop G, Tole L (2004) Measuring the health effects of air pollution: To what extent can we really say that people are dying from bad air? *Journal of Environmental Economics and Management*, 47, 30–54.

Koopman S (1993) Disturbance smoother for state space models. *Biometrika*, 80, 117–26.

Koopman S, Durbin J (2000) Fast filtering and smoothing for multivariate state space models. *Journal of Time Series Analysis*, 21, 281–96.

Koopman S, Shephard N, Doornik J (1999) Statistical algorithms for models in state space form using SsfPack 2.2. *Econometrics Journal*, 2, 113–66.

Korn E, Whittemore A (1979) Methods for analyzing panel studies of acute health effects of air pollution. *Biometrics*, 35, 795–802.

Kostaki A, Panousis V (2001) Expanding an abridged life table. *Demographic Research*, 5, 1, http://www.demographicresearch.org/Volumes/vol5/1/5-1.pdf.

Kottas A, Müller P, Quintana F (2005) A nonparametric Bayesian model for multivariate ordinal data. *Journal of Computational and Graphical Statistics*, 14, 610–25.

Koul H, Schick A (1996) Adaptive estimation in a random coefficient autoregressive model. *The Annals of Statistics*, 24, 1025–52.

Kozumi H (2004) Posterior analysis of latent competing risk models by parallel tempering. *Computational Statistics & Data Analysis*, 46, 441–58.

Kruijer W, Stein A, Schaafsma W, Heijting S (2007) Analyzing spatial count data, with an application to weed counts. *Environmental and Ecological Statistics*, 14, 399–410.

Kuhn E, Lavielle M (2005) Maximum likelihood estimation in nonlinear mixed effects models. *Computational Statistics & Data Analysis*, 49, 1020–38.

Kuhn I (2007) Incorporating spatial autocorrelation may invert observed patterns. *Diversity and Distributions*, 13, 66–69.

Kuo L, Mallick B (1997) Bayesian semiparametric inference for the accelerated failure-time model. *Canadian Journal of Statistics*, 25, 457–72.

Kuo L, Mallick B (1998) Variable selection for regression models. *Sankhya B*, 60, 65–81.

Lagazio C, Biggeri A, Dreassi E (2003) Age-period-cohort models and disease mapping. *Environmetrics*, 14, 475–90.

Laird N, Louis T (1989) Empirical Bayes ranking methods. *Journal of Educational Statistics*, 14, 29–46.

Lambert D (1992) Zero-inflated Poisson regression, with an application to defects in manufacturing. *Technometrics*, 34, 1–14.

Lambert P, Sutton A, Burton P, Abrams K, Jones D (2005) How vague is vague? A simulation study of the impact of the use of vague prior distributions in MCMC using WinBUGS. *Statistics in Medicine*, 24, 2401–28.

Lambert, P (2006) Comment on article by Browne and Draper. *Bayesian Analysis*, 1, 543–46.

Lancaster T (1990) *The econometric analysis of transition data.* Cambridge: Cambridge University Press.

Lancaster T (2002) Orthogonal parameters and panel data. *Review of Economic Studies*, 69, 647–66.

Lang S, Brezger A (2004) Bayesian P-splines. *Journal of Computational and Graphical Statistics*, 13, 183–212.

Lang S, Fronk E, Fahrmeir L (2002) Function estimation with locally adaptive dynamic models. *Computational Statistics*, 17, 479–500.

Lange K, Little R, Taylor M (1989) Robust statistical modeling using the t distribution. *Journal of the American Statistical Association*, 84, 881–96.

Langford I, Lewis T (1998) Outliers in multilevel data. *Journal of the Royal Statistical Society: Series A*, 161, 121–60.

Larch M, Walde J (2008) Lag or error – detecting the nature of spatial correlation. In C Preisach, H Burkhardt, L Schmidt-Thieme (eds), *Studies in classification, data analysis, and knowledge organization.* New York: Springer, 301–308.

Larson J, Soule S (2006) Sector Level Dynamics and Collective Action in the United States, 1965–1975. Working Paper, Department of Sociology, University of Arizona.

Laud P, Ibrahim J (1995) Predictive model selection. *Journal of the Royal Statistical Society: Series B*, 57, 247–62.

Lavine M (1999) Another look at conditionally Gaussian Markov random fields. In J Bernardo, J Berger, P Dawid, A Smith (eds), *Bayesian Statistics 6.* Oxford: Oxford University Press, 371–87.

Lavori P, Dawson R, Shera D (1995) A multiple imputation strategy for clinical trials with truncation of patient data. *Statistics in Medicine*, 14, 1913–25.

Lawless J, Crowder M (2004) Covariates and random effects in a gamma process model with application to degradation and failure. *Lifetime Data Analysis*, 10, 213–27.

Lawson A (2008) *Bayesian disease mapping: Hierarchical modeling in spatial epidemiology.* Boca Raton, FL: CRC Press.

Lawson A, Clark A (2002) Spatial mixture relative risk models applied to disease mapping. *Statistics in Medicine*, 21, 359–70.

LeSage J (1997) Bayesian estimation of spatial autoregressive models. *International Regional Science Review*, 20, 113–29.

LeSage J (1999) Spatial Econometrics. In RW Jackson (ed.), *The web book of regional science* (www.rri.wvu.edu/regscweb.htm), Morgantown, WV: Regional Research Institute, West Virginia University.

LeSage J (1999a) Econometrics toolbox for Matlab. www.spatial-econometrics .com/.

LeSage J (2004) A family of geographically weighted regression models. In L Anselin, R Florax, S Rey (eds), *Advances in spatial econometrics. Methodology, tools and applications.* New York: Springer, 241–64.

LeSage J, Kelley Pace R (2009) *Introduction to spatial econometrics.* CRC Press, Boca Raton, FL: Taylor & Francis.

Lee D, Shaddick G (2005) Time-varying coefficient models for the analysis of air pollution and health outcome data. University of Bath, Dept of Statistics, Working Paper 05/09.

Lee H, Higdon D, Calder C, Holloman C (2005) Efficient models for correlated data via convolutions of intrinsic processes. *Statistical Modelling*, 5, 53–74.

Lee J, Hwang R (2000) On estimation and prediction for temporally correlated longitudinal data. *Journal of Statistical Planning and Inference*, 87, 87–104.

Lee J, Sabavala D (1987) Bayesian estimation and prediction for the beta binomial model. *Journal of Business and Economic Statistics*, 5, 357–67.

Lee K, Thompson S (2008) Flexible parametric models for random-effects distributions. *Statistics in Medicine*, 27, 418–34.

Lee P (2004) *Bayesian statistics: An Introduction.* 3rd ed. London: Edward Arnold.

Lee R, Carter L (1992) Modeling and forecasting U.S. mortality. *Journal of the American Statistical Association*, 87, 659–71.

Lee S (1998) Coefficient constancy test in a random coefficient autoregressive model. *Journal of Statistical Planning and Inference*, 74, 93–101.

Lee S-Y (2007) *Structural equation modelling: A Bayesian approach.* New York: Wiley.

Lee S-Y, Shi J (2000) Joint Bayesian analysis of factor score and structural parameters in the factor analysis models. *Annals of the Institute of Statistical Mathematics*, 52, 722–36.

Lee S-Y, Song X-Y (2003) Bayesian model selection for mixtures of structural equation models with an unknown number of components. *British Journal of Mathematical and Statistical Psychology*, 56, 145–65.

Lee S-Y, Song X-Y (2004) Bayesian model comparison of nonlinear structural equation models with missing continuous and ordinal categorical data. *British Journal of Mathematical and Statistical Psychology*, 57, 131–50.

Lee S-Y, Song X-Y (2008) Bayesian model comparison of structural equation models. In D Dunson (ed.), *Random effect and latent variable model selection.* New York: Springer, 121–49.

Lee S-Y, Tang N (2006) Bayesian analysis of structural equation models with mixed exponential family and ordered categorical data. *British Journal of Mathematical and Statistical Psychology*, 59, 151–72.

Lee Y, Nelder J (2000) Two ways of modelling overdispersion in non-normal data. *Applied Statistics*, 49, 591–98.

Lee Y, Nelder J (2001) Modelling and analysing correlated non-normal data. *Statistical Modelling*, 1, 3–16.

Lee Y, Nelder J (2004) Conditional and marginal models: Another view. *Statistical Science*, 19, 219–38.

Leisch F (2004) FlexMix: A general framework for finite mixture models and latent class regression. *R. Journal of Statistical Software*, 11 (8), http://www.jstatsoft.org/v11/i08

Lenk P (1988) The logistic normal distribution for Bayesian nonparametric predictive densities. *Journal of the American Statistical Association*, 83, 509–16.

Lenk P (1999) Bayesian inference for semiparametric regression using a Fourier representation. *Journal of the Royal Statistical Society: Series B*, 61, 863–79.

Lenk P, DeSarbo W (2000) Bayesian inference for finite mixture models of generalized linear models with random effects. *Psychometrika*, 65, 475–96.

Leonard T (1973) A Bayesian method for histograms. *Biometrika*, 60, 297–308.

Leonte D, Nott D, Dunsmuir W (2003) Smoothing and change point detection for gamma ray count data. *Mathematical Geology*, 35, 175–94.

Leroux B, Lei X, Breslow N (1999) Estimation of disease rates in small areas: a new mixed model for spatial dependence. In M Halloran, D Berry (eds), *Statistical models in epidemiology, the environment and clinical trials*. New York: Springer-Verlag, 135–78.

Lesaffre E, Spiessens B (2001) On the effect of the number of quadrature points in a logistic random-effects model: An example. *Applied Statistics*, 50, 325–35.

Leslie D, Kohn R, Nott D (2007) A general approach to heteroscedastic linear regression. *Statistics and Computing*, 17, 131–46.

Leung Y, Mei C-L, Zhang W-X (2000) Statistical tests for spatial non-stationarity based on the geographically weighted regression model. *Environment and Planning A*, 32 (1), 9–32.

Levy J, Chemerynski S, Sarnat J (2005) Ozone exposure and mortality: An empiric Bayes metaregression analysis. *Epidemiology*, 16, 458–68.

Li K (1999) Bayesian analysis of duration models: An application to Chapter 11 bankruptcy. *Economics Letters*, 63, 305–12.

Li J, Yang X, Wu Y, Shoptaw S (2007) A random-effects Markov transition model for Poisson-distributed repeated measures with non-ignorable missing values. *Statistics in Medicine*, 26, 2519–32.

Li M (2007) Bayesian proportional hazard analysis of the timing of high school dropout decisions. *Econometric Reviews*, 26, 529–56.

Li Q, Stengos T (1994) Adaptive estimation in the panel data error component model with heteroskedasticity of unknown form. *International Economic Review*, 35, 981–1000.

Li W (1994) Time series models based on generalized linear models: Some further results. *Biometrics*, 50, 506–11.

Liang F, Paulo R, Molina G, Clyde M, Berger J (2008) Mixtures of g priors for Bayesian variable selection. *Journal of the American Statistical Association*, 103, 410–23.

Liang, K, Zeger S (1986) Longitudinal data analysis using generalized linear models. *Biometrika*, 73, 13–22.

Lichstein J, Simons T, Shriner S, Franzreb K (2002) Spatial autocorrelation and autoregressive models in ecology. *Ecological Monographs*, 72, 445–63.

Liechty J, Liechty M, Muller P (2004) Bayesian correlation estimation. *Biometrika*, 91, 1–14.

Lillard L, Willis R (1978) Dynamic aspects of earning mobility. *Econometrica*, 46, 985–1012.

Lin H, McCulloch C, Rosenheck R (2004) Latent pattern mixture models for informative intermittent missing data in longitudinal studies. *Biometrics*, 60, 295–305.

Lin T, Lee J, Hsieh W (2007b) Robust mixture modeling using the skew t distribution. *Statistics and Computing*, 17, 81–92.

Lin T, Lee J, Ni H (2004) Bayesian analysis of mixture modelling using the multivariate t distribution. *Statistics and Computing*, 14, 119–30.

Lin T, Lee J (2006) A robust approach to t linear mixed models applied to multiple sclerosis data. *Statistics in Medicine*, 25, 1397–1412.

Lin T, Lee J, Yen S (2007a) Finite mixture modelling using the skew normal distribution. *Statistica Sinica*, 17, 909–27.

Lindgren F, Rue H (2005) A note on the second order random walk model for irregular locations. Preprint Statistics 6/2005. Norges Teknisk-Naturvitenskapelige Universite.

Lindley D, Smith A (1972) Bayes estimates for the linear model. *Journal of the Royal Statistical Society: Series B*, 34, 1–41.

Lindsey J (1993) *Models for repeated measurements.* New York: Oxford University Press.

Lindstrom M, Bates D (1990) Nonlinear mixed effects models for repeated measures data. *Biometrics*, 46, 673–87.

Ling S (2004) Estimation and testing stationarity for double-autoregressive models. *Journal of the Royal Statistical Society: Series B*, 66, 63–78.

Litterman R (1986) Forecasting with Bayesian vector autoregressions—five years of experience. *Journal of Business & Economic Statistics*, 4, 25–38.

Little R (1993) Pattern-mixture models for multivariate incomplete data. *Journal of the American Statistical Association*, 88, 125–34.

Little R (1995) Modeling the drop-out mechanism in repeated-measures studies. *Journal of the American Statistical Association*, 90, 1112–21.

Little R, Rubin D (2002) *Statistical analysis with missing data*, 2nd edn. Hoboken, NJ: Wiley-Interscience.

Liu L, Hedeker, D (2006) A mixed-effects regression model for longitudinal multivariate ordinal data. *Biometrics*, 62, 261–68.

Liu N, Dey D (2007) Hierarchical overdispersed Poisson model with macrolevel autocorrelation. *Statistical Methodology*, doi:10.1016/i.stamet .2006.11.006.

Liu X, Wall M, Hodges J (2005) Generalized spatial structural equation modeling. *Biostatistics*, 6, 539–57.

Lockwood J, Doran H, McCaffrey D (2003) Using R for estimating longitudinal student achievement models. *R Newsletter*, 3, 17–23.

Lopes H, Muller P, Ravishanker N (2007) Bayesian computational methods in biomedical research. In R Khattree, D Naik (eds), *Computational methods in biomedical research.* New York: Deccer, 211–59.

Lopes H, West M (2004) Bayesian model assessment in factor analysis. *Statistica Sinica*, 14, 41–67.

Lubke G, Muthen B (2005) Investigating population heterogeneity with factor mixture models. *Psychological Methods*, 10, 21–39.

Lubrano M (1995) Testing for unit root in a Bayesian framework. *Journal of Econometrics*, 69, 81–109.

Lunn D, Spiegelhalter D, Thomas A, Best N (2009) The BUGS project: Evolution, critique and future directions. *Statistics in Medicine*, 28, 3049–67.

Müller P, Rosner G (1997) A Bayesian population model with hierarchical mixture priors applied to blood count data. *Journal of the American Statistical Association*, 92, 1279–92.

Ma G, Troxel A, Heitjan D (2005) An index of local sensitivity to nonignorable drop-out in longitudinal modelling. *Statistics in Medicine*, 24, 2129–50.

Ma Y, Genton M, Davidian M (2004) Linear mixed effects models with semiparametric generalized skew elliptical random effects. In M Genton (ed.), *Skew-Elliptical distributions and their applications: A journey beyond normality*. Boca Raton, FL: Chapman & Hall/CRC, 339–58.

MacCurdy T (1982) The use of time series processes to model the error structure of earnings in longitudinal data analysis. *Journal of Econometrics*, 18, 83–114.

MacNab Y (2007) Mapping disability-adjusted life years: A Bayesian hierarchical model framework for burden of disease and injury assessment. *Statistics in Medicine* 26, 4746–69.

MacNab Y, Gustafson P (2007) Regression B-spline smoothing in Bayesian disease mapping: With an application to patient safety surveillance, *Statistics in Medicine*, 26, 4455–74.

MacNab Y, Kmetic A, Gustafson P, Shaps S (2006) An innovative application of Bayesian disease mapping methods to patient safety research. *Statistics in Medicine*, 25, 3960–80.

MacNab Y, Qiu Z, Gustafson P, Dean C, Ohlsson A, Lee S (2004) Hierarchical Bayes analysis of multilevel health services data: A Canadian neonatal mortality study. *Health Services and Outcomes Research Methodology*, 5, 5–26.

Madsen L, Dalthorp D (2007) Simulating correlated count data. *Environmental and Ecological Statistics*, 14, 129–48.

Makuch R, Stephens M, Escobar M (1989) Generalized binomial models to examine the historical control assumption in active control equivalence studies. *The Statistician*, 38, 61–70.

Malaeb Z, Summers K, Pugesek B (2000) Using structural equation modeling to investigate relationships among ecological variables. *Environmental and Ecological Statistics*, 7, 93–111.

Malchow-Moller N, Svarer M (2003) Estimation of the multinomial logit model with random effects. *Applied Economics Letters*, 10, 389–92.

Manda S, Gilthorpe M, Tu Y, Blance A, Mayhew M (2005) A Bayesian analysis of amalgam restorations in the Royal Air Force using the counting process approach with nested frailty effects, *Statistical Methods in Medical Research*, 14, 567–78.

Mantel N, Hankey B (1978) A logistic regression analysis of response time data where the hazard function is time dependent. *Communications in Statistics A*, 7, 333–47.

Mardia K (1988) Multi-dimensional multivariate Gaussian Markov random fields with application to image processing. *Journal of Multivariate Analysis*, 24, 265–84.

Mardia K, Watkins A (1989) On multimodality of the likelihood in the spatial linear model. *Biometrika*, 76, 289–95.

Marin J, Mengersen K, Robert C (2005) Bayesian modelling and inference on mixtures of distributions. In D Dey, C Rao (eds), *Handbook of Statistics 25*. Amsterdam: Elsevier, 15840–45.

Marriott J, Naylor J, Tremayne A (2003) Exploring economic time series: A Bayesian graphical approach. *Econometrics Journal*, 6, 124–45.

Marriott J, Ravishanker N, Gelfand A, Pai J (1996) Bayesian analysis of ARMA processes: Complete sampling based inference under full likelihoods. In D Barry, K Chaloner, J Geweke (eds), *Bayesian analysis in statistics and econometrics.* New York: Wiley, 243–56.

Marsh H, Grayson D (1994) Longitudinal confirmatory factor analysis: Common, time-specific, item-specific, and residual-error components of variance. *Structural Equation Modeling*, 1, 116–45.

Marshall C, Best N, Bottle A, Aylin P (2004) Statistical issues in the prospective monitoring of health outcomes across multiple units. *Journal of the Royal Statistical Society: Series A*, 167, 541–59.

Marshall C, Spiegelhalter D (2003) Approximate cross-validatory predictive checks in disease mapping models. *Statistics in Medicine*, 22, 1649–60.

Marshall C, Spiegelhalter D (2007) Identifying outliers in Bayesian hierarchical models: A simulation-based approach. *Bayesian Analysis*, 2, 1–33.

Marshall E, Spiegelhalter D (1998) Comparing institutional performance using Markov Chain Monte Carlo methods. In B Everitt, G Dunn (eds), *Statistical analysis of medical data: New developments.* London: Arnold, 229–49.

Marshall E, Spiegelhalter D (2003) Approximate cross-validatory predictive checks in disease mapping models. *Statistics in Medicine*, 22, 1649–60.

Marshall E, Spiegelhalter D (2007) Simulation-based tests for divergent behaviour in hierarchical models. *Bayesian Analysis*, 2, 409–44.

Martin D, Raftery A (1987) Non-Gaussian state-space modeling of non-stationary time series: Robustness, computation, and non-Euclidean models. *Journal of the American Statistical Association*, 82, 1044–50.

Martin T, Wintle B, Rhodes J, Kuhnert P, Field S, Low-Choy S, Tyre A, Possingham H (2005) Zero tolerance ecology: Improving ecological inference by modelling the source of zero observations. *Ecology Letters*, 811, 1235–46.

Martinez-Beneito M, Lopez-Quilez A, Botella-Rocamora P (2008) An autoregressive approach to spatio-temporal disease mapping. *Statistics in Medicine*, 27, 2874–89.

Mazumdar S, Tang G, Houck P, Dew M, Begley A, Scott J, Mulsant B, Reynolds C (2007). Statistical analysis of longitudinal psychiatric data with dropouts. *Journal of Psychiatric Research*, 41, 1032–41.

McAdam D, Su Y (2002) The war at home: Antiwar protests and congressional voting, 1965 to 1973. *American Sociological Review*, 67, 696–721.

McCarthy M (2007) *Bayesian methods for ecology.* Cambridge: Cambridge University Press.

McCullagh P, Nelder J (1989) *Generalized linear models.* 2nd ed. Boca Raton, FL: Chapman & Hall/CRC.

McCulloch R, Polson N, Rossi P (2000) A Bayesian analysis of the multinomial probit model with fully identified parameters. *Journal of Econometrics*, 99, 173–93.

McCulloch R, Rossi P (1994) An exact likelihood analysis of the multinomial probit model. *Journal of Econometrics*, 64, 207–40.

McCulloch R, Tsay R (1993) Bayesian inference and prediction for mean and variance shifts in autoregressive time series. *Journal of the American Statistical Association*, 88, 968–78.

McCulloch R, Tsay R (1994) Bayesian analysis of autoregressive time series via the Gibbs sampler. *Journal of Time Series Analysis*, 15, 235–50.

McFadden D (1974) Conditional logit analysis of qualitative choice behaviour. In P Zarembka (ed.), *Frontiers in econometrics*. New York: Academic Press, 105–42.

McMillen D (1992) Probit with spatial autocorrelation. *Journal of Regional Science*, 32, 335–48.

Mebane W, Sekhon J (2004) Robust estimation and outlier detection for overdispersed multinomial models of count data. *American Journal of Political Science*, 48, 391–410.

Mehnert W, Smans M, Muir C, Mohner M, Schon D (1992) *Atlas of cancer incidence in the former German Democratic Republic 1978–1982*. New York: Oxford University Press.

Menard S (2002) *Longitudinal research*, 2nd edn. London: Sage.

Meng X (1994) Posterior predictive p-values. *The Annals of Statistics*, 22, 1142–60.

Meng X, Wong H (1996) Simulating ratios of normalizing constants using a simple identity: A theoretical exploration. *Statistica Sinica*, 6, 831–60.

Mengersen K, Tweedie R (1996) Rates of convergence of the Hastings and Metropolis algorithms. *The Annals of Statistics*, 24, 101–21.

Metropolis N, Rosenbluth A, Teller A, Teller E (1953) Equations of state calculations by fast computing machines. *Journal of Chemical Physics*, 21, 1087–92.

Meyer K (2005) Random regression analyses using B-splines to model growth of Australian Angus cattle. *Genetics Selection Evolution*, 37, 473–500.

Meyer M, Laud P (2002) Predictive variable selection in generalized linear models. *Journal of the American Statistical Association*, 97, 859–71.

Meyer R, Millar B (1998) Bayesian stock assessment using a nonlinear state-space model. In B Marx, H Friedl (eds), *Statistical Modeling. Proceedings, of the 13th International Workshop on Statistical Modelling*. Thousand Oaks, California: Sage, 284–91.

Meyer R, Yu J (2000) BUGS for a Bayesian analysis of stochastic volatility models. *Econometrics Journal*, 3, 198–215.

Mezzetti M (2006) Bayesian correlated factor analysis for spatial data. Proceedings Compstat 2006, In A Rizzi, M Vichi (eds), International Association for Statistical Computing. New York: Springer.

Migon H, Gamerman D, Lopes H, Ferreira M (2005) Dynamic models. In D Dey, C Rao (eds), *Handbook of statistics, volume 25: Bayesian thinking, modeling and computation*, Chapter 19, Amsterdam: Elsevier, 553–88.

Migon H, Moreira A (2004) Core inflation: Robust common trend model forecasting. *Brazilian Review of Econometrics*, 24, 1–19.

Militino A, Ugarte M, Dean C (2001) The use of mixture models for identifying high risks in disease mapping. *Statistics in Medicine*, 20, 2035–49.

Millar R (2004) Sensitivity of Bayes estimators to hyper-parameters, with an application to maximum yield from fisheries. *Biometrics*, 60, 536–42.

Millar R, Meyer R (2000) State-space modeling of nonlinear fisheries biomass dynamics using the Gibbs sampler. *Applied Statistics*, 49, 327–42.

Mira A, Petrone S (1996) Bayesian hierarchical nonparametric inference for change point problems. In J Bernardo, J Berger, A Dawid, A Smith (eds), *Bayesian Statistics 5*. Oxford: Oxford University Press, 693–703.

Mitchell T, Beauchamp J (1988) Bayesian variable selection in linear regression. *Journal of the American Statistical Association*, 83, 1023–36.

Moauro P, Savio G (2005) Temporal disaggregation using multivariate structural time series models. *Journal of Econometrics*, 8, 214–34.

Mohr D (2006) Bayesian identification of clustered outliers in multiple regression. *Computational Statistics and Data Analysis*, 51, 3955–67.

Molenberghs G, Verbeke G (2004) An introduction to (generalized) (non)linear mixed models. In P de Boeck (ed.), *Explanatory item response models: A generalized linear and nonlinear approach.* New York: Springer, 111–53.

Molenberghs G, Verbeke G (2006) *Models for discrete longitudinal data.* New York: Springer.

Molenberghs G, Verbeke G, Demetrio C (2007) An extended random-effects approach to modelling repeated, overdispersed count data. *Lifetime Data Analysis*, 13, 513–31.

Mollié A (1996) Bayesian mapping of disease. In W Gilks, S Richardson, D Spiegelhalter (eds), *Markov Chain Monte Carlo in practice*, Chapter 20. London: Chapman and Hall, 359–79.

Mollie A, Richardson S (1991) Empirical Bayes estimates of cancer mortality rates using spatial models. *Statistics in Medicine*, 10, 95–112.

Morgan B (1988) Extended models for quantal response data. *Statistica Neerlandica*, 42, 253–72.

Morris C, Norton E, Zhou X (1994) Parametric duration analysis of nursing home usage, Chapter 12 in Case Studies. In N Lange, L Ryan, L Billard, D Brillinger, L Conquest, J Greenhouse (eds), *Biometry.* New York: John Wiley, 231–48.

Mosler K (2003) Mixture models in econometric duration analysis. *Applied Stochastic Models in Business and Industry*, 19, 91–104.

Moustaki I (2000) A review of exploratory factor analysis for ordinal categorical data. In R Cudeck, S du Toit, D Sorbom (eds), *Structural equation models: Present and future.* Lincolnwood, IL: Scientific Software International, 461–80.

Muller, P, Erkanli, A, West, M (1996) Bayesian curve fitting using multivariate normal mixtures. *Biometrika*, 83, 67–79.

Muthén B, Asparouhov T (2009) Multilevel regression mixture analysis. *Journal of the Royal Statistical Society: Series A*, 172, 639–57.

Muthén B, Brown C, Masyn K, Jo B, Khoo S, Yang C, Wang C, Kellam S, Carlin J, Liao J (2002) General growth mixture modeling for randomized preventive interventions. *Biostatistics*, 3, 459–75.

Muthen B (1984) A general structural equation model with dichotomous, ordered categorical, and continuous latent variable indicators. *Psychometrika*, 49, 115–32.

Muthen B, Masyn K (2005) Discrete-time survival mixture analysis. *Journal of Educational and Behavioral Statistics*, 30, 27–58.

Namboodiri K, Suchindran C (1987) *Life table techniques and their applications.* New York: Academic Press.

Nandram B, Kim H (2002) Marginal likelihoods for a class of Bayesian generalized linear models. *Journal of Statistical Computation and Simulation*, 73, 319–40.

Natarajan R, Kass R (2000) Reference Bayesian methods for generalized linear mixed models. *Journal of the American Statistical Association*, 95, 227–37.

Natarajan R, McCulloch C (1995) A note on the existence of the posterior distribution for a class of mixed models. *Biometrika*, 82, 639–64.

Natarajan R, McCulloch C (1998) Gibbs sampling with diffuse proper priors: A valid approach to data-driven inference? *Journal of Computational and Graphical Statistics*, 7, 267–77.

Natario I, Knorr-Held L (2003) Non-parametric ecological regression and spatial variation. *Biometrical Journal*, 45, 670–88.

Neal, R (2000) Markov chain sampling methods for Dirichlet process mixture models. *Journal of Computational and Graphical Statistics*, 9, 249–65.

Nelson K, Leroux B (2006) Statistical models for autocorrelated count data. *Statistics in Medicine*, 25, 1413–30.

Nerlove M (2002) Essays in panel data econometrics. New York: Cambridge University Press.

Neuhaus J, Hauck W, Kalbfleisch J (1992) The effects of mixture distribution misspecification when fitting mixed-effects logistic models. *Biometrika*, 79, 755–62.

Neuhaus J, Kalbfleisch J, Hauck W (1991) A comparison of cluster-specific and population-averaged approaches for analyzing correlated binary data. *International Statistical Review*, B59, 25–35.

Neuhaus J, McCulloch C (2006) Separating between- and within-cluster covariate effects using conditional and partitioning methods. *Journal of the Royal Statistical Society B*, 68, 859–72.

Neves C, Migon H (2007) Bayesian graduation of mortality rates: An application to reserve evaluation, Insurance. *Mathematics and Economics*, 40, 424–34.

Ngo L,Wand M (2004) Smoothing with mixed model software. *Journal of Statistical Software*, 9(1), http://www.jstatsoft.org/v09/i01

Nicholls D, Quinn B (1980) The estimation of random coefficient autoregressive models I. *Journal of Time Series*, 1, 37–115.

Nicholls D, Quinn B (1982) *Random coefficient autoregressive models: An introduction.* New York: Springer-Verlag.

Niedermeier K, Von Eye A (1999) *Statistical analysis of longitudinal categorical data in the social and behavioral sciences: An introduction with computer illustrations.* Mahwah, NJ: Lawrence Erlbaum Associates.

Nikolov M, Coull B, Catalano P (2007) An informative Bayesian structural equation model to assess source-specific health effects of air pollution. *Biostatistics*, 8, 609–24.

Norton J, Niu X (2009) Intrinsically autoregressive spatiotemporal models with application to aggregated birth outcomes. *Journal of the American Statistical Association*, 104, 638–49.

Ntzoufras I (2009) *Bayesian modeling using Winbugs.* Chichester, Sussex: Wiley.

Nunez-Anton V, Zimmerman D (2000) Modeling non-stationary longitudinal data. *Biometrics*, 56, 699–705.

O'Brien S, Dunson D (2004) Bayesian multivariate logistic regression. *Biometrics*, 60, 739–46.

O'Sullivan D, Unwin D (2002) *Geographic information analysis.* Chichester, Sussex: Wiley.

Oh M-S, Lim Y (2001) Bayesian analysis of time series Poisson data. *Journal of Applied Statistics*, 28, 259–71.

Ohlssen D, Sharples L, Spiegelhalter D (2007) Flexible random-effects models using Bayesian semi-parametric models: Applications to institutional comparisons. *Statistics in Medicine*, 26, 2088–2112.

Oman S, Meir N, Halm N (1999) Comparing two measures of creatinine clearance: An application of errors-in-variables and bootstrap techniques. *Applied Statistics*, 48, 39–52.

Omori Y (2003) Discrete duration model having autoregressive random effects with application to Japanese diffusion index. *Journal of the Japan Statistical Society*, 33, 1–22.

Opsomer J, Claeskens G, Ranalli M, Kauermann G, Breidt F (2008) Non-parametric small area estimation using penalized spline regression. *Journal of the Royal Statistical Society: Series B*, 70, 265–86.

Orbe J, Núñez-Antón V (2006) Alternative approaches to study lifetime data under different scenarios: From the PH to the modified semiparametric AFT model. *Computational Statistics & Data Analysis*, 50, 1565–82.

Ord J, Snyder R, Koehler A, Hyndman R, Leeds M (2005) Time series forecasting: the case for the single source of error state space approach. Working Paper 7/05, Department of Econometrics and Business Statistics, Monash University.

Osborn J (1975) A multiplicative model for the analysis of vital statistics rates. *Applied Statistics*, 24, 75–84.

Osborne P, Foody G, Suárez-Seoane S (2007) Non-stationarity and local approaches to modelling the distributions of wildlife. *Diversity and Distributions*, 13, 313–23.

Oud J, Folmer H (2008) A structural equation approach to models with spatial dependence. *Geographical Analysis*, 40, 152–66.

Paap R (2002) What are the advantages of MCMC based inference in latent variable models? *Statistica Neerlandica*, 56, 2–22.

Paap R, van Dijk H (2003) Bayes estimation of Markov trends in possibly cointegrated series: An application to U.S. consumption and income. *Journal of Business & Economic Statistics*, 21, 547–63.

Paddock S (2007) Bayesian variable selection for longitudinal substance abuse treatment data subject to informative censoring. *Journal of the Royal Statistical Society: Series C*, 56, 293–311.

Palmer J, Pettit L (1996) Risks of using improper priors with Gibbs sampling and autocorrelated errors. *Journal of Computational and Graphical Statistics*, 5, 245–49.

Palomo J, Dunson D, Bollen K (2007) Bayesian structural equation modeling. In S-Y Lee (ed.), *Handbook of latent variable and related models.* Amsterdam: Elsevier, 163–68.

Panagiotelis A, Smith M (2008) Bayesian identification, selection and estimation of functions in high-dimensional additive models. *Journal of Econometrics*, 143, 291–316.

Papaspiliopoulos O, Roberts G, Skold M (2003) Non-centered parameterisations for hierarchical models and data augmentation. In J Bernardo, S Bayarri, J Berger, A Dawid, D Heckerman, A Smith, M West (eds), *Bayesian statistics 7.* Oxford: Oxford University Press, 307–26.

Parmigiani G (2002) *Modeling in medical decision making: A Bayesian approach.* New York: Wiley.

Parsons N, Edmondson R, Gilmour S (2006) A generalized estimating equation method for fitting autocorrelated ordinal score data with an application in horticultural research. *Journal of the Royal Statistical Society: Series C*, 55, 507–24.

Pastor N (2003) Methods for the analysis of explanatory linear regression models with missing data not at random, *Quality and Quantity*, 37, 363–376.

Patwardhad A, Small M (1992) Bayesian methods for model uncertainty analysis with application to future sea level rise. *Risk Analysis*, 12, 513–23.

Pauler D, Wakefield J (2000) Modeling and implementation issues in Bayesian meta-analysis. In D Stangl, D Berry (eds), *Bayesian meta-analysis.* New York: Marcel Dekker, 205–30.

Pearlman J (1980) An algorithm for the exact likelihood of a high-order autoregressive-moving average process. *Biometrika*, 67, 232–33.

Pedroza C (2006) A Bayesian forecasting model: predicting U.S. male mortality. *Biostatistics*, 7, 530–50.

Peng F, Jacobs R, Tanner M (1996) Bayesian inference in mixtures-of-experts and hierarchical mixtures-of-experts models with an application to speech recognition. *Journal of the American Statistical Association*, 91, 953–60.

Penzer J (2006) Diagnosing seasonal shifts in time series using state space models. *Statistical Methodology*, 3, 193–210.

Perperoglou A, van Houwelingen H, Henderson R (2006) A relaxation of the gamma frailty (Burr) model 2006. *Statistics in Medicine*, 25, 4253–66.

Perreault L, Berniera J, Bobéeb B, Parent E (2000) Bayesian change-point analysis in hydrometeorological time series; comparison of change-point models and forecasting. *Journal of Hydrology*, 235, 242–63.

Pettit L, Young K (1990) Measuring the effect of observations on Bayes factors. *Biometrika*, 77, 455–66.

Pettitt A, Tran T, Haynes M, Hay J (2006) A Bayesian hierarchical model for categorical longitudinal data from a social survey of immigrants. *Journal of the Royal Statistical Society: Series A*, 127, 97–114.

Pettitt A, Weir I, Hart A (2002) A conditional autoregressive Gaussian process for irregularly spaced multivariate data with application to modelling large sets of binary data. *Statistics and Computing*, 12, 353–67.

Phillips P, Durlauf S (1986) Multiple time series regression with integrated processes. *The Review of Economic Studies*, 53, 473–95.

Phillips P (1991) To criticize the critics: An objective Bayesian analysis of stochastic trends. *Journal of Applied Econometrics*, 6, 333–64.

Pickles A, Crouchley R (1995) A comparison of frailty models for multivariate survival data. *Statistics in Medicine*, 14, 1447–61.

Piegorsch W, Bailer J (2005) *Analyzing environmental data*. Chichester, UK: Wiley.

Pinheiro J, Liu C, Wu Y (2001) Efficient algorithms for robust estimation in linear mixed-effects models using the multivariate t distribution. *Journal of Computational and Graphical Statistics*, 10, 249–76.

Pitman J, Yor M (1997) The two-parameter Poisson-Dirichlet distribution derived from a stable subordinator. *Annals of Probability*, 25, 855–900.

Pitt M, Shephard N (1999) Time varying covariances: A factor stochastic volatility approach. In J Bernardo, J Berger, A Dawid, A Smith (eds), *Bayesian statistics 6*. Oxford: Oxford University Press, 547–70.

Plummer M (2008) Penalized loss functions for Bayesian model comparison. *Biostatistics*, 9, 523–39.

Podlich H, Faddy M, Smyth G (2004) Semi-parametric extended Poisson process models for count data. *Statistics and Computing*, 14, 311–21.

Pourahmadi M (1999) Joint mean-covariance models with applications to longitudinal data: Unconstrained parameterisation. *Biometrika*, 86, 677–90.

Pourahmadi M (2000) Maximum likelihood estimation of generalized linear models for multivariate normal covariance matrix. *Biometrika*, 87, 425–35.

Pourahmadi M (2002) Graphical diagnostics for modeling unstructured covariance matrices. *International Statistical Review*, 70, 395–417.

Pourahmadi M, Daniels M (2002) Dynamic conditionally linear mixed models for longitudinal data. *Biometrics*, 58, 225–31.

Prado R, Huerta G, West M (2000) Bayesian time-varying autoregressions: Theory, methods and applications. *Journal of the Institute of Mathematics and Statistics of the University of Sao Paolo*, 4, 405–22.

Prado R, West M (1997) Exploratory modelling of multiple non-stationary time series: Latent process structure and decompositions. In T Gregoire (ed.), *Modelling longitudinal and spatially correlated data.* New York: Springer-Verlag.

Press S, Shigemasu K (1989) Bayesian inference in factor analysis. In L Gleser, M Perleman, S J Press, A Sampson (eds), *Contributions to probability and statistics.* New York: Springer-Verlag, 271–87.

Prevost T, Abrams K, Jones D (2000) Hierarchical models in generalized synthesis of evidence: An example based on studies of breast cancer screening. *Statistics in Medicine*, 19, 3359–76.

Proietti T (2006) Measuring core inflation by multivariate structural time series models. Research Paper Series 83, Tor Vergata University, CEIS.

Qian S, Reckhow K, Zhai J, McMahon G (2005) Nonlinear regression modeling of nutrient loads in streams: A Bayesian approach. *Water Resources Research*, 41, W07012, doi:10.1029/2005WR003986.

Qin Z, Damien P, Walker S (2000) Uniform scale mixture models with applications to variance regression. Working Paper 14, Univ Michigan Business School.

Qiu Z, Song P, Tan M (2002) Bayesian hierarchical models for multi-level repeated ordinal data using WinBUGS. *Journal of Biopharmaceutical Statistics*, 12, 121–35.

Quinn K (2004) Bayesian factor analysis for mixed ordinal and continuous responses. *Political Analysis*, 12, 338–53.

Quintana F, Tam W (1996) Bayesian estimation of beta-binomial models by simulating posterior densities. *Revista de la Sociedad Chilena de Estadística*, 13, 43–56.

Quintana, F, Müller P, Rosner G (2008) A semiparametric Bayesian model for repeated binary measurements. *Applied Statistics*, 57, 419–31.

Rabe-Hesketh S, Skrondal A, Pickles A (2004) *GLLAMM manual.* U.C. Berkeley Division of Biostatistics Working Paper 160.

Raftery A (1996) Approximate Bayes factors and accounting for model uncertainty in generalized linear models. *Biometrika*, 83, 251–66.

Raftery A, Hout M (1993) Maximally maintained inequality: Expansion, reform and opportunity in Irish schools. *Sociology of Education*, 66, 41–62.

Raftery A, Lewis S (1992). One long run with diagnostics: Implementation strategies for Markov Chain Monte Carlo. *Statistical Science*, 7, 493–97.

Raftery A, Lewis S (1996) The number of iterations, convergence diagnostics and generic Metropolis algorithms. In W Gilks, D Spiegelhalter, S Richardson (eds), *Practical Markov Chain Monte Carlo.* London: Chapman & Hall.

Rao J (2003) *Small area estimation.* New York: Wiley.

Rattanasiri S, Bohning D, Roianavipart P, Athipanyakom S (2004) A mixture model application in disease mapping of malaria. *Southeast Asian Journal of Tropical Medicine and Public Health*, 35, 38–47.

Raudenbush S (1993) A crossed random effects model for unbalanced data with applications in cross-sectional and longitudinal research. *Journal of Educational Statistics*, 18, 321–49.

Ravishanker N, Dey D (2000) Multivariate survival models with a mixture of positive stable frailties. *Methodology and Computing in Applied Probability*, 2, 293–308.

Redner R, Walker H (1984) Mixture densities, maximum likelihood, and the EM algorithm. *SIAM Review*, 26, 195–239.

Reich B, Fuentes M (2007) A multivariate semiparametric Bayesian spatial modeling framework for hurricane surface wind fields. *Annals of Applied Statistics*, 1, 249–64.

Reich B, Hodges J (2008) Modeling longitudinal spatial periodontal data: A spatially-adaptive model with tools for specifying priors and checking fit. *Biometrics*, 64, 790–99.

Reis E, Salazar E, Gamerman D (2006) Comparison of sampling schemes for dynamic linear models. *International Statistical Review*, 74, 203–14.

Ribeiro P, Diggle P (2001) geoR: A package from geostatistical analysis. *R-NEWS*, 1 (2), 15–18 June. ISSN 1609-3631. URL http://cran.R-project .org/doc/Rnews.

Rice K (2005) Bayesian measures of goodness of fit. In P Armitage, T Colton (eds), *Encyclopedia of biostatistics*. Chichester: John Wiley.

Rice K, Spiegelhalter D (2006) A simple diagnostic plot connecting robust estimation, outlier detection, and false discovery rates. *Journal of Applied Statistics*, 33, 1131–47.

Richards H, Barry, R (1998) U.S. Life Tables for 1990 by sex, race, and education. *Journal of Forensic Economics*, 11, 9–26.

Richardson S, Green P (1997) On Bayesian analysis of mixtures with an unknown number of components. *Journal of the Royal Statistical Society: Series B*, 59, 731–58.

Richardson S, Guihenneuc C, Lasserre V (1992) Spatial linear models with autocorrelated error structure. *The Statistician*, 41, 539–57.

Richardson S, Monfort C (2000) Ecological correlation studies. In P Elliott, J Wakefield, N Best and D Briggs (eds), *Spatial epidemiology methods and applications*. Oxford: Oxford University Press, 205–20.

Richardson S, Thomson A, Best N, Elliott, P (2004) Interpreting posterior relative risk estimates in disease-mapping studies. *Environmental Health Perspectives*, 112, 1016–25.

Rigby R (1997) Bayesian discrimination between two multivariate normal populations with equal covariance matrices. *Journal of the American Statistical Association*, 92, 1151–54.

Riggan W, Manton K, Creason J, Woodbury M, Stallard E (1991) Assessment of spatial variation of risks in small populations. *Environmental Health Perspectives*, 96, 223–38.

Robert C (1996) Mixtures of distributions: Inferences and estimation. In W Gilks, S Richardson, D Spiegelhalter (eds), *Markov Chain Monte Carlo in Practice*, Chapter 24. Boca Raton, FL: Chapman and Hall/CRC, 441–64.

Robert C, Mengersen K (1999) Reparameterisation issues in mixture modelling and their bearing on the Gibbs sampler. *Computational Statistics and Data Analysis*, 29, 325–43.

Robert C, Titterington D (1998) On perfect simulation for some mixtures of distributions. *Statistics and Computing*, 8, 145–58.

Roberts G, Gelman A, Gilks W (1997) Weak convergence and optimal scaling of random walk metropolis algorithms. *The Annals of Applied Probability*, 7, 110–20.

Roberts G, Rosenthal J (2004) General state space Markov chains and MCMC algorithms. *Probability Surveys*, 1, 20–71.

Roberts G, Sahu S (2001) Approximate predetermined convergence properties of the Gibbs sampler. *Journal of Computational and Graphical Statistics*, 10, 216–29.

Roberts G, Tweedie R (1996) Geometric convergence and central limit theorems for multidimensional Hastings and Metropolis algorithms. *Biometrika*, 83, 95–110.

Roberts G, Sahu S (1997) Updating schemes, correlation structures, blocking and parameterization of the Gibbs sampler. *Journal of the Royal Statistical Society: Series B*, 59, 291–317.

Roberts S, Husmeier D, Rezek I, Penny W (1998) Bayesian approaches to Gaussian mixture modeling. *IEEE Transactions on Pattern Analysis and Machine Intelligence*, 20, 1133–42.

Robertson C, Fryer J (1968) Some descriptive properties of normal mixtures. *Skand Aktuar Tidskr*, 52, 137–46.

Rodríguez-Avi J, Conde-Sánchez A, Sáez-Castillo A, Olmo-Jimé nez M (2007) A generalization of the beta–binomial distribution. *Applied Statistics*, 56, 51–61.

Rodrigues A, Assuncao R (2008) Propriety of posterior in Bayesian space varying parameter models with normal data. *Statistics & Probability Letters*, 78, 2408–11.

Rodrigues-Motta R, Gianola D, Heringstad B, Rosa G, Chang Y (2007) A zero-inflated Poisson model for genetic analysis of the number of mastitis cases in Norwegian red cows. *Journal of Dairy Science*, 90, 5306–15.

Rodriguez G, Goldman N (2001) Improved estimation procedures for multilevel models with binary response: A case study. *Journal of the Royal Statistical Society: Series A*, 164, 339–55.

Roeder K, Wasserman L (1997) Practical Bayesian density estimation using mixtures of normals. *Journal of the American Statistical Association*, 92, 894–902.

Rosenbaum P, Rubin D (1983) The central role of the propensity score in observational studies for causal effects. *Biometrika*, 70, 41–55.

Rossi P, Allenby G, McCulloch R (2005) *Bayesian statistics and marketing.* Chichester, England: Wiley.

Rousseeuw P (1984) Least median of squares regression. *Journal of the American Statistical Association*, 79, 871–80.

Roy J, Lin X (2000) Latent variable models for longitudinal data with multiple continuous outcomes. *Biometrics*, 56, 1047–54.

Roy J, Lin, X (2002) Analysis of multivariate longitudinal outcomes with non-ignorable dropouts and missing covariates: Changes in methadone treatment practices. *Journal of the American Statistical Association*, 97, 40–52.

Roy N (2002) Is adaptive estimation useful for panel models with heteroskedasticity in the individual specific error component? Some Monte Carlo evidence. *Econometric Reviews*, 21, 189–203.

Royston P (1993) A toolkit for testing for non-normality in complete and censored samples. *The Statistician*, 42, 37–43.

Rubin D (1976) Inference and missing data. *Biometrika*, 63, 581–92.

Rubin D (1987) *Multiple imputation for nonresponse in surveys.* New York: Wiley.

Rubin D, Schenker N (1986) Multiple imputation for interval estimation from simple random samples with ignorable nonresponse. *Journal of the American Statistical Association*, 81, 366–74.

Rue H (2001) Fast sampling of Gaussian Markov random fields. *Journal of the Royal Statistical Society: Series B*, 63, 325–38.

Rue H, Held L (2005) *Gaussian Markov Random Fields: Theory and applications.* London: Chapman & Hall.

Rue H, Tjelmeland H (2002) Fitting Gaussian Markov random fields to Gaussian fields. *Scandinavian Journal of Statistics*, 29, 31–49.

Rupp A, Dey D, Zumbo B (2004) To Bayes or not to Bayes, from whether to when: Applications of Bayesian methodology to modeling. *Structural Equation Modeling*, 11, 424–51.

Ruppert D, Carroll R (2000) Spatially-adaptive penalties for spline fitting. *Australian & New Zealand Journal of Statistics*, 42, 205–23.

Ruppert D, Wand M, Carroll R (2003) *Semiparametric regression.* Cambridge: Cambridge University Press.

Ruspini E (1999) Longitudinal research and the analysis of social change. *Quality and Quantity*, 33, 219–27.

Ryu D, Sinha D, Mallick B, Lipsitz S, Lipshultz S (2007) Longitudinal studies with outcome-dependent follow-up: Models and Bayesian regression. *Journal of the American Statistical Association*, 102, 952–61.

Sahu S (2002) Bayesian estimation and model choice in item response models. *Journal of Statistical Computation and Simulation*, 72, 217–32.

Sahu S, Dey D (2000) A comparison of frailty and other models for bivariate survival data. *Lifetime Data Analysis*, 6, 207–28.

Sahu S, Dey D (2004) On a Bayesian multivariate survival model with skewed frailty. In M Genton (ed.), *Skew-elliptical distributions and their applications: A journey beyond normality.* Boca Raton, FL: CRC/Chapman & Hall, 321–38.

Sahu S, Dey D, Aslanidou H, Sinha D (1997) A Weibull regression model with gamma frailties for multivariate survival data. *Lifetime Data Analysis*, 3, 123–37.

Sahu S, Dey D, Branco M (2003) A new class of multivariate skew distributions with applications to Bayesian regression models. *The Canadian Journal of Statistics*, 31, 129–50.

Sain S, Cressie N (2007) A spatial model for multivariate lattice data. *Journal of Econometrics*, 140, 226–59.

Sala-i-Martin X, Doppelhofer G, Miller R (2004) Determinants of long-term growth: A Bayesian averaging of classical estimates (BACE) approach. *American Economic Review*, 94, 813–35.

Sanchez B, Butdz-Jorgensen E, Ryan L, Hu H (2005) Structural equation models: A review with applications to environmental epidemiology. *Journal of the American Statistical Association*, 100, 1443–55.

Sargent D (1997) A flexible approach to time-varying coefficients in the Cox regression setting. *Lifetime Data Analysis*, 3, 13–25.

Schabenberger O, Gotway C (2004) *Statistical methods for spatial data analysis*. Boca Raton, FL: Chapman & Hall/CRC.

Schafer J (1997) Imputation of missing covariates under a multivariate linear mixed model. Technical report, Department of Statistics, The Pennsylvania State University.

Schafer J, Graham J (2002) Missing data: Our view of the state of the art. *Psychological Methods*, 7, 147–77.

Scheines R, Hoijtink H, Boomsma A (1999) Bayesian estimation and testing of structural equation models. *Psychometrika*, 64, 37–52.

Schmid V, Held L (2004) Bayesian extrapolation of space-time trends for cancer registry data. *Biometrics*, 60, 1034–42.

Schmid V, Held L (2007) Bayesian age-period-cohort modeling and prediction—BAMP. *Journal of Statistical Software*, 21 (8), http://www.jstatsoft.org/.

Schmidt P, Witte A (1989) Predicting criminal recidivism using 'split population' survival time models. *Journal of Econometrics*, 40, 141–59.

Schoen R, Weinick R (1993) The slowing metabolism of marriage: Figures from 1988 U.S. marital status life tables. *Demography*, 30, 734–46.

Schotman P, Van Dijk H (1991) On Bayesian routes to unit roots. *Journal of Applied Econometrics*, 6, 387–401.

Schwartz J (1993) Air pollution and daily mortality in Birmingham, Alabama. *American Journal of Epidemiology*, 137, 1136–47.

Schwarz G (1978) Estimating the dimension of a model. *Annals of Statistics*, 6, 461–64.

Scollnik D (1995) Bayesian analysis of two overdispersed Poisson models. *Biometrics*, 51, 1117–26.

Scollnik D (2002) Implementation of four models for outstanding liabilities in WinBUGS: A discussion of a paper by Ntzoufras and Dellaportas. *North American Actuarial Journal*, 6, 128–36.

Scott S (2002) Bayesian methods for hidden Markov models: Recursive computing in the 21st century, *Journal of the American Statistical Association*, 97, 337–51.

Scott S (2003) Data augmentation for the Bayesian analysis of multinomial logit model. *Proceedings of American Statistical Association Section on Bayesian Statistical Science*. Alexandria, VA: American Statistical Association.

Seltzer M (1993) Sensitivity analysis for fixed effects in the hierarchical model: A Gibbs sampling approach. *Journal of Educational Statistics*, 18, 207–35.

Seltzer M, Novak J, Choi K, Lim N (2002) Sensitivity analysis for hierarchical models employing t level-1 assumptions. *Journal of Educational and Behavioral Statistics*, 27, 181–222.

Seltzer M, Wong W, Bryk A (1996) Bayesian inference in applications of hierarchical models: Issues and methods. *Journal of Educational and Behavioral Statistics*, 21, 131–67.

Sethuraman J (1994) A constructive definition of Dirichlet priors. *Statistica Sinica*, 4, 639–50.

Shao Q, Zhou X (2004) A new parametric model for survival data with long-term survivors. *Statistics in Medicine*, 23, 3525–43.

Sharples L (1990) Identification and accommodation of outliers in general hierarchical models. *Biometrika*, 77, 445–53.

Shen W, Louis T (1998) Triple-goal estimates in two-stage hierarchical models. *Journal of the Royal Statistical Society: Series B*, 60, 455–71.

Shephard N, Pitt M (1997) Likelihood analysis of non-Gaussian measurement time series. *Biometrika*, 84, 653–67.

Shi J, Lee S-Y (2000) Latent variable models with mixed continuous and polytomous data. *Journal of the Royal Statistical Society B*, 62, 77–87.

Shively T, Kohn R, Wood S (1999) Variable selection and function estimation in additive nonparametric regression using a data-based prior. *Journal of the American Statistical Association*, 94, 777–807.

Shoemaker J, Painter I, We B (1999) Bayesian statistics in genetics: a guide for the uninitiated. *Trends in Genetics*, 15, 354–58.

Shumway R, Stoffer D (2006) *Time series analysis and its applications; with R examples*. New York: Springer.

Silliman N (1997) Hierarchical selection models with applications in meta-analysis. *Journal of the American Statistical Association*, 92, 926–36.

Silva G, Dean C, Niyonsenga T, Vanasse A (2008) Hierarchical Bayesian spatiotemporal analysis of revascularization odds using smoothing splines. *Statistics in Medicine*, 27, 2381–401.

Silva R, Lopes H, Migon H (2006) The extended generalized inverse Gaussian distribution for log-linear and stochastic volatility models. *Brazilian Journal of Probability and Statistics*, 20, 67–91.

Sims C, Zha T (1998) Bayesian methods for dynamic multivariate models. *International Economic Review*, 39, 949–68.

Sinha D (1993) Semiparametric Bayesian analysis of multiple event time data. *Journal of the American Statistical Association*, 88, 979–83.

Sinha D, Chen M-H, Ghosh S (1999) Bayesian analysis and model selection for interval-censored survival data. *Biometrics*, 55, 585–90.

Sinha D, Ghosh S (2006) Multiple events time data: A Bayesian recourse. In D Dey, C Rao (eds), Bayesian thinking: Modeling and computation, Handbook of Statistics, Vol 25. Amsterdam: Elsevier, 891–906.

Sinha D, Patra K, Dey, D (2003) Modelling accelerated life test data by using a Bayesian approach. *Journal of the Royal Statistical Society C*, 52, 249–59.

Sinharay S (2004) Experiences with Markov Chain Monte Carlo convergence assessment in two psychometric examples. *Journal of Educational and Behavioral Statistics*, 29, 461–88.

Sinharay S, Stern H (2003) Posterior predictive model checking in hierarchical models. *Journal of Statistical Planning and Inference*, 111, 209–21.

Sinharay S, Stern H (2005) An empirical comparison of methods for computing Bayes factors in generalized linear mixed models. *Journal of Computational and Graphical Statistics*, 14, 415–35.

Siu N, Kelly D (1998) Bayesian parameter estimation in probabilistic risk assessment. *Reliability Engineering and System Safety*, 62, 89–116.

Skrondal A, Rabe-Hesketh S (2004) *Generalized latent variable modeling: Multilevel, longitudinal and structural equation models.* Boca Raton, FL: Chapman & Hall/CRC.

Skrondal A, Rabe-Hesketh S (2007) Latent variable modelling: A survey. *Scandinavian Journal of Statistics*, 34, 712–45.

Smith A (1973a) A general Bayesian linear model. *Journal of the Royal Statistical Society: Series B*, 35, 67–75.

Smith A (1973b) Bayes estimates in one-way and two-way models. *Biometrika*, 60, 319–29.

Smith A, Gelfand A (1992) Bayesian statistics without tears: A sampling-resampling perspective. *The American Statistician*, 46, 84–88.

Smith A, Roberts G (1993) Bayesian computation via the Gibbs sampler and related Markov chain Monte Carlo methods. *Journal of the Royal Statistical Society: Series B*, 55, 3–23.

Smith M, Kohn R (1996) Nonparametric regression using Bayesian variable selection. *Journal of Econometrics*, 75, 317–44.

Smith M, Kohn R (1997) A Bayesian approach to nonparametric bivariate regression. *Journal of the American Statistical Association*, 92, 1522–35.

Smith M, Kohn R (2002) Parsimonious covariance matrix estimation for longitudinal data. *Journal of the American Statistical Association*, 97, 1141–53.

Smith M, Wong C-M, Kohn R (1998) Additive nonparametric regression with autocorrelated errors. *Journal of the Royal Statistical Society: Series B*, 60, 311–31.

Smith R, Davis J, Sacks J (2000) Regression models for air pollution and daily mortality: Analysis of data from Birmingham, Alabama. *Environmetrics*, 11, 719–43.

Smith T, LeSage J (2004) A Bayesian probit model with spatial dependencies. In J LeSage, R Kelley Pace (eds), *Advances in Econometrics*: Vol 18: *Spatial and spatiotemporal econometrics.* Amsterdam: Elsevier Science, 127–60.

Smith T, Spiegelhalter D, Thomas A (1995) Bayesian approaches to random-effects meta-analysis: A comparative study. *Statistics in Medicine*, 14, 2685–99.

Snijders T, Berkhof J (2002) Diagnostic checks for multilevel models. In J de Leeuw, I Kreft (eds), *Handbook of quantitative multilevel analysis.* Boston: Kluwer, 139–73.

Snijders T, Bosker R (1999) *Multilevel analysis. An introduction to basic and advanced multilevel modelling.* London: Sage.

Sohn Y, Chang I, Moon T (2007) Random effects Weibull regression model for occupational lifetime. *European Journal of Operational Research*, 179, 124–31.

Song J, Belin T (2004) Imputation for incomplete high-dimensional multivariate normal data using a common factor model. *Statistics in Medicine*, 2004, 23, 2827–43.

Song J, Ghosh M, Miaou S, Mallick B (2005) Bayesian multivariate spatial models for roadway traffic crash mapping. *Journal of Multivariate Analysis*, 97, 246–73.

Song X, Lee S (2004) A Bayesian model selection method with applications. *Computational Statistics & Data Analysis*, 40, 539–57.

Song X-Y, Lee S-Y, Ng M, So W-Y, Chan J (2006) Bayesian analysis of structural equation models with multinomial variables and an application to type 2 diabetic nephropathy. *Statistics in Medicine*, 26, 2348–69.

Speed T, Kiiveri H (1986) Gaussian distributions over finite graphs. *Annals of Statistics*, 14, 138–50.

Spiegelhalter D (1998) Bayesian graphical modelling: A case-study in monitoring health outcomes. *Applied Statistics*, 47, 115–33.

Spiegelhalter D (1999) Surgical audit: Statistical lessons from Nightingale and Codman. *Journal of the Royal Statistical Society: Series A*, 162, 45–58.

Spiegelhalter D (2004) Incorporating Bayesian ideas into health-care evaluation. *Statistical Science*, 19, 156–74.

Spiegelhalter D (2005) Handling over-dispersion of performance indicators. *Quality and Safety in Health Care*, 14, 347–51.

Spiegelhalter D (2006) Two brief topics on modelling with WinBUGS. Presented at ICEBUGS Conference, Helsinki 2006 (available from http://mathstat.helsinki.fi/openbugs/IceBUGS/IceBUGSTimetable.html).

Spiegelhalter D, Abrams K, Myles J (2004) *Bayesian approaches to clinical trials and health-care evaluation.* New York: Wiley.

Spiegelhalter D, Best N, Carlin B, van der Linde A (2002) Bayesian measures of model complexity and fit. *Journal of the Royal Statistical Society: Series B*, 64, 583–639.

Spiegelhalter D, Freedman L, Parmar M (1994) Bayesian approaches to randomised trials. *Journal of the Royal Statistical Society: Series A*, 157, 357–87.

Spiess M (2006) Estimation of a two-equation panel model with mixed continuous and ordered categorical outcomes and missing data. *Journal of the Royal Statistical Society: Series C*, 55, 525–38.

Staudenmayer J, Lake E, Wand M (2009) Robustness for general design mixed models using the t distribution. *Statistical Modelling*, 9, 235–55.

Stephens M (2000) Bayesian analysis of mixture models with an unknown number of components – An alternative to reversible iump methods. *Annals of Statistics*, 28 (1), 40–74.

Stern H, Sinharay S (2005) Bayesian model checking and model diagnostics. In D Dey, C Rao (eds), *Bayesian thinking: Modeling and computation, handbook of statistics.* vol. 25. Amsterdam: Elsevier, 171–92.

Stock J, Watson M (1998) Median unbiased estimation of coefficient variance in a time-varying parameter model. *Journal of the American Statistical Association*, 93, 349–58.

Strickland C, Turner I, Denham R, Mengersen K (2008) Efficient Bayesian estimation of multivariate state space models. Source: http://eprints .qut.edu.au.

Sturtz S, Ligges U, Gelman A (2005) R2WinBUGS: A package for running WinBUGS from R. *Journal of Statistical Software*, 12 (3), 1–16.

Subramanian S (2004) The relevance of multilevel statistical methods for identifying causal neighborhood effects. *Social Science & Medicine*, 58, 1961–67.

Subramanian S, Jones K, Duncan C (2003) Multilevel methods for public health research. In I Kawachi, L Berkman (eds), *Neighborhoods and health.* New York: Oxford University Press, 65–111.

Sun D, Speckman P, Tsutakawa R (2000) Random effects in generalized linear mixed models (GLMMs). In D Dey, S Ghosh, B Mallick (eds), *Generalized linear models: A Bayesian perspective.* Dekker: New York, 23–39.

Sun D, Tsutakawa R, Kim H, He Z (2000) Spatio-temporal interaction with disease mapping. *Statistics in Medicine*, 19, 2015–35.

Sun D, Tsutakawa R, Speckman P (1999) Posterior distribution of hierarchical models using CAR(1) distributions. *Biometrika*, 86, 341–50.

Swaminathan H, Rogers H (1990) Detecting differential item functioning using logistic regression procedures. *Journal of Educational Measurement*, 27, 361–70.

Swamy P (1971) *Statistical inference in random coefficient regression models.* New York: Springer-Verlag.

Swamy P, Mehta J (1975) Bayesian and non-Bayesian analysis of switching regressions and random coefficient regression. *Journal of the American Statistical Association*, 70, 593–602.

Swanson D, Clauser B, Case S, Nungester R, Featherman C (2002) Analysis of differential item functioning (DIF) using hierarchical logistic regression models. *Journal of Educational and Behavioral Statistics*, 27, 53–75.

Symanski E, Sallsten G, Chan W (2001) Heterogeneity in sources of exposure variability among groups of workers exposed to inorganic mercury. *Annals of Occupational Hygiene*, 45, 677–87.

Tan M, Qu Y, Mascha E, Schubert A (1999) A Bayesian hierarchical model for multi-level repeated ordinal data: Analysis of oral practice examinations in a large anaesthesiology training programme. *Statistics in Medicine*, 18, 1983–92.

Tanizaki H (2003) Nonlinear and non-Gaussian state-space modeling with Monte Carlo techniques: A survey and comparative study. In C Rao, D Shanbhag (eds), *Handbook of Statistics.* Vol. 21. *Stochastic processes: Modeling and simulation*, Amsterdam: Elsevier, Chap. 22, 871–929.

Tanner M (1996) Tools for statistical inference: Methods for the exploration of posterior distributions and likelihood functions. 3rd edition. New York: Springer.

Teather D (1984) The estimation of exchangeable binomial parameters. *Communications in Statistics, Part A*, 13, 671–80.

Ten Have T, Kunselman A, Pulkstenis E, Landis R (1998) Mixed effects logistic regression models for longitudinal binary response data with informative drop-out. *Biometrics*, 54, 367–83.

Thall P, Vail S (1990) Some covariance models for longitudinal count data with overdispersion. *Biometrics*, 46, 657–71.

Tiao G, Tsay R (1989) Model specification in multivariate time series. *Journal of the Royal Statistical Society*, 51B, 157–213.

Tierney L (1994) Markov chains for exploring posterior distributions. *Annals of Statistics*, 21, 1701–62.

Tierney L, Kadane J (1986) Accurate approximations for posterior moments and marginal densities. *Journal of the American Statistical Association*, 81, 82–86.

Toft N, Innocent G, Gettinby G, Reid S (2007) Assessing the convergence of Markov Chain Monte Carlo methods: An example from evaluation of diagnostic tests in absence of a gold standard. *Preventive Veterinary Medicine*, 79, 244–56.

Tosch T, Holmes P (1980) A bivariate failure model. *Journal of the American Statistical Association*, 75, 415–17.

Troughton P, Godsill S (1998) Bayesian model selection for linear and non-linear time series using the Gibbs sampler. In J Fine McWhirter, I Prouder (eds), *Mathematics in Signal Processing IV*. Oxford: Oxford University Press, 249–61.

Troxel A, Harrington D, Lipsitz S (1998) Analysis of longitudinal data with non-ignorable non-monotone missing values. *Applied Statistics*, 47, 425–38.

Troxel A, Ma G, Heitjan D (2004) An index of local sensitivity to nonignorability. *Statistica Sinica*, 14, 1221–37.

Tsai M-Y, Hsiao C (2008) Computation of reference Bayesian inference for variance components in longitudinal studies. *Computational Statistics*, 23, 587–604.

Tsay R (1986) Time series model specification in the presence of outliers. *Journal of the American Statistical Association*, 81, 132–41.

Tsodikov A, Ibrahim J, Yakovlev A (2003) Estimating cure rates from survival data: An alternative to two-component mixture models. *Journal of the American Statistical Association*, 98, 1063–78.

Tutz G, Kauermann G (2003) Generalized linear random effects models with varying coefficients. *Computational Statistics & Data Analysis*, 43 (1), 13–28.

Tutz G, Reithinger F (2007) A boosting approach to flexible semiparametric mixed models. *Statistics in Medicine*, 26, 2872–900.

Tzala E, Best N (2007) Bayesian latent variable modelling of multivariate spatio-temporal variation in cancer mortality. *Statistical Methods in Medical Research.* Sep 13 (epub).

Tzala E, Best N (2008) Bayesian latent variable modelling of multivariate spatio-temporal variation in cancer mortality. *Statistical Methods in Medical Research*, 17, 97-118.

Ulrick S (2007) *Using semiparametric methods in an analysis of earnings mobility.* Available at SSRN: http://ssrn.com/abstract=654741.

van Dongen S (2006) Prior specification in Bayesian statistics: Three cautionary tales. *Journal of Theoretical Biology*, 242, 90–100.

van Duijn M, Jansen M (1995) Modelling repeated count data: Some extensions of the Rasch Poisson counts model. *Journal of Educational and Behavioral Statistics*, 20, 241–58.

van Dyk D (2003) Hierarchical models, data augmentation, and Markov Chain Monte Carlo. In G Babu, E Feigelson (eds), *Statistical challenges in modern astronomy III.* New York: Springer, 41–56.

van Dyk D, Meng X-L (2001) The art of data augmentation. *Journal of Computational and Graphical Statistics*, 10, 1–111.

van Houwelingen H, Arends L, Stiinen T (2002) Advanced methods in meta-analysis: Multivariate approach and meta-regression. *Statistics in Medicine*, 21, 589–624.

Van den Berg, G (2001) Duration models: Specification, identification, and multiple durations. In J Heckman, E Leamer (eds), *Handbook of Econometrics 5*, Amsterdam: North Holland.

Vanhonacker W (1990) On Bayesian estimation of model parameters. *Marketing Science*, 9, 54–56.

Vannucci M (2000) Matlab code for Bayesian variable selection. *ISBA Bulletin*, 7 (3), 12–13.

Vehtari A, Lampinen J (2002) Expected utility estimation via cross-validation. In J Bernardo, M Bayarri, J Berger, A Dawid, D Heckerman, A Smith, M West (eds), *Bayesian Statistics 7.* Clarendon Press, 701–10.

Venables W, Ripley B (1994) *Modern applied statistics with S-plus.* New-York: Springer-Verlag.

Verbeke G, Lesaffre E (1996) A linear mixed-effects model with heterogeneity in the random-effects population. *Journal of the American Statistical Association*, 91, 217–21.

Verdinelli I, Wasserman L (1991) Bayesian analysis of outlier problems using the Gibbs sampler. *Statistics and Computing*, 1, 105–17.

Viele K, Tong B (2002) Modeling with mixtures of linear regressions. *Statistics and Computing*, 12, 315–30.

Vines S, Gilks W, Wild P (1996) Fitting Bayesian multiple random effects models. *Statistics and Computing*, 6, 337–46.

Viswanathan B, Manatunga A (2001) Diagnostic plots for assessing the frailty distribution in multivariate survival data. *Lifetime Data Analysis*, 7, 143–55.

Wahba G (1983) Bayesian confidence intervals for the cross validated smoothing spline. *Journal of the Royal Statistical Society: Series B*, 45, 133–50.

Wakefield J (2007) Disease mapping and spatial regression with count data. *Biostatistics*, 8, 158–83.

Wakefield J, Gelfand A, Smith A (1991) Efficient generation of random variates via the ratio-of-uniforms method. *Statistics and Computing*, 1, 129–33.

Wakefield J, Smith A, Racine-Poon A, Gelfand A (1994) Bayesian analysis of linear and non-linear population models using the Gibbs sampler. *Applied Statistics*, 43, 201–21.

Wakefield, J, Walker, S (1997) Bayesian nonparametric population models: Formulation and comparison with likelihood approaches. *Journal of Pharmacokinetics and Biopharmaceutics*, 25, 235–53.

Walker S, Damien P, Laud P, Smith A (1999) Bayesian nonparametric inference for random distributions and related functions. *Journal of the Royal Statistical Society: Series B*, 61, 485–527.

Walker S, Mallick B (1999) A Bayesian semiparametric accelerated failure time model. *Biometrics*, 55, 477–83.

Wall M (2004) A close look at the spatial structure implied by the CAR and SAR models. *Journal of Statistical Planning and Inference*, 121, 311–24.

Waller L (2002) Hierarchical models for disease mapping. In A El-Shaarawi, W Piegorsch (eds), *Encyclopedia of Environmetrics.* Chichester: Wiley, 1004–1007.

Waller L, Gotway C (2004) *Applied spatial statistics for public health data.* Chichester: Wiley.

Wand J, Shotts K, Sekhon J, Mebane J, Herron M, Brady H (2001) The butterfly did it: The aberrant vote for Buchanan in Palm Beach County, Florida. *American Political Science Review*, 95, 793–810.

Wand M (2003) Smoothing and mixed models. *Computational Statistics*, 18, 223–49.

Wang D, Lin J-Y, Yu T (2006) A MIMIC approach to modeling the underground economy in Taiwan. *Physica A*, 371, 536–42.

Wang F, Wall M (2003) Generalized common spatial factor model. *Biostatistics*, 4, 569–82.

Wang P, Puterman M (1999a) Markov Poisson regression models for discrete time series, part 1: Methodology. *Journal of Applied Statistics*, 26, 855–69.

Wang P, Puterman M (1999b) Markov Poisson regression models, part 2: Applications. *Journal of Applied Statistics*, 26, 871–82.

Wang P, Puterman M, Cockburn I, Le N (1996) Mixed Poisson regression models with covariate dependent rates. *Biometrics*, 52, 381–400.

Warn D, Thompson S, Spiegelhalter D (2002) Bayesian random effects meta-analysis of trials with binary outcomes: Methods for the absolute risk difference and relative risk scales. *Statistics in Medicine*, 21, 1601–23.

Warton D (2005) Many zeros does not mean zero inflation: Comparing the goodness-of-fit of parametric models to multivariate abundance data. *Environmetrics*, 16, 275–89.

Wasserman L (2000) Asymptotic inference for mixture models using data-dependent priors. *Journal of the Royal Statistical Society: Series B*, 62, 159–80.

Watson T, Christian C, Mason A, Smith M, Meyer R (2002) Bayesian-based decision support system for water distribution systems. In 5th International Conference on Hydroinformatics. Cardiff University, UK.

Webster R, Oliver M, Muir K, Mann J (1994) Kriging the local risk of a rare disease from a register of diagnoses. *Geographical Analysis*, 26, 168–85.

Wedel M, Bockenholt U, Kamakura W (2003) Factor models for multivariate count data. *Journal of Multivariate Analysis*, 87, 356–69.

Wedel M, Desarbo W, Bult J, Ramaswamy V (1993) A latent class Poisson regression model for heterogeneous count data. *Journal of Applied Econometrics*, 8, 397–411.

Weems K, Smith P (2004) On robustness of maximum likelihood estimates for Poisson-lognormal models. *Statistics & Probability Letters*, 66, 189–96.

Wei L (1992) The accelerated failure time model: A useful alternative to the Cox regression model in survival analysis. *Statistics in Medicine*, 11, 1871–79.

Weir I, Pettitt N (1999) Spatial modelling for binary data using a hidden conditional autoregressive Gaussian process: A multivariate extension of the probit model. *Statistics and Computing*, 9, 77–86.

Weiss R (1994) Pediatric pain, predictive inference and sensitivity analysis. *Evaluation Review*, 18, 651–78.

Weiss R (2005) *Modelling longitudinal data*. New York: Springer.

Weiss R, Cho M, Yanuzzi M (1999) On Bayesian calculations for mixture priors and likelihoods. *Statistics in Medicine*, 18, 1555–70.

West M (1984) Outlier models and prior distributions in Bayesian linear regression. *Journal of the Royal Statistical Society: Series B*, 46, 431–39.

West M (1992) Modelling with mixtures. In J Bernardo, J Berger, A Dawid, A Smith (eds), *Bayesian Statistics 4*. New York: Oxford University Press, 503–24.

West M (1996) Bayesian time series: models and computations for the analysis of time series in the physical sciences. In J Skilling (ed.), *Maximum entropy and bayesian methods*, Dordrecht: Kluwer Academic, 23–34.

West M (1998) Bayesian forecasting. In S Kotz, C Read, D Banks (eds), *Encyclopedia of Statistical Sciences*. Chichester: Wiley.

West M, Harrison J (1997) *Bayesian forecasting and dynamic models*. New York: Springer.

West M, Harrison P, Migon H (1985) Dynamic generalised linear models and Bayesian forecasting. *Journal of the American Statistical Association*, 80, 73–97.

West M, Harrison P (1997) *Bayesian forecasting and dynamic models*, 2nd edn. New York: Springer-Verlag.

West M, Muller P, Escobar M (1994) Hierarchical priors and mixture models, with application in regression and density estimation. In P Freeman, A Smith (eds), *Aspects of uncertainty: A tribute to D.V. Lindley*. Chichester: Wiley, 363–86.

Wheeler D, Calder C (2006) Bayesian spatially varying coefficient models in the presence of collinearity. *Proceedings of the Joint Statistical Meetings.* Seattle, WA. August 6–10, 2006.

Wheeler D, Calder C (2007) An assessment of coefficient accuracy in linear regression models with spatially varying coefficients. *Journal of Geographical Systems*, 9, 145–66.

Wheeler D, Tiefelsdorf M (2005) Multicollinearity and correlation among local regression coefficients in geographically weighted regression. *Journal of Geographical Systems*, 7, 161–87.

Wikle C (2003) Hierarchical models in environmental science. *International Statistical Review*, 71, 181–99.

Willink R, Lira I (2005) A united interpretation of different uncertainty intervals. *Measurement*, 38, 61–66.

Winkelmann R, Zimmermann K (1995) Recent developments in count data modelling: Theory and application. *Journal of Economic Surveys*, 9, 1–24.

Winkelmann, R, Boes S (2005) *Analysis of microdata.* New York: Springer-Verlag.

Witkovsky V (1996) On variance-covariance components estimation in linear models with AR(1) disturbances. *Acta Mathematicae Universitatis Comenianae*, 65, 129–39.

Wong O (1977) A competing-risk model based on the life table procedure in epidemiologic studies. *International Journal of Epidemiology*, 6, 153–59.

Wood S, Jiang W, Tanner M (2002) Bayesian mixture of splines for spatially adaptive nonparametric regression. *Biometrika*, 89, 513–28.

Wood S, Kohn R (1998) A Bayesian approach to robust nonparametric binary regression. *Journal of the American Statistical Association*, 93, 203–13.

Woodworth G (2004) *Biostatistics: A Bayesian introduction.* Chichester: John Wiley & Sons.

Wooldridge J (2005) Simple solutions to the initial conditions problem in dynamic, nonlinear panel data models with unobserved heterogeneity. *Journal of Applied Econometrics*, 20, 39–54.

Wu H, Zhang J-T (2006) *Nonparametric regression methods for longitudinal data analysis.* Chichester: John Wiley.

Wu M, Bailey K (1989) Estimation and comparison of changes in the presence of informative right censoring: Conditional linear model. *Biometrics*, 45, 939–55.

Xia Y, Weng S, Zhang C, Li S (2005) Mixture random effect model based meta-analysis for medical data mining. *Lecture Notes in Artificial Intelligence*, 3587, 630–40.

Xie M, He B, Goh T (2001) Zero-inflated Poisson model in statistical process control. *Computational Statistics & Data Analysis*, 38, 191–201.

Xu S, Jones R, Grunwald G (2007) Analysis of longitudinal count data with serial correlation. *Biometrical Journal*, 49, 416–28.

Yang R , Chen M-H (1995) Bayesian analysis for random coefficient regression models using noninformative priors. *Journal of Multivariate Analysis*, 55, 283–311.

Yang X, Shoptaw S (2005) Assessing missing data assumptions in longitudinal studies: An example using a smoking cessation trial. *Drug and Alcohol Dependence*, 77, 213–25.

Yanli Z, Wall M (2004) Investigating the use of the variogram for lattice data. *Journal of Computational and Graphical Statistics*, 13, 719–38.

Yashin A, Iachine I, Begun A, Vaupel J (2001) Hidden frailty: Myths and reality. Department of Statistics and Demography, SDU-Odense University, Research Report 34.

Yau K (2001) Multilevel models for survival analysis with random effects. *Biometrics*, 57, 96–102.

Yau K, Kuk A (2002) Robust estimation in generalized linear mixed models. *Journal of the Royal Statistical Society B*, 64, 101–17.

Yau K, Lee A, Ng A (2003) Finite mixture regression model with random effects: Application to neonatal hospital length of stay. *Computational Statistics and Data Analysis*, 41, 359–66.

Yau P, Kohn R (2003) Estimation and variable selection in nonparametric heteroscedastic regression. *Statistics and Computing*, 13, 191–208.

Yau P, Kohn R, Wood S (2003) Bayesian variable selection and model averaging in high-dimensional multinomial nonparametric regression. *Journal of Computational and Graphical Statistics*, 12, 23–54.

Yin G (2005) Bayesian cure rate frailty models with application to a root canal therapy study. *Biometrics*, 61, 552–58.

Yin G, Ibrahim J (2005) A class of Bayesian shared gamma frailty models with multivariate failure time data. *Biometrics*, 61, 208–16.

Yin G, Ibrahim J (2006) Bayesian transformation hazard models. In *IMS monograph series*, Vol. 49. Beachwood, OH: Institute of Mathematical Statistics, 170–82.

Yu J, Meyer A (2006) Multivariate stochastic volatility models: Bayesian estimation and model comparison. *Econometric Reviews*, 25, 361–84.

Yuan C, Druzdzel M (2005) How heavy should the tails be? In I Russell, Z Markov (eds), *Proceedings of the Eighteenth International FLAIRS Conference (FLAIRS-05).* Menlo Park, CA: AAAI Press/The MIT Press, 799–804.

Yuan K-H, Bentler P, Chan W (2004) Structural equation modeling with heavy tailed distributions. *Psychometrika*, 69, 421–36.

Zayeri F, Kazemnejad A, Khanafshar N, Nayeri F (2005) Modeling repeated ordinal responses using a family of power transformations: Application to neonatal hypothermia data. *BMC Medical Research Methodology*, 5, 29, http://www.biomedcentral.com/1471-2288/5/29

Zellner A (1983) Applications of Bayesian analysis in econometrics. *The Statistician*, 132, 23–34.

Zellner A (1986) On assessing prior distributions and Bayesian regression analysis with g-prior distributions. In P Goel, A Zellner (eds), *Bayesian inference and decision techniques: Essays in honor of Bruno de Finetti.* Amsterdam: Elsevier, 233–43.

Zhang H (2002) On estimation and prediction for spatial generalized linear mixed model. *Biometrics*, 58, 129–36.

Zhang D, Davidian M (2001) Linear mixed models with flexible distributions of random effects for longitudinal data. *Biometrics*, 57, 795–802.

Zhao Y, Staudenmayer J, Coull B, Wand M (2006) General design Bayesian generalized linear mixed models. *Statistical Science* 21, 35–51.

Zhu H, Lee S (2001) A Bayesian analysis of finite mixtures in the LISREL model. *Psychometrika*, 66, 133–52.

Zhu L, Gorman D, Horel S (2006) Hierarchical Bayesian spatial models for alcohol availability, drug "hot spots" and violent crime. *International Journal of Health Geographics*, 5, 54.

Zhu M, Lu A (2004) The counter-intuitive non-informative prior for the Bernoulli family. *Journal of Statistics Education [Online]*, 12 (2), http://www.amstat.org/publications/jsc/v12n2/zhu.pdf

Zivot E, Wang J (2006) *Modeling financial time Series with S-PLUS*. Berlin: Springer.

Zuur G, Garthwaite P, Fryer R (2002) Practical use of MCMC methods: Lessons from a case study. *Biometrical Journal*, 44, 433–55.

Index

E

H

M

N

T

Z